CLIMATE CHANGE ADAPTATION and MITIGATION MANAGEMENT OPTIONS

A Guide for Natural Resource Managers in Southern Forest Ecosystems

CLIMATE CHANGE ADAPTATION and MITIGATION MANAGEMENT OPTIONS

A Guide for Natural Resource Managers in Southern Forest Ecosystems

Edited by

JAMES M. VOSE

KIER D. KLEPZIG

CRC Press is an imprint of the
Taylor & Francis Group, an **informa** business

CRC Press
Taylor & Francis Group
6000 Broken Sound Parkway NW, Suite 300
Boca Raton, FL 33487-2742

Printed on acid-free paper
Version Date: 20130923

International Standard Book Number-13: 978-1-4665-7275-1 (Hardback)

Visit the Taylor & Francis Web site at
http://www.taylorandfrancis.com

and the CRC Press Web site at
http://www.crcpress.com

Contents

Preface

The rapid pace of climate change and its direct and indirect effects on forest ecosystems present a pressing need for better scientific understanding and the development of new science-management partnerships. Understanding the effects of stressors and disturbances (including climatic variability), and developing and testing science-based management options to deal with them, have been core research tasks for researchers in the southern United States for decades. Climate change adds a new dimension to this task because it can directly impact forest ecosystems over large spatial scales and interact with other stressors and disturbances to create stress complexes that may have an even greater impact than any single stressor. In addition, the large spatial scale and complex interactions that occur with climate change make traditional experimental approaches and direct application of existing scientific studies difficult. Despite these challenges, climate change science is progressing rapidly, and new insights from recent syntheses, models, experiments, and observations provide enough information to begin taking action now.

One of the purposes of this book was to begin engaging scientists from a range of disciplines in the process of thinking about and addressing the implications of climate change in their areas of disciplinary expertise: water, wildlife, biodiversity, forest productivity, recreation, wildfire; and insects, diseases, and pathogens. In this, we believe we undoubtedly succeeded. While some of the authors have extensive expertise in climate science, others do not. As a result, many of the chapter authors extended their knowledge of climate models (general circulation models; GCMs) and climate change terminology; and synthesized new and disparate information. While we provided authors with a common "scenario-driven" conceptual framework and overarching goals, they were also given full freedom to use other data and develop their own approaches and methods as needed. As a result, the chapters (all of which were peer reviewed and revised in response to reviewer comments) reflect a blend of data syntheses; conceptual, simple, and complex models; and general versus specific approaches and examples. A combination of all of these approaches, and the ongoing dialogue this volume will hopefully inspire, will be required to address the complex and dynamic problem of managing southern forest ecosystems in response to climate change.

James M. Vose and Kier D. Klepzig

Editors

James M. Vose is a research ecologist and project leader of the USDA Forest Service, Southern Research Station, Center for Integrated Forest Science (CIFS) at North Carolina State University in Raleigh, NC. Prior to his current appointment as project leader of CIFS, he spent 25 years at the Coweeta Hydrologic Laboratory studying watershed ecosystem responses to disturbances and forest management. He received degrees in forestry from Southern Illinois University (BS), Northern Arizona University (MS), and North Carolina State University (PhD). He has authored over 170 scientific papers and serves as adjunct faculty at the University of Georgia, Virginia Tech, and North Carolina State University. He recently served as a coeditor and coauthor for the USDA National Climate Assessment Forest Sector Report published in 2012.

Kier D. Klepzig is an entomologist and assistant director of the USDA Forest Service, Southern Research Station in Asheville, NC. Prior to his current appointment, he spent 14 years as a research entomologist and project leader of the Insects, Diseases, and Invasive Plants Research Unit in Pineville, LA. He studied bark beetle fungal interactions and symbiosis. He received degrees in biology and reclamation (double major) from the University of Wisconsin-Platteville (BS), and in entomology and plant pathology (double major) from the University of Wisconsin-Madison (MS, PhD). He has authored over 90 scientific papers and serves as an editor for *Environmental Entomology*. He recently served as a coeditor and coauthor of the landmark book *SPBII*. He is the most recent winner of the A.D. Hopkins Award for excellence in forest entomology.

Contributors

Devendra M. Amatya
Center for Watershed Science Research
Southern Research Station
USDA Forest Service
Cordesville, South Carolina

Ashley E. Askew
Department of Statistics
University of Georgia
Athens, Georgia

J.D. Austin
Department of Wildlife Ecology and Conservation
University of Florida
Gainesville, Florida

J.M. Bowker
Integrating Human and Natural Systems in Urban and Urbanizing Environments
Southern Research Station
USDA Forest Service
Athens, Georgia

Megan L. Buchanan
Department of Geography
University of Alabama
Tuscaloosa, Alabama

John R. Butnor
Forest Genetics and Ecosystems Biology
Southern Research Station
USDA Forest Service
South Burlington, Vermont

Peter V. Caldwell
Eastern Forest Environmental Threat Assessment Center
Southern Research Station
USDA Forest Service
Asheville, North Carolina

Zanethia D. Choice
Center for Bottomland Hardwoods Research
Southern Research Station
USDA Forest Service
Asheville, North Carolina

Stacy L. Clark
Upland Hardwood Ecology and Management
Southern Research Station
USDA Forest Service
Knoxville, Tennessee

J. Alan Clingenpeel
Ouachita National Forest
USDA Forest Service
Hot Springs, Arkansas

Barton D. Clinton (Retired)
Center for Watershed Science Research
Southern Research Station
USDA Forest Service
Cordesville, South Carolina

Paul A. Conrads
South Carolina Water Science Center
U.S. Geological Survey
Columbia, South Carolina

H. Ken Cordell
Southern Research Station
USDA Forest Service
Athens, Georgia

Zhaohua Dai
University of New Hampshire
Durham, New Hampshire

Songlin Fei
Department of Forestry and Natural Resources
Purdue University
West Lafayette, Indiana

Kathleen E. Franzreb (Retired)
Upland Hardwood Ecology and Management
Southern Research Station
USDA Forest Service
Knoxville, Tennessee

E.M. Fucik
Department of Biology
Stephen F. Austin University
Nacogdoches, Texas

Carlos A. Gonzalez-Benecke
School of Forest Resources and Conservation
University of Florida
Gainesville, Florida

Scott L. Goodrick
Center for Forest Disturbance Science
Southern Research Station
USDA Forest Service
Athens, Georgia

Cathryn H. Greenberg
Upland Hardwood Ecology and Management
Southern Research Station
USDA Forest Service
Asheville, North Carolina

James M. Guldin
Ecology and Management of Southern Pines
Southern Research Station
USDA Forest Service
Hot Springs, Arkansas

Qinfeng Guo
Eastern Forest Environmental Threat Assessment Center
Southern Research Station
USDA Forest Service
Asheville, North Carolina

Justin L. Hart
Department of Geography
University of Alabama
Tuscaloosa, Alabama

James D. Haywood
Restoring Longleaf Pine Ecosystems
Southern Research Station
USDA Forest Service
Pineville, Louisiana

Thomas P. Holmes
Forest Economics and Policy
Southern Research Station
USDA Forest Service
Asheville, North Carolina

Kurt H. Johnsen
Forest Genetics and Ecosystems Biology
Southern Research Station
USDA Forest Service
Asheville, North Carolina

Cassandra Johnson-Gaither
Integrating Human and Natural Systems in Urban and Urbanizing Environments
Southern Research Station
USDA Forest Service
Athens, Georgia

Binita K.C.
Department of Geography
University of Georgia
Athens, Georgia

Donald J. Kaczmarek
Forest Watershed Science
Southern Research Station
USDA Forest Service
Columbia, South Carolina

Tara L. Keyser
Upland Hardwood Ecology and Management
Southern Research Station
USDA Forest Service
Asheville, North Carolina

Kier D. Klepzig
Southern Research Station
USDA Forest Service
Asheville, North Carolina

M.A. Kwiatkowski
Department of Biology
Stephen F. Austin University
Nacogdoches, Texas

Shelby Gull Laird
Charles Sturt University
New South Wales, Australia

Harbin Li
Rangeland Management
USDA Forest Service
Cordesville, South Carolina

Yongqiang Liu
Center for Forest Disturbance Science
Southern Research Station
USDA Forest Service
Athens, Georgia

Susan C. Loeb
Upland Hardwood Ecology and Management
Southern Research Station
USDA Forest Service
Clemson, South Carolina

Chris A. Maier
Forest Genetics and Biological Foundations
Southern Research Station
USDA Forest Service
Asheville, North Carolina

Daniel A. Marion
Center for Bottomland Hardwoods Research
Southern Research Station
USDA Forest Service
Asheville, North Carolina

Heather R. McCarthy
Department of Microbiology and Plant Biology
University of Oklahoma
Norman, Oklahoma

W. Henry McNab
Upland Hardwood Ecology and Management
Southern Research Station
USDA Forest Service
Asheville, North Carolina

Steve McNulty
Eastern Forest Environmental Threat Assessment Center
Southern Research Station
USDA Forest Service
Asheville, North Carolina

Jennifer A. Moore Meyers
Eastern Forest Environmental Threat Assessment Center
Southern Research Station
USDA Forest Service
Asheville, North Carolina

Chelcy F. Miniat
Center for Watershed Science Research
Southern Research Station
USDA Forest Service
Cordesville, South Carolina

Thomas L. Mote
Department of Geography
University of Georgia
Athens, Georgia

Rabiu Olatinwo
Department of Entomology
Louisiana State University AgCenter
Baton Rouge, Louisiana

William Otrosina
Insects, Diseases, and Invasive Plants
Southern Research Station
USDA Forest Service
Athens, Georgia

Ying Ouyang
Center for Bottomland Hardwoods Research
Southern Research Station
USDA Forest Service
Asheville, North Carolina

B.R. Parresol
Pacific Northwest Research Station
USDA Forest Service
Vancouver, Washington

Roger W. Perry
Southern Pine Ecology and Management
Southern Research Station
USDA Forest Service
Hot Springs, Arkansas

Neelam Poudyal
Warnell School of Forestry and Natural Resources
University of Georgia
Athens, Georgia

Jeffrey P. Prestemon
Forest Economics and Policy
Southern Research Station
USDA Forest Service
Asheville, North Carolina

Edwin A. Roehl Jr.
Advanced Data Mining International

D. Craig Rudolph
Southern Pine Ecology
Southern Research Station
USDA Forest Service
Nacogdoches, Texas

Daniel Saenz
Southern Pine Ecology
Southern Research Station
USDA Forest Service
Nacogdoches, Texas

John Schelhas
Integrating Human and Natural Systems in Urban and Urbanizing Environments
Southern Research Station
USDA Forest Service
Athens, Georgia

Lynne Seymour
Department of Statistics
University of Georgia
Athens, Georgia

J. Marshall Shepherd
Department of Geography
University of Georgia
Athens, Georgia

Martin A. Spetich
Upland Hardwood Ecology and Management
Southern Research Station
USDA Forest Service
Hot Springs, Arkansas

John A. Stanturf
Center for Forest Disturbance Science
Southern Research Station
USDA Forest Service
Athens, Georgia

Douglas Streett
Insects, Diseases, and Invasive Plants
Southern Research Station
USDA Forest Service
Asheville, North Carolina

Christopher W. Strother
Department of Geography
University of Georgia
Athens, Georgia

Ge Sun
Eastern Forest Environmental Threat Assessment Center
Southern Research Station
USDA Forest Service
Asheville, North Carolina

G.W. Tanner (Retired)
Department of Wildlife Ecology and Conservation
University of Florida
Gainesville, Florida

Scott J. Torreano
Department of Forestry and Geology
University of the South
Sewanee, Tennessee

Carl Trettin
Center for Watershed Science Research
Southern Research Station
USDA Forest Service
Cordesville, South Carolina

James M. Vose
Center for Integrated Forest Science and Synthesis
Southern Research Station
USDA Forest Service
Asheville, North Carolina

David N. Wear
Center for Integrated Forest Science and Synthesis
Southern Research Station
USDA Forest Service
Asheville, North Carolina

Eric Winters
Upland Hardwood Ecology and Management
Southern Research Station
USDA Forest Service
Clemson, South Carolina

Stanley J. Zarnoch
Southern Research Station
USDA Forest Service
Asheville, North Carolina

Wayne Zipperer
Integrating Human and Natural Systems in Urban and Urbanizing Environments
Southern Research Station
USDA Forest Service
Gainesville, Florida

1 Introduction to Climate Change Adaptation and Mitigation Management Options

James M. Vose and Kier D. Klepzig

CONTENTS

Climate is a critical factor shaping the structure and function of forest ecosystems in the Southern United States. Human induced changes in climate systems have resulted in an increase in the global average air temperature of about 0.8°C since the 1900s (Pachuri and Reisinger 2007). Data from long-term weather stations show that overall, the continental United States has warmed during the past century, but that the magnitude and direction of change vary by geographic area (Backlund et al. 2008). The primary driving force behind this overall warming is an increase in carbon dioxide (CO_2) and other greenhouse gas emissions, a trend that is likely to continue over the next century (Karl et al. 2009). For example, by 2100, further warming in the United States is expected to range from 2.5°C to 5.3°C relative to the 1971 to 2000 time period (Kunkel et al. 2011).

While warming and elevated CO_2 are important aspects of climate change, projections of increased climate variability and extreme weather events—such as drought, heat waves, heavy rains, tornados, and hurricanes (Easterling et al. 2000; Huntington 2006)—are expected to have an even greater impact on forest ecosystems than increases in CO_2 and temperature alone (Dale et al. 2001; Kunkel et al. 2011; Vose et al. 2012). Indirect effects may be equally or more significant, as the frequency, magnitude, and severity of wildfires, insect and pathogen outbreaks, and the spread of nonnative invasive species may be amplified by climate change (Vose et al. 2012). Combined, these direct and indirect effects of climate change are likely to create conditions that have not as yet been observed and may shape ecosystems in ways that have no historical analog (Williams and Jackson 2007). Some of these effects may be already occurring (Parmesan and Yohe 2003). For example, forest dieback, large insect outbreaks, and large wildfires during the past decade (Bentz et al. 2009; 2010; Breshears et al. 2005, 2009; Turetsky et al. 2010) may be signals of the potential effects of a rapidly changing climate on forest ecosystems (Vose et al. 2012).

The history of forestry in the South is one of managing disturbances, whether from early unsustainable logging practices or from wildfires, insects, and diseases. Almost a century of federal, university, and private industry research has produced an abundance of silvicultural studies, long-term data on trends in forest conditions and environmental changes, and expertise in modeling the effects of disturbances ranging from wildfires to insects to climate change. This science served to reforest the almost completely cutover landscapes at the beginning of the twentieth century and to establish the most productive forest region in the world (Prestemon and Abt 2002). Land managers are now being challenged to sustainably manage forest ecosystems in an increasingly uncertain, but likely very different, set of future climate conditions and disturbance regimes. The task becomes even more difficult because of co-occurring increases in landscape fragmentation, greater numbers of invasive species, changing social and economic conditions, and greater demands for ecosystem services from a growing population (Wear and Greis in press). Hence, land managers will need to consider multiple risks as they make decisions about activities on forest lands.

Land managers often look to past experiences and well-established scientific knowledge for guidance before deciding how to manage their forests. For example, restoration activities are often guided by the structure and function of historical stand types and conditions such as longleaf pine (*Pinus palustris*). Commercial forest management proceeds from an understanding of historical growth and yield coupled with an understanding of the risks of forest damage or mortality. In some circumstances, these historical conditions may be useful analogs for contemporary and future management; in others, change occurs at such a fast pace and broad scale that historical analogs provide poor guidance (Hobbs et al. 2011). In these circumstances, how does a land manager develop, evaluate, and implement the appropriate management actions? Scientists and managers have begun to develop some general guidelines, principles, and tools (e.g., Millar et al. 2007; Peterson et al. 2011) to help land managers begin incorporating climate change considerations into planning, management, and decision making. Management responses to climate change can generally be classified into three categories: adaptation, mitigation, and no action. Mitigation, which reduces or offsets CO_2 emissions, includes increasing storage (in forest systems or long-lived wood-based products) or offsetting fossil fuel use. Adaptation includes activities that help ecosystems resist the effects of climate change, be more resilient to the effects of climate change, or facilitate the transition to a new state after ecosystems have been subjected to the effects of climate change (Millar et al. 2007). The longevity of forests creates challenges not shared by other managed ecosystems such as agricultural crops, where changing species or genotypes and other management practices can be quickly modified when conditions change, new technologies are introduced, and new best management practices are adopted. Instead, forest management is a long-term investment—management actions implemented today can greatly constrain management possibilities over the next several decades. Although this long-term aspect of forest management does not preclude change, it does impart restrictions on "nimbleness," increasing the urgency for basing today's decisions on the best available information.

Now more than ever, land managers need credible and concrete examples of how to blend expert knowledge, science, and on-the-ground experience (Vose et al. 2012) to address climate change. Some information can be extracted directly from the large body of science that focuses on forest ecosystem responses to disturbance and environmental variability; however, conducting controlled studies of how changes in temperature and other co-occurring factors, such as precipitation amount and variability, impact ecosystem structure and function is extremely difficult, especially at the temporal and spatial scales needed to understand the consequences and implications for forest management. As a result, some of the science and approaches used to develop our current understanding of forests and forest management in southern forests may not fully apply to the future complexities of climate change. To address this knowledge deficit, scientists can combine synthesis of current knowledge, retrospective analyses of long-term data, modeling, and their own experiential knowledge to formulate working hypotheses about climate-change impacts on ecosystems and to help managers develop options to mitigate change or help forests adapt. These working hypotheses can also guide future forest science. That is the approach used in this book.

The objective of this book is to synthesize the best available expert knowledge by combining scientific literature, tools and models, and the experiences of scientists and land managers in the South to answer the question: *Can forest management enhance the sustainability of southern forest ecosystems and their values under climate change?* It provides a comprehensive analysis of the management options that could be used by natural resource managers to help southern forest ecosystems adapt to the impacts of climate change or manage forests to help mitigate climate change.

APPROACH

Providing a scientific framework for managing forests in the face of climate change has been the focus of several recent papers and overviews, including Baron et al. (2008), Joyce et al. (2009), Millar et al. (2007), and Peterson et al. (2011). They have provided the foundation for many of the concepts and definitions used by the authors in this book; however, they provide little direct guidance specific to land managers in the South. To help address this need, we assembled a team of scientists that represented multiple disciplines and many years of experience studying southern forest ecosystems and developing management options for protecting or enhancing their values. Our goal was to develop a comprehensive, cohesive, and integrated analysis of potential climate change impacts and management options to address those impacts. Doing so required developing a common conceptual framework, consistent definitions and data sources, and an overall organization and analytical structure that could span scientific disciplines. Data and knowledge limitations usually required that each of the chapter authors adapt and modify aspects the common conceptual framework to meet their particular needs.

THREATS AND VALUES

We organized the chapters by the values that southern forest ecosystems provide and the threats to the sustainability of those values (Table 1.1), each of which was identified and refined at a stakeholder workshop conducted early in the process. Workshop attendees included scientists, public and private land managers, and nongovernment organizations (NGOs). The list was not intended to be comprehensive of all of the values provided by (or threats to) southern forests; rather, they were collectively identified as critically important, and as areas where sufficient science and expertise were available to characterize risks and vulnerabilities and to develop potential management options.

Decisions about areas of focus were also guided by results from the Southern Forest Futures Project (SFFP) (www.srs.fs.usda.gov/futures/[Date accessed: October 25, 2012]; Wear and Greis in press), which provides a science-based "futuring" analysis for the forests of the 13 southern states. Organized by a set of scenarios (described fully in Chapter 2) and using a combination of computer

TABLE 1.1
Threats and Values of Forest Ecosystems in the Southern United States

Threats	Values
Insects, disease, invasives	Timber, fiber, and carbon sequestration
Wildfire	Water quality and quantity
	Species and habitats
	Plants
	Wildlife
	Recreation

models and science synthesis, the SFFP examines a variety of possible futures and how they could shape forests and their ecosystem services and values. Its ultimate goal was to translate a vast array of science and modeling results into useable information by government, the natural resource community, and other key stakeholders for southern forest management and policy analysis. Indeed, many of the analyses, databases, and findings from the SFFP serve as a foundation for several of the chapters in this book.

Ecological Subregions

The South is a region of highly complex landscapes, ranging from the mountainous areas in the Southern Appalachians and Cumberland, to the Piedmont, the Mississippi Alluvial Valley, the Mid-South, and the flat landscape of the lower Coastal Plain. These subregions vary considerably in biophysical characteristics such as climate, soils, and vegetation; in processes such as water, carbon, and nutrient cycling; and in socioeconomic conditions, land use patterns, and forest management opportunities. We used the subregion classification system (Figure 1.1) developed for the SFFP (Wear and Greis in press) as a framework both for describing this variation in our analyses of risks, vulnerabilities, and impacts, and for developing management options for the chapters in this book. We recognized that a significant level of variation in biophysical and socioeconomic conditions also exists *within* the subregions, but addressing specific site/stand condition or socioeconomic/management constraints was beyond the scope of this book. Also, our experience has been that within the context of climate change, the use of general analyses or simple model predictions at the stand or site level cannot produce meaningful specific recommendations. This means that land managers must continue to blend the best and most appropriately synthesized science with their own experience and professional judgment in making stand-level decisions. To provide the best available science and account for some of this variation among subregions, we described any key vulnerabilities (Schneider et al. 2007) and potential impacts that we were able to identify as specific to each subregion. Authors were encouraged to use case studies to provide specific examples and demonstrate how the concepts, analyses, and potential management options could be applied within the subregions.

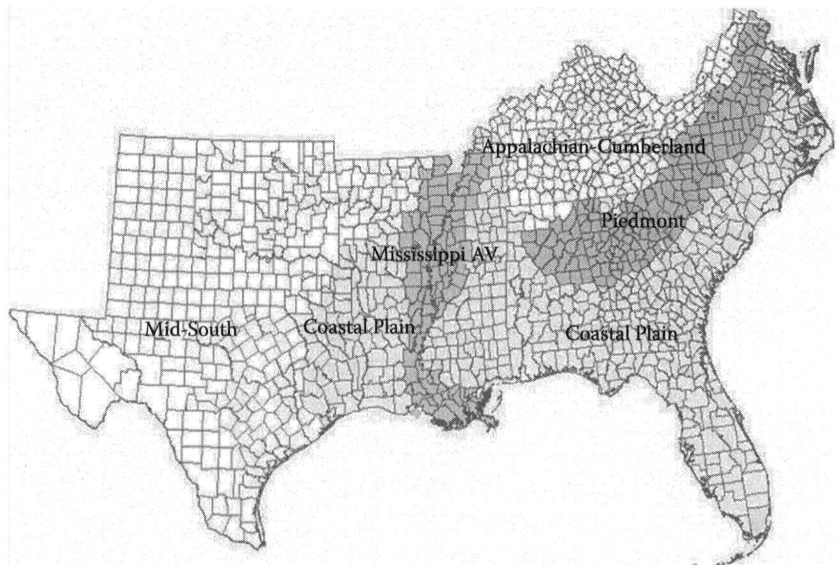

FIGURE 1.1 Ecological subregions of the Southern United States. (Wear, D.N. and J.G. Greis, eds. In press. The Southern Forest Futures Project: Technical Report.)

OVERALL CONCEPTUAL FRAMEWORK

Forest ecosystems are unique among all land uses because of their longevity and relative stability; even so, each has a structure and function that can be substantially altered by forest management. Both managed and unmanaged forest ecosystems frequently experience variability in weather systems (with or without climate change), as well as a myriad of other co-occurring disturbances and stressors. As a result, natural resources managers already have extensive experience managing forests, both to increase resistance and resiliency to historical disturbances and stressors and to restore desired conditions. However, the rate and magnitude of changes in biophysical conditions, as well as in the forest ecosystems that these biophysical conditions influence, will likely test the efficacy of current management approaches and guidelines. Nevertheless, this experience should help enable them to understand and implement "climate smart" management practices.

Despite these challenges, managing in the face of climate change is imperative for ensuring the sustainability of ecosystem services in southern forests. The response of forests to changing biophysical conditions will result from choices to: (1) respond to anticipated changes, (2) react to observed changes, (3) take no action in response to anticipated or observed changes, or (4) combine several of the above options. Indeed, any of these four approaches may be most appropriate based on assessments of critical vulnerabilities and risks. Proactive or reactive approaches to forest management can influence ecosystem responses to climate change by altering structural and functional attributes that determine response thresholds in either a positive manner (undesirable changes are less likely to occur) or a negative manner (undesirable change are more likely to occur).

Structural and functional attributes of forest ecosystems determine their resistance and resilience to historic patterns of climatic variability, and can facilitate long-term persistence of species and community types. As such, large and rapid changes in the structure and function of forest ecosystems usually only occur at the extremes of climate and other physical conditions. The threshold for these rapid, and sometimes permanent, changes varies considerably by ecosystem type and condition. For example, ecosystems that have been substantially degraded may have a narrower threshold of response to climate change than ecosystems that have been subject to less degradation. Conversely, heavily degraded ecosystems may have reached a new level of stability (e.g., they may now have a higher proportion of disturbance tolerant species) that reduces vulnerability to extreme disturbances. More extreme disturbances (such as hurricanes, catastrophic wildfires, and extreme drought) are difficult to predict and are often (but not always) localized. For example, an increase in the number of extended droughts may lead to more frequent and intense ecological disturbances, which in turn would lead to rapid changes in forest composition and dynamics (McKenzie et al. 2004). As a result, managing forests for frequent, unpredictable, and localized events can be very difficult, and in many situations, the severity of the disturbance is so great that preventive management activities are futile. In these situations, postdisturbance restoration or "facilitated transitions" are the only possible responses. Responses to longer-term climate change, such as a gradual warming, can be viewed in the same conceptual framework (see Chapter 3). Gradual changes in average climate or atmospheric environment produce gradual changes in ecosystems. For example, forest species distribution and abundance have shifted over long time scales as individuals respond to variability in temperature and precipitation, and to climatic-induced changes in wildfire and other disturbance regimes (Whitlock et al. 2008; Anderson et al. 2009).

COMMON DATABASES

One of the challenges of making "climate smart" management decisions is the variation (in space and over time) in projections of climatic future conditions. Uncertainty poses challenges to decision making in any context, but land managers will need to make decisions and management choices within the context of largely uncertain projected future climate conditions, as well as uncertain anticipated impacts. This variation in future projections is driven by a variety of factors, such as the choice of general circulation models, assumptions about future carbon dioxide emissions, and differences in approaches used to downscale coarse scale models to finer spatial scales (statistical vs. dynamic downscaling). To reduce the variability associated with model choice, emissions futures, or downscaling approaches across chapters, authors were provided with common databases and projections of future climatic conditions to 2060. We used the climate futures developed by the SFFP, which included four general circulation models (CSIROMK2, CSIROMK3.5, HadCM3, and MIROC3.2) and two emissions storylines (A1B and B2) from the Intergovernmental Panel on Climate Change (2007). These scenarios resulted in a range of possible futures derived from both model performance and emissions scenarios, and were intended to bracket "low" and "high" projections of future temperature and precipitation across the South. In addition, we developed an "ensemble"-based approach to develop temperature and precipitation projections using four general circulation models (CGCM3, CCSM3, HadCM3, and GFDLCM2.1) and three IPCC emission scenarios (B1, A1B, and B2). This ensemble approach averages variability in projections associated with differences in model performance, but maintains variation associated with emissions scenarios (Chapter 2).

ORGANIZATION

We used a series of workshops to develop a conceptual framework and identify key databases, afterward establishing a common framework for each chapter organized around the following questions:

1. In the Southern United States, where are social or biological systems most vulnerable to climate change?
2. Where will the consequences of climate change be the greatest (what areas have the greatest risk)?
3. What management options can be implemented to reduce vulnerability and risk (how can we manage to increase resistance and resilience to climate change)?
4. What are the key unknowns and uncertainties?

Teams of expert scientists for the threats and values identified in Table 1.1 were asked to synthesize the best available science, implement new syntheses and analyses, and use their best professional judgments to answer these questions (Chapters 3 through 12). In some cases, chapter authors used case studies to highlight specific geographic areas or species of concern. There is considerable overlap among many of these threats and values, as well as interdependencies and positive and negative interactions. For example, a management recommendation intended to mitigate a threat to one value could also have either a positive or negative impact on another value. Implications for some of the most significant interactions are discussed in Chapter 13.

CONCLUSIONS

Ensuring the sustainability of southern forest ecosystems and their values under climate change will require management decisions informed by the best available science. Our hope is that this book will serve as a valuable resource for both long-term planning and day-to-day forest management activities. Managing southern forests in response to or in anticipation of disturbance is nothing new for the southern natural resources community. Although uncertainty exists about the location, magnitude, and timing of climate-change effects on southern forest ecosystems, sufficient scientific information and tools are available to begin taking action now. The anticipated rapid pace of climate change, coupled with changes in disturbance regimes and other co-occurring stressors, will challenge the applicability of present-day best-management practices. The authors of this book have attempted to provide a linkage between current management actions and future management options that would anticipate a changing climate. Establishing this foundation of knowledge now could ensure a broader range of options for managing southern forests and protecting their values in the future. This approach requires strong partnerships between land managers and the science community, with successes and failures shared and evaluated, new science rapidly translated into management implications, and adaptive management embraced and implemented.

REFERENCES

Anderson, T., Carstensen, J., Hernandez-Garcia, E., and Duarte, C. 2009. Ecological thresholds and regime shifts: Approaches to identification. *Trends in Ecology and Evolution*. 24:49–57.

Backlund, P., Janetos, A., Schimel, D. et al. 2008. *The Effects of Climate Change on Agriculture, Land Resources, Water Resources, and Biodiversity in the United States*. Washington, DC: U.S. Department of Agriculture. 362 p.

Baron, J.S., Allen, C.D., Fleishman, E. et al. 2008. National parks. In: Julius, S.H., West, J.M., eds. *Preliminary Review of Adaptation Options for Climate-Sensitive Ecosystems and Responses: A Report by the U.S. Climate Change Science Program and the Subcommittee on Climate Research*. Washington, DC: U.S. Environmental Protection Agency: 4–1 to 4–68.

Bentz, B., Logan, J., MacMahon, J. et al. 2009. *Bark Beetle Outbreaks in Western North America: Causes and Consequences*. Salt Lake City, UT: University of Utah Press. 44 p.

Bentz, B.J., Régnière, J., Fettig, C.J. et al. 2010. Climate change and bark beetles of the western United States and Canada: Direct and indirect effects. *BioScience*. 60:602–613.

Breshears, D.D., Cobb, N.S., Rich, P.M. et al. 2005. Regional vegetation die-off in response to global-change-type drought. *Proceedings of the National Academy of Sciences of the United States of America*. 102: 15144–15148.

Breshears, D.D., Myers, O.B., Myer, C.W. et al. 2009. Tree die-off in response to global change-type drought: Mortality insights from a decade of plant water potential measurements. *Frontiers in Ecology and the Environment*. 7:185–189.

Dale, V.H., Joyce, L.A., McNulty, S. et al. 2001. Climate change and forest disturbances. *BioScience*. 51: 723–734.

Easterling, D.R., Meehl, G.A., Parmesan, C. et al. 2000. Climate extremes: observations, modeling, and impacts. *Science*. 289:2068–2074.

Hobbs, R.J., Hallett, L.M., Ehrlich, P.R., and Mooney, H.A. 2011. Intervention ecology: Applying ecological science in the twenty-first century. *Bioscience*. 61(6):442–450.

Huntington, T.G. 2006. Evidence for intensification of the global water cycle: Review and synthesis. *Journal of Hydrology*. 319:83–95.

Intergovernmental Panel on Climate Change. 2007. Climate change 2007: Synthesis report. Contribution of working groups I, II, and III to the Fourth Assessment Report of the Intergovernmental Panel on Climate Change. IPCC, Geneva, Switzerland.

Joyce, L.A., Blate, G.M., McNulty, S.G. et al. 2009. Managing for multiple resources under climate change. *Environmental Management*. 44:1022–1032.

Karl, T.R., Melillo, J.M., and Peterson, T.C., eds. 2009. *Global Climate Change Impacts in the United States: A State of Knowledge Report from the U.S. Global Change Research Program*. New York, NY: Cambridge University Press. 192 p.

Kunkel, K.E., Stevens, L., and Stevens, S. 2011. *Continental United States Climate Outlook*. Asheville, NC: Cooperative Institute for Climate and Satellites: National Climatic Data Center. Unpublished Draft.

McKenzie, D., Gedalof, Z., Peterson, D.L., and Mote, P. 2004. Climatic change, wildfire and conservation. *Conservation Biology*. 18:890–902.

Millar, C.I., Stephenson, N.L., and Stephens, S.L. 2007. Climate change and forests of the future: Managing in the face of uncertainty. *Ecological Applications*. 17(8):2145–2151. http://www.sgcp.ncsu.edu:8090/ [Date accessed: October 25, 2012]

Pachuri, R.K. and Reisinger, A., eds. 2007. Climate change 2007: Synthesis report: Contribution of Working Groups I, II and III to the fourth assessment report of the Intergovernmental Panel on Climate Change, Geneva, Switzerland: Intergovernmental Panel on Climate Change. 104 p.

Parmesan, C. and Yohe, G. 2003. A globally coherent fingerprint of climate change across natural systems. *Nature*. 421:37–42.

Peterson, D.L., Millar, C.I., Joyce, L.A. et al. 2011. *Responding to Climate Change on National Forests: A Guidebook for Developing Management Options*. Gen. Tech. Rep. PNW-GTR-855. Portland, OR: U.S. Department of Agriculture, Forest Service, Pacific Northwest Research Station. 109 p.

Prestemon, J.P. and Abt, R.C. 2002. The southern timber market to 2040. *Journal of Forestry*. 100(7):16–22.

Schneider, S.H., Semenov, S., and Patwardhan, A. 2007. Assessing key vulnerabilities and the risk from climate change. In: Parry, M.L., Canziani, O.F., Palutikof, P.J. et al., eds. *Climate Change 2007: Impacts, Adaptation and Vulnerability—Contribution of Working Group II to the Fourth Assessment Report of the Intergovernmental Panel on Climate Change*. Cambridge, United Kingdom: Cambridge University Press: 779–810. Chapter 19.

Turetsky, M.R., Kane, E.S., Harden, J.W. et al. 2010. Recent acceleration of biomass burning and carbon losses in Alaskan forests and peatlands. *Nature Geoscience*. 4:27–31.

Vose, J.M., Peterson, D.L., and Patel-Weynand, T., eds. 2012. *Effects of Climatic Variability and Change on Forest Ecosystems: A Comprehensive Science Synthesis for the U.S. Forest Sector*. Gen. Rech. Rep. PNW-GTR-870. Portland, OR: U.S. Department of Agriculture, Forest Service, Pacific Northwest Research Station. 265 p.

Wear, D.N. and Greis, J.G., eds. In press. The Southern Forest Futures Project: Technical Report.

Whitlock, C., Marlon, J., Briles, C., Brunelle, A., Long. C., and Bartlein, P.J. 2008. Long-term relations among fire, fuels, and climate in the northwestern U.S. based on lake-sediment studies. *Journal of International Wildfire Research*. 17(1):72–83.

Williams, J.W. and Jackson, S.T. 2007. Novel climates, no-analog communities, and ecological surprises. *Frontiers in Ecology and Environment*. 5(9):475–482.

2 Framing the Future in the Southern United States

Climate, Land Use, and Forest Conditions

David N. Wear, Thomas L. Mote, J. Marshall Shepherd, Binita K.C., and Christopher W. Strother

CONTENTS

The Intergovernmental Panel on Climate Change (IPCC) has concluded, with 90% certainty, that human or "anthropogenic" activities (emissions of greenhouse gases, aerosols and pollution, land-use/land-cover change) have altered global temperature patterns over the past 100–150 years (IPCC 2007a). Such temperature changes have a set of cascading, and sometimes amplifying, effects on the entire global climate system, including the water cycle, cryosphere, hurricane intensity, and sea level. This chapter develops a set of scenarios for exploring potential climate and resource effects futures in the Southern United States.

The term global warming refers to an increase in average atmospheric temperature, but this definition masks much seasonal and regional variability, such as that which has been observed in the Southern United States. Temperature observations in the Southern United States have fluctuated, but generally indicate a cooling trend in the second half of the twentieth century, especially in the late spring–early summer period (Portmann et al. 2009). Many studies have hypothesized the reasons for this cooling trend (Misra et al. 2012), including pollution increases, reforestation, irrigation, urban land cover variations, and the use of pesticides. Nevertheless, recent climatological analysis suggests that the cooling trend has abated and that an upward temperature trend is emerging (Tebaldi et al. 2012). Precipitation has also shown wide year-to-year variability with rainfall increasing about 10% for Georgia, Alabama, and Florida (Karl et al. 2009). Indeed, rainfall is quite variable across the region and is very much influenced by large-scale driving variables such as hurricanes and El Niño and La Niña events.

Concerns about climate change have prompted the development of general circulation models (also known as climate models) to predict how changes in atmospheric greenhouse gases (e.g., carbon dioxide and methane), solar output, orbital parameters, vegetation, and other factors might affect climate variables such as temperature, cloud formation, and precipitation. The climate models have limitations, one of which is the inability to project alternative futures, compromising their usefulness as a tool for examining future climates given their economic, social, technology drivers—factors that may hold substantial influence over how climate is expressed in ecological systems and may directly influence the structure of future forests.

These issues led the IPCC to develop additional models (under different greenhouse gas and societal scenarios) using a range of physical assumptions and parameterizations. The IPCC commissioned a special report to generate storylines of greenhouse gas emissions, set within four broad descriptions of future economic, demographic, political, environmental, and technological change (Nakicenovic et al. 2000). The *A1 storyline* describes a future of very rapid economic growth with a global population that peaks in mid-century, and then decreases. Modifications to A1 include the *A1B storyline*, which assumes future technology provides a balance in energy sources. The *A2 storyline* describes a continuously increasing global population and economic growth that is more regionally oriented; that is, with less market integration and trade among countries. Population growth for the *B1 storyline* is the same as A1, but the economic future for B1 is rapid change toward a service and information economy, with a strong emphasis on clean and resource-efficient technologies. The *B2 storyline* describes a growing population and intermediate economic growth, but a preference for national solutions over global integration. Another limitation of these climate models is that their scale is on the order of 50–150 km, making "downscaling" a temperature or precipitation projection to local and regional areas neither trivial nor error free; some of these issues have been addressed by ensembling techniques and various statistical or dynamic downscaling approaches (Winkler et al. 2011a).

In this chapter, we describe and provide land-use and climate-change projections using methods developed for the Southern Forest Futures Project (SFFP; Wear and Greis 2012, in press), as well as climate change projections using a climate model ensemble and a statistical downscaling approach. Most subsequent chapters utilize a set of discrete futures described by the SFFP, some use model averages provided by the ensembles, while a few use both sources. Projections served two purposes: (1) they characterize a range of possible land use and climate futures for the Southern United States; and (2) they provide a common frame of reference and common variables for chapter authors to use in their interpretations, modeling, and other analyses.

DEVELOPING THE ENSEMBLE OF CLIMATE MODELS

Monthly temperature and precipitation predictions for 2000–2060 were created for each county in 17 states (Alabama, Arkansas, Delaware, Florida, Georgia, Kentucky, Louisiana, Maryland, Missouri, Mississippi, North Carolina, Oklahoma, South Carolina, Tennessee, Texas, Virginia, and West Virginia). Each monthly prediction is an ensemble, or average, of multiple simulations from four climate models—a subset of models described in Maurer et al. (2007): CGCM3 (Flato and Boer 2001), CCSM3 (Collins et al. 2006), HadCM3 (Gordon et al. 2000), and GFDLCM2.1 (Delworth et al. 2006). Individual ensembles were created for three IPCC greenhouse gas emissions storylines: B1, A1B, and A2.

Global climate model output from the CMIP3 multi-model dataset (Meehl et al. 2007) were downscaled as described by Maurer et al. (2007). The downscaling used a bias-correction/spatial-downscaling method (Wood et al. 2004), which generated gridded fields of precipitation and surface air temperature over the conterminous United States and portions of Canada and Mexico using quantile-mapping of observations from 1950 to 1999 (Adam and Lettenmaier 2003) and an interpolation of monthly bias-corrected climate-model anomalies onto a fine-scale grid of historical climate data that produced a monthly time series at each 1/8-degree grid cell (~140 km^2 per grid cell). Quantile mapping corrects for model bias using the empirical transformation of Panofsky and Brier (1968). A correction function is calculated by matching the distributions of the simulated data with the observed data by quantile. This correction function is then used to correct the data in the validation period (e.g., Boé et al. 2007). The 1/8-degree downscaled output was averaged for all simulations for each combination of a given climate model and emissions storyline. The resulting averages for the four climate models were then averaged for each emissions storyline. Finally, the resultant ensemble 1/8-degree fields were used to generate county-level data using nearest-neighbor interpolation. Changes in monthly temperature and precipitation were calculated by differencing the 2001 to 2010 period from the 2051 to 2060 period.

Statistical downscaling allows production of long periods of climate scenarios for multiple locations in a manner much more computationally efficient than using nested high-resolution models (i.e., dynamical downscaling). However, statistical downscaling approaches have several common limitations. These limitations include: (1) underestimation of extremes unless approaches are used to adjust the variance, (2) the assumption of stationarity between large-scale and local climate in the future, and (3) the creation of individual climate predictants (i.e., temperature, humidity, precipitation) that may be physically inconsistent with one another (Winkler et al. 2011b). The users must remain vigilant to the potential effects of these limitations on the results of statistical downscaling.

DEVELOPING THE CORNERSTONE FUTURES

We examined alternative scenarios affecting southern forests developed for the SFFP (Wear and Greis 2012, in press) to provide a mechanism for considering and preparing for changes in southern forests and the benefits they provide. Because any single projection of global (or regional) biological, physical, and social systems has a high probability of being incorrect, the SFFP constructed a range of possibilities that describe the forces influencing forests. This set of six future scenarios, labeled "Cornerstone Futures," provides a foundation for many of the analyses in this book. Each Cornerstone Future is a comprehensive and coherent (internally consistent) combination of varying climatic, demographic, and economic changes in the South; simulating these changes allowed us to forecast likely impacts on the amount and characteristics of forests. In this section, we describe the land-use and forest forecasts that are coupled with climate, demographic, and economic changes as a foundation for analysis. Forecasts were derived from simulations using the assessment system (USFAS) developed originally for the Resources Planning Act (RPA) 2010 Assessment (U.S. Department of Agriculture Forest Service 2012b) and provided forecasts based on broad economic, timber market, demographic, and climate futures (Wear and Greis in press).

The USFAS provided the modeling framework for our analysis and we considered driving factors in terms of the inputs from this model (Figure 2.1). Market futures were driven by price forecasts, with prices increasing as timber products became scarcer and decreasing as supply increased. Land-use models within the USFAS were driven by population and income projections. The effects of climate variables on projections of forest conditions were manifest in forest-type distributions and forest productivity. Although based on IPCC storylines, the RPA scenarios also contain data and detail relevant to conditions in the United States—specifically climate and socioeconomic

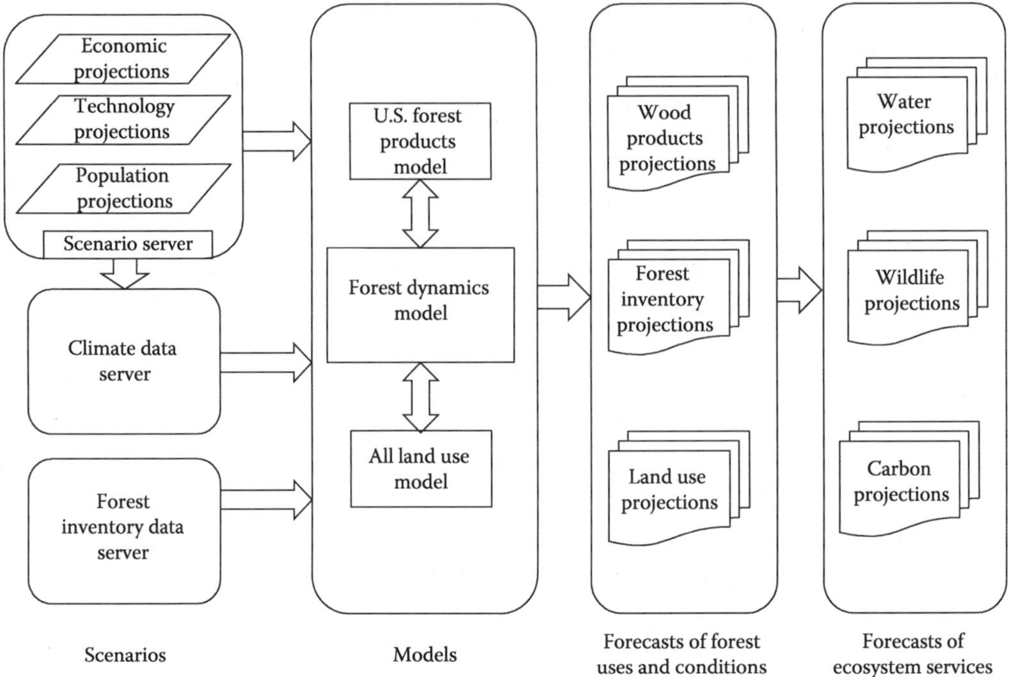

FIGURE 2.1 General schematic of the U.S. Forest Assessment System (USFAS).

projections, downscaled to the county level (U.S. Department of Agriculture Forest Service 2012a). Because the USFAS had originally been designed to develop the RPA assessment, we began by evaluating whether the existing RPA scenarios were adequate for addressing the issues of the SFFP. Note that the RPA adopted a version of A1 (A1B) because of its assumptions about energy futures (high demand for woody biomass) and dropped B1 because of data compatibility issues; A1B, A2, and B2 were evaluated.

For each of the IPCC storylines, the 2010 RPA scenarios provide unique forecasts of population and economic growth, downscaled to U.S. counties through 2060. The storylines were used as input to emissions modeling systems; their outputs were used as inputs to run models that generated alternative climate forecasts. The RPA analyses paired the results from three climate models with the storylines, resulting in nine potential climate futures. The climate-model data were downscaled using a statistical approach to the 0.5 arc minute and then aggregated to the county scale (Coulson et al. 2010). For the A1B and A2 storylines, the climate models are MIROC3.2, CSIRO MK3.5, and CGCM3; for the B2 storylines, they are the HadCM3, CSIROMK2, and CGCM2.

The climate-model projections were generated as changes from simulated historical monthly averages (1961 to 1990) for each climate model (Joyce et al. in press; Price et al. 2011). Basing forecasts on the simulated values was intended to correct for differential biases of the climate models at the fine scale used for the analysis. Change forecasts were then applied to historical data to generate projections. Monthly average daily maximum and minimum temperatures, precipitation, along with other variables not considered in the SFFP analysis were generated to 2100 (compared to 2060 for the SFFP).

To summarize, the RPA scenarios provided a set of climate futures defined by three general circulation models for each of three storylines. Within USFAS, land-use models defined changes in forest area (and other uses) produced by the economic variables from the storylines as well as timber and crop prices (Wear 2011). Forest-dynamics models forecasted changes in forest conditions produced by harvesting, which in turn was influenced by the economic variables (Polyakov et al. 2010), and changes associated with aging, disturbance, and climate (Wear et al. 2013).

Two other important forces of change—forest harvesting and the potential impacts of wood-based energy on other wood products markets—were also evaluated. The USFAS addressed alternative scenarios for wood production in two ways. The first approach was to apply alternative projections of prices for forest products to forecast changes in harvesting. This "price exogenous" approach, although simple, allowed the USFAS to simulate increasing and decreasing scarcity in markets without specifically addressing market dynamics and wood-products demand structures. The second approach was to incorporate explicit models of market demands for various forest products, which for the RPA scenarios was the forest-products model that was developed by Ince et al. (2011) to simulate specific market futures associated with bioenergy/wood products markets for the A1B, A2, B2, and other storylines. The forest-products model was chosen for the RPA scenarios because it can incorporate demands for all classes of U.S. wood products within a global marketing framework (Raunikar et al. 2010) and can be driven by variables taken from the same storylines.

In developing the Cornerstone Futures, we used exogenous price forecasts, in particular, three "price exogenous" scenarios: constant timber prices, increasing prices (plus 1% per year), and decreasing prices (minus 1% per year). Increasing prices reflect an increasing scarcity of timber products and therefore can be applied to two possible futures: a shortage in available timber supplies or an increased demand to satisfy existing uses or emerging uses such as bioenergy. Decreasing prices reflect decreasing scarcity consistent with a contraction in demands for products (such as pulpwood for paper production) or a rapid expansion in supplies derived from intensive management. We used the 1% increase and decrease rates (Figure 2.2) to bookend the analysis of markets because they are consistent with real price growth over the expansionary phase of southern timber markets from the 1980s through the 1990s. This approach allowed us to examine a broad range of market futures without explicit consideration of sources of demand growth or decline.

For our initial set of Cornerstone Futures, therefore, we started with the nine RPA scenarios for forecasts of climatic and socioeconomic conditions, and then applied the three alternative timber market scenarios. None of the resulting 27 initial possibilities was considered more likely than the other—they all were considered plausible. Each likely contained some unique insights into future resource uses and conditions; nevertheless, in the interest of practicality and clarity, we selected a smaller subset for detailed analysis and discussion as well as for use in the chapters of this book.

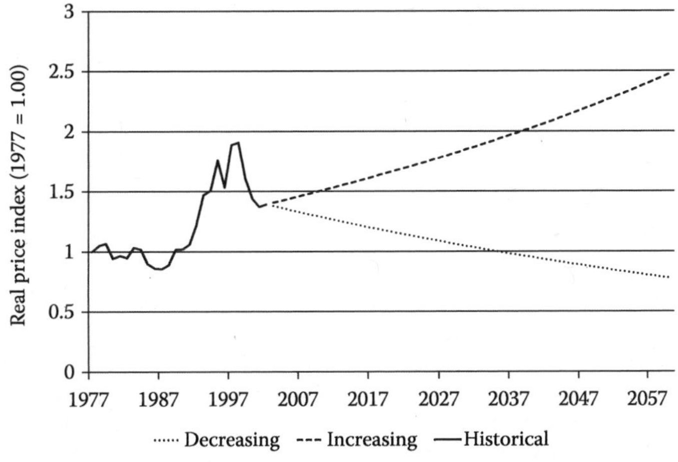

FIGURE 2.2 Historical and projected real price index for timber products in the southern United States. (From Timber Mart South for historical data, 2010 Resources Planning Act assessment total product output charts.)

Final Set of Futures

To begin defining the set of Cornerstone Futures to be used for detailed analysis, we conducted USFAS land-use, forest-condition, and timber-harvesting simulations for the initial set of 27 alternative futures and applied measures of total forest land, biomass in forests, and other variables to compare the resulting forecasts. This process was complicated because the alternative futures had different rankings based on which variable was used to construct the ranking—for example, the same future might forecast the biggest loss of forest land and only a moderate level of future biomass loss (estimated as growing stock volumes).

Timber prices—We started by dropping the set of futures with constant prices. In every comparison, these futures yielded forecasts that were intermediate between futures with increasing and decreasing prices, meaning that the increasing and decreasing price futures bracketed the constant price futures for all variables evaluated. This step reduced the number of alternative futures to 18.

Biomass volume and land use—We next used two highly aggregate metrics to compare forecasts across the remaining futures: total volume of biomass by broad forest types and total area of forest land (Wear et al. in press).

Total inventory volume follows a broad range of trajectories across futures. Increases are only expected for the B2 storyline with low prices (resulting in lower harvesting), which predicts the lowest urbanization (relatively low population growth and moderate income growth). All other futures would result in biomass expansion through 2030 or 2040, followed by decreases. The future with the lowest biomass in 2060 is predicted using the A1B storyline (moderate population growth and high income growth) combined with high timber prices. Softwood volumes would increase to 2030 but then either level off (A1B/high prices) or increase through 2060, with the highest rate of increase for B2/low prices. Hardwood volumes are projected to decrease after 2030 for all futures except B2/low prices, with the largest decrease for A1B/high prices.

Forest area is forecasted to decrease in response to the economic/population forecasts from all the storylines and both the timber price futures (by construction, land use is not directly responsive to climate). Low population and income growth reduces urbanization and conversion of forest land. In addition, high timber prices discourage deforestation. Therefore, the B2 storyline (moderate income growth) coupled with high prices would yield the smallest loss of forest land by 2060, and the A1B storyline (rapid economic growth) coupled with low prices would yield the largest loss. With the storyline held constant, low prices would yield more forest loss than high prices. Because the A2 storyline is intermediate to the A1B/high forest loss and B2/low forest loss and is also intermediate in the biomass volume forecasts (for all climate projections), it was dropped from the analysis.

The range of outcomes for these two variables suggested inclusion of four futures for consideration. A high economic growth/increasing timber price future (A1B/high price) and a low growth/decreasing price future (B2/low price) bracket the projections of total forest biomass. For forest area change projections, the brackets are a low economic growth/increasing timber price future (B2/high price), which could reflect less globalization (more isolated nation economies) and increasing U.S. scarcity of wood products in the face of less trade; and a high growth/decreasing price (A1B/low price) future, which could reflect a shift in timber production offshore to support global economic growth (or simply a decline in the demand for forest products). These became the initial set of four Cornerstone Futures.

Although the timber-price and storyline effects overshadowed the effects of climate variation (Wear et al. in press), we decided not to eliminate climate variation from consideration because we wanted to account for spatial variations that may be masked by the aggregate outcomes. Accordingly, we introduced climate variation by assigning a climate model to each of the four cornerstones identified previously: MIROC3.2 to A1B/high prices, HadCM3 to the B2/low prices, CSIROMK3.5 to the A1B/low prices, and CSIROMK2 to the B2/high prices. These climate models were selected to provide a variety of spatial expressions of the climate projections (MIROC3.2 A1B, for example, is generally warmer than the others, but these differences do not significantly affect the projections of aggregate forest conditions).

A review of these four Cornerstone Futures indicated that forest-investment forecasts, a key element in the development of southern forests since the 1950s, failed to respond with fluctuations in market futures. We therefore expanded our scope to address the effects of forest planting in a way that was consistent with our modeled changes in timber markets. To accomplish this, we augmented USFAS with a simple model that assumes replanting of all current plantations as well as a specified portion of other harvested forests. These assumptions were derived from historical rates of planting for each of the 13 states and from consultations with professional foresters on the likely path of future planting. The planting rates adopted for these baseline assumptions are more moderate than the aggressive expansion of plantations in the 1990s, and more closely reflect the economic conditions and trends of the 2000s. Because planting rates are tied to harvesting (which controls the availability of forests for planting) and to land-use changes, the area projected for planting was assumed to vary somewhat across simulated futures. We adjusted this baseline projection approach to introduce broader variation in the planting rate and to reflect the assumptions about future timber markets. The result was two additional futures. For the first, we increased planting rates in harvested areas by 50% from base rates for the MIROC3.2-A1B/high price future, therefore yielding higher planting rates than the other Cornerstone Futures but not as high as was experienced in the 1990s—we concluded that this would be plausible in light of observed nursery capacity and forest management. For the second, we decreased planting rates by 50% from base rates for the HadCM3-B2/low price future, therefore yielding a very moderate increase in forest plantations that would level off after about 2030.

Summary

Cornerstones A through D are defined by the matrix formed by intersecting low and high population and income forecasts with increasing and decreasing timber price futures as described in the preceding:

- *Cornerstone A:* High population/income growth along with increasing timber prices and baseline rates of post-harvest tree planting.
- *Cornerstone B:* High population/income growth along with decreasing timber prices and baseline rates of post-harvest tree planting.
- *Cornerstone C:* Low population/income growth along with increasing timber prices and baseline rates of post-harvest tree planting.
- *Cornerstone D:* Low population/income growth along with decreasing timber prices and baseline rates of post-harvest tree planting.

In developing these four Cornerstone Futures, we used baseline rates of tree planting from data collected by the Forest Inventory and Analysis (FIA) program of the U.S. Forest Service. Two other Cornerstones were developed that either augment planting rates by 50% of Cornerstone A (Cornerstone E), where economic growth is strong and timber markets are expanding; or by decreasing rates by 50% of Cornerstone D (Cornerstone F), where economic growth is reduced and timber markets are declining.

- *Cornerstone E:* High population/income growth along with increasing timber prices and high tree planting rates.
- *Cornerstone F:* Low population/income growth along with decreasing timber prices and low tree planting rates.

The six Cornerstone Futures define a broad range of potential future conditions within which forests might develop. They describe a range of futures identified as important driving factors—wood products markets, bioenergy, land uses, and climate changes—in a process that began with solicitation of input in a series of public meetings and ended with refinement by a panel of recognized natural resource experts. They address bioenergy and wood-products markets in a qualitative fashion—through exogenously defined trajectories of timber prices—that captures a broad range of market conditions. And they address land use and climate change in a detailed and spatially explicitly way through projections of population, income, temperatures, and precipitation downscaled to the county level. The next step in the analysis was to evaluate these Cornerstone Futures using the models contained in the U.S. Forest Assessment System.

LAND-USE MODELS

Land-use patterns define both the extent of human presence on a landscape and the ability of land to provide a full range of ecosystem services. The future sustainability of forests in the South has been and will continue to be largely influenced by the dynamics of land use. And as the regional population grows, so too will the area of developed uses. The pattern of these developments, returns from the various products of rural land, and the inherent productivity of the land will determine the distribution of forest, crop, and other rural land uses—therein lies the structure and function of terrestrial ecosystems (Chen et al. 2006; Wear 2002).

The USFAS forecasts land use based on the RPA econometric land-use models developed by Wear (2011) to reflect variations in land-use patterns and biophysical capabilities across the U.S. regions. The land-use model for the South addresses changes in all uses of land for all of the 13 states in the SFFP analysis area (with the exception of central and western Texas and Oklahoma, where the land-use model developed for the Rocky Mountain/Great Plains region was applied) and is driven by population and income growth along with the prices of timber products.

Each land-use model has two major components: (1) changes in county-level population and personal income, which are used to simulate future urbanization; and (2) allocations of rural land among competing uses, which likely result from predicted urbanization and rural land rents. Output from both components is based on land-use data from the 1987 and 1997 National Resource Inventories (U.S. Department of Agriculture Natural Resources Conservation Service 2009) to ensure that forecasted land-use changes are generally consistent with observed urbanization intensities and rural-land-use changes (see Wear 2011).

We examine how land use could respond to the economic and population forecasts associated with the Cornerstone Futures. Our forecasts use empirical models to address the Cornerstone Futures and to examine some specific questions about alternative land-use futures. Land-use forecasts play a central role in USFAS (Wear in press), with the information developed in this chapter providing one of the inputs to the USFAS forest-dynamics model, which in turn generates forecasts of southern forest conditions, which are described subsequently.

FOREST CONDITIONS MODELS

Within the forest-dynamics model of the USFAS (Figure 2.1), the future of every plot in the FIA inventory was projected in a multiple stage process (Wear et al. 2013). The harvest-choice model assigned a management intensity choice (no harvest, partial harvest, or final harvest) based on timber prices (from the forest products module) and then on condition of the plot (Polyakov et al. 2010). The age of each plot was determined for the next period, and if harvested, the plot was determined to be naturally regenerated or planted. Forecasted climate (including temperature and precipitation) was assigned and forest conditions on the plot were inferred based on the harvest/no harvest decision, age, and climate selection.

Planting probabilities were derived from the frequency of observed planting on harvested plots, calculated for each forest type in each state using the most recent inventory period and were adjusted to reflect the assumptions of each Cornerstone Future as described earlier.

To incorporate forest market data, we adopted a simplified process that specifies a price trajectory for each Cornerstone Future and provides input both on individual plot-harvesting decisions and on the overall supply of timber. Consistent with theory, higher prices yielded more harvesting and larger timber supplies; lower prices yielded smaller timber supplies. Harvest-choice models, based on empirical analysis, were consistent with harvesting behavior observed in the late 1990s and early 2000s.

Projections of forest area from the land-use module also fed into the forest dynamics module and the projection of future market conditions. Changes forecasted at the county level for nonfederal land were based on National Resource Inventory data (U.S. Department of Agriculture 2009) and were used to rescale the area represented by each plot in the county.

The assignment of future plot conditions involved a resampling of historical plot records called whole-plot imputation (Wear et al. 2013): selecting a historical plot with comparable conditions to represent the conditions of the future plot location. The selected historical plot was as close as possible to the original plot location to allow for orderly changes in conditions. For example, if plot conditions were forecasted to be warmer, the resampling algorithm would first look within the same survey unit to find a historical plot with similar temperature increases. Finding none, the algorithm would extend the search to adjacent units until an appropriate match is found. This process was repeated for every time step (or interval between measurements or projections) to generate plot forecasts over time. The inventory forecast was completed by coupling the plot forecast with the land-use forecasts, which were applied to adjust the area represented by each nonfederal plot within each county (through the plot expansion factor described in the previous paragraph).

The forest-dynamics model was based on several probabilities, including probabilistic harvest-choice, forest-investment, and forest-transition models that were implemented with random draws from probability distributions, and a whole-plot imputation that was based on a random selection of a subset of historical plots (with replacement). This meant that forecasts could vary from one run of the forest-dynamics model to the next. The forecasts for the 50-year simulation in this chapter are based on 26 runs, one of which was selected as representative based on the central tendency across several variables. The full suite of 26 runs, providing information about the uncertainty of the forecasts, was used whenever confidence intervals were needed.

The time step of the simulation depended on the FIA inventories that underlay much of the modeling (Huggett et al. in press). Because starting years and time step varied from state to state, the forecast periods were staggered across the region; for example, a time step of 6 years starting in 2007 for one state compared to 6 years starting in 2008 for its neighbor. For reporting across the region, we selected 10-year intervals, each beginning with a zero-ending year (e.g., 1990), and then attached forecasts from the nearest year to the referenced decade—identical to how FIA assigns years for aggregate inventories (Smith et al. 2009).

VARIATION ACROSS THE CORNERSTONE FUTURES

Figure 2.3 shows the six Cornerstone Futures in a diagram that emphasizes their key variables. Cornerstones A through D are defined by the matrix formed by intersecting RPA/IPCC storylines A1B and B2 with increasing and decreasing timber price futures. Cornerstones E and F depart from these four by either augmenting the planting rates in Cornerstone A (E) or by decreasing the planting rates in Cornerstone D (F).

Storylines vary in their projections of population density. A2, the storyline not used within the Cornerstone Futures, yields the highest population growth with an 80% increase from 2006 to 2060. The lowest population growth is associated with B2 (40%) and A1B is bracketed by the two (60%). Because urbanization is also fueled by income levels, A1B, with its strong economic growth, actually results in the most urbanization and highest losses of forest area (described in the following); B2 results in the lowest urbanization and forest losses.

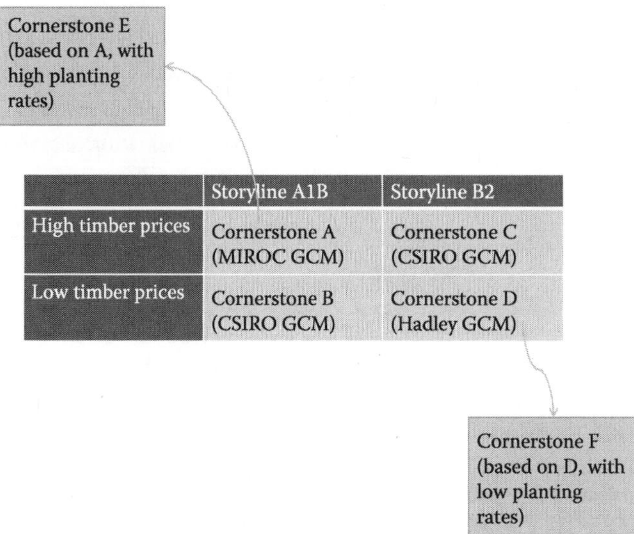

FIGURE 2.3 The six Cornerstone Futures defined by permutations of storylines from the 2010 Resources Planning Act (RPA) assessment, three general circulation models (GCMs), and two timber price futures; and then expanded by evaluating increased and decreased forest planting rates.

Historically, population density has varied across the five subregions of the South (Figure 2.4). In 2006, the Piedmont had the highest population density (about 250 people per square mile or ppsm), followed by the Coastal Plain and Appalachian-Cumberland highlands with intermediate densities (100–150 ppsm) and the Mid-South and Mississippi Alluvial Valley with the lowest densities (75 ppsm). This general trend would continue over the A1B storyline projection period, with growth

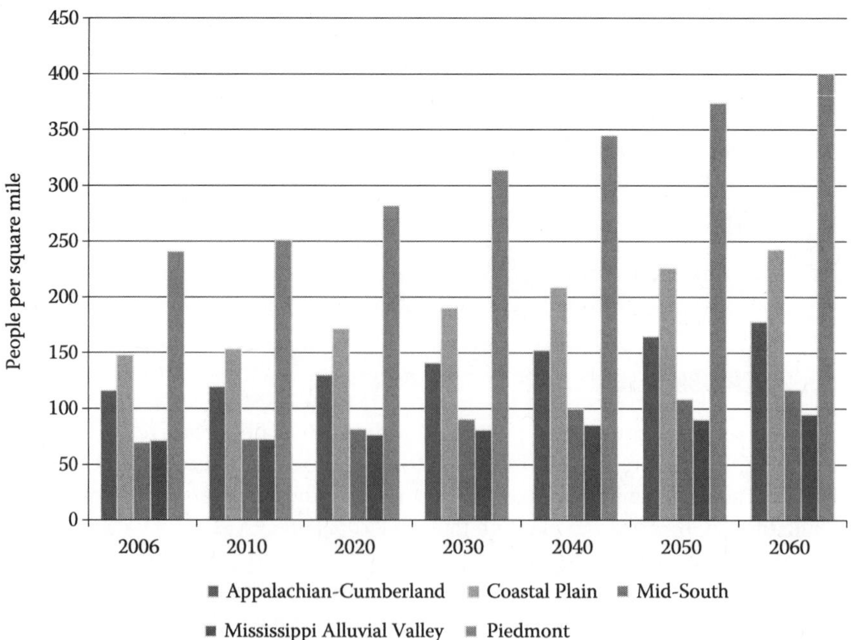

FIGURE 2.4 Projections of population density, 2006–2060, for the five subregions of the Southern United States under storyline A1B (low population growth, high economic growth, high energy use) from the 2010 Resources Planning Act (RPA) assessment.

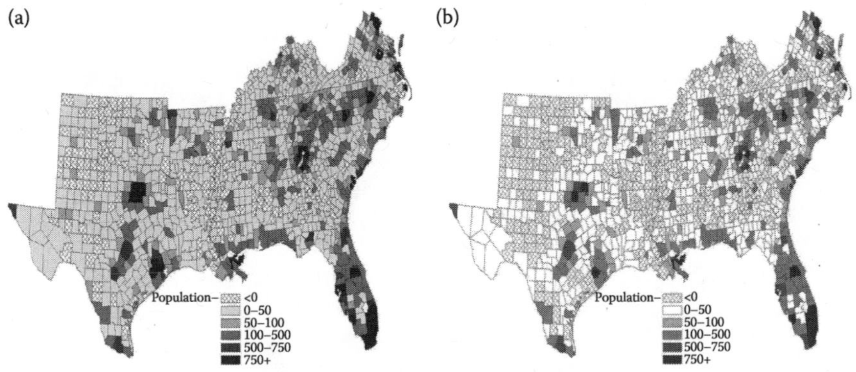

FIGURE 2.5 Projection of population change in the Southern United States, 2010–2060, assuming the (a) A1B storyline of low population growth, high economic growth, high energy use; and (b) B2 storyline of moderate growth and energy use from the 2010 Resources Planning Act (RPA) assessment. Note that counties marked with a cross-hatch have forecasted population losses.

strongest on the Piedmont (an additional 150 ppsm). By 2060, the population density in the Coastal Plain would be as high as current population densities in the Piedmont.

Even within the subregions, population change would not be evenly spread. Forecasted population growth from 2006 and 2060 (Figure 2.5a,b) shows that several areas are expected to experience population decreases. This includes parts of the High Plains in Texas and Oklahoma, much of the Mississippi Alluvial Valley, and parts of southern Alabama and Mississippi. Population growth in the South would continue to be organized around major metropolitan centers, especially Atlanta, Miami, Houston, Dallas, Washington, Nashville (TN), Charlotte (NC), and Raleigh (NC).

Cornerstone Futures were also framed by timber market projections over the next 50 years. These projections did not account for short-run business cycles or the pattern of economic recovery from the recent recession, instead attempting to capture some broader long-run potentials for market development. Price forecasts defined by the Cornerstone Futures anticipated an orderly progression, either increasing or decreasing in real terms at 1% per year from a 2005 base. That year, prices were below their peak values from the late 1990s, especially for pulpwood-sized material.

We also held the real returns to agricultural crops constant through the period. Because future markets could depart from these assumptions, we used additional analyses to examine the sensitivity of future forest conditions to general market conditions (described in the preceding), providing a framework for evaluating forest product/bioenergy market possibilities.

OUTLOOK FOR CLIMATE

Output from Ensembles

Climate projections showed that January temperatures would increase by no more than 0.5°C under the B1 ensemble, but by more than 2°C under A1B and A2. January precipitation would decrease by up to 1 mm/day in the lower Mississippi Alluvial Valley and Tennessee Valley under the B1 and A1B ensembles, with increased precipitation to the west and east. Under the A2 ensemble, the largest increase in precipitation would occur over Kentucky and Tennessee, with reduced precipitation to the west and south.

Under the A2 ensemble, July temperatures would increase as much as 3°C from western Tennessee to Oklahoma and western Texas (Figure 2.6a–c). The A1B ensemble shows temperature increases in excess of 3.5°C in the Mississippi Alluvial Valley (Figure 2.6a–c), compared to >2°C

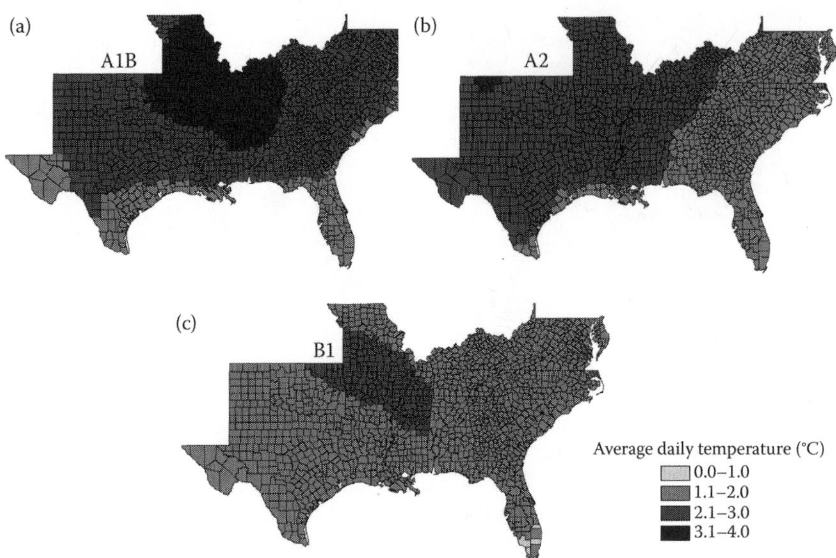

FIGURE 2.6 Difference in July temperature in the Southern United States, comparing the 2051–2060 period to the 2001–2010 period under the (a) A1B storyline of low population growth, high economic growth, high-energy use; (b) A2 storyline of continuous population growth and high economic growth; and (c) B1 storyline of low population growth, high economic growth, and shift to a service/information economy. (Adapted from Intergovernmental Panel on Climate Change 2007b. Climate Change 2007: The Physical Science Basis. Contribution of Working Group I to the Fourth Assessment Report of the Intergovernmental Panel on Climate Change. Cambridge University Press, Cambridge, UK.)

temperature increases under the B1 ensemble (13°C). Under all three storylines, nearly all areas of the South are expected to increase by ≥1°C.

July precipitation is projected to increase from Georgia northward to Virginia, Maryland, and Delaware, ranging from 0.6 mm/day under the A2 ensemble (Figure 2.7a–c) to 0.8 mm/day under the A1B ensemble (Figure 2.7a–c). From 2051 to 2060, the ensembles predict drier conditions from the Appalachians westward and in southern Florida (≥0.8 mm/day).

OUTPUT FROM CORNERSTONE FUTURES

Figures 2.8 through 2.10 present graphic and map displays of precipitation data, and Figures 2.11 through 2.13 present graphic and map displays of temperature data from the Southern Forest Futures Project (McNulty et al. in press). Characterized by low population growth and high energy use/economic growth (MIROC3.2 A1B), Cornerstone A is forecasted to be dry and hot, with average annual precipitation of 912 mm and average annual temperature of 20.22°C. Annual precipitation expected for any southern county ranges from 103 to 4999 mm, and temperature ranges from –12.01°C to 50.24°C. Also characterized by low population growth and high energy use/economic growth (CSIROMK3.5 A1B), Cornerstone B is forecasted to be wet and warm, with average annual precipitation of 1167 mm and average temperature of 19.06°C. Annual precipitation expected for southern counties ranges from 93 to 3912 mm, and temperature ranges from –11.21°C to 44.24°C. Characterized by moderate population/income growth and energy use (CSIROMK2 B2), Cornerstone C is forecasted to be moderate and warm, with average annual precipitation of 1083 mm and average annual temperature of 19.45°C. Annual precipitation expected for any southern county ranges from 35 to 2641 mm, which would break the 1956 regional low of 42 mm in Texas (Burt 2007). Temperature is expected to range from –19.73°C to 45.39°C. Also characterized by moderate population/income growth and energy use (HadCM3 B2), Cornerstone D is also forecasted to be moderate and warm, with average

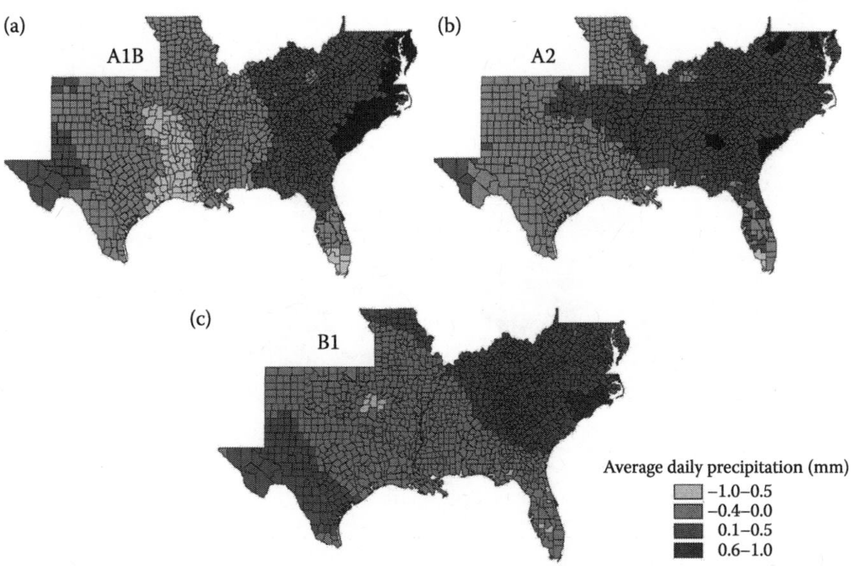

FIGURE 2.7 Difference in July precipitation in the Southern United States, comparing the 2051–2060 period to the 2001–2010 period under the (a) A1B storyline of low population growth, high economic growth, high energy use; (b) A2 storyline of continuous population growth and high economic growth; and (c) B1 storyline of low population growth, high economic growth, and shift to a service/information economy. (Adapted from Intergovernmental Panel on Climate Change 2007b. Climate Change 2007: The Physical Science Basis. Contribution of Working Group I to the Fourth Assessment Report of the Intergovernmental Panel on Climate Change. Cambridge University Press, Cambridge, UK.)

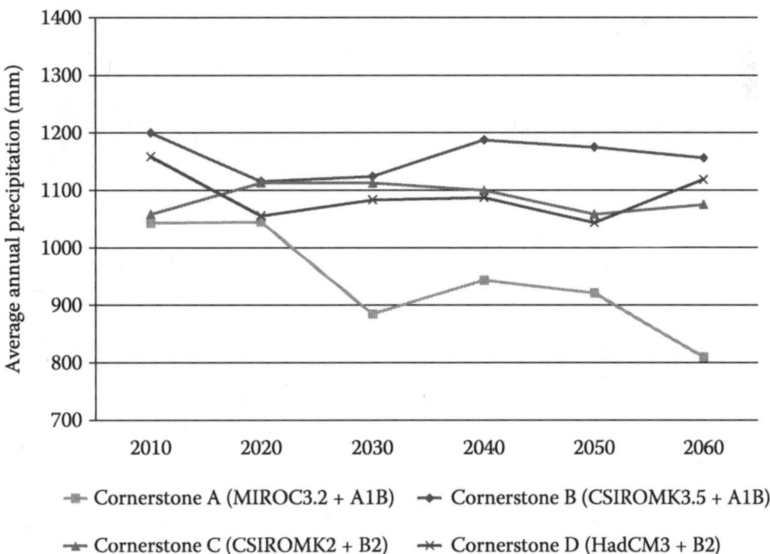

FIGURE 2.8 Predicted annual precipitation (2010, 2020, 2030, 2040, 2050, and 2060) for the Southern United States as forecasted by four Cornerstone Futures (A through D), each of which represents a general circulation model paired with one of two emissions storylines—A1B representing low population/high economic growth, high energy use, and B2 representing moderate growth and use. (Adapted from Intergovernmental Panel on Climate Change 2007b. Climate Change 2007: The Physical Science Basis. Contribution of Working Group I to the Fourth Assessment Report of the Intergovernmental Panel on Climate Change. Cambridge University Press, Cambridge, UK.)

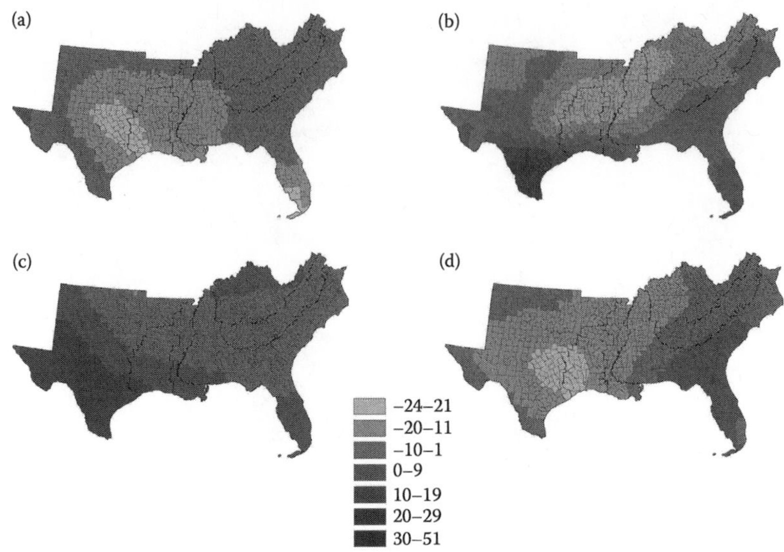

FIGURE 2.9 Predicted change in precipitation from 2010 to 2050 for the Southern United States as forecasted by four Cornerstone Futures, each of which represents a general circulation model paired with one of two emissions storylines—A1B representing low population/high economic growth, high energy use, and B2 representing moderate growth and use: (a) MIROC3.2 A1B, (b) CSIROMK3.5 A1B, (c) CSIROMK2 B2, and (d) HadCM3 B2. (Adapted from Intergovernmental Panel on Climate Change 2007b. Climate Change 2007: The Physical Science Basis. Contribution of Working Group I to the Fourth Assessment Report of the Intergovernmental Panel on Climate Change. Cambridge University Press, Cambridge, UK.)

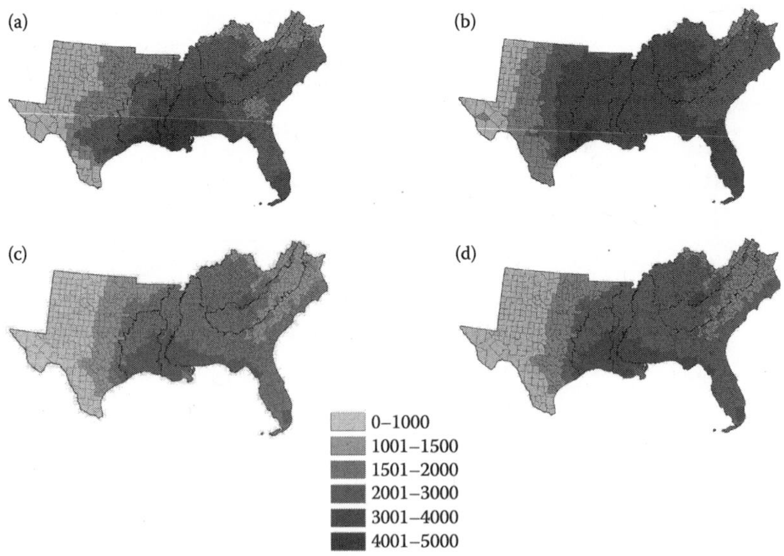

FIGURE 2.10 Maximum precipitation from 2010 to 2060 for the Southern United States as forecasted by four Cornerstone Futures, each of which represents a general circulation model paired with one of two emissions storylines—A1B representing low population/high economic growth, high energy use, and B2 representing moderate growth and use: (a) MIROC3.2 A1B, (b) CSIROMK3.5 A1B, (c) CSIROMK2 B2, and (d) HadCM3 B2. (Adapted from Intergovernmental Panel on Climate Change 2007b. Climate Change 2007: The Physical Science Basis. Contribution of Working Group I to the Fourth Assessment Report of the Intergovernmental Panel on Climate Change. Cambridge University Press, Cambridge, UK.)

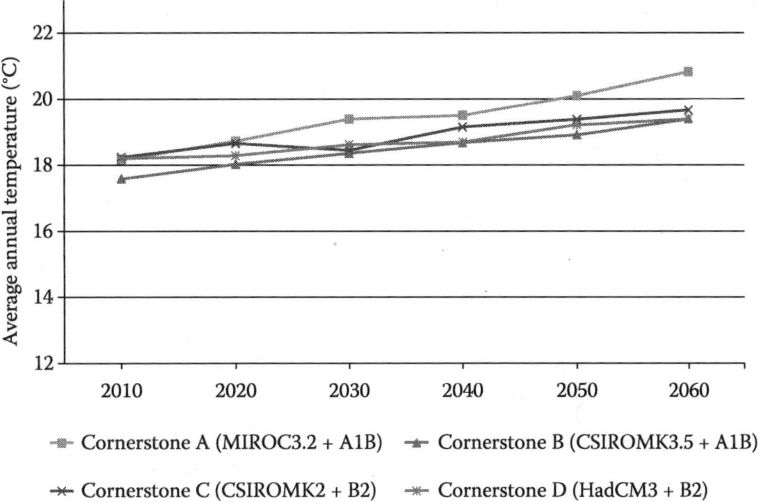

FIGURE 2.11 Predicted annual air temperature (2010, 2020, 2030, 2040, 2050, and 2060) for the Southern United States as forecasted by four Cornerstone Futures (A through D), each of which represents a general circulation model paired with one of two emissions storylines—A1B representing low population/high economic growth, high energy use, and B2 representing moderate growth and use. (Adapted from Intergovernmental Panel on Climate Change 2007b. Climate Change 2007: The Physical Science Basis. Contribution of Working Group I to the Fourth Assessment Report of the Intergovernmental Panel on Climate Change. Cambridge University Press, Cambridge, UK.)

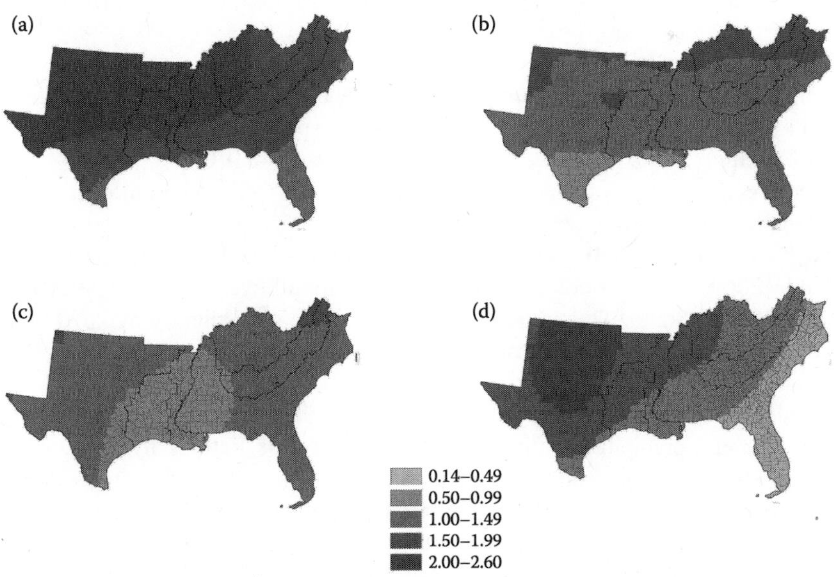

FIGURE 2.12 Predicted change in air temperature from 2010 to 2050 for the Southern United States as forecasted by four Cornerstone Futures, each of which represents a general circulation model paired with one of two emissions storylines—A1B representing low population/high economic growth, high energy use, and B2 representing moderate growth and use: (a) MIROC3.2 A1B, (b) CSIROMK3.5 A1B, (c) CSIROMK2 B2, and (d) HadCM3 B2. (Adapted from Intergovernmental Panel on Climate Change 2007b. Climate Change 2007: The Physical Science Basis. Contribution of Working Group I to the Fourth Assessment Report of the Intergovernmental Panel on Climate Change. Cambridge University Press, Cambridge, UK.)

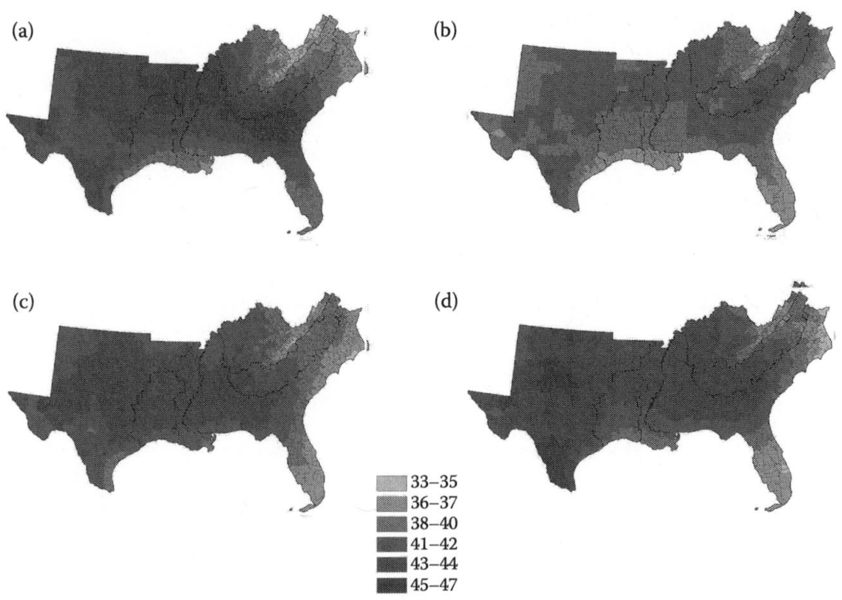

FIGURE 2.13 Maximum air temperature from 2010 to 2060 for the Southern United States as forecasted by four Cornerstone Futures, each of which represents a general circulation model paired with one of two emissions storylines—A1B representing low population/high economic growth, high energy use, and B2 representing moderate growth and use: (a) MIROC3.2 A1B, (b) CSIROMK3.5 A1B, (c) CSIROMK2 B2, and (d) HadCM3 B2. (Adapted from Intergovernmental Panel on Climate Change 2007b. Climate Change 2007: The Physical Science Basis. Contribution of Working Group I to the Fourth Assessment Report of the Intergovernmental Panel on Climate Change. Cambridge University Press, Cambridge, UK.)

annual precipitation of 1106 mm (higher than Cornerstone C) and average annual temperature of 19.3°C (lower than Cornerstone C). Annual precipitation expected for any southern county ranges from 102 to 2708 mm, and temperature ranges from –18.7°C to 48.0°C.

Forecasted precipitation and temperature averages are not likely to be uniform throughout the South, and significant variations expected across the five subregions. The high energy use/economic growth (MIROC3.2 A1B) under Cornerstone A is predicted to result in the least decadal precipitation by 2060, with an overall average of 810 mm for all five southern subregions and a low of 525 mm in the Mid-South. This trend is expected to abate only slightly by 2090 to an average of 858 mm for all subregions and 535 mm for the Mid-South; still much drier than the historical overall average of 1136 mm.

Although also based on high-energy-use/economic-growth, Cornerstone B (CSIROMK3.5 A1B) predicts more decadal precipitation than the other Cornerstone Futures by 2060, with an overall average of 1156 mm. This trend continues into 2090, with an overall average predicted to be 1223 mm. Cornerstone B also predicts cooler decadal temperatures than the other Cornerstone Futures by 2060—with an overall average of 19.4°C—for every subregion except the Mid-South. This trend continues into 2090, with an overall average of 20.1°C for Cornerstone B, lower than all the other cornerstones for all subregions.

Cornerstone A predicts warmer decadal temperatures than the other Cornerstone Futures by 2060, with an overall average of 20.8°C for all five southern subregions. This trend continues into 2090, with an overall average of 21.8°C, ahead of all the other cornerstones for all subregions.

Electing to use discrete Cornerstone Futures with different climate futures prevented us from isolating the effects of climate versus population, land use, or other driving forces behind the scenario. Evaluating these secondary and tertiary effects using the same factorial analysis that went into the development of the Cornerstone Futures, which were based on emissions storylines and climate models, was beyond the capacity of the Futures Project. However, forest forecasts for the full factorial were the basis for the selection of the Cornerstone Futures and are described in detail in Wear et al. (in press).

OUTLOOK FOR LAND USE

Land-use forecasts indicate a range of results for the Cornerstone Futures (Figure 2.14a–d). Urbanization would add between 29 and 42 million acres of developed uses by 2060, with losses of varying degrees accruing for all other land uses. The Cornerstone Futures are in general agreement about predicted changes for range and pasture use, but not for cropland and forest area. Predicted losses range from about 11 million acres (6.5%) to about 22 million acres (13.1%) for forest uses, and from about 5 million acres (6%) to about 16 million acres (19%) for cropland uses.

URBAN LAND USES

By model construction, urbanization forecasts are driven exclusively by population and income and are not influenced by the future trajectory of timber or agricultural prices. Cornerstones A and B (with the A1B storyline) predict the same higher rates of income and population growth. The result would be an expansion in urban uses of about 43 million acres (about 143%) by 2060 from the 1997

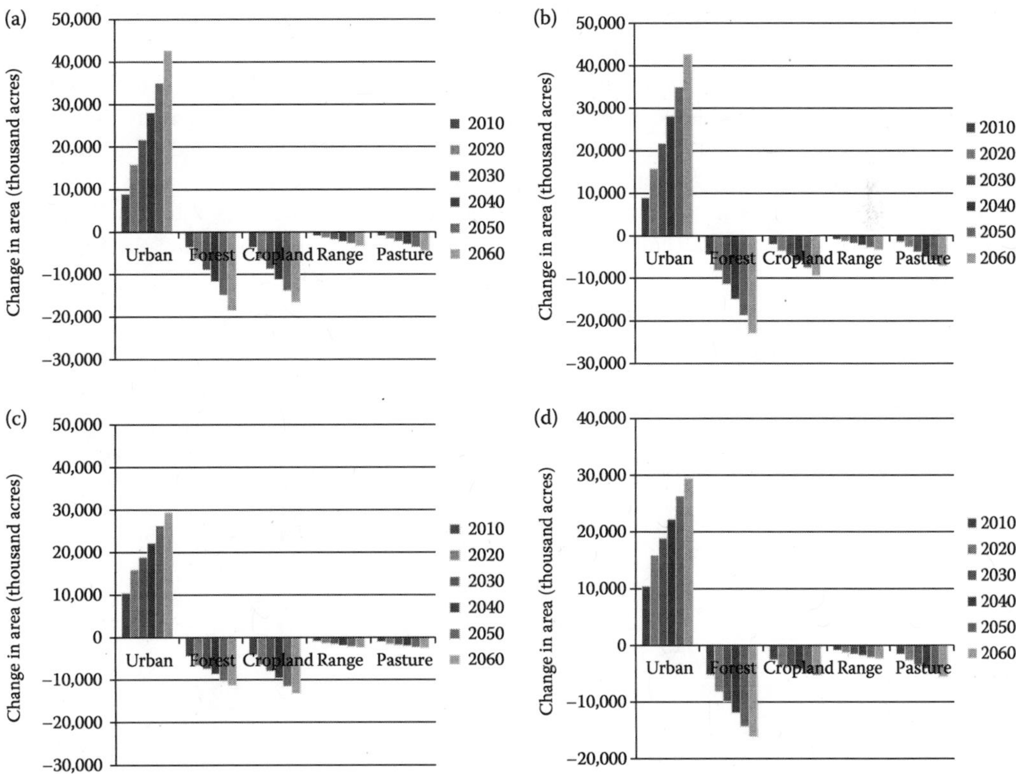

FIGURE 2.14 Changes in urban, forest, cropland, range, and pasture land uses in the Southern United States under each of four Cornerstone Futures: (a) large urbanization gains with increasing timber prices, (b) large urbanization gains with decreasing timber prices, (c) moderate urbanization gains with increasing timber prices, and (d) moderate urbanization gains with decreasing timber prices.

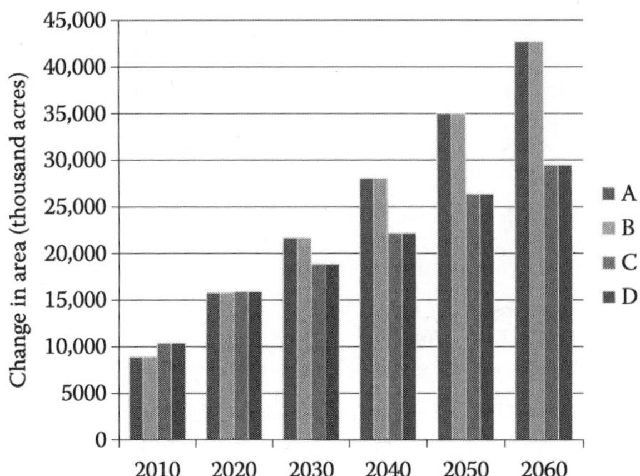

FIGURE 2.15 Change in urban land uses for the Southern United States, 1997–2060, under four Cornerstone Futures: (A) large urbanization gains with increasing timber prices; (B) large urbanization gains with decreasing timber prices; (C) moderate urbanization gains with increasing timber prices; and (D) moderate urbanization gains with decreasing timber prices.

base of about 30 million acres (Figure 2.15). Cornerstones C and D (with the B2 storyline) predict lower rates of income and population growth, with a resulting gain in urban uses of about 30 million acres (98%).

Urbanization is highest in areas experiencing the highest population growth; for the South, this is at the periphery of urban centers (Figure 2.16). For Cornerstones C and D, gains in urban uses are expected to be widespread, with the exception of a few areas of decreasing populations (such as the Mississippi Alluvial Valley and southwestern Alabama). For Cornerstones A and B, urbanization would spread out across an even broader area, highlighting the linkage to income.

The amount of urban growth is expected to vary across the five subregions of the South (Figure 2.17 for Cornerstones A and B). Under Cornerstones A and B, almost 18 million of the 43 million

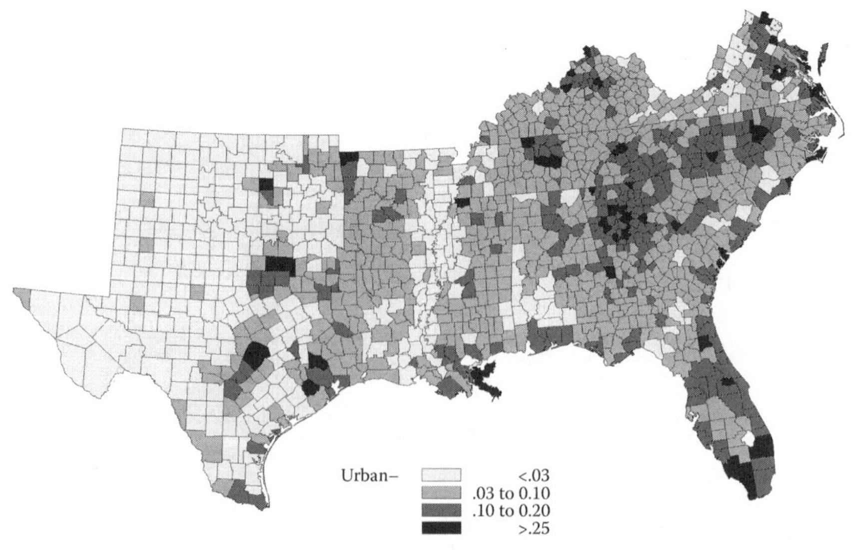

FIGURE 2.16 Changing urban land uses in the Southern United States, 1997–2060, based on an expectation of large urbanization gains with decreasing timber prices (Cornerstone C).

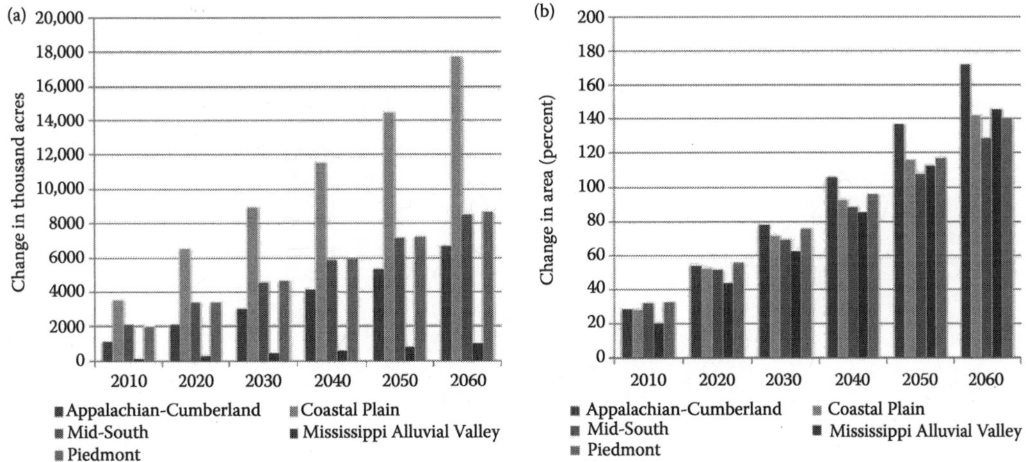

FIGURE 2.17 Changing urban area in the five subregions of the Southern United States, 1997–2060, expressed in (a) acres and (b) percent; and based on an expectation of large urbanization gains, either with increasing timber prices (Cornerstone A) or with decreasing timber prices (Cornerstone B).

acres of additional urban area would be on the Coastal Plain. The Piedmont and Mid-South would add about 9 million acres each, the Appalachian-Cumberland highlands would add about 7 million acres, and the Mississippi Alluvial Valley would have a comparatively small increase. The rate of growth would be highest for the Appalachian-Cumberland highlands (175%) because of its relatively small 1997 urban base; the fastest growing areas are central-northern Kentucky (an area bordered by Lexington, Louisville, and Cincinnati), and, in Tennessee, near Nashville and Knoxville. Growth rates for the other four subregions range from 125% to 140%.

Forest Land Uses

Unlike urban land uses, the amount of forest land in the South depends on timber prices as well as the more dominant population- and income-growth drivers of urbanization. All Cornerstone Futures predict losses, but with varying degrees of loss. The biggest loss is projected to be 23 million acres (13%) by 2060 for Cornerstone B, which is based on high economic growth (storyline A1B) and decreasing timber prices (Figure 2.18). At the other end of the spectrum is a projected loss of about 11 million acres (7%) for Cornerstone C, which is based on low economic growth (storyline B2) and increasing timber prices. Comparing forecasts for Cornerstones A and B with those for Cornerstones C and D shows a 5-million acre difference between a future of increasing timber prices and a future of decreasing prices, confirming that the effects of the economic/population storyline dominate the effects of timber prices.

Forest losses would be especially high in a few areas. For all Cornerstone Futures, losses are expected to be concentrated in the Piedmont from northern Georgia through North Carolina and into parts of Virginia, as Figure 2.19 shows for Cornerstone C (selected because it is bracketed by the other Cornerstone Futures). Other areas of concentrated forest losses would be the Atlantic Coast, along the Gulf of Mexico, and in parts of eastern Texas outside of Houston. The income-fueled development in Cornerstones A and B would spread low-intensity forest losses across a broader area (Figure 2.20).

Under Cornerstone B, forest losses are forecasted to be highest in the Coastal Plain, at about 12 million acres, and lowest in the Mississippi Alluvial Valley and the Mid-South (Figure 2.21a,b). Percentage losses would be highest in the Piedmont (21%), followed by the Appalachian-Cumberland highlands (13%) and the Coastal Plain (11%).

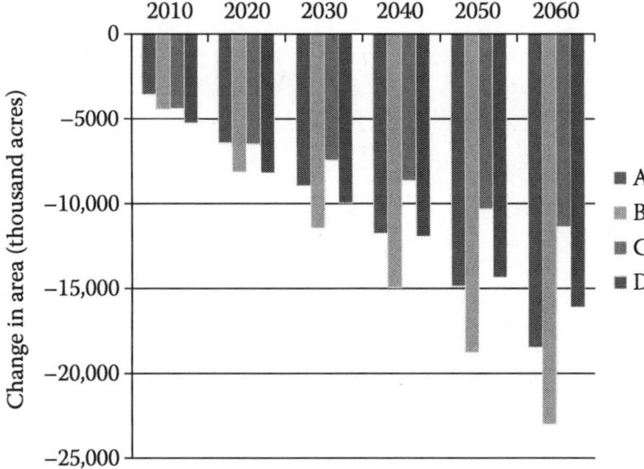

FIGURE 2.18 Changing forest land uses for the Southern United States, 1997 to 2060, under four Cornerstone Futures: A, large urbanization gains with increasing timber prices; B, large urbanization gains with decreasing timber prices; C, moderate urbanization gains with increasing timber prices; and D, moderate urbanization gains with decreasing timber prices.

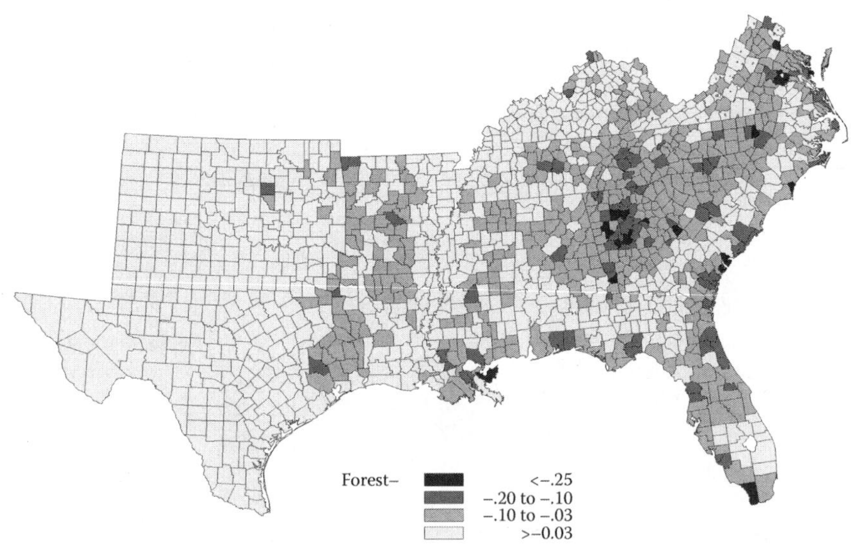

FIGURE 2.19 Changing forest land uses in the Southern United States, 1997 to 2060, based on an expectation of large urbanization gains and decreasing timber prices (Cornerstone B).

Other Land Uses

Cropland—As with forest area, the change in cropland area would depend on the economic conditions defined by each alternative future. However, unlike forest area, which is dominated by urbanization patterns (driven by the A1B storyline), cropland is more heavily influenced by timber prices. Losses would range from about 16 million acres under the high economic growth (A1B) of Cornerstone A with increasing timber prices, to only about 5 million acres under the lower economic growth (B2) of Cornerstone D with decreasing timber prices. The difference in crop loss

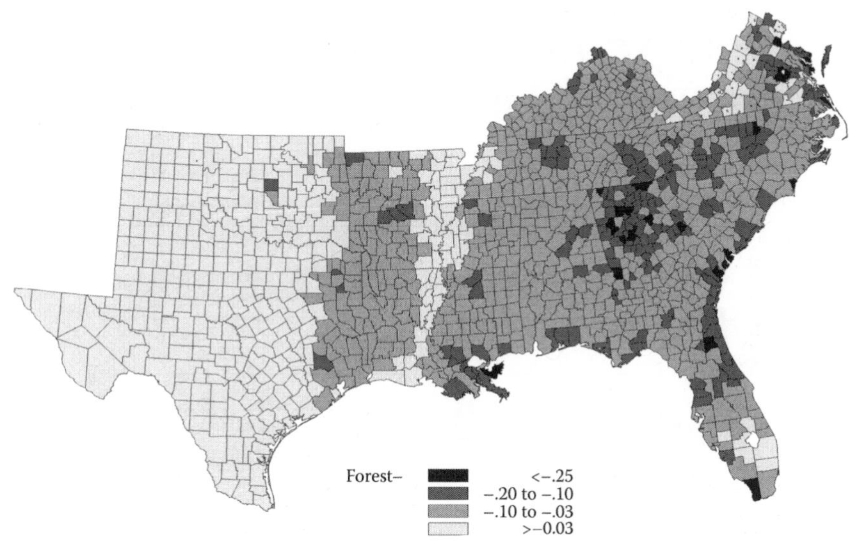

FIGURE 2.20 Changing forest land uses in the Southern United States, 1997 to 2060, based on an expectation of large urbanization gains and decreasing timber prices (Cornerstone B).

between storylines A1B and B2 (holding price futures constant) would be about 3 million acres. The difference between increasing and decreasing price futures (holding storylines constant) would be about 8 million acres.

Cornerstone D, which predicts the lowest levels of cropland loss, shows especially high levels of loss in North Carolina, southern Florida, central Kentucky and Tennessee, and the area in Texas bordered by Dallas, Houston, and Austin. Under Cornerstone A, crop losses would be highest, spreading across broader areas of North Carolina, Tennessee, and Kentucky; and with additional losses in southeastern Georgia and the coastal areas of Texas and Louisiana. Among the five southern subregions, the highest percentage loss of cropland would be in the Piedmont (28% under Cornerstone B and 51% under Cornerstone A), followed by large areas in the Coastal Plain and the Appalachian-Cumberland highlands.

Pasture—The pattern of pasture losses across the Cornerstone Futures would be similar to forests, with the highest losses forecasted under Cornerstone B (about 7 million acres) and the lowest under Cornerstone C. Similar to the pattern of cropland forecasts, pasture area change is more heavily affected by timber price projections than by the economic growth forecasts.

Pasture losses for all the Cornerstone Futures would be concentrated in three broad zones: the first stretching from northern Georgia to northern Kentucky and including a large area of Tennessee, the second in Peninsular Florida, and the third including the Ozark-Ouachita Highlands and the Cross Timbers section of eastern Texas and Oklahoma. Variations across the five southern subregions would be substantial. As with forests and crops, the Piedmont would experience the largest percentage loss, about 25% for Cornerstone B, followed by the Appalachian-Cumberland highlands (15%), Coastal Plain (11%), and Mid-South (9%).

Rangeland—By construction, forecasts of change in range area are limited to Texas and Oklahoma and only reflect the effects of urbanization (not being sensitive to fluctuations in timber prices). Rangeland would decrease by about 2.5 million acres under Cornerstone Futures C and D and about 3.2 million acres under Cornerstones A and B, with losses concentrated in the urbanizing Cross Timbers section of eastern Texas and Oklahoma—especially around Dallas and Austin—and along the border with Mexico.

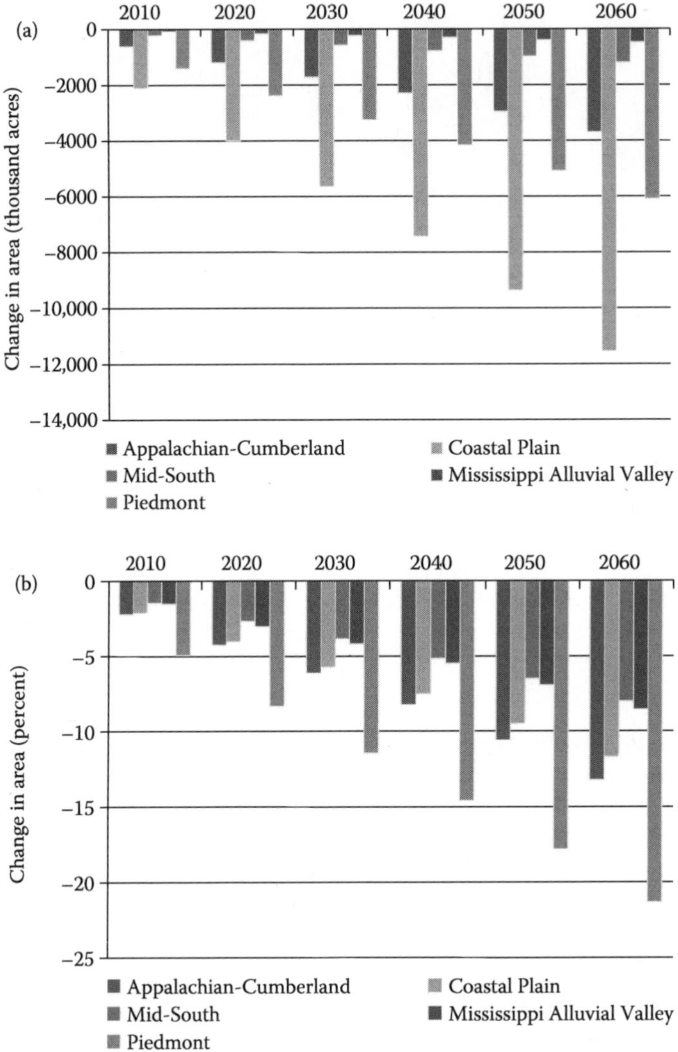

FIGURE 2.21 Changing forest area by southern subregion, 1997 to 2060, expressed in (a) acres and (b) percent; and based on an expectation of large urbanization gains with decreasing timber prices (Cornerstone B).

OUTLOOK FOR FOREST CONDITIONS

Forest Area

Forecasts of forest area change were derived from the land-use analysis described in the preceding. The land-use analysis (Wear, in press) only considered nonfederal land consistent with the National Resources Inventory design and started from a baseline of 1997; we updated the basis to 2010 and extended the analysis to all forested land based on FIA data.

All Cornerstone Futures predict decreases in forest area with losses ranging from 4 to 21 million acres (2–10%) by 2060, the result of population- and income-driven urbanization and of changes in the relative price of timber products (Figure 2.22). The smallest loss of forest area is forecasted for Cornerstone C, which would have the lowest population and income growth resulting in lowest urbanization, and increasing timber prices resulting in shifts of some rural land toward forest uses. The largest loss of forest area is forecasted for Cornerstone B, where population growth would be

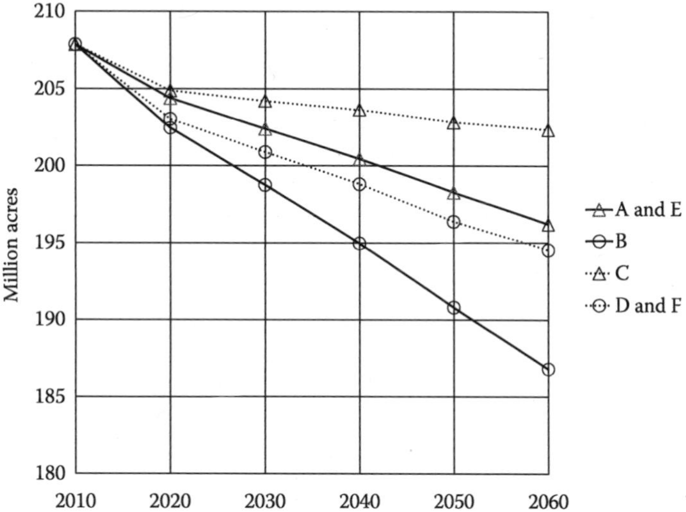

FIGURE 2.22 Forecasted total forest area in the Southern United States, 2010–2060, by Cornerstone Future (A representing large urbanization gains with increasing timber prices, B representing large urbanization gains with decreasing timber prices, C representing moderate urbanization gains with increasing timber prices, D representing moderate urbanization gains with decreasing timber prices, E representing large urbanization gains with increasing timber prices and increased planting, and F representing moderate urbanization gains with decreasing timber prices and decreased planting). Cornerstones A and E share a land-use model, as do Cornerstones D and F. The figure shows the overlapping trajectories for these.

moderate but income growth would be strong (resulting in high urbanization), and timber prices would decrease (resulting in shifts of forest land to agricultural uses). Figure 2.22 also shows that price effects dominate the projection of forest area, with the highest forest loss predicted for the Cornerstone Futures associated with decreasing prices (B, D, and F); the three Cornerstone Futures with the lowest forest loss would have increasing prices (A, C, and E).

Forest area change would also vary across the five management types: planted pine (*Pinus* spp.), natural pine, mixtures of oak (*Quercus* spp.) and pine, upland hardwood, and lowland hardwood. The upland and lowland hardwood types are forecasted to make up 51–53% of all forests in 2060, a decrease from about 54% in 2010 (Figure 2.23a,b). Changes, however, would be most profound among the softwood types (planted pine, natural pine, and oak-pine). These forest dynamics are heavily influenced by management for forest products, which in turn is driven by timber market conditions and by the rate of forest planting.

Planted pine—Of the five management types in the South, only planted pine is forecasted to increase in spite of overall decreases in forested area. The South now contains 39 million acres of planted pine (about 19% of total forest area), the culmination of an upward trend that started in the 1950s. Our projections of planted pine were driven by urbanization, timber prices, and planting rates (Figure 2.24a).

Cornerstone E, characterized by a relatively high level of urbanization as well as high timber prices and planting rates, would produce the largest expansion in planted pine (though at rates lower than those experienced in the 1990s) and yield an increase of 28.2 million acres by 2060 (about 560,000 acres per year). Under this Cornerstone, 34% of forests would be planted pine in 2060. Conversely, Cornerstone F, characterized by a relatively low level of urbanization as well as low timber prices and planting rates, would yield the smallest gain in planted pine area with an increase of 7.8 million acres by 2060 (24% of forest area). The remaining Cornerstone Futures have projections that are intermediate to these results. They cluster around the forecast for Cornerstone D with

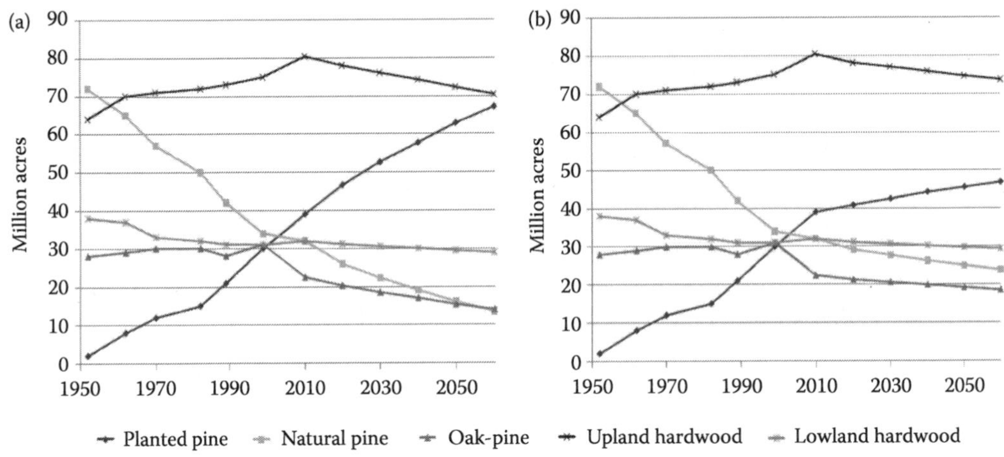

FIGURE 2.23 Forecasted forest area in the Southern United States by management type, 1952–2060, for (a) Cornerstone E, which is characterized by high urbanization, high timber prices, and more planting; and (b) Cornerstone F, which is characterized by low urbanization, low timber prices, and less planting.

its lower timber prices and slower urbanization: a gain of 16.8 million acres (about 0.3 million acres per year) with planted pine comprising 28% of the forest acreage in 2060.

Natural pine—Forecasted losses in the area of naturally regenerated pine forest types mirror the gains in planted pine forests and are therefore related, albeit inversely, to the condition of forest products markets (Figure 2.24b). The largest predicted decrease in natural pine—a loss of 58% from 31.5 million acres in 2010 to 13.5 million acres in 2060—is associated with Cornerstone E, which would have the highest planting rates. The smallest decrease would occur with Cornerstone F, with its lower timber prices and planting rates, but losses would still be substantial: 7.6 million acres or 25%. Regardless of the Cornerstone Future evaluated, naturally regenerated pine types are forecasted to decrease, continuing a trend that has dominated forest type dynamics since the 1960s.

Oak-pine—The area of the oak-pine management type would also decrease for all Cornerstone Futures, with a similar pattern of change but smaller acreage and percent changes than is forecasted for natural pines (Figure 2.24c). As with natural pine, oak-pine is more heavily influenced by timber market conditions than by urbanization rates. Oak-pine decreases would range from 8.5 million acres (38%) for Cornerstone E to 3.9 million acres (17%) for Cornerstone F.

Upland hardwood—At more than 80 million acres in 2010, upland hardwoods are the predominant management type in the South, more than double the area of the next largest type. Upland hardwoods are forecasted to decrease for all Cornerstone Futures (Figure 2.24d), and variations in forecasts are associated more with the rate of urbanization than with timber market futures. The three Cornerstone Futures that predict the lowest upland forest loss are associated with lower urbanization (C, D, and F), and the three that predict the highest loss are associated with the higher urbanization forecasts (A, B, and E). Loss of upland hardwood forests would range from 5.9 million acres (about 8%) for Cornerstone C to 11.2 million acres (14%) for Cornerstone B. For Cornerstone E, which is also characterized by high timber prices but with even higher rates of planting, the projected total area of planted pine forest would be nearly equal to upland hardwood forests in 2060, as the stimulating effects of price and planting on the pine type combine with the depressing effects of urbanization on the hardwood type.

Lowland hardwood—The area of lowland hardwood forests is also more sensitive to the rate of urbanization and less sensitive to forest products markets than the softwood types. For this management type (Figure 2.24e), forecasts indicate losses ranging from 1.7 (5%) to 4 million acres (12%) from a base of 32 million acres in 2010. The relative ranking of predicted change across

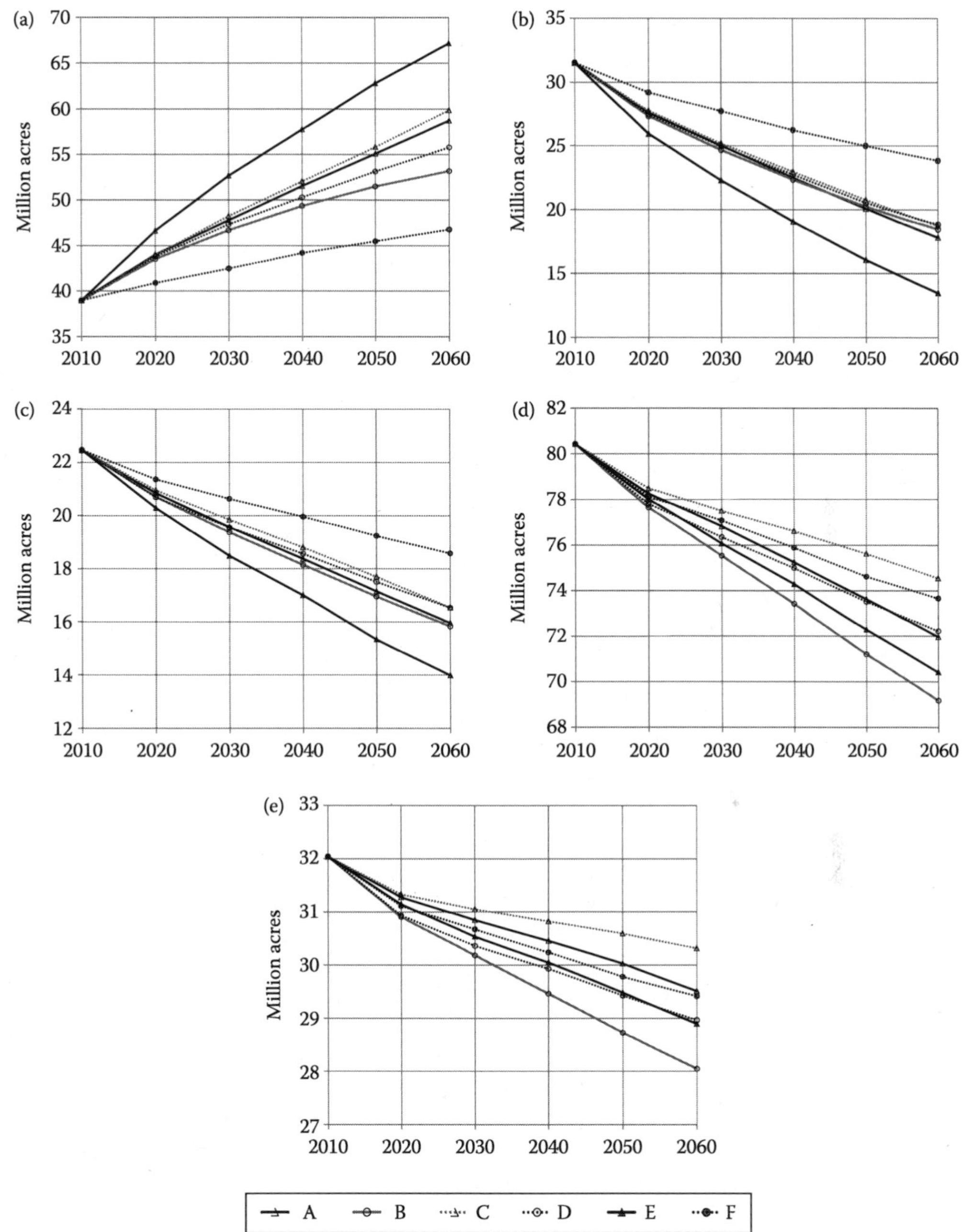

FIGURE 2.24 Forecasted forest area in the Southern United States under six Cornerstone Futures, 2010–2060, for (a) planted pine, (b) natural pine, (c) oak-pine, (d) upland hardwood, and (e) lowland hardwood management types.

Cornerstone Futures is identical to the forecasts for upland hardwood types. Lowland forests would lose proportionally less area than the other natural pine, oak-pine, and upland hardwood.

Regional variations—The forecasts of forest-type dynamics vary across the subregions (Figure 2.25a–e). Forest losses would be especially concentrated in the Piedmont and Appalachian-Cumberland highlands and intensive management in the Coastal Plain would influence management

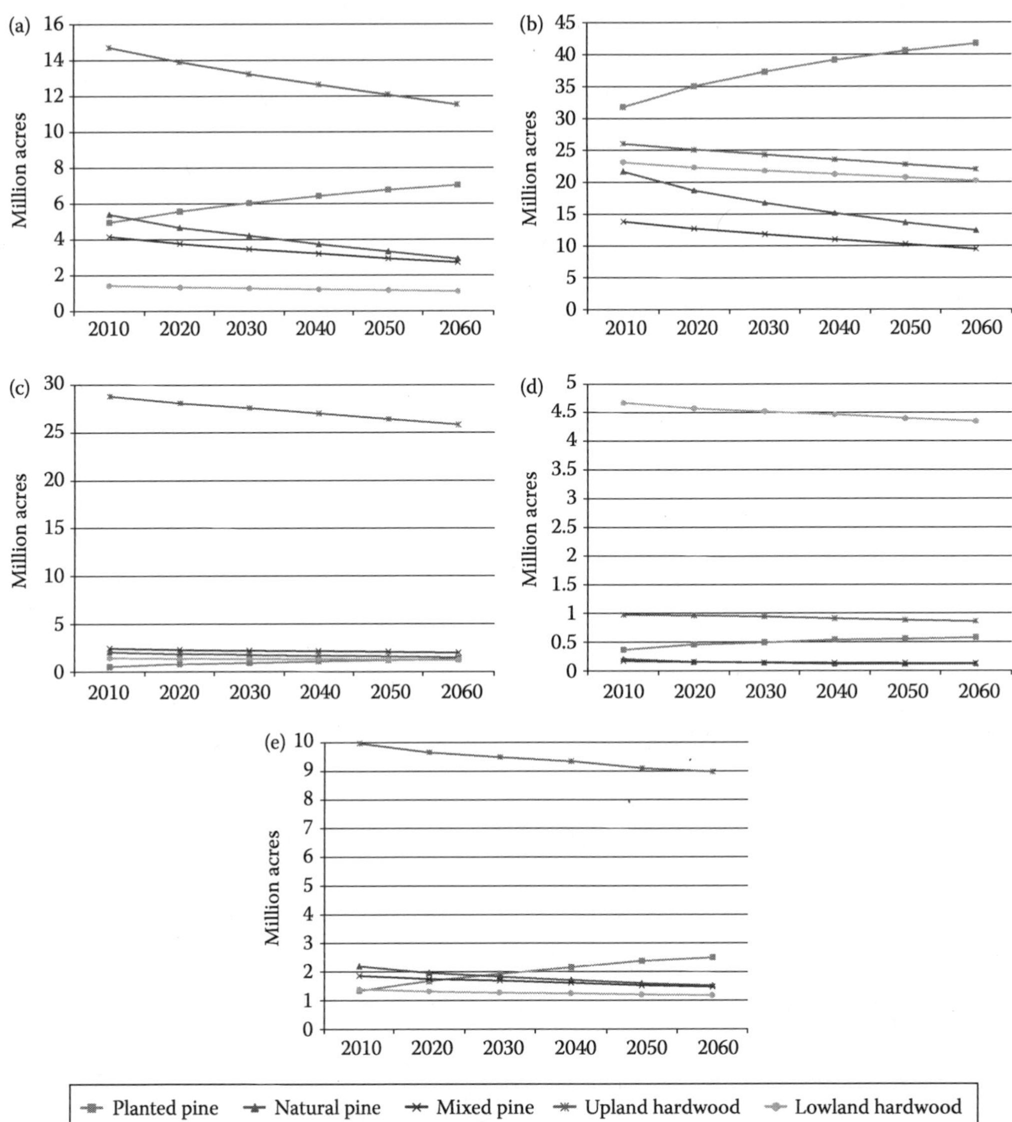

FIGURE 2.25 Forecasted forest area in the Southern United States by management type, 2010–2060, for (a) Piedmont, (b) Coastal Plain, (c) Appalachian-Cumberland highlands, (d) Mississippi Alluvial Valley, and (e) Mid-South under Cornerstone B, which is characterized by high urbanization and low timber prices.

types differently. The Coastal Plain, with 82% of planted pine in 2010, would experience the largest growth in planted pine area, from 32 to 43 million acres. Decreases in naturally regenerated pine would be largest in the Coastal Plain as well. Upland hardwood losses would be largest in the Piedmont and in the Appalachian-Cumberland highlands, reflecting the influence of urbanization on these types. Changes in lowland hardwood types would be more evenly spread across the South.

Forest Biomass

Of the various metrics available for measuring biomass changes for a site in a forest inventory, we opted to focus on the volume of growing stock because it is a useful index both for timber analysis

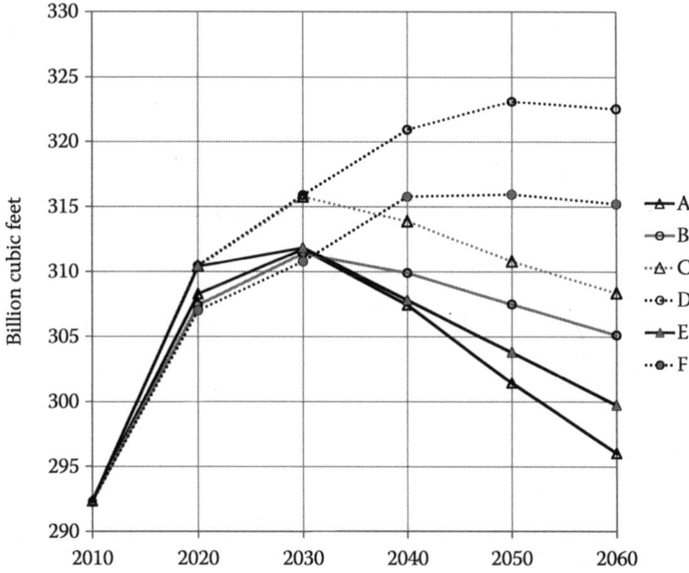

FIGURE 2.26 Total growing stock volume in the Southern United States, 2010–2060, under six Cornerstone Futures.

and for measuring other ecosystem services. Total growing stock volumes are forecasted to change in response to both land-use changes and timber harvesting levels (Figure 2.26). From a base of about 292 billion cubic feet in 2010, inventories would increase at most by about 11% in 2060 under the low urbanization/low timber price Cornerstone D. The smallest increase in total growing stock inventories would be found with the high urbanization/high timber price Cornerstone A, with an increase in volume to 2030 and then a decrease over the remainder of the forecast period. Under this cornerstone, the volume in 2060 would be about 1% higher than values observed in 2010.

Differing patterns of change between hardwood and softwood components of the inventory are generally countervailing in their effects. Under all Cornerstone Futures, softwood growing stock inventories would increase (Figure 2.27). For the low urbanization/low timber price Cornerstone D,

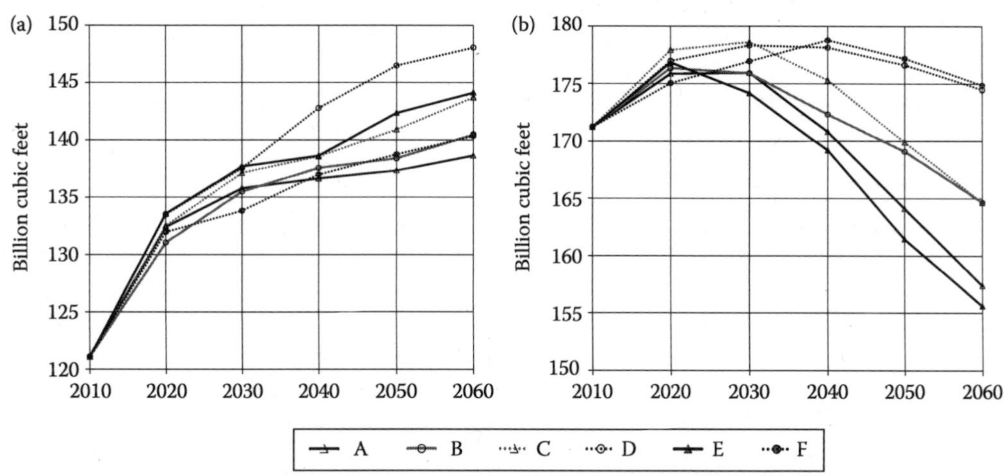

FIGURE 2.27 Total growing stock volume in the Southern United States, 2010–2060, under six Cornerstone Futures for (a) softwoods and (b) hardwoods.

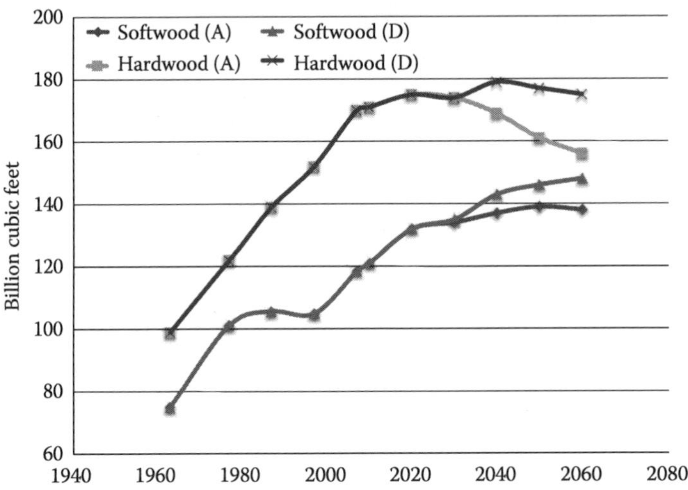

FIGURE 2.28 Historical (1962–1999) and forecasted (2020–2060) hardwood and softwood growing stock inventories in the Southern United States, for Cornerstones A (large urbanization gains with increasing timber prices) and D (moderate urbanization gains with decreasing timber prices).

softwood inventories would increase from a base of about 121 billion cubic feet in 2010 to ≤148 billion cubic feet (37 billion cubic feet or 22%). The smallest increase would be 15% (18 billion cubic feet) for the high urbanization/high timber prices Cornerstone A.

Starting from about 171 billion cubic feet in 2010, hardwood growing stock volumes would peak somewhere between 2020 and 2040 for all Cornerstone Futures and then decrease to the year 2060 (Figure 2.27b). The most pronounced decreases would be for Cornerstones A and E, both of which have high rates of urbanization and high timber prices. Predicted declines for these Cornerstone Futures from 2010 to 2060 are in the range of 15 billion cubic feet (9%), compared to 3 billion cubic feet (2%) for Cornerstones D and F.

These predicted changes in growing stock volume depart from historical patterns of volume accumulation in the South. From 1963 to 2010 southern forests accumulated about 2.5 billion cubic feet per year or roughly 70%, with hardwoods accounting for most of this biomass accumulation (61%). Although growth is projected to continue over at least the next 10 years, growing stock volume would reach a maximum and then decrease somewhat over the following 40 years (Figure 2.28), with hardwood growing stocks decreasing, especially in response to urbanization (Cornerstone A). Softwood volumes would increase only slightly.

Forest Carbon

We estimated that the carbon in southern forests in 2010 was about 12.4 billion tons stored in eight pools: down trees, standing dead trees, litter, soil organic carbon, live trees aboveground and belowground, and understory plants aboveground and belowground. Aboveground live trees and soil organic material made up 80% of the total carbon stock. Forecasts of future forest carbon stocks reflect changes in the amount of forest area and the composition of the forest inventory. However, because the model we used tracks only the carbon pool in forests and does not account for carbon transfers to agricultural and other land-use pools, our estimates are not comprehensive. Likewise, the model does not account for carbon that leaves forests as products and may remain sequestered for long periods of time in housing or other end uses (Heath et al. 2011).

Changes in forest carbon pools reflect both changes in growing stock volumes and changes in forest area (Figure 2.29). Under most Cornerstone Futures, tree carbon would peak in 2020 and then

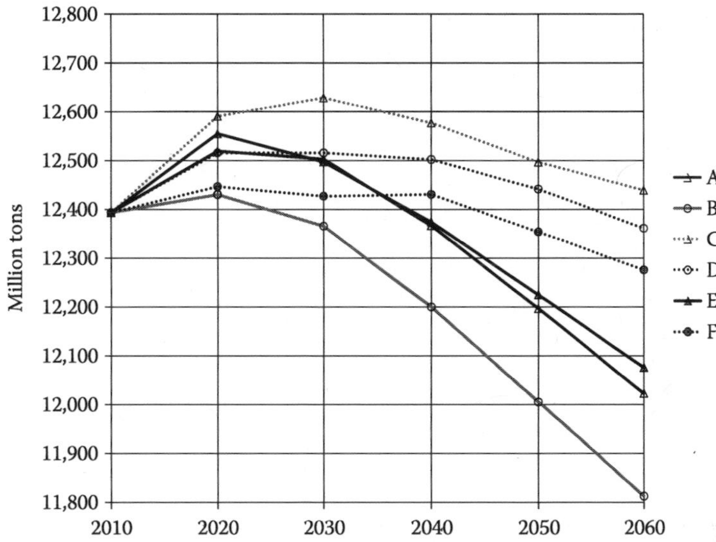

FIGURE 2.29 Total forest carbon stock in the Southern United States, 2010–2060, under six Cornerstone Futures.

level off or decrease; the exception would be the low urbanization/high timber prices Cornerstone C, which would peak in 2030. At the extreme, the forest carbon pool in 2060 would be 5% smaller than the pool in 2010 (a net emission of about 600 million tons). Carbon would accumulate as a result of net biomass growth on forested lands, but these gains would be offset by the loss of forested land through urbanization.

The clustering of carbon forecasts in Figure 2.29 reveals the interplay of urbanization and timber prices. The forecasts with the highest amount of carbon predicted for 2060 are associated with the low urbanization Cornerstone Futures (C, D, and F). The lowest carbon forecasted is for the high-urbanization Cornerstone Futures (A, B, and E). However, within each of these triplets, the lowest carbon levels are associated with lower timber prices (Cornerstones B and F); higher timber prices and resulting intensive forest management would lead to higher carbon sequestered in the forest pool. This suggests that urbanization patterns dominate the forecasts of carbon storage and that stronger forest product markets can ameliorate carbon losses.

Forest Composition and Age Classes

The USFAS was designed to replicate inventories across broad areas—for example, aggregates of at least several counties—and for broad species groupings. However, the resampling/imputation approach does provide some insights into how drivers of change may alter forest composition at a finer scale. For example, Figure 2.30 shows forest forecasts of hardwood groups for the high urbanization/low timber price Cornerstone B, which leads the other Cornerstone Futures in predicted decreases of forest area. For this cornerstone, all hardwood groups are forecasted to lose 16 million acres (14%) but not all would lose area at the same rate. Oak/hickory (*Carya* spp.) would lose about 1% compared to 10% or more for all other groups. The most substantial losses (26%) would be for the yellow poplars (*Liriodendron tulipifera*) at 4 million acres and the other hardwoods at 6 million acres. Digging a bit deeper, loses of the yellow poplar forest type are expected to be about 25% in the Coastal Plain (1.4 million acres) and the Appalachian-Cumberland highlands (1.0 million acres), compared to a more substantial loss of 34% in the Piedmont (1.6 million acres).

The forecasted area of softwood species groups is shown in Figure 2.31. Loblolly/shortleaf pine (*P. echinata*/*P. taeda*) is forecasted to remain dominant, with its area roughly level through the

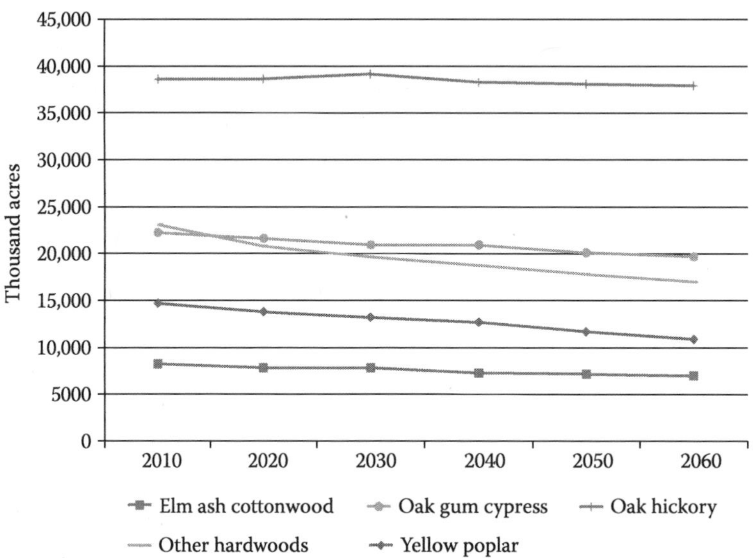

FIGURE 2.30 Forecasts of the area of hardwood species and groups in the Southern United States, 2010–2060, under Cornerstone B, which is characterized by high urbanization and low timber prices.

forecast period. Although overall forest area is expected to decrease and softwood removals are forecasted to increase, continuation of investment in plantations would enable this group to maintain its area. Longleaf/slash pine (*P. palustris/P. elliottii*) is forecasted to increase slightly and oak/pine is expected to decrease. All other groups would exhibit very little change.

Another element of forest conditions that may be especially important for wildlife is the age class distribution of management types (Chapter 11). Figure 2.32a–e shows forecasts of age classes for the high urbanization/high timber prices/more planting Cornerstone E. Early-age forests are <20 years, mid-age forests are 20–70 years, and old-age forests are >70 years. Because Cornerstone E

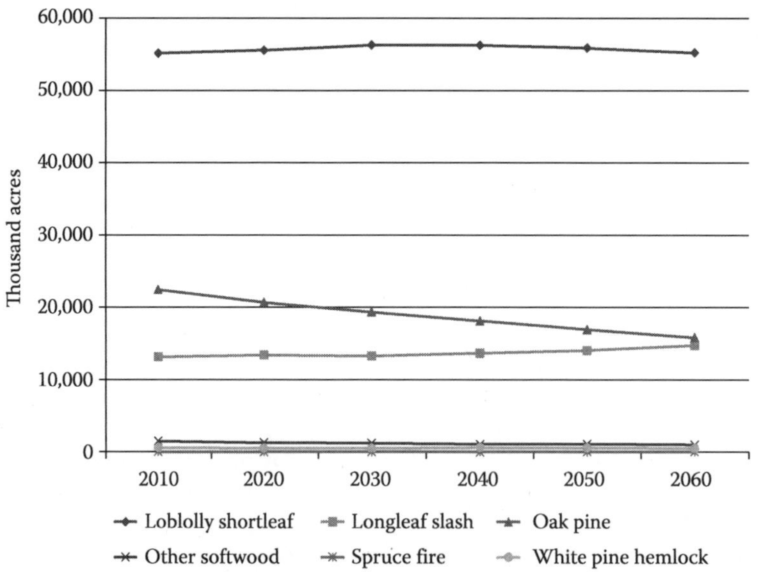

FIGURE 2.31 Forecasts of the area of softwood species groups in the Southern United States, 2010–2060, under Cornerstone B, which is characterized by high urbanization and low timber prices.

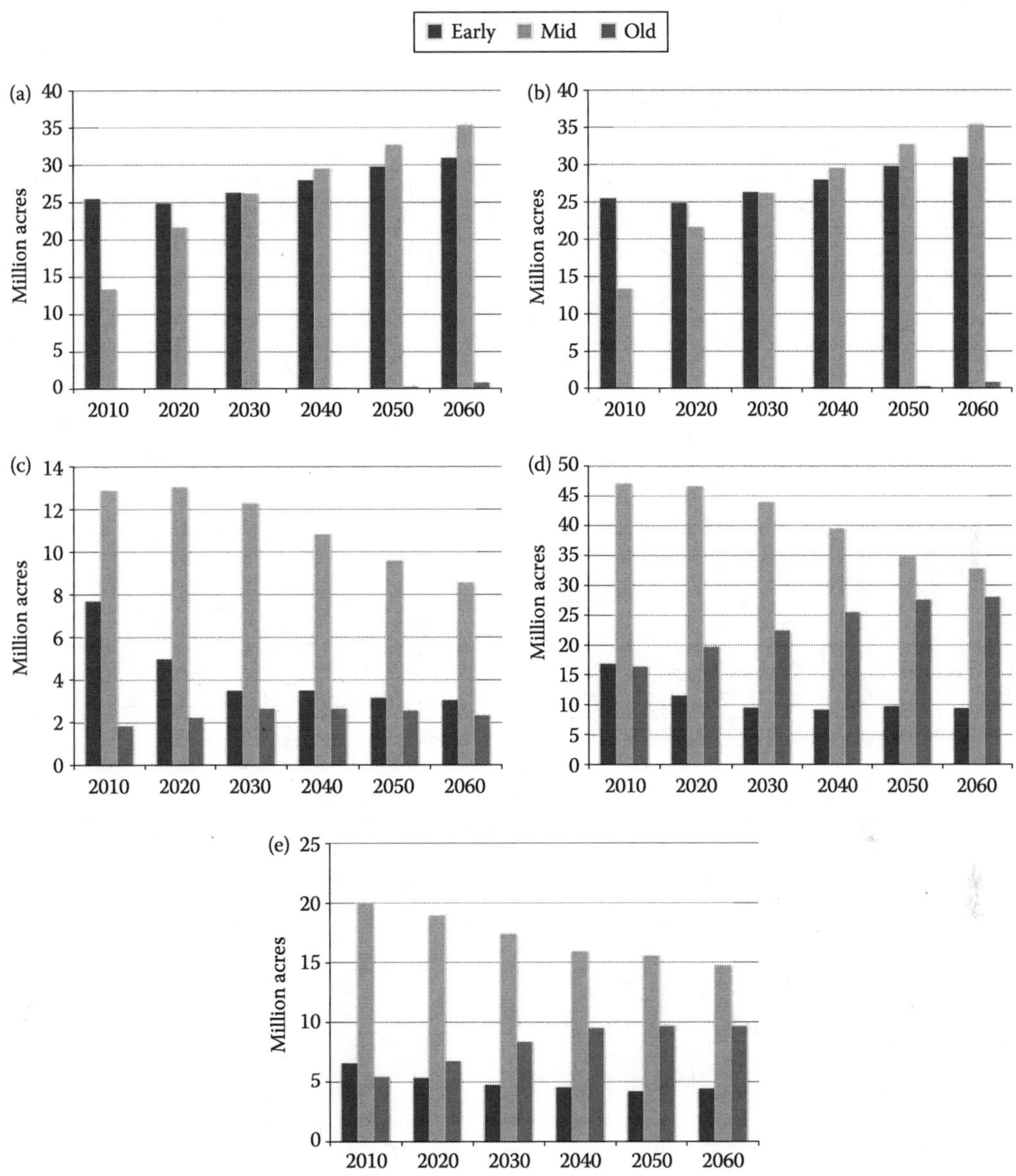

FIGURE 2.32 Forecasts of forest age classes in the Southern United States, 2010–2060, for (a) planted pine, (b) natural pine, (c) oak-pine, (d) upland hardwood, and (e) lowland hardwood management types under Cornerstone E, which is characterized by high urbanization, high timber prices, and increased planting.

predicts the largest change in management types and the most harvesting, it is a "best" case for the production of early-age forests.

With the exception of planted pine, all management types are forecasted to experience a decrease in early-age forests throughout the forecast period for Cornerstone E. Planted pine forests are forecasted to shift their distributions toward mid-age as the rate of planting decreases from its peak in the late 1990s. Mid-age forest area would increase substantially, and early-age forest area would increase at a more moderate rate. Practically no acres of old-age forest would remain in the planted pine management type.

High harvest rates and conversion to planted pine in Cornerstone E would shift the age class distribution of the naturally regenerated pine types. The area of old-age natural pine would stay relatively constant but mid-age forests would decrease by 13 million acres (about 64%) and early-age forests would decrease by about 4.5 million acres (58%). Oak-pine would experience a similar pattern of age class changes.

In contrast to the softwoods, hardwood forecasts would experience an increase in old-age forests. For upland hardwoods, the area of mid-age forests is forecasted to decrease by 14 million acres (down from 59% in 2010 to 47% in 2060). Over this same period, old-age forests are forecasted to increase by 12 million acres (up from 20% to 40%) and early-age forests would decrease by 8

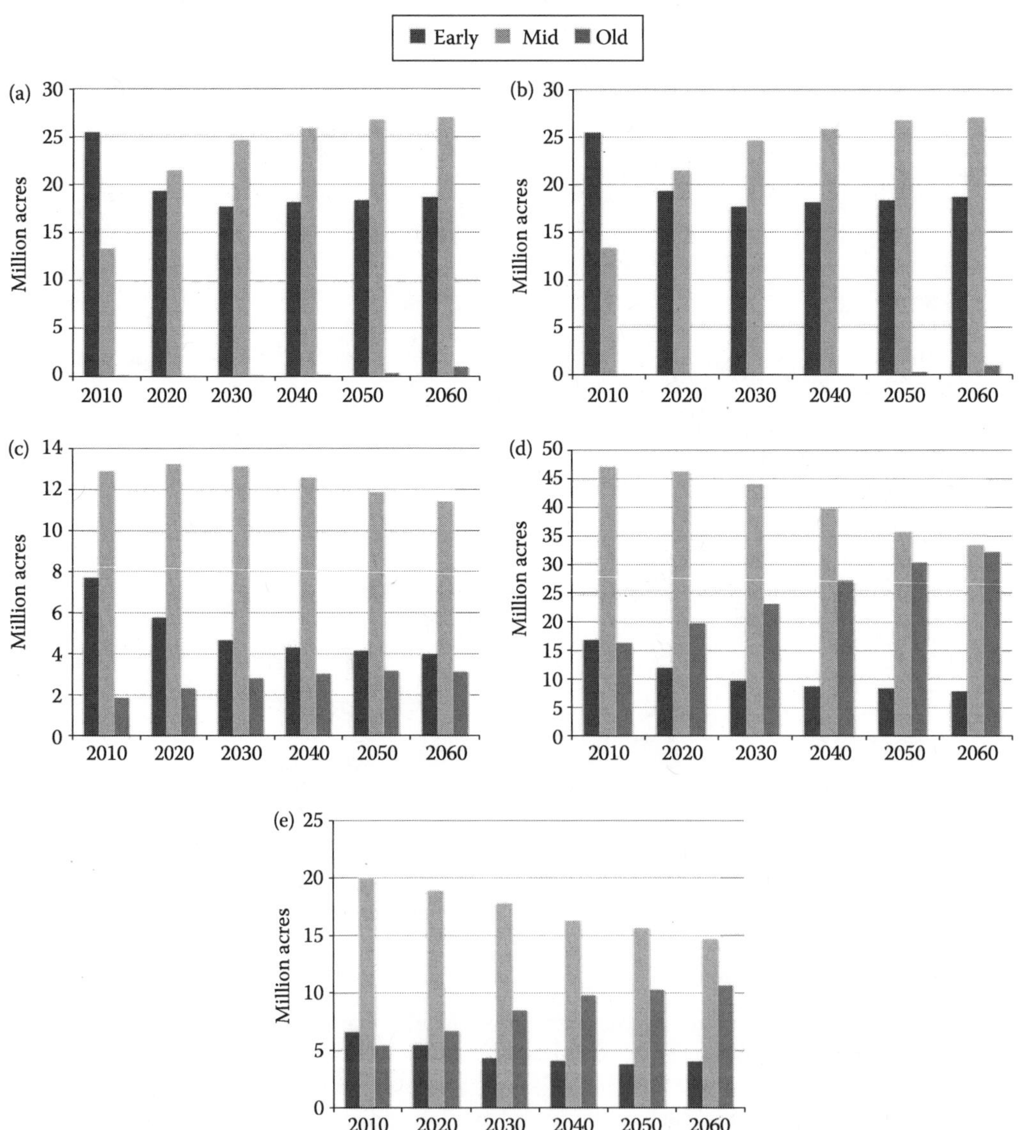

FIGURE 2.33 Forecasts of forest age classes in the Southern United States, 2010–2060, for (a) planted pine, (b) natural pine, (c) oak-pine, (d) upland hardwood, and (e) lowland hardwood management types under Cornerstone F, which is characterized by low urbanization, low timber prices, and decreased planting.

million acres (down from 21% to 13%). Overall, the shift among age classes reveals that the early-age component of the upland hardwood type would decrease by 44% at the same time that the old-age component would increase by 71%.

The pattern of change for the lowland hardwood management type would be similar to changes in upland hardwoods, but the changes would occur at different rates. The mid-age component of lowland hardwood forests would decrease by about 5 million acres (26%), compared to an increase of about 4 million acres (77%) for the old-age component, and a decrease of about 2 million acres (33%) for the early-age component. As with the upland management type, the total area would decrease but average age would increase.

Contrary to Cornerstone E, Cornerstone F (Figure 2.33a–e) is characterized by less planting and a lower harvest rate. The forecasts for Cornerstones E and F therefore bracket all the Cornerstone Futures. For Cornerstone F, the area of early-age planted pine would decrease over time (in contrast to increases simulated under Cornerstone E), and the age class distribution would approach a stasis.

Cornerstone F would also result in less dramatic changes in area of natural pine when compared to Cornerstone E, with more stability in the early-age class, increases in the old-age class, and smaller decreases of the mid-aged class (33% less). This pattern is mirrored in the oak-pine age classes. For hardwood management types, however, the differences between Cornerstones E and F were minor. This indicates that management changes strongly influence the age structures of softwoods but have little influence on hardwoods.

CONCLUSIONS

The SFFP describes a set of integrated social, economic, climate, and resource futures for exploring adaptation and mitigation strategies for the region's forests. Projections indicate relevant variability across the scenarios but also across space within each scenario. The six discrete Cornerstone Futures together form a broad range of plausible future conditions for the analysis of specific issues in subsequent chapters in this book. They allow for simultaneous consideration of the economic/social driving variables and the climate effect variables thereby supporting integrated analysis across various phenomena.

In contrast, some questions demanded averaged climate projections, without concurrent consideration of the socioeconomic drivers. These questions were best served by the ensembles of climate projections created for the IPCC storylines. These ensembles also demonstrate important variability across storylines and across space within storylines.

REFERENCES

Adam, J.C. and D.P. Lettenmaier 2003. Adjustment of global gridded precipitation for systematic bias. *Journal of Geophysical Research*, 108, 1–143.

Boé, J., L. Terray, F. Habets, and E. Martin 2007. Statistical and dynamical downscaling of the Seine basin climate for hydro-meteorological studies. *International Journal of Climatology*, 27, 1643–1655.

Burt, C.C. 2007. *Extreme Weather: A Guide and Record Book.* New York: W.W. Norton and Co. 303 p.

Chen, H., H.Q. Tian, M.L. Liu, J. Melillo, S.F. Pan, and C. Zhang 2006. Effect of land-cover change on terrestrial carbon dynamics in the southern United States. *Journal of Environmental Quality*, 35, 1533–1547.

Collins, W.D., P.J. Rasch, B.A. Boville et al. 2006. The community climate system model version 3 (CCSM3). *Journal of Climate*, 19, 2122–2143.

Coulson, D.P., L.A. Joyce, D.T. Price et al. 2010. Climate Scenarios for the conterminous United States at the county spatial scale using SRES scenarios A1B and A2 and PRISM climatology. Fort Collins, CO: U.S. Department of Agriculture Forest Service, Rocky Mountain Research Station. http://www.fs.fed.us/rm/data_archive/dataaccess/US_ClimateScenarios_county_A1B_A2_PRISM.shtml. [Date accessed: November 6, 2012].

Delworth, T.L., A.J. Broccoli, A. Rosati et al. 2006. GFDL's CM2 global coupled climate models—Part 1: Formulation and simulation characteristics. *Journal of Climate*, 19, 643–674.

Flato, G.M. and G.J. Boer 2001. Warming asymmetry in climate change simulations. *Geophysical Research Letters*, 28, 195–198.

Gordon, C., C. Cooper, C.A. Senior et al. 2000. The simulation of SST, sea ice extents and ocean heat transports in a version of the Hadley Centre coupled model without flux adjustments. *Climate Dynamics*, 16, 147–168.

Heath, L.S., J.E. Smith, K.E. Skog, D.J., Nowak, and C.W. Woodall 2011. Managed forest carbon estimates for the US greenhouse gas inventory, 1990–2008. *Journal of Forestry*, 109(3), 167–173.

Huggett, R., D.N. Wear et al. In press. Forest forecasts. In: D.N. Wear, and J.G. Greis, eds. The Southern Forest Futures Project: Technical Report.

Ince, P.J., A.D. Kramp, K.E., Skog, H.N. Spelter, and D.N. Wear 2011. US Forest Products Module for RPA Forest Assessment. Research Paper FPL-RP-662. Asheville, NC: U.S. Department of Agriculture Forest Service, Southern Research Station. 61 p.

Intergovernmental Panel on Climate Change 2007a. Climate change 2007: Synthesis report. Geneva, Switzerland. http://www.ipcc.ch/publications_and_data/ar4/syr/en/contents.html. [Date accessed: November 6, 2012].

Intergovernmental Panel on Climate Change 2007b. Climate Change 2007: The Physical Science Basis. Contribution of Working Group I to the Fourth Assessment Report of the Intergovernmental Panel on Climate Change. S. Solomon, D. Qin, M. Manning, Z. Chen, M. Marquis, K.B. Averyt, M. Tignor, and H.L. Miller, eds. Cambridge University Press, Cambridge, United Kingdom and New York, NY, USA.

Joyce, L.A., D.T. Price, D.P. Coulson, D.W. McKenney, R.M. Siltanen, P. Papadopol, and D. Lawrence In press. *Projecting climate change in the United States: A technical document supporting the Forest Service RPA 2010 Assessment.* Gen. Tech. Rept. RMRS-GTR-XXX. Fort Collins, CO: U.S. Department of Agriculture, Forest Service, Rocky Mountain Research Station. xx p.

Karl, T.R., J.M. Melillo, and T.C. Peterson 2009. *Global Climate Change Impacts in the United States*. Cambridge University Press, Cambridge, United Kingdom and New York, NY, USA.

Maurer, E.P., L. Brekke, T. Pruitt, and P.B. Duffy 2007. Fine-resolution climate projections enhance regional climate change impact studies. *Eos Trans. AGU*, 88(47), 504.

McNulty, S., M.J. Moore, P. Caldwell, G. Sun In press. Climate change. In: D.N. Wear, and J.G. Greis, eds. The Southern Forest Futures Project: Technical Report.

Meehl, G.A., C. Covey, T. Delworth et al. 2007. The WCRP CMIP3 multi-model dataset: A new era in climate change research. *Bulletin of the American Meteorological Society*, 88, 1383–1394.

Misra, V., J.-P. Michael, R. Boyles et al. 2012. Reconciling the spatial distribution of the surface temperature trends in the Southeastern United States. *Journal of Climate*, 25, 3610–3618.

Nakicenovic, N., J. Alcamo, G. Davis et al. 2000. *Special Report on Emissions Scenarios: A Special Report of Working Group III of the Intergovernmental Panel on Climate Change*, Cambridge University Press, Cambridge, U.K., 599 pp. http://www.grida.no/climate/ipcc/emission/index.htm http://www.climatecentral.org/wgts/heat-is-on/HeatIsOnReport.pdf [Date accessed: November 6, 2012].

Panofsky, H. and G. Brier 1968. *Some Applications of Statistics to Meteorology*. The Pennsylvania State University Press: University Park, 224 pp.

Polyakov, M., D.N. Wear, and R. Huggett 2010. Harvest choice and timber supply models for forest forecasting. *Forest Science* 56(4), 344–355.

Portmann, R.W., S. Solomon, and G.C. Hegel 2009. Spatial and seasonal patterns in climate change, temperatures, and precipitation across the United States. *Proceedings of the National Academy of Sciences*, 106, 7324–7329.

Price, D.T., D.W. McKenney, L.A. Joyce et al. 2011. *High-Resolution Interpolation of Climate Scenarios for Canada Derived from General Circulation Model Simulations.* Information Report NOR-X-421. Edmonton, AB: Natural Resources Canada, Canadian Forest Service, Northern Forestry Centre. 104 p.

Raunikar R., J. Buongiorno, J.A. Turner, and S. Zhu 2010. Global outlook for wood and forests with the bioenergy demand implied by scenarios of the Intergovernmental Panel on Climate Change. *Forest Policy and Economics*, 12, 48–56.

Shem, W., T.L. Mote, and J.M. Shepherd 2012. Validation of NARCCAP temperature data for some forest sites in the southeast United States. *Atmospheric Science Letters* 13, 275–282.

Smith, W.B., P.D. Miles, C.H. Perry, and S.A. Pugh 2009. *Forest Resources of the United States, 2007*. Gen. Tech. Rep. WO-78. Washington, DC: U.S. Department of Agriculture, Forest Service, Washington Office. 336 p.

Tebaldi, C., D. Smith, and N. Heller 2012: The heat is on: U.S. temperature trends. http://www.climatecentral.org/wgts/heat-is-on/HeatIsOnReport.pdf [Date accessed: November 6, 2012].

U.S. Department of Agriculture Natural Resources Conservation Service 2009. Summary Report: 2007 *National Resources Inventory.* U.S. Department of Agriculture Natural Resources Conservation Service, Washington, DC, and Center for Survey Statistics and Methodology, Iowa State University, Ames, Iowa. 123pp. http://www.nrcs.usda.gov/Internet/FSE_DOCUMENTS//stelprdb1041379.pdf [Date accessed: January 24, 2012].

U.S. Department of Agriculture Forest Service 2012a. Future of America's forests and rangelands: 2010 Resources Planning Act Assessment. Gen. Tech. Rep. WO-87. Washington, DC: Department of Agriculture, Forest Service. 198 p.

U.S. Department of Agriculture Forest Service 2012b. Future scenarios: A technical document supporting the Forest Service 2010 RPA assessment. Gen. Tech. Rep. RMRS-GTR-272. Fort Collins, CO: U.S. Department of Agriculture Forest Service, Rocky Mountain Research Station. 34 p.

Wear, D.N. 2002. Land Use. In: D.N. Wear and J.G. Greis, eds. Southern Forest Resource Assessment. Gen. Tech. Rep. SRS-53. Asheville, NC: U.S. Department of Agriculture, Forest Service, Southern Research Station, pp. 153–173.

Wear, D.N. 2011. Forecasts of County-level Land uses under Three Future Scenarios: A Technical Document Supporting the Forest Service 2010 RPA Assessment. Gen. Tech. Rep. SRS-141. Asheville, NC: U.S. Department of Agriculture Forest Service, Southern Research Station. 41 p.

Wear, D.N. In press. Land use forecasts. In: D.N. Wear and J.G. Greis, eds. The Southern Forest Futures Project: Technical Report.

Wear, D.N. and J.G. Greis. 2012. The Southern Forest Futures Project: Summary Report. Gen. Tech. Rep. SRS-168. Asheville, NC: U.S. Department of Agriculture Forest Service, Southern Research Station. 54 p.

Wear, D.N. and J.G. Greis, eds. In press. The Southern Forest Futures Project: Technical Report.

Wear, D.N., R. Huggett, and J.G. Greis In press. Alternative Futures. In: D.N. Wear and J.G. Greis, eds. The Southern Forest Futures Project: Technical Report.

Wear, D.N., R. Huggett, R. Li et al. 2013. Forecasts of forest conditions in regions of the United States under future scenarios: A technical document supporting the Forest Service 2010 RPA Assessment. Gen. Tech. Rep. SRS-170. Asheville, NC: U.S. Department of Agriculture Forest Service, Southern Research Station. 101 p.

Winkler, J.A., G.S. Guentchev, M. Liszewska, A. Perdinan, and P.-N. Tan 2011a. Climate scenario development and applications for local/regional climate change impact assessments: An overview for the non-climate scientist. Part II: Considerations when using climate change scenarios. *Geography Compass*, 5(6), 301–328, doi: 10.1111/j.1749-8198.2011.00426.x.

Winkler, J.A., G.S. Guentchev, A. Perdinan, P.-N. Tan, S. Zhong, M. Liszewska, Z. Abraham, T. Niedźwiedź, and Z. Ustrnul 2011b. Climate scenario development and applications for local/regional climate change impact assessments: An overview for the non-climate scientist. *Geography Compass*, 5, 275–300. doi: 10.1111/j.1749-8198.2011.00425.x

Wood, A.W., L.R. Leung, V. Sridhar, and D.P. Lettenmaier 2004. Hydrologic implications of dynamical and statistical approaches to downscaling climate model outputs. *Climatic Change*, 62, 189–216.

3 A Conceptual Framework for Adaptive Forest Management under Climate Change

Thomas P. Holmes, Steve McNulty, James M. Vose, Jeffrey P. Prestemon, and Harbin Li

CONTENTS

The consensus among most scientists is that the global climate is changing in response to a rapid increase in greenhouse gas emissions over the past 150 years. This perspective has prompted research on potential changes in future forest conditions so that management interventions might be developed to protect desired ecosystem services. Some of the most significant forest trends expected in response to climate change are: a shift in tree habitats and ranges perhaps in surprising directions (Iverson et al. 2008, 2011; Zhu et al. 2011); an increase in the rate and severity of disturbances such as pest outbreaks, wildfires, acidic deposition, drought, and storms (Allen et al. 2010; Ayres and Lombadero 2000; Breshears et al. 2005; Emanuel et al. 2008; Klos et al. 2009; McNulty and Boggs 2010; Raffa et al. 2008; Soja et al. 2007; Vose et al. 2012; Westerling et al. 2006, 2011); and a reduced ability of some forests to recover from forest disturbances (Thompson et al. 2009). Although these trends are projected to materialize over the next several decades, the precise timing, location, and intensity of changing climatic effects on forests are uncertain.

Because forests are long-lived, current management decisions carry a legacy that will affect their responses to climatic and biotic conditions far into the future. Understandably, many forest managers are reluctant to change their practices in the near term without persuasive evidence that consequential changes in forest health or productivity are underway. For example, research has shown that decision makers, in many different contexts, tend to favor the status quo (do nothing different) over other alternatives (Samuelson and Zeckhauser 1988), especially in situations where payoffs are uncertain (Kahneman and Tversky 1979). Consequently, forest management that seeks to maintain current forest conditions may dominate decision making, resulting in forests that are poorly adapted to future climate. Although sustainable forest management in the twenty-first century will require a willingness to experiment, learn, and adapt management strategies to changing conditions (Blate et al. 2009; Millar et al. 2007; Seastadt et al. 2008), the inducements needed to alter management regimes are poorly understood. Managers may be more inclined to implement "no regrets" strategies whereby benefits accrue with or without changes in climate (such as fuel load reduction); however, these strategies may be too conservative to increase adaptive capacity over large spatial scales.

The premise of this chapter is that better information leads to better decision making. Our goal, therefore, was to describe a general conceptual framework for an iterative decision-making process based on experimentation and scientific learning (adaptive forest management) that can help practitioners identify their best options when faced with uncertainty about future climate changes and their impacts on forest ecosystems. In addition, this conceptual framework and associated terminology served as guidance to ensure consistency in approach for the subsequent chapters in this book.

SUSTAINING ECOSYSTEM SERVICES

Natural systems are increasingly viewed as critical capital assets that provide a broad suite of ecosystem services valued by people (Daily et al. 2009; Mäler et al. 2008). The Millennium Ecosystem Assessment (2003) listed four categories of ecosystem services: provisioning (such as food, water, and timber), regulating (such as flood control), cultural (such as recreation), and supporting (such as nutrient cycling). Ecosystem services are valued by people because they help to sustain and protect human life as well as improve the quality of life. They derive from ecosystem processes that transform structural and functional ecosystem inputs (such as tree abundance and rates of evapotranspiration) into the outputs that people desire (such as clean water). When the values of ecosystem services are not considered by decision makers, they will be provided at suboptimal levels or not at all.

Climate change is anticipated to alter the amounts and kinds of ecosystem services forests provide (Figure 3.1) through mechanisms we describe in the following as slow or fast ecosystem disturbance processes.

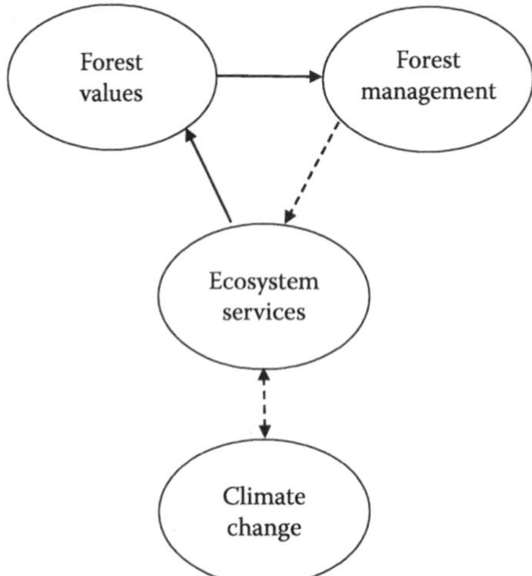

FIGURE 3.1 Diagram showing the linkages among forest values, forest management, ecosystem services, and climate change, with solid arrows indicating the direction of causality and broken arrows indicating biophysical production functions. Note that the downward pointing arrow linking ecosystem services to climate change represents mitigation through carbon sequestration.

Forest management can be used to modify the delivery of ecosystem services by adding or subtracting biophysical inputs. If the provision of ecosystem services is not consistent with the goals and desires of landowners or other stakeholders, the level of management inputs can be changed. The relatively long time lag between the implementation of forest management and the period during which desired ecosystem services are provided greatly complicates forest management decision making.

Adaptation and mitigation are two types of forest management strategies that can reduce the anticipated impacts of climate change. Adaptation—which is the focus of this chapter—refers to actions taken with the goal of decreasing the perceived negative impacts of climate change or increasing the perceived positive impacts (Millar et al. 2007). Investments in adaptation alter the ecosystem production function (the arrow linking forest management and ecosystem services in Figure 3.1) but do not directly influence the rate or magnitude of climate change. In contrast, investments in mitigation have a goal of reducing the anticipated severity of climate change. Within the forestry context, mitigation principally includes forest management activities that sequester carbon (Liao et al. 2010; Pan et al. 2011; Stoy et al. 2008) (the downward arrow linking ecosystem services (carbon sequestration) and climate change in Figure 3.1). To be effective, investments in mitigation and adaptation activities must recognize both the potential trade-offs and the complementarities between these activities (Callaway 2004).

Sustainable forest management means different things to different people. In this chapter, we define sustainable management as any management regime that maintains the productive capacity of forests so that the level of well-being available for future generations does not decline (see the Glossary at the end of the chapter). Mäler et al. (2008) proposed measures of societal well-being based on economic indicators that account for the value of ecosystem services provided by natural systems. Within the forestry context, Adamowicz (2003) recommended that levels of societal well-being provided by forest ecosystem services be assessed by a system that accounts for the value—expressed in units of currency—of forest ecosystem services that enter markets (such as timber) as well as the services that are not marketed (such as subsistence harvesting of mushrooms or aesthetic views). Market and nonmarket economic indicators, measures of market and nonmarket service values, offer a means of comparing the costs and benefits of changes in ecological production functions, thereby providing a basis for managerial decision making.

FOREST DISTURBANCES AND CLIMATE CHANGE

Although forests are owned and managed for a variety of goals and objectives, owners and managers generally seek to create, sustain, and enhance desirable conditions (or states) and avoid undesirable ones. These pursuits reflect the view that: (1) alternative biotic and abiotic conditions can produce alternative forest states, and (2) not all forest states are equally desired. For example, a plantation manager may hold a very different view of the desirability of wildfire than a manager who seeks to maintain and restore natural processes.

In this section, we provide an overview of two classes of disturbances that can alter the structure and function of forest ecosystems. We also review basic concepts describing the ability of ecological and economic systems to persist in the face of disturbances.

Pulse and Press Disturbances

Forest disturbances, by definition, are forces that alter the preexisting state of a forest ecosystem. Within the ecological literature, they have been classified as either press or pulse disturbances, depending on the ecosystem and the response variable of interest (Bender et al. 1984; Glasby and Underwood 1996). A press disturbance (also known as a stress) is a continuing disturbance, such as a gradual warming or gradual changes in other climate variables, that slowly alters the state of an ecosystem. A pulse disturbance (also known as a perturbation) is a short-term, high-intensity disturbance that causes sudden change and rapidly alters the state of an ecosystem. The ecological literature has additionally recognized that slowly changing, underlying control variables can act as triggers for rapid changes when controls exceed critical thresholds (Carpenter and Turner 2001; McNulty and Boggs 2010; Rinaldi and Scheffer 2000).

Increases in temperature and changes in patterns of precipitation can have direct impacts on the delivery of forest ecosystem services. For example, ecosystems (and species) at their range limits, such as native brook trout (*Salvelinus fontinalis*) and spruce–fir (*Picea rubens–Abies fraseri*) species in the Southern Appalachian Mountains, may be highly sensitive to short-term temperature extremes (pulse disturbances), especially when combined with other stressors (McNulty and Boggs 2010). Responses to longer-term climate change (press disturbances), such as rising mean annual temperatures, may take longer to manifest and yet may still produce long-term shifts in plant and insect phenology and other changes in ecological production functions.

Changes in temperature and precipitation from historical averages can also induce indirect (or secondary) impacts on ecological production functions. Perhaps the most dramatic indirect impacts of climate change on forests result from an increase in the frequency and intensity of pulse disturbances, such as wildfires and insect and disease outbreaks (Dale et al. 2001).

Managing Forests Using Resistance and Resilience Options

Organisms and ecosystems must develop physical or biological defenses to protect themselves against disturbances. Scientists have articulated several interrelated concepts that characterize the ability of an organism or ecosystem to survive, return to or maintain the same state when subjected to press or pulse disturbances. Two broad categories of persistence mechanisms have been defined. The first is often referred to as "engineering resilience" (Pimm 1984), or simply "resilience" in this chapter, and describes the length of time required for a disturbed system to return to some initial functional state. The second, "ecological resilience" (Holling 1973), or simply "resistance" in this chapter, is the magnitude of disturbance that can be absorbed before an ecosystem is significantly altered.

Resilience and resistance have been intuitively described using the example of a ball in a landscape of hills and valleys, with valleys of different sizes and shapes representing different "attractors" (organizing forces) into which an ecosystem (the ball) is drawn, and the neighboring hills defining the barriers between attractors (DeAngelis and Waterhouse 1987). Some valleys may be relatively shallow, with low, gradually sloping hills defining their boundaries; when the ball (ecosystem) is displaced from equilibrium by a disturbance, it will return to equilibrium relatively slowly, consistent with low resilience. In contrast, if a valley is relatively narrow and located between hills with steep sides, the ball will return to equilibrium quickly, consistent with high resilience. However, if a disturbance is large enough so that a ball is bumped from a valley over a neighboring hill, the ecosystem will move into a new valley with a different set of organizing forces and attain a new equilibrium (e.g., a shift from forest to shrub-land). The height of a hill describes the degree of ecosystem resistance: a valley located between tall hills provides a lot of resistance to major structural and functional changes resulting from a press or pulse disturbance. In contrast, a valley located between low hills offers little resistance to major changes in ecosystem structure or function that may result from a disturbance.

Within the forestry context, climate change is anticipated to create novel environmental conditions (Seastadt et al. 2008; Thompson et al. 2009) and alter the set of organizing forces affecting forest structure and function. Therefore, management efforts seeking to restore forest conditions to a state that reflects a pre-Columbian climatic regime would be less likely to succeed in providing desired ecosystem services than management based on resistance and resilience concepts. Millar et al. (2007, p. 2145) describe forest resistance management options as actions that "... forestall impacts and protect highly valued resources" particularly at local scales over the near term. Examples of forest resistance management (increasing the height of the neighboring "hills") include creating fuel breaks around high-risk, high-value forests or thinning high-value forests to protect against insect outbreaks. In contrast, forest resilience management options (increasing the steepness of the neighboring "hills") are those that "... improve the capacity of ecosystems to return to desired conditions after disturbance" (Millar et al. 2007, p. 2145). Examples of forest resilience management include activities that ensure adequate regeneration of desired species after a disturbance, for example, surplus seed-banking and focused revegetation (Spittlehouse and Stewart 2003). Finally, Millar et al. (2007, p. 2145) argue for response options that "... facilitate transition of ecosystems from current to new conditions." Intuitively, response options facilitate the movement of a ball across the stability landscape to a new (and desirable) valley. Examples include planting alternate genotypes or new species and aiding species range shifts through assisted species migration or maintenance of continuous transition pathways.

Economic–Ecologic Stability

Although stability concepts such as resistance and resilience initially were applied to natural systems, they have also been adopted by economists studying the impacts of ecological shocks on the stability of economic systems. Economic modeling has begun to recognize the interdependence of economic and ecological systems and has focused attention on the economic consequences of crossing critical thresholds in natural systems. Emerging from these models is a recommendation to use risk management approaches to reduce the probability of natural and economic systems shifting from desirable to undesirable states. For example, Perrings (1998) proposed that economic–ecologic systems be modeled as Markov processes, meaning that the resistance of systems in desirable states be measured by the probability that they will shift to undesirable states. Further, he considered transition probabilities to be functions of policies and management strategies designed to sustain desirable states and avoid undesirable states.

The second approach to risk management that specifically addresses climate thresholds is hedging. With hedging strategies, investments are made in the near term to protect against uncertain but intolerable impacts that may occur if a climate threshold is crossed (Keller et al. 2008; Yohe 1996; Yohe et al. 2004). The concept of a climate threshold (or tipping point) refers to "... a threshold above which damages caused by gradual climate change would climb dramatically" (Yohe et al. 2004, p. 416). Examples include the dieback of the Amazon rainforest (Kriegler et al. 2009), a rapid shortening of wildfire return intervals in the Greater Yellowstone ecosystem (Westerling et al. 2011), and a large-scale change from tundra to boreal forest in Alaska (Chapin and Starfield 1997), thereby changing albedo.

The third approach to risk management in economic–ecologic systems is based on understanding the likelihood of a "fat tail" in the extreme range of economic damage functions. Fat-tailed distributions of future economic losses that could result from climate extremes have much higher than historical probabilities of catastrophic economic damages as critical ecosystem processes are disrupted (Weitzman 2009, 2011). This line of thinking argues that more effort should be directed at understanding the extreme tails of climate-related probability functions rather than focusing on what is thought to be most likely.

The fourth risk management approach to sustaining economic–ecologic systems is based on the view that social well-being ultimately derives from the stock of manufactured capital (such as energy and transportation infrastructure, factories, and offices), natural capital (such as forests, clean air and water, living organisms, and the ozone layer), and human capital (such as health, knowledge, and skills). These three capital stocks constitute the wealth of a society (Arrow et al. 2003; Dasgupta 2008). Recognizing that some ecosystems can shift from a desired state to an undesired state when their resistance is exceeded, Mäler (2008) proposed that resistance ("ecological resilience" in his terminology) of natural capital to sudden, disruptive change should be considered a productive asset. Resistance provides insurance against undesired degradation and loss of natural capital, and the value of ecosystem resistance can be measured by the probability that a system will suddenly shift to an alternative state, multiplied by the difference in the value of ecosystem services provided by the states before and after the shift.

APPROACHING FORESTRY USING RISK MANAGEMENT TOOLS

Available Decision-Making Tools

Within the decision sciences, several approaches have been developed that provide guidance to decision makers faced with uncertainty about the future. These approaches can guide decision making within the context of climate change (Polasky et al. 2011). Each approach provides an opportunity to evaluate future consequences of current decisions and to learn from past decisions.

A standard approach to decision making under uncertainty, largely developed within the economics literature, is the expected utility model. This model uses information on probabilities of different futures, in combination with a set of possible management actions, to identify actions that maximize expected benefits (Shoemaker 1982). A second model to aid decision making under uncertainty is the maxi–min model (von Neumann and Morganstern 1947), which was developed to help identify the management action that produces the "least bad" outcome. A third approach is Bayesian updating, which combines a prior probability distribution of future conditions with newly collected data to estimate an updated probability distribution. This approach may be particularly useful for adaptive management under climate change (Prato 2005). A fourth approach, based on the concept of "safety first" (Roy 1952), is to make decisions that minimize the probability of losses exceeding some given critical threshold.

Finally, scenario planning is an approach to decision making that helps characterize hard-to-quantify uncertainties through construction of scenarios about how the future may unfold (Polasky et al. 2011). Scenario planning can be accomplished without assigning probabilities to future forest conditions that may result from a changing climate. The key to scenario planning is to define future threats and then identify management activities that would mitigate the impacts of those threats.

Implementing a Conceptual Framework for Adaptive Forest Management

This section proposes a risk management framework for forestry that integrates the concepts of ecosystem resilience and resistance with an economic perspective of social well-being. For purposes of this chapter, we defined risk as the functional relationship between a set of ecological or economic outcomes and the probability that each will occur. Uncertainty is the lack of knowledge about the parameters of probability distributions. Our framework reorients the problem of selecting preferred management strategies into a search for management options that shift anticipated distributions of future forest conditions (the structure and function of the assemblage of dominant plant species found at a given location) toward desirable outcomes using resilience, resistance, and response options. We illustrate our framework with a graphical depiction and explain how the framework could be implemented using several steps. Throughout our description of the framework, we explain the concepts using an example drawn from the Southern Appalachian Mountains, the high-elevation spruce–fir ecosystem that is found on ridges and mountaintops in North Carolina, Virginia, and Tennessee. The future of this ecosystem is threatened by a warmer climate (including hot spells), drought, and more severe storms (North Carolina Department of Environment and Natural Resources 2010). In addition to these direct climatic threats, acid deposition and insect outbreaks also pose threats to this ecosystem (Boggs et al. 2005; McNulty and Boggs 2010).

Step 1: Identify indicators of forest conditions—The first step in the development of a conceptual framework for forest management under climate change is to identify one or more indicators of forest condition that are thought to be critical to the provision of desired ecosystem services and are sensitive to a changing climate. Although we use a two-dimensional figure to illustrate a probability function associated with a single measure of forest condition, a three-dimensional figure could be used for two measures of forest condition, and so forth. More climate change variables may increase the complexity but not the basic form of the interactions (Tian et al. 2012).

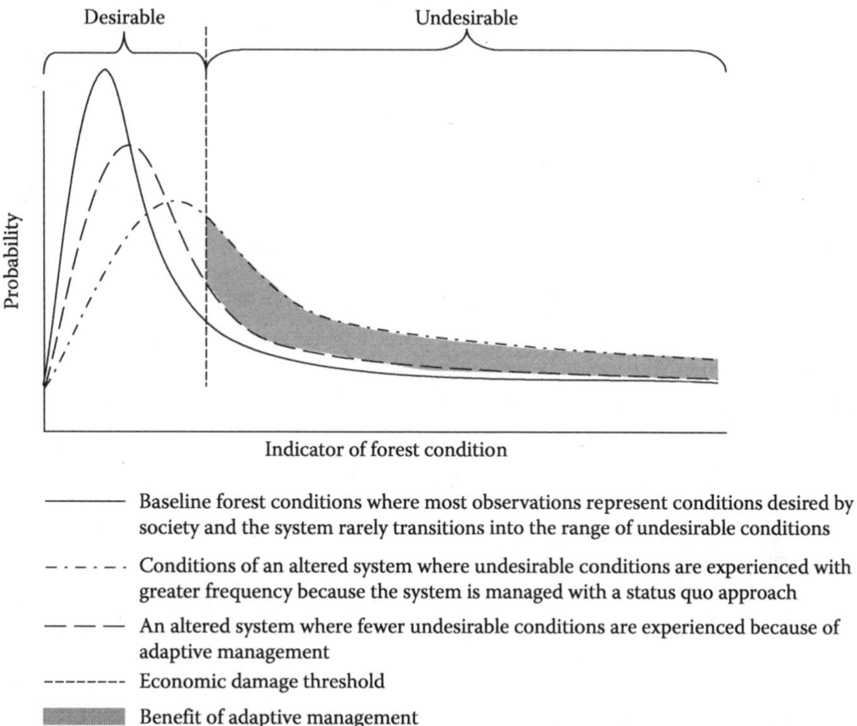
Desirable
Undesirable
Probability
Indicator of forest condition
Baseline forest conditions where most observations represent conditions desired by society and the system rarely transitions into the range of undesirable conditions
Conditions of an altered system where undesirable conditions are experienced with greater frequency because the system is managed with a status quo approach
An altered system where fewer undesirable conditions are experienced because of adaptive management
Economic damage threshold
Benefit of adaptive management

FIGURE 3.2 Schematic showing how future climate may change the probability density function of forest conditions in a given location.

Although undesirable forest conditions are always possible (Figure 3.2), forest ecosystems generally persist in the range of forest conditions desired by forest managers under current (baseline) climatic conditions and management regimes. This is because current forest management strategies have developed under a range of observed climatic conditions and disturbance regimes, and are implemented to prevent or minimize impacts or to facilitate postdisturbance recovery. The range of forest conditions depicted in Figure 3.2 represents factors such as forest structure and function and rates of carbon sequestration/net primary production (Kurz et al. 2008; Running 2008) that are sensitive to climate variables. For an example of an indicator of forest condition, we suggest the (live) basal area of red spruce and Fraser fir in the Southern Appalachian Mountains. This indicator is sensitive to both direct (e.g., drought) and indirect (e.g., insect outbreaks) impacts of climate change.

Step 2: Develop a baseline—After selecting one or more indicators of forest condition, the next step is to estimate the relative frequency (relative amount of time) associated with each level of the indicator under baseline conditions. This could be accomplished using historical data, preferably spanning many years (or decades) to provide a reasonable depiction of the proportion of years the indicator was observed in alternative conditions. Although we illustrate relative frequency (probability) concepts using smooth functions, the indicators of forest conditions could be represented by other, less smooth types of functions such as a triangular distribution that only requires estimates of the minimum, maximum, and most common values (Prato 2008).

Under baseline climate and management (Figure 3.2), most of the area under the relative frequency function is located within the desirable range of conditions, wherein valued ecosystem services and consumption goods are being produced. Equilibrium is found in the neighborhood of conditions where the relative frequency function is the greatest (because that is the most frequent level for the indicator). Even under baseline climate and management, an indicator of forest condition can shift from a desirable equilibrium into the undesirable range due to a pulse disturbance (such as a wildfire, insect outbreak, or storm). The transition from desirable to undesirable occurs when valued ecosystem services are either degraded or no longer being produced, resulting in economic losses (discussed further in the following). If an indicator of forest condition has low (high) resilience to a disturbance, the frequency distribution in the undesirable range will have a relatively fat (thin) tail because it takes a long (short) time to recover to equilibrium conditions.

Continuing with the spruce–fir ecosystem example, socially undesirable forest conditions could be depicted by the advanced mortality and loss of Fraser fir basal area after the ecosystem was invaded by the nonnative insect, balsam woolly adelgid (*Adelges tsugae*). Owing to the continued persistence of the adelgid, the system has been unable to fully recover and return to the pre-adelgid equilibrium. Consequently, the structure of the ecosystem has fundamentally changed and the entire relative frequency function has shifted toward the range of undesirable forest conditions (lower live basal area). This shift has increased the relative frequency of the system in the undesirable range and has made the ecosystem less resistant to other disturbances that could continue to fundamentally alter the system, including climate change.

Step 3: Identify consequences of no action—The third step in the development of our framework is to forecast a probability distribution for the chosen indicator(s) of future forest conditions. This step could be accomplished using ecosystem simulation models or expert opinion (Prato 2008). As already described, climate change is anticipated to increase both pulse and press disturbances in forests. These impacts can be illustrated by shifting average or typical future forest conditions toward the range of undesirable conditions, as well as by increasing the relative frequency of undesirable forest conditions. If forest managers decide to maintain the status quo and continue using historical management, not anticipating the change in the distribution of future forest conditions, they run the risk of misallocating management inputs. Misallocation of inputs would occur when managers believe that the probability distribution of forest conditions will not change in the future and they are, therefore, managing with the wrong probability function in mind. The ultimate outcome of misallocation could be a large increase in the area of the right-hand side tail of the probability function. Consequently, expected future economic damages would increase.

Step 4: Design actions (with consequences)—The fourth step in the development of our framework is to identify resilience, resistance, and response options that could shift the probability distribution of anticipated future forest conditions under status quo management back toward the desirable range of conditions (Figure 3.2). Reducing the probability of experiencing undesirable future forest conditions represents the benefit (reduction in future economic damages) of forward-looking adaptive management. The reduction in the probability of experiencing undesirable future forest conditions by shifting from status quo to adaptive management strategies is represented by the shaded area in Figure 3.2.

In our example of the Southern Appalachian spruce–fir ecosystem, adaptive management options could include restoring spruce and fir canopy, increasing connectivity between existing spruce–fir stands, reintroduction of rare species endemic to these forests, and wildfire exclusion (North Carolina Department of Environment and Natural Resources 2010). These activities can increase both the resilience and resistance of this ecosystem to pulse and press disturbances caused by climate change.

Step 5: Identify indicators of economic loss—In the preceding discussion, we have referred to economic damages as indicators of change in social well-being, and now we include them explicitly in our framework. The fifth step is to identify one or more indicators of economic damage that are directly related to the chosen indicator(s) of forest conditions. Economic damage is often measured by the sum of economic costs and losses—with costs incurred to prevent economic loss or to restore systems after a pulse or press disturbance, and losses incurred by the degradation or destruction of a valued ecosystem service. While the measurement of economic costs is relatively straightforward, the measurement of economic damages relies on the implementation of "welfare" economic methods that estimate the loss in market and nonmarket values to a broad spectrum of producers and consumers.

Continuing with our spruce–fir ecosystem example, cultural ecosystem services provided by this high-elevation ecosystem include the provision of recreational opportunities that may be diminished by advanced mortality and decline of these tree species. An economic indicator of the value of this ecosystem service can be measured using nonmarket valuation methods, as has been previously demonstrated (Haefele et al. 1991; Holmes and Kramer 1996; Kramer et al. 2003). As these authors note, in addition to the loss in recreational value incurred by the degradation of this ecosystem by the balsam woolly adelgid, substantial economic losses were incurred by people who do not intend to use the ecosystem, but simply value knowing that a healthy high-elevation ecosystem exists in the Southern Appalachian Mountains.

Step 6: Link economic impacts to forest conditions—After an indicator of economic damages has been selected, the next step in the development of our framework is to quantify how the economic indicator responds to changes in the chosen indicator of forest condition; for example, an increase in economic damages resulting from changes in the area of a spruce–fir ecosystem that had been subjected to insect-induced mortality (Holmes and Kramer 1996). The relationship between the indicator of forest condition and the indicator of economic damage for the forest ecosystem is then plotted (Figure 3.3):

- Quadrant I depicts the probability (relative frequency) function of a forest condition indictor with more frequent undesirable conditions resulting from status quo management under a changing climate (derived from Figure 3.2).
- Quadrant II traces the relationship between the forest conditions and economic damages. It shows that economic damage does not occur until the threshold of economic damage separating the desirable and undesirable forest conditions is crossed. Alterations in forest condition up to that threshold are too small to cause measurable economic damage. However, once that threshold is crossed, economic damages may rise at an increasing rate (faster than the rate that the indictor of forest condition moves into the range of undesirable conditions) as forest conditions are degraded.
- Quadrant III simply translates economic damages on the vertical axis to the horizontal axis, done to enable the plotting in the next quadrant.
- Quadrant IV shows the level of economic damage associated with the probability of each forest condition (the probabilistic damage function) given status quo management.

Two points are worth noting. First, mathematical integration of the area beneath the probabilistic damage function in quadrant IV will provide the expected value of economic damages given the distribution of forest conditions shown in quadrant I. Second, the shape of the probability function of economic damages shown in quadrant IV is determined both by the shape of the distribution of forest conditions shown in quadrant I and the shape of the economic damage function shown in quadrant II. Therefore, the expected value of economic damages, as an indicator of social well-being, is affected both by the distribution of anticipated future forest conditions and the related economic values.

Step 7: Describe probable outcomes of actions—The final step in developing the framework is to replace the probability function of future forest conditions under status quo management with alternative probability functions of forest conditions anticipated using different resilience, resistance, and response options. This step allows a manager to compare the expected economic losses under various management alternatives. Different management options will incur different management costs. Therefore, a forest manager can evaluate alternative strategies by adding the costs and expected economic losses associated with each management option. One criterion for selecting a preferred strategy is to choose the option that minimizes

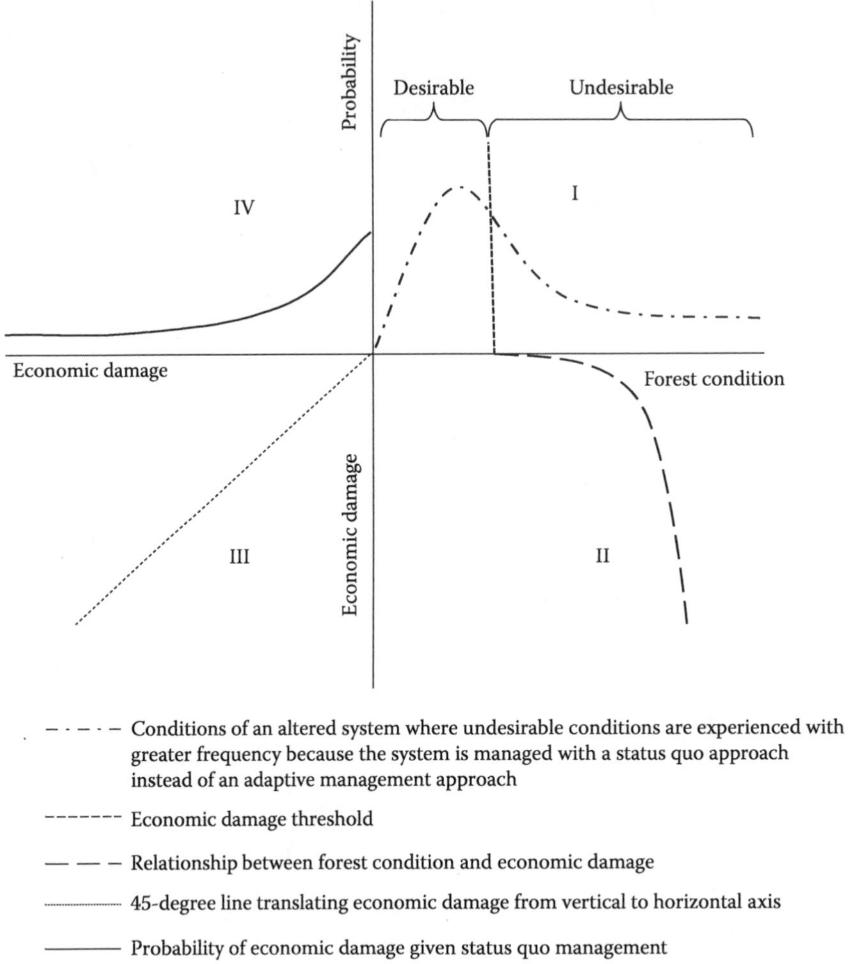

FIGURE 3.3 Four-quadrant diagram showing a conceptual model of the relationship between an identified indicator of forest condition and an identified indicator of economic damage for a forest ecosystem facing climate change. Note that the line in quadrant III merely represents a translation of economic damages from a vertical (quadrant II) to a horizontal plane.

the sum of costs plus expected losses. A second criterion is to choose the option that minimizes the probability of costs plus expected losses exceeding an intolerable level.

Over time, as new information about forest conditions and economic values becomes available, forest managers can update their assessments of how well resilience, resistance, and response options are providing sought-after ecosystem services. New information provides the basis for adaptive management, and by adapting to changing conditions (represented by updated probability and economic damage functions), managers can increase the probability of maintaining and enhancing desirable forest conditions.

FURTHER CONSIDERATIONS

Identifying Ecological and Economic Vulnerabilities

Vulnerability can be defined as the sensitivity of a system, subsystem, or system component to damage or harm resulting from exposure to a disturbance (Turner et al. 2003). It is, generally speaking, the inverse of resilience and resistance. That is, the most vulnerable systems (both ecologically and economically) have low resistance and low resilience. They are sensitive to even minor press or pulse disturbances, the magnitude of the impacts are large, and the duration of recovery is long.

For this chapter we defined three categories of vulnerabilities: vulnerability under current climate (VCC), vulnerability under future climate (VFC), and vulnerability under future climate with adaptive management (VFCAM). Combining the first two vulnerabilities with predicted climate change and the risk of incurring economic losses provides a template for identifying locations or ecosystems with the highest priority for management action; for example, where climate is likely to change the most, where low resistance and resilience causes a high probability of ecological or economic damage (VFC-VCC), and where the effects of management to reduce vulnerability (VFCAM-VCC < VFC-VCC) are large.

Partnerships Are Crucial

Adaptive management requires institutions that are flexible, willing to experiment with new (and perhaps unconventional) approaches, and able to incorporate learning from experience (Holling 2001; Seastadt et al. 2008). Within the United States, the national forests and experimental forests of the U.S. Forest Service are ideally poised for development into widely distributed working laboratories (both geographically and ecologically) to test hypotheses about adaptive management under climate change. Using these forests as laboratories would require close collaboration between the agency's management and research staffs. Such collaboration could lead to the identification of potential resilience, resistance, and response management strategies and to the creation of needed monitoring systems. The researchers could also provide the analytical resources needed to evaluate the alternative management strategies.

To further broaden the usefulness of this climate change-focused forest science, collaborative experiments could also be established across a spectrum of forest ownerships—from universities, to states, to forest industry and nonindustrial private owners. Such broadening would enable the testing of management strategies that recognize a wide range of potential ownership objectives and the diversity of ecosystem services that forests provide.

CONCLUSIONS

In this chapter, we have presented a conceptual framework for managing forests to adapt to climate change within the context of maintaining or enhancing the values that forests produce. Climate-induced changes in forest structure, composition, and function are likely to lead to overall negative economic impacts (costs plus losses). Although climate change could increase the benefits for some forest owners—perhaps by enhancing timber growth rates and timber growing opportunities—it is also likely to result in species range shifts and altered rates of press and pulse disturbances. Such changes would have negative impacts on both the commodity and the noncommodity values derived from public and private forests. Forest management, however, can alter ecological production functions in ways that reduce the overall negative impacts of climate change on ecosystems and the services these forests provide.

Forest management strategies designed to adapt to climate change by maintaining and enhancing desirable forest conditions will be made under conditions of risk and uncertainty. If probability functions can be estimated for indicators of forest conditions and for the parameters of economic damage functions, preferred actions could be selected based on criteria such as minimizing expected cost-plus-loss or minimizing the probability of economic damages exceeding an intolerable level. In situations where information is insufficient to estimate probability functions, other decision tools, such as scenario analysis, would be useful.

Whether conducting a risk analysis as described in our conceptual model is possible or not, forest managers can still consider what resilience, resistance, and response options may be available for managing their forests. The choice of preferred options will likely be very different across the suite of forest ownerships and types, because of varying ownership objectives and the diversity of ecosystem services each forest can provide. Perhaps the greatest challenges to maintaining desired forest conditions under future climates are faced by the owners of nonindustrial private forests. Because much of their forest acreage is passively managed, these owners may be reluctant to invest in enhancing resistance and resilience against future, uncertain levels of disturbances and species alteration. The result would be a large portion of the southern forested landscape that is highly vulnerable to climate change.

Scientific assessments can play a crucial role in designing adaptive forest management strategies. Research conducted across the entire spectrum of state, national, university, industrial, and private nonindustrial forests can be designed to test new strategies for adapting to climate change. An essential ingredient in sustainable forest management will be for all forest stakeholders to engage in maintaining the productivity of forest ecosystems and to develop a shared understanding of the social and economic values derived from productive forests.

GLOSSARY

Adaptation: Actions taken to decrease undesirable impacts of climate change, or that increase positive impacts.

Adaptive management: Iterative decision making based upon experimentation and scientific learning.

Ecosystem production functions: Processes that transform structural and functional ecosystem inputs into the services desired by people.

Ecosystem services: Outputs produced by ecosystems that are valued by people and contribute to their well-being.

Forest condition/state: The structure and function of the assemblage of dominant plant species found at a given location.

Mitigation: Actions taken to reduce the severity of potential climate change, such as sequestering carbon.

Press disturbance: A continuing disturbance, or stress, that slowly alters the state of an ecosystem.

Pulse disturbance: A short-term, high-intensity disturbance, or perturbation that rapidly alters the state of a system.

Resilience: The length of time required for a disturbed system to return to some initial functional state.

Resistance: The magnitude of disturbance that can be absorbed before an ecosystem state is significantly altered.

Risk: The functional relationship between a range of undesired ecological or economic conditions and their probability.

Sustainable management: Management that maintains the productive capacity of forests so that nondeclining levels of well-being are available for future generations.

Uncertainty: A lack of knowledge about the parameters of probability distributions.

Vulnerability: The sensitivity of a system, subsystem, or system component to damage or harm resulting from exposure to a disturbance.

REFERENCES

Adamowicz, W. 2003. Economic indicators of sustainable forest management: Theory versus practice. *Journal of Forest Economics* 9: 27–40.

Allen, C.D., A.K. Macalady, H. Chenchouni, D. Bachelet, N. McDowell, M. Vennetier, T. Kitzburger et al. 2010. A global overview of drought and heat induced mortality reveals emerging climate change risks for forests. *Forest Ecology and Management* 259(4): 660–684.

Arrow, K.J., P. Dasgupta, and K.-G. Mäler. 2003. Evaluating projects and assessing sustainable development in imperfect economies. *Environmental and Resource Economics* 26: 647–685.

Ayres, M.P. and M.J. Lombadero. 2000. Assessing the consequences of global change for forest disturbances from herbivores and pathogens. *The Science of the Total Environment* 262: 263–286.

Bender, E.A., T.J. Case, and M.E. Gilpin. 1984. Perturbation experiments in community ecology: Theory and practice. *Ecology* 65(1): 1–13.

Blate, G.M., L.A. Joyce, J.S. Little, S.G. McNulty, C.I. Millar, S.C. Moser, R.P. Neilson, K. O'Halloran, and D.L. Peterson. 2009. Adapting to climate change in United States national forests. *Unasylva* 60: 57–62.

Boggs, J.L., S.G. McNulty, M.J. Gavazzi, and J. Moore Myers. 2005. Tree growth, foliar chemistry, and nitrogen cycling across a nitrogen deposition gradient in southern Appalachian deciduous forests. *Can. J. For. Res.* 35: 1901–1913.

Breshears, D.D., N.S. Cobb, P.M. Rich, K.P. Price, C.D. Allen, R.G. Balice, W.H. Romme et al. 2005. Regional vegetation die-off in response to global-change-type drought. *Proceedings of the National Academy of Science* 102(42): 15144–15148.

Callaway, J.M. 2004. Adaptation benefits and costs: Are they important in the global policy picture and how can we estimate them? *Global Environmental Change* 14: 273–282.

Carpenter, S.R. and M.G. Turner. 2001. Hares and tortoises: Interactions of slow and fast variables in ecosystems. *Ecosystems* 3: 495–497.

Chapin, F.S., III and A.M. Starfield. 1997. Time lags and novel ecosystems in response to transient climatic change in arctic Alaska. *Climatic Change* 35: 449–461.

Daily, G.C., S. Polasky, J. Goldstein, P.M. Kareiva, H.A. Mooney, L. Pejchar, T.H. Ricketts, J. Salzman, and R. Shallenberger. 2009. Ecosystem services in decision making: Time to deliver. *Frontiers in Ecology and the Environment* 7(1): 21–28.

Dale, V.H., L.A. Joyce, S. McNulty, R.P. Neilson, M.P. Ayres, M.D. Flannigan, P.J. Hanson et al. 2001. Climate change and forest disturbances. *BioScience* 51(9): 723–734.

Dasgupta, P. 2008. Nature in economics. *Environmental and Resource Economics* 39: 1–7.

DeAngelis, D.L. and J.C. Waterhouse. 1987. Equilibrium and nonequilibrium concepts in ecological models. *Ecological Monographs* 57(1): 1–21.

Emanuel, K., R. Sundararajan, and J. Williams. 2008. Hurricanes and global warming: Results from downscaling IPCC AR4 simulations. *Bulletin of the American Meteorological Society* 89(3): 347–367.

Glasby, T.M. and A.J. Underwood. 1996. Sampling to differentiate between pulse and press perturbations. *Environmental Monitoring and Assessment* 42: 241–252.

Haefele, M., R.A. Kramer, and T.P. Holmes. 1991. Estimating the total value of forest quality in high-elevation spruce–fir forests. In: *The Economic Value of Wilderness: Proceedings of the Conference.* General Technical Report SE-78, Southern Forest Experiment Station, USDA Forest Service, Asheville, NC, pp. 91–96.

Holling, C.S. 1973. Resilience and stability of ecological systems. *Annual Review of Ecology and Systematics* 4: 1–23.

Holling, C.S. 2001. Understanding the complexity of economic, ecological, and social systems. *Ecosystems* 4: 390–405.

Holmes, T.P. and R.A. Kramer. 1996. Contingent valuation of ecosystem health. *Ecosystem Health* 2: 56–60.

Iverson, L.R., A.M. Prasad, S.N. Matthews, and M. Peters. 2008. Estimating potential habitat for 134 eastern tree species under six climatic scenarios. *Forest Ecology and Management* 254: 390–406.

Iverson, L., A.M. Prasad, S. Matthews, and M. Peters. 2011. Lessons learned while integrating habitat, dispersal, disturbance, and life-history traits into species habitat models under climate change. *Ecosystems* 14: 1005–1020.

Kahneman, D. and A. Tversky. 1979. Prospect theory: An analysis of decision under risk. *Econometrica* 47: 263–292.

Keller, K., G. Yohe, and M. Schlesinger. 2008. Managing the risks of climate thresholds: Uncertainties and information needs. *Climatic Change* 91: 5–10.

Klos, R.J., G.G. Wang, W.L. Bauerle, and J.R. Rieck. 2009. Drought impact on forest growth and mortality in southeast USA: An analysis using Forest Health and Monitoring data. *Ecological Applications* 19(3): 699–708.

Kramer, R.A., T.P. Holmes, and M. Haefele. 2003. Contingent valuation of forest ecosystem protection. In: *Forests in a Market Economy* (Eds. E.O. Sills and K.L. Abt). Kluwer Academic Publishers: Dordrecht, the Netherlands, pp. 303–320.

Kriegler, E., J.W. Hall, H. Held, R. Dawson, and H.J. Schnellnhuber. 2009. Imprecise probability assessment of tipping point in the climate system. *Proceedings of the National Academy of Sciences* 106(13): 5041–5046.

Kurz, W.A., C.C. Dymond, G. Stinson, G.J. Rampley, E.T. Neilson, A.L. Carroll, T. Ebata, and L. Safranyik. 2008. Mountain pine beetle and forest carbon feedback to climate change. *Nature* 452: 987–990.

Liao, C., Y. Luo, C. Fang, and B. Li. 2010. Ecosystem carbon stock influenced by plantation practice: Implications for planting forests as a measure of climate change mitigation. *PLoS ONE* 5(5): e10867.

Mäler, K.-G. 2008. Sustainable development and resilience in ecosystems. *Environmental and Resource Economics* 39: 17–24.

Mäler, K.-G., S. Aniyar, and Å. Jansson. 2008. Accounting for ecosystem services as a way to understand the requirements for sustainable development. *Proceedings of the National Academy of Sciences* 105: 9501–9506.

McNulty, S.G. and J.L. Boggs. 2010. A conceptual framework: Redefining forest soil's critical acid loads under a changing climate. *Environmental Pollution* 158: 2053–2058.

Millar, C.I., N.L. Stephenson, and S.L. Stephens. 2007. Climate change and forests of the future: Managing in the face of uncertainty. *Ecological Applications* 17(8): 2145–2151.

Millenium Ecosystem Assessment. 2003. *Ecosystems and Human Well-Being: A Framework for Assessment.* Island Press, Washington, DC.

North Carolina Department of Environment and Natural Resources. 2010. North Carolina ecosystem response to climate changes: DENR assessment of effects and adaptation measures, spruce–fir forests. http://portal.ncdenr.org/c/document_library/get_file?uuid=c180b2d8-8135-4b86-a9f0-e1cb344b3604&groupId=61587 [Date accessed: December 5, 2012].

Pan, Y., R.A. Birdsey, J. Fang, R. Houghton, P.E. Kauppi, W.A. Kurz, O.L. Phillips et al. 2011. A large and persistent carbon sink in the world's forests. *Science* 333: 988–993.

Perrings, C. 1998. Resilience in the dynamics of economy–environment systems. *Environmental and Resource Economics* 11: 503–520.

Pimm, S.L. 1984. The complexity and stability of ecosystems. *Nature* 307: 321–326.

Polasky, S., S.R. Carpenter, C. Folke, and B. Keeler. 2011. Decision-making under uncertainty: Environmental management in an era of global change. *Trends in Ecology and Evolution* 26(8): 398–404.

Prato, T. 2005. Bayesian adaptive management of ecosystem. *Ecological Modelling* 18: 147–156.

Prato, T. 2008. Conceptual framework for assessment and management of ecosystem impacts of climate change. *Ecological Complexity* 5: 329–338.

Raffa, K.F., B.H. Aukema, B.J. Bentz, A.L. Carroll, J.A. Hicke, M.G. Turner, and W.H. Romme. 2008. Cross-scale drivers of natural disturbance prone to anthropogenic amplification: The dynamics of bark beetle eruptions. *BioScience* 58(6): 501–517.

Rinaldi, S. and M. Scheffer. 2000. Geometric analysis of ecological models with slow and fast processes. *Ecology* 3: 507–521.

Roy, A.D. 1952. Safety first and the holding of assets. *Econometrica* 20(3): 431–449.

Running, S.W. 2008. Ecosystem disturbance, carbon, and climate. *Science* 321: 652–653.

Samuelson, W. and R. Zeckhauser. 1988. Status quo bias in decision making. *Journal of Risk and Uncertainty* 1(1): 7–59.

Seastadt, T.R., R.J. Hobbs, and K.N. Suding. 2008. Management of novel ecosystems: Are novel approaches required? *Frontiers in Ecology and the Environment* 6(10): 547–553.

Shoemaker, P.J.H. 1982. The expected utility model: Its variants, purposes, evidence and limitations. *Journal of Economic Literature* 20(2): 529–563.

Soja, A.J., N.M. Tchebakova, N.H.F. French, M.D. Flannigan, H.H. Shugart, B.J. Stocks, A.I. Sikhinin, E.I. Parfenova, F.S. Chapin III, and P.W. Stackhouse Jr. 2007. Climate-induced boreal forest change: Predictions versus current observations. *Global and Planetary Change* 56: 274–296.

Spittlehouse, D.L. and R.B. Stewart. 2003. Adaptation to climate change in forest management. *BC Journal of Ecosystems and Management* 4(1): 1–10.

Stoy, P.C., G.C. Katul, M.B.S. Siqueira, J-H. Juang, K.A. Novick, H.R. McCarthy, A.C. Oishi, and R. Oren. 2008. Role of vegetation in determining carbon sequestration along ecological succession in southeastern United States. *Global Change Biology* 14: 1–19.

Thompson, I., B. Mackey, S. McNulty, and A. Mossler. 2009. *Forest Resilience, Biodiversity, and Climate Change: A synthesis of the biodiversity/resilience/stability relationship in forest ecosystems.* Secretariat of the Convention on Biological Diversity, Montreal. Technical Series no. 43, 67 pages.

Tian, H., G. Chen, C. Zhang, M. Liu, G. Sun, A. Chappelka, W. Ren et al. 2012. Century-scale responses of ecosystem carbon storage and flux to multiple environmental changes in the southern United States. *Ecosystems* 15: 674–694.

Turner, B.L. II, R.E. Kasperson, P.A. Matson, J.J. McCarthy, R.W. Corell, L. Christensen, N. Eckley et al. 2003. A framework for vulnerability analysis in sustainability science. *Proceedings of the National Academy of Science* 100(14): 8074–8079.

Von Neumann, J. and O. Morganstern. 1947. *Theory of Games and Economic Behavior*. Princeton University Press, Princeton, NJ.

Vose, J.M., D.L., Peterson, and T. Patel-Weynand, eds. 2012. Effects of climatic variability and change on forest ecosystems: A comprehensive science synthesis for the U.S. forest sector. Gen. tech. Rep. PNW-GTR-870. Portland, OR: U.S. Department of Agriculture, Forest Service, Pacific Northwest Research Station. 265 p.

Weitzman, M.L. 2009. On modeling and interpreting the economics of catastrophic climate change. *The Review of Economics and Statistics* 91: 1–19.

Weitzman, M.L. 2011. Fat-tailed uncertainty in the economics of catastrophic climate change. *Review of Economics and Policy* 5: 275–292.

Westerling, A.L., H.G. Hildago, D.R. Cayan, and T.W. Swetnam. 2006. Warming and earlier spring increase western U.S. forest wildfire activity. *Science* 313: 940–943.

Westerling, A.L., M.G. Turner, E.A.H. Smithwick, W.H. Romme, and M.G. Ryan. 2011. Continued warming could transform Greater Yellowstone fire regimes by mid-21st century. *Proceedings of the National Academy of Science*, Early Edition, www.pnas.org/cgi/doi/10.1073/pnas.1110199108.

Yohe, G. 1996. Exercises in hedging against extreme consequences of global change and the expected value of information. *Global Environmental Change* 6: 87–101.

Yohe, G., N. Andronova, and M. Schlesinger. 2004. To hedge or not against an uncertain climate future? *Science* 306: 416–417.

Zhu, K., C.W. Woodall, and J.C. Clark. 2012. Failure to migrate: Lack of tree range expansion in response to climate change. *Global Change Biology* 18: 1042–1052.

4 Water Stress and Social Vulnerability in the Southern United States, 2010–2040

Cassandra Johnson-Gaither, John Schelhas, Wayne Zipperer, Ge Sun, Peter V. Caldwell, and Neelam Poudyal

CONTENTS

Water scarcities are striking in semiarid, subregions of the Southern United States such as Oklahoma and western Texas (Glennon 2009; Sabo et al. 2010). In Texas, water stress has been a constant concern since the 1950s when the state experienced severe drought conditions (Moore 2005). The nearly 2000-mile Rio Grande River, which forms part of the Texas–Mexico border, is the sole source of water for residents and businesses at the far western part of the state. The river also provides water for Ciudad Juarez in Mexico. Because of its strategic location, the Rio Grande is a site of water contestation between United States and Mexican concerns (Szydlowski 2007). To the east, a three-year, state-level dispute over water rights involved North and South Carolina, as each state struggled to assert rights to the waters of the 225-mile Catawba River, which flows through the two states. The Catawba is a source of domestic water for more than one million residents and provides electricity to nearly double that number of consumers. The dispute was settled amicably in 2010, but these examples illustrate that the oftentimes, taken-for-granted assumptions of freshwater access in industrialized nations are being challenged by decreasing supplies, resulting from both existing and projected climatic changes.

This chapter focuses on the social vulnerability of human communities to water stress in the U.S. South, commencing in 2010 and extending through 2040. One of the primary purposes for including it in this chapter is to highlight the interactions between climate change (in this case, through impacts on water availability) and humans, both in the context of social vulnerability and response drivers. To do so, we examine the intersection of social vulnerability and stresses on freshwater supply at 10-year intervals in the 13 southern states—Alabama, Arkansas, Florida, Georgia, Kentucky, Louisiana, Mississippi, North Carolina, Oklahoma, South Carolina, Texas, Tennessee, and Virginia. Using exploratory spatial-data analysis, we assess the spatial association between social vulnerability and water stress at the county level in each of the five subregions in the South— Appalachian-Cumberland highlands, Coastal Plain, Mississippi Alluvial Valley, Mid-South, and Piedmont (Eastern Forest Threat Assessment Center 2011).

We defined socially vulnerable populations as marginal groups with respect to social and human capital and entitlements/endowments—conditions that compromise a population's ability to anticipate, cope with, or recover from environmental hazards (Downing et al. 2005; Intergovernmental Panel on Climate Change 2007; Kelly and Adger 2000). Social vulnerability accounts for expected social inequities among classes, races, ethnic groups, and genders, and also includes processes and movements at scales beyond the local, for instance, immigration/migration and the built environment as they relate to human communities (level of urbanization, growth rate, and economic robustness).

Three primary factors are converging to threaten water resources in the South: increasing risks of water degradation from an expanding human population base, conversion of forests to alternative uses, and climate change (Lockaby et al. 2011; Sun et al. 2008). Climate change effects are expected to intensify impacts associated with land cover and population changes in the region over the next century (Stone 2012). Such changes include possible decreases in precipitation and increases in air temperatures, resulting in higher rates of evapotranspiration and increasing demand for crop irrigation (Sun et al. 2008). On average, water stress across the South has been relatively low since European colonization but varies spatially and has been appreciably higher in western Texas (Lockaby et al. 2011). In contrast to the westernmost subregions of the South, the humid, subtropical areas farther east have had a relatively abundant water supply. Again, however, population growth and urbanization, continuing forest losses, increasing drought frequencies, and the absence of mechanisms to address future water shortages could lead to a vulnerable situation vis-à-vis water availability.

We take an initial step toward examining the association between water stress and social vulnerability at the county level. As stated, a number of factors are expected to affect water stress over the next several decades. Because of data limitations, however, we are able to examine only the climate-related drivers of water stress for our analysis period (2010–2040). Our water stress variable reflects climate change only; therefore, the analysis should be taken as a screening-level assessment to identify places for further inquiry. For more detailed analysis of the effects of land-use change and population change on water stress, see Chapter 9.

Climate change effects are expected to manifest differently, depending on geography and the ability of populations to anticipate, withstand, and recover from or adapt to the resulting biophysical changes, as reported by the Intergovernmental Panel on Climate Change (2007):

> There are sharp differences across regions and those in the weakest economic position are often the most vulnerable to climate change and are frequently the most susceptible to climate-related damages.... There is increasing evidence of greater vulnerability of specific groups such as the poor and elderly not only in developing but also in developed countries....

The panel concluded that water resources will be among the most "climate-sensitive" systems that are likely to be affected by climate change.

A key factor in social vulnerability is what Adger and Kelly (1999) describe as the "architecture of entitlements"—an individual's access (by rights, knowledge, or both) to the complex web of material, cultural, social, and economic resources. Because entitlements are often disputed and challenged, they are embedded within the political economy both locally and over great distances. Entitlements are controlled primarily at the state level, although local custom can also play a role. Entitlements are typically distributed along sociodemographic lines, for example, according to wealth, caste (race/ethnicity), age, and physical and mental abilities.

The detection of inequitable access to freshwater sources is much more evident in developing countries where the divide between affluent and impoverished population sectors has been documented at length (Bakker 2007; Mukheibir 2010; Rockström 2003; Vincent 2004; Vörösmarty et al. 2005). Limited access to natural and cultural resources and technology, repression of basic freedoms and human rights, generational warfare, and environmental degradation are but a few of the factors that cause or contribute to extreme social vulnerabilities for the poor in the developing world (Wisner et al. 2004).

McCarthy (2002), however, proposes that inequities such as limited access to natural and cultural resources, marginalization of resource-dependent communities, and environmental degradation can also be found in the developed world, albeit at less-intense scales. Central to these inequities is the socially vulnerable or marginal status of resource users whose access is blocked by more powerful, external agents or by competition from those who have stronger connections to resource providers. For instance, the highly controversial "Water Wars" involves demands by Alabama, Florida, and Georgia for freshwater contained in the Apalachicola–Chattahoochee–Flint River Basin. In 1989, the U.S. Army Corps of Engineers announced plans to redistribute 20% of the hydroelectric water supply from Lake Sidney Lanier (headwaters for the basin) to metropolitan Atlanta until 2010 (Feldman 2008). Alabama and Florida responded with lawsuits in 1990. Competing demands actually stem from several sources: (1) the large, urban centers in Georgia (Atlanta) and Alabama (Birmingham) that require more water for domestic consumption; (2) agricultural uses by farmers in southwestern Georgia; (3) Lake Lanier recreation in northern Georgia; and (4) freshwater needs for ecological balance in the Apalachicola Bay estuary, which feeds the economically important oyster industry in the Florida panhandle (Feldman 2008). Perhaps, the most tension has been between Georgia and Florida, specifically, metropolitan Atlanta's demand for increased water retention in northern Georgia versus the need for freshwater flows to Florida's Apalachicola Bay.

This conflict was exacerbated in the summer of 2007 when a severe drought brought water demand to the forefront of southern politics. As the final recipient of water flows from the Apalachicola–Chattahoochee–Flint River Basin, Florida argued that stable flow regimes need to be maintained for ecosystem balance in the Apalachicola River and for species preservation—the freshwater fat three-ridge (*Amblema neislerii*) and the purple bankclimber (*Elliptoideus sloatianus*) are federally protected mollusks found only in the Apalachicola River—and maintenance of crucial saltwater-to-freshwater proportions in the ecologically and economically important Apalachicola Bay, which provides 90% of the oysters harvested in Florida and 10% of those supplied in the United States (Chapin 2003). Importantly, oyster harvesters and others claim that disruptions to the saline–freshwater balance in the bay would also disrupt traditional livelihoods and cultural practices.

The tristate Water Wars have been framed by the media as a contestation between rich, powerful interests in Georgia and poor fishermen in Florida (Williams 2007). This may be hyperbolic, but the strife highlights the potential for uneven access to resources as well the ability of large, urban population centers to control resources that are vital to the livelihood of smaller, resource-dependent communities.

Absent to new water sources or reductions in the rate of population growth, we may infer that disputes over water supplies will increase in coming decades and that socially marginal groups worldwide will be less able to afford this increasingly scarce commodity, whether publicly or privately supplied, or to influence the politics of water allocation (Dellapenna 2005). A U.S. Senate report states that water scarcity in Central Asia is now a national security concern because of the potential for armed conflict over water in the region, and, as economic growth intensifies in India, Africa, and the Middle East, so too does the demand for water and associated conflicts (British Broadcasting Corporation 2008; U.S. Senate Committee on Foreign Relations 2011).

The question then becomes: precisely how would social vulnerabilities exacerbate water stress in the U.S. South into the foreseeable future? There are not likely to be instances in which poor or socially vulnerable households are unable to acquire the resource. Rather, the amplification of water stresses by social vulnerability would be most likely in places where agriculture and other water-dependent economies predominate. For instance, in two Arizona valleys, Vásquez-León (2009) found that Hispanics had less access to institutional safeguards against drought than other farmers and workers, which perpetuated Hispanics' marginal position in agriculture. The study showed an unevenness of social capital and networks linking farmers to agricultural assistance programs (such as crop subsidies, federal disaster relief, and both public and private crop insurance) that had been designed to buffer against the full impact of seasonal floods and droughts. Because Hispanic farmers in the study typically did not produce commodity crops, such as cotton and corn, that qualify for government assistance, they relied on less-formal social networks involving family and friendship circles to help ensure the continuity of their farming operations. Also, their productivity was limited by a combination of factors including social connections, amount of land owned, English language fluency, technological literacy, and selective issuance of federal loans and water distribution. The study suggested that water scarcity, when compounded by social vulnerability can exacerbate the effects of climate vulnerability for a population that already experiences a greater degree of social marginalization and susceptibility than others.

In urban settings, water stress is also compounded by concentrations of socially vulnerable populations, many of whom are likely to be members of racial or ethnic minority groups but also the elderly and very young. But the water stress effect in urban areas is not on water scarcity but extreme precipitation events, for instance floods. Projected climate changes in precipitation and temperature will probably raise the risk of contamination of water used for drinking and recreational purposes (Patz et al. 2001). For instance, high levels of precipitation and runoff can greatly elevate counts of fecal bacteria in local streams and coastal waters (Paul and Meyer 2001). Similarly, increased runoff from extreme rain events can be a concern for cities with combined storm water drainage and sewage systems such as Atlanta, Georgia. Under normal conditions, wastewater treatment facilities handle combined sewage and storm water drainage. Under an extreme event, however, incoming water can exceed the capacity of the facility with overflows being discharged into surface waters. Such events have been linked to incidences of West Nile Virus in Atlanta (Chaves et al. 2009).

With respect to another type of environmental risk (wildfire) and socially vulnerable populations, Collins (2008a,b) found that in the White Mountains of Arizona, state and market institutions such as local fire-protection services and insurance insulated affluent homeowners from potential losses from wildfire. Lower socioeconomic groups, on the other hand, had to absorb the increased wildfire risk created by increasing migration to ecologically vulnerable places. Those with less income had fewer means to purchase or command the type of insulation readily available to higher income groups.

Given these examples from the agricultural, urban ecology, and wildfire risk literature, we suggest that the exacerbation of environmental risk, when social vulnerability is present, depends on the extent to which buffers or insulation from environmental stresses can be employed. Thus, it is the access to various forms of capital (human, social, physical, financial, or natural) that largely determines the extent to which vulnerabilities will intensify environmental stress (Wisner et al. 2004).

CONCEPTUALIZATIONS OF SOCIAL VULNERABILITY

Birkmann (2006) found at least 25 definitions of the term "vulnerability" and six "schools of thought" or analytic models of vulnerability in the literature. Although these definitions may differ with respect to scale and linkages to other crucial concepts such as sustainability, they are similar in that each acknowledges that both endemic and exterior conditions influence the extent of vulnerability. The endemic or interior basis of social vulnerability derives from research on famine and food insecurities (Bohle et al. 1994); in contrast, external vulnerability derives from the hazards literature, which focuses on quantifying risks associated with natural hazards and disaster.

The interior dimension of vulnerability is expressed as the "endpoint" or "outcome" perspective and the external dimension is expressed as the "starting point" or "contextual" framing (Brooks 2003; Kelly and Adger 2000; O'Brien et al. 2007). The outcome perspective constructs vulnerability as a result, as a state of being that arises after exposure and adaptation to some natural disaster or to more incremental forces such as climate change. This perspective makes no assumptions about the temporal or spatial variability of vulnerability.

The definition of vulnerability given by the Intergovernmental Panel on Climate Change (2001)—exposure of a system to climate change and variability, the sensitivity of a system to climate changes, and the ability of the system to adapt to the same—is an example of an outcome perspective on vulnerability. Exposure to hazard refers to the risk of danger (from water stress, wildfires, droughts, or hurricanes) to a population, sensitivity measures the sociodemographic factors associated with place that can either exacerbate or mitigate risk, and adaptive capacity is the ability of a community or population to reach a new equilibrium after all exposures and sensitivities have been accounted for (Polsky et al. 2007; Turner et al. 2003).

Alternatively, contextual vulnerability emphasizes the social conditions of place, identifying the political and cultural context of where environmental threats or risk exists. The focus rests on the larger institutions of place and how communities have greater or lesser access to the system's entitlement structure. Vulnerability is assumed to exist *a priori* to hazard exposure. The socioeconomic, political, and cultural fabric of place acts as a sieve through which biophysical hazards and scarcities are mediated or experienced (Pelling 2001).

Contextual vulnerabilities lie within the social and institutional processes that distribute entitlements across society, with entitlements understood as the differential access to and use of resources. Viewed through this lens, social vulnerability is an inherent or endemic quality of a system that may exist regardless of whether a natural hazard impacts a given locale. Kelly and Adger (2000) use the "wounded soldier" analogy to illustrate this framing: Injuries already existing or sustained in previous battles circumscribe a soldier's ability to withstand any present or future assaults, thus rendering him or her vulnerable. Similarly, a place or system with existing sociodemographic deficiencies would exist in a socially vulnerable state.

However, because human societies are not static, even seemingly immutable inabilities or vulnerabilities can be minimized with a collective effort to be resolute or adapt to environmental stresses over time. Importantly, the vulnerabilities of people and place change continually in response to feedback from the encounters of human and natural systems to physical and social processes (Birkmann 2006; Turner et al. 2003).

Cutter et al. (2003) developed a hazards-of-place model that extends the contextualization of vulnerability beyond sociodemographic variables (those associated with people) to other dimensions of vulnerability—urbanization, geography, growth rates, economic activity of place, and biophysical environment. The authors view social vulnerability in a broader context than poverty or residence in socially marginal geographies. When disaster strikes, wealthier, urbanized areas are considered as vulnerable as poorer places because of high wealth concentrations and the likelihood of large economic losses. Similarly, Brooks (2003) views social vulnerability not only in terms of socioeconomic characteristics but also as a function of the "physical environment as they relate to human systems," using as an example the engineering processes that change the direction and flow of water.

Turner et al. (2003) elaborate on the holistic nature of place by drawing attention to the way that sustainability contributes to vulnerability assessments. Important here is the inevitable interaction between biophysical and social systems (Figure 4.1). Their emphasis is on the vulnerabilities of the "coupled" system rather than looking at the vulnerability of social or biophysical systems in isolation. The model combines elements of both the outcome perspective and contextual framing: vulnerabilities are triggered by exposure to hazards or stressors (outcome), but the ability of the system to cope with the disturbance depends on conditions particular to the human and the natural environments. This means that the responses of systems are not uniform across place but vary according to the capabilities of each subsystem, whether human or physical, and that factors beyond

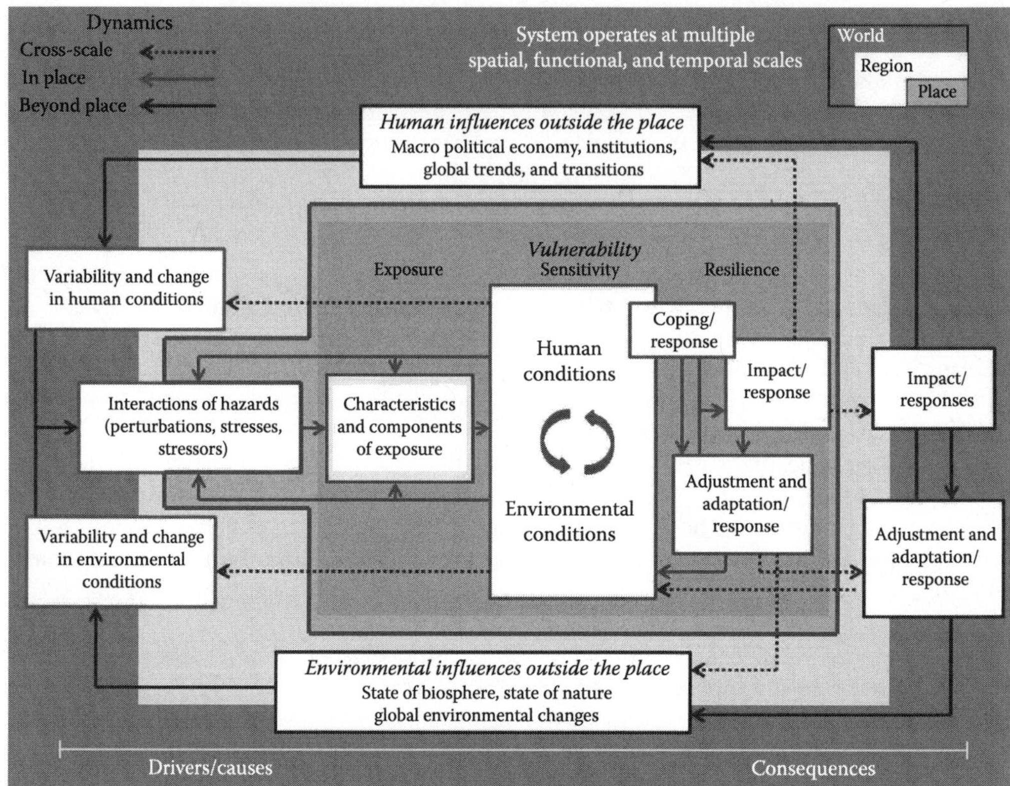

FIGURE 4.1 Schematic showing the components of social vulnerability and how they interact at various scales. (Adapted from Turner, B.L. et al., 2003. *Proceedings of the National Academy of Sciences*, July 8(100), 14:8074–8079.)

place influence the responses of systems and subsystems. Finally, the impacts of hazards are seen in how social and biophysical functions can withstand such impacts. Adaptations in the form of new policies and programs may be offered as a method of resilience; or the impact of the hazards may simply trigger adjustments and adaptations.

The next section describes the methodology used to examine the linkage between social vulnerability and water stress. Subsequently, we discuss findings and offer an alternative method for examining social vulnerability at the community level.

METHODOLOGY

Water Stress

We calculated a water stress value for each decade, 2010 through 2040; first averaging data by decade (2001–2010, 2011–2020, 2021–2030, and 2031–2040) and then averaging across four climate scenarios to develop an ensemble average for each hydrologic unit code (HUC) and decade. HUC codes are eight-digit numbers that identify watersheds addresses (nwis.waterdata.usgs.gov/tutorial/huc_def.html). The climate scenarios were developed by combining four commonly used general circulation models (MIROC3.2, CSIROMK2, CSIROMK3.5, and HadCM3) with two emissions storylines (Intergovernmental Panel on Climate Change 2007): A1B representing low population/high economic growth and high energy use, and B2 representing moderate growth and low energy use. We assumed constant land cover from 1997 levels developed by the U.S. Forest Service for the

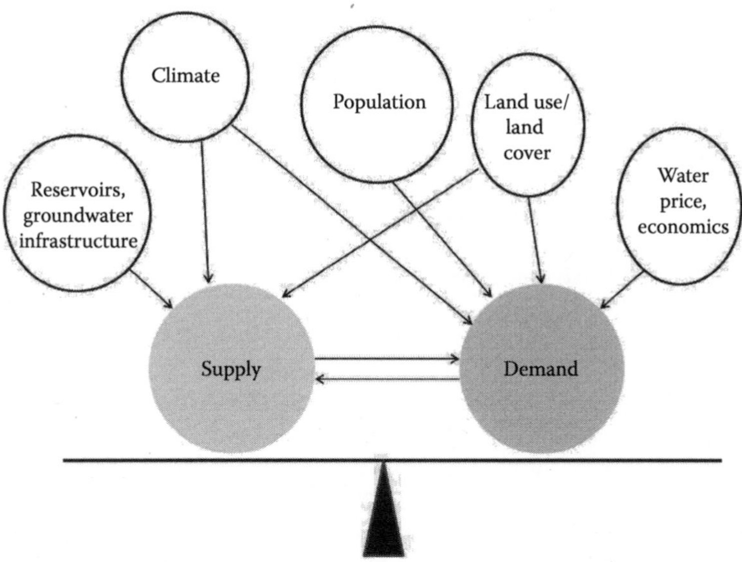

FIGURE 4.2 Factors affecting water supply demand. (Adapted from Sun, G. et al., 2008. *Journal of American Water Resources Association,* 44(6):1441–1457.)

2010 Resources Planning Act Assessment and constant water use from 2005 levels published by the U.S. Geological Survey (U.S. Forest Service 2012). All data were scaled to the county level.

Water stress is measured in terms of a water-accounting model or water supply stress index (WaSSI) (Sun et al. 2008), calculated as water demand divided by water supply. This straightforward explanation conceals the many complex interactions among the variables that constitute the index (Figure 4.2).

Lockaby et al. (2011) elaborate on the model:

> The scale of the water stress model can encompass an entire system from watershed to basin or any portion thereof, depending on the research question and availability of data to examine human water use and demand. The model simulates full monthly water balance, including evapotranspiration, soil moisture content, and water yield. Within each basin, spatially explicit land cover and soil data are used to account for evapotranspiration, infiltration, soil storage, snow accumulation and melt, surface runoff, and baseflow processes; and discharge is routed through the stream network from upstream to downstream watersheds.

Social Vulnerability

As with conceptualizations of social vulnerability, various approaches are available for measuring the construct, including quantitative indices and ethnographic methodologies that provide nuanced explanations of key concepts such as adaptive capacity and resilience (Füssel 2009; Nelson and Finan 2009). Numerous indicators have been developed (Cutter et al. 2003; Eriksen and Kelly 2007; Füssel 2009; Oxfam 2009; Polsky et al. 2007; Rygel, O'Sullivan, and Yarnal 2006). Many of these operationalize the definition of social vulnerability given by the third Intergovernmental Panel on Climate Change (2001):

$$\text{Social vulnerability} = \text{exposure} + \text{sensitivity} + \text{adaptive capacity} \tag{4.1}$$

We modified Equation 4.1 to develop a vulnerability equation that keeps the exposure variable but combines sensitivity and adaptive capacity into a *social vulnerability* index. In our model, the hazard occurrence is separated from social vulnerability, and the net effect of sensitivity and adaptive capacity gives social vulnerability. So, *exposure* indicates the degree of hazard occurrence (in the present case, water stress) in Equation 4.1, *sensitivity* measures sociodemographic characteristics at the individual and community scale, and *adaptive capacity* measures sociodemographic characteristics (or resistance) to those variables that would make a population or place more susceptible to environmental stress.

The sensitivity component of social vulnerability was measured with 10 sociodemographic variables from projected decadal estimates provided by Woods and Poole (2009) and from the 2000 U.S. Census. These are the proportion of county residents (1) more than 65 years old, (2) less than 5 years old, (3) female, (4) African American, (5) Hispanic, (6) Native American, (7) employed in farming, (8) employed in "forestry, fishing, related activities and other," (9) below the poverty level, and (10) aged 25 years or older without a high-school diploma. Of these, the first eight were projected to the decades 2010 through 2040 by Woods and Poole (2009). Poverty and education data are from the U.S. Census (2000); these remain constant from 2010 to 2040. U.S. Census 2010 data were not used because they had not yet been scaled to the county level at the time of writing. The adaptive capacity component of social vulnerability was measured by a single variable, the Woods & Poole Wealth Index (Woods and Poole 2009), 2010 through 2040. This index is an indicator of total personal income per capita. It is a summation of three weighted averages: regional per capita income divided by U.S. per capita income (contributing to 80% of the index); regional income from dividends, interest payments, and rent divided by the U.S. income from these sources (10% of index); and U.S. transfer payments divided by regional transfer payments (10% of index).

Each of the social vulnerability variables can have a direct bearing on social vulnerability for both individuals and communities. Age is a critical factor in vulnerability assessments because the elderly and very young are less able to garner resources in the event these become scarce. The same argument can be made for women relative to men. Racial and ethnic minority status, wealth, poverty status, and educational attainment can also play a role in resource acquisition (Collins 2008a,b; Taylor et al. 2007). Employment in agriculture or large-scale resource extraction (such as timber and mining) also bears directly on water stress, as these pursuits require abundant water resources.

All variables contributing to social vulnerability were combined additively to arrive at the social vulnerability index. The index was weighted by a county's reliance on activities relating to farming and fisheries. If 25% or more of the population were employed in farming or 15% or more were employed in fishing, each of these variables (percent employed in farming, percent employed in fishing) was assigned a weight of 0.60 and the remaining variables, except wealth, were assigned a weight of 0.40. If these conditions were not met, percent farming and fishing were assigned a weight of 0.20 and the other variables 0.80. We developed weights to distinguish counties that had stronger ties economically to natural resources, the idea being that resource-based economies would be more sensitive than others to water stresses. Weights were determined based on expert judgment of the impact of farming and fishing on county economies.

We tested for multicollinearity for variables comprising the social vulnerability component of water stress by examining a regression model where water stress was the dependent variable and the social vulnerability variables were predictors. Multicollinearity was detected by indications of inflated standard errors of the predictors. We examined the variance inflation factor (VIF) to detect fluctuating standard errors. Generally, VIF values greater than 10 suggest multicollinearity. VIF values for all predictors were less than 3.5, suggesting low or moderate multicollinearity. Proportions for each social vulnerability indicator were summed to produce the social vulnerability value for a given county in each ecoregion, per decade.

Bivariate Analysis

We examined the association between the two components of vulnerability (water stress and social vulnerability) by using the Local Indicator of Spatial Association (LISA statistic) (Anselin 1995). The LISA statistic indicates how well observations of one variable for a county (i) are associated with observations of a different variable in neighboring or adjacent counties. For our analysis, the variables were water stress for county *i* and social vulnerability in the cluster of counties adjacent to county *i*. A county was defined as being adjacent to county *i* based on a first-order, queen-contiguity weight matrix. Counties adjacent to county *i* that share a common border length or vertex were included in county *i*'s "neighborhood." The LISA statistic compares the average value for water stress in county *i* to the social vulnerability average for counties in the neighborhood of county *i*.

Bivariate LISA statistics were computed using GeoDa™ 0.9.5-I. Our analysis mapped four distinct spatial clusters pairing water stress and social vulnerability at the county level. The clusters, defined in relation to the average value for water stress and social vulnerability, are (1) high water stress–high social vulnerability, defined as a high water stress county adjacent to high social vulnerability neighboring counties; (2) low water stress–low social vulnerability, defined as a low water stress county adjacent to low social vulnerability neighboring counties; (3) low water stress–high social vulnerability, defined as a low water stress county adjacent to high social vulnerability neighboring counties; and (4) high water stress–low social vulnerability, defined as a high water stress county adjacent to low social vulnerability neighboring counties.

We defined high water stress–high social vulnerability counties as "hot spots." Given that we used actual population data rather than a sample, all associations and related LISA statistics were assumed to be significant. However, in keeping with conventional reporting of values significant at $p \leq 0.05$, we mapped only those clusters that met this criterion. Pseudo-*p* values were generated for LISA statistics using the 999-permutation criteria available in GeoDa0.9.5-I.

The calculation of the LISA statistic (Sunderlin et al. 2008) is

$$I_l = z_{xi} \sum_{j=1, j \neq i}^{N} w_{ij} z_{yj} \tag{4.2}$$

where I_l is the LISA statistic, x is water stress, and y is social vulnerability for county i and neighborhood j. Similarly, z_x represents the standardized value for water stress and z_y represents the standardized value for social vulnerability. The variable w_{ij} is the weight matrix that defines those counties comprising neighborhood j (also generated using GeoDa 0.9.5-I).

RESULTS OF UNIVARIATE ANALYSIS

Table 4.1 shows that water stress averages steadily increased in each subregion for the period 2010 through 2040. Table 4.2 shows similar increases for social vulnerability. Figure 4.3 maps water stress and Figure 4.4 maps social vulnerability in terms of its standard deviations from the average for the Coastal Plain in 2010. Figure 4.3 shows that the greatest water stress is in central and southern Florida counties, eastern South Carolina, and eastern Texas. Relatively, little is seen in southern Georgia and Alabama. Figure 4.4 shows that social vulnerability is highest in an arc-like pattern extending from eastern Virginia through the eastern Carolinas and into southern Georgia, Alabama, much of Mississippi, Louisiana, and eastern Texas. Compared to inland counties, coastal counties show less social vulnerability, as do all the counties in Florida.

SUBREGIONAL RESULTS OF BIVARIATE ANALYSIS

Appalachian-Cumberland Highlands and Piedmont

The Appalachian-Cumberland and Piedmont subregions are both represented in the Southern Appalachian landscape. This landscape encompasses one of the most diverse ecosystems that can be found in a temperate zone (Southern Appalachian Man and the Biosphere 1996). The Appalachian-Cumberland highlands cover approximately 99 million square miles and the Piedmont

TABLE 4.1
Current and Predicted Average WaSSI for the Five Subregions of the Southern United States, 2010 through 2040

Subregion	N	2010	2020	2030	2040
Appalachian-Cumberland highlands	252	0.011	0.012	0.013	0.014
		s.d. = 0.018	s.d. = 0.019	s.d. = 0.020	s.d. = 0.021
		max = 0.144	max = 0.148	max = 0.152	max = 0.154
		min = 0.000	min = 0.000	min = 0.000	min = 0.000
Piedmont	177	0.028	0.030	0.031	0.032
		s.d = 0.035	s.d. = 0.037	s.d. = 0.037	s.d. = 0.039
		max = 0.220	max = 0.226	max = 0.225	max = 0.232
		min = 0.001	min = 0.001	min = 0.001	min = 0.001
Coastal Plain	488	0.021	0.023	0.025	0.028
		s.d. = 0.030	s.d. = 0.034	s.d. = 0.036	s.d. = 0.040
		max = 0.231	max = 0.287	max = 0.317	max = 0.322
		min = 0.000	min = 0.000	min = 0.001	min = 0.001
Mississippi Alluvial Valley	64	0.037	0.039	0.042	0.042
		s.d. = 0.033	s.d. = 0.034	s.d. = 0.034	s.d. = 0.037
		max = 0.166	max = 0.169	max = 0.160	max = 0.187
		min = 0.000	min = 0.000	min = 0.000	min = 0.000
Mid-South	330	0.079	0.091	0.100	0.094
		s.d. = 0.098	s.d. = 0.107	s.d. = 0.116	s.d. = 0.106
		max = 0.60	max = 0.737	max = 0.840	max = 0.652
		min = 0.001	min = 0.000	min = 0.001	min = 0.000

Source: Adapted from Vörösmarty, C. J. et al., 2000. *Science,* 289:284–288.

Note: WaSSI is calculated as demand divided by supply: low is >0.1, medium is 0.1–0.2, medium–high is 0.2–0.4, high is ≥0.04. *N* is the number of measurements taken; s.d. is the standard deviation from the mean; max is the highest WaSSI for the subregion; min is the lowest WaSSI for the subregion.

approximately 79 million square miles. These two subregions share similar climatic characteristics and, to some extent, comparable social characteristics. The larger cities and metropolitan areas of the Appalachian-Cumberland highlands and Piedmont have been the epicenters of southern population expansion, economic growth, and cultural diversification over the past 30 years; however, some of the more rural and mountain communities continue to struggle with generational poverty and economic decline.

Figure 4.5a–d shows bivariate LISA clusters for the Appalachian-Cumberland highlands from 2010 to 2040. Three counties were consistently classed as hot spots across the four decades: Macon (North Carolina), Greene (Tennessee), and Washington (Virginia). In terms of social vulnerability, these are all counties with relatively low poverty and small minority populations, relative to their state averages; for Macon and Washington counties, the main drivers of vulnerability center around their low wealth and education indices. A large swath of counties with low water stress–high social vulnerability is located in southeastern Kentucky. These findings are consistent with other mappings of high-poverty counties in this part of Kentucky. Table 4.3 shows the distribution of counties according to cluster type by decade. In each decade, the percentage of each cluster type remained nearly constant. Only in 2040 did the number of high water stress–high social vulnerability clusters decrease to three from a high of six in 2030. Of the significant counties, the low water stress–high social vulnerability cluster predominated and the hot spots were least prevalent across all decades.

TABLE 4.2
Current and Predicted Average Social Vulnerability (SOVU) for the Five Subregions of the Southern United States, 2010 through 2040

Subregion	N	2010	2020	2030	2040
Appalachian-Cumberland highlands	252	0.317	0.342	0.376	0.408
		s.d. = 0.238	s.d. = 0.246	s.d. = 0.254	s.d. = 0.258
		max = 0.862	max = 0.882	max = 0.917	max = 0.959
		min = –0.630	min = –0.761	min = –0.938	min = –0.992
Piedmont	177	0.305	0.348	0.391	0.425
		s.d. = 0.348	s.d. = 0.337	s.d. = 0.335	s.d. = 0.339
		max = 1.166	max = 1.189	max = 1.20	max = 1.21
		min = –0.962	min = –0.948	min = –0.934	min = –0.891
Coastal Plain	488	0.508	0.544	0.580	0.606
		s.d. = 0.338	s.d. = 0.334	s.d. = 0.332	s.d = 0.335
		max = 1.407	max = 1.44	max = 1.48	max = 1.48
		min = –0.817	min = –0.735	min = –0.653	min = –0.723
Mississippi Alluvial Valley	64	0.667	0.716	0.746	0.761
		s.d. = 0.351	s.d. = 0.341	s.d. = 0.333	s.d. = 0.329
		max = 1.39	max = 1.41	max = 1.42	max = 1.42
		min = 0.058	min = 0.107	min = 0.176	min = 0.188
Mid-South	330	0.386	0.422	0.460	0.478
		s.d. = 0.419	s.d. = 0.419	s.d. = 0.418	s.d. = 0.415
		max = 1.93	max = 1.93	max = 1.95	max = 1.952
		min = –2.85	min = –2.80	min = –2.72	min = –2.357

Source: Adapted from Intergovernmental Panel on Climate Change. 2001. Contribution of Working Groups I, II, and III to the Third Assessment Report of the Intergovernmental Panel on Climate Change. Cambridge University Press.

Note: SOVU is calculated as the sum of exposure, sensitivity, and adaptive capacity. *N* is the number of measurements taken; s.d. is the standard deviation from the mean; max is the highest WaSSI for the subregion; min is the lowest WaSSI for the subregion.

Figure 4.6a–d shows cluster types for the Piedmont, 2010 through 2040. Counties with the-most consistent high water stress–high social vulnerability clustering are Henry (Virginia), Stokes, Rockingham, Cleveland, and Union (North Carolina); and York and Lancaster (South Carolina). Table 4.4 shows that again for the Piedmont, both the absolute number and percentage of cluster types remained essentially constant over the 30-year span.

Coastal Plain

The Coastal Plain encompasses land area extending from southern Virginia, down to southern Georgia and Alabama, and into Louisiana and eastern Texas—more than 79 million square miles. The Coastal Plain has also been one of the higher growth areas of the South, including metropolitan areas such as Miami and Houston. A mild climate, proximity to coastal areas, and lower cost of living in many of the rural areas of the Coastal Plain contribute to its attractiveness for migration. Again, however, persistently poor communities also characterize rural Black Belt areas that run through the deep South states of Arkansas, Georgia, Alabama, and Mississippi. The southern Black Belt is comprised of 623 counties spanning from southern Virginia through east Texas that have higher-than-average Black populations. Eighteen percent of the U.S. population is contained in the Black Belt (Allen-Smith et al. 2000; Johnson-Gaither et al. 2011). These counties are mostly adjacent although they span several states.

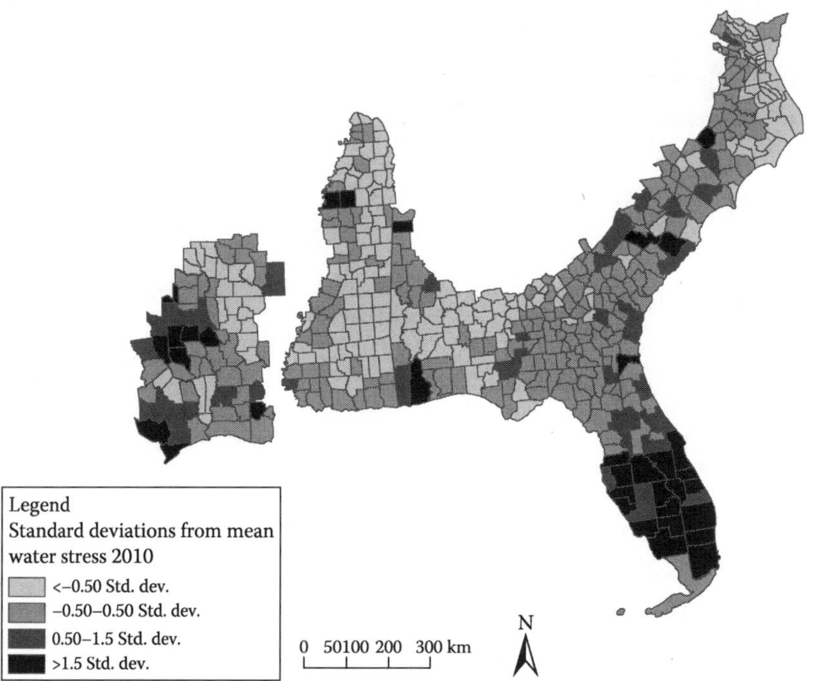

FIGURE 4.3 Water stress, 2010, on the Coastal Plain of the Southern United States, expressed as standard deviations from the mean (water stress is measured by the WaSSI).

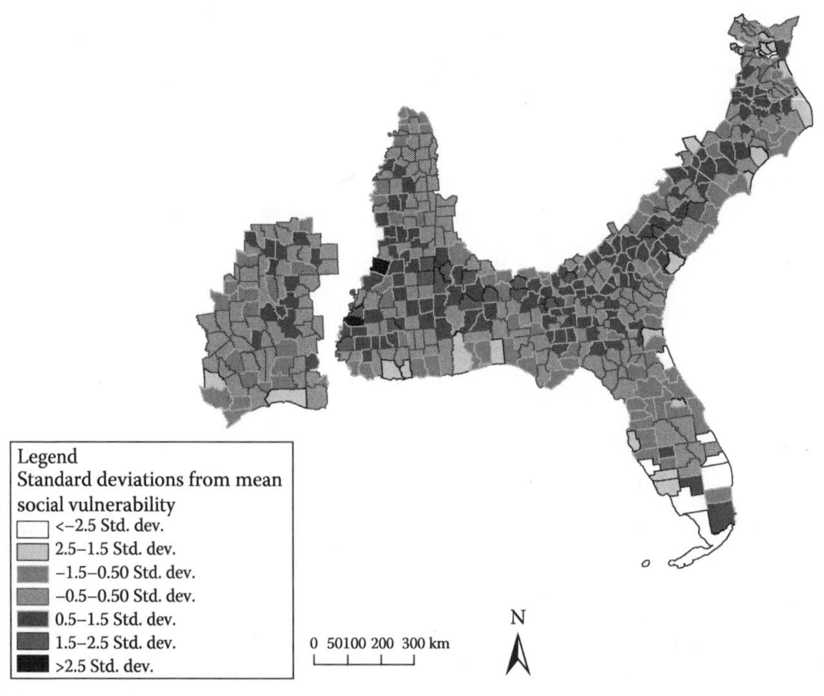

FIGURE 4.4 Social vulnerability, 2010, on the Coastal Plain of the Southern United States, expressed as standard deviations from the mean.

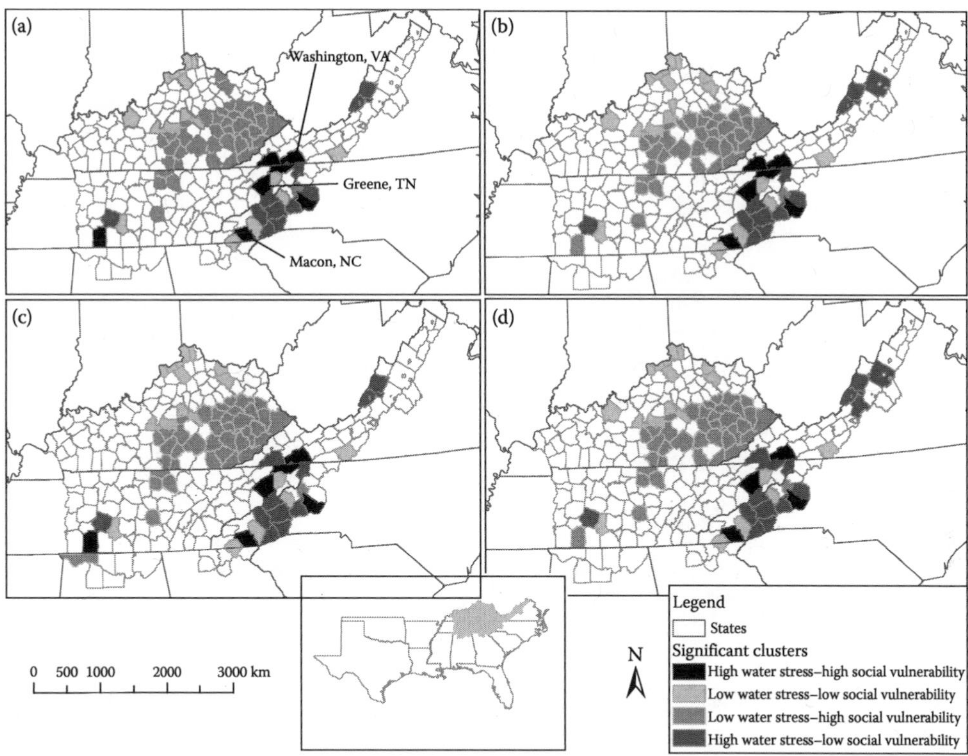

FIGURE 4.5 Convergence of water stress and social vulnerability in Appalachian-Cumberland counties of the Southern United States (a) in 2010 and predicted for (b) 2020, (c) 2030, and (d) 2040.

Figure 4.7a–d depicts clusters for the Coastal Plain. High water stress is indicated in central and southern Florida, accompanied by lower social vulnerability; in southern Alabama and Mississippi, higher levels of social vulnerability are prevalent but with little water stress. Hot-spot clusters appear only sporadically for each decade (Table 4.5).

TABLE 4.3
Current and Predicted Distribution of Counties for the Appalachian-Cumberland Highlands of the Southern United States, 2010–2040

	2010		2020		2030		2040	
Condition	**Number**	**%**	**Number**	**%**	**Number**	**%**	**Number**	**%**
High–high[a]	6	2.38	6	2.38	6	2.38	4	1.58
Low–low[b]	19	7.54	20	7.94	19	7.54	20	7.94
Low–high[c]	40	15.87	38	15.08	39	15.48	38	15.08
High–low[d]	12	4.76	12	4.76	11	4.37	15	5.96
Not significant	175	49.72	176	69.84	177	70.24	175	69.44
Total	252	100	252	100	252	100	252	100

[a] High water stress, high social vulnerability.
[b] Low water stress, low social vulnerability.
[c] Low water stress, high social vulnerability.
[d] High water stress, low social vulnerability.

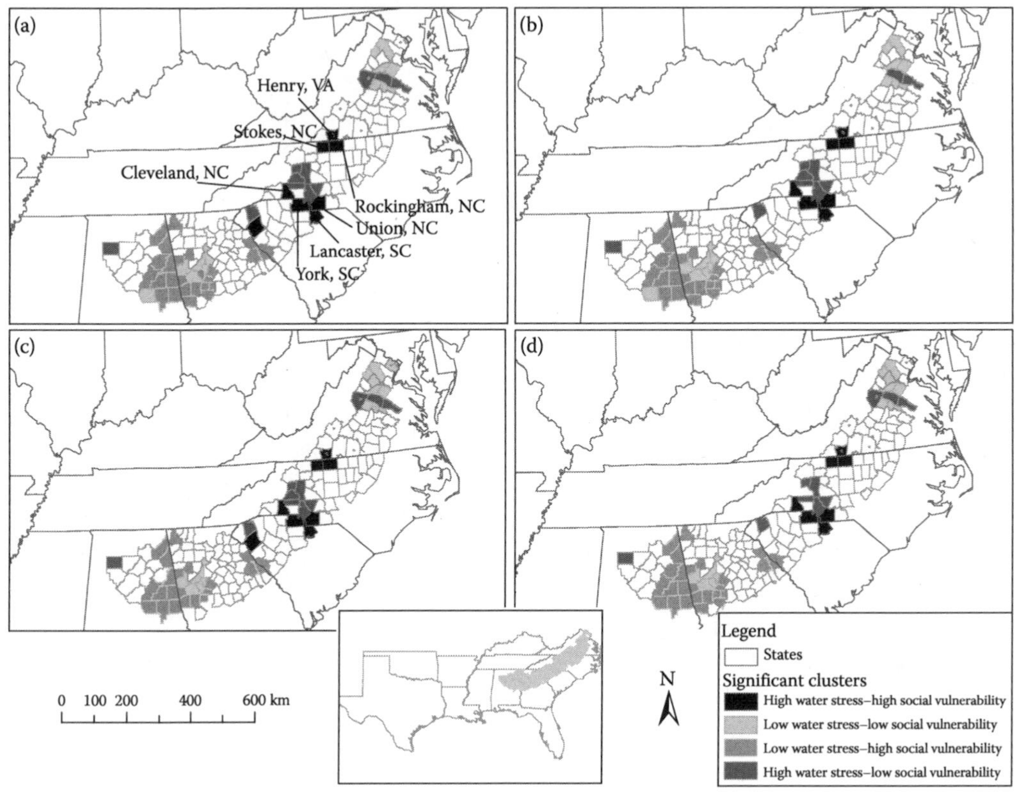

FIGURE 4.6 Convergence of water stress and social vulnerability in Piedmont counties of the Southern United States (a) in 2010 and predicted for (b) 2020, (c) 2030, and (d) 2040.

TABLE 4.4
Current and Predicted Distribution of Counties for the Piedmont of the Southern United States, 2010–2040

	2010		2020		2030		2040	
Condition	**Number**	**%**	**Number**	**%**	**Number**	**%**	**Number**	**%**
High–high[a]	8	4.51	7	3.95	8	4.52	7	3.95
Low–low[b]	12	6.78	13	7.34	13	8.47	12	6.78
Low–high[c]	26	14.69	25	14.12	22	14.69	26	14.69
High–low[d]	11	6.21	10	6.21	11	5.08	10	5.65
Not significant	120	67.80	122	69.49	123	69.49	122	68.93
Total	177	100	177	100	177	100	177	100

[a] High water stress, high social vulnerability.
[b] Low water stress, low social vulnerability.
[c] Low water stress, high social vulnerability.
[d] High water stress, low social vulnerability.

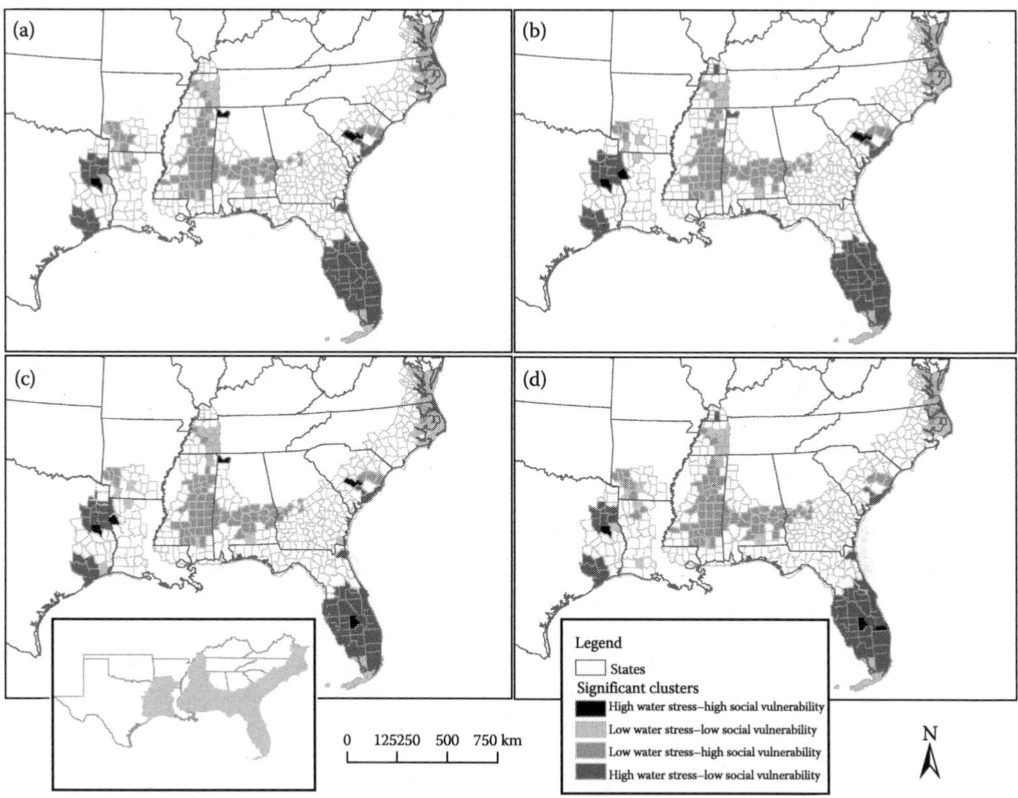

FIGURE 4.7 Convergence of water stress and social vulnerability in Coastal Plain counties of the Southern United States (a) in 2010 and predicted for (b) 2020, (c) 2030, and (d) 2040.

Mississippi Alluvial Valley

The Mississippi Alluvial Valley has its origins at the meeting of the Ohio and Mississippi Rivers in southern Illinois and extends to the Gulf of Mexico. Figure 4.8a–d shows no hot-spot clusters; overall, 11 or fewer counties were found to have a significant association between water stress and social vulnerability (Table 4.6). These results are not surprising given that the Mississippi Alluvial Valley is a riverine ecosystem that has a low-lying physiography and the largest coterminous wetlands system in North America (Omernik and Griffith 2008). Fertile soils make it especially suitable for farming, although flooding is common.

Mid-South

The Mid-South encompasses most of Texas, all of Oklahoma, and much of northwestern Arkansas. Table 4.7 shows that both the absolute number and percentage of cluster types remained virtually constantover the 30-year span. Figure 4.9a–d shows hot-spot clusters mainly in western and southern Texas.

Results are consistent with expectations for these parts of Texas, given that nearly all these hot-spot counties have high water stress (Texas Water Development Board 2008), relatively lower education rates, majority Hispanic populations, and sometimes higher-than-average poverty rates. For the seven counties comprising western Texas (El Paso, Hudspeth, Culberson, Jeff Davis, Presidio, Brewster, and Terrell), overall water demand is expected to increase by only 9% from 2010 to

TABLE 4.5
Current and Predicted Distribution of Counties for the Coastal Plain of the Southern United States, 2010–2040

	2010		2020		2030		2040	
Condition	**Number**	**%**	**Number**	**%**	**Number**	**%**	**Number**	**%**
High–high[a]	3	1.00	3	1.00	5	1.00	3	1.00
Low–low[b]	42	8.62	40	8.21	45	9.24	43	8.83
Low–high[c]	80	16.43	77	15.81	72	14.78	83	17.04
High–low[d]	43	8.83	47	9.65	46	9.45	43	8.83
Not significant	319	82.43	320	65.71	319	65.50	315	64.68
Total	487	100	487	100	487	100	487	100

[a] High water stress, high social vulnerability.
[b] Low water stress, low social vulnerability.
[c] Low water stress, high social vulnerability.
[d] High water stress, low social vulnerability.

2060, but municipal demand is expected to increase by 51% (Texas Water Development Board 2008). High water stress–low social vulnerability counties cluster in northwestern Texas. Low water stress–high social vulnerability counties cluster in eastern Oklahoma and western Arkansas and in some central and southern Texas areas; one extreme example is Cherokee County in Oklahoma, where roughly a third of the population is projected to be Native American in 2020 and the wealth index is only two-thirds the national average.

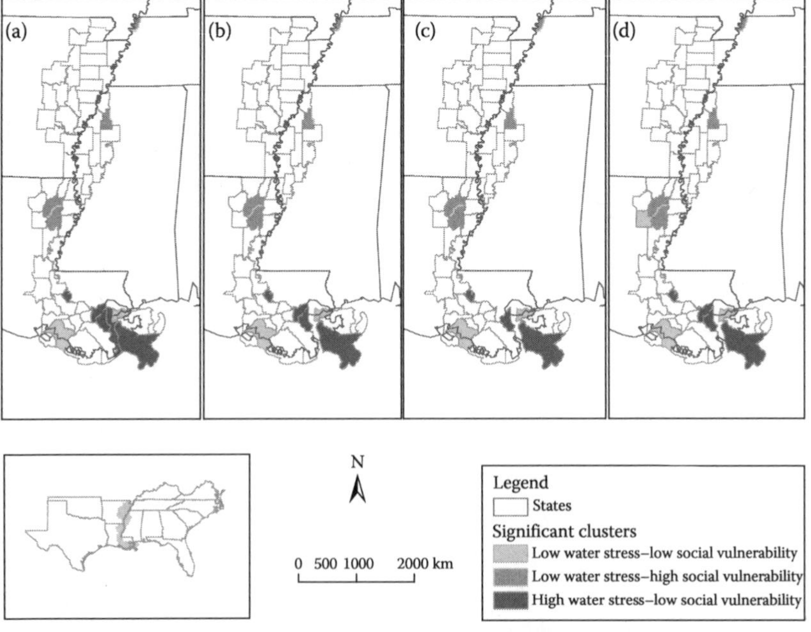

FIGURE 4.8 Convergence of water stress and social vulnerability in Mississippi Alluvial Valley counties of the Southern United States (a) in 2010 and predicted for (b) 2020, (c) 2030, and (d) 2040.

TABLE 4.6
Current and Predicted Distribution of Counties for Mississippi Alluvial Valley of the Southern United States, 2010–2040

	2010		2020		2030		2040	
Condition	**Number**	**%**	**Number**	**%**	**Number**	**%**	**Number**	**%**
High–high[a]	—	—	—	—	—	—	—	—
Low–low[b]	2	3.13	2	3.13	2	3.13	3	4.69
Low–high[c]	4	6.25	4	6.25	4	6.25	4	6.25
High–low[d]	4	6.25	4	7.82	3	4.69	4	6.25
Not significant	54	84.38	54	83.38	55	85.94	53	82.81
Total	64	100	64	100	64	100	64	100

Note: — means zero.
[a] High water stress, high social vulnerability.
[b] Low water stress, low social vulnerability.
[c] Low water stress, high social vulnerability.
[d] High water stress, low social vulnerability.

TABLE 4.7
Current and Predicted Distribution of Counties for the Mid-South of the Southern United States, 2010–2040

	2010		2020		2030		2040	
Condition	**Number**	**%**	**Number**	**%**	**Number**	**%**	**Number**	**%**
High–high[a]	26	7.88	22	6.67	27	8.18	26	7.88
Low–low[b]	56	16.97	55	16.67	53	16.06	54	16.36
Low–high[c]	43	13.03	43	13.03	44	13.33	45	13.63
High–low[d]	23	6.97	27	8.18	19	5.76	24	7.27
Not significant	182	55.15	183	55.45	187	56.67	181	54.85
Total	330	100	330	100	330	100	330	100

[a] High water stress, high social vulnerability.
[b] Low water stress, low social vulnerability.
[c] Low water stress, high social vulnerability.
[d] High water stress, low social vulnerability.

DISCUSSION

Returning to the question posed earlier concerning real-world interplays of social vulnerability and water stress, we offer further discussion along the lines of the relative lack of buffers or insulating factors that act to shield socially vulnerable groups from environmental stresses impacting landownership and improvement. We suggested that the vulnerability of a place is compounded when there are simultaneously environmental stresses (e.g., water scarcity) and socially vulnerable populations. Socially vulnerable populations with significant landownership or whose income depends to a large extent on water availability are typically less likely than those not in socially precarious conditions to be able to protect their ownership interests from environmental stresses

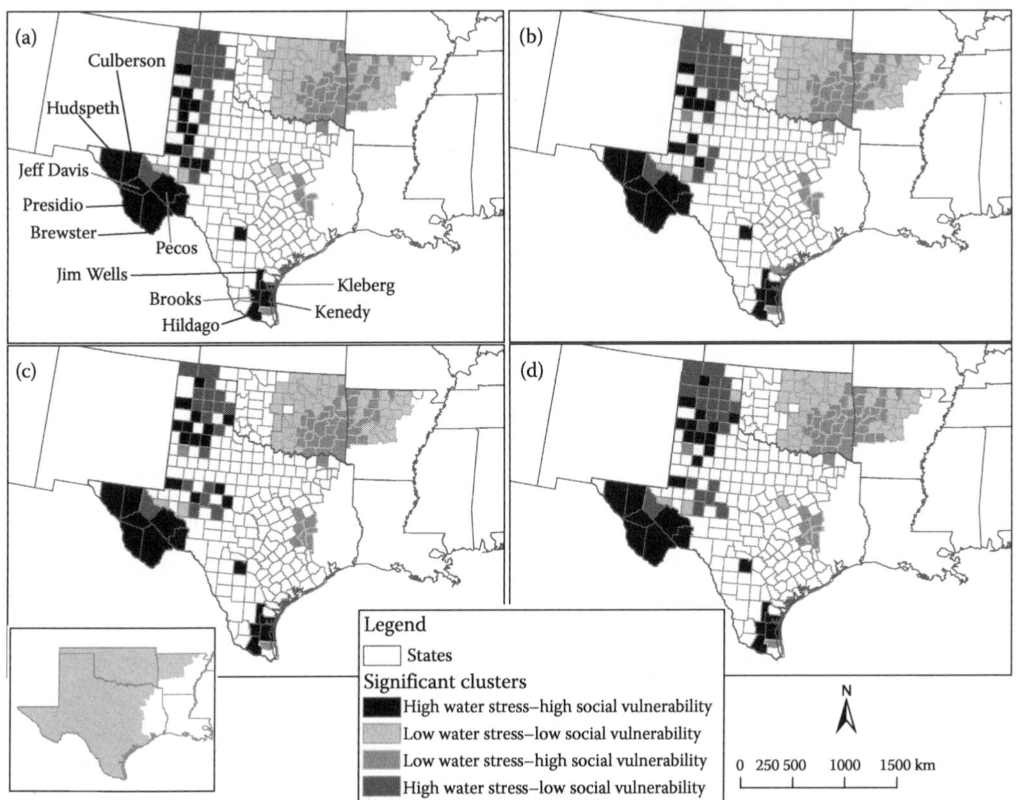

FIGURE 4.9 Convergence of water stress and social vulnerability in Mid-South counties of the Southern United States (a) in 2010 and predicted for (b) 2020, (c) 2030, and (d) 2040.

such as drought or wildfire (Collins 2008a,b; Johnson-Gaither et al. 2011). A crucial factor in such protection involves the legal status of the land in question. This question also references the problem of entitlements discussed earlier.

"Heir property," fractionation, or tenancy in common is a typical form of landownership in the South, particularly among Black-Belt African Americans but is also prevalent among Appalachian Whites and Native Americans (Deaton et al. 2009; Dyer et al. 2009; Mitchell 2000, 2001; Shoemaker 2003). Heir property describes undivided, real property owned simultaneously by a potentially unlimited number of heirs. No single owner has clear title to land. This type of land acquisition usually comes about in cases where a family member dies intestate or without a will specifying exact amounts and location of ownership accruing to descendants.

In 1978, Graber estimated that one-third of all Black-held land in the rural South could be classed as heir property. Graber (1978, p. 276) details that "heirs [sic] property is most acute in rural areas and in counties that are still untouched by heavy industry, by suburban home building, by oil exploration or by extensive resort development." Further, "[t]here is an absence of black real estate (and heir property) in those counties where large tracts of land are owned by timber or pulp companies, or where a large part of the land mass is in public ownership (i.e., national forests)" (p. 276). Later, Mitchell (2000) offered that 41% of African American-owned land in the southeast could be classed as heir property. Shoemaker (2003, p. 729) also asserts that in "Indian Country today, fractionation has reached crisis proportions," resulting in millions of actual heirs to small land parcels. Today, rights to Native Lands are recognized; however, Shoemaker (2003, p. 730) writes that "Other potential benefits" of the land are lost; and that "[h]omelessness and poverty persist within the traditional homeland, and Indian Country continues to face critical obstacles to true economic, cultural, and political security."

The common denominator among these groups with respect to heir property is lack of wealth and education. Socially vulnerable populations are more likely than middle-class or affluent groups to not have wills; and the lack of wills only serves to perpetuate social vulnerabilities. Although Dyer and Bailey (2008) call attention to the cultural significance of heir property for southern African Americans, Dyer et al. (2009) also argue that such landownership is also rife with a number of legal impediments that inhibit the effective use of such properties as collateral for home mortgages, any USDA Rural Development loans available for home building, or the array of farm programs offering protection from environmental hazards, including crop insurance (Dyer et al. 2009; Graber 1978; Shoemaker 2003). Moreover, Deaton et al. (2009) stress that heir property acts as a deterrent to wealth accumulation for individual families because intergenerational transfers of property and income are key to wealth accumulation in the United States.

Specifically, Deaton et al. (2009) write that heir property presents a problem for landowners because of "efficiency and displacement." Efficiency has to do with the underutilization of land and displacement with land loss. With reference to the efficiency difficulty, Deaton et al. (2009, p. 2347) describe heir property as a "tragedy of the anti-commons," meaning that the existence of multiple landowners can result in an apathetic engagement with privately held resources because of disincentives for land improvement. For example, in three case studies from Appalachia, Deaton et al. (2009) found that the lack of agreement on property management or selling price to an outside party resulted in nothing being done either to utilize resources or improve land, for a significant amount of time. Another problem with heir property is that any cotenant who wishes to realize the cash value of his or her land share can force a court sale of an entire property; the land is then sold to the highest bidder, resulting in the possible dispossession and displacement of actual family members. Graber (1978, p. 278) writes that heir property is land with "greatly diminished value" because of such restrictions; further, heir property "will not finance a home or farm equipment or serve as collateral for an emergency loan." A limited-resource landowner in a water-stressed location in need of a deeply drilled well would find difficulties in obtaining the loan if the intestate properties were intended for collateral. Given the prevalence of heir property among socially vulnerable populations, this form of landownership represents a hindrance to the acquisition of buffers such as property insurance and loans that could protect land from environmental stressors.

Policy and Management Implications

Mapping social vulnerability to environmental change and hazards, as we have done in this chapter, is useful in a number of ways. Social vulnerability indicators that draw from census data enable comprehensive mapping of social vulnerability over large geographic areas with a modest investment of time and little fieldwork. By highlighting vulnerable areas, maps help identify priority places for planning and for focusing responses to sudden environmental disasters, and for targeting mitigation efforts, such as planting trees to reduce urban heat island effects (Cox et al. 2006). However, social vulnerability indicators have limitations, many of which result from using census measures as proxy variables for more complicated social and economic processes that cannot be easily measured and are generally unavailable across large geographic areas.

Although indicators are useful for telling where to target interventions, Ribot (2011) suggests that developing effective interventions also requires a causal analysis of why people are at risk and identification of the potential entry points for risk reduction. As discussed, this would entail an examination of the architecture of entitlements as these relate to climate mitigation. Place-centered causal analysis starts with the unit at risk and the assets or entitlements that are (or are not) present, and then traces the causes of these conditions outward to the larger physical, social, and political–economic environment to understand why and when particular vulnerabilities occur at certain times

(Ribot 2009). This requires detailed, site-specific work for which ethnographic research and participatory planning techniques are well suited.

Ethnographic research can bring in important elements of human vulnerability and adaptive capacity that are difficult or impossible to understand from census data or to measure through surveys. Roncoli et al. (2009) propose that subjective judgment, cultural meanings, and political agendas shape both the awareness of potential climate changes and the responses to climate-related events. Ethnographic methods are needed to understand these patterns, as well as to help identify the processes that shift risks and costs to minority groups as the larger society plans and adapts to climate change.

The mapping presented in this chapter can be used as a first-level indicator of water stress vulnerability. Using the "hot spots" as a guide, research can be directed to a fine-tuned analysis of the contextual vulnerabilities and adaptive capacities within these counties. This step would necessitate collaboration between researchers and county managers (elected officials, extension agents, or human welfare agents) to explore the informal social networks that might operate among limited-resource landowners or farmers, for instance, the goal of which is to understand better how public resources may reinforce water stress alleviation provided by these networks. Informal networks supporting economic activities can include collectives based on friendship, kin, or religious ties. It is these unofficial, invisible webs that can provide the most aid to socially marginal populations in agricultural economies because, again, these populations typically make less use of formal-support networks. Yet, Vásquez-León (2009, p. 291) writes that "Climate-vulnerability studies that focus on the agricultural sector in the United States and Canada rarely look at informal networks or local-level collective strategies." This exploration would provide information on how groups acquire and share information about risk and disaster preparedness and management; further, the analysis would uncover factors impacting competition among groups for resource access and identify specific actions that could be undertaken by county managers to reduce tensions among various groups. While county agents may not be able to address long-standing inequities such as poverty, they may be able to discover why certain groups are routinely excluded from the benefit of state-sponsored buffers such as information delivery and other extension activities. We offer the current work as starting point for looking at the intersection of water stress and social vulnerability in the U.S. South.

REFERENCES

Adger, W.N. and Kelly, P.M. 1999. Social vulnerability to climate change and the architecture of entitlements. *Mitigation and adaptation strategies for global change*, 4:253–266.

Allen-Smith, J.E., Wimberley, R.C., and Morris, L.V. 2000. America's forgotten people and places: Ending the legacy of poverty in the rural south. *Journal of Agricultural and Applied Economics*, 32:319–329.

Anselin, L.1995. Local indicators of spatial association-LISA. *Geographical Analysis*, 27(2):93–115.

Bakker, K. 2007. Trickle down? Private sector participation and the pro-poor water supply debate in Jakarta, Indonesia. *Geoforum*, 38:855–868.

Birkmann, J. 2006. Measuring vulnerability to promote disaster-resilient societies: Conceptual frameworks and definitions. In: Birkmann, J. ed., *Measuring Vulnerability to Natural Hazard*, pp. 1–54. New York: United Nations University Press, pp. 524.

Bohle, H.G., Downing, T.E., and Watts, M.J. 1994. Climate change and social vulnerability: Toward a sociology and geography of food insecurity. *Global Environmental Change*, 4(1):37–48.

British Broadcasting Corporation. 2008. Worldwide battle for water. Available online at http://news.bbc.co.uk/2/hi/science/nature/7569453.stm.

Brooks, N. 2003. Vulnerability, Risk and adaptation: A conceptual framework. Working Paper 38. Tyndall Centre for Climate Change Research. Norwich, UK.

Chapin, T. 2003. A population and employment forecast for Franklin County. The Department of Urban and Regional Planning, Florida State University. http://mailer.fsu.edu/~tchapin/garnettchapin/urp5261/FC%20Final%20Report.pdf.

Chaves, L.F., Keogh, C.L., Vazquez-Prokopec, G.M., and Kitron. U.D. 2009. Combined sewage overflow enhances ovipositon of *Culex quinquefasciatus* (Diptera: Culicidae) in urban areas. *Journal of Medical Entomology,* 46:220–226.

Collins, T.W. 2008a. The political ecology of hazard vulnerability: Marginalization, facilitation and the production of differential risk to urban wildfires in Arizona's White mountains. *Journal of Political Ecology*, 15:21–43.

Collins, T. 2008b. What influences hazard mitigation? Household decision making about wildfire risks in Arizona's White Mountains. *The Professional Geographer*, 60:508–526.

Cox, J.R., Rosenzweig, C., Solecki, W., Goldberg, R., and Kinney, P. 2006. *Social Vulnerability to Climate Change: A Neighborhood Analysis of the Northeast U.S. Megaregion. Northeast Climate Impacts Assessment (NECIA).* Cambridge, MA: Union of Concerned Scientists.

Cutter, S., Boruff, B., and Shirley, W. 2003. Social vulnerability to environmental hazards. *Social Science Quarterly,* 84(1):242–261.

Deaton, B.J., Baxter, J., and Bratt, C.S. 2009. Examining the consequences and character of "heir property." *Ecological Economics*, 68:2344–2353.

Dellapenna, J.W. 2005. Markets for water: Time to put the myth to rest? *Journal of Contemporary Water Research and Education*, 131(1):33–41.

Downing, T., Aerts, J., Soussan, J., Bharwani, S., Ionescu, C., Hinkel, J., Klein, R. et al. 2005. Integrating Social Vulnerability into Water Management, NeWater Working Paper No. 5. New Approaches to Adaptive Water Management under Uncertainty. Available online at http://www.pik-potsdam.de/research/research-domains/transdisciplinary-concepts-and-methods/project-archive/favaia/pubs/downing_etal_2005.pdf. Date accessed 16 June 2011.

Dyer, J.F. and Bailey, C. 2008. A place to call home: Cultural understanding of heir property among rural African Americans. *Rural Sociology*, 73(3):317–338.

Dyer, J.F., Bailey, C., and Van Tran, N. 2009. Ownership characteristics of heir property in a black belt county: A quantitative approach. *Southern Rural Sociology*, 24(2):192–217.

Eastern Forest Environmental Threat Assessment Center. 2011. CCAMMO Ecoregions. Available online at: http://www.forestthreats.org/current-projects/project-summaries/CCAMMO_ecoregions.png/view. Date accessed 15 November 2011.

Eriksen, S.H. and Kelly, P.M. 2007. Developing credible vulnerability indicators for climate adaptation policy assessment. *Mitigation and Adaptation Strategies for Global Change*, 12:495–524.

Feldman, D.L. 2008. Barriers to adaptive management: Lessons from the Apalachicola-Chattahoochee-Flint compact. *Society and Natural Resources*, 21(6):512–525.

Füssel, H.M. 2009. Review and quantitative analysis of indices of climate change exposure, adaptive capacity, sensitivity, and impacts. Potsdam Institute for Climate Impact Research. Available online at: http://siteresources.worldbank.org/INTWDR2010/Resources/5287678-1255547194560/WDR2010_BG_Note_Fussel.pdf. Date accessed 9 September 2011.

Glennon, R.J. 2009. *Unquenchable: American's Water Crisis and What to Do About It.* Island Press, Washington, D.C.

Graber, C.S. 1978. Heirs property: The problems and possible solutions. *Clearinghouse Review*, 12:273–284.

Intergovernmental Panel on Climate Change. 2001. Contribution of Working Groups I, II, and III to the Third Assessment Report of the Intergovernmental Panel on Climate Change, Cambridge University Press.

Intergovernmental Panel on Climate Change. 2007. Intergovernmental Panel on Climate Change Synthesis Report. Available online at http://www.ipcc.ch/publications_and_data/ar4/syr/en/mains5-2.html. Date accessed 12 September 2011.

Johnson-Gaither, C., Gan, J., Jarret, A., Wyman, M.S., Malone, S., Adams, K.J., Bowker, J.M., and Stein, T.V. 2011. Black Belt Landowners Respond to State-sponsored Wildland Fire Mitigation Policies and Programs. General Technical Report, SRS-139. Asheville, NC: U.S. Department of Agriculture Forest Service, Southern Research Station, pp.25.

Kelly, P.M. and Adger, W.N. 2000. Theory and practice in assessing vulnerability to climate change and facilitating adaptation. *Climatic Change*, 47:325–352.

Lockaby, G., Nagy, C., Vose, J.M., Ford, C.R., Sun, G., McNulty, S., Caldwell, P., Cohen, E., and Moore Myers, J. 2011. Water and forests. In: *Southern Forest Futures Project*. Southern Research Station, Asheville, NC. Southern Forest Futures Project. Chapter 13. Available online at: http://www.srs.fs.usda.gov/futures/reports/draft/pdf/Chapter%2013.pdf. Date accessed 2 July 2013.

McCarthy. J. 2002. First world political ecology: Lessons from the wise use movement. *Environment and Planning A*, 34:1281–1302.

Mitchell, T.W. 2000. From Reconstruction to Deconstruction: Undermining Black Landownership, Political Independence, and Community through Partition Sales of Tenancies in Common. Land Tenure Center Research Paper 132. Land Tenure Center, University of Wisconsin-Madison. Available online at: http://papers.ssrn.com/sol3/papers.cfm?abstract_id=1544380. Date accessed unknown.

Mitchell, T.W. 2001. From reconstruction to deconstruction: Undermining black landownership, political independence, and community through partition sales of tenancies in common. *Northwestern University Law Review*, 95:505–580.

Moore, J.G. 2005. A half century of water resource planning and policy, 1950–2000. In: Norwine, J., Giardino, J.R., and Krishnamurthy, S. eds., *Water for Texas*, pp. 5–16. Moore Publication, College Station, Texas A & M University Agricultural Series, no. 6.

Mukheibir, P. 2010. Water access, water scarcity, and climate change. *Environmental Management*, 45: 1027–1039.

Nelson, D.R. and Finan, T.J. 2009. Praying for drought: Persistent vulnerability and the politics of patronage in Ceará, Brazil. *American Anthropologist*, 111(3):302–316.

O'Brien, K., Eriksen, S., Nygaard, L.P., and Schjolden, A. 2007. Why different interpretations of vulnerability matter in climate change discourse. *Climate Policy*, 7:73–88.

Omernik, J.M. and Griffith, G.E. 2008. Ecoregions of the Mississippi Alluvial Plain. *The Encyclopedia of the Earth*. Available at: http://www.eoearth.org/article/Ecoregions_of_the_Mississippi_Alluvial_Plain_(EPA). Date downloaded: 13 October 2011.

Oxfam. 2009. Exposed: Social vulnerability and climate change in the US southeast. Available online at: http://adapt.oxfamamerica.org/resources/Exposed_Report.pdf. Date accessed 8 December 2009.

Patz, J.A., McGeehin, M.A., Bernard, S.M., Ebi, K.L., Epstein, P.R., Grambsch, A., Gubler, D.J. et al. 2001. Potential consequences of climate variability and change for human health in the United States. In: N.A.S. Team, (ed.). *Climate Change Impacts on the United States: The Potential Consequences of Climate Variability and Change*, pp. 437–458. Cambridge: Cambridge University Press.

Paul, M.J. and Meyer, J.L. 2001. Streams in the urban landscape. *Annual Review of Ecology and Systematics*, 32:333–365.

Pelling, M. 2001. Natural disasters?. In: Castree, N. and Braun, B. eds., *Social Nature: Theory, Practice and Politics*, pp. 170–188. Oxford: Blackwell.

Polsky, C., Neff, R., and Yarnal, B. 2007. Building comparable global change vulnerability assessments: The vulnerability coping diagram. *Global Environmental Change*, 17(3–4):472–485.

Ribot, J. 2011. Vulnerability before adaptation: Toward transformative climate action. *Global Environmental Change,* 21:1160–1162.

Ribot, J.C. 2009. Vulnerability does not just fall from the sky: Toward multi-scale pro-poor climate policy. In: R. Mearns and A. Norton, eds. *Social Dimensions of Climate Change: Equity and Vulnerability in a Warming World*, pp. 47–74. Washington, DC: The World Bank.

Rockström, J. 2003. Resilience building and water demand management for drought mitigation. *Physics and Chemistry of the Earth,* 28:869–877.

Roncoli, C., T. Crane, and B. Orlove. 2009. Fielding climate change in cultural anthropology. In: S.A. Crate and M. Nuttall eds., *Anthropology and Climate Change*, pp. 87–115. Walnut Creek, CA; Left Coast Press.

Rygel, L., O'Sullivan, D., and Yarnal, B. 2006. A method for constructing a social vulnerability index: An application to hurricane storm surges in a developed country. *Mitigation and Adaptation Strategies for Global Change*, 11:741–764.

Sabo, J.L., Sinha, T., Bowling, L.C., Schoups, G.H.W., Wallender, W.W., Campana, M.E., Cherkauer, K.A. et al., 2010. Reclaiming freshwater sustainability in the Cadillac Desert. *Proceedings of the National Academy of Sciences of the United States of America*. Available online at http://www.pnas.org/content/107/50/21263.full.pdf+html. Date accessed 7 September 2011.

Science Daily. 2010. Southeastern U.S., with exception of Florida, likely to have serious water scarcity issues. John Kominoski quoted in *Science Daily*, 13 December 2010. Available online at http://www.sciencedaily.com/releases/2010/12/101213184436.htm. Date accessed 21 November 2011.

Shoemaker, J.A. 2003. Like snow in the spring time: Allotment, fractionation, and the Indian land tenure problem. *Wisconsin Law Review*, 729–788.

Southern Appalachian Man and the Biosphere. 1996. The southern Appalachian assessment terrestrial technical report. Atlanta: U.S. Department of Agriculture, Forest Service, Southern Region. 248 p.

Stone, B. 2012. *The City and the Coming Climate Change*. New York: Cambridge University Press.

Sun, G., McNulty, S.G., Moore Myers, J.A., and Cohen, E.C. 2008. Impacts of multiple stresses on water demand and supply across the southeastern United States. *Journal of American Water Resources Association,* 44(6):1441–1457.

Sunderlin, W.D., Dewi, S., Puntodewo, A., Muller, D., Angelsen, A., and Epprecht, M. 2008. Why forests are important for global poverty alleviation: A spatial explanation. *Ecology and Society*, 13(2):24.

Szydlowski, G.F. 2007. Commoditization of water: A look at Canadian bulk water exports, the Texas water dispute, and the ongoing battle under NAFTA for control of water resources. Notes and Comments. *Colorado Journal of Environmental Law and Policy*, 18(3):665–686.

Taylor, W., Floyd, M.F., Whitt-Glover, M., and Brooks, J. 2007. Environmental justice: A framework for collaboration between the public health and recreation and parks fields to study disparities and physical activity. *Journal of Physical Activity Health*, 4, S20–S29.

Texas Water Development Board. 2008. *Far West Texas Climate Change Conference: Study Findings and Conference Proceedings*. Available online at http://www.twdb.state.tx.us/publications/reports/climatechange.pdf. Date accessed 24 September 2012.

Turner, B.L., Kasperson, R.E., Matson, P.A., McCarthy, J.J., Corell, R.W., Christensen, L., Eckley, N. et al., 2003. A analysis in sustainability science framework for vulnerability. *Proceedings of the National Academy of Sciences*, July 8(100), 14:8074–8079.

U.S. Census Bureau. 2000. Summary File 3. Detailed Tables for Population, Poverty, Education (which tables are these). Universe Total Population. P6. Data Set: Census 2000 Summary File 3 (SF 3) Sample Data. Available online at http://factfinder.census.gov. Date accessed 7 December 2009.

U.S. Forest Service. 2012. Future of America's Forests and Rangelands: Forest Service 2010 Resources Planning Act Assessment. General Technical Report. WO-87. Washington, DC: 198 p.

U.S. Senate Committee on Foreign Relations. 2011. Avoiding water wars: Water scarcity and Central Asia's growing importance for stability in Afghanistan and Pakistan. Available online at: www.foreign.senate.gov/.

Vásquez-León, M. 2009. Hispanic farmers and farmworkers: Social networks, institutional exclusion, and climate vulnerability in Southeastern Arizona. *American Anthropologist,* 111(3):289–301.

Vincent, K. 2004. Creating an Index of Social Vulnerability to Climate Change for Africa. Tyndall Centre for Climate Change Research. Working Paper 56.

Vörösmarty, C.J., Douglas, E.M., Green, P.A., and Revenga, C. 2005. Geospatial indicators of emerging water stress: An application to Africa. *Royal Swedish Academy of Sciences*, 34(3):230–236.

Vörösmarty, C.J., Green, P., Salisbury, J., and Lammers, R.B. 2000. Global water resources: Vulnerability from climate change and population growth, *Science*, 289:284–288.

Williams, M. 2007. Proud, humble oystermen fear loosing a way of life, Cox News Services, Available online at www.lexisnexis.com.proxy-remote.galib.uga.edu/us. Date accessed, unknown.

Wisner, B., Blaikie, P., Cannon, T., and Davis, I. 2004. *At Risk: Natural Hazards, People's Vulnerability and Disasters*. New York: Routledge.

Woods and Poole. 2009. *The 2010. Complete Economic and Demographic Data Source (CEDDS)*. Washington, DC: Woods & Poole Economics, Inc. pp. 106.

5 Future Wildfire Trends, Impacts, and Mitigation Options in the Southern United States

Yongqiang Liu, Jeffrey P. Prestemon, Scott L. Goodrick, Thomas P. Holmes, John A. Stanturf, James M. Vose, and Ge Sun

CONTENTS

Wildfire is among the most common forest disturbances, affecting the structure, composition, and functions of many ecosystems. The complex role that wildfire plays in shaping forests has been described in terms of vegetation responses, which are characterized as dependent on, sensitive to, independent of, or influenced by fire (Myers 2006). Fire is essential in areas where species have evolved to withstand burning and facilitate the spread of combustion, such as the *Pinus* spp. found in the Coastal Plain of the Southern United States. Notable fire-dependent ecosystems include many boreal, temperate, and tropical coniferous forests, eucalyptus forests, most vegetation assemblages in Mediterranean-type climates, some forests dominated by oaks (*Quercus* spp.), grasslands, savannas, and marshes, and palm forests. At the other extreme, fire is largely absent where cold, wet, or dry conditions prevail (such as tundra landscapes, some rain forests, and deserts). Fire-sensitive ecosystems that have evolved without fire as a significant process have become more vulnerable to human activities such as stand fragmentation, alteration of fuels, and increased ignitions. Fire-influenced ecosystems generally are adjacent to areas where fire-dependent vegetation facilitates ignition and spreading of wildfires.

Wildfire, meanwhile, can be a major natural disaster. From 1992 to 2001, almost 2 million ha of U.S. forests and other ecosystems were burned by hundreds of thousands of wildfires annually, costing billions (U.S. Department of Agriculture Forest Service 2005). The 1997/1998 wildfires in Indonesia burned 8 million ha (Cochrane 2003). In catastrophic wildfires in January and February of 2009 in Victoria, Australia, the greatest damages were visited on "Black Saturday" (February 7, 2009) when over 2000 homes were destroyed or damaged and 173 people were killed; 430,000 ha were burned in the region in early 2009 (Teague et al. 2010). It is notable that, as in the case of the Black Saturday fire and many other large fires, some fires are simply beyond our control, regardless of the type, kind, or number of firefighting resources deployed. In the United States, large fires and the uncontrollable "mega-fires" of the kind cited by Williams (2004) account for 90% of the area burned and 80% of suppression costs, but together represent less than 1% of all wildfires (Williams 2004).

Wildfires can also produce severe environmental consequences. Smoke particles are a source of atmospheric aerosols, which affect atmospheric radiative transfer through the scattering and absorption of solar radiation and through the modification of cloud microphysics (Charlson et al. 1992). These processes can further modify clouds, precipitation, and atmospheric circulation (Ackerman et al. 2000; Liu 2005a,b). The particulates and other air pollutants from wildfires can degrade air quality (Riebau and Fox 2001), resulting in significant human health consequences (Rittmaster et al. 2006). Wildland fires contribute an estimated 15% of total particulate matter and 8% of carbon monoxide emissions over the Southern United States (Barnard and Sabo 2003). Burned areas are prone to severe soil erosion due to loss of ground vegetation and litter cover and an accelerated overland flow. Stormflow volume and peakflow rates increase dramatically in response to reduced soil infiltration rates and soil water storage because forest evapotranspiration rates are reduced. Increases of stormflow and soil erosion carry with them the potential to degrade water quality after severe wildfire events.

Weather and climate are determinants for wildfire characteristics along with fuel properties and topography (Pyne et al. 1996). Fire activity varies from one fire season to the next. For example, the burned area in the United States increased from 0.5 million ha in 1998 to 2.3 million ha in 1999 (National Interagency Fire Center 2010). Fire weather and climate also influence wildfire behavior and account for fire variability at various time scales. Under warm and dry conditions, a fire season becomes longer, and fires are easier to ignite and spread. The interannual variability in the atmospheric circulation patterns that brought drought conditions in the past are still a driving force in the variability of fire season severity (Westerling and Swetnam 2003). Contemporary observational data indicate statistically significant relations among wildfires, atmospheric conditions, and ocean conditions (Swetnam and Betancourt 1990; Brenner 1991; Prestemon et al. 2002; Skinner et al. 2002; Liu 2004, 2006; Dixon et al. 2008; Goodrick and Hanley 2009; Hoinka et al. 2009). Research shows that wildfires, especially catastrophic wildfires, have increased in recent decades in both the

United States and other parts of the world (Piñol et al. 1998; Food and Agriculture Organization of United Nations 2001; Gillett et al. 2004; Reinhard et al. 2005; Westerling et al. 2006). Among the converging factors were extreme weather events, such as extended drought, and climate change (Goldhammer and Price 1998; Stocks et al. 2002).

A new and challenging wildfire issue is the potential increase in occurrence of wildfires due to a changing climate brought on by the greenhouse effect. Many climate models have projected that the greenhouse effect will result in significant climate change by the end of this century (Intergovernmental Panel on Climate Change 2007), including an overall increase in temperature worldwide and a drying trend in many subtropical and mid-latitude areas. Thus, it is likely that wildfires will increase in these areas. One effect could be more fires that burn more intensely and spread faster in northern California (Fried et al. 2004). A 50% increase in fire occurrence is projected in boreal forests by the end of the century (Flannigan et al. 2009). Fire potential may increase significantly in several global geographic areas, including some in the United States (Liu et al. 2009b, 2012).

The South is one of the most productive forested regions in the country, with 81 million ha or 40% of the nation's forests in an area occupying only 24% of the nation's land area (Burkett et al. 1996). Furthermore, southern forests are dynamic ecosystems characterized by rapid growth—and hence rapid accumulation of fuels within a favorable climate—and a high fire-return rate of 3–5 years (Stanturf et al. 2002). The South leads the nation in annual wildfires, averaging approximately 45,000 fires a year from 1997 to 2003 (Gramley 2005). The region is also experiencing increased droughts. For example, during the 2007 drought in the southeastern United States, the worst drought in more than a century, severe wildfires in the spring in and around the Okefenokee National Wildlife Refuge on the southern-Georgia/northern-Florida border burned approximately 243,000 ha. The frequency of droughts across the region appears to be changing and the period from the mid- to late-1990s may have been wetter than the long-term average (Seager et al. 2009).

Like many other geographic areas in the nation and in the world, the South faces the challenge of potentially increased wildfires this century, resulting from the projected warmer temperatures and more frequent droughts that would occur in response to climate change. Increased wildfire activity would have some specific ecologic, environmental, social, and economic consequences. Continued population growth increases the potential threat these fires would pose to life and property. In addition, forestry and forestry-related industry represent a significant portion of the region's economy, making each fire a potential loss to a local economy. Also, the increases in wildfire potential would require increased future resources and management efforts for disaster prevention and recovery. Projections of future wildfire trends in the South under a changing climate are essential for accurately assessing possible impacts of climate-related future wildfire trends, including human health and safety impacts and environmental impacts, and are critical to designing and implementing necessary measures to mitigate impacts.

Wildfire in the South has been identified as a priority for many research programs including those funded through the National Fire Plan, the U.S. Environmental Protection Agency Star Program, and the U.S. Departments of Agriculture and the Interior's mutually sponsored Joint Fire Science Program. The objectives of these research programs included investigating and synthesizing the current status of wildfires, projecting future trends, assessing impacts of changes in wildfires on other ecosystem processes, and providing management options to mitigate the impacts, particularly in places where wildfire activity is projected to increase. This chapter presents the findings from these studies. Background information, including climate and vegetation, wildfire, fire–weather and fire–climate interactions, fire and climate change, and research and mitigation issues, is first provided. Future fire and fuel conditions (including projection approaches, climate change scenarios, and results), the impacts of future fire changes (including impacts on emissions, smoke and air quality, forest ecosystems, socioeconomics, hydrology, and regional climate), and management options for impact mitigation are then described. Finally, major findings and knowledge gaps are summarized together with suggested future research needs.

BASICS OF WILDFIRE AND CLIMATE IN THE SOUTH

Climate

Consisting of the 13 states roughly south of the Ohio River and from Texas to the Atlantic Coast, the South can be classified by topography and ecological features into the following level II eco-regions (U.S. Environmental Protection Agency 2008): (1) Coastal Plain, consisting of the coastlines along the Atlantic Ocean and the Gulf of Mexico, including the Florida peninsula and the Mississippi Alluvial Valley; (2) Piedmont and Southern Appalachian Mountains, including the Appalachian plateaus and mountain ranges; (3) Interior Highlands, consisting of the Interior Low Plateaus of Kentucky and Tennessee and the Ozark-Ouachita Highlands; and (4) western Ranges and Plains, consisting of central and western Texas and Oklahoma. The first region of these classifications roughly corresponds to the eastern Coastal Plain, the western Coastal Plain, and the Mississippi Alluvial Valley eco-regions, while the three others roughly correspond to the Piedmont, Appalachian-Cumberland, and Mid-South eco-regions (Wear and Greis 2012).

The region primarily has a humid subtropical climate except for a tropical climate in southern Florida and a semi-arid climate in western Texas and Oklahoma [*Times* (UK) 1993]. Annual daily temperature averages range from greater than 21°C in southern Florida and Texas to 13–16°C in northern areas. Annual precipitation is 1270–1780 mm in the Mid-South including Louisiana, Mississippi, Alabama, and Tennessee, areas of Georgia and Florida, and areas along the Atlantic coastline. Precipitation reduces to 1015–1270 mm toward the Atlantic coastal areas and northern areas of the region, and to 300–500 mm toward western Texas and Oklahoma.* Seasonal variability is significant in most of the region, characterized by hot, humid summers and mild to cool winters. The major weather and climate extremes include tornados, hurricanes, excessive lightning, and drought, with drought the largest contributor to large wildfires.

Fuel

In the vegetation types defined by the National Fire Danger Rating System fuel models, the Coastal Plain is dominated by open pine (*Pinus* spp.) stands—with perennial grasses and forbs as the primary ground fuel—in the coastal area along the Gulf of Mexico and hardwoods in the coastal area along the Atlantic Ocean. Major pine species are longleaf (*P. palustris*), slash (*P. elliottii*), and loblolly (*P. taeda*) pines (Wade et al. 2000). Florida has a mixture of dense live brush, agriculture, and sawgrass (*Cladium* spp.). The western Coastal Plain is dominated by natural pine stands, southern pine plantations, and hardwoods as the primary fuel. The Mississippi Alluvial Valley is dominated by agriculture, and the Piedmont by southern pine plantations and natural pine stands. Shortleaf pine (*P. echinata*) is more widespread in the Piedmont and mountains than in the Coastal Plain. The Appalachian-Cumberland highlands are dominated by pine with some perennial grasses. The central Texas and Oklahoma areas of the Mid-South are dominated by intermediate brush to the south and agriculture to the north, compared to grasses, a mixture of sagebrush (*Artemisia* spp.) and grasses, and some agriculture in western Texas.

Wildfire

The characteristics of wildfires in a geographic area are usually described in terms of fire regime and fire history. Fire regime describes the long-term presence of fire in an ecosystem (Brown 2000), mainly characterized by fire frequency (or fire return interval) and fire severity. Fire regimes can be classified as understory, stand-replacement, or mixed (Brown and Smith 2000). Understory-regime fires generally do not kill the dominant vegetation or substantially change its structure.

* See www.hurricane.ncdc.noaa.gov/climaps

A stand-replacement fire kills the aboveground parts of the dominant vegetation, changing the aboveground structure substantially. Mixed-regime fires can either cause selective mortality in dominant vegetation—depending on a species' susceptibility to fire—or can at some times limit effects to the understory and at other times to the dominant vegetation in a stand replacement. Fire severity depends on the type of fire regimes and is defined as the "degree to which a site has been altered or disrupted by fire; loosely, a product of fire intensity and residence time" by the National Wildfire Coordinating Group (2005). Fire severity is also often partially measured by the amount of total area burned. Burned area does not reflect the exact features of fire intensity and period. However, burned area is closely related to how intense and long-lasting a fire is: an intense and long-lasting fire usually leads to a large burned area.

The fire regimes of southern ecosystems have been described in detail (Wade et al. 2000; Stanturf et al. 2002; Fowler and Konopik 2007). Intervals between fires are primarily determined by vegetation species, which in turn depend on the physiographic characteristics of the eco-region. Intervals between fires may be as short as a year or as long as centuries. Before European settlement, frequent low severity understory fires characterized most Coastal Plain ecosystems with a return interval of 1–4 years (Table 5.1). Blowdowns and droughts led to occasional severe fires (Myers and Van Lear 1997). Mixed stands of oak and hickory (*Carya* spp.) in the Piedmont had a return interval of less than 35 years, compared to less than 200 years for Table Mountain pine (*P. pungens*). The return interval of mixed mesophytic species depended on whether they grew on the eastern or the western side of the Southern Appalachians.

The long history of fire since humans arrived in the South can be divided into five periods (Stanturf et al. 2002). (1) The pre-Columbian period, more than 500 years ago, in which Native Americans used fires extensively as a landscape management tool. (2) The early European settlement period, from 500 to about 110 years ago, in which European settlers likewise used fire

TABLE 5.1
Occurrence and Frequency of Pre-Settlement Fire Regime Types in the Southern United States, by the Society of American Foresters

	Frequency (Years) by Fire Regime		
Vegetation	**Understory**	**Mixed**	**Stand Replacement**
Longleaf pine	1–4		
Slash pine	1–4		
Loblolly pine	1–4		
Shortleaf pine	2–15		
Oak-hickory	<35		
Pond pine		6–25	
Pitch and Virginia pines		10–35	
Table Mountain pine		<200	
Mixed mesophytic		10–35 or >200	
Bottomland hardwoods		<200	
Sand pine			20–60
Bay forests			20–100
Atlantic white cedar			35–200
Northern hardwoods			300–500

Source: Modified from Wade, D.D. et al. 2000. *Wildland Fire in Ecosystems: Effects of Fire on Flora.* GTR RMRS-42. Ogden, UT: USDA Forest Service, Rocky Mountain Research Station: 53–96.

culturally but also introduced livestock and new farming practices involving widespread land clearing. (3) From the late 1890s to the 1920s, in which the remaining southern forests were extensively logged to support economic and population expansion and in which wildfires were common due to logging slash accumulations. (4) In reaction to these widespread and destructive wildfires, the fourth period, characterized by fire suppression, began in the 1920s and extended to the 1980s. (5) The current period, in which the natural role of fire is increasingly recognized and incorporated into forest management.

FIRE WEATHER AND CLIMATE

Fire weather and climate describe the atmospheric conditions that influence fires (Flannigan and Wotton 2001). Weather refers to atmospheric elements (such as temperature, humidity, pressure, winds, and precipitation) and the related processes or systems (such as fronts, cyclonic and anticyclonic circulations, troughs and ridges, and jet streams) on time scales of hours to weeks. Conversely, climate describes the statistics of weather over a long period (multiple decades). In fire research and management, however, weather conditions for fire (fire weather) often refer to atmospheric conditions and processes for individual fires on specific days and months, but climate conditions for fire (fire climate) refer to conditions during an entire fire season, inter-fire season variability, and long-term trends.

The relationship between weather and fire is often expressed in the fire environment triangle (Figure 5.1), with fire behavior in the center reflecting the degree of fire suppression difficulty based on ignition, spread, and intensity. Ignition is the process of increasing fuel temperature—often by external or internal heat energy or lightning—to a critical value (ignition temperature); that is, a temperature at which combustion starts. Heat sources can be natural (radiation, sensitive heat, chemical energy) or related to human activities (such as arson, equipment sparks, or arcing power lines). Lightning, which is of special concern when occurring in the absence of rain, initiates a series of chain reactions that generate the needed heat energy for ignition. Fire spread is the process of igniting new fuels from a single burn point. The rate of fire spread varies with time. The rate is controlled by ambient and fire-induced winds and relative humidity, which in turn determine fuel moisture. Fire intensity,* is proportional to fire spread rate, flame residence time, and reaction intensity. It is also sometimes measured by flame length.

Weather, fuel, and topography form the sides of the fire environment triangle; the role of each is described in Table 5.2. Models are available to predict the probability of ignition through heating or lightning (Latham and Williams 2001) and to calculate fire spread and intensity as a function of fuels, weather, and topography (Rothermel 1972; Finney 1998; Keane et al. 2003).

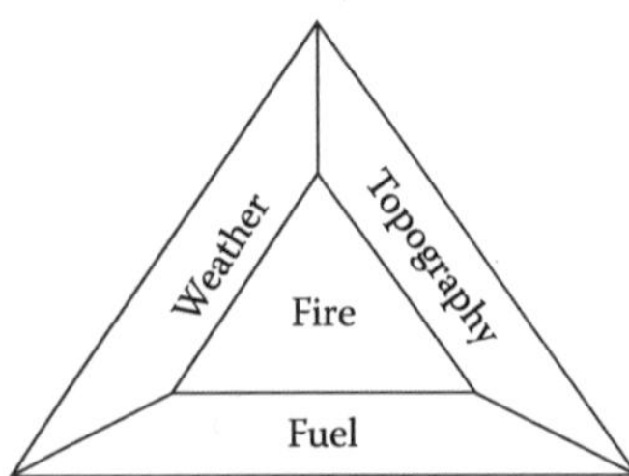

FIGURE 5.1 Fire environmental triangle. (Adapted from na.fs.fed.us/fire_poster/science_of_fire.htm)

* The literature presents a confusing array of definitions for fire intensity. According to Keeley (2009), "fire intensity represents the energy released during various phases of the fire and no single metric captures all of the relevant aspects of fire energy."

TABLE 5.2
Fire Environmental Factors, Their Elements, and Their Impacts on Fire Behavior

Factor	Parameters	Roles
Weather	Wind, temperature, relative humidity, air pressure, precipitation	High temperature reduces fuel moisture. Wind pushes a fire along. Low relative humidity dries out fuels causing them to ignite more easily. Precipitation puts out a fire and conversely a lack of precipitation can make fire more likely by drying out the fuels.
Fuels	Density (light or heavy), arrangement, moisture	The drier and lighter the fuels the more easily they will ignite. A continuous layer of fuels on the forest floor can aid in the spread of a fire.
Topography	Flat or sloped, aspect	Fire moves more rapidly up hills than down hills or over flat surfaces. Fire is more likely on southern and western aspects, which are drier.

Source: Adapted from na.fs.fed.us/fire_poster/science_of_fire.htm

Fire weather and climate conditions only provide necessary conditions rather than sufficient conditions for fire occurrence. In other words, certain weather patterns such as dry and hot weather do not guarantee the occurrence of a fire at a specific location. Rather, such conditions may confer a higher probability for a fire or fires at a location within an area of similar weather, fuel, and topography, assuming the same ignition probability. In addition, when looking at fire-climate relationships across a range of time scales (seasonal, interannual, decadal, or longer scales), the focus is often on total area burned rather than fire ignition and spread processes.

Fire potential—the measure of the chance that a fire of a certain severity will occur in an area—is often used as a surrogate for real fires. It is often estimated as a fire danger rating that is based on weather and fuel conditions. Different from fire behavior prediction, which is a property of an individual fire, fire danger rating (or fire potential) focuses on the fire situation over a geographic area. Furthermore, fire behavior prediction estimates what a fire will do, while fire danger rating is typically an ordinal index relating the probability that fires of a certain severity level will occur. Many rating systems have been developed, including the National Fire Danger Rating System. These systems often consist of a number of indices expressing fuel conditions (fuel moisture levels and energy release), weather conditions as expressed by the Keetch–Byram Drought Index (KBDI) or Fire Weather Index, and potential fire behavior (spread component and burning index).

Fire–weather and fire–climate relations have been studied extensively, including a review of studies in the last century by Flannigan and Wotton (2001) and a systematic study for the South by Heilman et al. (1998). Cold frontal passage, dry spells, and low relative humidity were found to be the most important weather determinants of area burned. These elements influence fuel moisture and associated fire danger components. Fire season variability is mainly governed by the interannual variability of atmospheric conditions, with fires often occurring during periods of drought, abnormal ocean conditions, and other anomalies of weather and climate. The anomalies in the sea-surface temperature (SST) of the equatorial eastern Pacific Ocean such as El Niño and La Niña can change the atmospheric circulation patterns in the Southern United States through atmospheric teleconnection and can modify fire weather in this region.

FIRE AND CLIMATE CHANGE

The features of fire regimes in the South described above evolved based on the climate and fuel conditions in the past. Because fire environmental conditions are expected to change this century, fire regimes could change as well. Environmental changes will occur in response to changes in climate, land use, and socioeconomic and environmental variables affecting ignitions by people, as well as change in wildfire management approaches. Understanding the possible change in fire regimes is essential to assessing the potential impact of future wildfire trends.

One among many variables affecting fire regimes, a changing climate could have various impacts on fires in the South (Table 5.3), but the relationships are complex. Projected temperature increases across the South would contribute to longer fire seasons and increases in fire frequency, intensity, and total burned area. Temperature change also could indirectly impact fires by changing fuel conditions. Increased temperature would likely increase evaporation, thereby reducing fuel moisture and increasing fire occurrence. Higher temperatures also could affect ignition rates. The impact on fuel loading is more complex. Longer growing seasons associated with temperature increases can increase fuel loading by increasing productivity, but temperature increases can also decrease vegetation growth and thereby fuel loading because of reduced water availability due to increased rates of evaporation and transpiration.

The contributions of precipitation and humidity are also complex. Precipitation is projected to decrease in many subtropical and mid-latitude areas, reducing fuel moisture and increasing fire potential. At the same time, a reduction in precipitation would reduce water availability for plant growth, leading to less fuel and lower fire potential. Clearly, projecting the effects of altered precipitation is accompanied by more uncertainty than projecting the effects of increased air temperature. Additional uncertainty comes from the atmospheric models: projected precipitation change often shows no clear trends, even over large areas. Along with changes in average precipitation, most general circulation models (GCMs) also project more frequent precipitation anomalies, such as drought, that would increase fire activity. Although increased temperature would reduce relative humidity locally, it would also increase evaporation from ocean and land surfaces, thereby producing an overall increase in relative humidity. Effects on relative humidity are also difficult to predict because of the dependence of atmospheric humidity on precipitation, which removes water vapor from the atmosphere.

Surface wind is determined by surface roughness and spatial differences in atmospheric heating, which in turn are influenced by complex thermal and dynamic processes in the atmosphere. The strong winds that have the biggest impact on fires are related to cold fronts and other weather systems that are expected to change in frequency and intensity in the future. Thus, winds and their fire impacts will likely change as well, although projections of changes in these features of climate are even more uncertain than projections of changes in precipitation, especially in areas with complex topography.

TABLE 5.3

Response of Fire and Fuel Properties to Possible Changes in Various Atmospheric Elements and Processes

	Prediction	Fire Response			Fuel Response		
Change	**Confidence**	**Frequency**	**Intensity**	**Season**	**Area**	**Loading**	**Moisture**
Increased temperature	High	+	+	+	+	+/–	–
Decreased precipitation	Low	+	+	+	+	–	–
Increased drought	High	+	+	+	+	–	–
Changed relative humidity	Low	+/–	+/–	+/–	+/–	+/–	+/–
Increased wind strength	Low	+	+	+	+	No change	No change
Increased lightning	Low	+	+	+	+	No change	No change

Lightning currently accounts for a small share of wildfire ignitions in the South compared to human ignitions but played an important ecological role in the past (Myers and Van Lear 1997; Outcalt 2008) and could play an increasingly important role in the future. Lightning is another complex process that could become more frequent due to warming and increased trends in atmospheric instability, despite the projected precipitation decreases in many places of subtropical and mid-latitude areas (Shankar et al. 2009).

ROLE OF MITIGATION

The role of forest management is illustrated in Figure 5.2 using an example of prescribed burning to reduce frequency of wildfire occurrence by removing accumulated understory fuels. Assuming that the current fire potential is at a moderate level of the KBDI of 250 and that prescribed burning is conducted every four years, the corresponding wildfire frequency is assumed to be twice every 100 years. Under a changing climate, fire potential is projected to increase to a higher level (KBDI of 350). If prescribed burning remains once every four years, wildfire frequency would increase to three times every 100 years. One of the mitigation options could be to double the rate of prescribed burning, to every two years. As a result, wildfire frequency would remain at twice per 100 years.

PROJECTIONS OF FUTURE FIRE AND FUELS

Use and Limitations of KBDI

The KBDI is used by fire managers in the South and modelers as an indicator of current and future fire potential. A detailed description of the development and application of this index was presented in Keetch and Byram (1968) and summarized in Liu et al. (2009b). The maximum value for KBDI is 800. KBDI is classified into eight drought stages by increments of 100 (Keetch and Byram 1968). Two adjacent stages represent a change of one level in fire severity potential (Table 5.4). The KBDI depends on its historical values; in other words, it has "memory" in the sense that current values depend on previous values. For example, if a drought has occurred for one month, reduced rainfall

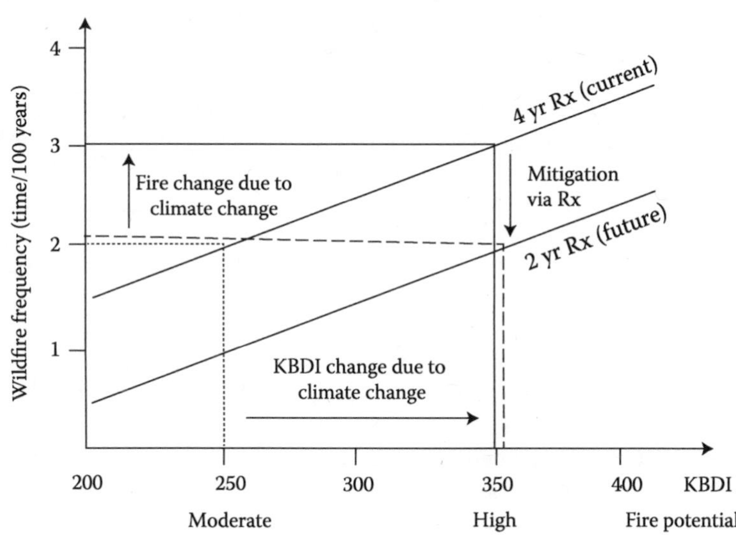

FIGURE 5.2 Schematic showing how prescribed burning can mitigate the impacts of climate change on wildfire.

TABLE 5.4
Fire Potential Classifications, Based on KBDI

Level	KBDI	Condition	Typical Period
Low	0–200	Soil moisture and large class fuel moistures are high and do not contribute much to fire intensity.	Spring dormant season following winter precipitation.
Moderate	200–400	Lower litter and duff layers are drying and beginning to contribute to fire intensity.	Late spring, early growing season.
High	400–600	Lower litter and duff layers actively contribute to fire intensity and will burn actively.	Late summer, early fall.
Extreme	600–800	Intense, deep burning fires with significant downwind spotting can be expected. Live fuels can also be expected to burn actively at these levels.	Often associated with periods of severe drought.

Source: Reorganized from U.S. Department of Agriculture Forest Service, 2005. Wildland Fire Assessment System www.wfas.net/index.php/keetch-byram-index-moisture-drought-49.

will impact the KBDI in future months. For this reason, KBDI is a fire index more suitable than most other indices to measure long-term fire potential.

The KBDI has some limitations. First, the peak values of KBDI often lag the peak of the fire season in the South. Second, the range of a specific fire potential level could vary slightly with area and season (Goodrick 1999) and fire type (Melton 1989, 1996). And third, specific KBDI values cannot be directly compared between locations with different climates because the drying rate in the index is a function of the average annual precipitation for a given location. Despite the potential limitations of the functional form used in the KBDI to parameterize evapotranspiration, the index is still a viable means of assessing the potential impacts of a changing climate on fire potential because it focuses on the relative changes produced by changes in temperature and precipitation.

STATISTICAL DOWNSCALING OF CLIMATE CHANGE SCENARIOS

To make wildfire projections based on climate change projections at a spatial scale of relevance to management (e.g., 30 × 30 km), the coarse spatial and temporal scales of projections produced by GCMs need to be downscaled to the spatial scale of inference. One option for obtaining fine scale projections from coarser scale GCMs is statistical downscaling. Statistical downscaling requires the estimation of statistical relationships among observational data and the coarse model data, and then combining these relationships using spatial interpolation. Although limited by the assumption that the factors influencing the finer spatial scale climate will remain constant throughout the projection period, these techniques are able to provide a first approximation of regional climatic conditions without the computational expense of higher resolution physical modeling.

We employed county-level temperature and precipitation data derived from an ensemble average of four GCMs [Canada General Circulation Model, version 3 (CGCM3), Geophysical Fluid Dynamic Laboratory (GFDL) model, Community Climate System Model Version 3 (CCSM3), and Hadley Center Climate Model, version 3 (HadCM3)] for three greenhouse gas emissions scenarios used by the Intergovernmental Panel on Climate Change (A1B, A2, and B1) to produce values for every month from 2010 to 2060 (Intergovernmental Panel on Climate Change 2007). Using the ensemble average of four climate models limits the impact of bias from individual models. Making projections under three different emissions scenarios allows us to sample a range of potential future conditions, possibly revealing how sensitive simulated futures are to the emissions scenario used.

Emissions scenarios combine two sets of divergent tendencies: one set varies between strong economic values and strong environmental values, and the other set between increasing globalization and increasing regionalization (Nakicenovic et al. 2000). The A1 scenario family describes a future of very rapid economic growth, global population that peaks in mid-century and declines thereafter, and entails the supposed rapid introduction of new and more efficient technologies. Within that family, A1B represents a balance between fossil fuels and alternative energy sources. The A2 scenario differs in that population is assumed to continuously increase, economic development is more regionally focused, and the introduction of new technologies is slower and more fragmented, limiting the adoption of alternative fuels. The B1 scenario is similar to the A1 family but describes a more integrated world characterized by an emphasis on global approaches to economic, social, and environmental stability. The atmospheric CO_2 concentration in 2008 was about 360 ppmv* for all scenarios. It would increase to about 700 for A1B (950 for A1F1), 850 for A2, 550 for B1, and 600 for B2 by 2100. For the A1B scenario, GCMs estimate the global average temperature to rise by approximately 2.8°C by the end of this century, the A2 by 3.4°C, and the B1 by 1.8°C.

The temperature and precipitation data for our analysis were derived from the bias-corrected and spatially downscaled climate projections originally derived from the Climate Model Intercomparison Program (CMIP3) data by Maurer et al. (2007).† To achieve county level detail, the data were resampled using a nearest-neighbor approach: the value for each county was assigned based on the grid point nearest the county centroid. Temperature information was in the form of average daily temperature, and precipitation values reflected the average rainfall per day for each month. These values were not ideal for calculating KBDI, which normally uses daily maximum temperatures and total rainfall. This data limitation was overcome by assuming that the daily maximum temperature is 15% higher than the daily average temperature, that rain falls every two days, and that soil moisture begins each month at saturated conditions (zero KBDI). The use of a 15% increase from daily average temperature to achieve the daily high temperature is arbitrary, but it provided a generally good approximation in the South based on an application of the model to historical data. The assumption of rain every other day maximized the daily drying to align with the KBDI assumption that the first 6.5 mm of rainfall is insufficient to lower the drought index. Starting the KBDI calculations from zero each month provided an indicator of how quickly the soil could dry out each month based solely on the meteorological conditions of that month and not on any residual dryness.

In analyzing how climate change could impact the KBDI for all three emission scenarios across the South, we first established data for the baseline decade of 2000–2009 by examining KBDI patterns in the months of January, March, May, July, September, and November for those years. Next, we examined departures from these baseline patterns for 2010–2019, 2030–2039, and 2050–2059. Our goals were to identify eco-regions where fire potential is changing substantially from current conditions, gauge the level of uncertainty in the projections by noting differences among emissions scenarios, and translate these changes into impacts on fire season duration and severity.

For January during the baseline period, conditions are consistent across the scenarios (Figure 5.3). Cold temperatures throughout most of the region strongly limit drying except in the Florida peninsula and along the Texas coastline. Precipitation is more than sufficient to counter drying and keep the soil near saturation. By the 2010 to 2019 period, drying is expected to begin spreading up the Florida peninsula and into the Eastern Atlantic and Southern Gulf sections of the Coastal Plain as well as from the Texas coastline to the Western Gulf section of the Coastal Plain. Few differences are expected among the scenarios, with the A2 scenario projecting the most drying followed by A1B, and B1 projecting the least. By the 2030 decade, drying is expected to spread along the Gulf of Mexico with the A1B scenario projecting strong drying along southern areas of the Deltaic Plain section of the Mississippi Alluvial Valley. Overall, the A1B scenario projects the most drying along the Gulf of Mexico, while the B1 scenario projects the least. In summary, the changes are expected to be relatively minor because the most severe drying, found in the A2 scenario, would only change the KBDI by about 40 units by the 2050 decade.

* ppmv = parts per million by volume.

† Data available at gdo-dcp.ucllnl.org/downscaled_cmip3_projections/

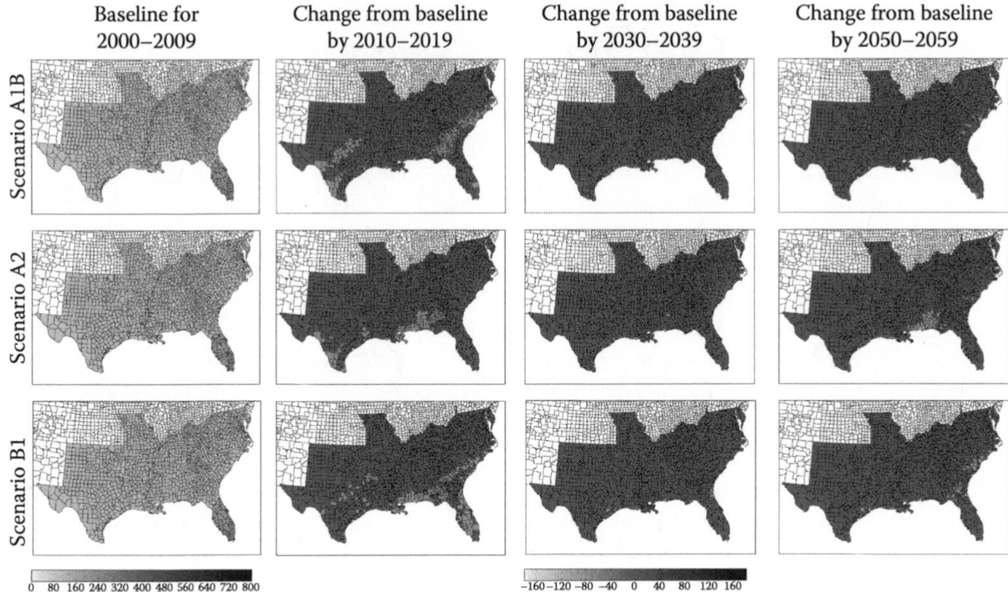

FIGURE 5.3 January fire potential for the baseline decade followed by ensemble, statistically downscaled projections of changes in fire potential for three subsequent decades assuming different Intergovernmental Panel on Climate Change (2007) emissions storylines: A1B, A2, and B1.

By March of the baseline period, dryness has begun to spread northward from the Florida peninsula and Texas coastline (Figure 5.4) with decadal averages of KBDI rising to 250–300 in some areas. As with the January baseline period, the emission scenarios have little impact on baseline conditions. By 2010, however, differences are expected among the emissions scenarios. Although projecting overall drying across most of the South, the B1 scenario projects that drying will be most severe in the Deltaic Plain section of the Mississippi Alluvial Valley, will spread across the Southern Gulf section of the Coastal Plain by 2030, and then expand into the Eastern Atlantic section by 2050. In the Blue Ridge, Northern/Southern Ridge Valley, and into the Cumberland Plateau regions, few changes in KBDI are expected, with perhaps slightly wetter conditions regardless of scenario or decade—suggesting a slight preference of the Ohio River Valley for storm tracks in the models.

In May, very slight differences among the scenarios are evident for the baseline period (Figure 5.5). Drying spreads throughout the South, except for the northernmost areas. May typically marks the peak of the spring fire season, particularly with regard to area burned. The May peak is likely to continue in the future, with all scenarios indicating that late spring will become even drier in most areas. As early as 2010, significant differences among the scenarios are expected. A1B and B1 project the most significant drying, particularly in the Texas/Oklahoma Cross Timbers section of the Mid-South and the western and eastern Middle Gulf sections of the Coastal Plain. All three scenarios project a tendency for wetter conditions in the Southern Gulf and Eastern Atlantic sections of the Coastal Plain in 2010. However, these wet areas are not expected to persist through the 2030 and 2050 decades. By 2050, the A1B scenario projects more substantial drying, an increase in KBDI of over 150 points, with the highest values centered over Louisiana and spreading throughout the Western and Middle Gulf sections of the Coastal Plain. Even the scenario with the smallest projected changes for May, those of the A2 scenario, indicates significant drying and hence longer spring fire seasons by 2050.

During the baseline period for July, high temperatures and limited rainfall result in dry conditions across much of the South (Figure 5.6). The most severe drying is projected to be centered

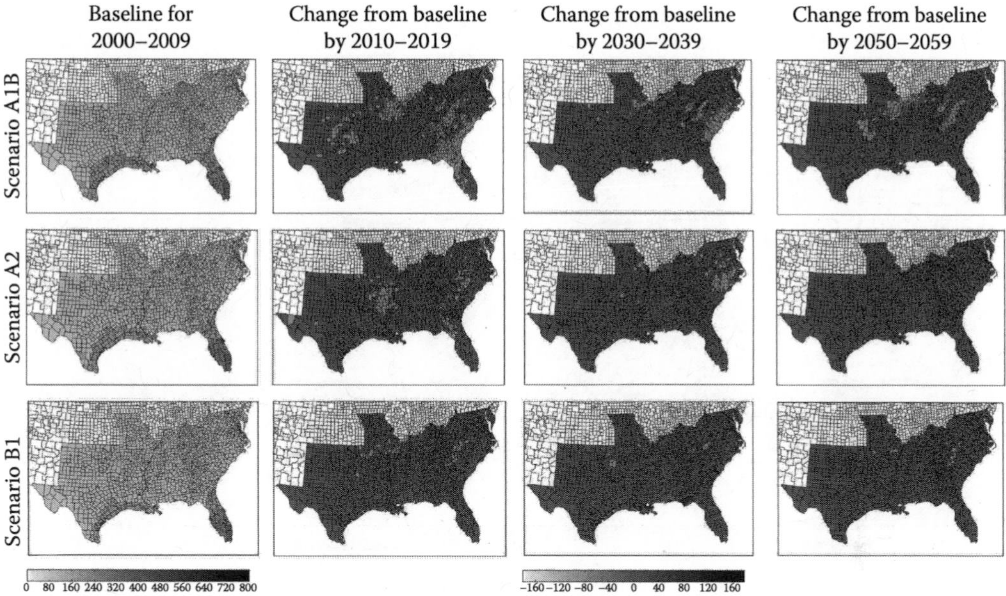

FIGURE 5.4 March fire potential for the baseline decade followed by ensemble, statistically downscaled projections of changes in fire potential for three subsequent decades assuming different Intergovernmental Panel on Climate Change (2007) emissions storylines: A1B, A2, and B1.

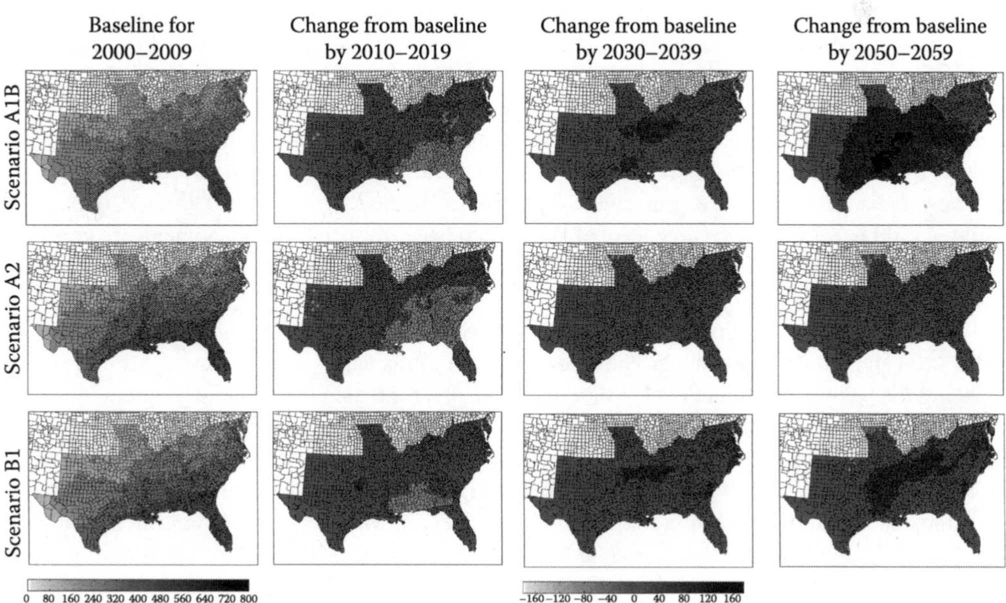

FIGURE 5.5 May fire potential for the baseline decade followed by ensemble, statistically downscaled projections of changes in fire potential for three subsequent decades assuming different Intergovernmental Panel on Climate Change (2007) emissions storylines: A1B, A2, and B1.

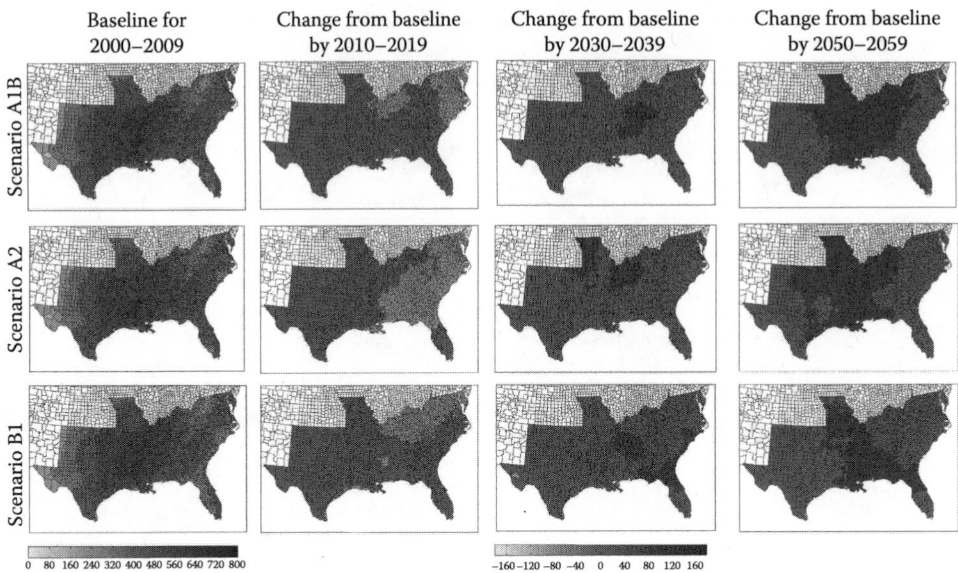

FIGURE 5.6 July fire potential for the baseline decade followed by ensemble, statistically downscaled projections of changes in fire potential for three subsequent decades assuming different Intergovernmental Panel on Climate Change (2007) emissions storylines: A1B, A2, and B1.

over Louisiana, Mississippi, and Arkansas, but with most coastal areas noticeably wetter from sea-breeze-induced thunderstorms. As with projections of baseline conditions for early months of the year, only subtle differences separate the three scenarios. Differences are projected to begin appearing as soon as 2010, when the A2 scenario indicates wetter conditions for the Southern Appalachians, Piedmont, Southern Gulf, and Eastern Atlantic. The A1B and B1 scenario projections are very similar across the region, except for the northern half of the Florida peninsula, where B1 projects more pronounced drying. By 2030, no areas of increased moisture are projected to remain, and all scenarios indicate conditions becoming drier, although the spatial pattern varies by scenario; for example, the A2 projects the lowest increase in dryness. By 2050, the A1B and A2 scenarios indicate KBDI increases of about 100 units across the majority of the South. The A1B projects the most intense drying in northern Alabama and Mississippi. Even the coastal areas with relatively moist conditions are projected to experience intense drying by 2030 and 2050. These results should be viewed with some level of skepticism because such local phenomena are generated based only on statistical downscaling of the coarser scale GCM projections, which do not show such small scale variations.

The area of dry conditions begins to contract by September for the baseline period (Figure 5.7) and is still centered over Louisiana. The extent of the dry area varies slightly among scenarios, with A2 being the wettest. September is the month that displays the greatest variability among the emissions scenarios for all future time periods. For 2010, the B1 scenario projects increased moisture along the Texas coastline and across the Florida peninsula as well as across the northern half of the region, with only slight drying in other areas. A1B introduces strong drying along the Atlantic coastline and the Southern Gulf section of the Coastal Plain, with wetter conditions along the Texas coastline and slightly wetter conditions along the northern half of the region, but not to the same spatial extent as in B1. The A2 projects a combination of features from the other two scenarios, but overall is drier for much of the region, with drying along the Atlantic coastline (although more spatially limited than in A1B) and wetter conditions in the Florida peninsula and along the Texas coastline. By 2030, all emissions scenarios project similar spatial patterns but vary in their intensity. The North Atlantic section of the Coastal Plain and the Central Appalachian Piedmont are projected to

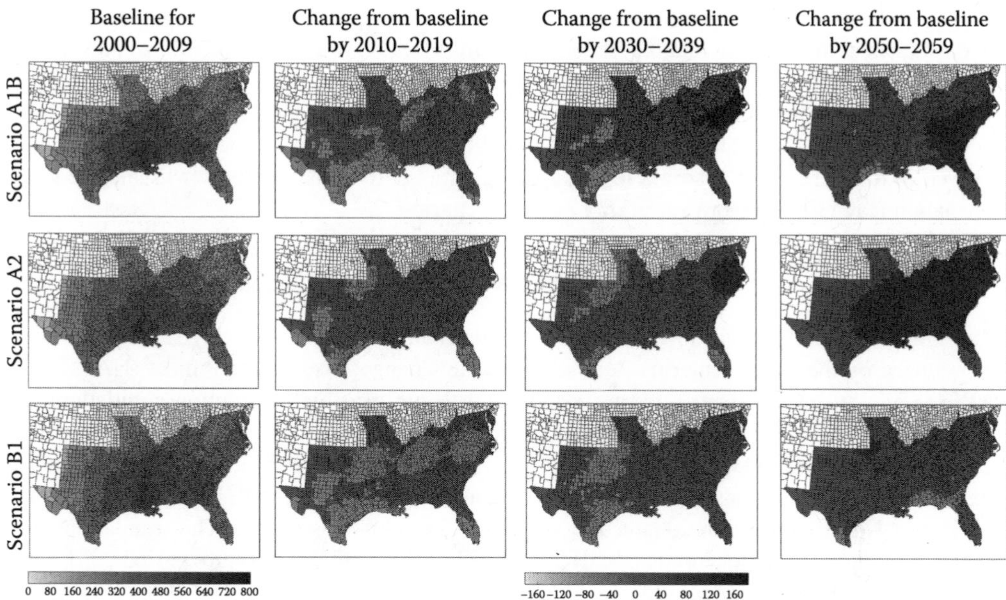

FIGURE 5.7 September fire potential for the baseline decade followed by ensemble, statistically downscaled projections of changes in fire potential for three subsequent decades assuming different Intergovernmental Panel on Climate Change (2007) emissions storylines: A1B, A2, and B1.

become a center of drying that varies in intensity from weak for B1 to strong for A1B. Although A2 does not project as high a peak in KBDI as does A1B, the spatial extent of the strong drying is larger. The Texas coastline and an area along the Texas–Oklahoma border are expected to become a center of wetter conditions, with a 40-unit decrease in KBDI. By 2050, the A2 scenario projects significant drying; much of the region experiences increases of at least 80 units in KBDI and most of the Northern Atlantic section of the Coastal Plain increasing by approximately 160 units. The A1B scenario projects a similar, although less extreme, drying trend that intensifies the dryness in eastern areas and largely eliminates the wetter conditions in the western half of the region. By 2050, the B1 scenario projects a spatial pattern that is different from the other scenarios, with region-wide drying that is centered in the Ozark-Ouachita Highlands. For those areas of the region that typically experience an autumn fire season, these results suggest an earlier start of the fire season and more severe conditions during the fire season.

The overall impact of climate change on fire potential in the South, as reflected by projected changes in the KBDI, is a gradual shift toward more severe fire conditions. The lengths of the spring and autumn fire seasons are projected to increase, and the extent of the drying is likely to be more severe. The early spring fire season is projected to be concentrated in eastern reaches of the Gulf of Mexico coast; however, the increase in severity is likely to be more widespread in coming decades. The projected dryness during the summer may mean new areas of the South that are subject to a summer fire season (or at least a later end to the spring season and earlier start of the autumn season).

The impact of the different emissions scenarios on fire potential is not large through 2060; the differences among model results are generally small. However, the consistency in emissions impacts shows that a dramatic decrease in fire potential is not likely, based on scenarios that all suggest an increase in fire potential in the coming decades. Although the models are in agreement, several factors give reason to advise caution in interpreting and applying these results, especially for any type of regional assessment. Large-scale global models are not currently run at a resolution capable of resolving all important weather phenomena. Features such as sea breezes—a mechanism for significant rainfall in coastal areas—and topographic modification of frontal systems by the Appalachian

Mountains are not adequately represented. Instead, the regional projections presented in this section rely on statistical information relating global model information to local observations; they assume that these relationships will remain constant in the future. The choice of emissions scenario drives the general circulation model, but the same statistical relations are used to translate output from all models to the region level. The statistical commonalities may limit the degree of variability that we observed in the projections. Using a regional climate model to dynamically downscale the global information may result in more variability among scenarios.

DYNAMICAL DOWNSCALING OF CLIMATE CHANGE SCENARIOS

An alternative to the statistical downscaling of climate change scenarios is the dynamical downscaling produced by the North America Regional Climate Change Assessment Project (NARCCAP). NARCCAP is an international program established to produce high-resolution simulations that describe uncertainties in regional scale projections of future climate and generate scenarios for use in impacts research (Mearns et al. 2009). The NARCCAP regional climate change scenarios were obtained by running a set of regional climate models that were driven by general circulation models over North America in conjunction with the A2 emissions scenario. The simulations were conducted for the current (recent historical) period 1971–2000 and for the future period 2041–2070, and the spatial resolution was 50 km.

These scenarios have several features different from the statistical downscaling approach. First, they provide daily output. Second, maximum temperature data are available, which is one of the variables needed for KBDI calculations. Third, other variables such as relative humidity and wind are also available, which can be used together with temperature and precipitation to calculate other fuel and fire indices in addition to the KBDI. These additional variables and the calculated indices are useful when projecting weather conditions critical to prescribed burning (see Management Options for Mitigation).

We used downscaling of HadCM3 with the Hadley Regional Model, version 3 (HRM3), which has been used for projecting fire potential trends in North America (Liu et al. 2010, 2012). The spatial scope of NARCCAP is North America. For this chapter, however, we only used the data at grid points within the South. Figures 5.8 through 5.10 show averages of temperature, precipitation, and KBDI. The simulated current maximum temperature shows a clear seasonal cycle for the region as

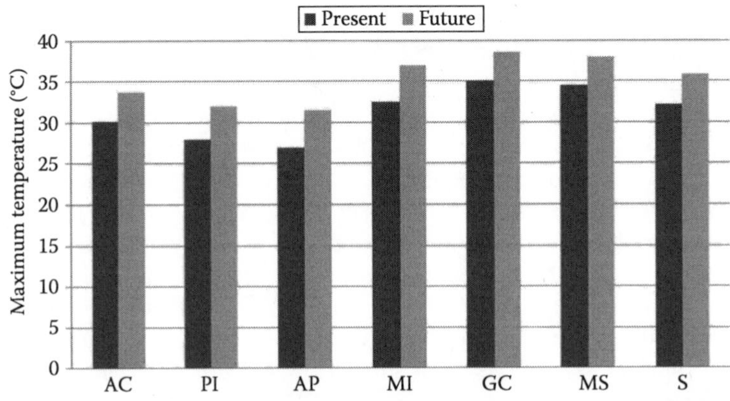

FIGURE 5.8 Current (1971–2000, left bar) and future (2041–2070, each right bar) change in seasonal temperature averaged over the Atlantic Coast (AC), Piedmont (PI), Appalachian Mountains (AP), Mississippi Alluvial Valley (MI), Gulf Coast (GC), Mid-South (MS), and entire Southern United States (S). (Adapted from Liu, Y.-Q., Goodrick, S.L., Stanturf, J.A. 2012. *Forest Ecology and Management.* 294: 120–155. doi: 10.1016/j.foreco.2012.06.049.)

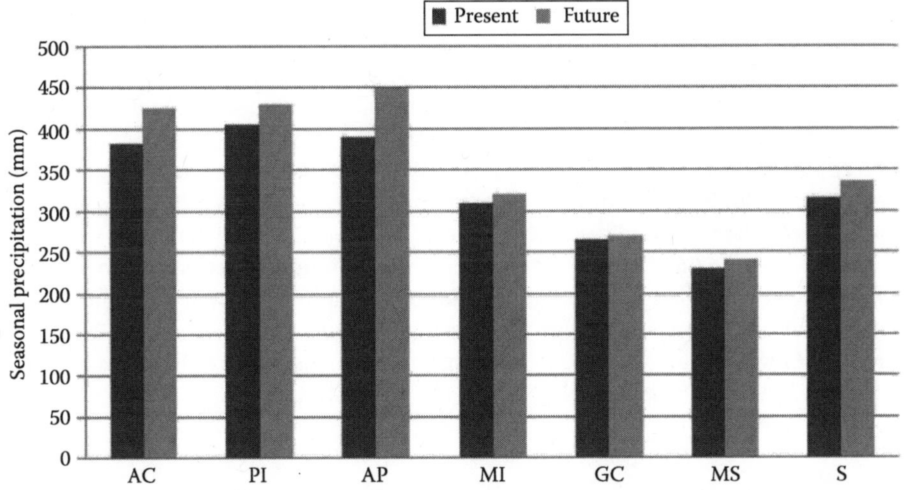

FIGURE 5.9 Same as Figure 5.8 except for precipitation.

a whole (Figure 5.8), increasing from about 10°C in winter to 20°C in spring, and 32°C in summer, and then decreasing to 22°C in autumn. Temperature decreases from the Atlantic Coastal Plain to the Piedmont and to Appalachian-Cumberland highlands for all four seasons; it generally increases from the Mississippi Alluvial Valley and southern Coastal Plain to the Mid-South. Regional maximum temperatures are projected by the middle of this century to increase by about 3–4°C during all seasons, with the largest increases projected for summer. There are no significant intraregional differences projected. The HadCM3 and HRM3 project the same spatial pattern but the HRM3 projections are slightly lower, especially in the western areas of the South.

The simulated current precipitation in the South (Figure 5.9) also shows the same seasonal cycle as that of temperature, increasing from winter (240 mm) to spring and summer (320 mm), and then decreasing in autumn (200 mm). In the western areas of the region, however, precipitation peaks in spring rather than summer. Precipitation is projected to increase for the region, by greater than 50 mm during summer in some areas. HRM3 precipitation projections for western areas are

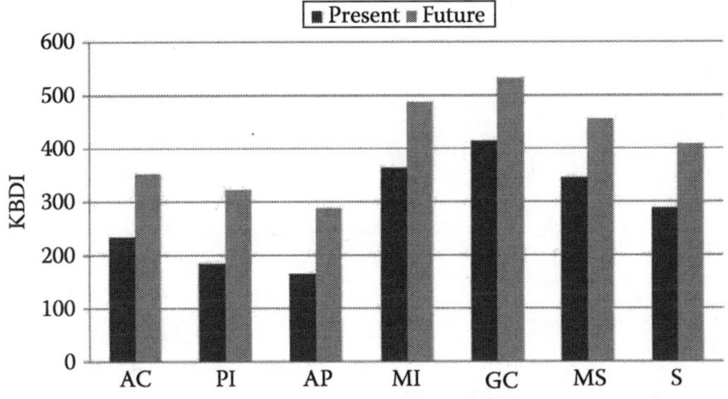

FIGURE 5.10 Summer KBDI in the Southern U.S. eco-regions for the present period of 1971–2000 and future change by 2041–2070. The regions below each panel are Atlantic Coast (AP), Piedmont (PI), Appalachian (AP), Mississippi (MI), Gulf Coast (GC), Mid-South (MS), and entire Southern U.S. (S). (Adapted from Liu, Y.-Q., Goodrick, S.L., Stanturf, J.A. 2012. *Forest Ecology and Management.* 294: 120–155. doi: 10.1016/j.foreco.2012.06.049.)

substantially different from the decrease projected by HadCM3. This divergence in precipitation projections should have a significant effect on the future KBDI calculations.

Current KBDI values are usually small in winter and spring and large in summer and autumn in all eco-regions (Figure 5.10). Present summer and autumn KBDI values are around 200 (considered in the upper KBDI range for low fire potential or the lower range for moderate fire potential) in the three eastern eco-regions, and future values rise to 350–400 (the upper KBDI range for moderate fire potential or the lower range for high fire potential). Meanwhile, present summer and autumn KBDI values are around 400 (the upper KBDI range for moderate fire potential or the lower range for high fire potential) in the three western eco-regions, and future values change to about 500 (the middle KBDI range for high fire potential). For the entire South, summer and autumn fire potential changes from moderate at present to high fire potential in future.

Figure 5.11 shows monthly variations of current and projected future KBDI for the South.

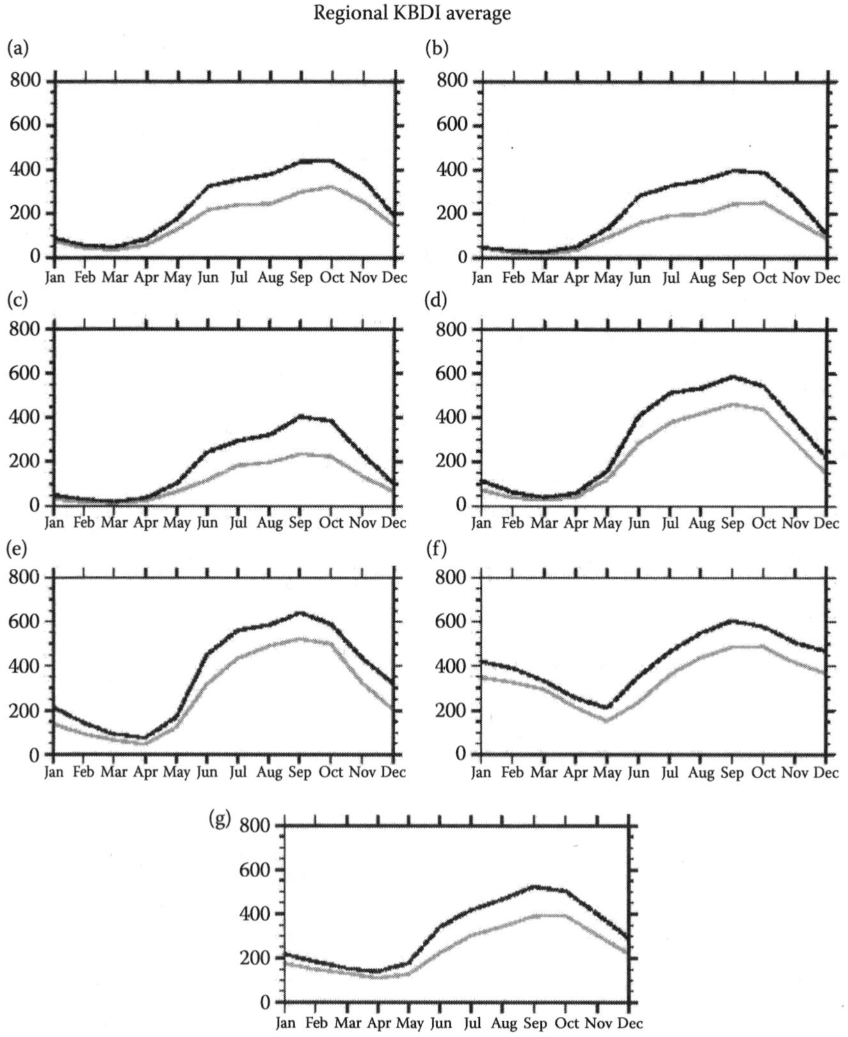

FIGURE 5.11 Monthly current and predicted fire potential, based on KBDI, averaged over the (a) Atlantic Coast, (b) Piedmont, (c) Appalachian Mountains, (d) Mississippi Alluvial Valley, (e) Gulf Coast, (f) Mid-South, and (g) entire Southern United States.

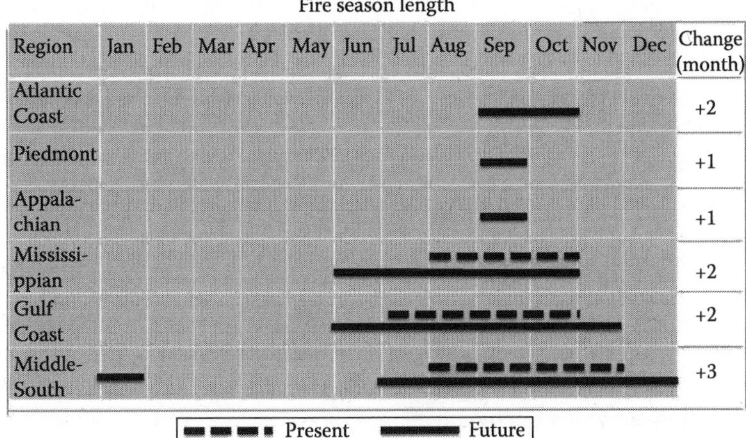

Fire season length
Region
Jan
Feb
Mar
Apr
May
Jun
Jul
Aug
Sep
Oct
Nov
Dec
Change (month)
Atlantic Coast
+2
Piedmont
+1
Appala-chian
+1
Mississi-ppian
+2
Gulf Coast
+2
Middle-South
+3
Present
Future

FIGURE 5.12 Current and predicted length of fire season within the Southern United States, based on KBDI ratings of moderate or high fire potential. (Adapted from Liu, Y.-Q., Goodrick, S.L., Stanturf, J.A. 2012. *Forest Ecology and Management*. 294: 120–155. doi: 10.1016/j.foreco.2012.06.049.)

Current KBDI starts with a low value of about 170 in January. It gradually decreases to less than 100 units in April, increases to about 350 in September and October, and then decreases again to about 190 in December. The corresponding fire potential is low from December to May and moderate from June to November. The future KBDI is projected to increase for all months, substantially after May, and with the largest increase of about 150 in September. Future fire potential is expected to remain low from January to May but to change from moderate to high during the July to October period. Intraregional changes in future fire potential are similar, although actual KBDI values vary.

The projected increase in fire potential suggests that fire seasons might become longer in the future (Figure 5.12). The length of a fire season, measured by the number of the months with high or extreme fire potential level, might increase by 1–3 months.

TRENDS AND PROJECTIONS FOR FIRE EXTENT

Wildfire in a Changing Climate and Societal Environment

If climate changes as projected and society continues to develop in the coming decades, then wildfire activity in the South will also change. Research has shown that wildfire responds to weather, fuels, and inputs from people. Human inputs have included intentional activities, such as prevention, fuels management, and suppression, designed to reduce wildfire occurrence and spread. They also include intentional and unintentional human fire starts and land use changes that can influence the frequency, location, and size of wildfires.

Recent research has shown that fire prevention efforts can be effective (Butry et al. 2010; Prestemon et al. 2010), that law enforcement can reduce the frequency of arson wildfires (Prestemon and Butry 2005), and that residential and commercial development patterns affect the severity and extent of wildfires in the South (Prestemon et al. 2002; Mercer and Prestemon 2005; Mercer et al. 2007). With respect to development patterns, research has shown that accumulations of forest fuels from fire suppression encourage the spread and increase the intensity of wildfires when they occur; wildfires burning in heavy fuels tend to be harder to extinguish. Human populations themselves seem to present positive risk factors for wildfire (Donoghue and Main 1985); human-ignited wildfires tend to be clustered around places with human populations (Zhai et al. 2003; Genton et al. 2006), confirming that as human populations grow, wildfire ignitions by people are more frequent, all other factors considered.

Because of the links between fuels and fuel conditions, weather and climate, as climate in the South warms or dries, wildfires could become larger and more intense—again, all other factors considered. The conclusion by Westerling et al. (2006) that wetter weather, which may be experienced in some places with climate change, would result in less frequent, smaller, and less intense wildfires is supported by much research in the South for both human- and lightning-caused fires (Donoghue and Main 1985; Prestemon et al. 2002; Mercer et al. 2007; Prestemon and Butry 2005; Butry et al. 2010).

Many natural resource scientists, land managers, and policy makers have expressed concern about the implications of climate change on wildfire activity. Other sections of this chapter document how climate change may lead to higher fire potential due to drying, warming, and longer fire seasons. Complicating questions of climate change, however, is the likelihood that society is also projected to change significantly in the coming decades. Human populations are growing, including in the South, and most economists predict continued economic expansion. Therefore, projections of future wildfire activity would be incomplete unless they considered societal change. Because of the link between greenhouse gas emissions and economic activity, climate can be said to partly depend on how society changes; Nakicenovic et al. (2000) provide a number of scenarios that describe this kind of dependence.

Results from the various scenarios described by Nakicenovic et al. (2000) and later projections in the Intergovernmental Panel on Climate Change (2007), fourth assessment, indicate that climate in the South will be warmer and, in many areas, drier. The effects of those changes on wildfire, however, are likely to be complex. Humans affect ignition processes, spread processes, and land uses—all of which have a bearing on wildfire projections. In an initial attempt to understand the effects of such changes, we developed statistical models of wildfire in the South, based on historical data, that relate wildfire activity to fuel conditions, broad descriptions of ecological conditions (ecological classification), weather, human populations, and economic activity. Climate, land use, and socioeconomic variables were then projected. This effort was part of larger effort to understand the air quality implications of altered fire activity in the South as a result of climate change (Shankar et al. 2009). An A1B scenario projection paired with the CESM3 (Community Earth System Model 2011) was used as the basis for projections of weather, fire, demographics, timber harvesting activity, and land use. Projections ran from 2002 (the base year) to 2050 and for two intervening years: 2020 and 2030.

Fire Projection Methods

We projected fire area burned on grid cells measuring 12 × 12 km for most of the South (southern Kentucky to southern Virginia, Florida to Texas). The base year for wildfire data—distinguished by cause (human or lightning)—and associated population, economic, fuels, and weather data was 2002. Cross-sectional sample selection models (Greene 1997) by cause were estimated for human-caused wildfire and lightning-caused wildfire. The modeling occurred in two stages: the first stage used a Probit model to predict whether fire occurred in the spatial unit of observation during 2002, and the second stage made a least-squares estimate of area burned using only the observations of fires recorded in 2002. In this second stage equation, the amount of fire recorded in the grid cell was regressed on a factor that measured the probability of having fire as well as a set of other predictors. The threshold used in this model for determining whether fire occurred in the grid cell was whether at least 5 ha burned in the cell. The model domain included 13,956 grid cells; hence 13,956 observations in both sample selection models. Of these, 450 grid cells had at least 5 ha burned by lightning-caused wildfires and 3882 cells had at least 5 ha burned by human-ignited wildfires.

In the first stage of the selection model, the fire occurrence was expressed as a function of income (economic output per unit area), population per unit area, forest land per unit area, fuel levels, and average wind speeds. Intercept shifting dummies reflecting ecological and states' boundaries were included to allow for absolute fire probability differences across these geographical units. The second stage of the model was expressed as a function of the same variables, predicting burned area given that a fire occurred in that grid cell in 2002. Detailed results are available in Shankar et al. (2009).

The statistical models' predicted area burned for 2002 was calibrated to match the region-wide total of area burned for 2002 by cause. The calibration factors for lightning- and human-ignited wildfires were then used in the projection years of 2020, 2030, and 2050. Projections of area burned in these future years were done by applying the estimated statistical models to the predictors. Projections of county and state level variables (forest, income, population, climate variables) were derived from the 2010 Resources Planning Act Assessment (U.S. Department of Agriculture Forest Service 2012). Forest area projections were based on work by Wear (2011), but adjusted for the base year 2002 using the National Land Cover Data from the U.S. Department of the Interior, U.S. Geological Survey. Assignment of grid cells within the region was based on work by Rudis (1999).

Fire Projections Results

Results of the wildfire projections in the South are illustrated in Figure 5.13. Lightning fires are projected to rise from base year (2002) levels. The base-year burned area levels for lightning might have been usually low in 2002, but the projections clearly show an increase, from about 18,000 ha in 2002 to 51,000 ha by 2050, with slightly higher levels expected in 2030 than 2050. Conversely, area burned in human-ignited wildfires by 2050 is projected to decrease by 35%, from about 141,000 ha in 2002 to 92,000 ha in 2050. In aggregate, the decrease in area burned by human-ignited wildfires outweighs the increase in area burned by lightning-ignited wildfires, producing a projected total

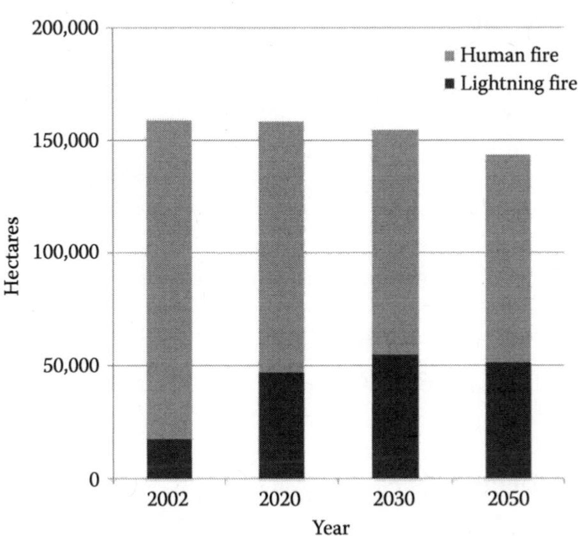

FIGURE 5.13 Area affected by human- versus lightning-caused wildfire in the Southern United States, 2002 (baseline) to 2050 (projected) assuming the Intergovernmental Panel on Climate Change (2007) A1B emissions storyline of rapid economic growth, global population that peaks in mid-century and declines thereafter, rapid introduction of new and more efficient technologies, and a balance between fossil fuels and alternative energy sources.

decrease in area burned of about 10% by 2050 (about 144,000 ha in 2050, compared to 159,000 ha in 2002).

The effects and importance of the variables used in the projection can be appreciated by examining the statistical modeling results (Shankar et al. 2009). One general result of the modeling is that, in aggregate, forest area is positively related to the area burned by both lightning fires and human-ignited fires. As forest area increases, the number of wildfires and area burned would increase. Land use projections used in this study show an aggregate loss in forest land, a change that therefore would be expected to lower overall wildfire activity. Furthermore, income increases have negative effects on both the occurrences of human-ignited wildfires in our statistical models and on the area burned by such wildfires. Therefore, as incomes rise in aggregate across the South, human-ignited wildfires would be expected to decline in both frequency and size. This effect would be expected: as values at risk increase, communities have been shown to devote greater resources toward both preventing fires (Prestemon et al. 2010) and extinguishing them more quickly. Conversely, climate changes projected under scenario A1B with the CESM3 generally show higher winter and summer temperatures and lower overall humidity in the South by 2050. These trends are likely to lead to higher overall burn probabilities and areas burned. These effects, even adding in the effects of higher populations, apparently are less important for human-caused fires compared to changes in forest area and income.

For lightning fires, however, wildfire frequencies are more heavily influenced by trends in temperatures (upward) and fuel moisture. Although the size of such fires would be expected to decline because of greater suppression efforts enabled by higher incomes (attempts to protect values at risk), the tendencies for lightning fires to be larger under a warmer and drier climate outweigh the efforts to control their extent.

TRENDS IN FUEL LOADING

To generate high-resolution datasets that reflect the spatial–temporal dynamics of fuel loads from 2002 to 2050; Zhang et al. (2010) used a global dynamic vegetation model to incorporate simulated ecosystem dynamics into the default (contemporary) fuel loading map developed by the Fuel Characteristic Classification System. Current fuel loading increases from about 6.7 Mg/ha in the coastal areas along the Atlantic Ocean and Gulf of Mexico to 11 Mg/ha or larger in the Appalachian-Cumberland and Ozark-Ouachita highlands; and is less than 2.2 Mg/ha in western Texas. Future fuel loading is expected to be reduced from current level in the central areas of the region, with the largest reduction of about 3.4 Mg/ha in the northern areas. In contrast, the projected fuel loading would increase in Atlantic coastal areas and the Piedmont, with the largest increase of about 3.4 Mg/ha. Fuel loading is expected to be slightly reduced in the central area of western Texas and Oklahoma and increased in the northern and southern areas.

IMPACTS OF PROJECTED FIRE CHANGES

FIRE EMISSIONS

Understanding the potential impact of climate change on fire emissions requires an analysis of the weather and the fuel components of the fire triangle, because both contribute to the total amount of fuel consumed. The most critical factors in determining fuel consumption are the initial amount of available fuel and its moisture content. Climate change could alter fuel loading by changing plant productivity and decomposition rates, as well as by causing shifts in species distribution. Warmer and drier conditions would result in more fuel being consumed. For a more complete picture of climate impacts on fire emissions in the South, we must expand our scope beyond wildfires to also include prescribed fires—the primary tool for preventing wildfires. Although climate may shift toward warmer and drier conditions, these conditions may not be acceptable for prescribed fires because such conditions may be outside the parameters under which land managers can safely conduct a prescribed burn to accomplish their management objectives.

On an annual basis across the South, generally more area is burned due to prescribed fires than wildfires (Figure 5.14). Although the prescribed fire data shown are limited to what was reported for the 2002 emissions inventory, prescribed fire acreage generally shows less interannual variability than wildfires. The wildfire data shown reflect a 5-year average for 1997 to 2002 (Southern Group of State Foresters 2010). Most prescribed burns are accomplished during the first half of the year before the spring wildfire season peaks. For this assessment of the impact of climate change on fire emissions, we assumed that this annual distribution of area burned remains constant.

Climate information is supplied by two models: the Model for Interdisciplinary Research on Climate (the MIROC3.2 model), obtained from the Center for Climate System Research, University of Japan (K-1 Model Developers 2004); and the Commonwealth Scientific and Industrial Research Organization Mk3.5 (the CSIROMK3.5 model), provided by the Commonwealth Scientific and Industrial Research Organization in Australia (Gordon et al. 2002)—both forced by the A1B emissions scenario. These model/scenario combinations were selected from among those used in the Southern Forest Futures Project because CSIROMK3.5 model reflects well the ensemble average that we used in our statistical downscaling analysis, while the MIROC3.2 model projects a future that is among the driest. The average KBDI projected by the CSIROMK3.5 model show modest changes between 2010 and 2060 (Figure 5.15a). The principal distinction of the CSIROMK3.5 compared to the MIROC3.2 is that the former projects a drying that begins in the spring and extends through the summer. In contrast, the MIROC3.2 model exhibits a much stronger summer drying that begins later in the year and persists until later in autumn (Figure 5.15b).

Climate models that provide only maximum temperature and average daily precipitation information are of limited use in supplying the fuel moisture information required for most methods of calculating fuel consumption; and KBDI by itself is not directly useful. To circumvent this limitation, we used a simple equation based on the National Fire Danger Rating System burning index to calculate fuel consumption (Goodrick et al. 2010). Burning index values were not developed by direct calculation (because, as stated above, we did not have all the information required for such calculations). Instead, observed burning index values calculated for weather stations across the South were used to create a set of burning index distributions as a function of KBDI.

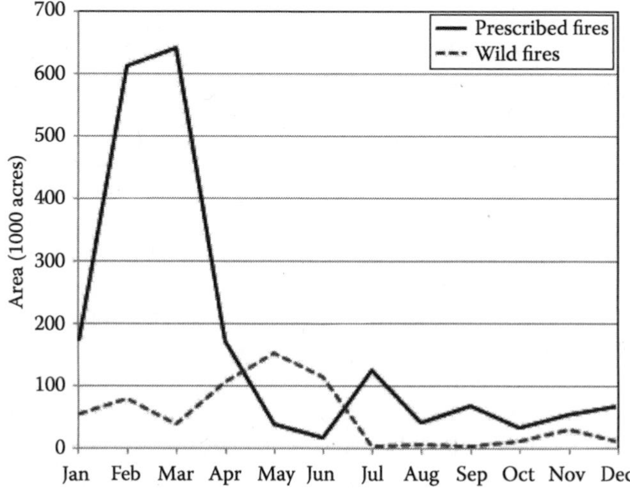

FIGURE 5.14 Average area burned by month for prescribed fires and wildfires from 1997 to 2002 in the Southern United States.

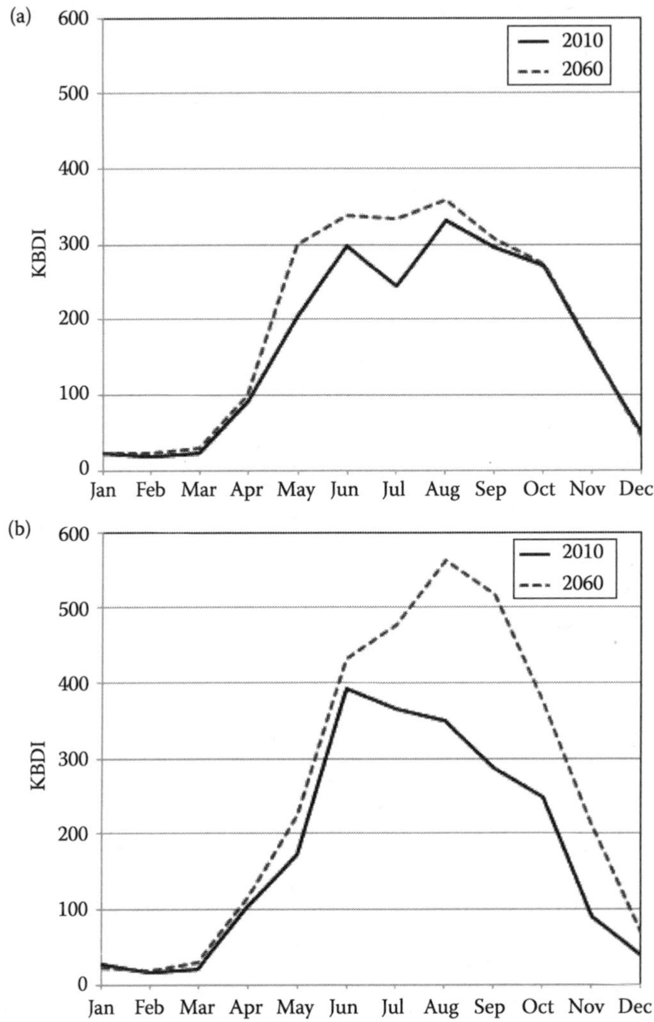

FIGURE 5.15 Monthly KBDI averages for the Southern United States in 2010 and projected for 2060, assuming the Intergovernmental Panel on Climate Change (2007) A1B emissions storyline (a) CSIROMK3.5 and (b) MIROC3.2.

For each month, the burned area of prescribed fires and wildfires was divided into 500-acre (approximately 202 ha*) fires labeled *p* for prescribed fires and *w* for wildfires. Fuel consumption was calculated following Goodrick et al. (2010) using a (1) fuel loading assigned by random draw from the spatially weighted distribution of southern fuel types from the Fuel Characteristic Classification System (with nonflammable types excluded), and (2) burning index value assigned by random draw from the observed burning index distribution that corresponded to the projected KBDI for that month. The primary difference between wildfires and prescribed fires in the analysis was that wildfires are allowed to occur for any burning index value, but prescribed fires are restricted to burning index values less than 35 points. The threshold of 35 was chosen because it reflects a flame length of 1.07 m, which is below the upper limit for which hand crews can safely work a fire-line (1.22 m). Carbon dioxide emissions were determined by simply multiplying the total fuel consumed by an emissions factor.

* The original English units were used in the calculation to avoid introducing artifacts into the data.

For the CSIROMK3.5 model, predicted changes in average fuel consumption (expressed as tons per acre) between 2010 and 2060 only occur from April through July (Figure 5.16a).

Although the MIROC3.2-based predictions are quite similar to the CSIROMK3.5 values, averaging around 6 tons/acre (approximately 13.4 Mg/ha; Figure 5.16b), the changes occur throughout the summer and into autumn.

The monthly carbon dioxide emissions for the CSIROMK3.5 projections for prescribed fires and wildfires in 2010 and 2060 are shown in Figure 5.17a. Although prescribed fire emissions change only slightly, the springtime peak in dryness coinciding with the peak in wildfire activity is expected to result in increased wildfire emissions. This increase is solely caused by the change in climate, with changes in fuel loading not included. The MIROC3.2 projections produce very little change in overall fire emissions because predicted drying occurs in summer, the time of historically low fire activity (Figure 5.17b). These results suggest that the timing of increased drying is potentially more important than the amount of drying, and that drying at times of peak fire occurrence will have a greater impact than drying at other times of the year.

In addition to the influence of fire properties, a change in fuel loading would also influence the effect of climate change on fire emissions (Liu et al. 2011). The predicted changes in fuel loading would lead to an increase in prescribed fire emissions of 500 tons (~454 Mg) from 2002 to 2050 in the eastern areas of the region, compared to a decrease in the central areas, assuming that total area burned remains unchanged.

Land cover change, another cause for changes in future fuel loading, is also expected to contribute to emissions. For prescribed fire emissions when fuel loading is combined with the land cover changes resulting from scenario A1B, instead of increasing the effect of fuel loading alone, emissions in central and northwest Florida would decrease—indicating that the amount of fuel loading reduced by urbanization would outweigh the amount of fuel loading increased by climate change.

Smoke and Air Quality

Future increases in fire activity would produce more smoke and lead to severe air quality impacts, a more far-reaching problem than just changes in fire occurrence and severity. Smoke is produced when wood and other organic material combust, producing a mixture of gases, solid particles, and droplets. Smoke impacts can generally be characterized into two classes—visibility related and health related. Visibility impacts range from regional haze that obscures general visibility and degrades scenic vistas, to dramatic visibility reductions that create a hazard to air and ground transportation. Smoke can cause safety problems when it impedes visibility, a motor vehicle hazard.

Health-related impacts negatively change or limit human habitation or activity (Achtemeier et al. 2001), which is of special concern for those with respiratory problems and other smoke-sensitive illnesses (Naeher et al. 2007). Health-related impacts are regulated through the National Ambient Air Quality Standards (NAAQS) outlined in the Clean Air Act. Wildfire emissions are important sources for particulate matter sizes above 2.5 μm ($PM_{2.5}$) and are precursors of ozone, both of which are subject to monitoring. One recent example is the smoke plume that was transported to Atlanta and other metropolitan areas during the 2007 Okefenokee wildfires that straddled the Georgia-Florida border (Odman et al. 2007). The resulting concentrations of particulate matter exceeded the danger threshold for 2.5 μm and caused significant human health problems in those areas. Prescribed fire emissions are also the source of air pollutants, as Liu et al. (2009a) found in a study on smoke incursion into urban areas of a prescribed burn in central Georgia, USA, on February 28, 2007. Using a smoke simulation system, model results indicated that the smoke invaded metropolitan Atlanta during the evening rush hour. The plumes caused severe air quality problems in Atlanta. Some hourly ground $PM_{2.5}$ concentrations at three metropolitan Atlanta locations were three to four times as high as the daily (24-h) NAAQS level.

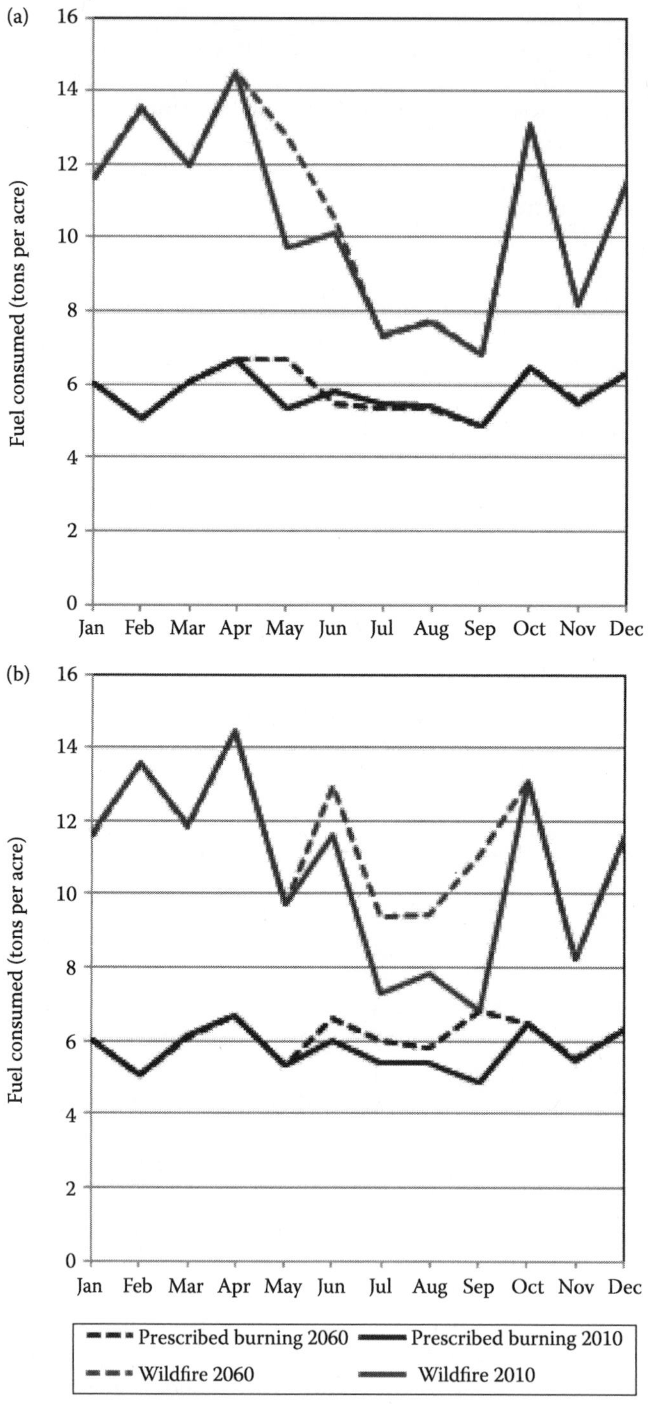

FIGURE 5.16 Average fuel consumption by month for prescribed fires and wildfires in the Southern United States in 2010 and projected for 2060, assuming the Intergovernmental Panel on Climate Change (2007) A1B emissions storyline (a) CSIROMK3.5 and (b) MIROC3.2.

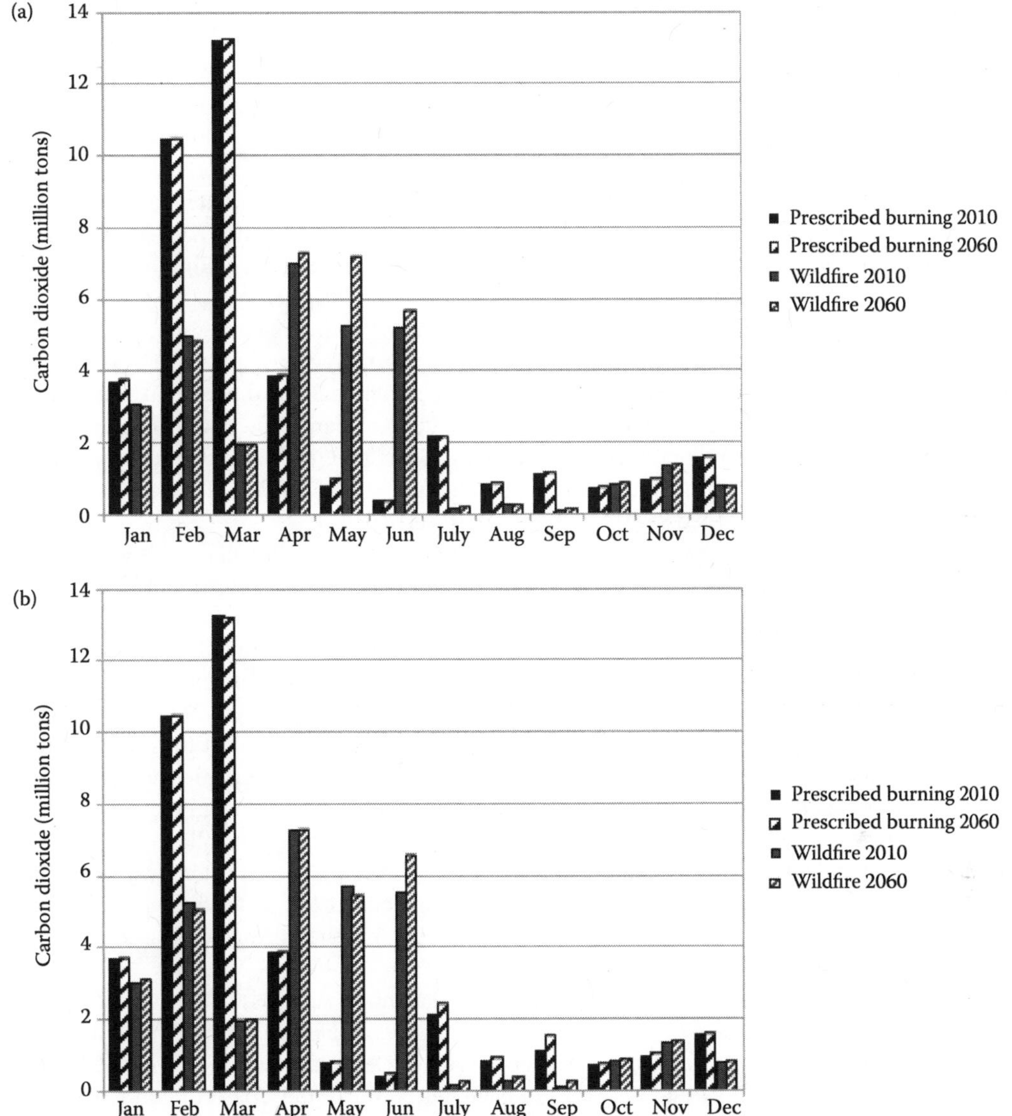

FIGURE 5.17 Monthly carbon dioxide emissions for prescribed fires and wildfires in the Southern United States in 2010 and projected for 2060, assuming the Intergovernmental Panel on Climate Change (2007) A1B emissions storyline (a) CSIROMK3.5 and (b) MIROC3.2.

Public tolerance for smoke has diminished over time, increasing the frequency of complaints about smoke impacts from prescribed burning, fire used to manage wildlands, and wildfires (National Science and Analysis Team 2012). In some situations, lawsuits have affected regional prescribed burning programs, prompting greater emphasis to be placed on smoke impacts when management options are considered (Stanturf and Goodrick 2011).

Water and Soil

Although both wildfires and prescribed burning can have negative impacts on water and soil (Neary et al. 2005), most studies have focused on prescribed burning because of ease in comparing changes before and after treatment. Opportunities to monitor large fire impacts have been rare in the Southern United States.

The impacts of fires on water quality and nutrient cycling include a reduction of total ecosystem nitrogen availability as a result of volatilization and leaching (Knoepp and Swank 1993) and increased sediment loading (Knoepp and Swank 1993; Vose et al. 1999). The magnitude of the effects varies greatly and depends on fuels, soil properties, topography, climate, weather, and fire frequency and intensity (Richter et al. 1982). A single fire occurring after an extended period of fire suppression and fuel accumulation may have a greater impact on water quality, nutrient cycling, and sedimentation than would multiple fires occurring at more frequent intervals. Fire impacts on sediment movement are also more pronounced on sloping than level landscapes.

Ursic (1969) described the effects of prescribed burning on hydrology and water quality at two abandoned fields in the Gulf Coastal Plain in Mississippi. Stormflow during the first year increased 48% in one catchment, and continued to increase in the second and third years. Treatment of the second catchment, which had a fragipan that impeded deep recharge, did not change the volume of stormflow but significantly increased peak discharges and overland flow. Sediment production increased from 0.11 to 1.9 Mg/ha in the first catchment and by 7.5 Mg/ha in the second catchment during the first year, but dropped to <0.56 Mg/ha the third year.

Douglass and Van Lear (1983) reported responses of nutrient and sediment exports to prescribed burning for a Piedmont site at the Clemson Experimental Forest in South Carolina. Four loblolly pine watersheds were burned twice at an 18-month interval. The first burn took place in March and the second in September. The prescribed burns did not change water quality of the streams.

Clinton et al. (2000) summarized the results of four experiments that examined stream nitrate (NO_3^-) responses to forest fires at the Nantahala National Forest in western North Carolina: an autumn fell-and-burn fire (Jacob's Branch) and two spring stand-replacement fires (Wine Spring Creek and Hickory Branch) implemented to improve degraded xeric oak-pine forests, and an autumn arson-related wildfire (Joyce Kilmer) that burned the understory in an old-growth mesic and xeric forest. Stream nitrate was elevated by 0.03 mg/L for eight months following the burn on Jacob's Branch and by 0.06 mg/L for six weeks following Joyce Kilmer fire. The two spring burn sites experienced no change in stream nitrate. The authors concluded that nitrogen released during the spring burns was immobilized by vegetation uptake, but that nitrogen released during the autumn burns was not.

Vose et al. (2005) used a combination of field studies and modeling to assess the impacts of varying fire regimes on water quality across a geographic gradient in North Carolina. Field study sites were located in the Nantahala National Forest in the Southern Appalachians, the Uwharrie National Forest in the Piedmont region, and the Croatan National Forest in the Coastal Plain region. This study suggested that nitrogen (NO_3^-, NH_4^+) was not affected by prescribed fires of any intensity or severity.

Neary and Currier (1982) monitored stream chemistry (nitrate nitrogen, ammoniacal nitrogen, orthophosphate, sodium, potassium, calcium, and magnesium) and total suspended solids for five streams burned by wildfires in the Blue Ridge Mountains of South Carolina. Increases in stream water nitrate were attributed to fertilizer applications. Elevated concentrations of nitrate and orthophosphate in stream water occurred mostly during stormflow events, and average concentrations were not significantly higher than those observed on undisturbed watersheds. Concentrations of anions (sodium, potassium, calcium, and magnesium) ranged from 12% to 82% above background levels during the monitoring period.

Forest fires can burn significant amounts of forest understory canopy, litter, and duff layers, leaving soils unprotected against raindrop impact. The combustion of forest litter and plants in high-intensity forest fires can create and concentrate long-chained organic compounds that induce water repellency in soils (Doerr et al. 2000). This combination of factors reduces infiltration and increases runoff and soil erosion, especially in the Western United States (Tiedemann et al. 1979; Wright and Bailey 1982; Wolgemuth 2001). However, Wohlgemuth (2001) found that during fire events on southern California chaparral watersheds, forests that had been treated with prescribed fires had erosion rates lower than previously unburned forests. Water repellency has not been found to occur in soils of the Southern United States.

Literature suggests that fire generally has less effect on sediment loading in the South than in the West (Goebell et al. 1967; Van Lear and Waldrop 1986; Van Lear and Danielovich 1988; Shahlee et al. 1991; Marion and Ursic 1992; Swift et al. 1993). Increased soil erosion following fires is frequently associated with forest floor disturbances caused by mechanical site preparation during fire controlling activities, and least with direct fire influences. Similarly, operationally disturbed sites and especially skid trails have been found to be more susceptible to postfire erosion (Ursic 1970; Van Lear et al. 1985). However, because most fire research in the Southern Appalachian Mountains has involved fires of low to moderate intensity (Van Lear and Waldrop 1989; Swift et al. 1993), their results have limited applications.

Socioeconomic Impacts

Future wildfires may induce a variety of socioeconomic consequences for people living in fire-prone areas, including loss of life, increased morbidity, loss of property, and the necessity of making investments to reduce fire-related risks. Although socioeconomic impacts are likely to occur principally in communities located in the wildland–urban interface (WUI; the area where residential development is in close proximity to private and public wildlands), urban populations will not be immune to impacts, particularly smoke-related impacts. At the national scale, the area in the WUI increased by over 50% between 1970 and 2000, and is anticipated to increase another 10% by 2030 (Theobald and Romme 2007). Similar to national trends, the WUI area in the South (roughly 88 million acres) is growing rapidly (Southern Group of State Foresters 2008).

Wildfires emit fine particulate matter in smoke, and epidemiological studies have shown that high levels of particulate matter can adversely affect human health. Some evidence exists that high particulate matter levels produced by wildfires increase mortality risk, especially for elderly populations (Sastry 2002; Kochi et al. 2012). Additional research demonstrates a strong association between particulate matter generated by wildfire and various sources of morbidity such as asthma and general respiratory effects (Rittmaster et al. 2006; Kochi et al. 2010).

Health-related impacts of wildfires generate economic impacts through losses in productivity, defensive expenditures taken to lessen health impacts, and a general loss of well-being (utility). Although only a few studies have attempted to estimate the economic losses associated with the health effects of wildfires, these studies indicate that economic impacts can be substantial. For example, a study conducted in Alberta, Canada, reported that the loss in utility associated with smoke from a large wildfire that impacted people living in Edmonton caused economic losses that were only second to timber losses associated with the fire (Rittmaster et al. 2006). Kochi et al. (2012) calculated that excess cardiorespiratory deaths due to smoke exposure during the severe 2003 wildfire season in California in 2003 was approximately $1 billion. Further, Richardson et al. (2012) estimated that people exposed to the Station Fire of 2009 in Los Angeles County, California, spent about $85.00 per day for defensive expenditures (such as wearing a face mask, running the air conditioner more than usual, or taking medications). As citizens in five California cities were exposed to smoke from that fire, which lasted several weeks, it is clear that the economic costs associated with defensive expenditures can be very large.

In addition to health impacts, wildfires pose a direct threat to lives and property. An extreme example is provided by the California fires of late October 2003, which burned over 300,000 ha in one week, destroyed over 3000 homes, and killed 26 people (Keely et al. 2004). Kochi et al. (2012) estimated that 133 additional deaths from wildfire smoke exposure occurred in southern California as a result of these 2003 wildfires. Although the impacts of the 1998 wildfires in Florida were less extreme, these wildfires nonetheless destroyed 336 homes, 33 businesses, and several cars and boats (Butry et al. 2001).

The increasing frequency and severity of wildfires in forested residential neighborhoods in the United States has caused fire managers and policymakers to emphasize the role of homeowner and community mitigation activities to reduce the hazards associated with wildfires. The Firewise Communities program was initiated in 2002 to respond to this need. During 2011, 713 communities were active in this program, having invested over $103 million in wildfire risk mitigation activities (Firewise Communities/USA 2012). The popularity of these programs is growing rapidly, with nearly one-third of the investment made in 2010 alone. More than half of the active communities are located in the Southern United States.

Within the United States, nearly one-third of the WUI occurs in either loblolly-shortleaf pine or longleaf-slash pine forest types, in which wildfires often burn at high intensity and are difficult to control (Theobald and Romme 2007). Although it is logical that communities located in these forest types might be interested in investing in Firewise Communities or similar activities, recent research suggests that poorer communities living in and near high fire risk landscapes are less likely to invest in fire risk mitigation programs (Gaither et al. 2011). These communities appear to be especially vulnerable to potential changes in fire regimes due to climate change.

Regional Climate

Carbon dioxide and aerosol particles emitted into the atmosphere during wildfires can alter climate—an effect that would increase with increased fire activity. The greenhouse effect from increased carbon dioxide gases in the atmosphere is one of the major contributors for climate change at long-term (decade and century) scales and is among the most important and challenging environmental issues facing world leaders. Greenhouse gases in the atmosphere can absorb long-wave radiation emitted from the ground, which reduces heat energy lost into space. As a result, the temperature of the earth–atmosphere system increases and the water cycle is accelerated. Many atmospheric general circulation models have projected an increase in global temperature by 4°C to about 6°C and significant changes in precipitation by the end of this century.

Smoke particles from wildland fires can affect climate by scattering and absorbing short-wave (solar) radiation (direct radiative forcing) and by modifying cloud microphysics (indirect radiative forcing), with further consequences to cloud formation and precipitation processes and atmospheric circulation (Ackerman et al. 2000; Liu 2005a). In contrast, smoke aerosols have much shorter life spans, but much larger spatial variabilities. Thus, they mainly affect short-term (daily, monthly, seasonal) regional climate variability. For example, smoke aerosols from the Yellowstone National Park wildfires may have played a role in a drought experienced later in 1988 in the Northern United States. Simulations showed the Northwest experiencing the most widespread precipitation decreases in response to radiative forcing of smoke aerosols, with large reductions (about 30 mm) in the northeastern Midwest (Liu 2005b). Meanwhile, precipitation increased in the Southwest, the southeastern Midwest, and the Northeast, but it decreased in the South. The simulated spatial pattern of precipitation anomalies was similar to the observed pattern, suggesting that the smoke particles from the wildfire may have enhanced the drought.

Carbon emitted from biomass burning from all causes contributes significantly to global carbon emissions. Average annual global fire carbon emissions were estimated at about 2 Pg in the 1990s, about one-third of the total global carbon emissions. This contribution could be exceptionally large over a short period of time, before being offset by carbon uptake from vegetation regrowth in burned areas. For example, carbon emissions during the 1997–1998 Indonesian wildfires were the equivalent to the total global carbon uptake by the terrestrial biosphere in a typical year (Page et al. 2002; Tacconi et al. 2007). Although wildfires in many forest ecosystems occur naturally, large amounts of carbon stored in forest ecosystems are lost permanently by deforestation in many regions such as the Amazon, where forest biomass is burned in the conversion to agriculture or pasture (e.g., Van der Werf et al. 2010).

MANAGEMENT OPTIONS FOR MITIGATION

Prescribed Burning

Prescribed burning is among the set of critical forest management tools that can be used to mitigate the impacts of climate change on wildfires in the South, but it has many potential benefits. Aside from potentially reducing the overall economic and negative ecological impacts of wildfires, prescribed fire can be used to enhance forage opportunities for wildlife and livestock (Waldrop and Goodrick 2012) and help to restore and maintain the fire-adapted ecosystems that many species depend on (e.g., Brockway et al. 2005). In recognition of these potential benefits, land managers in the South prescribe burn approximately 3.2 million ha annually—more than in all other regions of the United States combined (Wade et al. 2000). Broad evidence of prescribed fire's effects at reducing understory fuels and therefore wildfire risks in the South is revealed by the fact that the total prescribed area burned in the South is about 60% of the prescribed area burned for the United States, much larger than its 20% share of wildfire area burned (Figure 5.18).

Prescribed burning can also be a management option for reducing the impacts of any future increases in wildfire potential emanating from climate change. Wildfires typically occur under drier conditions that favor higher-intensity and more complete fuel consumption. Instead, prescribed burning in the South is conducted at higher fuel moistures under meteorological conditions that favor low-intensity fires with lower fuel consumption. Therefore, prescribed burning potentially results in lower emissions than wildfires (Urbanski et al. 2009). Recent studies have provided quantitative estimates of the role of prescribe fire in other U.S. regions. For example, Wiedinmyer and Hurteau (2010) used a regional fire emissions model to estimate daily carbon dioxide fire emissions for 2001–2008 for the West and found that wide-scale prescribed fire application reduces carbon dioxide fire emissions by 18 to 25% generally and by as much as 60% in specific forest systems. Robertson (2007) pointed out that managers could choose certain weather conditions to conduct prescribed burning under which fire emissions could be minimal. Narayan (2007) showed that prescribed burning can significantly reduce carbon dioxide emissions in European countries that experience high fire occurrence. This author estimated that wildfire emissions were about 11 million Mg/year over a 5-year period, compared to about 6 million Mg/year for prescribed burning.

Wildfire potential is projected to rise under some climate scenarios (perhaps especially from lightning), so more frequent prescribed burning could be used in the future to mitigate some

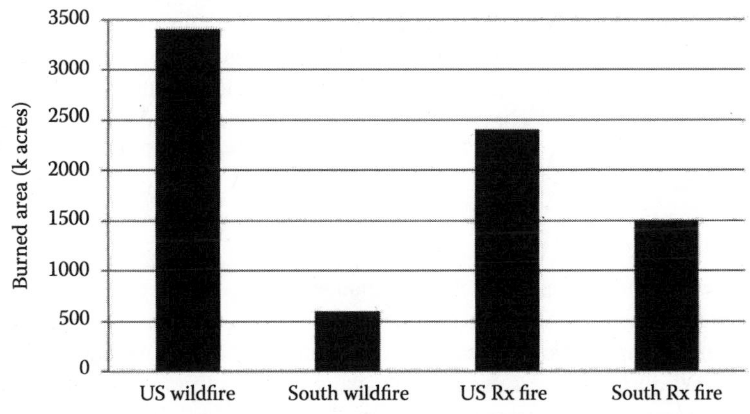

FIGURE 5.18 Burned areas in 2010 for the South and the United States by wildfire and prescribed fire.

of the negative impacts of wildfires. The challenges that southern fire and land managers may face in the future include a reduced burning window caused by changes in weather conditions (Liu 2012), higher risk of prescribed fire escape into the wildland–urban interface and other developed landscapes, and greater constraints in use because of air quality concerns. These constraints to action by managers imply that aggressive fuel load management will be needed where and when possible if future wildfire risks are to be reduced. The size of prescribed burns may be increased so that more area can be burned during the smaller number of favorable days. Burning fewer larger areas will require somewhat fewer resources than more, smaller fires would require. Greater preplanning and coordination of resources will be needed so that open windows for prescribed burning are efficiently utilized. Maintaining or increasing first attack response times will be required, posing logistical challenges in the face of more frequent ignitions and a longer fire season. Fire suppression efforts will have to adjust in order to maintain or increase firefighter safety. Fuel loads may need to be managed at reduced levels, requiring more frequent prescribed burning.

Growing urbanization of the landscape will increase the area of land at the interface with wildlands, indicating that greater engagement between fire managers and communities would be another strategy for mitigating the effects of higher wildfire potential. Prescribed burning can be conducted safely in the interface zone, but more skilled personnel will be needed. Policy and regulatory changes may be needed to provide liability protection for burners. Effective communication with the public will be critical to gaining acceptance for prescribed burning (e.g., Shindler et al. 2009), and personnel may require enhanced communication training. Educational efforts aimed at reducing arson and accidental ignitions as well as fire-wise landscaping around residences could receive increased support as prevention measures (Butry et al. 2010). Support could be provided by state forestry agencies as technical assistance or subsidies. Alternatively, regulatory approaches to risk reduction, including mandated insurance coverage for wildfires, could be another avenue toward reducing some of the negative impacts of future wildfires (Haines et al. 2008). Mechanical and chemical fuel reduction techniques may become more economically feasible under certain conditions (Prestemon et al. 2012), at least in creating buffers between structures and forests in the wildland–urban interface.

Forest Restoration

Another climate change-related wildfire mitigation option is to convert vegetation to more fire-tolerant species (e.g., replacing loblolly pine with longleaf pine or with broadleaved species). Altering forest stand structure, especially by planting at wide spacing to reduce fuel mass per unit area, may alter fire behavior in a way that produces several net benefits for ecosystems, landowners, and society. Fire behavior models may need revision to account for novel fuel types if vegetation changes significantly (e.g., planting *Eucalyptus* species; Goodrick and Stanturf 2012). This strategy, however, would not be without risks or controversy (e.g., McLachlan et al. 2007), but its viability as a wildfire risk mitigation measure merits additional study (Millar et al. 2007).

Smoke Management

Finally, as wildfire potential rises with climate change, managers, modelers, and planners could work jointly to reduce the many negative impacts from wildfire and prescribed fire-related smoke. Effective smoke management would entail the development and deployment of advanced smoke transport models. Wildfire managers seeking to expand the use of prescribed fire will need to pay increasing attention to air quality constraints, particularly if larger areas are burned within an airshed on a given day. This more concentrated prescribed fire activity will need careful management to avoid impacting urban areas.

DISCUSSION

Information on southern wildfire, potential future wildfire trends under a changing climate; the ecological, environmental, and socioeconomic impacts of southern wildland fires; and the forest management mitigation options available to fire managers and policy makers in the South have been presented in this chapter. Major findings include:

Future fire potential: Fire potential is expected to increase in the South in this century as a result of increased concentrations of greenhouse gases.

- The hotspot for the largest projected increase in future fire potential is the western coastal area, along the Gulf of Mexico, in the early spring. This hotspot extends to the central areas of the region in late spring, and spreads farther east, to the Atlantic coastline in summer and early autumn.
- The most significant increase in future fire potential is projected to occur during summer and autumn, when fire potential would increase roughly from the current low level to the moderate level in eastern areas, and from the current moderate level to the high level in western areas of the South.
- The length of fire seasons in the South will likely increase by a few months.
- Projected fire potential is unlikely to increase significantly until 2030–2040.

Burned area: Actual burned areas for a specific landscape would not necessarily increase even with a projected increase in fire potential. A decline in human-ignited wildfire area and an increase in lightning-ignited wildfire area from 2002 to 2050 are projected under a single climate change scenario. Reductions in forest area and changes in other societal factors linked to wildfire activity result in a net reduction in overall fire activity for the region as a whole.

Fuel loading: Fuels may increase or decrease through the effects on productivity and decomposition rates; higher precipitation will increase productivity and thus fuel loading. Conversely, higher temperatures will increase respiration and possibly decomposition rates, thereby lowering fuel loading. Such changes will depend upon vegetation type, soils, and anomalies from current conditions. Future fuel loading is projected to decrease in the western areas of the region and increase in the eastern areas.

Impacts: The projected changes in future wildfire occurrences are expected to have some substantial impacts on forest ecosystems, the environment, and society in the South.

- Future emissions from wildfires could increase in forested areas, especially during late spring and early summer; emissions from prescribed burning, assuming that the current level continues, are likely to decrease in most areas because of projected reductions in fuel loading.
- Increased wildfire emissions could have important smoke and air quality impacts at regional and local scales.
- With the projected increases in wildland–urban interface, wildfire and smoke are expected to result in increased wildfire economic damages and higher wildfire management costs.
- Increased wildfire occurrence in forested areas could reduce total ecosystem nitrogen and increase sediment loading.
- Increased emissions of smoke particles could reduce solar radiation absorbed by forest plants and soils, leading to a cooling effect, changes in heat and water fluxes, and potential changes in regional circulations and precipitation.
- Increased productivity and higher fuel loading, combined with fire weather changes, will favor longer wildfire seasons, increasing the need for prescribed burning to reduce hazardous fuels.

- Longer wildfire seasons may decrease the number of days when prescribed burning may be safely conducted; thus, a greater area will require prescribed burning within a shorter season.
- Increasing urbanization will expand the wildland–urban interface zone but may reduce the opportunities for conducting prescribed burning.
- Regulatory constraints on smoke driven by air quality concerns could decrease opportunities for prescribed burning.

Mitigation:

- Prescribed burning is a forest management tool that has been used extensively in the South to reduce wildfire risks by reducing understory fuel accumulation. It may be among the most useful options for mitigating the impacts of potential increases in wildfire under a changing climate.
- Higher rates of prescribed burning would bring new challenges to fire and land managers, whose use of the tool is restricted by many factors, including weather conditions, risk of escape into developed areas, and smoke-related safety and air quality regulations. Because weather, ambient air quality conditions, and values at risk are likely to change with an enhanced greenhouse effect and population and economic growth in the region, understanding how prescribed fire options may change along with them is critical.
- Altering forest stand structure, especially by planting at wide spacing to reduce fuel mass per unit area, may alter fire behavior in a way that produces several net benefits for ecosystems, landowners, and society.
- As wildfire potential rises with climate change, managers, modelers, and planners could work jointly to reduce the many negative impacts from wildfire and prescribed fire related smoke.

FUTURE RESEARCH NEEDS

The results provided in this chapter are largely preliminary and further research on the issues below is needed to improve our understanding of future fire trends, their impacts, and mitigation options that might be available:

1. The advantages, disadvantages, strengths, and weaknesses of statistical versus dynamic downscaling of projected climate variables need to be clarified with respect to wildfire. Regional climate change scenarios require comparison and interpretation. Two types of downscaled climate change scenarios were presented in this chapter. Statistical downscaling, which includes projections from multiple models and ensemble projections with multiple emissions scenarios, has a higher spatial resolution and provides outputs over various projection periods but is limited to monthly temperature and precipitation projections. Conversely, dynamical downscaling provides daily values for more variables, but has a lower spatial resolution and, for our analysis, was limited to a single general circulation model and emissions scenario for projections of fire potential and total burned area. Comparing the impacts of the differences between the two types of downscaling techniques and using multiple general circulation models and emissions scenarios for dynamic downscaling would improve projections.
2. It is critically important to develop long, consistent historical wildfire datasets for the South that can be used to develop robust statistical models. Very limited fire occurrence and burned area data were used in the analyses reported here: wildfire data for only 5 years for the fire emissions calculation and a single year for projections of future burned areas

and prescribed burning. Wildfires have significant interannual variability. Thus, using the limited fire data increases uncertainty and may obscure significant trends.

3. Alternative statistical models are needed for fire projections, including alternative functional forms, which would enable identification of the modeling framework most likely to accurately predict wildfire given a climate change scenario and expected changes in society. Also, in light of limitations with the KBDI, applications and comparisons of other fire indices, especially those that include wind and humidity factors, would be useful for a more complete understanding of future wildfire potential but would also require improved climate modeling capability.
4. Scientific understanding of fire emissions effects on climate change merits significant advancement, in multiple dimensions. Although some quantitative estimates of the impacts of future fires on emissions were made, other impacts of wildfires were only approached by synthesizing existing studies. For more extensive and reliable quantitative estimates of fire impacts, additional work is needed in a number of areas. The first is to develop more detailed spatial projections of future fires, including frequency, intensity, and burned areas in specific landscapes and ecosystems. Although finer scale resolution of climate variables and fire present statistical challenges associated with false precision, especially at the finest time and spatial scales, it is hoped that more scientific study can provide analysts with real improvements in the reliability of such fine scale projections. The second is to develop more complete datasets of ecological, environmental, and socioeconomic processes along with their interactions with wildfires. The third is to improve our capacity in data processing and computation, which is especially important to modeling the regional air quality impacts of future wildfires.
5. Fuel loading under alternative climate change scenarios needs additional study. In addition to weather and climate, fuel is a critical element in understanding wildfire trends and the potential impacts on emissions and air quality. The projections of future changes in fuel loading described in this chapter were made using a single climate change scenario. This projection approach could be improved by using multiple scenarios. A more refined understanding of fuel loading changes would be enabled by understanding better how climate change would affect vegetation types, representing a challenging but important task for the vegetation and fire modeling communities.
6. More specific understanding is needed of the dynamics of fuel loading under altered climate, productivity, and vegetation types and of the changes needed in burning frequency. Much of our exploration of climate change related wildfire mitigation options was focused on managing fuel loads by prescribed burning. But climate change's likely impacts on burning windows and smoke production and transport may limit its viability under many circumstances. Given these potential limitations, more science is needed on how other fuel reduction strategies could be affordably and effectively implemented. Included in this science would be enhanced understanding of the trade-offs between prescribed fire, nonfire fuel management, provision of ecosystem services, and wildfire risk mitigation. Wildfire in forest ecosystems contributes significantly to carbon emissions. Hence, an elucidation of the trade-offs among carbon emitted from periodic low-intensity prescribed fire, other kinds of fuel reduction treatments, and infrequent but high-intensity wildfire would improve analysts' and managers' ability to recommend and implement the combination of management actions that yielding the greatest possible benefits.

ACKNOWLEDGMENTS

Discussion with and suggestions from Dave Meriwether of Forest Service Region 8, Beth Buchanan of Forest Service Southern Research Station, and James Rogers of North Carolina Department of Environment and Natural Resources are appreciated. The research on fire potential projections using the KBDI was partially supported by the U.S. Joint Fire Science Program under Agreement No. JFSP11172. The dynamical downscaled regional climate change scenario was produced by the North America Regional Climate Change Assessment Program (NARCCAP).

REFERENCES

Achtemeier G.L., Jackson, B., Brenner, J.D. 2001. Problem and nuisance smoke. In: Hardy, C., Ottmar, R.D., Peterson, J.L. et al. (eds.) *Smoke Management Guide for Prescribed and Wildland Fire 2001 Edition. National Wildfire Coordination Group*, PMS 420-2 and NFES 1279. http://www.nwcg.gov/pms/pubs/SMG/SMG-72.pdf (accessed June 14, 2012).

Ackerman, A.S., Toon, O.B., Stevens, D.E., Heyms, A.J., Ramanathan, V., Welton, E.J. 2000. Reduction of tropical cloudiness by soot. *Science* 288: 1042–1047.

Barnard, W., Sabo, W. 2003. Review of 1999 NEI (version 2, final) and recommendations for developing the 2002 VISTAS inventory for regional haze modeling (area and point sources). Prepared for VISTAS, Asheville, NC, USA.

Brenner, J. 1991. Southern oscillation anomalies and their relationship to wildfire activity in Florida. *International Journal of Wildland Fire* 1: 73–78.

Brockway, D.G., Outcalt, K.W., Tomczak, D., Johnson, E.E. 2005. Restoring longleaf pine forest ecosystems in the southern U.S. Chapter 32 In: Stanturf, J.A., Madsen, P. (eds.), *Restoration of Boreal and Temperate Forests*. CRC Press, Boca Raton, FL, pp. 501–519.

Brown, J.K. 2000. Introduction and fire regimes. In: Brown, J.K.; Smith, J.K., (eds.), *Wildland Fire in Ecosystems: Effects of Fire on Flora*. Gen. Tech. Rep. RMRS–42. Ogden, UT: U.S. Department of Agriculture, Forest Service, Rocky Mountain Research Station, pp. 1–8. Vol. 2.

Brown, J.K., Smith, J.K. (eds.). 2000. *Wildland Fire in Ecosystems: Effects of Fire on Flora*. Gen. Tech. Rep. RMRS–42. Ogden, UT: U.S. Department of Agriculture Forest Service, Rocky Mountain Research Station, Vol. 2, 257 p.

Burkett, V.R., Beasley, R.S., Roussopoulos, P., Barnett, J.P. 1996. Toward southern forest sustainability: A science agenda. Southern Region Forest Research Report to the Seventh American Forest Congress, February 20–24, 1996. Office of the Seventh American Forest Congress, Yale School of Forestry, New Haven, CT, 27 pp.

Butry, D.T., Mercer, D.E., Prestemon, J.P., Pye, J.M., Holmes, T.P. 2001. What is the price of catastrophic wildfire? *Journal of Forestry* 99: 9–17.

Butry, D.T., Prestemon, J.P., Abt, K.L., Sutphen, R. 2010. Economic optimisation of wildfire intervention activities. *International Journal of Wildland Fire* 19: 659–672.

Charlson, R.J., Schwartz, S.E., Hales, J.M., Cess, R.D., Coakley Jr., J.A., Hansen, J.E., Hoffman, D.J. 1992. Climate forcing by anthropogenic sulfate aerosols. *Science* 255: 423–430.

Clinton, B.D., Vose, J.M., Knoepp, J.D., Elliot, K.J. 2000. Stream nitrate response to different burning treatments in Southern Appalachian forests. In: *Proceedings: Fire Conference 2000: First National Conference on Fire Ecology, Prevention and Management*, San Diego, CA, Nov 27–Dec 1.

Cochrane, M.A. 2003. Fire science for rainforests. *Nature* 421: 913–919.

Community Earth System Model. 2011. http://www.cesm.ucar.edu/. (accessed June 14, 2012).

Dixon, P.G., Goodrich, G.B., Cooke, W.H. 2008. Using teleconnections to predict wildfires in Mississippi. *Monthly Weather Review* 136: 2804–2811.

Doerr, S.H., Shakesby, R.A., Walsh, R.P.D. 2000. Soil water repellency: Its causes, characteristics and hydrogeomorphological significance. *Earth-Science Reviews* 51: 33–65.

Donoghue, L.R., Main, W.A. 1985. Some factors influencing wildfire occurrence and measurement of fire prevention effectiveness. *Journal of Environmental Management* 20: 87–96.

Douglass, J.E., Van Lear, D.H. 1983. Prescribed burning and water quality of ephemeral streams in Piedmont of South Carolina. *Forest Science* 29: 181–189.

Finney, M.A. 1998. FARSITE: Fire area simulation model development and evaluation Research Paper RMRS-4. USDA Forest Service, Rocky Mountain Research Station, Fort Collins, CO.

Firewise Communities/USA. 2012. Firewise Communities/USA Project Report. www.firewise.org/Communities/USA-Recognition-Program.aspx

Flannigan, M.D., Wotton, B.M. 2001. Climate, weather, and area burned, in *Forest Fires: Behavior and Ecological Effects,* E.A. Johnson and K. Miyanishi (eds.), Academic Press, San Diego, pp. 351–375.

Flannigan, M., Stocks, B., Turetsky, M., Wotton, W. 2009. Impacts of climate change on fire activity and fire management in the circumboreal forest. *Global Change Biology* 15: 549–560.

Food and Agriculture Organization of United Nations. 2001. Global forest fire assessment 990–2000 (Forest Resources Assessment - WP 55).

Fowler, C., Konopik, E. 2007. The history of fire in the southern United States, *Human Ecology Review* 14: 165–176.

Fried, J.S., Torn, M.S., Mills, E. 2004. The impact of climate change on wildfire severity: A regional forecast for Northern California. *Climatic Change* 64: 169–191.

Gaither, C.J., Poudyal, N.C., Goodrick, S., Bowker, J.M., Malone, S., Gan, J. 2011. Wildland fire risk and social vulnerability in the Southeastern United States: An exploratory spatial data analysis approach. *Forest Policy and Economics* 13: 24–36.

Genton, M.G., Butry, D.T., Gumpertz, M.L., Prestemon, J.P. 2006. Spatio-temporal analysis of wildfire ignitions in the St. Johns River Water Management District, Florida. *International Journal of Wildland Fire* 15: 87–97.

Gillett, N.P., Weaver, A.J., Zwiers, F.W., Flannigan, M.D. 2004. Detecting the effect of climate change on Canadian forest fires. *Geophysical Research Letters* 31, L18211, doi: 10.1029/2004GL020876.

Goebell, N.B., Brender, E.V., Cooper, R.W. 1967. Prescribed burning of pine-hardwood stands in the upper Piedmont of South Carolina. Res. Ser. 16. Clemson, SC: Clemson University, Department of Forestry.

Goldhammer, J.G., Price, C. 1998. Potential impacts of climate change on fire regimes in the tropics based on Magicc and a GISS GCM-Derived Lightning Model. *Climatic Change* 39: 273–296.

Goodrick, S.L. 1999. Regional/Seasonal KBDI Classification, Florida Department of Agriculture and Consumer Services, Division of Forestry. www. fl-dof.com/fire_weather/information/kbdi_seasonal.pdf.

Goodrick, S.L., Hanley, D.E. 2009. Florida wildfire activity and atmospheric teleconnections. *International Journal of Wildland Fire* 18: 476–482.

Goodrick, S.L., Stanturf, J.A. 2012. Evaluating potential changes in fire risk from *Eucalyptus* plantings in the Southern United States. *International Journal of Forestry Research* 2012, Article ID 680246, nine pages, 2012. doi:10.1155/2012/680246.

Goodrick, S.L., Shea, D., Blake, J. 2010. Estimating fuel consumption for the upper coastal plain of South Carolina. *Southern Journal Applied Forestry* 34: 5–12.

Gordon, H.B., Rotstayn, L.D., McGregor, J.L., Dix, M.R., Kowalczyk, E.A., O'Farrell, S.P., Waterman, L.J. et al. 2002: The CSIRO Mk3 Climate System Model [Electronic publication]. Aspendale: CSIRO Atmospheric Research. (CSIRO Atmospheric Research technical paper; no. 60). 130 pp. Available at http://www.dar.csiro.au/publications/gordon_2002a.pdf

Gramley, M. 2005. Fire in the South: A Report by the Southern Group of State Foresters. Winder GA: Southern Group of State Foresters, http://216.226.177.78/PDFs/fire_in_the_south.pdf (accessed June 14, 2012).

Greene, W.H. 1997. *Econometric Analysis,* 3rd ed. MacMillan Publishing Company, New York, NY. 1075 p.

Haines, T.K., Renner, C.R., Reams, M.A. 2008. A review of state and local regulation for wildfire mitigation. In: Holmes, T.P., Prestemon, J.P., Abt, K.L. (eds.), *The Economics of Forest Disturbances: Wildfires, Storms, and Invasive Species*. Springer: Dordrecht, The Netherlands. pp. 273–293.

Heilman, W.E., Potter, B.E., Zerbe, J.I. 1998. Regional climate change in the Southern United States: The implications for wildfire occurrence, In: Mickler, Fox (eds.), *The Productivity & Sustainability of Southern Forest Ecosystems in a Changing Environment*, Springer-Verlag, New York, Inc., pp. 683–699.

Hoinka, K.P., Carvalho, A., Miranda, A. 2009. Regional-scale weather patterns and wildland fires in central Portugal. *International Journal of Wildland Fire* 18: 36–49. DOI: 10.1071/WF07045.

Intergovernmental Panel on Climate Change. 2007. *Climate Change 2007.* Cambridge University Press, Cambridge, UK.

K-1 Model Developers. 2004. K-1 coupled model (MIROC) description, K-1 technical report, 1, H. Hasumi and S. Emori (eds.), Center for Climate System Research, University of Tokyo, 34pp. (http://www.ccsr.u-tokyo.ac.jp/kyosei/hasumi/MIROC/tech-repo.pdf).

Keane, R.E., Rollins, M., Parsons, R. 2003. Developing the spatial programs and models needed for implementation of the LANDFIRE Project. In: *2nd International Wildland Fire Ecology and Fire Management Congress*. Orlando, FL.

Keeley, J.E. 2009. Fire intensity, fire severity and burn severity: A brief review and suggested usage. *International Journal of Wildland Fire* 18: 116–126.

Keeley, J.E., Fotheringham, C.J., Moritz, M.A. 2004. Lessons from the October 2003 wildfires in Southern California. *Journal of Forestry* 102(7): 26–31.

Keetch, J.J., Byram, G.M. 1968. A drought index for forest fire control. USDA Forest Service, Southeast Forest Experiment Station Research Paper SE-38. Asheville, NC. 32 pp.

Knoepp, J.D., Swank, W.T. 1993. Site preparation burning to improve Southern Appalachian pine-hardwood stands: Nitrogen responses in soil, water and streams. *Canadian Journal of Forest Research* 23: 2263–2270.

Kochi, I., Donovan, G.H., Champ, P.A., Loomis, J.B. 2010. The economic cost of adverse health effects from wildfire-smoke exposure: A review. *International Journal of Wildland Fire* 19: 803–817.

Kochi, I., Champ, P.A., Loomis, J.B., Donovan, G.H. 2012. Valuing mortality impacts of smoke exposure from major southern California wildfires. *Journal of Forest Economics* 18: 61–75.

Latham, D., Williams, E. 2001. Lightning and forest fires, in *Forest Fires—Behavior and Ecological Effects* E.A. Johnson and K. Miyanishi (eds.), Academic Press, San Diego, pp. 376–418.

Liu, Y.-Q. 2004. Variability of wildland fire emissions across the contiguous United States. *Atmospheric Environment* 38: 3489–3499.

Liu, Y.-Q. 2005a. Atmospheric response and feedback to radiative forcing from biomass burning in tropical South America. *Agricultural and Forest Meteorology* 133: 40–57.

Liu, Y.-Q. 2005b. Enhancement of the 1988 Northern U.S. drought due to wildfires. *Geophysical Research Letters* 32: L1080610.1029/2005GL022411.

Liu, Y.-Q. 2006. Northwest Pacific warming and intense Northwestern U.S. wildfire, *Geophys. Res. Let.* 33(21), L21710, 10.1029/GL027442.

Liu, Y.-Q. 2012. Reduced time windows for prescribed burning in continental United States under a changing climate, *Fourth International Conference on Climate Change University of Washington*, Seattle, USA, July 12–13, 2012.

Liu, Y.-Q., Goodrick, S.L., Stanturf, J.A. 2012. Future U.S. wildfire potential trends projected using a dynamically downscaled climate change scenario. *Forest Ecology and Management.* 294: 120–155. doi: 10.1016/j.foreco.2012.06.049.

Liu Y.-Q., Stanturf, J., Goodrick, S. 2009b. Trends in global wildfire potential in a changing climate. *Forest Ecology and Management* 259: 685–697. doi:10.1016/j.foreco.2009.

Liu, Y.-Q., Stanturf, J.A., Goodrick, S.L. 2010. Future trends in wildfire potential in North America. *XXIII UFRO World Congress*, Seoul, Republic of Korea, August 23–28, 2010. International Forestry Review 12(5): 57.

Liu Y.-Q., Goodrick S., Achtemeier G., Jackson W., Qu, J., Wang W. 2009a. Smoke incursions into urban areas: Simulation of a Georgia prescribed burn. *International Journal of Wildland Fire* 18: 336–348.

Liu, Y.-Q., Stanturf, J.A., Goodrick, S.L., Tian, H., Zhang, C., Wang, Y., Zeng, T. 2011. Changes in carbon emissions from prescribed burning in the southeast due to climate change induced fuel disturbances. *Oral presentation at the 16th Biennial Southern Silvicultural Research Conference*, Charleston, SC, Feb. 15–17.

Marion, D.A., Ursic, S.J. 1992. Sediment production in forests of the Coastal Plain, Piedmont and Interior Highlands: Technical workshop on sediment. *Proceedings of the EPA/Forest Service Workshop,* February 3–7. Corvallis, OR; Washington, DC: Terrene Institute: 19–27. http://www.srs.fs.usda.gov/pubs/2827#sthash.iOOiIFVk.dpuf.

Maurer, E.P., Brekke, L., Pruitt, T., Duffy, P.B. 2007. Fine-resolution climate projections enhance regional climate change impact studies. *EOS* 88(47): 504.

McLachlan, J.S., Hellmann, J.J., Schwartz, M.W. 2007. A framework for debate of assisted migration in an era of climate change. *Conservation Biology* 21(2): 297–302.

Mearns, L.O., Gutowski, W.J., Jones, R., Leung, L.-Y., McGinnis, S., Nunes, A.M.B., Qian, Y. 2009. A regional climate change assessment program for North America. *EOS* 90(36): 311–312.

Melton, M. 1989. Keetch-Byram Drought Index: A guide to fire conditions and suppression problems. *Fire Management Notes* 50(4): 30–34.

Melton, M. 1996. Keetch-Byram drought index revisited: Prescribed fire applications. *Fire Management Notes*, 56(4): 7–11. http://www.fs.fed.us/fire/fmt/index.html.

Mercer, D.E., Prestemon, J.P. 2005. Comparing production function models for wildfire risk analysis in the Wildland–Urban Interface. *Forest Policy and Economics* 7(5): 782–795.

Mercer, D.E., Prestemon, J.P., Butry, D.T., Pye, J.M. 2007. Evaluating alternative prescribed burning policies to reduce net economic damages from wildfire. *American Journal of Agricultural Economics* 89(1): 63–77.

Millar, C.I., Stephenson, N.L., Stephens, S.L. 2007. Climate change and forests of the future: Managing in the face of uncertainty. *Ecological Applications* 17(8): 2145–2151.

Myers, R.K., Van Lear, D.H. 1997. Hurricane-fire interactions in coastal forests of the South: A review and hypothesis. *Forest Ecology and Management* 103: 265–276.

Myers, R.L. 2006. Living with fire—Sustaining ecosystems & livelihoods through integrated fire management. Global Fire Initiative, The Nature Conservancy, Tallahassee, FL, 28 pp. www.nature.org/initiatives/fire/files/integrated_fire_management_myers_2006.pdf

Naeher, L.P., Brauer, M., Lipsett, M., Zelikoff, J.T., Simpson, C.D., Koenig, J.Q., Smith, K.R. 2007. Woodsmoke health effects: A review. *Inhalation Technology* 19: 67–106.

Nakicenovic, N., Alcamo, J., Davis, G. et al. 2000. Special report on emissions scenarios: A special report of working group III of the Intergovernmental Panel on Climate Change. Cambridge, UK: Cambridge University Press. 599 p. www.grida.no/climate/ipcc/emission/index.htm. (accessed December 9, 2010).

Narayan, C. 2007. Review of CO_2 emissions mitigation through prescribed burning, EFI Technical Report 25, 1–58.

National Interagency Fire Center. 2010. Fire Information—Wildland Fire Statistics, http://www.nifc.gov/fire-Info/fireInfo_statistics.html

National Science and Analysis Team. 2012. Scientific basis for modeling wildland fire management. 71pp. Available at http://www.forestthreats.org/products/publications/NSAT_Phase_2_Summary_Report.pdf

National Wildfire Coordinating Group. 2005. Glossary of wildland fire terminology. http://www.nwcg.gov/.

Neary, D.G., Currier, J.B. 1982. Impact of wildfire and watershed restoration on water quality in South Carolina's Blue Ridge Mountains. *Southern Journal of Applied Forestry* 6(2): 81–90.

Neary, D.G., Ryan, K.C., DeBano, L.F., (eds.). 2005. (revised 2008). Wildland fire in ecosystems: Effects of fire on soils and water. Gen. Tech. Rep. RMRS-GTR-42-vol.4. Ogden, UT: U.S. Department of Agriculture, Forest Service, Rocky Mountain Research Station. 250 p.

Odman, T., Hu, Y., Russell, A. 2007. Forecasting the impacts of wild fires. 2007 CMAQ User Workshop, Chapel Hill, NC. (http://www.cmascenter.org/conference/2007/blank.cfm?CONF_PRES_ID=339)

Outcalt, K.W. 2008. Lightning, fire and longleaf pine: Using natural disturbance to guide management. *Forest Ecology and Management* 255: 3351–3359.

Page, S.E., Siegert, F., Rieley, J.O., Boehm, H-.D.V., Jaya, A., Limin, S. 2002. The amount of carbon released from peat and forest fires in Indonesia during 1977. *Nature* 420: 61–65.

Piñol, J., Terradas, J., Lloret, F. 1998. Climate warming, wildfire hazard, and wildfire occurrence in coastal eastern Spain. *Climatic Change* 38: 345–357.

Prestemon, J.P., Abt, K.L. Barbour, R.J. 2012. Quantifying the net economic benefits of mechanical wildfire hazard treatments on timberlands of the western United States. *Forest Policy and Economics* 21: 44–53.

Prestemon, J.P., Butry, D.T. 2005. Time to burn: Modeling wildland arson as an autoregressive crime function. *American Journal of Agricultural Economics* 87(3): 756–770.

Prestemon, J.P., Butry, D.T., Abt, K.L., Sutphen, R. 2010. Net benefits of wildfire prevention education efforts. *Forest Science* 56(2): 181–192.

Prestemon, J.P., Pye, J.M., Butry, D.T., Holmes, T.P., Mercer, D.E. 2002. Understanding broad scale wildfire risks in a human-dominated landscape. *Forest Science* 48(4): 685–693.

Pyne, S., Andrews, P.L., Laven, R.D., 1996. *Introduction to Wildland Fire*, John Wiley & Sons, New York, NY, 769pp.

Reinhard, M., Rebetez, M., Schlaepfer, R. 2005. Recent climate change: Rethinking drought in the context of forest fire research in Ticino, south of Switzerland. *Theoretical and Applied Climatology* 82: 17–25.

Richardson, L.A., Champ, P.A., Loomis, J.B. 2012. The hidden cost of wildfires: Economic valuation of health effects of smoke exposure in Southern California. *Journal of Forest Economics* 18: 14–35.

Richter, D.D., Ralston, C.W., Harms, W.R. 1982. Prescribed fire: Effects of water quality and forest nutrient cycling. *Science* 215: 661–663.

Riebau, A.R., Fox, D. 2001. The new smoke management. *International Journal of Wildland Fire* 10: 415–427.

Rittmaster, R., Adamowicz, W.L., Amiro, B., Pelletier, R.B. 2006. Economic analysis of health effects from forest fires. *Canadian Journal of Forest Research* 36: 868–877.

Robertson, K. 2007. Wildfire, prescribed fire, and climate change in Florida, www.talltimbers.org/images/fireecology/Wildfire-PF-ClimateChange6–07.pdf

Rothermel, R.C. 1972. A mathematical model for predicting fire spread in wildland fuels. USDA For. Sen. Res. Pap. INT-115.

Rudis, V. 1999. Ecological eco-region codes by county, coterminous United States. USDA Forest Service General Technical Report SRS-36. 95 p.

Sastry, N. 2002. Forest fires, air pollution, and mortality in Southeast Asia. *Demography* 39: 1–23.

Seager, R., Tzanova, A., Nakamura, J. 2009. Drought in the Southeastern United States: Causes, variability over the last millennium and the potential for future hydroclimate change. *Journal of Climate* 22: 5021–5045.

Shahlee, A.K., Nutter, W.L., Morris, L.A., Robichaud, P.R. 1991. Erosion studies in burned forest sites of Georgia. *Proceedings of the 1991 Georgia Water Resources Conference* (Ed. Kathryn J. Hatcher), March 19–20. University of Georgia, Athens, GA.

Shankar, U., Fox, D.G., Xiu, A. 2009. NCER Assistance Agreement Final Project Report Investigation of the Interactions between Climate Change, Biomass, Forest Fires, and Air Quality with an Integrated Modeling Approach. Final Report prepared for the Environmental Protection Agency National Center for Environmental Research. 13 p. www.ie.unc.edu/cempd/projects/FIRE/documents.cfm (accessed June 14, 2012).

Shindler, B.A., Toman, E., McCaffrey, S.M. 2009. Public perspectives of fire, fuels and the Forest Service in the Great lakes Region: A survey of citizen-agency communication and trust. *International Journal of Wildland Fire* 18(2): 157–164.

Skinner, W.R., Flannigan, M.D., Stocks, B.J. 2002. A 500 hPa synoptic wildland fire climatology for large Canadian forest fires, 1959–1996. *Theoretical and Applied Climatology* 71: 157–169.

Southern Group of State Foresters. 2008. Fire in the South 2. www.southernwildfirerisk.com/reports/FireintheSouth2.pdf.

Southern Group of State Foresters. 2010. Southern Wildfire Risk Assessment. Available at www.southernwildfirerisk.com

Stanturf, J.A., Goodrick, S.L. 2011. Chapter 17: Fire. In: *Forest Futures Technical Report*. D.N. Wear and J.G. Greis (eds). http://www.srs.fs.fed.usda.gov/futures/.

Stanturf J.A., Wade, D.D., Waldrop, T.A., Kennard, D.K., Achtemeier, G.L. 2002. Fires in southern forest landscapes, Chapter 25 . In: Wear, D.M. and Greis, J. (eds.), *The Southern Forest Resource Assessment*, Gen. Tech. Rep. SRS-53 USDA Forest Service, 607–630.

Stocks, B.J., Mason, J.A., Todd, J.B. 2002. Large forest fires in Canada, 1959–1997. *Journal of Geophysical Research (D)* 107, doi:10.1029/2001JD000484.

Swetnam, T.W., Betancourt, J.L. 1990. Fire-Southern Oscillation relations in the southwestern United States. *Science* 249: 1017–1020.

Swift, L.W. Jr., Elliot, K.J., Ottmar, R.D., Vihnanek, R.E. 1993. Site preparation burning to improve Southern Appalachian pine-hardwood stands: Fire characteristics and soil erosion, moisture and temperature. *Canadian Journal of Forest Research*. 23: 2242–2254.

Tacconi, L., Moore, P.F., Kaimowitz, D. 2007. Fires in tropical forests—What is really the problem? Lessons from Indonesia. Mitigation Adaptation Strategies. *Global Change* 12: 55–66.

Teague, B., McLeod, R., Pascoe, S. 2010. 2009 Victorian Bushfires Royal Commission, Final Report: Volume 1—The Fires and the Fire-Related Deaths. (Government of Victoria: Melbourne, Australia). www.royalcommission.vic.gov.au/Commission-Reports (accessed 7 June 2012).

Theobald, D.M., Romme, W.H. 2007. Expansion of the US wildland–urban interface. *Landscape and Urban Planning* 83: 340–354.

Tiedemann, A.R., Conrad, C.E., Dietrich, J.H. et al. 1979. Effects of fire on water: A state of the knowledge review. Gen. Tech. Rep. WO–10. Washington, DC: [U.S. Department of Agriculture], Forest Service. 28 p.

Times (UK), 1993, *The Times Atlas of the World* 1993. Times Books, ISBN 0–7230–0492–7. Available at en.wikipedia.org/wiki/Humid_subtropical_climate

Urbanski, S.P., Hao, W.M., Baker, S. 2009. Chemical composition of wildland fire emissions. Chapter 4. In: Bytnerowicz, A., Arbaugh, M., Riebau, A., Andersen, C. (eds.). *Wildland Fires and Air Pollution*. Developments in Environmental Science Volume 8. Elsevier, Amsterdam.

Ursic, S.J. 1969. Hydrologic effects of prescribed burning on abandoned fields in northern Mississippi. Res. Pap. SO–46. New Orleans: U.S. Department of Agriculture, Forest Service, Southern Forest Experiment Station.

Ursic, S.J. 1970. Hydrologic effects of prescribed burning and deadening upland hardwoods in northern Mississippi. U.S. For. Serv. South. Region Manage. Bull. R8–MB 4.

U.S. Department of Agriculture Forest Service. 2005. Fire statistics. http://www.usfa.fema.gov/wildfire/ (accessed June 14, 2012).

U.S. Department of Agriculture, Forest Service, Rocky Mountain Research Station. 2005. Fire statistics. http://www.usfa.fema.gov/wildfire/. (accessed June 14, 2012).

U.S. Environmental Protection Agency. 2008. Ecoregion Maps and GIS Resources. Available at http://www.epa.gov/wed/pages/ecoregions.htm (accessed April 10, 2008).

USDA Forest Service. 2012. Future scenarios: A technical document supporting the Forest Service 2010 RPA Assessment. Gen. Tech. Rep. RMRS-GTR-272. Fort Collins, CO: U.S. Department of Agriculture, Forest Service Rocky Mountain Research Station, 34 p.

Van der Werf, G.R., Randerson, J.T., Giglio, L., Collatz, G.J., Mu, M., Kasibhatla, P.S., Morton, D.C., DeFries, R.S., Jin, Y., van Leeuwen, T.T. 2010. Global fire emissions and the contribution of deforestation, savanna, forest, agricultural, and peat fires (1997–2009). *Atmospheric Chemistry and Physics Discussions* 10:16153–16230.

Van Lear, D.H., Danielovich, S.J. 1988. Soil movement after broadcast burning in the Southern Appalachians. *Southern Journal of Applied Forestry*. 12: 49–53.

Van Lear, D.H., Douglass, J.E., Cox, S.K., Augspuger, M.K. 1985. Sediment and nutrient export in runoff from burned and harvested pine watershed in the South Carolina Piedmont. *Journal of Environmental Quality* 14: 169–174.

Van Lear, D.H., Waldrop, T.A. 1986. Current practices and recent advances in prescribed burning. In: Carpenter, S.B. (ed.), *Proceedings: Southern Forestry Symposium*. Oklahoma State University, Stillwater, OK:, pp. 69–83.

Van Lear, D.H., Waldrop, T.A. 1989. History, uses, and effects of fire in the Appalachians. Gen. Tech. Rep. SE–54. Asheville, NC: U.S. Department of Agriculture, Forest Service, Southeastern Forest Experiment Station. 20 p.

Vose, J.M., Swank, W.T., Clinton, B.D. et al. 1999. Using stand replacement fires to restore Southern Appalachian pine-hardwood ecosystems: Effects on mass, carbon, and nutrient pools. *Forest Ecology and Management* 114: 215–226.

Vose, J.M., Laseter, S.H., Sun, G., McNulty, S.G. 2005. Stream nitrogen response to fire in the southeastern U.S. In *3rd International Nitrogen Conference*, 12–16 October.

Wade, D.D., Brock, B.L., Brose, P.H., Grace, J.B., Hoch, G.A., Patterson III, W.A. 2000. Fire in eastern ecosystems. In: Brown, J.K., Smith, J.K., (eds.), *Wildland Fire in Ecosystems: Effects of Fire on Flora*. GTR RMRS-42. Ogden, UT: USDA Forest Service, Rocky Mountain Research Station: 53–96.

Waldrop, T.A., Goodrick, S.L. 2012. Introduction to prescribed fires in Southern ecosystems. Science Update SRS-054. USDA Forest Service, Asheville, NC.

Wear, D.N. 2011. Forecasts of county-level land uses under three future scenarios: A technical document supporting the Forest Service 2010 RPA Assessment. Gen. Tech. Rep. SRS-141. Asheville, NC: U.S. Department of Agriculture Forest Service, Southern Research Station. 41 p.

Wear, D.N., Greis, J.G. (eds.) 2012. The Southern Forest Futures Project: Summary Report. Gen. Tech. Rep. SRS-168. Asheville, NC: U.S. Department of Agriculture Forest Service, Southern Research Station. 54 p.

Westerling, A.L., Swetnam, T. 2003. Interannual to decadal drought and wildfire in the western United States, *EOS* 84(49), 545: 554–555.

Westerling, A.L., Hidalgo, H.G., Cayan, D.R., Swetnam, T.W. 2006. Warming and earlier spring increase Western U.S. wildfire activity. *Science* 313(5789): 940–943.

Wiedinmyer, C. and Hurteau, M. 2010. Prescribed fire as a means of reducing forest carbon emissions in the Western United States. *Environmental Science Technology* 44: 1926–1932.

Williams, J.T. 2004. Managing fire-dependent ecosystems: We need a public lands policy debate. *Fire Management Today* 64(2): 6–11.

Wohlgemuth, P.M. 2001. Prescribed fire as a watershed sediment management tool: An example from Southern California. *Watershed Management and Operations Management* 2000: 1-10.

Wright, H.A., Bailey, A.W. 1982. *Fire Ecology*. New York: John Wiley.

Zhai, Y., Munn, I.A., Evans, I.A. 2003. Modeling forest fire probabilities in the South Central United States. *Southern Journal of Applied Forestry* 27(1): 11–17.

Zhang, C., Tian, H.Q., Wang, Y.H., Zeng, T., Liu, Y.Q. 2010. Predicting response of fuel load to future changes in climate and atmospheric composition in the southeastern United States. *Forest Ecology and Management* 260: 556–564.

6 Climate-Induced Changes in Vulnerability to Biological Threats in the Southern United States

Rabiu Olatinwo, Qinfeng Guo, Songlin Fei, William Otrosina, Kier D. Klepzig, and Douglas Streett

CONTENTS

The South is characterized by hot summers, mild winter temperatures, and high humidity compared to the other regions of the continental United States. During the past three decades, the annual average temperature has risen in the South, particularly during the winter months (Karl et al. 2009), and the occurrence of temperatures below freezing has declined by 4–7 days per year in most of the region. Some areas have experienced >20 fewer days below freezing, and the average autumn precipitation in other areas has increased approximately 30% since 1901 (Karl et al. 2009). Although many areas have reported increases in heavy rainfall events (Karl and Knight 1998; Keim 1997), others have experienced moderate-to-severe increases in drought over the last three decades.

Globally, growing atmospheric greenhouse gas concentrations emitted from human activities could cause annual temperatures to increase between 1.8°C and 4.0°C by the end of the twenty-first century (Intergovernmental Panel on Climate Change 2007). Projected future climatic changes are expected to affect the condition, composition, distribution, and productivity of southern forests by the end of this century (www.srs.fs.usda.gov/futures/date accessed: August 13, 2012). At high latitudes and elevations in North America, the rise in temperatures could exceed global averages (Intergovernmental Panel on Climate Change 2007). Changes in environmental factors including temperature, precipitation, and associated factors could affect the occurrence and impacts of forest diseases, native and nonnative pest insects, and nonnative plant species in several ways, some of which are difficult to predict. Understanding the potential impacts of climate change on these biological threats is therefore of great importance to the future of southern forests. However, many uncertainties and gaps still exist in our current knowledge about interactions among these organisms and southern forests.

FOREST DISEASE AND CLIMATE CHANGE

Relatively few studies address the direct effects of climate change on forest diseases. As drought and other weather-driven abiotic stressors increase or are amplified in a changing climate, some scientists have predicted increases in the frequency and severity of forest diseases (e.g., Sturrock et al. 2011). Nevertheless, the examples and scenarios they provided have a significant degree of uncertainty. Although other scientists have reached similar conclusions for organisms such as the fungus (*Heterobasidion irregulare*) that causes annosus root disease (Kliejunas et al. 2009; Otrosina and Garbelotto 2010), these studies are also couched in terms of significant uncertainty and scant empirical evidence.

One major limitation in such prediction is the biotrophic (host dependent) or at least hemibiotrophic nature (ability to continue survival on dead tissue) of the causal organisms behind most forest diseases. As such, host distributions would be the starting point for discussion of how climate change could influence forest diseases. Although well-reasoned and comprehensive predictions of host distributions under varying climate regimes exist, the complex and co-evolved host–pathogen interactions must also be considered, as these interactions add a considerable degree of uncertainty to forecasting efforts. Fungal pathogens are highly adaptable and their reproductive systems are geared for coping with highly variable climatic regimes, host genetics, and host physiology. A case in point is the annosus root disease pathogen that causes significant mortality of southern conifers. Gonthier et al. (2007) discovered that this species is causing extensive mortality of Corsican pine (*Pinus nigra* var. maritime) in the central western coast of Italy, where the climate is warm but drier than in the Southern United States; however, the fungus is apparently spreading and causing higher mortality as it invades newly forested areas, possibly supplanting or hybridizing the native variety. Evidence from genetic analyses of its origins and spread over geological time scales also suggests a high degree of adaptability and mobility. In fact, changing climates may have played a key role in the evolution and speciation of this fungus, affecting host specialization (Linzer et al. 2008; Otrosina et al. 1993). The question remains, therefore, as to the degree to which this pathogen will respond to changes over the relatively short time scale of 50 years.

Climate can affect host genetics. Mitton and Duran (2004) found genetic variation in populations of two needle pinyon pine (*Pinus edulis*) that they associated with summer rainfall patterns—the selection for certain alleles correlated with sites having more rainfall during summer. Similar mechanisms may exist in host species from the Southern United States where a warmer, drier climate could cause the selection of certain genotypes in Southern pine hosts. Any such changes in hosts could, of course, affect interactions with pathogens.

Many soil processes that directly affect forest tree productivity are impacted by drought. Long-term drought events can result in increased host stress and lead to an increase in the severity and mortality caused by root disease. Conversely, tree mortality associated with drought could reduce stand density, thus reducing the opportunity for root contact, contagion, and demand for nitrogen. Nitrogen losses during a return to normal precipitation could affect soil nutrient status and further complicate potential host–pathogen interactions for a time (McNulty and Boggs 2010).

Increased air temperature can decrease the incidence of stem rust pathogens, which have exacting requirements for infection and spore survival on plant surfaces. Another complicating factor, especially with fusiform rust (*Cronartium quercuum* f. sp. *fusiforme*), is the requirement of an alternate host, principally water oak (*Quercus nigra*). The differing requirements for infection of oaks would also complicate any model that attempts to superimpose temperature and moisture regime layers on incidence and severity of this disease.

Increasing atmospheric carbon dioxide can also affect forest diseases by influencing the physiology of their hosts. For root disease, increasing atmospheric carbon dioxide can lead to increased root growth, and by extension, increased probability of infection by pathogens such as annosus root disease (Kliejunas et al. 2009). Although no empirical evidence supports this notion, elevated carbon dioxide concentrations could increase the amount of susceptible pine tissue available for infection by fusiform rust or by pitch canker (*Fusarium circinatum*), increasing the likelihood that disease incidence will increase. For example, increased atmospheric carbon dioxide concentrations may increase growth or amount of juvenile tissue available to infect. Juvenile tissue favors these pathogens. However, some experimental data suggest the opposite; that is, that these pathogens are seemingly unaffected by elevated carbon dioxide (Runion et al. 2010) and that changes in host defensive compounds could result in no increase in disease incidence.

These are only examples of the results of climate change—forest pathogen interactions are only a partial illustration of possible ramifications and complexities. In general, data are extremely rare or nonexistent for most forest diseases.

FOREST INSECTS AND CLIMATE CHANGE

Changes in climate have already contributed to various insect outbreaks over the past decades (Karl et al. 2009) and more outbreaks are expected in the future (Bale et al. 2002; Klapwijk et al. 2012) partly because of the direct link between insect population growth and seasonal temperature (Danks 1987). A warming climate would likely hasten proliferation of potentially devastating insect species such as bark beetles, defoliators, sap-suckers, borers, and weevils. Any future increase in outbreaks and severity of insect pests could increase host-tree mortality, heighten forest disturbance, and reduce tree vitality (Pimentel et al. 2004; Stadler et al. 2005; Volney and Fleming 2000). Recent years have seen increases in outbreaks of both native species, such as southern pine beetles (*Dendroctonus frontalis*) and pine root collar weevils (*Hylobius radicis*), and nonnative species, such as emerald ash borers (*Agrilus planipennis*), Sirex woodwasps (*Sirex noctilio*), gypsy moths (*Lymantria dispar*), hemlock woolly adelgids (*Adelges tsugae*), and redbay ambrosia beetles (*Xyleborus glabratus*).

Temperature plays a decisive role in the survival, growth, distribution, and diversity of many organisms. Increasing temperatures could expand the northern ranges of many nonnative insect species in the United States. Insect developmental thresholds, the rate of multiple life stages (generations), adult longevity, prolonged adult emergence, and insect flight are all affected directly by temperature (Bentz et al. 2010). However, the spread of an individual species is also influenced by its adaptive capability and biological properties, by the diversity of potential host species, and by the prevailing ecosystem disturbance (Mack et al. 2000). Shifts in temperature and precipitation have already shown some correlations with outbreak of bark beetles, which are highly sensitive to changes in temperature and water stress (Berg et al. 2006; Powell and Logan 2005).

From 1960 to 2004, minimum temperatures have increased by approximately 3.3°C in the South. Over the same period, outbreaks of southern pine beetles have extended northward, matching physiological model predictions of cold tolerance for the species (Trân et al. 2007). Similarly, warming trends have been associated with shifts in the duration of generations for spruce beetles (*Dendroctonus rufipennis*) in the Western United States (Hansen et al. 2001; Werner et al. 2006) and of mountain pine beetles (*Dendroctonus ponderosae*) in high-elevation forests (Bentz and Schen-Langenheim 2007). The expansion in tree host range could lead to the dispersal of many insect species and expansion of communities that are associated with them (symbionts). Climate change may promote rapid adaptation of nonnative species to seasonal temperature changes (Balanyá et al. 2006; Bradshaw and Holzapfel 2006) and expansion of their ranges into the new niches created by increased temperatures (Battistia et al. 2006; Nealis and Peter 2009).

Warmer winter seasons could increase survival rates of insects throughout the winter, allow more generations per season, and possibly expand their ranges northward (Karl et al. 2009). Insects generally grow under medium thermal conditions ranging from 5°C to 37°C, with optimal growth occurring at 20–30°C and thermal death usually occurring at 46°C (Dwinell 2001). Bentz et al. (2010) identified direct correlations between increased minimum temperatures and reduced cold-induced beetle mortality, while Ungerer et al. (1999) described the development rate of southern pine beetles as a function of temperature, where air temperatures of –16°C result in almost 100% mortality. Hence, the range of the southern pine beetle is more likely dependent on winter temperatures. Therefore, warmer winters would likely benefit the southern pine beetle to expand its range northward.

In addition to changing developmental rates, the metabolic rates of insect species almost double with a 10°C increase in temperature (Clarke and Fraser 2004; Gillooly et al. 2001), a clear indication that insect physiology is highly sensitive to changes in temperature. Therefore, a warmer climate would likely accelerate insect feeding, development, and movements with possible effects on the rate of fecundity, survival, generation time, and dispersal (Bale et al. 2002). In addition to increasing insect metabolism during the growing season, warmer temperatures can also reduce insect mortality from extreme cold in winter (Ayres and Lombardero 2000; Bale et al. 2002), resulting in thriving insect populations through the spring and into summer.

Climate change could also indirectly affect insect populations through impacts on natural enemies (Burnett 1949), important insect symbionts (Lombardero et al. 2003; Six and Bentz 2007), host physiology, and host range distributions (Bentz et al. 2010). Bark beetles and redbay ambrosia beetles typically rely on a complex relationship with fungal symbionts to survive and complete their life cycles. However, these symbionts could have different thermal ranges for optimal growth (Rice et al. 2008). Therefore, a significant shift in thermal range could influence the fungal symbionts that are vectored by dispersing beetles (Six and Bentz 2007), and the physiological responses of hosts to environmental changes could also lead to shifts in the insects that feed on them (Seybold et al. 1995). Shugart (2003) noted that an approximately 2°C increase in temperature has led to significant changes in the composition and formation of new forest communities, most of which no longer exist today, over several thousand years.

As warming trends progress, insect distributions are extending northward (Parmesan 2006; Ungerer et al. 1999). However, some species could be limited to a specific area because of unfavorable climatic conditions rather than unavailability of tree hosts (Salinas-Moreno et al. 2004). Future warmer winter temperatures could remove existing range barriers, creating conditions that are conducive for nonnative insects to spread into places where hosts are currently abundant. The resulting influx could result in competition with native species.

Although the complexity of insect species' responses to climate change results in uncertainties in predicting the extent and nature of impacts, predicted changes in temperature from various general circulation models serve as an important basis for estimating biological responses to changing conditions (Millar et al. 2007). The general circulation models and quantitative models based on documented biological parameters of individual species should provide some insight into how southern forest ecosystems would likely respond to climate change. Warming would only translate to insect population growth or range expansions if the insect species can adapt to changing forest ecosystems (Bentz et al. 2010; Bradshaw and Holzapfel 2006). Overall, the potential threat of future insect attacks (native and nonnative species), will depend on how well they response to shifts in environmental factors such as temperature or precipitation ranges, and on their interactions with host plant species and other associated organisms within the forest ecosystem.

NONNATIVE PLANT SPECIES AND CLIMATE CHANGE

In general, nonnative plant species will likely face similar effects from climate change as their native counterparts (Parmesan 2006), perhaps with varying level of impacts. Global climate change, along with other natural and human-caused factors, could lead to a major conservation threat, which could change species geographic range, community composition, and structure and function of ecosystems (Mack et al. 2000). Similar to native plants, many introduced species, especially those that are highly invasive, could shift their distributional ranges accordingly, albeit with a time lag in response. For example, under warming conditions, including those seen in the last few decades, invasives of all kinds are moving higher in latitude and elevation (Guo 2010), threatening those native species that have small populations and small distribution ranges (e.g., the species on mountaintops). In a recent study of 825 birds, mammals, and plants introduced between continents and from eastern Asia to North America, Guo et al. (2012b) found strong correlations between the latitudinal distributions in their native lands and the lands they invade; nevertheless, relatively more species occur at latitudes higher than those in their native lands, and fewer occur at latitudes lower than those in their native lands.

In addition, climate change will likely influence the establishment of new species and the effectiveness of control strategies. Nonnative plants that are fast growing and responsive to resources would be favored by environmental changes that increase resource availability. Many endangered species already have limited habitat on mountaintops and fragmented landscapes and no viable escape routes. A warming climate, upward movement of nonnatives, and increasing urbanization and fragmentation would further jeopardize their further existence.

Current forest management practices have given scant consideration to the possible consequences associated with climate change, even though most—if not all—simulations and models have projected significant shifts in species ranges after introduction of nonnative plants. The resulting alterations in ecosystems would greatly affect how we manage our forests and the future management policy (Guo et al. 2012b).

This chapter describes the interrelationships among insects, diseases, and nonnative plants, and assesses the vulnerability of the southern forest ecosystem within the context of future climate conditions. Specifically, we discuss the impacts of future climate change on selected insects, diseases, and plants species; along with projections and distributions of host species or the local conditions that affect host species at the subregional level and critical areas within subregions. Our goals are to benefit stakeholders including land managers and private landowners and to assist the Forest Service in future strategic planning of resource allocations to mitigate the effects of climate change on southern forests.

METHODS

FUTURE CLIMATE

One climate projection for 2080 (Karl et al. 2009) indicates that average temperatures in the South will rise by 2.5°C under lower emissions projections to 5.0°C under higher emissions projections, with an increase of approximately 5.8°C in summer. The states that border the Gulf of Mexico are expected to have less rainfall in winter and spring, compared to states farther north. The frequency, duration, and intensity of droughts will most likely increase with higher temperatures, more evaporation from soils, and more water loss from plants. The number of days per year with peak temperatures >32.2°C is expected to rise significantly, especially under the higher emissions projection. The largest temperature increase can be expected in summer, when increases in the number of very hot days could mean more and more-intense droughts and wildfires. Increases in temperatures on the very hot days are projected to outpace increases in average temperature (Karl et al. 2009).

To develop estimates of carbon dioxide that humans could produce over the next 100 years, the Intergovernmental Panel on Climate Change (2007) developed a set of greenhouse-gas emissions storylines, based on economic, environmental, global, and regional considerations.

In this chapter, we evaluate the impacts of future climate change by 2020 and 2060 on selected invasive insects, diseases, and nonnative plants (Table 6.1) by examining an ensemble of four general circulation models (CGCM3, GFDL, CCSM3, and HadCM3, described in Chapter 2), anticipated to be moderate and warm, in combination with three of the Intergovernmental Panel on Climate Change storylines (A1B, A2, and B1). A storyline is a narrative description of a scenario (or a family of scenarios), highlighting the main scenario characteristics and dynamics, and the relationships between key driving forces (Nakićenović et al. 2000): A1B representing moderate population growth and high-energy use; A2 representing continuously increasing population, regionally focused economic growth, and slower introduction of alternative fuels technology; and the B1 representing moderate population growth and an emphasis on global approaches to economic and environmental stability.

INSECT INFESTATIONS AND DISEASES

We examined representative species from five classes of insects (Table 6.1): borer (emerald ash borer and Sirex woodwasp), bark beetle (southern pine beetle), defoliator (gypsy moth), sapsucker (hemlock woolly adelgid), and weevil (pine root collar weevil).

The three representative diseases of major significance to the southern forests selected were fusiform rust, annosus root disease, and laurel wilt caused by a fungus (*Raffaelea lauricola*) carried by redbay ambrosia beetles. All plant diseases and insect species selected have unique biological and environmental requirements (Table 6.1) and distribution patterns, which depend on host species distributions across the United States (Figure 6.1). Temperature and precipitation are the two main factors represented in the models presented in this chapter. Optimum temperature range is critical for individual representative species to establish and thrive in a given environment. Hence, in this study, maps of suitable habitat were produced by using (1) individual species temperature range (i.e., lower and upper limits required by the organism at any given time through the season), (2) the corresponding host winter temperature range (upper and lower limits), and (3) the host water availability requirements (i.e., annual minimum precipitation required for growth).

Availability of water is perhaps the most important environmental factor affecting the geographic distribution and structure of vegetation (Woodward 1987). The northern range of distribution for many species is usually limited by minimum annual average temperature; minimum temperature also influences the growing degree-days (the average of the daily maximum and minimum temperatures minus a base temperature of 10°C) that are necessary for physiological processes, such as spring budburst, flowering, pollen production, and seed production (Karlsson 2002; Linkosalo

TABLE 6.1
Temperature and Precipitation Requirements for Several U.S. Invasive Insects and Diseases and Their Hosts

Invasive Species	Temperature (°C)			Preferred Host		
	Extreme	Range	Summer Extreme	Species	Precipitation Range (mm)[a]	January Temperature Range (°C)[a]
Insects						
Redbay ambrosia beetle (*Xyleborus glabratus*)	<–27[b]	10.0–25.0	—	Redbay (*Persea borbonia*)	1020–1630	3–20
Emerald ash borer (*Agrilus planipennis*)	<–30[c]	13.4–25.0[d,e,f]	>32[g]	Green ash (*Fraxinus pennsylvanica*)	380–1520	–18 to 13
Gypsy moth (*Lymantria dispar*)	<–20[h]	3.0–28.0[h,i]	>41[g,h]	Northern red oak (*Quercus rubra*)	760–2030	4–16
Hemlock woolly adelgid (*Adelges tsugae*)	<–30[j]	3.9–25.0[k]	—	Eastern hemlock (*Tsuga canadensis*)	740–1270	–12 to 6
Pine root collar weevil (*Hylobius radicis*)	—	8.5–28.0[l]	>40	Red pine (*Pinus resinosa*)	510–1010	–18 to –4
Sirex woodwasp (*Sirex noctilio*)	—	6.8–25.0[m]	>35	Loblolly pine (*Pinus taeda*)	1020–1520	4–16
Southern pine beetle (*Dendroctonus frontalis*)	<–16	6.7–36.7[n]	>37	Loblolly pine (*Pinus taeda*)	1020–1520	4–16
Diseases						
Annosum root rot (*Heterobasidion irregulare*)	—	5.0–32.0[o]	>35	Loblolly pine (*Pinus taeda*)	1020–1520	4–16
Fusiform rust (*Cronartium fusiforme*)	—	16.0–25.0	—	Slash pine (*Pinus elliottii*)	1020–1520	–18 to 17
Laurel wilt (*Raffaelea lauricola*)	—	10.0–30.0[p]	>34	Redbay (*Persea borbonia*)	1020–1630	3–20

Note: — means no information available.

[a] Burns and Honkala, 1990 (*Source:* www.na.fs.fed.us/spfo/pubs/silvics_manual/table_of_contents.htm, date accessed: August 13, 2012.)

[b] www.forestthreats.org/publications/fhm/frank-koch/Spatio-temporal_analysis_of_Xyleborus_glabratus_invasion_in_eastern_US_forests.pdf, date accessed: August 13, 2012.

[c] www.fs.usda.gov/Internet/FSE_DOCUMENTS/stelprdb5191794.pdf, date accessed: August 13, 2012.

[d] www.insectscience.org/10.128/i1536-2442-10-128.pdf, date accessed: August 13, 2012.

[e] nrs.fs.fed.us/pubs/gtr/gtr-nrs-p-51papers/54keena-p-51.pdf, date accessed: August 13, 2012.

[f] esameetings.allenpress.com/2009/Paper17566.html, date accessed: August 13, 2012.

[g] pestthreats.umd.edu/content/documents/EABBulletin.pdf, date accessed: August 13, 2012.

[h] www.nappfast.org/caps_pests/maps/Documentation/Lymantriamathuradocumentation.pdf, date accessed: August 13, 2012.

continued

TABLE 6.1 (continued)
Temperature and Precipitation Requirements for Several U.S. Invasive Insects and Diseases and Their Hosts

[i] www.ipm.ucdavis.edu/PHENOLOGY/ma-gypsy_moth.html, date accessed: August 13, 2012.

[j] www.biocap.ca/rif/report/Hunt_S.pdf, date accessed: August 13, 2012.

[k] www.insectscience.org/10.62/i1536-2442-10-62.pdf, date accessed: August 13, 2012.

[l] www.na.fs.fed.us/pubs/misc/prc_weevil/prc_weevil.pdf, date accessed: August 13, 2012.

[m] www.aphis.usda.gov/plant_health/plant_pest_info/sirex/downloads/sirex-pra.pdf.

[n] repository.tamu.edu/bitstream/handle/1969.1/ETD-TAMU-1360/GRAY-DISSERTATION.pdf?sequence=1, date accessed: August 13, 2012.

[o] www.springerlink.com/content/06640483645677p1/fulltext.pdf, date accessed: August 13, 2012.

[p] www.public.iastate.edu/~tcharrin/399harrington8-53.pdf, date accessed: August 13, 2012.

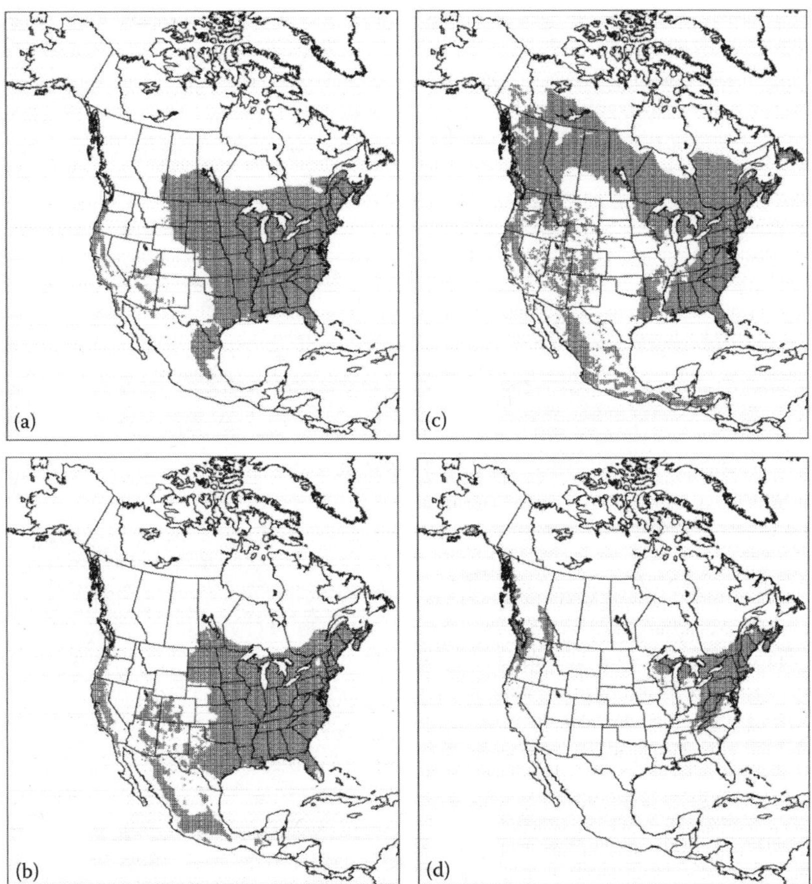

FIGURE 6.1 Distributions of conifer and hardwood hosts for several important invasive insects and diseases of the United States: (a) ash, (b) oak, (c) pine, and (d) hemlock. (Adapted from pubs.usgs.gov/pp/p1650-a/, date accessed: August 13, 2012.)

et al. 2000). In this chapter, insect/disease, host species, and climate scenarios were represented in models examined to obtain long-term projections (i.e., 2020 and 2060). We used annual minimum precipitation and the winter (i.e., January) temperature to produce a climate envelop, that is, a map of likely suitable habitats for individual host plants under different climate scenarios (A1B, A2, and B1 storylines for 2020 and 2060). First, we developed a map of potential suitable habitats for each disease/insect species using the temperature requirements for the insect/disease development in January and July (Table 6.1), which represents suitable habitat in the winter and in the summer, and also indicates the north and south limits for the species. Second, we produced a map of suitable habitat for the corresponding host plant species using the temperature range required by host plant species in January (upper and lower limits), and the annual minimum precipitation required for development of host plant (Thompson et al. 1999) (pubs.usgs.gov/pp/p1650-a/). The climate envelopes for representative hosts were comparable to distributions of major conifer and hardwood hosts (Thompson et al. 1999). Third, we overlaid host and insect/disease maps in DIVA-GIS version 7.5.0 (http://www.diva-gis.org) to produce a suitable habitat map for each representative species (Table 6.1).

To highlight the need for daily weather data in future research and the importance of accumulated degree-days, we used a simple regression analysis to establish relationships between accumulated degree-days from the dynamic downscaling daily temperature data (Mearns et al. 2009) and the monthly average temperature ensemble general circulation model data (A2 storyline). The resulting regression equation was applied to the monthly average temperature from the ensemble data to calculate the degree-days accumulation from March to August, which represents the period between spring and summer. The degree-days data were used to produce maps of suitable habitat under the A1B, A2, and B1 storylines for 2020 and 2060.

Nonnative Plants

To study the potential impact of climate change on nonnative invasive plant species, we analyzed five representative species of different growth forms (Table 6.2): garlic mustard (*Alliaria petiolata*) for forbs, cogongrass (*Imperata cylindrical)* for grasses and canes, Chinese privet (*Ligustrum sinense*) for shrubs, kudzu (*Pueraria montana* var. *lobata*) for vines, and tallowtree (*Triadica sebifera*) for trees. To reconstruct the spatial distribution of each species, we used a spatial distribution database developed by the Eastern Forest Environmental Threat Assessment Center of the Forest Service (www.forestthreats.org, date accessed: August 13, 2012).

Because many species have been introduced to the United States over a relatively short period of time and are thus still spreading, their distribution boundaries (such as their northern and southern latitudinal limits) are difficult to define. For this reason, we used their native ranges as references to determine their warmest and coldest climatic limits (Guo 2006). We projected future distributions at the county level for the year of 2020 and 2060 (10-year and 50-year projections).

The geographic distribution of a species is the product of complex interactions among environmental adaptations, ecological and evolutionary processes, and alterations by humans (Guo et al. 2012a). Climatic variables are the first-order constraints of species distribution, as evidenced in many range-equilibrium studies (Geber and Eckhart 2005; Griffith and Watson 2006). In this

TABLE 6.2
Characteristics and Climatic Limits for Five Nonnative Plant Species of the Southern United States

Species	Growth Form	Native Region(s)	Years Since Introduction (Estimated)	Southern Limit Average Temperature (°C)		Northern Limit Average Temperature (°C)		Annual Precipitation (mm)
				January	July	January	July	
Garlic mustard (*Alliaria petiolata*)	Forb	Europe	200	9	28	−17	3.9	≥500
Cogongrass (*Imperata cylindrical*)	Grass	Asia	100	15	35	−14	21	1000–2500
Chinese privet (*Ligustrum sinense*)	Shrub	Asia	150	13.3	28.5	−10	24.7	700–1600
Kudzu (*Pueraria montana* var. *lobata*)	Vine	Eastern, southern Asia	100	>14	30	−14	16	>1000
Chinese tallowtree (*Triadica sebifera*)	Tree	Eastern Asia	300	10	30	−2	27.5	700–2800

analysis, we employed a bioclimate modeling approach to study the potential habitats, and a set of temperature and precipitation variables to represent the limits in native ranges (Table 6.2). The underlying assumption of bioclimate-based invasive species distribution modeling is that a species can only survive in environments with conditions that match those of its native range (Peterson 2003; Peterson and Vieglais 2001). This general consistency in environmental requirements after a species is introduced into a new area allowed us to ascertain the maximum extent of invasion and identify the geographic areas with native resources at risk. Bioclimate-based models are useful and, to some extent, effective tools in predicting the impacts of climate change on the distribution of species, at least as a first approximation (Pearson and Dawson 2003).

OUTLOOK FOR FOREST PESTS

It is important to note that temperature and precipitation were the only weather factors represented in the models used for producing the 2020 and 2060 projections in this chapter. However, references were made to other ecological factors that might be important, but not represented in the models (e.g., wet springs) in discussing on each representative species.

Annosus Root Disease

Annosus root disease is a deadly disease of forest trees that is often associated with thinning susceptible hosts such as intensively planted loblolly, longleaf, shortleaf, slash, and white pines. The disease has resulted in significant losses of conifers across the South, where slash and loblolly pines are common susceptible hosts (Robbins 1984; Stambaugh 1989). Infection is particularly common in freshly cut stump surfaces and is often transmitted through root grafts that function as conduits to nearby susceptible hosts.

Impacts of climate change—Spores are disseminated throughout the year, making annosus root disease highly dependent on favorable environmental conditions for successful germination and establishment. The fungus does not do as well in cooler, wetter climates; the optimum temperature is approximately 24°C. Ambient temperatures of 45°C can render the spores inactive, and mycelia in wood can die from exposure to 40°C for an hour (Duerr and Mistretta in press). The range of annosus root disease extends throughout most southern forests with few areas remaining for further spread. Changes in environmental factors including increased temperatures, reduced rainfall, and increased host growth could result in increased activity attributable to increased host susceptibility. A drier/warmer climate could also increase pine susceptibility on well-drained sandy sites and on reforested farms where pines are already highly susceptible to the disease (Duerr and Mistretta in press), causing reductions in the distribution range of the host species and ultimately limiting the overall negative impact of this disease by blocking infestations in new locations.

By 2020 and 2060, suitable July habitat for annosus root disease is expected to cover most of the South, with the exception of southern Florida, some parts of Oklahoma, and a large part of western Texas. Under the A1B storyline, the northern limit could shrink by 2060. The pattern of suitable July habitat will likely be similar for the three storylines, with slightly larger January habitable areas under A2 storyline (Figure 6.2). A significant retreat of the range's southern edge is unlikely.

Although the disease is found throughout much of the United States, it is more common and severe in the South. Therefore, a warmer climate and a shift in host ranges could mean that the disease will become more common in latitudes farther north.

Management—As temperatures increase, thermal treatment of the stumps by sunlight could become even more effective in preventing infection. Other efforts, such as restoration of the less susceptible longleaf pine to its previous range, could further decrease the range of annosus root disease. Using borax as a stump treatment in uninfected stands is expected to continue to be an effective tool for managing the disease (Duerr and Mistretta in press).

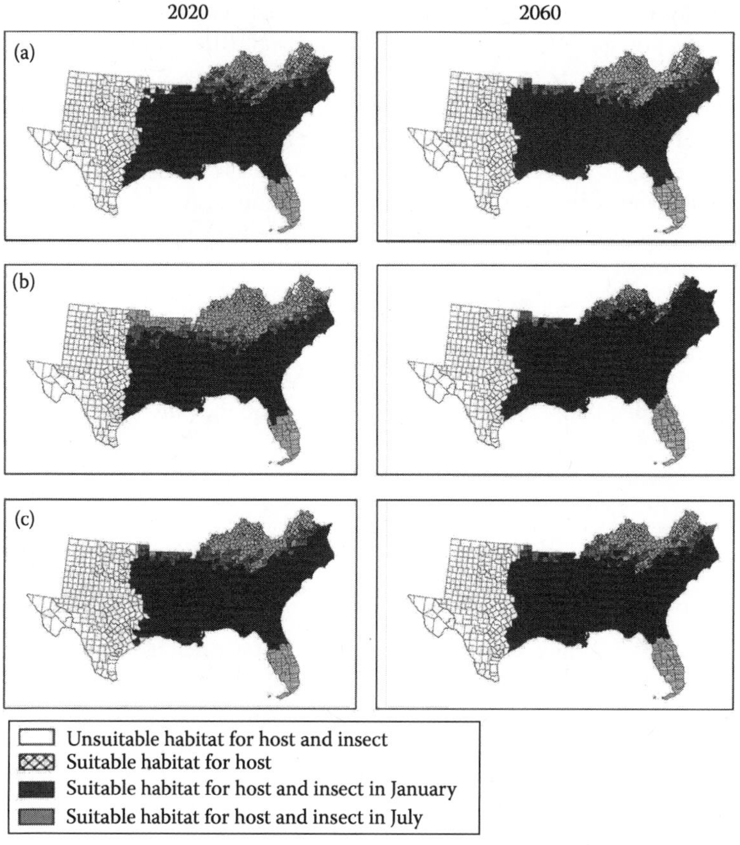

FIGURE 6.2 Suitable habitat for annosus root disease in January and July (2020 and 2060) in the Southern United States, projected using an ensemble of selected general circulation models (CGCM3, GFDL, CCSM3, and HadCM3) and assuming three Intergovernmental Panel on Climate Change (2007) emissions storylines; (a) A1B, (b) A2, and (c) B1 (see text and Chapter 2 for storyline descriptions).

Fusiform Rust

Fusiform rust is an important destructive disease of southern pines that primarily occurs on susceptible species such as loblolly and slash pines, where it causes the formation of fusiform-shaped galls that are fatal to host trees if they occur on the main stem (Anderson et al. 1980; Czabator 1971; Duerr and Mistretta in press). Among the most important factors affecting the incidence of fusiform rust in forest stands are weather, fire, soil cultivation, tree species, genetic strain, stand age, rate of height growth, and density of stocking (Goggans 1957). In the 1930s, extensive planting of susceptible slash and loblolly pines later led to the spread of fusiform rust throughout the host range, with infected trees found in all southern pine ecosystems (Duerr and Mistretta in press; Ward and Mistretta 2002). Heavy losses were observed in Coastal Plain sites from Louisiana to southeastern South Carolina (Duerr and Mistretta in press). Phelps and Czabator (1978) reported that the highest incidences of fusiform rust were in the loblolly and slash pine stands of Georgia, Alabama, South Carolina, and Mississippi. In Louisiana, incidence was low for loblolly and high for slash, whereas Florida had a higher incidence for loblolly and a lower incidence for slash pine. Incidence in Arkansas, Virginia, Texas, and North Carolina was generally low.

Impacts of climate change—Temperature (15–27°C) and humidity (97–100%) range requirements for infection (Phelps and Czabator 1978) and the late winter and early spring weather affect the extent and severity of infection during a given year (Goggans 1957). In the next 50 years, the range of fusiform rust is expected to expand concomitantly as the ranges of its pine and (and alternatively, oak) hosts expand with migration from coastal areas upward into the Appalachian Mountains. Already, the disease is widely distributed within its geographical host range in the South; therefore, a warmer/drier climate could be of less concern than the possibility that a virulent strain will emerge on current hosts. Duerr and Mistretta (in press) and Chakraborty et al. (1998) predict that any losses of pine hosts in coastal areas (with warmer, drier climate regimes) would be offset by gains in the Piedmont and in the lower reaches of the Appalachian Mountains. Any northward shift in planting loblolly pine prompted by warmer climate regimes would likely be matched by a rapid movement of fusiform rust into the new planting areas.

By 2020 and 2060, the potential suitable habitat for fusiform rust will likely extend throughout most of the region with a few unsuitable habitats below the south limit in southern Florida, western Oklahoma, and parts of central and western Texas. The suitable July habitat under the A1B, A2, and B1 storylines in 2020 is expected to shrink by 2060 (Figure 6.3).

Management—The amount of fungal inoculum available, the susceptibility of individual pine species, the abundance of the alternate susceptible hosts (oaks), and favorable climatic conditions are critical for the spread of fusiform rust. Longleaf pines are moderately resistant and shortleaf pines are highly resistant to the fungus (Anonymous 1989). Although several uncertainties still exist, planting of resistant seedlings in anticipation of shifts in host geographic range and changes in

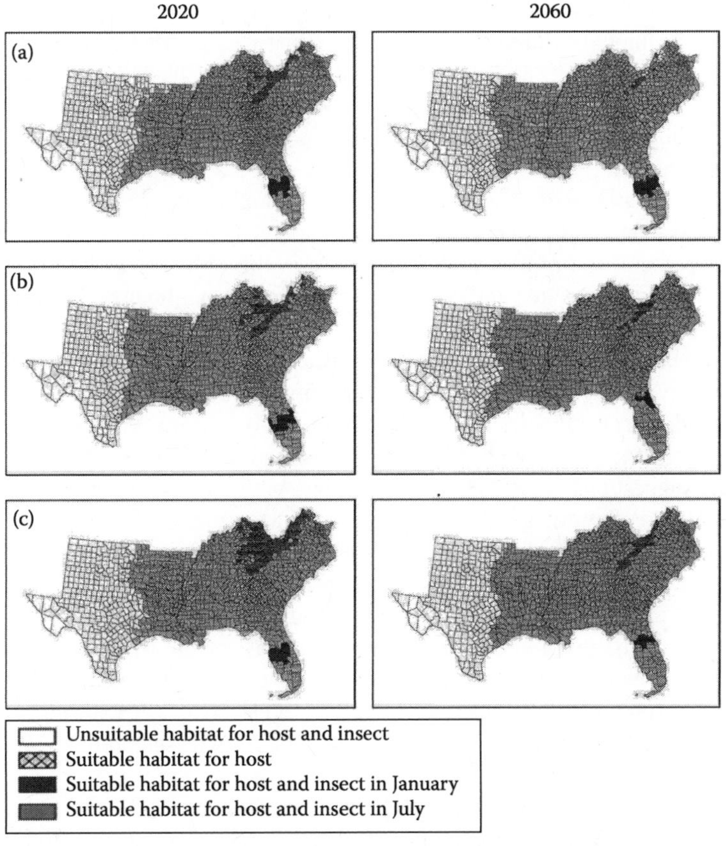

FIGURE 6.3 Suitable habitat for fusiform rust in January and July (2020 and 2060) in the Southern United States, projected using an ensemble of selected general circulation models (CGCM3, GFDL, CCSM3, and HadCM3) and assuming three Intergovernmental Panel on Climate Change (2007) emissions storylines; (a) A1B, (b) A2, and (c) B1 (see text and Chapter 2 for storyline descriptions).

climate could be a useful management strategy in reducing the impact of the disease. Other strategies for plantations and forests include avoiding over-fertilization of seedlings in nurseries, using effective systemic fungicides to protect seedlings in nurseries, applying silvicultural manipulations in young stands to favor healthy saplings, and favoring the deployment of genetically screened resistant seedlings (Duerr and Mistretta in press).

Laurel Wilt/Redbay Ambrosia Beetle Complex

The redbay ambrosia beetle was first reported in the United States in 2002 at the Port Wentworth in Georgia (Rabaglia 2005) and was later found to be consistently associated with unusual mortality of redbay (*Persea borbonia*) and sassafras (*Sassafras albidum*) in coastal areas of Georgia and South Carolina (Fraedrich et al. 2008). Most ambrosia beetles attack stressed, dead, or dying woody plants but the redbay species is unusual in that it attacks healthy trees. Initial attacks occur on healthy trees followed by reproduction on dead tissue. The beetle feeds on plants in the *Lauraceae*. It is thought to be attracted to host plant volatiles, which presumably trigger attacks (Harrington and Fraedrich 2010; Hanula et al. 2008). A single attack by a redbay ambrosia beetle is sufficient to transmit the fungal pathogen that causes laurel wilt, which quickly causes mortality (Fraedrich et al. 2008; Harrington et al. 2008).

Redbay, the favored host of the redbay ambrosia beetle (Hanula and Sullivan 2008), grows in a climate ranging from warm–temperate along the Atlantic Coast to semitropical in southern Florida and the coast of Texas (Brendemuehl 1990). According to the Global Invasive Database provided by the International Union for Conservation of Nature Invasive Species Group (2007a) (http://www.issg.org/database/species/ecology.asp?si=1536&fr=1sts=sss&lang=EN, date accessed: June 26, 2013), other hosts of the redbay ambrosia beetle include swamp bay (*Persea palustris*), avocado (*Persea americana*), camphortree (*Cinnamomum camphora*), pondberry (*Lindera melissifolia*), and pondspice (*Litsea aestivalis*).

The laurel wilt fungus, a food source of the redbay ambrosia beetle, causes vascular damage in redbay, sassafras, and other southern laurel species (Fraedrich et al. 2008; Harrington et al. 2008). At present, the redbay ambrosia beetle is the only known vector. The disease is widespread in six southern states and continues to spread throughout the ranges of redbay and sassafras.

Impacts of climate change on the redbay ambrosia beetle—The redbay ambrosia beetle has a much narrower native geographic range than other ambrosia beetle species and is limited to subtropical or warm temperate climates that are characteristic of eastern Asia (Koch and Smith 2008; Schiefer and Bright 2004). Although flight activity is typically highest in the summer (Hanula et al. 2008), data suggest that adults can also fly during winter and summer months (Mayfield and Thomas 2006). Unlike any other parts of the United States, the southern Coastal Plain has sufficient rainfall across the months of the growing season to provide the amount of moisture needed to sustain the beetle's fungal symbiont. Whether temperature is more of a constraint than moisture remains uncertain. The beetle-symbiont interaction could be constrained to the southern Coastal Plain, and therefore unlikely to spread into other areas east of the Mississippi River (Koch and Smith 2008; Rabaglia et al. 2006). Duerr and Mistretta (in press) indicate uncertainty as to whether both the beetle and fungus will be able to establish on other hosts as their ranges shift under the influence of climate change, making the likelihood of laurel wilt spreading beyond its projected range highly uncertain.

The beetle and the laurel wilt are already as far south as Homestead, Florida. The three emission storylines (A1B, A2, and B1) evaluated in this study indicate a likely decline in potential suitable habitats for the redbay ambrosia beetle by 2060. The B1 storyline will likely have larger areas of suitable habitat in July 2020 compared to the other storylines (Figure 6.4); by 2060, suitable habitat in southern Florida and southeastern Texas is likely to shift slightly southward in January for all storylines.

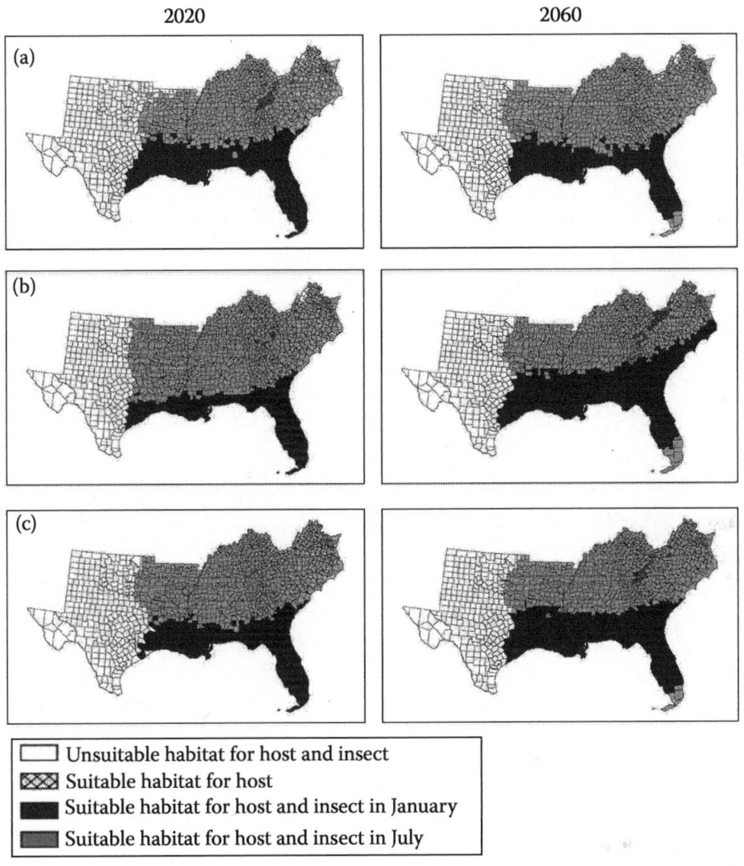

FIGURE 6.4 Suitable habitat for the redbay ambrosia beetle in January and July (2020 and 2060) in the Southern United States, projected using an ensemble of selected general circulation models (CGCM3, GFDL, CCSM3, and HadCM3) and assuming three Intergovernmental Panel on Climate Change (2007) emissions storylines; (a) A1B, (b) A2, and (c) B1 (see text and Chapter 2 for storyline descriptions).

Impacts of climate change on laurel wilt—Sustaining laurel wilt requires the presence of the redbay ambrosia beetle and viable host tissue, both of which need adequate water, although the fungus is not one that grows better under more moist or wet conditions. Koch and Smith (2008) projected that laurel wilt will reach its northern host range by 2020 and its western host range (eastern Texas) by 2040. This projection was only based on the natural range of redbay and on the climatic barriers that could limit the progress of the redbay ambrosia beetle and its fungal symbiont. In 2020, suitable July habitat is expected to be fairly similar for all three storylines, with few differences in the northern limit (Figure 6.5). Suitable July habitat is projected to shrink markedly by 2060. The projections indicate a northward shift in suitable January habitat in southern Florida and the lower Coastal Plain, where only few southern-most counties could become unsuitable in 2060. However, conditions will remain suitable for the hosts, redbay in the south and sassafras to the north.

Management—Because redbay ambrosia beetles can fly, laurel wilt can spread over short distances and perhaps longer distances when the beetles are carried by prevailing winds. The pattern of disease progression reported in Georgia and elsewhere strongly suggests that humans are aiding in its long distance spread. In one instance, an infection center was found adjacent to a campground, which most certainly occurred because a camper had brought in firewood that that had died from laurel wilt and was infested with the redbay ambrosia beetle (Cameron et al. 2008). Other than the use of integrated management approach, no single control measure is effective in protecting

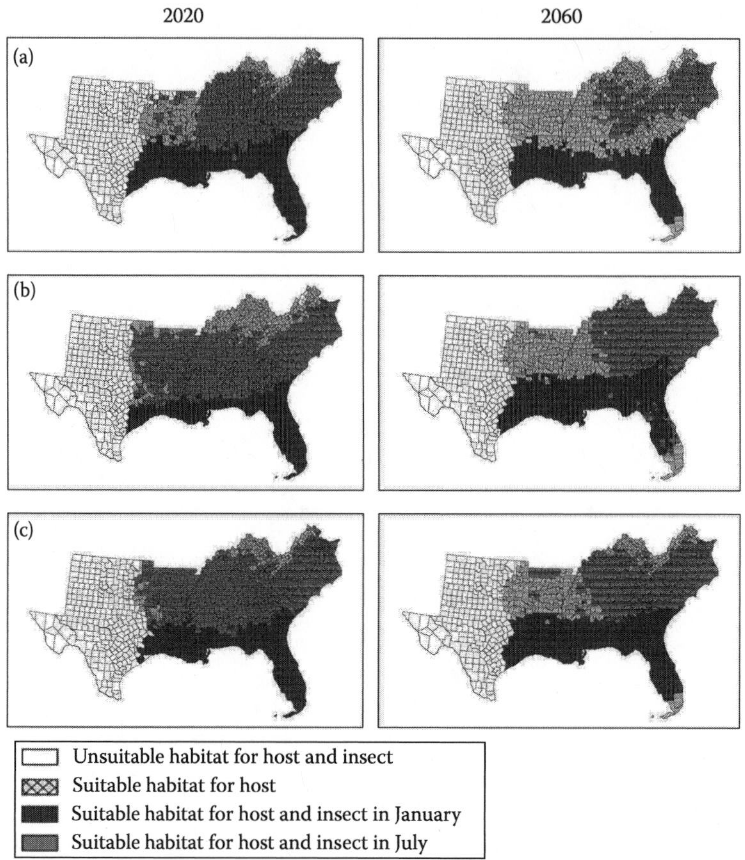

FIGURE 6.5 Suitable habitat for laurel wilt in January and July (2020 and 2060) in the Southern United States, projected using an ensemble of selected general circulation models (CGCM3, GFDL, CCSM3, and HadCM3) and assuming three Intergovernmental Panel on Climate Change (2007) emissions storylines; (a) A1B, (b) A2, and (c) B1 (see text and Chapter 2 for storyline descriptions).

forest and woodland trees against laurel wilt. A demethylation-inhibiting fungicide showed some promising results for preventing the disease in treated trees. However, the cost of treatments will likely limit its use in the future, except for high value trees (Mayfield and Thomas 2006). Therefore, adequate and effective dissemination of vital information and an increase in local awareness of the disease are critical in dealing with laurel wilt.

Local park-recreational activities could have played an important role in the spread of the redbay ambrosia beetle and host-tree mortality in Georgia. Therefore, effective phytosanitary practices to manage infested plant materials are critical, including monitoring, early detection, and effective disposal (chipping or burying) of infected materials (Mayfield and Thomas 2006).

Southern Pine Beetle

The southern pine beetle is a destructive insect pest of southern pine forests (Meeker et al. 2004; Thatcher and Conner 1985) where it has killed >1 million acres of pines, valued at >$1.5 billion, during outbreaks from 1999 to 2002 (Duerr and Mistretta in press). Southern pine beetle populations can increase rapidly, killing large numbers of trees, during periodic outbreaks; however, it could also go through periods of low activity when finding a single infested tree or capturing beetles in pheromone traps becomes rare (Billings and Upton 2010; Thatcher et al. 1980; Thatcher and

Barry 1982). Southern pine beetles attack all species of pines, including loblolly, shortleaf, slash, pitch, Virginia (*Pinus virginiana*), and pond (*Pinus serotina)*, although impacts on longleaf pine (*Pinus palustris*) are generally lower. Successful infestations have been observed in eastern white pines and Table Mountain pines (*Pinus pungens*), and an outbreak was recently observed in New Jersey pitch pines. Attacks on trees <5 years or <2 inches in diameter are very rare, but plantations with uniform-aged single species are susceptible as they age (Duerr and Mistretta in press). Mature trees in pure, dense stands or unthinned pine plantations are most susceptible to attack (Cameron and Billings 1988); therefore, large acreages of pine plantations established in the South over the past five decades could continue to be particularly vulnerable.

Impacts of climate change—Trân et al. (2007) reported that southern pine beetle populations decline when winter minimum temperatures drop below –14°C; and that developmental processes increase with temperatures ranging from approximately zero development at 5 to 10°C, and begin to decline when temperatures exceed 27–32°C (Wagner et al. 1981, 1984a). Extremely hot summer temperatures can be deadly to southern pine beetles (Wagner et al. 1984b) and could affect their interactions with associated forest communities (Hofstetter et al. 2006, 2007; Lombardero et al. 2003). Every stage of the southern pine beetle life cycle can be found in the South throughout all seasons (Thatcher and Pickard 1966, 1967), with a generation completed in about 50 days at 25°C, 100 days at 17°C, and 200 days at 12.5°C (Friedenberg et al. 2007; Wagner et al. 1984a). In the next 50 years, warmer winters could support more generations of southern pine beetles per year (Duerr and Mistretta in press, Gan 2004), but could also disrupt the timing of adult emergence with the onset of new infestations in spring (Billings and Kibbe 1978). Extremely hot summers could increase mortality, hinder flight, reduce physiological activities, and disrupt population growth. Pine stands in the northern edges of the region and even farther north than the historical southern pine beetle range will most likely experience significantly greater southern pine beetle activity and impacts than in the past (Duerr and Mistretta in press).

Under the three storylines, southern pine beetles would likely maintain a considerable suitable habitat from January to July based on 2020 and 2060 projections. Much of the South would provide suitable habitat, the exception being some southern Florida counties where environmental conditions could be unsuitable for either the host (loblolly pine) or the beetle. The pattern of suitable habitat in July will likely be similar for the three storylines, with a slight northern shift in under A2 by 2060 (Figure 6.6).

Management—Adequate management efforts, types of forest composition, direct suppression of new invasions, and other actions will play a significant role in determining southern pine beetle activity and future damages to hosts (Friedenberg et al. 2008). If, in the next 50 years, forest fragmentation increases, age distributions shift to predominantly younger classes, and thinning of plantations increases, the impacts of southern pine beetles could be lower, despite a warmer climate. Conversely, if the frequency of extreme precipitation events (drought and flooding) increases, pines could become stressed and vulnerable to increased southern pine beetle activity, resulting in significant damage (Duerr and Mistretta in press). Therefore, management options such as planting the appropriate species for a given site, lowering planting densities, and thinning pine stands (which increases stand vigor and resiliency) will likely reduce the risks of southern pine beetle outbreaks. Early detection and monitoring for infestation spots, followed by prompt direct suppression of active spots (Billings 1980), could help minimize damage when outbreaks do occur.

Gypsy Moth

Gypsy moth, a native of Europe and Asia, was accidentally introduced into the United States in 1869 (Lechowicz and Mauffelte 1986). Repeated defoliation by gypsy moth has significant effects on the health of oak forests (Campbell and Sloan 1977). Gypsy moth caterpillars feed on a wide range of trees and shrubs, but prefer oaks (Liebhold et al. 1995). Severe defoliation could leave oak trees vulnerable to secondary attacks by pests or pathogens such as the twolined chestnut borer (*Agrilus bilineatus*) or Armillaria root disease caused by a fungus (*Armillaria mellea*). Extended drought could also predispose oak trees to gypsy moth attack and increase oak tree mortality (Duerr

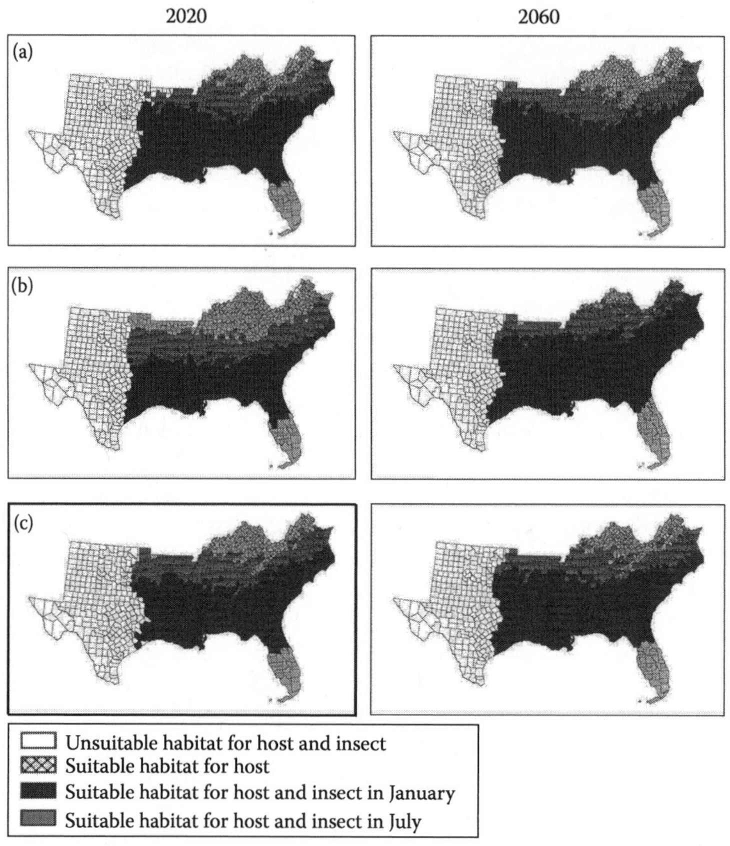

FIGURE 6.6 Suitable habitat for the southern pine beetle in January and July (2020 and 2060) in the Southern United States, projected using an ensemble of selected general circulation models (CGCM3, GFDL, CCSM3, and HadCM3) and assuming three Intergovernmental Panel on Climate Change (2007) emissions storylines; (a) A1B, (b) A2, and (c) B1 (see text and Chapter 2 for storyline descriptions).

and Mistretta in press). Highly favored host species include sweetgum (*Liquidambar styraciflua*), northern red oak (*Q. rubra*), and American basswood (*Tilia americana*). Species of limited suitability include pines, ashes, maples (*Acer* spp.), American beech (*Fagus grandifolia*), and black cherry (*Prunus serotina*). Species that are not favored or are avoided include blackgum (*Nyssa sylvatica*), yellow poplar (*Liriodendron tulipifera*), black locust (*Robinia pseudoacacia*), bald cypress (*Taxodium distichum*), southern magnolia (*Magnolia grandiflora*), and tupelo (Duerr and Mistretta 2011, in press).

Impacts of climate change—Duerr and Mistretta (in press) suggested that regardless of changes in climate, the range of gypsy moth infestation is expected to expand, and at a faster rate than can be attributed to any potential climate change. Logan et al. (2007) noticed that the climatic suitability for the gypsy moth in the Western United States could improve rapidly, due to a general warming trend that began few decades ago and still continues today. In the southeast, the gypsy moth will likely encounter lower concentrations of oak and cove hardwoods as it moves south and west, and forest susceptibility will decrease in some areas. Therefore, a significant portion of the South could experience widespread gypsy moth infestations over the next 50 years. Mixtures of oak and hickory (*Carya* spp.) could displace the boreal forests at higher elevations in the South, and a drier climate could increase host stress and reduce buildup of the gypsy moth's fungal predator, which thrives during wet spring weather. Future outbreaks will depend on the current management efforts to slow, suppress, and eradicate infestations; the abundance and health of hardwood forests that are subject to infestation; and the resilience of the gypsy moth, its natural enemies, and its host trees to changes in temperature and moisture (Duerr and Mistretta in press).

The pattern of potential suitable habitat for the gypsy moth in January would be similar for all three storylines in the 2020 projections. Much of the South is potentially suitable, with the exception of a few counties in southern Florida, western Oklahoma, and Texas. The suitable habitat in July could become bigger under the B1 storyline compared to A1B and A2. Reduction in July suitable habitat is likely by 2060 under the A1B storyline. The A2 storyline predicts a slight northward shift in 2060 (Figure 6.7).

Management—Human population growth could also increase the spread of gypsy moths through increases in recreation and transportation. Therefore, continued vigilance and the application of early detection methods are needed to reduce infestations and help prevent rapid spread in the South. Logan et al. (2007) demonstrated the use of a risk assessment system for predicting establishment of the gypsy moth. Several methods are available for managing the spread of the gypsy moth, including insecticides (*Bacillus thuringiensis* or diflubenzuron), or pheromone flakes to disrupt mating (Duerr and Mistretta in press). According to Sharov et al. (2002), the multi-agency pilot project called "Slow the Spread Program" reduced the rate of spread from approximately 25 miles a year to 7–10 miles per year. Continuation of these programs means that the gypsy moth would

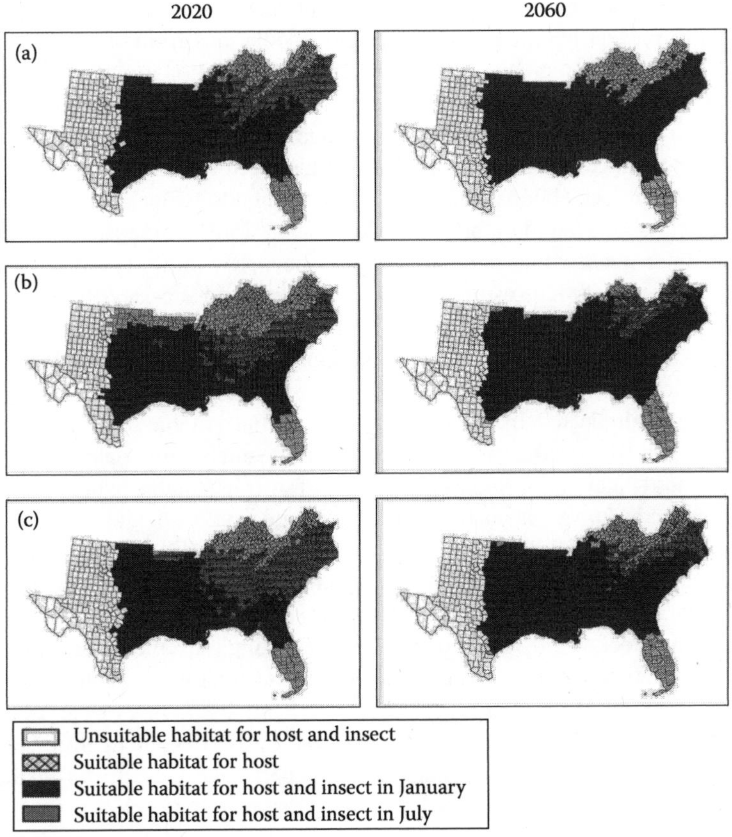

FIGURE 6.7 Suitable habitat for the gypsy moth in January and July (2020 and 2060) in the Southern United States, projected using an ensemble of selected general circulation models (CGCM3, GFDL, CCSM3, and HadCM3) and assuming three Intergovernmental Panel on Climate Change (2007) emissions storylines; (a) A1B, (b) A2, and (c) B1 (see text and Chapter 2 for storyline descriptions).

take the next 50 years to move 350–500 miles farther south, compared to 25–30 years without the program (Duerr and Mistretta in press).

Hemlock Woolly Adelgid

The hemlock woolly adelgid is native to Asia. In North America, this destructive invasive pest of ornamental and forest hemlock trees (*Tsuga canadensis*) was first reported in the western Canadian province of British Columbia in the 1920s (Canadian Food Inspection Agency 2011), and was initially thought to have crossed the U.S. border in Virginia years later. However, Havill et al. (2006) show they were very distinct genetically. Heavy infestations in the Eastern United States have killed host trees in as little as four years, although some trees have survived infestations for more than a decade (McClure et al. 2001). Hemlock mortality can occur quickly and uniformly throughout a stand (Fajvan and Wood 2010), and periods of low winter temperatures have been thought to cause high hemlock woolly adelgid mortality, thereby hindering population establishment and spread (Evans and Gregoire 2007; Morin et al. 2009; Parker et al. 1999; Shields and Cheah 2005). Hemlock woolly adelgid is widely distributed among hemlock growing areas, where minimum winter temperatures stay above −28.8°C. Its cold hardiness depends on geography and season, experiencing a gradual loss of cold tolerance as winter progresses (Skinner et al. 2003).

Impacts of climate change—McClure and Cheah (2002) indicated that the northerly spread and future range of hemlock woolly adelgid could be hindered by the duration, severity, and timing of minimum winter temperature. Paradis et al. (2008) found that an average winter temperature of −5°C is required to prevent hemlock woolly adelgid populations from expanding and spreading. Warmer winters would likely increase survival, fecundity, population growth, and extent of infestations (Pontius et al. 2002, 2006), which could lead to more outbreaks in its current range (Parker et al. 1997), as well as a northward expansion in range (Paradis et al. 2008). Morin et al. (2009) reported that the abundance of host hemlock was a major factor in the spread rate of the hemlock woolly adelgid with hemlock abundance negatively correlated to temperature. Widespread mortality of hemlocks following adelgid outbreaks would change forest composition, structure, nutrient cycling, surface water quality, and populations of associated wildlife (Ford et al. 2012; Jenkins et al. 1999; Kizlinski et al. 2002; Ross et al. 2003; Spaulding and Rieske 2010; Tingley et al. 2002; Vose et al. 2013; Webster et al. 2012). Historical spread of hemlock woolly adelgid was estimated at 3.6 km/year (±0.2 km) from east to west and 5.8 km/year (±0.28 km) from north to south (U.S. Department of Agriculture Forest Service 2011a).

The projection maps indicate a likely shrinkage in potential suitable July habitats under the A1B storyline by 2060. The B1 storyline will likely have larger suitable July habitat by 2020 compared to the other storylines (Figure 6.8). A marked northward shift in suitable habitat in January is likely under the A2 storyline by 2060. The distribution of hemlock woolly adelgid habitat could be limited to ecosystems in upper areas of the region.

Management—Hemlock woolly adelgid management involves the integrated use of cultural, chemical, biological, and genetic approaches to reduce adelgid populations. The cultural management approach for hemlock woolly adelgid includes preventing movement of infested plant materials to noninfested locations, especially between March and June when adelgid eggs and crawlers are abundant and seek out stress resistant cultivars of hemlock. Cowles et al. (2006) found that application of the systemic insecticide imidacloprid to soil was effective in eliminating the adelgid for a few years. Application of chemical insecticides is particularly useful for managing hemlock woolly adelgid in urban landscapes, and has been used extensively in select priority forest settings (McClure et al. 2001).

Hemlock woolly adelgids do not have host resistance or natural enemies in eastern areas of North America, although local arthropod predators help regulate their populations in their native ecosystem (McClure et al. 2001). *Laricobius nigrinus* Fender (Coleoptera: Derodontidae), native to western North America, has also been observed as a predator of hemlock woolly adelgids (Mausel et al.

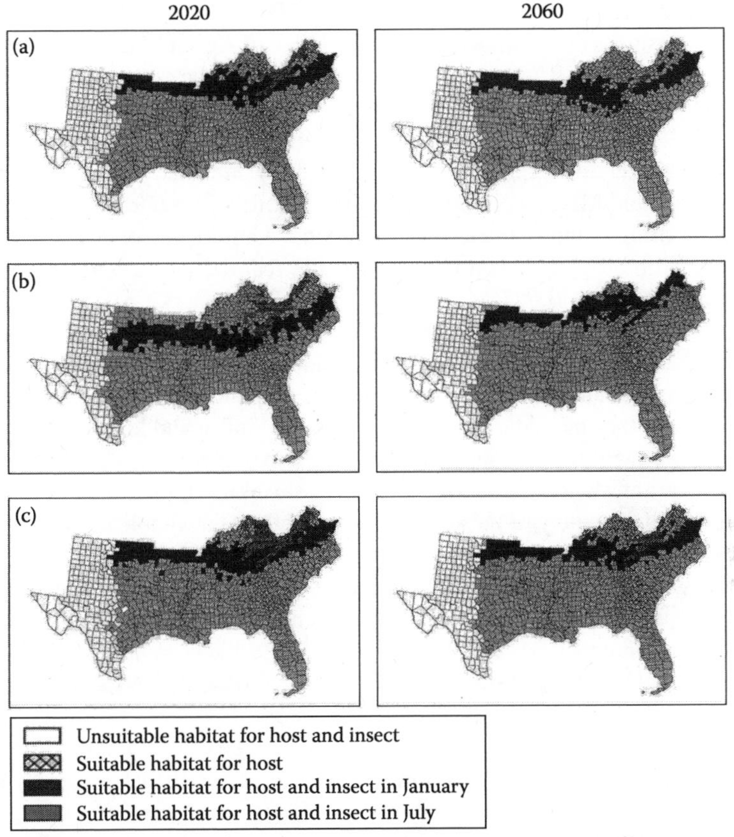

FIGURE 6.8 Suitable habitat for the hemlock woolly adelgid in January and July (2020 and 2060) in the Southern United States, projected using an ensemble of selected general circulation models (CGCM3, GFDL, CCSM3, and HadCM3) and assuming three Intergovernmental Panel on Climate Change (2007) emissions storylines; (a) A1B, (b) A2, and (c) B1 (see text and Chapter 2 for storyline descriptions).

2010). Predatory beetles imported from China and Japan have shown promising results (McClure et al. 2001; Onken and Reardon 2011). The widespread mortality seen in hemlock could mean replacement by American beech, northern red oak, sugar maple (*Acer saccharum*), yellow birch (*Betula alleghaniensis*), and red maple (*Acer rubrum*). Vose et al. (2013) identified needs for restoration efforts to include novel approaches, such as the introduction of nonnative species, facilitating movement of native species to new habitats (e.g., white pine), and aggressive management of existing species (e.g., rhododendron) with mechanical removal, fire, or chemicals. According to Jonas et al. (2012), in addition to encouraging natural regeneration of hemlock in damaged forest stands, growing offsite stocks of hemlock seedlings for replant, creating hemlock woolly adelgid-resistant hybrid hemlocks, and replanting using already resistant western hemlock or Chinese hemlock are useful genetic management strategies against hemlock woolly adelgid.

Emerald Ash Borer

The emerald ash borer is a devastating, wood-boring beetle first discovered in 2002 near Detroit and across the Canadian border in Windsor, Ontario (Kovacs et al. 2010). More than 50 million ash trees (*Fraxinus* spp.) have died, are dying, or are heavily infested in Michigan, Indiana, and Ohio (Smith et al. 2009), with additional infestations reported in 12 other states (Illinois, Iowa,

Kentucky, Maryland, Minnesota, Missouri, New York, Pennsylvania, Tennessee, Virginia, West Virginia, and Wisconsin) as of early 2011 (U.S. Department of Agriculture Animal and Plant Health Inspection Service 2011b). Ash trees are an important part of the rural and urban forests of the United States, valued at >$282 billion (U.S. Department of Agriculture Animal and Plant Health Inspection Service 2003). Unfortunately, all 16 species of ash tree in North America appeared to be susceptible to the emerald ash borer, which has significantly damaged the ecology and economy of infested areas (Duerr and Mistretta in press). Keena et al. (2009) reported that the optimum temperature for egg hatching is approximately 25°C, and that adult longevity decreases as temperature increases from 20°C to 30°C. Sobek et al. (2009) noted that the emerald ash borer larvae cannot survive temperatures below –30.6°C.

Impacts of climate change—Although susceptible species, such as green ash (*Fraxinus pennsylvanica*) and white ash (*Fraxinus americana*), have wide distributions in the South, they are not a dominant component of southern forests but a common component of riparian forests in this region. Over the next 50 years, the emerald ash borer could possibly infest and kill many, if not most, of the ash trees in the South. Millions of ashes could die from extreme drought, rapid insect population growth, and multiple infestation outbreaks, ultimately shrinking their southern range (Duerr and Mistretta in press). Unfortunately, at present, little information is available about the life cycle, flight capabilities, host preferences, natural enemies, or management methods.

The potential suitable habitat for the emerald ash borer in July is predicted to be significantly larger in 2020 compared to 2060 for all three emissions storylines. In 2020, the largest suitable habitat in July would be under B1, and A1B and A2 would likely be similar in distributions but smaller than B1. A slight northward shift in 2060 is possible under the A2 storyline at the southern edge of the range (Figure 6.9).

Management—The ability to detect early infestations will be crucial in preventing the spread of the emerald ash borer in the South (Duerr and Mistretta in press). Efforts currently geared toward controlling and eradicating the emerald ash borer—imposing quarantines, conducting surveys, drawing boundaries around areas that are adjacent to confirmed infested sites, removing ash trees, and developing information—would support management efforts. Lack of effective survey and control technology has made containment efforts challenging in some areas where the emerald ash borer has been found. A practical option could be to delineate and protect small pockets of exceptional ash resource as "ash conservation areas." Currently available chemical treatments are not cost effective for large-scale implementation. Mercader et al. (2011) identified management options with insecticides that significantly reduced the spread of *A. planipennis* when treatments were applied 1–4 years after infestations were initiated. Also, some parasitoids of larva and egg are being investigated for use as biological control agents (U.S. Department of Agriculture Animal and Plant Health Inspection Service 2011b). If effective, these control agents could help mitigate the spread of emerald ash borer populations, but they are not expected to completely stop the spread (U.S. Department of Agriculture Animal and Plant Health Inspection Service 2011b) with future climate change.

Pine Root Collar Weevil

The pine root collar weevil is a native insect pest of pine trees in north central and northeastern areas of the United States. The larval stage of the weevil causes major injury to host trees by feeding below ground in the root collar, root crown, and large roots (Eliason and McCullough 1995; Wilson and Schmiege 1975). Native and nonnative species of pines are susceptible to the weevil, which attacks both young and old pines. Young trees <4 inches in diameter are more vulnerable to attacks than larger trees, and can be killed in a single year by as few as two to five larvae. Attacks by weevils generally reduce growth rate, limit transport of nutrients, and increase susceptibility of the host trees to heavy wind, snow, and secondary pests. Adult feeding also girdles small shoots and branches, causing them to die and turn red. Infestations are more pronounced in pine

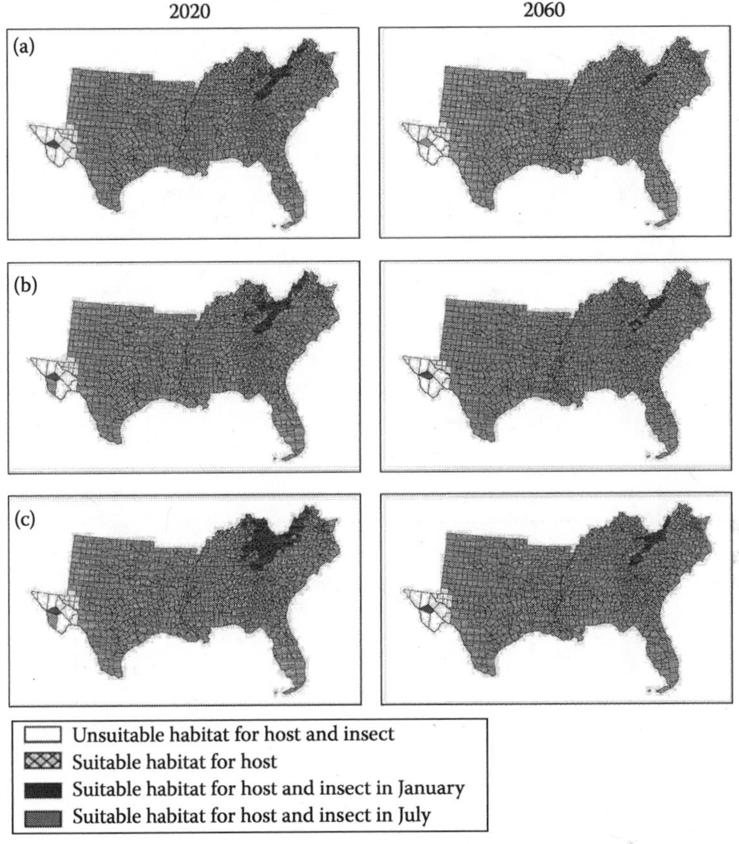

FIGURE 6.9 Suitable habitat for the emerald ash borer in January and July (2020 and 2060) in the Southern United States, projected using an ensemble of selected general circulation models (CGCM3, GFDL, CCSM3, and HadCM3) and assuming three Intergovernmental Panel on Climate Change (2007) emissions storylines; (a) A1B, (b) A2, and (c) B1 (see text and Chapter 2 for storyline descriptions).

plantations and windbreaks (Eliason and McCullough 1995; Wilson and Schmiege 1975). This weevil has also been implicated as a contributor to red pine (*Pinus resinosa*) decline, and was among other insects found to be significantly more abundant in declining stands than in healthy red pine (Klepzig et al. 1991).

Impacts of climate change—Excessive rainfall can cause mortality of pine root collar weevil pupae. Pruning lower branches of host trees can also expose larvae and adults to higher temperatures, causing disruptions in developmental activities (Wilson and Schmiege 1975). Temperatures <8.5°C during the egg developmental period are insufficient to support increases in weevil populations. Therefore, future change in temperatures will potentially influence the northern extent of the weevil's range. Egg development is very sensitive to temperature changes <12°C, but only slightly affected by temperature ≥18°C. A prolonged cooling of the root collar area can greatly influence egg development. At about 18°C, a temperature increase of 1°C decreases the incubation period by about a day (Wilson and Millers 1982). Hotter and wetter future conditions would disrupt egg development and spread of infestation. The projection maps indicate no suitable habitat for the pine root collar weevil because red-pine distribution in the South is limited to Virginia and North Carolina (Figure 6.10).

Management—Currently, several management practices can help reduce the likelihood of pine root collar weevil outbreaks. Some management practices suggested by Wilson and Millers (1982) are to (1) plant highly resistant pine species in areas with reported incidences of weevils, (2) avoid

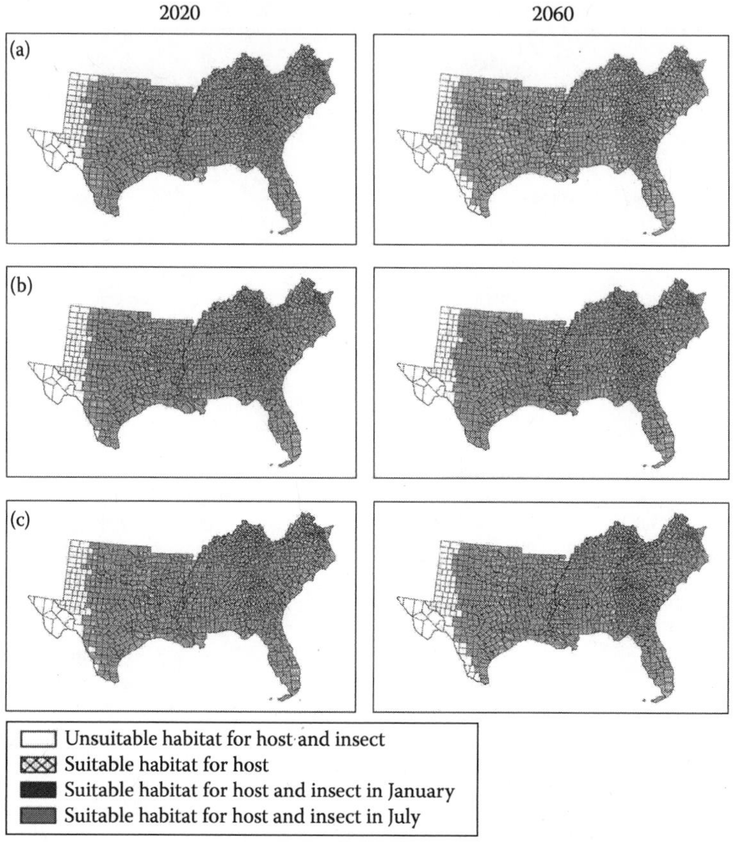

FIGURE 6.10 Suitable habitat for the pine root collar weevil in January and July (2020 and 2060) in the Southern United States, projected using an ensemble of selected general circulation models (CGCM3, GFDL, CCSM3, and HadCM3) and assuming three Intergovernmental Panel on Climate Change (2007) emissions storylines; (a) A1B, (b) A2, and (c) B1 (see text and Chapter 2 for storyline descriptions).

planting mixtures of susceptible and resistant species, (3) cut and destroy the root collars of open-grown, older susceptible pines before planting new ones, (4) maintain a fully stocked stand, because closed stands retard buildup of weevil populations, and (5) prune the lower 2–4 feet of branches from standing pines and rake away the duff beneath infested trees, thereby hindering the normal daytime activities of the adults by depriving them of a cool, shaded hiding place. Planting a resistant species is important for effective management. Pitch pine (*Pinus rigida*) is one of the least susceptible hosts. Eastern white pine (*Pinus strobus*) is rarely attacked unless adjacent to or planted with the highly susceptible species (Wilson and Schmiege 1975).

Sirex Woodwasp

The Sirex woodwasp is native to Europe, Asia, and northern Africa. It was first discovered in North America in 2004 (Hoebeke et al. 2005); its first known establishment in North America was reported in the Canadian province of Ontario and across the U.S. border in New York in late autumn of 2005 (Rabaglia and Lewis 2006). In the southern hemisphere, the species has caused significant mortality, especially in Monterey pine (*Pinus radiata*) and loblolly pine (*Pinus taeda*) plantations (Hopkins et al. 2008; Rabaglia and Lewis 2006). Slash pine (*Pinus elliottii*), shortleaf pine (*Pinus echinata*), ponderosa pine (*Pinus ponderosa*), lodgepole pine (*Pinus contorta*), and jack pine (*Pinus *banksiana*) are also susceptible (Haugen and Hoebeke 2005; Rabaglia and Lewis 2006). The estimated annual value of southern softwood logs, pulpwood, lumber, and veneer is >$8 billion. If the level of attacks by Sirex woodwasps is as aggressive as in South America and Australia, Duerr and Mistretta (in press) speculated that many important southern pine species could be vulnerable to high losses.

Sirex woodwasps usually complete a single generation per year; however, it could take up to two years for them to develop fully in the colder parts of their range. Adults emerge from July to September, with the peak around mid-August (Pollard et al. 2006). The symbiotic fungus (*Amylostereum areolatum*) that is vectored by Sirex grows rapidly and excretes wood-digesting enzymes into the sapwood (Haugen and Hoebeke 2005), causing foliage to wilt and yellow and disrupting water movement, which often results in the death of the host (Haugen and Hoebeke 2005; International Union for Conservation of Nature Invasive Species Group 2007b).

Impacts of climate change—Loblolly pine, a major pine species in the South, would be a suitable host if Sirex woodwasps were to move into the South (Duerr and Mistretta in press). Although human-aided infestations or natural spread could introduce Sirex woodwasps into southern forests within the next 50 years, potentially resulting in significant ecological and economic losses, the complexity of southern forests (their mixed stands, high biodiversity, and abundance of possible competitors, predators, and parasitoids) would tend to reduce the level of vulnerability compared to areas where damages have been reported in monoculture pine plantations (Dodds et al. 2007). However, the recent higher frequency of tropical storms and hurricanes in the South has left some forests damaged and vulnerable to Sirex infestations, increasing the risk to economic vitality of softwoods (Duerr and Mistretta in press). Rabaglia and Lewis (2006) predicted that the spread of Sirex could take 42 years to reach the Gulf of Mexico, but their prediction does not appear to have considered the impact of projected climate change.

Projections under the B1 storyline (Figure 6.11) indicate a larger suitable July habitat for Sirex woodwasp by 2020 compared to any other storylines and a significant reduction in potential suitable July habitat under A1B storyline by 2060. Expansion and northward shift of suitable January habitat are likely under the A2 storyline by 2060.

Management—The management approaches currently available for Sirex woodwasps include prevention and suppression techniques such as thinning to increase growth and vigor (Haugen et al. 1990). Destruction of infested trees can also lower abundance of Sirex woodwasps in newly infested areas (Hurley et al. 2007). Biological control agents, including a parasitic nematode (*Deladenus siricidicola*) that infests Sirex woodwasp larvae and ultimately sterilizes adult females, have been successful in other countries. Successful evaluation of this nematode for use in U.S. forests should provide a potential control option for southern landowners and land managers against future threats (Fernández-Arhex and Corley 2005; Haugen and Hoebeke 2005). However, the existence in the South of numerous competitors (such as bark beetles, wood borers, and fungi) for dead and dying pines, may very well complicate or even preclude the establishment of Sirex, the use of biological control to manage it, or both.

OUTLOOK FOR NONNATIVE PLANTS

Garlic Mustard

Introduced from Europe in the 1800s initially for medicinal purpose, garlic mustard is a cool-season biennial forb that grows often in colonies in a variety of habitats. Seeds can be dispersed by humans, animals, and water and can remain dormant up to six years. Garlic mustard produces allelopathic chemicals that can inhibit the regeneration and growth of nearby native plants.

Impacts of climate change—By 2020, the habitat range for garlic mustard in the South is projected to be largest under the A2 storyline, with ranges very similar for A1B and B1 (Figure 6.12). The projected habitat is expected to shift northward by 2060 under all three storylines, with the largest shift

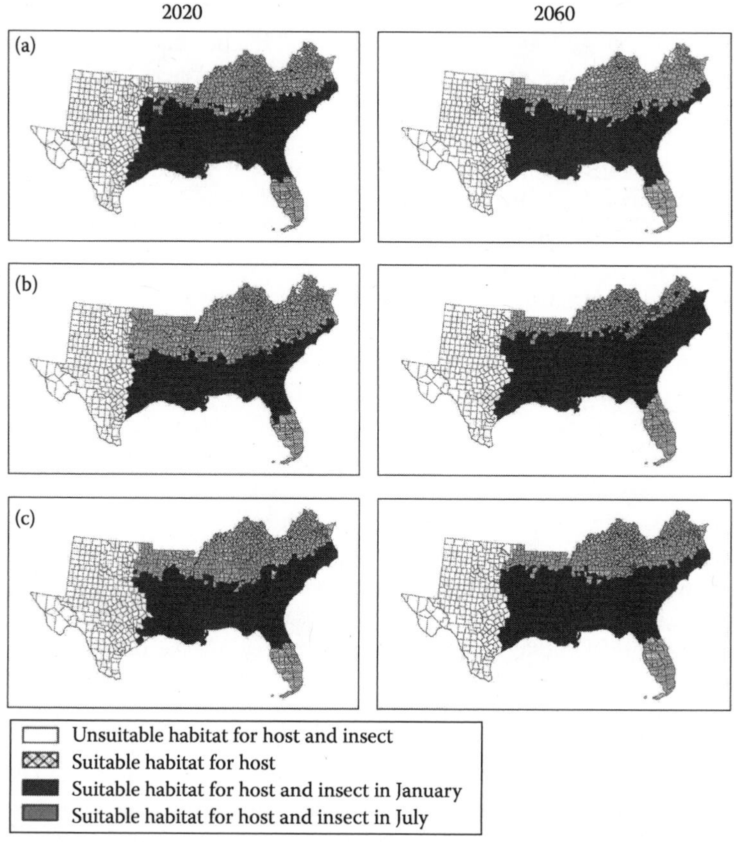

FIGURE 6.11 Suitable habitat for the Sirex woodwasp in January and July (2020 and 2060) in the Southern United States, projected using an ensemble of selected general circulation models (CGCM3, GFDL, CCSM3, and HadCM3) and assuming three Intergovernmental Panel on Climate Change (2007) emissions storylines; (a) A1B, (b) A2, and (c) B1 (see text and Chapter 2 for storyline descriptions).

under A1B and smallest under B1. Because garlic mustard is a cool-season forb, the northward shift is not surprising with warmer temperatures, especially at the southern edge of its range.

Management—The current distribution of garlic mustard is mainly concentrated in the northern portions (Kentucky, Tennessee, and Virginia) of the region. Based on the predicted major northward shift under future climate predictions, a regionally coordinated approach that includes early detection and rapid response to newly established populations could be the key to preventing the southward spread of this species. Management strategies such as minimizing disturbances near known infestations, removing newly infested populations, and minimizing seed dispersal by equipment and vehicles could greatly mitigate the spread of garlic mustard.

Cogongrass

Introduced from Asia in the early 1900s initially for soil erosion control, cogongrass is now widely distributed in Africa, Australia, southern Asia, the Pacific Islands, southern Europe, the Mediterranean, the Middle East, Argentina, Chile, Colombia, the Caribbean, and the Southern United States. It is a highly aggressive, colony-forming perennial grass that invades a wide range of habitats and can cause widespread fires. Worldwide, cogongrass is most invasive in wet tropical and subtropical areas that receive 750–5000 mm of rainfall annually. It spreads by both seed and rhizomes, can tolerate hot weather, but is sensitive to cold.

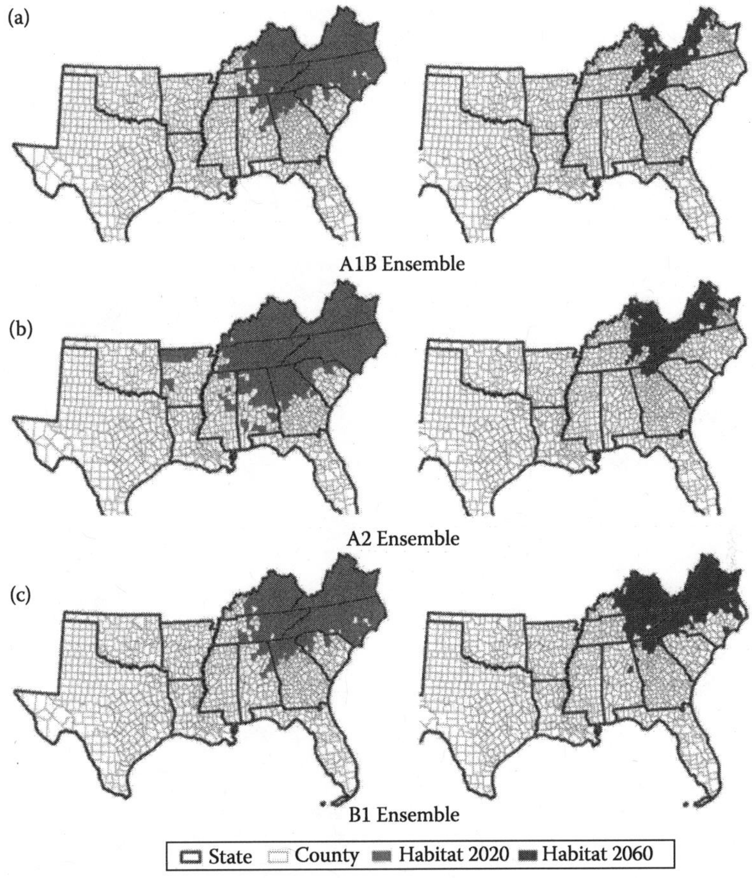

FIGURE 6.12 Suitable habitat for garlic mustard in 2020 and 2060 in the Southern United States, projected using an ensemble of selected general circulation models (CGCM3, GFDL, CCSM3, and HadCM3) and assuming three Intergovernmental Panel on Climate Change (2007) emissions storylines; (a) A1B, (b) A2, and (c) B1 (see text and Chapter 2 for storyline descriptions).

Impacts of climate change—In general, cogongrass is favored by a warming climate. All three climate change storylines predict that potential habitat for cogongrass will cover the majority of the South in 2020 (Figure 6.13); exceptions are western Texas (constrained by low precipitation) and southern Florida (constrained by high summer temperatures). Predicted potential cogongrass habitats in 2060 are very similar to the 2020 predictions, with only minor discrepancies at the western and southern edges of the range.

Management—Cogongrass is currently widely distributed in the lower portion of the South. The predicted climate change would favor its further invasion into the upper portion, making the management of this species more difficult. A complete eradication at this stage is nearly impossible. However, common management practices, such as practicing proper equipment sanitation and raising public awareness, would greatly assist the mitigation of this invasive grass species.

Chinese Privet

Introduced from China in the early 1800s as an ornamental plant, the Chinese privet reproduces and colonizes through both sprouts and seeds. It often grows to form dense thickets thus shading and outcompeting many native species with severe ecological and economic consequences.

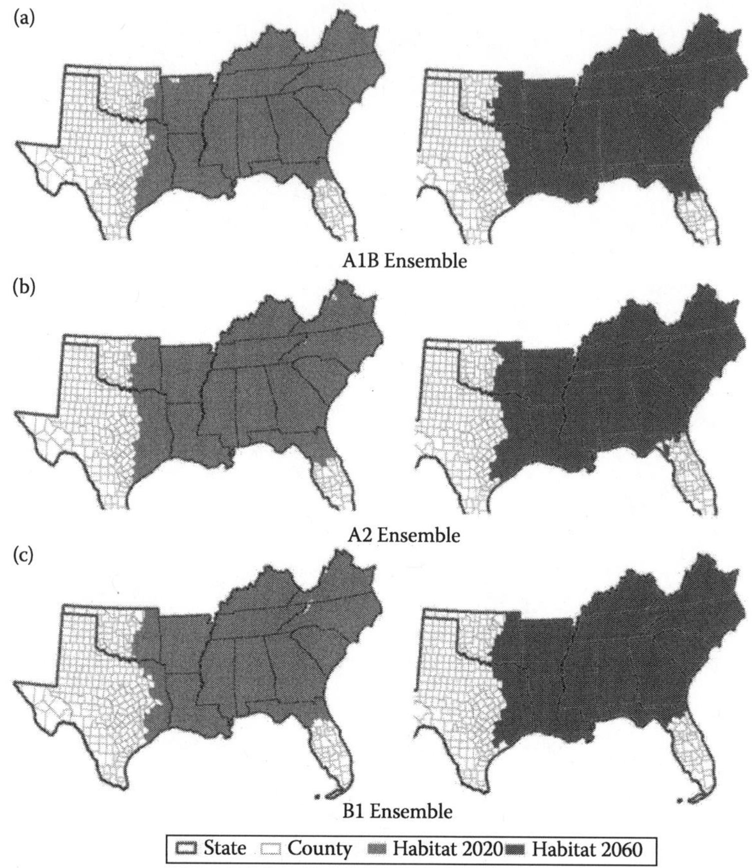

FIGURE 6.13 Suitable habitat for cogongrass in 2020 and 2060 in the Southern United States, projected using an ensemble of selected general circulation models (CGCM3, GFDL, CCSM3, and HadCM3) and assuming three Intergovernmental Panel on Climate Change (2007) emissions storylines; (a) A1B, (b) A2, and (c) B1 (see text and Chapter 2 for storyline descriptions).

Impacts of climate change—The potential habitat for Chinese privet is predicted for all of the southern states except Louisiana and Oklahoma in 2020 based on the A2 ensemble climate change storyline (Figure 6.14), and less so under B1. The A1B-based storyline predicts the fewest suitable areas. As with garlic mustard, prominent northward shift is predicted from 2020 to 2060. By 2060, the suitable Chinese privet habitat is predicted to be concentrated in the upper portions of the South (Kentucky, Virginia, Tennessee, and North Carolina).

Management—Chinese privet currently has a wide distribution across the entire South. Although higher-latitude regions may have increased the risk of Chinese privet invasion, the threat from this species could be alleviated with climate change, especially in the lower portions of the South. Management practices such as prevention of seed spread and dispersal and early detection/removal can mitigate the impacts of this invasive shrub species.

Kudzu

A high-climbing perennial vine initially introduced from Japan in 1876 (and later from China, possibly multiple times), kudzu has been a strong symbol of plant invasions in the United States, especially in the South. The species has been mostly used as a medicinal and is regarded as having

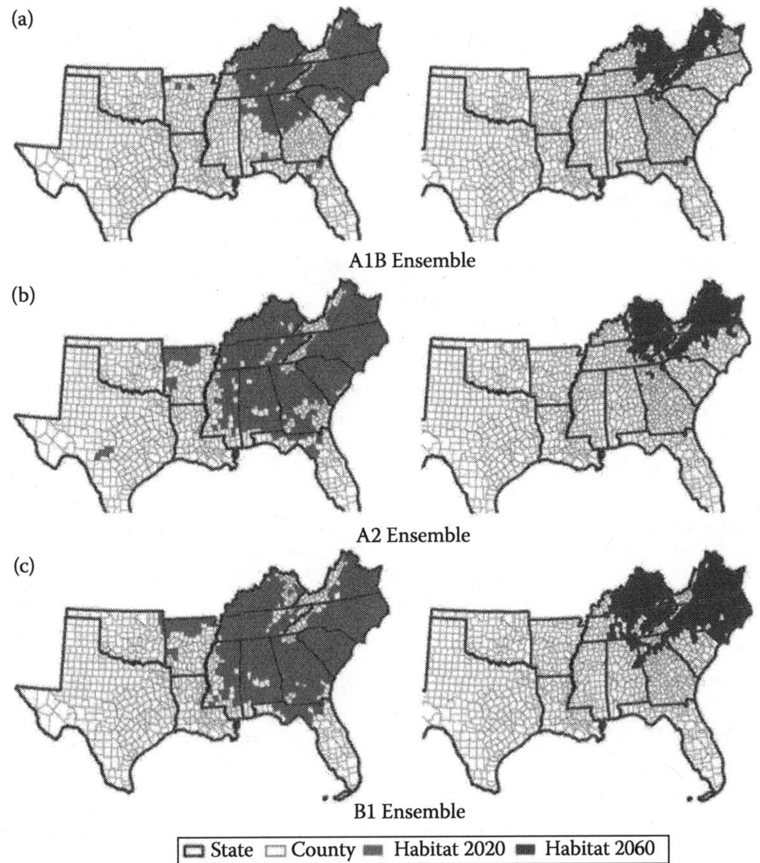

FIGURE 6.14 Suitable habitat for Chinese privet in 2020 and 2060 in the Southern United States, projected using an ensemble of selected general circulation models (CGCM3, GFDL, CCSM3, and HadCM3) and assuming three Intergovernmental Panel on Climate Change (2007) emissions storylines; (a) A1B, (b) A2, and (c) B1 (see text and Chapter 2 for storyline descriptions).

great economic values in its native lands (Li et al. 2011). In countries where it has been introduced, such as the United States, it was first promoted as an ornamental plant, then for erosion control, forage for domestic animals, and soil improvement. However, with time, the species has escaped and now grows in extensive habitats and high abundance. Although it grows over the tops of native forest species, thereby killing them through shading, it is mostly found along the roads, forest edges, and internal forests that have been disturbed. Many treatment methods have been attempted. Some control methods are available and effective but very costly.

Impacts of climate change—For 2020, potential kudzu habitats are predicted to be similar under the A2 and B1 storylines; the A1B predicts the least coverage, mainly along its western edge. A notable shift to the northeast is likely by 2060 with the smallest shift predicted under B1 (Figure 6.15).

Management—Kudzu is widely distributed across the entire South and in lower portions farther north, where the risk of invasion would increase as temperatures increase. In the South, the threat from this invasive vine at the western and southern edges of its range (especially in Texas, Oklahoma, Louisiana, and Florida) could be alleviated; for the western states, the main reason for kudzu decline would the predicted decrease in precipitation. Management practices such as herbicide applications, grazing, and prescribed fire can help eradicate and contain the kudzu invasion.

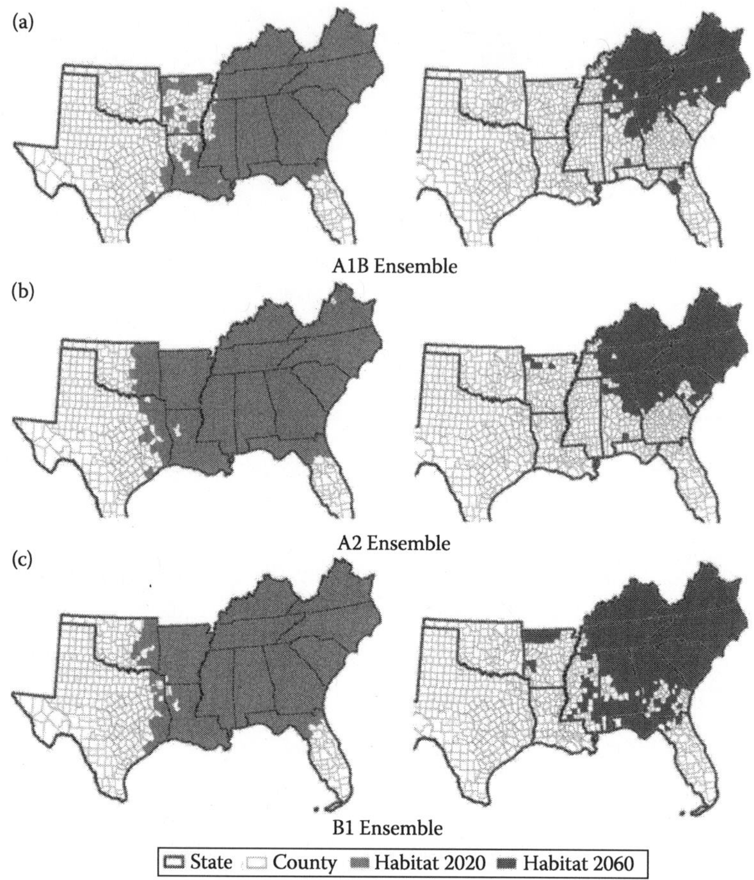

FIGURE 6.15 Suitable habitat for kudzu in 2020 and 2060 in the Southern United States, projected using an ensemble of selected general circulation models (CGCM3, GFDL, CCSM3, and HadCM3) and assuming three Intergovernmental Panel on Climate Change (2007) emissions storylines; (a) A1B, (b) A2, and (c) B1 (see text and Chapter 2 for storyline descriptions).

Chinese Tallowtree

Introduced from China in 1700s as an ornamental plant, Chinese tallowtree is found throughout southeastern Asia (Pattison and Mack 2009). As with kudzu, the species has always been considered as having high economic values for production of "tallow oil" and thus has been widely planted in its native lands (Zhang and Lin 1991). In the lands where it has been introduced, tallowtree grows in relatively wet areas although it can grow well in uplands if the infestation becomes large enough. It has also been shown to have allelopathic effects on many native species. Chinese tallowtree is a serious threat because of its ability to invade high quality, undisturbed forests.

Impacts of climate change—Unlike the other species discussed, the habitat for Chinese tallowtree is predicted to concentrate in the central areas of the South in 2020, with the widest coverage to be found under the A2 storyline. A significant northward shift is predicted from 2020 to 2060 (Figure 6.16), as well as a shift eastward, especially dramatic under the A2 storyline.

Management—The current distribution of Chinese tallowtree is mainly in the coastal areas of the South. Climate change would likely facilitate its range into the central portion by 2020 and into the central–northern portion by 2060, rendering the currently invaded areas as unfavorable. From a management prospective, climate change can effectively confine Chinese tallowtree to a

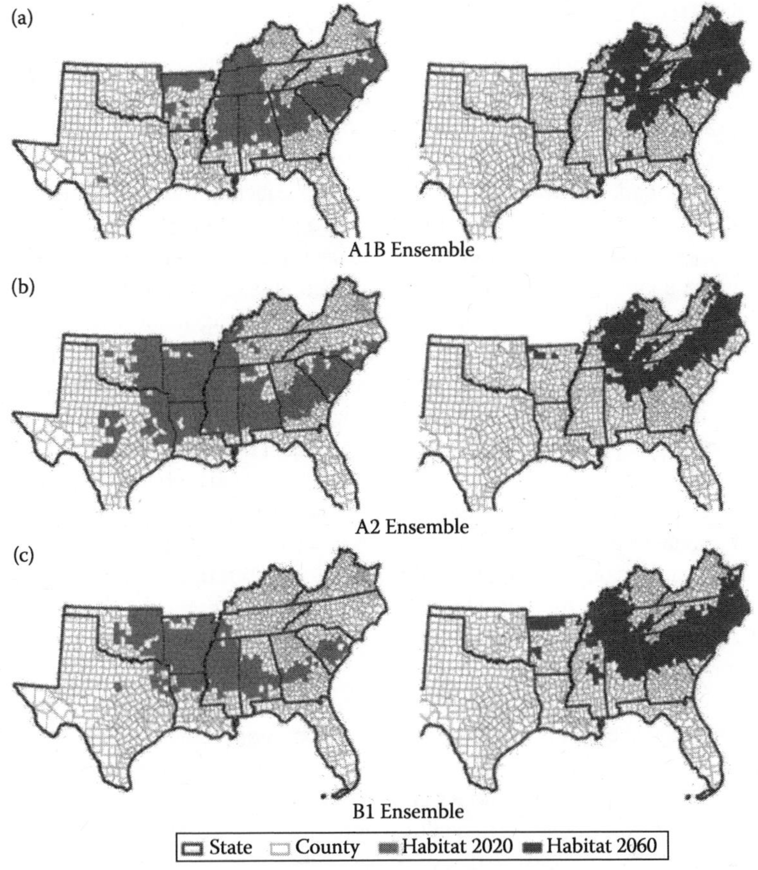

FIGURE 6.16 Suitable habitat for Chinese tallowtree in 2020 and 2060 in the Southern United States, projected using an ensemble of selected general circulation models (CGCM3, GFDL, CCSM3, and HadCM3) and assuming three Intergovernmental Panel on Climate Change (2007) emissions storylines; (a) A1B, (b) A2, and (c) B1 (see text and Chapter 2 for storyline descriptions).

small area if the species can be prevented from spreading northward. As with other species, removing new infestations and minimizing seed dispersal can greatly mitigate the spread of this invasive tree species.

Potential Habitats

Our projections show that all five species exhibited some degree of northward (or poleward) shift in latitude in 2020 and even more in 2060. Clear indications of sensitivity to climate change are the variability, both in the species' distribution shifts and in the two forecasting periods (ending in 2020 and 2060). For example, for projections in 2020, garlic mustard exhibited a small northward shift in its ranges under the A2 storyline; however, in the 2060 projections, there was a greater shift under A1B than under B1. In 2020, cogongrass also shifted northward under both A1B and B1 but the northward shift in 2060 was smaller. In 2020, Chinese privet showed a similar northward shift under both storylines; and, by 2060, it had a similar shift to garlic mustard. For kudzu, in 2020 and under A2, the northward shift was small but its western edge shifted eastward under A1B; similarly, in 2060, kudzu exhibited trends similar to garlic mustard. Of the five species, Chinese tallow tree exhibited the smallest northward shift in both 2020 and 2060 although under A1B; its western edge also shifted eastward.

MITIGATION STRATEGIES

In the past, short-term weather variability has created extreme events such as droughts, heavy storms, flooding, and wildfires with long-term devastating impacts on forest ecosystems, particularly in the United States. Extreme droughts often lead to forest fire, an important forest disturbance linked to a warming climate (Westerling et al. 2006), in a cycle of events that can reduce the resistance of surviving trees to attacks by bark beetles and other insects, subsequently increasing fuel loads from dead tree hosts and increasing the frequency of forest fires. Integrated management strategies are preferred for mitigating losses from insect attacks such as those caused by southern pine beetles (Billings et al. 2004). Typically, the approach involves prevention, detection, and control. Preventive measures may include thinning to reduce stand density; removing infested, damaged, and weakened trees; and harvesting before trees become overly mature. Once an outbreak begins, the attention usually shifts to prompt detection and suppression of individual infestations thereby substantially reducing loss of forest resources (Clarke and Billings 2003). Replanting using seedling from a variety of resistant families would be an important component for long-term management of negatively impacted forest ecosystems.

Some recommendations are already in place to help land managers effectively prepare for climate change: (1) reduce current environmental stresses and increase resistance against climate-exacerbated disturbances such as wildfire or insect outbreaks by thinning and fuel abatement treatments at landscape scales and strategically placed area treatments; (2) apply early detection and rapid response systems to identify plant, animal, and ecosystem responses to climate change; (3) plan for higher elevation insect outbreaks, species mortality events, and altered fire regimes; (4) facilitate natural selection by enhancing disturbances that initiate increased seedling development and genetic mixing; (5) reduce homogeneity of stand structure and synchrony of disturbance patterns by promoting diverse age classes, species mixes, and genetic diversity; (6) cultivate an increased understanding of climate change; and (7) expand partnerships and capitalize on potential new ecosystem services opportunities under climate change; and (8) increase collaboration among federal agencies (U.S. Department of Agriculture Forest Service 2009).

Furthermore, the Southern Forest Futures Project, initiated by the Forest Service, is an ongoing effort geared toward in-depth documentation of potential future threats to southern forests, with the goal of developing products that will guide land managers in implementing appropriate adaptation and mitigation management strategies.

What Can the Past Tell Us about Managing in the Future?

We expect that most of the measures mentioned above and other management strategies currently available for addressing short-term climate variability impacts will be useful in managing long-term climate change impacts, perhaps with some modifications. However, maintaining a healthy forest into the next century will require adaptation and mitigation strategies as well as adequate planning to identify vulnerabilities. Because the scope of likely impacts is so broad, managing for them all within the forest ecosystem would be impossible. However, implementing management options that enable natural adaptation of the forests to climate change will be critical.

Future adaptation strategies focus on reducing the risk of negative impacts while capitalizing on any potential opportunities brought by climate change. Mitigation strategies can focus on the pace and extent of climate change impacts and increasing carbon sequestration through healthy forests. The overall goal would involve short-term management options that forestall impacts and long-term options for anticipating and managing changing forest ecosystem.

What Can Be Done Now?

Some existing strategies could be effective for managing short-term, climate variability-related events and disturbances, while others would be more useful for managing long-term climate change effects. Overall, early planning of management options offers a good opportunity to identify vulnerabilities, the magnitude and timing (now or later) of potential impacts, the likelihood of impacts and confidence in estimates, and potential strategies for adaptation and mitigation (U.S. Department of Agriculture Forest Service 2009).

DISCUSSION

In this chapter, we defined vulnerability as the susceptibility of a forest ecosystem to stress or disturbance that invokes an undesirable change in structure or function. Vulnerability of an ecosystem to climate change is a function of (1) exposure, or the degree to which the ecosystem has been exposed to climate change and variability, (2) sensitivity to these changes, and (3) adaptive capacity of the ecosystem to respond to these changes. Therefore, developing strategies to help the ecosystem adapt is critically important (Intergovernmental Panel on Climate Change 2007; Williams et al. 2008).

The most vulnerable ecosystems will have low resistance and resilience to climate change, because they cannot cope with short- and long-term exposures to change, are highly sensitive to change, and lack adaptive capacity to withstand change. Vulnerability assessments generally focus on quantifying relative exposures (such as short- or long-term changes in precipitation and temperature) and sensitivity to climate changes (determined by ecological, genetic, and physiological traits), as well as the adaptive capacity of species or ecosystems, which can help direct or prioritize management efforts toward addressing future threats (DeWan et al. 2010).

Species or ecosystems with a high degree of adaptive capacity are subject to fewer risks from climate change impacts compared to species or ecosystems with relatively low adaptive capacity (DeWan et al. 2010). However, most species may lack the capacity to adapt given the rate and magnitude of projected changes (Bradshaw and Holzapfel 2006; Williams et al. 2008). Evidence suggests that entire species tend to shift their geographical distributions in response to climate change, rather than undergoing the major evolutionary adaptations needed for conservation of the original range (Parmesan 2006; DeWan et al. 2010).

As shown in Table 6.3, several physiological and life-history traits can influence species' vulnerability and response to climate change disturbances (DeWan et al. 2010). Therefore, the species most at risk tend to have a narrow range of physiological tolerance to factors such as temperature, water availability, and fire; low genetic variability; long generation times and long time to sexual maturity; specialized requirements (hosts, vector, symbionts, or habitat) that may be restricted; poor dispersal (adaptive capability); and a narrow geographic range.

Conversely, species that are least at risk to climate change are those with a broad range of physiological tolerance to environmental factors including temperature, precipitation, and fire. Species with a high degree of phenotypic plasticity and genetic variability, short generation time (rapid life cycles), and short time to reach sexual maturity, high fecundity, low specificity in habitat requirements (such as food sources and sites), good dispersal capability, and broad geographic ranges will be the least at risk (Steffen et al. 2009).

We anticipate an increase in the level of insect attacks (both native and nonnative) in response to climate change. We also anticipate that projected increases in temperature would increase the frequency of drought stress on host trees. Some southern tree species are likely to shift or shrink in geographical range. Cascading effects could include a more vulnerable forest ecosystem with fewer stands, causing reduced value and insufficient genetic diversity to withstand further changes. Current biological systems and social activities tied to the ecosystem could also become more vulnerable, and others could cease to exist in the future. Attacks of dead and dying host trees by invasive species have already resulted in the destruction of habitats, economic losses in the forest products sector, and the loss of trees in cities, towns, and natural areas that had once supported thriving forests.

TABLE 6.3
Parameters for Evaluating Adaptive Capability and Vulnerability of Species to Climate Change

Vulnerability Parameter	Least at Risk	Most at Risk
Physiological tolerance	Broad	Narrow
Genetic variability	High	Low
Generation/maturity	Short	Long
Fecundity/sporulation	High	Low
Specialized requirements	No	Yes
Dispersal capability	Good	Poor
Geographic range	Broad	Narrow
Host diversity	Broad	Narrow

Source: Adapted from Steffen, W. et al. 2009. Australia's biodiversity and climate change: A strategic assessment of the vulnerability of Australia's biodiversity to climate change. A report to the Natural Resource Management Ministerial Council commissioned by the Australian government. CSIRO Publishing.

Areas of Greatest Concern

Diseases and insects—Thus far, much of the emerald ash borer infestation has been concentrated in northeastern areas of the United States; however, in approximately 50 years (2060), southern forests are expected to lose millions of ash trees from emerald ash borer outbreaks triggered by projected extreme drought and rapid insect development. This could eventually shrink the range of ash trees in the region (Duerr and Mistretta in press).

The spread of the hemlock woolly adelgid (U.S. Department of Agriculture Forest Service 2011a) will likely continue from westward and southward. A warmer and drier climate would increase the likelihood of southern pine beetle activity across southern forests, especially in dense unthinned plantations and in pine acreage of highly susceptible species. A northward spread of southern pine beetles beyond the Appalachian-Cumberland highlands would likely increase and infestations would be more frequent.

Under a warmer and drier climate, the complex interaction between the redbay ambrosia beetle and fungal symbionts may be limited to the southern Coastal Plain (Koch and Smith 2008); however, it is possible that laurel wilt could spread to sassafras farther north since sassafras ranges extend to south Canada. Although redbay are restricted to certain parts of the southern geographic range, key areas with high concentrations of redbay have yet to be invaded, while others are imminently threatened (Koch and Smith 2008).

With the increasing possibility of Sirex woodwasp introduction in the South, the extent of climate change impacts on Sirex and host pine trees, which are uniformly distributed in the South, still remains uncertain. The likelihood that tropical storms and hurricanes will increase in the South may increase the likelihood of significant damage to southern forests, making them even more vulnerable to Sirex woodwasp attacks and infestations. According to Rabaglia and Lewis (2006), the spread of Sirex woodwasp would take 10 years to reach Virginia and 42 years to reach the Gulf of Mexico in Alabama, even without climate change.

With the current rate of spread, the gypsy moth could encounter lower concentrations of susceptible hosts as it moves south and west. Although overall forest susceptibility will likely decrease in most areas, the large distribution ranges of potential hosts could allow the spread to persist, raising the possibility of a significant widespread infestation in some southern landscapes within the next 50 years (Duerr and Mistretta in press).

Warmer or mild winters will likely extend the northward range of the pine root collar weevil. Because adults overwinter in the soil or in bark crevices before resuming activities in the spring, warmer winters would allow prolonged activity and an increase in the number of generations per year; at the same time, a decrease in precipitation may reduce the weevil's activity. Healthy forest stands are therefore critical for limiting infestations in the next 50 years. Inadequately managed forest stands will likely experience an increase in attacks and damage caused by the weevil.

Nonnative plants—Although the extent of northward (and sometimes eastward) shifts varies considerably among species and emissions storylines, all five species shift their distributions northward (poleward) in the 2020 and 2060 projections (Guo et al. 2012b). However, note that the vastly increased distributions and spread expected to result from climate change are not documented in this chapter because they are predicted to occur outside the Southern United States. In addition, the northward shifts of the five nonnative plants does not necessarily mean fewer threats to forest ecosystems, as other resident or newly introduced invasives would occupy these habitats.

Knowing where introduced species, especially those that are invasive, are spreading and at what pace is critical for invasive species management and conservation policy and planning. The five species used as examples of dominant invasives in the South for this chapter have continuously increased in occupied areas (Guo et al. 2006), causing drastic ecological and economical damage (Miller et al. in press). For forest and invasive species management, the areas that are subject to climate-induced introductions would experience fewer and less severe invasions if disturbances were kept to a minimum (Guo and Norman 2012). For the areas where a particular species retreats, the negative effects of the former resident could persist, thus requiring continued monitoring and management. Monitoring is also needed to update predictions about which species are likely to invade and at what rate, so that plans are in place to avoid potential damages.

Projections show that some species (such as cogongrass) that occur in much of Florida would have less coverage in 2020 and even less in 2060. In reality, speculating that an invasive species could completely emigrate from any area would be inaccurate, simply because many species are likely to find suitable microhabitats—somewhat similar to the northward migration during the glacier retreat when many species remained at lower latitudes but moved to higher elevations. This is a limitation of our modeling effort, as our projections do not capture the microenvironments that are still suitable for some species during their poleward migration or spread.

As with many studies using climate/land-use change projections for future species distribution forecasts, our projections also have several limitations. First, they do not account for possible biotic interactions (such as competition, predation, and transmission of diseases) among species in projected areas; that is, interactions with resident species and other new species that may invade. In reality, most species could actually spread to smaller areas than those projected because factors other than climate are also at work. Second, the climate variables used for native species are not definitive, simply because of the difficulty in identifying (without experimental support) which climate factor is critical (for temperature alone, options include average annual temperature, warmest/coldest monthly temperature, days when temperature is below a certain level, or the highest or lowest temperature). Third, the five species represent only a small fraction of broadly defined life/growth forms.

For these reasons, caution is advised when interpreting these results. What is needed is more information regarding how species might respond to climate change, in addition to information on how species respond to environmental factors such as temperature and moisture. This information combined with emerging techniques in modeling/simulation would be useful for evaluating responses of individual species (Albright et al. 2010) in an environment or interactions with other species in a given environment.

Management Options for Reducing Vulnerability

Diseases and insects—An integrated approach for managing anticipated spread of invasive species (such as the redbay ambrosia beetle) would necessarily consider trends in human population growth and recreational activities in areas where the beetles have recently become a major problem. A combination of elimination, proper disposal, and quarantine of infested plant materials is an effective management strategy, because employing most existing measures individually is inadequate against laurel wilt infections—though it might slow down the spread, it is impossible to cover such a large area. Results from fungicide applications have been promising, although they are limited to prevention of laurel wilt in high-value trees.

Most management options currently available for emerald ash borer are not very effective. However, future management efforts in response to climate change should include survey and quarantine of infested plant materials and sites, removal of weak and vulnerable ash trees, and establishment of "ash conservation areas" to protect small pockets of exceptional ash resource. Biological control methods using parasitoids of larvae and eggs combined with limited chemical treatments in conserved areas could play a useful role in reducing the spread of the emerald ash borer in the South.

Increased human populations could increase the spread of gypsy moth by increasing the number of parks and recreational activities. Therefore, continued vigilance and the use of early detection methods will be vital in reducing the risk of gypsy moth epidemics in the South. A combination of aerial applications of biological control agents, treatment with pesticides, deployment of pheromones traps, and continuation of programs such as "Slow the Spread" would help protect southern landscapes for several years.

Future management options for the hemlock woolly adelgid include restricting movement of infested plant materials to noninfested locations and planting new stands of stress-resistant cultivars. Hardwoods that are not susceptible to the hemlock woolly adelgid could replace hemlocks in places where the adelgid has caused widespread mortality.

Future outbreaks of southern pine beetles could be minimized through early detection and monitoring of new infestation spots, followed by timely control measures to suppress establishment of active spots. Lowering stand densities and thinning vulnerable pine stands would increase vigor and resiliency to southern pine beetle attacks.

Implementation of phytosanitary management practices would reduce the risk of pine root collar weevil infestations in vulnerable host stands. Measures include avoiding planting mixtures of susceptible and resistant species together, destroying the root collars of older susceptible pines, maintaining a fully stocked stand, pruning lower branches, and clearing beneath infested trees. These practices will likely reduce further spread of the pine collar weevil.

In addition to prevention and suppression management techniques to combat the threats posed by Sirex woodwasp, introducing biological enemies (such as parasitic nematodes) may become an effective tool for lowering the population below damaging threshold levels. Biological control methods will provide added options to landowners and land managers against future threats. Prevention measures such as thinning of susceptible or stressed pines would also increase overall stand vigor, and elimination of infested trees would help keep Sirex woodwasp populations low.

Restoring the less susceptible longleaf pine should be considered in areas where annosus root disease has caused significant damage to the forest stands. In response to likely shifts in geographic range from climate change, management options for fusiform rust would include deploying screened, resistant seedlings in areas with a history of high rust incidence, avoiding over-fertilization of seedlings in nurseries, and maintaining healthy saplings through effective silvicultural methods.

Some adaptive management strategies may be useful; examples include prescribed burning, restoring tree species, enabling migration of new tree species, instituting measures to enable migration for species that cannot migrate fast enough to keep up with climate change, and managing invasive species. However, these strategies are sometimes difficult to implement and limited in effectiveness (Frelich and Reich 2009). Overall, the forecasts and management options discussed in this chapter would help strengthen the resilience of southern forests in withstanding possible adverse impacts of future climate change. Although several uncertainties are inherent in models and projections used in this chapter, we believe that optimizing current models and taking advantage of additional management options as new technologies and adaptive strategies emerge would do much to address the threats of the future.

Nonnative plants—Options to reduce the vulnerability of forests to nonnative invasive plants include implementing management strategies that minimize disturbances near known infestations, removing newly infested populations, and minimizing seed dispersal by equipment and vehicles. These strategies could greatly mitigate the spread of garlic mustard. Although a complete eradication of cogongrass is nearly impossible at this stage, common phytosanitary management practices and greater public awareness would significantly enhance mitigation efforts. Other management practices such as preventing seed spread and dispersal (e.g., by monitoring possible seed-carrying vehicles and travelers) and implementing early detection and removal measures can mitigate the impacts of invasive shrub species. Management practices that involve herbicides, grazing, and prescribed fire can help eradicate and contain kudzu invasions. Interestingly, climate change may serve as a natural management option by confining the spread of Chinese tallowtree to a smaller area and preventing or slowing its spread northward. As with other nonnative plant species, removing newly infested populations and the reducing seed dispersal would greatly mitigate the spread of this invasive tree species.

Current Knowledge Gaps/Areas of Future Research

The suitable habitat projections that were developed using the ensemble of selected general circulation models and the three emissions storylines (A1B, A2, and B1) have limitations. They lack several environmental parameters that could be critical in biological modeling: daily temperature parameters required for monitoring insect development or relative humidity for monitoring diseases such as fusiform rust.

Figure 6.17 shows the importance of daily data, including average, minimum, and maximum temperatures—in examining the potential number of insect generations (populations) through the season under different climate change storylines. Produced from a regression relationship between accumulated degree-days from the daily dynamic downscaling data and the monthly ensemble general circulation model data, this analysis emphasizes both the need for daily weather data and the importance of accumulated degree-days in predicting the potential suitable habitat for pests such as gypsy moths.

The critical gaps in our current knowledge of climate effects on various species that we have discussed in this chapter bring a high level of uncertainty to the projections of consequences for the incidence or severity of forest diseases, invasive insects, and nonnative plants. Often, limited information is available about the life cycle of an insect or pathogen, or about insect–fungal interactions, especially in situations where new interactions may develop.

Existing scenarios and attempts to model and predict outcomes that are based on a limited number of parameters (temperature or precipitation) are less valuable as decision-making tools than as a validation for continuing sound forest management practices to control the specific species discussed in this chapter, and as tools for exposing critical research needs. The information presented simply highlights the need for much more basic research on virtually all ramifications of climate change as it affects diseases, invasive insects, and nonnative plants, with the long-term goal of strengthening adaptation, restoration, and mitigation strategies, and incorporating them into management of southern forests (Guo and Norman 2012).

The management options presented in this chapter are specific to the individual representative species discussed. These options are not applicable to other species in the same class, because of differences in biology and unique interactions between species and their hosts. As such, they serve as examples of what is available and as guidance on what to consider when addressing nonnative and invasive species.

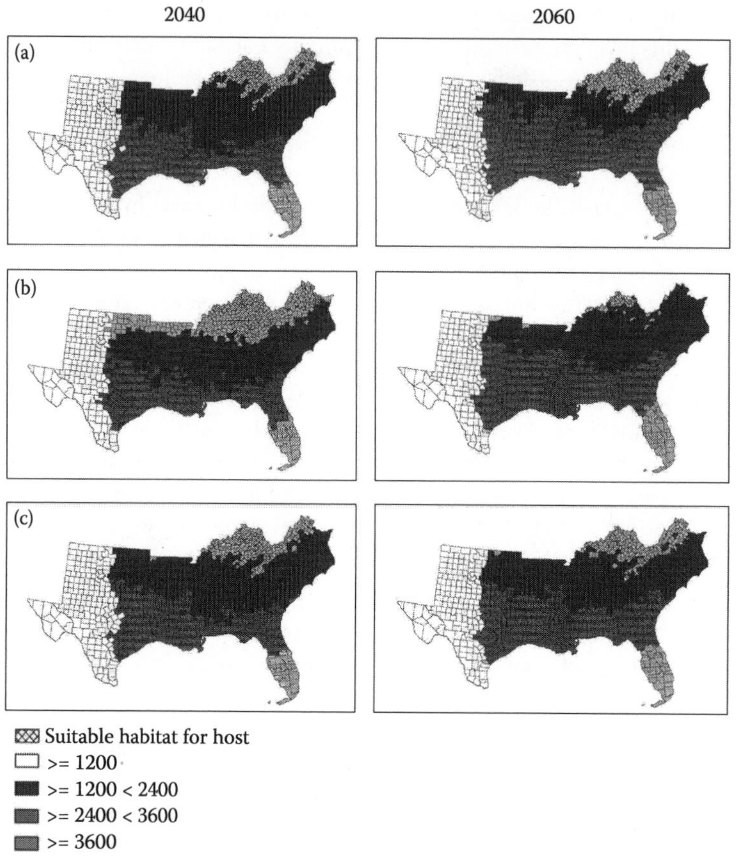

FIGURE 6.17 Accumulation of gypsy moths generations from March to August in 2020 and 2060, projected assuming a baseline count of 1200 per generation at 10°C and three Intergovernmental Panel on Climate Change (2007) emissions storylines; (a) A1B, (b) A2, and (c) B1 (see text and Chapter 2 for storyline descriptions). This figure was produced using a regression equation from the relationship between monthly average temperature and monthly accumulation of degree-days (days of sufficiently warm/cold temperatures needed for physiological processes such as egg hatching and first flight) based on the A2 (2010 data) and future temperature (2020 and 2060) projection data provided by Yongqiang Liu.

ACKNOWLEDGMENTS

We thank J. Hanula, A. E. Mayfield III, and Jeremy Allison for their helpful comments. Qinfeng Guo was supported by a National Science Foundation Grant #0640058. Our appreciation goes to contributing authors, Shelby Gull Laird, McNabb Henry, and Yongqiang Liu.

REFERENCES

Albright, T.P., Chen, H., Chen, L.J., and Q.F. Guo. 2010. The ecological niche and reciprocal prediction of the disjunct distribution of an invasive species: The example of *Ailanthus altissima*. *Biol. Invas.* 12:2413–2427.

Anderson, R.L., Powers, H., and G. Snow. 1980. How to identify fusiform rust and what to do about it. For. Bull. SA–FB/P24. Atlanta: U.S. Department of Agriculture Forest Service, Southern Region, State & Private Forestry, Forest Insect and Disease Management. 12 p.

Anonymous. 1989. Insects and Diseases of Trees in the South. USDA For. Serv. Protection Report R8-PR16.

Ayres, M.P. and M.J. Lombardero. 2000. Assessing the consequences of global change for forest disturbance from herbivores and pathogens. *Science of the Total Environment* 262:263–286.

Balanyá, J., Oller, J.M., Huey, R.B., Gilchrist, G.W., and L. Serra. 2006. Global genetic change tracks global climate warming in *Drosophila subobscura. Science* 313:1773–1775.

Bale, J.S., Masters, G.J., Hodkinson, I.D., Awmack, C., Bezemer, T.M., Brown, V.K., Butterfield, J. et al. 2002. Herbivory in global climate change research: Direct effects of rising temperature on insect herbivores. *Global Change Biology*, 8:1–16.

Battistia, A., Stastny, M., Buffo, E., and S. Larsson. 2006. A rapid altitudinal range expansion in the pine processionary moth produced by the 2003 climatic anomaly. *Global Change Biology* 12:662–671.

Bentz, B.J. and G. Schen-Langenheim. 2007. The mountain pine beetle and whitebark pine waltz: Has the music changed? *Proceedings of the Conference Whitebark Pine: A Pacific Coast Perspective* (16 June 2011; www.fs.fed.us/r6/nr/fid/wbpine/papers/2007-wbp-impacts-bentz.pdf).

Bentz, B.J., Régnière, J., Fettig, C.J. et al. 2010. Climate change and bark beetles of the western United States and Canada: Direct and indirect effects. *BioScience*, 60, 602–613.

Berg, E.E., Henry, J.D., Fastie, C.L., De Volder, A.D., and S.M. Matsuoka. 2006. Spruce beetle outbreaks on the Kenai Peninsula, Alaska, and Kluane National Park and Reserve, Yukon Territory: Relationship to summer temperatures and regional differences in disturbance regimes. *Forest Ecology and Management* 227:219–232.

Billings, R.F. 1980. Direct control. In: *The Southern Pine Beetle.* Thatcher, R.C., Searcy, J.L., Coster, J.E., and G.D., Hertel, eds. Sci. & Ed. Admin. Tech. Bull. 1631, Washington, DC: USDA Forest Service, Expanded Southern Pine Beetle Research and Applications Program: pp. 178–192.

Billings, R.F. and C.A. Kibbe. 1978. Seasonal relationships between southern pine beetle brood development and loblolly pine color in East Texas. *Southwestern Entomologist* 3(2):89–95.

Billings, R.F. and W.W. Upton. 2010. A methodology for assessing annual risk of southern pine beetle outbreaks across the southern region using pheromone traps. In: *Advances in Threat Assessment and Their Application to Forest and Rangeland Management.* Pye, J.M., Rauscher, H.M., Sands, Y., Lee, D.C., and Beatty, J.S., tech. eds. Gen. Tech. Rep. PNW-GTR-802. Portland, OR: U.S. Department of Agriculture Forest Service, Pacific Northwest and Southern Research Stations. Vol. 1:pp. 73–85.

Billings, R.F., Clarke, S.R., Espino Mendoza, V., Cordón Cabrera, P., Meléndez Figueroa, B., Ramón Campos, J., and G. Baeza. 2004. Bark beetle outbreaks and fire: A devastating combination for Central America's pine forests. *Unasylva*, 217:15–21.

Bradshaw, W.E. and C.M. Holzapfel. 2006. Evolutionary response to rapid climate change. *Science* 312:1477–1478.

Brendemuehl, R.H. 1990. *Persea borbonia* (L.) Spreng. Redbay. In: *Silvics of North America, Volume 2, Hardwoods.* Burns, R.M. and Honkala, B.H. eds. Agriculture Handbook 654, USDA Forest Service, Washington, DC, pp. 530–506. (http://www.na.fs.fed.us/pubs/silvics_manual/volume_2/persea/borbonia.htm, date accessed: August 13, 2012)

Burnett, T. 1949. The effect of temperature on an insect host-parasite population. *Ecology*, 30:113–134. doi:10.2307/1931181.

Burns, R.M. and B.H. Honkala. 1990. *Silvics of North America: 1. Conifers; 2. Hardwoods*. Agriculture Handbook 654. U.S. Department of Agriculture, Forest Service, Washington, DC. vol. 2, 877 p. (http://www.na.fs.fed.us/spfo/pubs/silvics_manual/table_of_contents.htm, date accessed: August 13, 2012)

Cameron, R.S. and R.F Billings. 1988. Southern pine beetle: Factors associated with spot occurrence and spread in young plantations. *Southern Journal of Applied Forestry.* 10:105–108.

Cameron, R.S., Bates, C., and J. Johnson. 2008. Distribution and Spread of Laurel Wilt Disease in Georgia: 2006–08 Survey and Field Observations. Georgia Forestry Commission report. September 2008. 28 p.

Campbell, R.W. and R.J. Sloan. 1977. Forest stand responses to defoliation by gypsy moth. *For. Sci. Monogr. 19. Forest Science*. 23(Suppl.):1–35.

Canadian Food Inspection Agency. 2011. *Adelges tsugae* (Annand)—Hemlock woolly adelgid. CFIA Pest Fact Sheet. *Adelges tsugae* (Annand)—Hemlock woolly adelgid (http://www.inspection.gc.ca/english/plaveg/pestrava/adetsu/tech/adetsue.shtml, accessed July 2011).

Chakraborty, S., Murray, G.M., Magarey, P.A., Yonow, T., O'Brien, R.G., Croft, B.J., Barbetti, M.J. et al. 1998. Potential impact of climate change on plant diseases of economic significance to Australia. *Australas. Plant Pathol.* 27:15–35.

Clarke, A. and K.P.P. Fraser. 2004. Why does metabolism scale with temperature? *Funct. Ecol.* 18:243–251. doi:10.1111/j.0269-8463.2004.00841.x.

Clarke, S.R. and R.F. Billings. 2003. Analysis of the southern pine beetle suppression program on the national forests in Texas in the 1990s. *Journal of Forestry*, 27(2):122–129.

Cowles, R.S., Montgomery, M.E., and C.A.S.J. Cheah. 2006. Activity and residues of imidacloprid applied to soil and tree trunks to control hemlock woolly adelgid (Hemiptera: Adelgidae) in forests. *Journal of Economic Entomology* 99:1258–1267.

Czabator, F.J. 1971. Fusiform rust of southern pines—A critical review. Res. Pap. SO–65. New Orleans: U.S. Department of Agriculture Forest Service, Southern Forest Experiment Station. 39 p.

Danks, H.V. 1987. Insect dormancy: An ecological perspective. Monograph Series no. 1. Biological Survey of Canada (Terrestrial Arthropods).

DeWan, A., Dubois, N., Theoharides, K., and J. Boshoven. 2010. *Understanding the Impacts of Climate Change on Fish and Wildlife in North Carolina*. Defenders of Wildlife, Washington, DC.

Dodds, K.J., Cooke, R.R., and D.W. Gilmore. 2007. Silvicultural options to reduce pine susceptibility to attack by a newly detected invasive species. *North. J. Appl. For.* 24(3):165–167.

Duerr, D.A. and P.A. Mistretta. (in press). Insect and disease pests of southern forests. In: *Southern Forest Futures Project* (http://www.srs.fs.usda.gov/futures/reports/draft/Frame.htm, date accessed: August 13, 2012).

Dwinell, L.D. 2001. Potential use of elevated temperatures to manage pests in transported wood. Exotic Forest Pests Online Symposium, April 16–29, 2001 (http://www.scientificsocieties.org/aps/proceedings/exoticpest/Papers/dwinell_temperature.htm, date accessed: August 13, 2012).

Eliason, E.A. and D. McCullough. 1995. *Life History and Control of Pine Root Collar Weevil in Christmas Tree Fields*. MSU Department of Forestry. Extention Bulletin E2560. (http://web2.msue.msu.edu/bulletins/Bulletin/PDF/E2560.pdf, accessed on July 16, 2011).

Evans, A.M. and T.G. Gregoire. 2007. A geographically variable model of hemlock woolly adelgid spread. *Biological Invasions*. 9:369–382.

Fajvan, M.A. and P.B. Wood. 2010. Maintenance of eastern hemlock forests: Factors associated with hemlock vulnerability to hemlock woolly adelgid. In: *Proceedings from the Conference on the Ecology and Management of High-Elevation Forests in the Central and Southern Appalachian Mountains*. Rentch, J.S., Schuler, T.M., eds. 2009 May 14–15; Slatyfork, WV. Gen. Tech. Rep. NRS-P-64. Newtown Square, PA: U.S. Department of Agriculture, Forest Service, Northern Research Station: pp. 31–38.

Fernández-Arhex, V. and J.C. Corley. 2005. The functional response of *Ibalia leucospoides* (Hymenoptera: Ibaliidae), a parasitoid of *Sirex noctilio* (Hymenoptera: Siricidae). *Biocontrol Science and Technology* 15(2):207–212.

Ford, C.R., Elliott, K.J., Clinton, B.D., Kloeppel, B.D., and J.M. Vose. 2012. Forest dynamics following eastern hemlock mortality in the southern Appalachians. *Oikos* 121:523–536.

Fraedrich, S.W., Harrington, T.C., Rabaglia, R.J., Ulyshen, M.D., Mayfield, A.E., Hanula, J.L., Eickwort, J.M., and D.R. Miller. 2008. A fungal symbiont of the redbay ambrosia beetle causes a lethal wilt in redbay and other Lauraceae in the southeastern United States. *Plant Disease* 92:215–224.

Frelich, L.E. and P.B. Reich. 2009. Wilderness conservation in an era of global warming and invasive species: A case study from Minnesota's Boundary Waters Canoe Area Wilderness. *Natural Areas Journal* 29:385–393.

Friedenberg, N.A., Powell, J.A., and M.P. Ayres. 2007. Synchrony's double edge: Transient dynamics and the Allee effect in stage-structured populations. *Ecology Letters* 10:564–573.

Friedenberg, N.A., Sarkar, S., Kouchoukos, N., Billings, R.F., and M.P. Ayres. 2008. Temperature extremes, density dependence, and southern pine beetle (Coleoptera: Curculionidae) population dynamics in east Texas. *Environmental Entomology* 37:650–659.

Gan, J. 2004. Risk and damage of southern pine beetle outbreaks under global climate change. *Forest Ecology and Management* 191:61–71.

Geber, M.A. and V.M. Eckhart. 2005. Experimental studies of adaptation in *Clarkia xantiana*. II. Fitness variation across a subspecies border. *Evolution* 59(3):521–531.

Gillooly, J.F., Brown, J.H., West, G.B., Savage, V.M., and E.L. Charnov. 2001. Effects of size and temperature on metabolic rate. *Science* 293:2248–2251. doi:10.1126/science. 1061967.

Goggans, J.F. 1957. Southern fusiform rust. Some factors affecting its incidence in Alabama's Coastal Plain Region. Alabama Agricultural Experiment Station. Bulletin 304. March 1957.

Gonthier, P., Nicoletti, G., Linzer, R., Gugliemo, F., and M. Garbelotto. 2007. Invasion of European pine stands by an North American forest pathogen and its hybridization with a native interfertile taxon. *Molecular Ecology* 16:1389–1400.

Griffith, T.M. and M.A. Watson. 2006. Is evolution necessary for range expansion? Manipulating reproductive timing of a weedy annual transplanted beyond its range. *American Naturalist* 167:153–164.

Guo, Q.F. 2006. Intercontinental biotic invasions: What can we learn from native populations and habitats? *Biol. Invas.* 8:1451–1459.

Guo, Q.F. 2010. Poleward shifts of introduced species under global climate warming. In: *Ecological Complexity and Ecological Vision.* Wu, Y. and Fan, J. eds. Higher Education Press and Springer, Beijing, pp. 140–146.

Guo, Q.F. and S.P. Norman. 2012. Improve restoration to control plant invasions under climate change. In: *Invasive Plant Ecology.* Jose, S., Singh, H., Batish, D., and Kohli, R. eds. CRC Press, Boca Raton, FL, pp. 203–216.

Guo, Q.F., Qian, H., Ricklefs, R.E., and W. Xi. 2006. Distributions of exotic plants in eastern Asia and North America. *Ecol. Lett.* 9:827–834.

Guo, Q.F., Rejmánek, M., and J. Wen. 2012a. Geographical, socioeconomic, and ecological determinants of exotic plant naturalization in the United States: Insights and updates from improved data. *NeoBiota* 12:41–55.

Guo, Q.F., Sax, D.F., Qian, H., and R. Early. 2012b. Latitudinal shifts of introduced species: Possible causes and implications. *Biol. Invas.* DOI 10.1007/s10530-011-0094-8.

Hansen, E.M., Bentz, B.J., and D.L. Turner. 2001. Temperature-based model for predicting univoltine brood proportions in spruce beetle (Coleoptera: Scolytidae). *Canadian Entomologist* 133:827–841.

Hanula, J.L. and B. Sullivan. 2008. Manuka oil and phoebe oil are attractive baits for redbay ambrosia beetle, *Xyleborus glabratus* (Coleoptera: Scolytinae), the vector of laurel wilt. *Environ. Entomol.* 37:1403–1409.

Hanula, J.L., Mayfield, A.E., Fraedrich, S.W., and R.J. Rabaglia. 2008. Biology and host associations of the redbay ambrosia beetle, *Xyleborus glabratus* (Coleoptera: Curculionidae: Scolytinae), exotic vector of laurel wilt killing redbay (*Persea borbonia*) trees in the Southeastern United States. *J. Econ. Entomol.* 101(4):1276–1286.

Harrington, T.C. and S.W. Fraedrich. 2010. Quantification of propagules of the laurel wilt fungus and other mycangial fungi from the redbay ambrosia beetle, *Xyleborus glabratus*. *Phytopathology* 100:1118–1123.

Harrington, T.C., Fraedrich, S.W., and D.N. Aghayeva. 2008. A new ambrosia beetle symbiont and pathogen on the Lauraceae. *Mycotaxon* 104:399–404.

Haugen, D.A. and E.R. Hoebeke. 2005. Sirex woodwasp—*Sirex noctilio* F. (Hymenoptera: Siricidae), USDA Forest Service Pest Alert NA-PR-07-05.

Haugen, D.A., Bedding, R.A., Underdown, M.G., and F.G. Neumann. 1990. National strategy for control of *Sirex noctilio* in Australia. Australian Forest Grower 13(2 Special Liftout Section No. 13).

Havill, N.P., Montgomery, M.E., Yu, G., Shiyake, S., and A. Caccone. 2006. Mitochondrial DNA from hemlock woolly adelgid (Hemiptera: Adelgidae) suggest cryptic speciation and pinpoints the source of the introduction to eastern North America. *Ann. Entomol. Soc. Am.* 99:195–203.

Hoebeke, E.R., Haugen, D.A., and R.A. Haack. 2005. *Sirex noctilio*: Discovery of a Palearctic siricid woodwasp in New York. *Newsletter of the Michigan Entomological Society* 50(1&2):2425.

Hofstetter, R.W., Dempsey, T.D., Klepzig, K.D., and M.P. Ayres. 2007. Temperature-dependent effects on mutualistic and phoretic associations. *Community Ecology* 8:47–56. IPCC Intergovernmental Panel on Climate Change. 2007. *Climate Change 2007: The Scientific Basis.* Cambridge University Press.

Hofstetter, R.W., Klepzig, K.D., Moser, J.C., and M.P. Ayres. 2006. Seasonal dynamics of mites and fungi and their interaction with southern pine beetle. *Environmental Entomology* 35:22–30.

Hopkins, J.D., Walkingstick, T., and J.L. Wallace. 2008. Invasive woodwasp, *Sirex noctilio*: A potential pest of pines in Arkansas, University of Arkansas Agriculture and Natural Resources Publication No. FSA7071 (FSA7071).

Hurley, B.P., Slippers, B., and M.J. Wingfield. 2007. A comparison of control results for the alien invasive woodwasp, *Sirex noctilio*, in the southern hemisphere. *Agricultural and Forest Entomology* 9(3):159–171.

Intergovernmental Panel on Climate Change. 2007. Climate Change 2007. Synthesis Report. Contribution of Working Groups I, II, and III to the Fourth Assessment Report of the Intergovernmental Panel on Climate Change. IPCC, Geneva, Switzerland. (http://www.ipcc.ch/publications_and_data/publications_and_data_reports.shtml) [Date accessed: July 22, 2011].

International Union for Conservation of Nature Invasive Species Group. 2007a. Xyleborus glabratus (insect). Global Invasive Species Database. http://www.issg.org/database/species/ecology.asp?si=1536&fr=1&sts=sss&lang=EN, date accessed: June 26, 2013 (Archived at PERAL).

International Union for Conservation of Nature Invasive Species Group. 2007b. *Sirex noctilio* (insect). Global Invasive Species Database. http://www.issg.org/database/species/ecology.asp?si=1211&fr=1&sts=&lang=EN, date accessed: August 13, 2012 (Archived at PERAL).

Jenkins, J.C., Aber, J.D., and C.D. Canham. 1999. Hemlock woolly adelgid impacts on community structure and N cycling rates in eastern hemlock forests. *Can. J. For. Res.* 29:630–645.

Jonas, S., Xi, W., Waldron, J., and R. Coulson. 2012. Ecological considerations for forest restoration following hemlock woolly adelgid outbreaks. *Tree and Forestry Science and Biotechnology* 6(Special Issue 1): 22–26.

Karl, T.R. and R.W. Knight. 1998. Secular trends of precipitation amount, frequency, and intensity in the United States. *Bulletin of the American Metrological Society*, 79(2):231–241.

Karl, T.R., Melillo, J.M., and T.C. Peterson (eds.). 2009. *Global Climate Change Impacts in the United States.* Cambridge University Press, New York.

Karlsson, M.G. 2002. Flower formation in *Primula vulgaris* is affectedby temperature, photoperiod and daily light integral. *Scientia Horticulturae* 95:99–110.

Keena, M.A., Gould, J., and L.S. Bauer. 2009. Factors that influence emerald ash borer (*Agrilus planipennis*) adult longevity and oviposition under laboratory conditions 2009 USDA Research Forum on Invasive Species (http://nrs.fs.fed.us/pubs/gtr/gtr-nrs-p-51papers/54keena-p-51.pdf, accessed June 30, 2011).

Keim, B.D. 1997. Preliminary analysis of the temporal patterns of heavy rainfall across the southeastern United States. *Professional Geographer* 49(1):94–104.

Kizlinski, M., Orwig, D.A., Cobb, R., and D. Foster. 2002. Direct and indirect ecosystem consequences of an invasive pest on forests dominated by eastern hemlock. *Journal of Biogeography* 29:1489–1503.

Klapwijk, M.J., Battisti, A., Ayres, M.P., and S. Larsson. 2012. Assessing the impact of climate change on outbreak potential. In: *Insect Outbreaks Revisited.* Barbosa, P., Schultz, J.C., Letourneau, D. eds. Blackwell Publishing Ltd, Oxford, UK.

Klepzig, K.D., Raffa, K.F., and E.B. Smalley. 1991. Association of an insect–fungal complex with red pine decline in Wisconsin. *Forest Science* 37:1119–1139.

Kliejunas, J.T., Geils, B.W., Glaeser, J.M., Goheen, E.M., Hennon, P., Kim, M., Kope, H., Stone, J., Sturrock, R., and S.J. Frankel. 2009. Review of literature on climate change and forest diseases of western North America. Gen. Tech. Rep. PSW-GTR-225. Albany, CA: U.S. Department of Agriculture, Forest Service, Pacific Southwest Research Station, 54 p.

Koch, F.H. and W.D. Smith. 2008. Spatio-temporal analysis of *Xyleborus glabratus* (Coleoptera: Circulionidae: Scolytinae) invasion in eastern U.S. forests. *Environmental Entomology* V37(2):442–452.

Kovacs, K., Haight, R.G., McCullough, D., Mercader, R., Siegert, N., and A.M. Liebhold. 2010. Cost of potential emerald ash borer damage in U.S. communities, 2009–2019. *Ecological Economics* 69(3):569–578.

Lechowicz, M.J. and Y. Mauffelte. 1986. Host preferences of the gypsy moth in eastern North America versus European forests. *Rev. Entomol. Queb.* 31:43–51.

Li, Z.Y., Dong, Q., Albright, T., and Q.F. Guo. 2011. Natural and human dimensions of a quasi-wild species: The case of kudzu. *Biological Invasions* 13:2167–2179.

Liebhold, A.M., Gottschalk, K.W., Muzika, R.M., Montgomery, M.E., Young, R., O'Day, K., and B. Kelley. 1995. Suitability of North American tree species to the gypsy moth: A summary of field and laboratory tests. USDA For. Serv. Gen. Tech. Rep. NE- 211.

Linkosalo, T., Carter, T.R., Hakkinen, R., and P. Hari. 2000. Predicting spring phenology and frost damage of *Betula* spp. under climatic warming: A comparison of two models. *Tree Physiology* 20:1175–1182.

Linzer, R.E., Otrosina, W.J., Gonthier, P., Bruhn, J., LaFlamme, G., Bussieres, G., and M. Garbelotto. 2008. Inferences on the phylogeography of the fungal pathogen *Heterobasidion annosum*, including evidence of interspecific horizontal genetic transfer and of human-mediated, long-range dispersal. *Molecular Phylogenetics and Evolution* 46:844–862.

Logan, J.A., Régnière, J., Gray, D.R., and A.S. Munson. 2007. Risk assessment in face of a changing environment: Gypsy moth and climate change in Utah. *Ecological Applications* 17:101–117.

Lombardero, M.J., Ayres, M.P., Hofstetter, R.W., Moser, J.C., and K.D. Klepzig. 2003. Strong indirect interactions of *Tarsonemus* mites (Acarina: Tarsonemidae) and *Dendroctonus frontalis* (Coleoptera: Scolytidae). *Oikos* 102:243–252.

Mack, R.N., Simberloff, D., Lonsdale, W.M., Evans, H., Clout, M., and F.A. Bazzaz. 2000. Biotic invasions: Causes, epidemiology, global consequences, and control. *Ecological Applications* (10) 3:689–710.

Mausel, D.L., Salom, S.M., Kok, L.T., and G.A. Davis. 2010. Establishment of the hemlock woolly adelgid predator, *Laricobius nigrinus* (Coleoptera: Derodontidae), in the Eastern United States. *Environmental Entomology* 39(2):440–448.

Mayfield III, A.E. and M.C. Thomas. 2006. *The redbay Ambrosia beetle, Xyleborus glabratus Eichhoff (Scolytinae: Curculionidae). FDACS Div.* of Plant Industry, Pest Alert, p.1–2.

McClure, M.S. and C.A.S.-J. Cheah. 2002. Important mortality factors in the life cycle of hemlock woolly adelgid. *Adelges tsugae* Annand (Homoptera: Aldelgid) in the Northeastern United States. *Proceedings: Hemlock Woolly Adelgid in the Eastern United States Symposium.* East Brunswick, New Jersey, February 5-7, pp. 13–22.

McClure, M.S., Salom, S., and K.S. Shields. 2001. *Hemlock Wooly Adelgid. Forest Health Technology Enterprise Team.* U.S. Forest Service Publication FHTET–2001–03. Morgantown, WV, 14 pp.

McNulty, S.G. and J.L. Boggs. 2010. A conceptual framework: Redefining forest soil's critical acid loads under a changing climate. *Environmental Pollution* 158:2053–2058.

Mearns, L.O., Gutowski, W.J., Jones, R., Leung, L-Y., McGinnis, S., Nunes, A.M.B., and Y. Qian. 2009. A regional climate change assessment program for North America. *EOS* 90(36):311–312.

Meeker, J.R., Dixon, W.N., Foltz J.L., and T.R Fasulo. 2004. Southern Pine Beetle, *Dendroctonus frontalis* Zimmermann (Insecta: Coleoptera: Scolytidae). EENY-176 Entomology and Nematology Department, Cooperative Extension Service, Institute of Food and Agricultural Sciences, University of Florida, 6 pp.

Mercader, R.J., Siegert, N.W., Liebhold, A.M., and D.G. McCullough. 2011. Simulating the effectiveness of three potential management options to slow the spread of emerald ash borer (*Agrilus planipennis*) populations in localized outlier sites. *Canadian Journal of Forest Research*, 41(2):254–264.

Millar, C.I., Stephenson, N.L., and S.L. Stephens. 2007. Climate change and forests of the future: Managing in the face of uncertainty. *Ecological Applications* 17:2145–2151.

Miller, J., Lemke, D., and J. Coulston. in press. *The Invasion of Southern Forests by Nonnative Plants: Current and Future Occupation with Impacts, Management Strategies, and Mitigation Approaches. Southern Forest Future Project.* USDA Forest Service, Southern Research Station, Asheville, NC.

Mitton, J.B. and K.L. Duran. 2004. Genetic variation in pinion pine, *Pinus edulis*, associated with summer precipitation. *Molecular Ecology* 13:1259–1264.

Morin, R., Liebhold, A., and K.W. Gottschalk. 2009. Ansiotrophic spread of hemlock woolly adelgid in the eastern United States. *Biological Invasions* 11:2341–2350.

Nakićenović, N., Alcamo, J., Davis, G., deVries, B., Fenhann, J., Gaffin, S., Gregory, K., Grübler, A. et al. 2000. *Emissions Scenarios, Special Report of Working Group III of the Intergovernmental Panel on Climate Change.* IPCC, Geneva, and Cambridge University Press, Cambridge, 595 pp. (ISBN 0-521-80081-1 hardback, ISBN 0-521-80493-0 paperback).

Nealis, V. and B. Peter. 2009. Risk assessment of the threat of mountain pine beetle to Canada's boreal and eastern pine forests. Natural Resources Canada, Canadian Forest Service. Information Report BC-X-417.

Onken, B.P. and R.C. Reardon. 2011. An overview and outlook for biological control of hemlock woolly adelgid. In: *Implementation and Status of Biological Control of the Hemlock Woolley Adelgid.* Onken, B. and Reardon, R. eds. USDA Forest Service, Forest Health Technology Enterprise Team, Publication FHTET- 2011-04, pp. 22–228.

Otrosina, W.J., Chase, T.E., Cobb, F.W. Jr., and K. Korhonen. 1993. Population structure of *Heterobasidion annosum* form North America and Europe. *Can. J. Botany.* 71:1064–1071.

Otrosina, W.L. and M. Garbelotto. 2010. *Heterobasidion occidentale* sp. nov. and *Heterobasidion irregulare* nom. nov.: A disposition of North American Heterobasidion species. *Fungal Biology* 114(1):16–25.

Paradis, A., Elkinton, J., Hayhoe, K., and J. Buonaccorsi. 2008. Role of winter temperature and climate change on the survival and future range expansion of the hemlock woolly adelgid (*Adelges tsugae*) in eastern North America. *Mitigation and Adaptation Strategies for Global Change*, 13(5–6):541–554.

Parker, B.L., Skinner, M., Gouli, S., Ashikaga, T., and H.B. Teillon. 1997. Survival of hemlock woolly adelgid at low temperatures. *Forest Science* 44:414–420.

Parker, B.L., Skinner, M., Gouli, S., Ashikaga, T., and H.B. Teillon. 1999. Low lethal temperatures for hemlock woolly adelgid (Homoptera: Adelgidae). *Environmental Entomology* 28:1085–1091.

Parmesan, C. 2006. Ecological and evolutionary responses to resent climatic change. *Annual Review of Ecology, Evolution and Systematics* 37:637–669. doi:10.1146/annurev.ecolsys.37.091305.110100.

Pattison, R.R. and R.N. Mack. 2009. Environmental constraints on the invasion of *Triadica sebifera* in the eastern United States: An experimental filed assessment. *Oecologia* 158:591–602.

Pearson, R.G. and T.P. Dawson. 2003. Predicting the impacts of climate change on the distribution of species: Are bioclimate envelope models useful? *Global Ecology and Biogeography* 12:361–371.

Peterson, A.T. 2003. Predicting the geography of species' invasion via ecological niche modeling. *Quarterly Review of Biology* 78:419–433.

Peterson, A.T. and D.A. Vieglais. 2001. Predicting species invasions using ecological niche modeling. *BioScience* 51:363–371.

Phelps, W.R. and F.L. Czabator. 1978. Fusiform rust of southern pines. Forest Insect and Disease Leaflet No. 2:7 pp. USDA Forest Service. www.na.fs.fed.us/spfo/pubs/fidls/fusiform/fidl-fusi.html

Pimentel, D., Zúñiga, R., and D. Morrison. 2004. Update on the environmental and economic costs associated with alien-invasive species in the United States. *Ecological Economics* 52:273–288.

Pollard, J., Czerwinski, E., and T. Scarr. 2006. Sirex woodwasp, *Sirex noctilio* (F.). Forest health alert. Ontario Ministry of Natural Resources, Forest Health and Silviculture Section. FHA-2-2006 https://ospace.scholarsportal.info/bitstream/1873/778/1/265664.pdf (Accessed July 07, 2011).

Pontius, J., Hallett, R., and J. Jenkins. 2006. Foliar chemistry linked to infestation and susceptibility to hemlock woolly adelgid (Homoptera: Adelgidae). *Environmental Entomology* 35:112–120.

Pontius, J., Hallett, R., and M. Martin. 2002. Examining the role of foliar chemistry in hemlock woolly adelgid infestation and hemlock decline. pp. 86–99, In: *Symposium on the Hemlock Woolly Adelgid in Eastern North America.* Reardon, R., Onken, B. and Lashomb, J. eds. US Forest Service, New Brunswick, NJ.

Powell, J.A. and J.A. Logan. 2005. Insect seasonality: Circle map analysis of temperature-driven life cycles. *Theoretical Population Biology* 67:161–179.

Rabaglia, R. 2005. Exotic forest pest information system for North America. *Xyleborus glabratus*. (http://spf-nic.fs.fed.us/exfor/data/pestreports.cfm?pestidval=148&langdisplay=english. Accessed July 6, 2011.)

Rabaglia, R. and J. Lewis. 2006. Economic analysis of the potential impact of *Sirex noctilio*, with emphasis on pines in the southeastern United States. U.S. Dep. Agric. Forest Service Forest Health Protection, Executive Summary, Arlington, VA, 6 pp.

Rabaglia, R.J., S.A. Dole, and A.I. Cognato. 2006. Review of American Xyleborina (Coleoptera: Curculionidae: Scolytinae) occurring North of Mexico, with an illustrated key. *Annals of the Entomological Society of America* 99:1034–1056.

Rice, A.V., Thormann, M.N., and D.W. Langor. 2008. Mountain pine beetle-associated blue-stain fungi are differentially adapted to boreal temperatures. *Forest Pathology* 38:113–123.

Robbins, K. 1984. *Annosus Root Rot in Eastern Conifers.* For. Insect and Dis. Leafl. 76. Washington, DC: U.S. Department of Agriculture Forest Service. 9 p. (http://www.fs.fed.us/r6/nr/fid/fidls/fidl-76.pdf). [Date accessed: September 28, 2011.]

Ross, R.M., Bennett, R.M., Snyder, C.D., Young, J.A., Smith, D.R., and D.P. Lemarie. 2003. Influence of eastern hemlock (*Tsuga canadensis* L.) on fish community structure and function in headwater streams of the Delaware River basin. *Ecology of Freshwater Fish* 12:60–65.

Runion, G.B., Prior, S.A., Rogers, H.H., and R.J. Mitchell. 2010. Effects of elevated atmospheric CO_2 on two southern forest diseases. *New Forests* 39:275–285.

Salinas-Moreno, Y., Mendoza, M.G., Barrios, M.A., Cisneros, R., Macías-Sámano, J., and G. Zúñiga. 2004. Areography of the genus *Dendroctonus* (Coleoptera: Curculionidae: Scolytinae) in Mexico. *Journal of Biogeography* 31:1163–1177.

Schiefer, T.L. and D.E. Bright. 2004. *Xylosandrus mutilatus* (Blandford), an exotic ambrosia beetle (Coleoptera: Curculionidae: Scolytinae: Xyleborini) new to North America. *Coleopts. Bull.* 58:431–438.

Seybold, S.J., Ohtsuka, T., Wood, D.L., and I. Kubo. 1995. The enantiomeric composition of ipsdienol: A chemotaxonomic character for North American populations of *Ips* spp. in the pini subgeneric group (Coleoptera: Scolytidae). *Journal of Chemical Ecology* 21:995–1016.

Sharov, A.A., Leonard, D., Liebhold, A.M., Roberts, E.A., and W. Dickerson. 2002. "Slow the Spread": A national program to contain the gypsy moth. *Journal of Forestry* 100:30–35.

Shields, K.S. and C.A.S.-J. Cheah. 2005. Winter mortality in *Adelges tsugae* populations in 2003 and 2004. In: *Proceedings of the 3rd Symposium on Hemlock Woolly Adelgid in the Eastern United States.* Onken, B., and Reardon, R. eds. FHTET-2005-01. Asheville, NC: U.S. Department of Agriculture, Forest Service, Forest Health and Technology Enterprise Team, pp. 354–356.

Shugart, H.H. 2003. *A Theory of Forest Dynamics: The Ecological Implications of Forest Succession Models.* The Blackburn Press, New Jersey.

Six, D.L. and B.J. Bentz. 2007. Temperature determines symbiont abundance in a multipartite bark beetle-fungus ectosymbiosis. *Microbial Ecology* 54:112–118.

Skinner, M., Parker, B.L., Gouli, S., and T. Ashikaga. 2003. Regional responses of hemlock woolly adelgid (Homoptera: Adelgida) to low temperatures. *Environmental Entomology* 32:523–528.

Smith, E.L., Storer, A.J., and B.K. Roosien. 2009. Emerald ash borer infestation rates in Michigan, Ohio, and Indiana [Abstract]. In: *Proceedings. 20th U.S. Department of Agriculture Interagency Research Forum on Invasive Species.* McManus, K.A. and Gottschalk, K.W. eds. Gen. Tech. Rep. NRS–P–51. Newtown Square, PA: U.S. Department of Agriculture Forest Service, Northern Research Station, p. 96.

Sobek, S., Crosthwaite, J.C., and B.J. Sinclair. 2009. Is overwintering biology of invasive insects affected by climate change? *Plasticity of cold tolerance in the emerald ash borer (Agrilus planipennis). 94th Annual Meeting of the Ecological Society of America.* August 2–7, 2009 (Abstract) (http://esameetings.allenpress.com/2009/Paper17566.html, date accessed: August 13, 2012).

Spaulding, H.L. and L.K. Rieske. 2010. The aftermath of an invasion: Structure and composition of Central Appalachian hemlock forests following establishment of the hemlock woolly adelgid, *Adelges tsugae*. *Biological Invasions* 12:3135–3143.

Stadler, B., Muller, T., Orwig, D., and R. Cobb. 2005. Hemlock woolly adelgid in New England forests: Canopy impacts transforming ecosystem processes and landscapes. *Ecosystems (N.Y., Print)*, 8:233–247. doi:10.1007/s10021-003-0092-5.

Stambaugh, W.J. 1989. Annosus root disease in Europe and the Southeastern United States: Occurrence, research, and historical perspective. In: Otrosina, William J.; Scharpf, Robert F., tech. coords. *Proceedings of the symposium on research and management of annosus root disease (Heterobasidion annosum) in Western North America*. Gen. Tech. Rep. PSW–116. Berkeley, CA: U.S. Department of Agriculture Forest Service, Pacific Southwest Forest and Range Experiment Station, pp. 3–9.

Steffen, W., Burbidge, A.A., Hughes, L., Kitching, R., Lindenmayer, D., Musgrave, W., Smith, M.S., and P. Werner. 2009. Australia's biodiversity and climate change: A strategic assessment of the vulnerability of Australia's biodiversity to climate change. A report to the Natural Resource Management Ministerial Council commissioned by the Australian government. CSIRO Publishing.

Sturrock, R.N., Frankel, S.J., Brown, A.V., Hennon, P.E., Kliejunas, J.T., Lewis, K.J., Worrall, J.J., and A.J. Woods. 2011. Climate change and forest disease. *Plant Pathology* 60:133–149.

Thatcher, R.C. and P.J. Barry. 1982. *Southern Pine Beetle.* USDA Forest Service, Washington, DC, Forest and Disease Leaflet No. 49, 7 p.

Thatcher, R.C. and M.D. Conner. 1985. *Identification and Biology of Southern Pine Bark Beetles.* USDA Forest Service, Washington, DC, Handbook No. 634, 14 p.

Thatcher, R.C. and L.S. Pickard. 1966. The clerid beetle, *Thanasimus dubius*, as a predator of the southern pine beetle. *Journal of Economic Entomology* 59:955–957.

Thatcher, R.C. and L.S. Pickard. 1967. Seasonal development of the southern pine beetle in East Texas. *Journal of Economic Entomology* 60:656–568.

Thatcher, R.C., Searcy, J.L., Coster, J.E., and G.D. Hertel, eds. 1980. The southern pine beetle. USDA, expanded southern pine beetle research and application program, forest service, science and education administration, pineville, LA. *Technical Bulletin* 1631, 265.

Thompson, R.S., Anderson, K.H., and P.J. Bartlein. 1999. Atlas of relations between climatic parameters and distributions of important trees and shrubs in North America. U.S. Geological Survey Professional Paper 1650 A&B Online Version 1.0 (http://pubs.usgs.gov/pp/p1650-a/).

Tingley, M.W., Orwig, D.A., Field, R., and G. Motzkin. 2002. Avian response to removal of a forest dominant: Consequences of hemlock woolly adelgid infestations. *Journal of Biogeography* 29:1505–1516.

Trân, J.K., Ylioja, T., Billings, R.F., Régnière, J., and M.P. Ayres. 2007. Impact of minimum winter temperatures on the population dynamics of *Dendroctonus frontalis*. *Ecological Applications* 17(3):882–899.

Ungerer, M.J., Ayers, M.P., and M.J. Lombardero. 1999. Climate and the northern distribution limits of *Dendroctonus frontalis* Zimmermann (Coeoptera: Scolytidae). *Journal of Biogeography* 26, 1133–1145.

U.S. Department of Agriculture Forest Service. 2011a. Forest health protection hemlock woolly adelgid [Online]. US Forest Service Northeastern Area. (Accessed on 19 July 2011 from http://na.fs.fed.us/fhp/hwa/, date accessed: August 13, 2012).

U.S. Department of Agriculture, Animal and Plant Health Inspection Service. 2011b. Cooperative emerald ash borer project: Emerald ash borer locations in Illinois, Indiana, Iowa, Kentucky, Maryland, Michigan, Minnesota, Missouri, New York, Ohio, Pennsylvania, Virginia, Wisconsin, West Virginia and Canada: February 1, 2011 (http://www.emeraldashborer.info/files/MultiState_EABpos.pdf, date accessed: March 1, 2011).

U.S. Department of Agriculture, Animal and Plant Health Inspection Service. 2003. Emerald ash borer; quarantine and regulations: Interim rule and request for comment. Federal Register. 68:59,082–59,091.

U.S. Department of Agricuture (USDA) Forest Service. 2009. *National Forest Management Options in Response to Climate Change.* Short Subjects from SAP 4.4 National Forests No 1. July 2009 (http://www.fs.fed.us/ccrc/files/NationalForestManagement_final.pdf, date accessed: August 13, 2012).

Volney, W.J.A. and R.A. Fleming. 2000. Climate change and impacts of boreal forest insects. *Agriculture, Ecosystems & Environment* 82:283–294. doi:10.1016/S0167-8809(00)00232-2.

Vose, J.M., Wear, D.N., Mayfield III, A.E., and C.D. Nelson. 2013. Hemlock woolly adelgid in the southern Appalachians: Control strategies, ecological impacts, and potential management responses. *Forest Ecology and Management* 291:209–219.

Wagner, T.L., Feldman, R.M., Gagne, J.A., Cover, J.D., Coulson, R.N., and R.M. Schoolfield. 1981. Factors affecting gallery construction, oviposition, and reemergence of *Dendroctronus frontalis* in the laboratory. *Annals of the Entomological Society of America* 74:255–273.

Wagner, T.L., Gagne, J.A., Sharpe, P.J.H., and R.N. Coulson. 1984a. A biophysical model of southern pine beetle *Dendroctonus frontalis* (Coleoptera: Scolytidae) development. *Ecological Modelling* 21:125–147.

Wagner, T.L., Gagne, J.A., Sharpe, P.J.H., and R.N. Coulson. 1984b. Effects of constant temperature on longevity of adult southern pine beetles *Dendroctonus frontalis* Coleoptera Scolytidae. *Environmental Entomology* 13:1125–1130.

Ward, J.D. and P.A. Mistretta. 2002. Impact of pests on forest health. In: Wear, D.N.; Greis, J.G. Southern forest resource assessment. Gen. Tech. Rep. SRS-53. Asheville, NC: U.S. Department of Agriculture Forest Service, Southern Research Station: pp. 403–428. [Chapter 17].

Webster, J.R., Morkeskia, K., Wojculewskia, C.A., Niederlehnera, B.R., Benfielda, E.F., and K.J. Elliott. 2012. Effects of hemlock mortality on streams in the Southern Appalachian Mountains. *The American Midland Naturalist* 168(1):112–131.

Werner, R.A., Holsten, E.H., Matsuoka, S.M., and R.E. Burnside. 2006. Spruce beetles and forest ecosystems in south-central Alaska: A review of 30 years of research. *Forest Ecology and Management* 227:195–206.

Westerling, A.L., Hidalgo, H.G., Cayan, D.R., and T.W. Swetnam. 2006. Warming and earlier spring increase western US forest wildfire activity. *Science* 313:940–943.

Williams, S.E., Shoo, L.P., Isaac, J.L., Hoffmann, A.A., and G. Langham. 2008. Towards an integrated framework for assessing the vulnerability of species to climate change. *PLoS Biology* 6:2621–2626.

Wilson, L.F. and I. Millers. 1982. Pine root collar weevil-its ecology and management. Tech. Bull. No. 1675. Washington, DC: U.S. Department of Agriculture; 1982, 34p.

Wilson, L.F. and D.C. Schmiege. 1975. *Pine Root Collar Weevil. Forest Pest Leaflet 39 (revised).* USDA Forest Service, Washington, DC, 8 p.

Woodward, F.I. 1987. *Climate and Plant Distribution.*: Cambridge University Press, New York, NY, 174 p.

Zhang, K.D. and Y.T. Lin. 1991. *Chinese Tallow Tree*. China Forestry Press, Beijing.

7 Adapting Silviculture to a Changing Climate in the Southern United States

James M. Guldin

CONTENTS

Questions about how forests might respond to climate change are often addressed through planning, prediction, and modeling at the landscape scale. A recent synthesis of climate-change impacts on forest management and policy found that the earth is warmer than it has been in the recent past, and that 11 of the last 12 years rank among the 12 warmest since 1850 (Solomon et al. 2007). Projections are that global average surface temperatures will be 3.25–7.2°F warmer at the end of the this century, and the concern is that this will lead to an increase in the frequency and severity of natural disturbances such as wildfires, insect outbreaks, and disease epidemics (Dale et al. 2001; Malmsheimer et al. 2008).

For many scientists, the question is not whether a changing climate affects forests, but how and when the effects might occur. Certainly, major regional shifts in climate have occurred before, but not since European settlement of North America and not with the world's population at 7 billion. Within a global context, questions persist about whether climate change is a naturally occurring event, caused by human activity, or a combination of both. That debate is best conducted in academic and legislative circles, provided that sufficient and credible information is available to inform the policy decisions being considered about the ecological, economic, and social conditions that a changing climate could bring to modern society.

On the other hand, from the perspective of foresters and natural resource managers working in the woods, deciding whether climate change is a natural or human-caused phenomenon does not help answer the question of what to do about it. One obvious response is that we must alter the species composition of our forests to account for the likelihood that the natural ranges of species could change. But practical issues are at play that will likely limit our ability to promote widespread changes in species composition at the landscape scale, especially in the forests of the Southern United States.

UNLIKELY PROSPECTS FOR AN EASY SOLUTION

A sense of the scale of the challenge inherent in making dramatic changes in species composition, and some perspective on what that might cost, can be appreciated by reviewing the South's experience with the rise of intensive plantation management of southern pines (*Pinus* spp.). There are about 200 million acres of timberland in the Southern United States (Wear and Greis 2002). From about 1950 to 2010, the active management of forest lands in the South led to the conversion of about 35 million acres of that timberland from naturally regenerated stands—pines, pines and hardwoods, and oaks (*Quercus* spp.) and pines—into pine plantations. Moreover, projections are that by 2050, an additional 15 million acres beyond that currently converted will also be in plantations (Wear and Greis 2002). Most plantations occupy land owned or managed by forest industry or acquired from forest industry owners by timber investment management organizations or real estate investment trusts (Binkley et al. 1996; Bliss et al. 2010). Virtually all are being managed primarily for industrial wood production.

In essence, by 2050, it will have taken 100 years to convert a quarter of the South's 200 million forested acres from natural mixed stands to pure or mostly pure pine plantations, or about 500,000 acres a year. Because conversions occurred primarily on privately owned land managed by forest industry and timberland investment owners, they were the easiest from an operational perspective; all that was needed was a corporate decision, with minimal complications socially and politically.

This conversion has not been inexpensive. In classical forest economics, the cost of establishing a new forest stand is independent of the proceeds from harvesting the previous stand. So the investment in a new pine plantation encompasses site preparation, disposal of slash, treatment of competing vegetation, treatment of the forest floor as needed, reforestation through planting genetically improved pines, and release of the pines through chemical applications and fertilization. The cost of stand establishment was an estimated $250 per acre for the 35 million acres established from 1950 through 2010 (Barlow et al. 2009), resulting in a nearly $9 billion investment over the 60 years. Converting another 15 million acres from 2010 to 2050 at the current rate of $400 per acre (Barlow et al. 2009) would cost another $6 billion. In short, the bill for converting a quarter of the South's forests from naturally regenerated stands to intensively managed plantations during the period from 1950 to 2050 would be roughly $15 billion (or $150 million annually), and 500,000 acres was converted annually to stay on schedule. Forest industry and timberland investment owners were more than willing to underwrite this expense, because doing so has guaranteed a continuous supply of wood and fiber (Figure 7.1).

Given the cost, the acreage involved, and the fact that both public and nonindustrial private forest land ownership comes into play, the likelihood of a "climate change conservation program" on a significant portion of the remaining 75% of southern forests is difficult to imagine. Suppose, for example, we knew with certainty that by 2060, another 25% of the southern forested land base would be subjected to conditions not conducive for sustainability of the forest types they currently support. Harvesting those 50 million acres and replacing them with forest types that are better adapted to the expected habitat conditions could become a regional or national priority. But over the next 50 years, this would require conversion of a million acres a year (twice the historical rate of plantation conversions) at an annual cost of $400 million, given an estimated cost of $400 an acre. To date, no federal, state, local, or private entity has given any indication to support a climate change conservation program remotely approaching this scale and cost.

Moreover, converting lands not held by industrial or investment owners prompts questions about who has the authority to make such decisions. On private lands, which are held by a range of corporate and individual owners including farmers, retirees, families, trusts, estates, partnerships, businesses, clubs, and tribes, any conversion would probably need to be voluntary, and would have complicated social, political, and legal ramifications. It is easy to imagine that

FIGURE 7.1 Two loblolly pine plantations in Ashley County (Arkansas) typify widespread forest type conversions across the Southern United States. (Photo by James M. Guldin.)

landowners currently active in forest management would be more comfortable investing in forest conversion for climate change, especially if issues of forest health were involved. Others, who have not been active forest managers in the past, might not be persuaded to be in the future, especially if out-of-pocket investment is required. On public lands such as National Forests, regulations are in place that might allow timber sale proceeds to help support the costs of conversion. But the more difficult challenge for National Forest land managers would be convincing the public to support this activity; traditional user groups and the public in general might raise substantial questions about managing National Forests in such an intensive program of widespread forest conversion.

Finally, all this speculation assumes that natural resource managers and scientists know how to manage forests for species movement and assisted migration, and whether this would be effective—but we do not. And even if we did, our understanding of interactions among tree species, and between tree species and less dominant shrub and herb species needed to properly relocate a forest ecosystem, are virtually nonexistent. Moreover, our technical ability to raising sufficient nursery stock for herbaceous annual and woody perennial plants is limited in scale and scope. In essence, it is easy to advocate the importance of region-wide forest conversion, but the practical implementation of it at an ecologically meaningful scale for landowners who may be unwilling or unable to commit the necessary resources would be simply impossible in the current economic and political climate.

The likely outcome of this speculation is fairly straightforward: for the immediate future, the best efforts to manage forests in ways that promote adaptation to climate change will be incorporated into existing forest management activities, most likely on land where active management already occurs—by forest industry, by private landowners taking advantage of stewardship or tree farm programs with the help of consulting foresters, and by natural resource managers on federal, state, and local government lands.

A potentially more interesting question from the perspective of stand-level silviculture can be stated simply: *What can be done during the course of active management on public and private lands that would increase the resistance and resilience of forest stands within the context of climate change?* Research is needed on what silvicultural practices would be appropriate to apply to a given stand in an environment of climate change, and how a forester might apply them. As with most decisions that affect forest management, the kinds of silvicultural prescriptions that foresters might consider under a changing climate will vary depending on land ownership objectives, forest types, and the geographic location. And perhaps most importantly, implementation requires a consistent landowner commitment to active forest management.

A SILVICULTURAL PERSPECTIVE

The practical questions that foresters face in managing forest stands revolve around predictions of how a given stand would be affected by changing climate conditions and determinations of the specific silvicultural practices that would be useful in that context. The silvicultural prescription must essentially carry a stand from its current condition to some desired future condition. Increasingly, natural resource professionals and technicians with tree-marking paint guns in the woods are confronted by the uncertainty of changing climate conditions and the uncertainty of how local conditions might change, but they must implement the silvicultural prescription in spite of these uncertain outcomes.

However, this is not a great philosophical leap for foresters to make. Natural disturbances often derail long-term management plans, and foresters are adept at modifying silvicultural plans to fit new local conditions. Some thought must be given to the specific practices that promote stand resistance to disturbance, as well as those that enable forests to recover or to be reestablished should disturbance events cause significant damage. In this chapter a broad range of silvicultural concepts is suggested within the context of adaptive management under changing climate conditions that would enable existing practices to become more robust in a changing environment. Perhaps the same arguments being made about adaptive response to changing climatic conditions per se should be expanded to address other ecological forcing factors likely to be important in the future—increasing concentration of atmospheric carbon dioxide, increasing effects of invasive species, and ongoing effects of air pollution, for example.

Adaptive management to address climate change depends on two concepts that govern the response of forests to any disturbance, resistance, and resilience (see definitions in Chapter 3). Briefly, resistance is the ability to withstand impacts, allowing a stand to maintain its structure and developmental trajectory under the influence of changes. Conversely, resilience is the ability of a stand to recover from damages after a disturbance, quickly recovering its structure and developmental trajectory or assuming a different structure and developmental dynamic in the aftermath of change events.

These are practical and current questions. Regarding national forests, demand is growing for management activities that reflect a consideration of climate change, and resource managers are struggling to respond. Issues of climate change have implications for stand-level decisions on intensively managed industrial and investment lands as well. And what does a consulting forester propose if a client asks for management advice that will preserve the family forest in a changing climate?

The three dominant silviculture textbooks (Daniel et al. 1979; Nyland 2002; Smith et al. 1997) define silviculture as an ecological science subject to economic and social constraints. But the questions surrounding climate change suggest a modification, or at least a more explicit understanding, that silviculture is an applied science that will be increasingly influenced by changing ecological conditions in the future.

This definition offers foresters more freedom to adapt silvicultural practices from the academic perspective of changing ecological conditions. Those practicing in southern forests have faced significant ecological changes in the past: the loss of the American chestnut (*Castanea dentata*), changes in forest conditions created by the gypsy moth (*Lymantria dispar*) and southern pine beetle (*Dendroctonus frontalis*), and most recently, the challenges associated with the loss of eastern hemlock (*Tsuga canadensis*) and Carolina hemlock (*Tsuga caroliniana*) from the hemlock woolly adelgid (*Adelges tsugae*). The profession itself developed in response to concerns about the sustainability of eastern forests during and after the wave of logging of the virgin forests more than a century ago.

From this historical perspective, acknowledging or refuting whether climate change is occurring is less relevant than understanding how the practice of silviculture would be modified if a given forest stand faced uncertain ecological changes over time. In addition, it might be unwise to propose widely different silvicultural practices or prescriptions than those currently being used; after all, the models of projected changes may be incorrect, and observed changes may not be as ecologically important in some areas versus others. Also, consideration must be given to the geographic range of species being managed because silvicultural practices proposed for a species in the heart of its natural range may be different from those that would be proposed for the same species near the limits of its range.

SILVICULTURAL SYSTEMS

The three silviculture textbooks cited above generally separate silvicultural systems into three stages: (1) establishment of regeneration and associated treatments, (2) treatments applied to improve immature stands, and (3) reproduction cutting applied from stand maturity through final harvesting. Opportunities to prescribe stand-level treatments that address climate change will vary among these three stages.

Twentieth-century science provided the tools needed for most of the managed forest types in the South. At the beginning of the century, cutting the virgin forest left southern forests in understocked and unproductive condition. The professional response was to ensure recovery with a host of silvicultural systems and regeneration methods (Dana 1951). The range of intensity was broad. Approaches that built on natural stand management were highlighted by the "manage what remains" approach that evolved into even-aged shelterwood and uneven-aged selection methods for southern pines in the Atlantic and western Gulf Coastal Plain from the 1930s to 1960 (Chapman 1942; Guldin and Baker 1998; Reynolds 1959). Approaches that relied on artificial regeneration included tremendous efforts at afforestation of cutover lands epitomized by the Yazoo-Little Tallahatchie Flood Control Project in the 1950s (Williston 1988), development of direct seeding technology to restore vast areas of denuded forests in the lower western Coastal Plain (Derr and Mann 1971), and, of course, the development of technology for planting pines that led directly to the tree improvement programs and revolutionized intensive forestry across the South (Fox et al. 2007; Wakeley 1954). Managing second-growth southern pines created a boom in industry towns such as Franklin, VA; St. Joe, FL; Mobile, AL; Crossett, AR; Bogalusa, LA; and Diboll, TX.

The second half of the twentieth century also saw outstanding application of science to forestry, including development of genetically improved planting stock and nursery practices to outplant superior seedlings of loblolly pine (*Pinus taeda*) and three other major pines species and hardwoods such as eastern cottonwood (*Populus deltoides*) for commercial timber and fiber production. The associated science to ensure establishment, survival, and growth of planted southern pines may be the greatest success story of southern forestry. But significant advances also occurred in the development of natural regeneration methods (Figure 7.2)—the shelterwood method for sustainability of longleaf pine (*Pinus palustris*) on the Coastal Plain (Croker and Boyer 1975) and on upland stands of oak and hickory (*Carya* spp.) in the Southern Appalachian Mountains (Loftis 1990); the irregular shelterwood method in naturally regenerated loblolly and shortleaf (*Pinus echinata*) pine stands in the upper western Coastal Plain championed by the Crossett Division of Georgia-Pacific LLC through the 1990s (Zeide and Sharer 2000); the continued work with low-cost natural regeneration alternatives in southern forests for public and private landowners in North Carolina, Georgia, and Arkansas (Guldin 2004); and restoration of pine-woodland communities on Red Hills hunting preserves (Masters et al. 2007) and national forests west of the Mississippi (Guldin 2008), both of which helped reverse declines in red-cockaded woodpecker populations (Figure 7.3).

These successes of the twentieth century have helped the natural resources community focus on the fundamental question of forest sustainability. Although people often think that forest

FIGURE 7.2 The shelterwood reproduction cutting method applied in longleaf pine stands on the Savannah River Forest Site (South Carolina). (Photo by David Wilson.)

sustainability means avoiding active management altogether, the reverse is actually true for the simple reason that forests continually grow and renew themselves in dynamic ways and that southern forests especially grow rapidly. In addition, human populations are dynamic and increasing, which places an ever-increasing strain on the timber, recreation, water, and wildlife resources that southern forests provide.

Across the South, forest sustainability can be defined as the ability of forests to meet the needs of society in a continual way over time, despite the increasing influence of human populations on

FIGURE 7.3 Pine-woodland restoration of longleaf pine stands on the lower western Coastal Plain, Sam Houston National Forest in Walker County (Texas). (Photo by James M. Guldin.)

forest cover and the encroachment of urban areas into wildlands—and quite possibly, amid a changing climate. But for the silviculturist, forest sustainability is achieved one stand at a time, and the question boils down to this: how to ensure the successful establishment and development of the desired species in a new age class or stand after harvesting. This is important, because if the plans of the silviculturist are properly made, over time the new age class will eventually consist of the desired primary and dominant species from seedlings to saplings, through pulpwood size classes, and eventually into sawtimber size classes.

Regeneration Establishment

The best single opportunity to influence the future direction of management in a forest stand is during the regeneration phase. The mode of origin of a new age cohort (whether planted or from natural regeneration, from seeds or sprouts), the variability of species composition, and the genetic identity and variability within the species in the new age cohort are elements that the silviculturist can influence. They have implications for the future of the stand regardless of forest landowner or management scenario.

Moreover, once a new regeneration cohort is established and has grown past the sapling stage, the ability to add new species or genotypes quickly diminishes. That cohort is off on its own, to grow within the weather events and climate conditions that will nurture or challenge the trees within it, to the point of maturity and harvesting. The only opportunity to modify the development of that original cohort is by removing some of it, establishing a new age cohort to co-occupy the site, or by removing that original cohort and starting anew. Thus, if the concern is about conditions that may change over time, the best advice for the forester is to choose the best species to plant, or to encourage the development of the desired species, density, and distribution during natural regeneration.

Essentially, the silviculturist considering the establishment of a new age cohort on a given site should assess the likelihood of changing climatic conditions on that site over the life of that new age cohort, and should consider whether to modify the new age cohort by adding species absent from the site, or perhaps using a genetically improved family line known to be successful under those climatic conditions that are expected, such as a drought-tolerant family of loblolly pine in the western part of the species' natural range, for example.

Planting is the most certain tool for successful stand establishment and development because it offers the forester the most control over seedling density and supplemental treatments. It is also an effective way to modify the genetics of a species currently found on the site that might be of questionable origin (e.g., survivors of previous mismanagement), and it is the most certain way to import or re-introduce a species that is absent from the existing stand. Conversely, natural regeneration depends upon existing seed sources in the stand. It can vary if for no other reason than the sporadic nature of seed crops, which can range from nearly every year in loblolly pine stands on the western Coastal Plain (Cain and Shelton 2001) to one year in three for shortleaf pine stands in the Interior Highlands (Shelton and Wittwer 1996) to one or two years a decade in longleaf pine stands along the Gulf of Mexico (Boyer 1991). Textbooks tell foresters to correlate harvesting and site preparation with anticipated cone crops when planning for natural regeneration, but this is difficult to do in practice. An easier and more effective approach is to establish advance growth of desired seedlings before harvesting, and then release the already-established age cohort during harvesting—a tactic that is critical to the success of the even-aged shelterwood and uneven-aged selection methods.

These decisions about the kind of regeneration to use are closely tied to the choice of reproduction cutting method to use, which in turn affects whether the new stand is naturally regenerated or planted; whether its origin is seeds, sprouts, or clones; and whether species composition is narrow or broad. These choices influence subsequent management under even-aged or uneven-aged systems as well as the duration of even-aged rotations or length of cutting-cycle harvests.

FIGURE 7.4 A mixed shortleaf pine-oak regeneration cohort after precommercial thinning and prescribed burning, established after the seed cut in an irregular shelterwood regeneration of a shortleaf pine stand on the Ouachita National Forest in Scott County (Arkansas). (Photo by James M. Guldin.)

The establishment of a new regeneration cohort (Figure 7.4) is the single-most important decision that guides how a stand will adapt to changing climate conditions, should they occur. The desired species or mixtures of species must be robust in the current climate, because it is in the current climate that the new age cohort must be successfully established. But the new age cohort must also be robust within the context of the future climate, at least for the length of time that the forester expects the new stand to occupy the site.

Unfortunately, the early hurdles are also the highest. For natural regeneration, the probability of mortality is highest in the first growing season; similarly, for planted seedlings, mortality is highest in the first season after outplanting. So even if the new age cohort is planned for a changing climatic condition that will be robust 30 or 40 years into the rotation, the most immediate concern is in the first year from threats such as drought-related mortality (e.g., Lambeth et al. 1984).

Genetic Variability of Regeneration

Genetic diversity is recognized as the best safeguard against ecological uncertainties such as climate change (Ledig and Kitzmiller 1992). Silviculturists have long observed a difference between methods in which stands regenerate from seed (high forest) and methods in which they regenerate from sprouts and stumps (low forest). The reality is that both forms of regeneration are often at work in combination, especially in naturally regenerated stands. In pine plantations, there are questions about whether to plant seedlings from one or more genetically improved families, which would provide some degree of genetic variability; plantations can also be established using clonal stock (pines) or clonal cuttings—cottonwood (*Populus* spp.), for example—where all of the trees in the new age cohort are genetically identical from one clone, or planted using a small number of clones, within each of which individuals are genetically identical.

This distinction is important for silviculturists to consider within the context of a changing climate. Conceptually, whether a regeneration cohort is dominated by trees grown from seeds or from sprouts or clones determines whether it will be genetically variable or genetically uniform, which brings concepts of fitness versus flexibility from the domain of forest genetics into play.

Regenerating for genetic fitness: Fitness is the domain of asexual reproduction, or sprouting. The traditional approach in this context is where new sprouts develop after the parent tree is cut or top-killed. If a tree has been previously successful on a given site or in a given stand, a sprout from that parent will have identical genetic composition, and should be adapted to the site and optimized for success in the future. The underlying assumption, of course, is that environmental conditions in the future will be identical to the conditions in the past, an assumption that comes into question within the context of climate change.

In southern forestry, sprouts are important in oak-dominated forests; also occasionally in shortleaf pines (Figure 7.5) where sprouting only occurs in seedlings and saplings as an adaptation to fire (Mattoon 1915). In hardwoods, especially oaks, sprouting is a more complicated process. Some sprouts develop as seedling or sapling sprouts in the well-known process of advance growth (Johnson et al. 2002); these sprouts originated from acorn germination, and thus represent new genotypes not genetically identical to the parent trees. When established beneath a fully stocked mature stand, these oak seedlings grow, die back, resprout, grow larger, die back, resprout, grow still larger, and so on for a number of iterations. All the while, the rootstock continues to develop in size and vigor. When the overstory is removed by harvesting or through another disturbance, these larger advance-growth saplings have a good chance to dominate in the new age cohort of seedlings. But it is often the case that this desirable advance growth of seedling and sapling sprouts is inadequate to

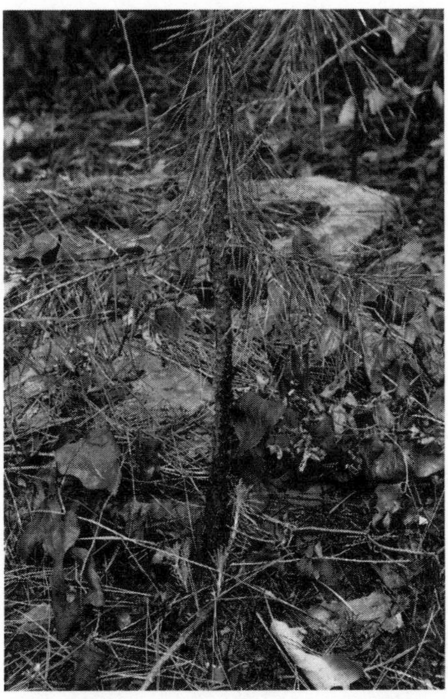

FIGURE 7.5 Shortleaf pine resprouting after being top-killed by summer prescribed fire on the Ouachita National Forest in Scott County (Arkansas). (Photo by Richard Straight.)

reforest the site. In this situation, sprouting from stumps or roots, which in fact are clones that are genetically identical to the parent tree, are used to supplement stocking from advance growth. This process has been well quantified, and predictive models are available to determine whether advance growth in a given stand is adequate or must be supplemented with stump or root sprouts (Johnson et al. 2002; Sander 1971; Sander et al. 1984). Thus, the regeneration cohort in most naturally regenerated oak-hickory stands includes a varying mixture of seed-origin and sprout-origin stock.

Sprouts also have a role in certain intensive forestry applications. Commercial production of cottonwood on short rotations along the Mississippi River uses planting stock consisting of 0.5-m sprouts cut from the roots or stumps of preferred clones. Each sprout is planted such that most is below ground, leaving only a small portion protruding above ground with buds oriented upward so they can sprout and develop into a new sapling. Clonal propagation of pines has developed in commercial plantations as well, such that a new stand can be established with one or more genetically identical clones selected for fast growth and optimum development. The primary disadvantage is the potential for reduced genetic diversity, especially if only one clone is used in the plantation.

Regenerating for genetic flexibility: Genetic flexibility is the domain of sexual reproduction, during which genetic recombination and gene flow result in outcrossed progeny that have different genetic traits from their parents. In periods of ecological uncertainty, a diverse age cohort with many individuals that collectively have a variable genetic base would be a good starting point for establishing a new stand. As weather and climate influence the stand, seedlings and saplings will express their genetic potential and develop within the environment in which they find themselves. Some individuals will develop more capably than others, based on their ability to express dominance, capture available resources, and compete successfully with their neighbors.

Over the past 70 years, genetic improvement programs have taken advantage of the inherent genetic variability in southern pines to identify families selected from the wild that have superior growth and form, evaluate controlled crosses of those selected families in progeny tests and outplanting, and identify families that far outperform average open-pollinated pines (Allen et al. 2005; Fox et al. 2007). Most of the pine plantations established in the South derive from several dozen common improved families whose geographic origin is well known by tree improvement cooperatives, but are perhaps less well known to forest managers and landowners. In seed orchards, mature trees from these families are managed for seed production using a variety of techniques, including bulk seed orchard collections where neither parent is known, open-pollinated collections where the maternal parent is known, and full-sib collections where both parents are known (McKeand et al. 2003). Most seed collected is open-pollinated, the hope being that the paternal parent is from another improved family within the seed orchard, but pollen might also come from unimproved wild trees in the area. Whether open-pollinated or full-sib origin, though, meiosis and gene recombination occur so that collected seeds have inherent genetic variability rather than being genetically identical.

A spectrum of tradeoffs: Based on this discussion, the tree breeder's dilemma is clearly one of gain versus risk (McKeand et al. 2003): gain in product volume and homogeneity versus risk of loss should the families that are outplanted suddenly encounter a disturbance event to which they are unable to adapt.

In the spectrum of genetic variability in regenerating stands, the greatest variability consists of a diversity of species that have established a new age cohort from seed origin. Intermediate genetic variability consists of a more narrow distribution of species with new propagules of seed (or seedling sprout) origin, possibly including some clonal progeny originating as sprouts from the previously harvested stand. Plantations established using orchard bulk lot, open-pollinated families, or full sib families still have a good deal of inherent genetic variability, especially if the families that are planted vary within a given ownership across stands. The least genetic variability is found in clonal plantings, especially if only one clone is used across a stand.

Within the context of climate change, the genetic variability of a given regeneration cohort in a stand under management will be an important consideration for silviculturists, and for landowners.

The decision space will include a projection of the gain-versus-risk tradeoff for the expected duration of the new age cohort. At the more variable end of this spectrum, the silvicultural approach would be to establish a wide variety of species of seed origin and then "let nature sort them out." At the less variable end, the gamble might be that the new cohort could reach maturity "before a changing environment wipes them out."

A few pine progeny tests in the South offer some guidance with respect to movement of seed sources. Schmidtling (1992, 1994) suggests that average monthly minimum temperature is a better predictor of successful seed movement in loblolly pine than average monthly temperature. Wells and Wakeley (1966) found that northward movement of seed sources results in improved growth over local sources, but movement too far north results in damage from early bud break.

In theory, a number of opportunities exist to make better decisions along the spectrum of genetic variability, but what is missing is research to support them. For example, one way that climate change is predicted to act in the western Coastal Plain is through increases in the mean minimum monthly temperatures, but scientists do not understand the performance of open-pollinated families, full-sib families, or clones with respect to this; if we did, we might be able to recommend a given family or clone for the expected change in mean minimum monthly temperature. If a stand under management is near the southern limit of its dominant species' range and just north of another species' range, the question that arises is whether to rely on native seed through natural regeneration, use planting to enrich or supplement the genetic variability of the dominant species, or diversify by planting some of the neighboring species. Research is needed to advance understanding of silvicultural tactics such as these, especially with respect to underplanting or interplanting, which is generally less successful than planting in the open (Jones 1975; Wakeley 1968).

There are also questions about how well our existing set of improved families and clones of loblolly pine will compete in conditions different from those in which their families were initially evaluated, or in an environment where interspecific competition early on is common. Schmidtling (1994) reported that native populations of loblolly pine west of the Mississippi River behaved differently from those on the eastern side because of the confounding effect of increasingly xeric conditions, which is exactly what part of the region may face according to some climate change models. Foresters do not have much experience with planting a minor component of a different species, either within or outside its natural range, that is absent from an existing regeneration cohort.

There is a practical question as well, similar to some of the challenges faced in longleaf pine restoration, and that is finding sufficient quantities of seed of proper genetic origin for restoration in a climate change context, and sufficient nursery capacity to grow the quantity of seedlings required. And finally, there are questions about quantifying a landscape model of genetic variability within the context of changing climatic conditions that would help evaluate the potential success of a new age cohort at a given location within that landscape. The answers to these and other questions—both basic and from applied science—will be important to those making silvicultural prescriptions for managed forests in the upcoming century.

Intermediate Treatments

Intermediate treatments are applied in immature stands to ensure that individual trees of the desired species will develop and to maintain the health, diversity, productivity, and sustainability of the stand. Practices include precommercial and commercial thinning (Figure 7.6), stand improvement treatments that remove undesirable individuals or species competing with the desired species, pruning to improve wood quality, and treatments such as prescribed burning that maintain desired structural conditions (Figure 7.7).

Commercial thinning is by far the most common intermediate treatment. Usually, the trees that are cut during thinning are removed so as to improve the growth and vigor of the trees that remain, and to provide some economic return to the landowner. Sometimes these priorities get reversed, so that the financial return from harvesting is of greater interest to the landowner than the silvicultural

FIGURE 7.6 The first commercial row-thinning in an overstocked, privately-owned loblolly pine plantation in Bradley County (Arkansas). (Photo by James M. Guldin.)

FIGURE 7.7 Prescribed burning on the Crossett Experimental Forest in Ashley County (Arkansas). (Photo by Virginia McDaniel.)

benefit, but that moves the treatment away from the realm of good silviculture to the less desirable practice of high-grading or selective cutting. A proper thinning has the goal of improving stem density and growth of the best trees in the stand.

A key element in the definition of intermediate treatments is the notion of stand immaturity. These treatments are not intended to secure a new cohort of natural regeneration for a mature stand, although that is the occasional, albeit accidental, outcome. Instead, the goal is to manipulate species composition and growth so that the stand will develop more rapidly and will support habitat conditions that better meet the needs of the landowner than if the treatment had not been applied. If natural regeneration happens to result from the thinning, the savvy silviculturist is one who focuses on the future development of dominant and codominant trees rather than on the happenstance new crop of regeneration.

The value of intermediate treatments in the face of disturbances was quantified in the pest management literature from the latter part of the last century, specifically with respect to research on southern pine beetle hazard and risk by Belanger (1980). He discovered that pine stands differ in their susceptibility based on site and stand conditions, and that thinning to a certain residual basal area on certain sites at a certain stand age reduces the likelihood of outbreaks. This phenomenon is probably associated with the beneficial effect of thinning on tree vigor, because vigorous trees produce higher levels of oleoresin that are thought to defeat the beetle's effort to infest the tree.

Many foresters assume that thinning and other intermediate treatments designed to maintain vigorous stands full of healthy trees are an important defense against changing climate conditions. But success will likely depend on the nature of the disturbance events associated with changing climate. For example, an increase in the extent and severity of summer drought might increase wildfire activity. In that event, maintaining thinned stands with an open understory through midstory removals and prescribed burning might be an effective tactic. This was found to be effective in limiting stand mortality, but not in halting overall wildfire spread, during the 2002 Hayman fire in Colorado (Graham 2003). Moreover, a failure to conduct a timely thinning or to control understory vegetation in an immature stand has been viewed more as a lost opportunity than a serious silvicultural mistake, but that failure to prescribe timely treatments may become more of a risk under changing climate and associated disturbance events.

Resilience depends on the kind of damage a stand suffers during a disturbance event, and the degree to which a manageable residual stand will survive. In the upper western Coastal Plain, research on rehabilitation of understocked stands (Baker and Shelton 1998a,b,c,d) shows that naturally regenerated loblolly-shortleaf stands with as low as 30% stocking become fully stocked more quickly through management than by removing the residual material and establishing a new stand. Decision support tactics such as this might become increasingly important to help in silvicultural triage after a major disturbance event such as the landfall of a hurricane. For stands that cannot recover from disturbance, decisions will have to be made about removing the remaining material from the site, and essentially treating the stand as newly harvested and subject to the establishment of a new regeneration cohort, if the landowner can afford to do so.

In practical terms, the choice between resistance and resilience versus rehabilitation depends on whether the silvicultural system that was in place before the disturbance can proceed, or whether revisions will be needed. Changes in the prevailing silvicultural system will not be required if the stand can resist the disturbance. Conversely, if the stand is dramatically altered but still able to recover, a revision in long-term silvicultural planning will most likely be required.

Reproduction Cutting Methods

The choice and timing of reproduction cutting is typically under the direction and control of forest landowners and the foresters who advise them. The decision might be based on a long-term management plan that has been written and approved by the appropriate parties, it might be made in response to disturbance events, or it might be at the sole discretion of a landowner, perhaps one who needs cash from a timber sale or one who wants to manage newly acquired forests differently. Within these boundaries, decisions are made about even-aged versus uneven-aged methods, the specific methods that will be used, and the manner in which harvesting will be conducted. These decisions influence the new stand that will develop after the cutting.

For example, the ultimate even-aged stand is a modern plantation that grows genetically improved trees, all often planted within the same week. These stands are established after clearcutting and site preparation; after planting they are often fertilized and treated with herbicides to control woody or herbaceous competition. In southern pines, this capital-intensive silvicultural system produces fast-growing stands with an expected rotation age of 25 years at most.

Foresters managing these stands have a choice about the genetic adaptation of different planting-stock families to the site conditions, and they may have a preference for the growth rates of one family or clone over another. In a scenario where climate change might affect stand establishment and development, foresters might also need to consider the survival of a given family or clone to increasing drought or temperature in the critical first year of outplanting, and also whether the trees will survive, and thrive, in the conditions expected over the 25-year life of the stand (Lambeth et al. 1984).

The choice of gain versus risk in this context is one related to genetic diversity. Greater plantation diversity, but less of a gain in volume growth, will be found with stock from bulk orchard seed or from open pollinated families; slightly less diversity and slightly more gain will be in full-sib planting stock. Clonal stock has no inherent diversity unless mixtures of clones are selected, which again would confer some additional diversity at the expense of the maximum gain in volume growth that single clone plantations provide. When the stand is harvested 25 years later, the forester in charge will have similar decisions to make using the available technologies of the future—both to forecast the climate conditions of the next rotation and to select the best planting stock, which will probably be improved beyond that which is currently available, for the expected climate at that time.

Private nonindustrial landowners are often unable or unwilling to practice intensive forest management, given the high cash outlay that is required early in the life of a stand. For such a landowner, low-cost alternatives include even-aged naturally regenerated stands using the shelterwood or seed-tree method; managing such stands to a 40- to 50-year rotation might better suit their financial condition and management objectives (Guldin 2004; Zeide and Sharer 2000). The disadvantage to long rotations is the length of time until a new regeneration cohort is obtained, during which changing conditions might be better suited to different species or different genotypes than the parents can provide.

The most extreme example of long-term, even-aged rotations being actively used in southern pine stands is on federal lands, where foresters manage for the endangered red-cockaded woodpecker (Figure 7.8) using a regional recovery plan that calls for 80- to 120-year rotations and the irregular shelterwood method of regeneration (Guldin 2004). These foresters can only hope that new age cohorts established during the shelterwood seed cut are sufficiently robust to adapt and survive through maturity in whatever climate conditions prevail during the 80- to 120-year duration of the rotation.

The uneven-aged selection method was used to rehabilitate cutover understocked southern pine stands into fully stocked sawtimber stands from the mid-1930s through the 1960s, as demonstrated by research at the Crossett Experimental Forest in Arkansas (Guldin and Baker 1998; Reynolds 1959). The Crossett stands were managed using annual cutting cycle harvests from 1937 to 1969, and periodic cutting cycle harvests since (Baker et al. 1996). Natural regeneration is the rule with the selection method (Figure 7.9), with cutting cycles varying from 5 to 7 years for loblolly and slash (*Pinus elliottii*) pines in the western Coastal Plain (Baker et al. 1996), to 10 years for longleaf pines (Farrar 1996), and to 20 years for upland oaks (Loewenstein et al. 2000).

Thus, choosing a reproduction cutting method is the first opportunity for a forester concerned about climate change to influence the frequency of origin of new age cohorts as well as their species

FIGURE 7.8 A shortleaf pine-dominated stand after thinning and cyclical prescribed burning in a shortleaf pine-bluestem management area of the Ouachita National Forest in Polk County (Arkansas). (Photo by James M. Guldin.)

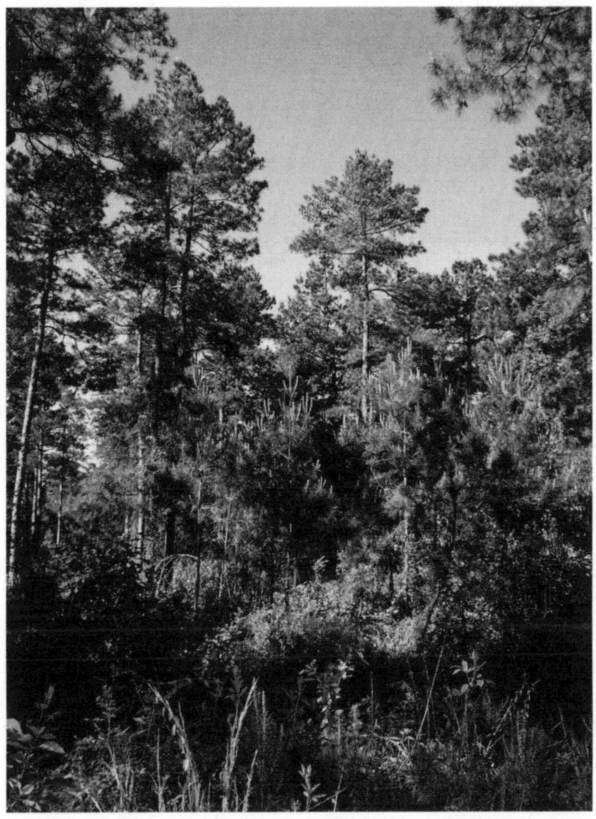

FIGURE 7.9 Uneven-aged stand structure on the Poor Farm Forestry Forty demonstration at the Crossett Experimental Forest in Ashley County (Arkansas). (Photo by James M. Guldin.)

composition. This may be the single most important way to affect how a stand will respond to changing climate conditions, should they occur. The selected species or mixtures of species must be robust in the current climate, because it is in the current climate that the new age cohort must become successfully established. But the new age cohort must also be robust within the context of what the future climate is likely to be, at least during the period that the forester expects the new stand to occupy the site.

Finally, there may be a silvicultural case to justify heterogeneity at the landscape scale to a greater degree than we have done in the past, especially for landowners holding large blocks of forest land such as National Forests or timber investment organizations. Puettmann et al. (2009) make a compelling case to manage for heterogeneity and complexity to increase resiliency at the stand and landscape scale. The value of large blocks of land under one ownership being treated with similar silvicultural systems, such as pine plantations, even if established to ensure stand-level genetic diversity, may be less resilient than a landscape or ownership where heterogeneity of age, structure, and origin of regeneration among stands are promoted.

DISCUSSION

Barring huge shifts in public attitudes about fiscal and economic priorities, the establishment of a regional climate change adaptation program for the 200 million acres of southern forest land is unlikely any time soon. Forest ownership patterns, responsibility for management decisions on public and private lands, the capital required to convert existing forests to new species, and the source of that capital are major complications. No segment within society, government, or the natural-resources profession has the wherewithal to deliver the significant changes in forest types that would be needed for even a small portion of the southern landscape in the face of a changing climate.

Instead, progress in achieving resistance and resilience to climate change will probably become an outgrowth of the ongoing activities currently being practiced on forest industry, timberland investment, and real-estate investment ownerships, on nonindustrial private forest ownerships already in active management, and on government lands such as national and state forests. The foresters who work on these landscapes will be on the front lines in modifying silvicultural prescriptions to carry stands from their existing condition to a desired future condition that includes adaptation to climate change.

Thus, managing forest stands with attention to the possibility of climate change will not be much different in principle from the work that foresters have done over the past century, only with different treatment prescriptions based on stand management objectives that include the prospect of a changing climate. Essentially the process has three steps. The first is to identify the current condition of the stand. The second is to identify and quantify the desired future condition of the stand that not only meets the ownership objectives of the landowner, but also is ecologically and silviculturally robust in the context of changing climate in the locality. The third is to develop the detailed silvicultural prescription that carries the stand from its existing condition to its desired future condition. The first two steps are no different from standard silvicultural practice over the past century, but the third step will require innovative prescriptions based on the concepts discussed in this chapter.

Ownership objectives vary, as does the detail with which the forest landowner can express those objectives. On public lands such as national forests, the stand will have been assigned to a given management area under the prevailing land and resource management plan, which spells out management standards and guides, sometimes in considerable detail. On private lands managed for industry or investment purposes, the overarching goals of ownership will also be described in detail, and the silvicultural systems available for implementation may be carefully prescribed as well based on company philosophy. In contrast, nonindustrial private forest landowners may have in mind only rudimentary objectives and outcomes that are only vaguely quantified. Regardless of scale and ownership, what is needed are improved methods of identifying how the stand being studied currently meets the objectives, or can be managed to meet them in the future.

Descriptions of a stand's current ecological and silvicultural conditions are independent of the landowner's objectives, although they will certainly have been influenced by the vigor with which the stand has been managed in the past. Key variables include stand structure, species composition, stem density, basal area, volume, and other pertinent data that would be collected during a typical stand exam or timber cruise. The history of the stand should be reconstructed, with special attention to whether the dominant and codominant trees in the stand were established by natural or artificial regeneration, whether the stand is even-aged or uneven-aged, and what overarching silvicultural system (or lack thereof) has been employed. In some instances, this information will be readily available, especially on government and private industrial or investment lands; on nonindustrial private forest lands, these data may require some silvicultural detective work.

All of this information provides the groundwork for the important final step—developing a robust silvicultural prescription that meets the landowner's needs within the context of a changing climate, as well as one that can take the existing stand to its desired future condition. The first prescription to be developed will manage the existing stand through the point of reproduction cutting and will use appropriate silvicultural practices—such as the intermediate treatments outlined above—to improve the resistance of the new stand to the climate changes that will be expected for the duration of its rotation or cutting cycle. In other words, given that there is not much ability to influence the species composition of a fully stocked immature or mature stand, the preferred treatments will be those that optimize its ability to resist the disturbance events expected through maturity. These treatments are largely consistent with standard good forestry practices: maintaining individual tree vigor through timely thinning, using prescribed burning in ecosystems adapted to fire, and perhaps considering whether the stand should be harvested prematurely.

Subsequent prescriptions to establish an entirely new age cohort in the stand will rely upon reproduction cutting methods that enable the establishment of seedlings, saplings, and sprouts, whether planted, of natural origin, or both; this new age cohort will have the desired diversity of genetics, parental families, and species, thereby developing a species composition that will allow the stand to resist, or recover from, the climate changes expected as that age cohort matures over the rotation from sapling stage through maturity. Key decisions will be whether to rely on natural or artificial regeneration, and what source to use for seeds, sprouts, or clones. The mixture of species in natural regeneration, or of families used as planting stock in plantations, is critical. The reproduction cutting method used to regenerate the next stand will dictate the number of age cohorts, which in turn affects how frequently new cohorts of regeneration can be established. Finally, the expected age to maturity of dominant trees in the new age cohort will need to be decided, whether short (<25 years), moderate (25–50 years), long (50–100 years), or very long (>100 years). All of these decisions must be integrated into a silvicultural system that is also geared to meet the needs of the landowner.

Thus, the new or modified silvicultural treatments prescribed within the context of climate change will probably share a number of common attributes. Attention to the genetic diversity of new age cohorts will be increasingly important. For planted stands, this implies a diversity of genotypes in seedlings produced from existing orchards, and a diversity of clones planted in clonal forestry plantations. There might be reason to develop a genetic "line of custody" that allows landowners or the foresters who advise them to request known seed sources, or at least to know and quantify the diversity of the plantations that are being established, and also to encourage tree improvement cooperatives to develop new ways of labeling and handling seeds and seedlings. For naturally regenerated stands, enhanced genetic diversity would mean greater reliance on seed origin stock rather than sprouts, greater species diversity in new stands, and possibly enrichment planting at levels of 50–200 stems per acre to supplement natural regeneration with species that are currently absent from the site.

Another key element may be the frequency of age cohorts, where quicker turnover of dominant and codominant species can be a hedge against changing climate conditions. If changes are rapid, establishing new age cohorts more frequently may allow for quicker natural selection of individuals and species that are adapted to the changing site conditions. The result might be shorter even-aged rotations, especially on plantations, and broader use of uneven-aged silviculture or irregular stand structures—three or more age classes with 5- to 20-year cutting cycles—on public and nonindustrial private lands. This latter approach would have the added benefit of representing low-cost management with minimal out-of-pocket investment (Baker et al. 1996) that some landowners would find to be an appealing alternative to intensive plantation management.

Using prescribed burning or mechanical/manual "fire surrogate" treatments such as mulching will likely become more important for maintaining the resistance of stands in the face of increasing wildfire hazard under changing climatic conditions. Unfortunately, increasing the use of prescribed burning may be increasingly constrained by issues such as public resistance to smoke (see Chapter 5), and a declining and aging workforce on state and federal lands. But in an environment where surface fires may become increasingly frequent, maintaining stands in a condition that resists damage from wildfire, such as by applying restoration treatments that promote and maintain open understory conditions especially in long-rotation even-aged stands on federal lands, will become more important over time.

A final key descriptor will probably relate to forest health. The kinds of silvicultural practices that have effectively maintained forest health in the past, such as frequent thinning and management at low basal areas to maintain individual tree vigor, are probably a good place to begin preparing the forest stands of today for the changes expected over the next several decades.

SUMMARY

Climate change offers challenges and opportunities to the silviculturist, in the short- and long term. The short-term challenges are likely to be related to disturbance events that interfere with existing management plans, and that will require a response that ameliorates or rehabilitates affected stands through practices that promote continued resistance, resilience, or recovery from the disturbance depending upon the degree to which the current management program was disrupted. Challenges in the long term will focus on an enhanced understanding of the genetic diversity and species composition of the new regeneration cohorts that are established to replace the existing stand and that will be robust in future climatic conditions that are projected to occur, and the degree to which species not currently on the site but forecast to be suited to it can be successfully established. The likelihood of seeing a stand develop as planned from sapling stage to maturity will probably vary inversely with rotation age or the interval between new age cohorts. All in all, rather than developing overarching management plans for forested properties that are not likely to change over time, the role of silviculturists in the future will more closely resemble incident response, bringing adaptive management to bear in the face of changing ecological conditions over time.

REFERENCES

Allen, H.L., Fox, T.R., and Campbell, R.G. 2005. What is ahead for intensive pine plantation silviculture in the South? *Southern Journal of Applied Forestry*. 29, 62–69.

Baker, J.B. and Shelton, M.G. 1998a. Rehabilitation of understocked loblolly-shortleaf pine stands-I. Recently cutover natural stands. *Southern Journal of Applied Forestry*. 22, 35–40.

Baker, J.B. and Shelton, M.G. 1998b. Rehabilitation of understocked loblolly-shortleaf pine stands-III. Development of intermediate and suppressed trees following release in natural stands. *Southern Journal of Applied Forestry*. 22, 41–46.

Baker, J.B. and Shelton, M.G. 1998c. Rehabilitation of understocked loblolly-shortleaf pine stands-III. Natural stands cutover 15 years previously but unmanaged. *Southern Journal of Applied Forestry*. 22, 47–52.

Baker, J.B. and Shelton, M.G. 1998d. Rehabilitation of understocked loblolly-shortleaf pine stands-IV. Natural and planted seedling/sapling stands. *Southern Journal of Applied Forestry*. 22, 53–59.

Baker, J.B., Cain, M.D., Guldin, J.M., Murphy, P.A., and Shelton, M.G. 1996. Uneven-aged silviculture for the loblolly and shortleaf pine forest cover types. General Technical Report SO-118. Asheville, NC: U.S. Department of Agriculture, Forest Service, Southern Research Station. 65 p.

Barlow, R.J., Smidt, M.F., Morse, J.Z., and Dubois, M.R. 2009. Cost and cost trends for forestry practices in the South. *Forest Landowner*. 68(5), 5–12.

Belanger, R.P. 1980. Silvicultural guidelines for reducing losses to the southern pine beetle. In: Thatcher, R.C., Searcy, J.L., Coster, J.E., and Hertel, G.D. (eds.). *The Southern Pine Beetle*. Technical Bulletin 1631. Washington, DC: U.S. Department of Agriculture, Forest Service.

Binkley, C.S., Raper, C.F., and Washburn, C.L. 1996. Institutional ownership of US timberland—History, rationale, and implications for forest management. *Journal of Forestry*. 94, 21–28.

Bliss, J.C., Kelly, E.C., Abrams, J., Bailey, C., and Dyer, J. 2010. Disintegration of the U.S. industrial forest estate: Dynamics, trajectories, and questions. *Small-Scale Forestry*. 9, 53–60.

Boyer, W.D. 1991. Anticipating good longleaf pine cone crops—The key to successful natural regeneration. *Alabama's Treasured Forests*. Fall 1996: 24–26.

Cain, M.D. and Shelton, M.G. 2001. Twenty years of natural loblolly and shortleaf pine seed production on the Crossett Experimental Forest in southeastern Arkansas. *Southern Journal of Applied Forestry*. 25, 40–45.

Chapman, H.H. 1942. *Management of Loblolly Pine in the Pine-Hardwood Region in Arkansas and in Louisiana west of the Mississippi River*. Bulletin #49. New Haven, CT: Yale University, School of Forestry. 150 p.

Croker, T.C. Jr. and Boyer, W.D. 1975. Regenerating longleaf pine naturally. Research Paper SO-105. New Orleans, LA: U.S. Department of Agriculture, Forest Service, Southern Forest Experiment Station. 21 p.

Dale, V.H., Joyce, L.A., McNulty, S., Neilson, R.P., Ayres, M.P., Flannigan, M.D., Hanson, P.J. et al. 2001. Climate change and forest disturbances. *Bioscience*. 51, 723–734.

Dana, S.T. 1951. The growth of forestry in the past half century. *Journal of Forestry*. 49, 86–92.

Daniel, T.W., Helms, J.A., and Baker, F.S. 1979. *Principles of Silviculture*, 2nd ed. New York: McGraw-Hill 521 pp.

Derr, H.J., Mann, W.F. Jr. 1971. *Direct Seeding Pines in the South*. Agriculture Handbook 391. Washington, DC: U.S. Department of Agriculture.

Farrar, R.M. 1996. *Fundamentals of Uneven-Aged Management in Southern Pine*. Misc. Pub. 9, Tallahassee, FL: Tall Timbers Research Station. 52 p.

Fox, T.R., Jokela, E.J., and Allen, H.L. 2007. The development of pine plantation silviculture in the southern United States. *Journal of Forestry*. 105, 337–347.

Graham, R.T., tech. (ed.). 2003. *Hayman Fire case study: Summary*. Gen. Tech. Rep. RMRS-GTR-115. Ogden, UT: U.S. Department of Agriculture, Forest Service, Rocky Mountain Research Station. 32 p.

Guldin, J.M. 2004. Reproduction cutting methods for naturally-regenerated southern pine stands in the South. In: H.M. Rauscher and K. Johnsen (eds.), *Southern Forest Science: Past, Present, and Future*. General Technical Report SRS-75. USDA Forest Service, Asheville, NC, pp. 83–95.

Guldin, J.M. 2008. The silviculture of restoration: A historical perspective with contemporary application, p. 23–35. In: Deal, R.L., tech. (ed.), *Integrated Restoration of Forested Ecosystems to Achieve Multiresource Benefits: Proceedings of the 2007 National Silviculture Workshop*. Gen. Tech. Rep. PNW-GTR-733. Portland, OR: U.S. Department of Agriculture, Forest Service, Pacific Northwest Research Station. 306 p.

Guldin, J.M. and Baker, J.B. 1998. Uneven-aged silviculture, southern style. *Journal of Forestry*. 96(7), 22–26.

Guldin, J.M., Strom, J., Montague, W., and Hedrick, L.D. 2004. Shortleaf pine-bluestem habitat restoration in the Interior Highlands—Implications for stand growth and regeneration, pp. 182–190. In: Shepperd, W.D. and Eskew, L.D. (compilers). *Silviculture in special places: proceedings of the 2003 National Silviculture Workshop* Proceedings, RMRS-P-34. Fort Collins, CO: U.S. Department of Agriculture, Forest Service, Rocky Mountain Research Station. 255 p.

Johnson, P.S., Shifley, S.R., and Rogers, R. 2002. *The Ecology and Silviculture of Oaks*. New York, NY: CABI Publishing. 503 p.

Jones, E.P. Jr. 1975. Interplanting is futile in slash pine plantations. *Tree Planters' Notes* 25(1), 19–22.

Lambeth, C.C., Dougherty, P.M., Gladstone, W.T., McCullough, R.B., and Wells, O.O. 1984. Large-scale planting of North Carolina loblolly pine in Arkansas and Oklahoma: A case of gain versus risk. *Journal of Forestry* 82(12), 736–741.

Ledig, F.T. and Kitzmiller, J.H. 1992. Genetic strategies for reforestation in the face of global climate change. *Forest Ecology and Management* 50, 153–169.

Loewenstein, E.F., Johnson, P.S., and Garrett, H.E. 2000. Age and diameter structure of a managed uneven-aged oak forest. *Canadian Journal of Forest Research*. 30, 1060–1070.

Loftis, D.L. 1990. A shelterwood method for regenerating red oak in the southern Appalachians. *Forest Science*. 36(4), 917–929.

Malmsheimer, R.W., Heffernan, P., Brink, S., Crandall, D., Deneke, F., Galik, C., Gee, E. et al. 2008. Forest management solutions for mitigating climate change in the United States. *Journal of Forestry*. 106(3), 115–173.

Masters, R.E., Robertson, K., Palmer, B., Cox, J., McGorty, K., Green, L., and Ambrose, C. 2007. Red Hills forest stewardship guide. Misc. Pub. 12. Tallahassee, FL: Tall Timbers Research Station. 79 p.

Mattoon, W.B. 1915. *Life History of Shortleaf Pine.* Bulletin 244. Washington, DC: U.S. Department of Agriculture. 46 p.

McKeand, S., Mullin, T., Byram, T., and White, T. 2003. Deployment of genetically improved loblolly and slash pines in the South. *Journal of Forestry* 101(3), 32–37.

Nyland, R.D. 2002. *Silviculture: Concepts and Applications*, 2nd edition. The McGraw-Hill Companies, New York. 704 pp.

Puettmann, K.J., Coates, K.D., and Messier, C.C. 2009. *A Critique of Silviculture: Managing for Complexity.* Island Press, Washington, D.C. 188 p.

Reynolds, R.R. 1959. *Eighteen years of selection timber management on the Crossett Experimental Forest.* Tech. Bull. 1206. Washington, D.C.: U.S. Department of Agriculture, Forest Service. 68 p.

Sander, I.L. 1971. Height growth of new oak sprouts depends on size of advance reproduction. *Journal of Forestry* 69(11), 809–811.

Sander, I.L., Johnson, P.S., and Rogers, R. 1984. Evaluating oak advance reproduction in the Missouri Ozarks. Research Paper NC-251. St. Paul, MN: U.S. Dept. of Agriculture, Forest Service, North Central Forest Experiment Station. 16 p.

Schmidtling, R.C. 1992. A minimum-temperature model for racial variation in loblolly pine provenance tests. In: *Proceedings, 12th North American Forest Biology Workshop*, Sault Ste. Marie, Ont., Canada. 124 p.

Schmidtling, R.C. 1994. Use of provenance tests to predict response to climatic change: Loblolly pine and Norway spruce. *Tree Physiology*. 14, 805–817.

Shelton, M.G. and Wittwer, R.F. 1996. Shortleaf pine seed production in natural stands in the Ouachita and Ozark Mountains. *Southern Journal of Applied Forestry*. 20(2), 74–80.

Smith, D.M., Larson, B.C., Kelty, M.J., Ashton, P.M.S. 1997. *The Practice of Silviculture, Applied Forest Ecology*, 9th edition. John Wiley & Sons, New York. 560 pp.

Solomon, S., Qin, D., Manning, M., Chen, Z., Maquis, M., Averyt, K.B., Tignor, T.M., and Miller, H.L. (eds.), 2007. Summary for policymakers. Climate change 2007: The physical science basis. *Contribution of Working Group I to the Fourth Assessment Report of the Intergovernmental Panel on Climate Change.* Cambridge University Press, Cambridge, UK.

Wakeley, P.C. 1954. *Planting the Southern Pines.* Agriculture Monograph 18. Washington, DC: U.S. Department of Agriculture. 233 p.

Wakeley, P.C. 1968. Replacement planting of southern pines unsuccessful. USDA Forest Service, Research Note SO-85. New Orleans, LA: U.S. Department of Agriculture, Forest Service, Southern Forest Experiment Station. 4 p.

Wear, D.N. and Greis, J.G. 2002. *The Southern Forest Resource Assessment: Summary Report.* Gen. Tech. Rep. SRS-54. Asheville, NC: U.S. Department of Agriculture, Forest Service, Southern Research Station. 103 p.

Wells, O.O. and Wakeley, P.C. 1966. Geographic variation in survival, growth, and fusiform rust infection of planted loblolly pine. *Forest Science Monograph.* 11, 40 p.

Williston, H.L. 1988. The Yazoo-Little Tallahatchie Flood Prevention Project: A history of the Forest Service's role. Forestry report R8-FR. Atlanta, GA: U.S. Department of Agriculture, Forest Service, Southern Region. 63 p.

Zeide, B. and Sharer, D. 2000. Good forestry at a glance: A guide for managing even-aged loblolly pine stands. Arkansas Forest Resources Center Series 003. Division of Agriculture, Arkansas Agricultural Experiment Station, University of Arkansas, Fayetteville, AR, 19 pp.

8 Productivity and Carbon Sequestration of Forests in the Southern United States

Kurt H. Johnsen, Tara L. Keyser, John R. Butnor, Carlos A. Gonzalez-Benecke, Donald J. Kaczmarek, Chris A. Maier, Heather R. McCarthy, and Ge Sun

CONTENTS

Sixty percent of the Southern United States landscape is forested (Wear 2002). Forest types vary greatly among the five subregions of the South, which include the Coastal Plain, Piedmont, Appalachian-Cumberland, Mid-South, and the Mississippi Alluvial Valley. Current inventory data show upland hardwood forests being the predominant forest type in the South (>30 million ha) followed by planted pine (>15 million ha), natural pine and bottomland hardwoods (~13 million ha), and oak-pine (>3 million ha) forest types (Huggett et al. in press). These forest ecosystems provide a multitude of ecosystem goods and services including clean water and air, wildlife habitat, recreation and aesthetics, timber and fiber production, and CO_2 sequestration. Southern forests play an important role in meeting the current and future timber and fiber needs across the United States, as harvesting has substantially decreased, in other regions of the country. As a whole, the South's forest sector produces approximately 60% of the total U.S. wood production, more wood than any other single nation (Prestemon and Abt 2002).

The future of forests and forest management in the Southern United States face many uncertainties. Land-use change, population growth, urbanization, changing public values, unstable timber markets, and a changing climate are some of the factors that will influence forests and forest management in the South (Wear and Greis in press). Although forest management is unable to address the anticipated changes in forestland availability due to shifts in demography (e.g., increasing urbanization, land-use change, etc.), it can be used to increase the adaptability of forests to changing climate conditions. Proactive forest management that focuses on increasing resistance (i.e., reducing a stand's susceptibility to disturbance) and resilience (i.e., reducing the negative effects of disturbance and/or reducing recovery time following disturbance) may ameliorate some of the anticipated changes in forest structure, function, and productivity anticipated under a changing climate. Climate change is likely to affect forest structure and function through a variety of pathways, such as changing disturbance patterns, shifting species composition, and altering growth and productivity. Because the South produces the vast majority of wood-related raw materials, long-term trends in forest productivity are especially important in assessing whether southern forests can continue to meet the nation's timber and fiber demands in the context of a changing climate.

For large-scale planning purposes, models of forest growth and yield and/or forest stand dynamics have been traditionally used to project the growth and productivity of forest stands into the future. These models are a valuable tool in our conceptual understanding of how forests grow and of the products they can produce. However, existing growth and yield models or ecological process models essentially operate in a twentieth-century environment, because the long-term studies on which these models were based have largely been conducted, and algorithms developed from them, prior to the turn of the twenty-first century. Empirical forest growth models have been developed using common static measures of site productivity (i.e., site index [SI]) without direct consideration of fluctuating climatic variables. Consequently, parameter estimates in these models will not automatically adjust in response to the projected changes in climate, leaving the expected increases or decreases in forest productivity unaccounted for in model predictions. Current growth models, including empirical, process, and/or hybrid models (see Medlyn et al. 2011 for a review of models), should, therefore, be examined to determine how the interactions between climatic and nonclimatic factors limit stand-level productivity.

Since the 1950s forest productivity, as defined as the change in growing stock volume over time, in the South has steadily increased (Smith et al. 2007). During this period, net volume per hectare increased 95% in the Southern United States (Smith et al. 2007). In 2010, productivity of southern forests was estimated at 8.3 billion cubic meters, with softwood and hardwood growing stock approximating 3.4 and 4.8 billion cubic meters, respectively (Huggett et al. in press). Past increases in forest productivity can be attributed to a variety of factors, including a reduction in harvesting on National Forest System lands, afforestation activities, intensive forest management, an increase in the area converted from natural forest conditions (e.g., upland hardwood and/or natural pine forest types) to planted pine, and the age class distribution of forest stands across the region (Hicke et al. 2002).

The increase in forest productivity observed over the last four decades in the South has coincided with an identifiable change in climate. Since the 1960s, the average annual temperature across the Southern United States has increased (McNulty et al. in press). Although no significant trend in average annual precipitation has accompanied the increase in temperatures (McNulty et al. in press), patterns of precipitation have been altered, with areas experiencing moderate to severe spring and summer drought increasing across the South (Karl et al. 2009). Under future emission scenarios outlined in the Southern Forest Future's Project (SFFP), average annual temperatures in the Southern United States are expected to increase by an additional 2.5–3.5°C by 2060 (Wear and Greis 2012). Models predicting precipitation patterns are less consistent and show a high degree of spatial variability across the South's 13 states. Of the four climate scenarios reported on utilized in the SFFP, the MIROC3.2 + A1B scenario forecasts the most drastic decrease in precipitation, with up to a 24% reduction, relative to 2010 levels, in average annual precipitation forecasted by 2040 (McNulty et al. in press). The remaining climate change scenarios (CSIROMK3.5 + A1B, CISROMK2 + B2, HadCM3 + B2) predict that by 2040, average annual precipitation across the South may decrease (but to a lesser degree than the MIROC3.2 + A1B scenario), remain similar, or even increase relative to 2010 levels, depending on subregion (McNulty et al. in press). In addition to forecasted changes in total annual precipitation, the seasonality of precipitation may be altered, depending on climate change scenario and subregion.

Climate variables, including temperature and precipitation, exert a strong influence over site productivity. The expected trend of increased temperatures and decreased growing season precipitation under various climate change scenarios has implications for the future productivity of southern forestlands. Although factors controlling forest productivity in terms of net primary productivity (NPP) are numerous and include biotic factors such as leaf area, the efficiency of the foliage to absorb solar radiation [i.e., light use efficiency (ε)], soil nutrient availability, and species composition, abiotic factors that will be altered under a changing climate, including the amount of solar radiation, temperature, and available water (Churkina and Running 1998; Hicke et al. 2002; Running et al. 2004), have the greatest control over productivity. Global analyses indicate that NPP in the common forest types of the South is limited by available water and temperature (Churkina and Running 1998; Running et al. 2004). Solar radiation only represents a limiting factor in the tropics due to cloud cover (Running et al. 2004).

Therefore, climate change clearly has the potential to impact forest productivity and thus carbon (C) sequestration. Land occupied by southern forests represents 30% of the total forestland in the United States (Han et al. 2007). Owing to their extent and high productivity, southern forests have also been estimated to account for 36% of the C sequestered in the conterminous United States (Turner et al. 1995). Han et al. (2007) estimated forests in the South sequester 13% of regional greenhouse emissions.

Carbon can be sequestered via southern forests by two main routes: *in situ* and *ex situ* C sequestration (Figure 8.1, Marland and Marland 1992; Johnsen et al. 2001a, Gonzalez-Benecke et al. 2010). Carbon sequestered *in situ* is C that is tied up in aboveground and belowground biomass and necromass. Carbon is sequestered *ex situ* via incorporation into wood products that store C away from the atmosphere for different durations depending on the forest product type. As described below, gross primary productivity (GPP) is a critical component of *in situ* ecosystem C sequestration. GPP is the component of ecosystem C sequestration most amenable to influence by forest managers even in the event of climate change. Thus, forest managers can directly influence the rate of C sequestration (Johnsen et al. 2001a; Ryan et al. 2010; McKinley et al. 2011).

In this chapter, we review how forest productivity and C sequestration are related, discuss the impacts of selected press and pulse effects (Chapter 2 and briefly described below) on productivity

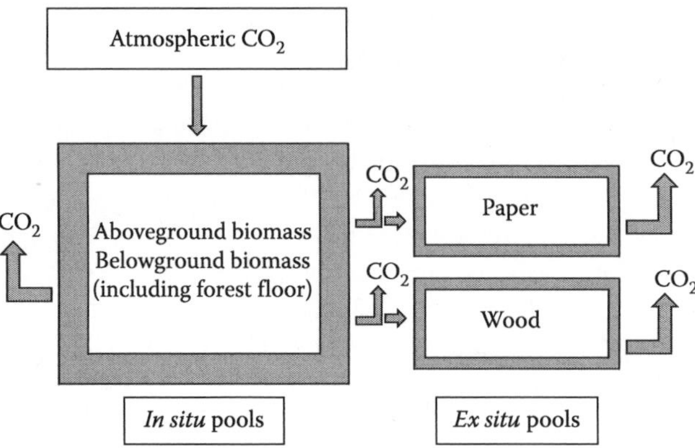

FIGURE 8.1 Conceptual model of carbon sequestration via southern forests. (Adapted from Johnsen, K.H. et al. 2001a. *Journal of Forestry* 99:14–21.)

and/or C sequestration, model the potential effects of climate change on forest productivity in the Southern United States, examine several case studies that highlight the potential impact of climate change and/or the impact of forest management on forest productivity and/or C sequestration, and, finally, consider general management options, including the potential of bioenergy production as they impact productivity and C sequestration options. We concentrate on biological climate change effects on forest productivity and C sequestration, excluding important impacts of land-use change, so that we can address strategies that land-use managers can use to ameliorate the impacts of climate change and/or maintain or increase C sequestration at the stand level. Impacts associated with land-use change including issues of leakage, permanence and disturbances are reviewed by McKinley et al. (2011).

IN SITU CARBON SEQUESTRATION

The Forest Carbon Cycle: The Relationship between Net Primary Production, Gross Primary Production, Net Ecosystem Production, and Ecosystem Respiration

Southern forests contain about 30% of the nation's C stock (Mickler et al. 2004) and play a prominent role in the regional and global C cycle (Turner et al. 1995). Forests exchange large amounts of C, as CO_2, with the atmosphere and store C in living plant biomass, detritus, and soil (Figure 8.2). Ecosystem C accumulation is essentially the balance between C gain and loss through photosynthesis and respiration. Carbon gain or GPP is the sum of individual leaf photosynthesis and represents the total C input into the ecosystem (Figure 8.2). GPP is integrated over space and time—typically one year—and is often expressed in terms of C (e.g., Mg C ha^{-1} $year^{-1}$). Annual Forest GPP is a function of leaf area index (LAI, leaf surface area per unit ground area) and the duration of display (Chapin et al. 2002). The biophysical mechanisms (e.g., light, temperature, and moisture) regulating annual GPP are well understood and can be modeled with reasonable accuracy (e.g., Landsberg and Waring 1997). Conversely, the mechanisms controlling C allocation to the growth and maintenance of different stand components are less well understood. About half of annual GPP is lost through autotrophic respiration (R_A), which is the total release of C, as CO_2, from all living primary producers, and represents the energy required for tissue growth, nutrient uptake and transport, and tissue maintenance. The balance between GPP and R_A is NPP ($NPP = GPP - R_A$); the net flux of C from the atmosphere into organic matter (i.e., foliage, branches, stems, reproductive organs, and roots). Forest NPP is a fundamental ecological variable because it measures the amount of energy input that drives ecosystem metabolism (Chapin et al. 2002). NPP is usually measured as the increment of new biomass or C equivalent (e.g., Mg C ha^{-1} $year^{-1}$), but also includes C loss in root exudates, herbivory,

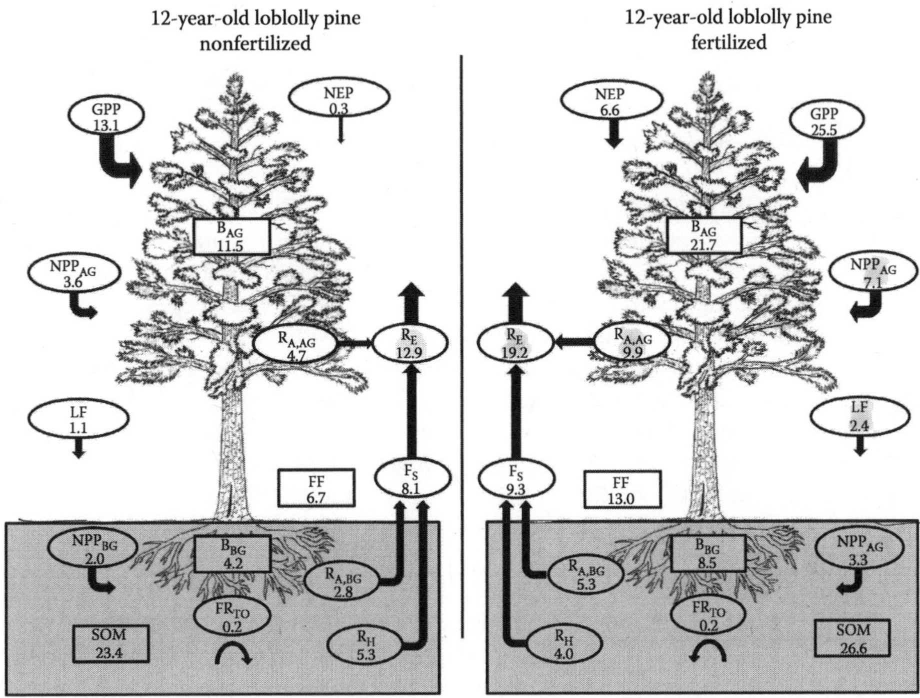

FIGURE 8.2 Major carbon pools (Mg C ha^{-1}; boxes: B_{AG}—aboveground biomass, B_{BG}—belowground biomass, FF—forest floor, SOM—soil organic matter) and fluxes (Mg C ha^{-1} $year^{-1}$; oval: NEP—net ecosystem productivity, GPP—gross primary productivity, NPP_{AG}—aboveground net primary productivity, NPP_{BG}—belowground net primary productivity, LF—litterfall, $R_{A,AG}$—aboveground autotrophic respiration, $R_{A,BG}$—belowground autotrophic respiration, R_H—heterotrophic respiration, F_S—soil CO_2 efflux, R_E—ecosystem respiration, FR_{TO}—fine root turnover) for a 12-year-old loblolly pine plantation (Maier et al. 2004). F_S is the combined respiration of heterotrophic soil microbes and plant roots. R_H was estimated as the difference between F_S and root respiration ($R_H = F_S - R_{A,BG}$) (Maier and Kress 2000). Estimate of the carbon flux in annual FR_{TO} (0.22 $year^{-1}$, Johnsen et al. 2005) are not included in NPP_{BG} pool.

and volatile emissions (Chapin et al. 2006). Annually these additional C losses are relatively small (<5%), but may be a substantial component of the C budget when assessed over many years or after large episodic disturbances (e.g., insect outbreaks). As the products of NPP die, dead organic matter or detritus accumulates on the forest floor and soil. Decomposition of detritus releases C to the atmosphere through heterotrophic respiration (R_H) and represents the second largest C flux from the ecosystem (Figure 8.2). Carbon not consumed in R_H accumulates in the soil, retained in chemically bound organomineral aggregates resistant to decomposition. This recalcitrant C pool can persist for thousands of years and accounts for most of the ecosystem carbon storage in some forest ecosystems.

Allocation of NPP into various ecosystem components (e.g., foliage, stems, and roots) determines ecosystem structure and can have a strong effect on other ecosystem processes such as biogeochemical cycling, water use, and C sequestration. The physiological and biochemical mechanisms that control allocation, particularly belowground, is only partially understood (Litton et al. 2007). Carbon allocated belowground supports root production and maintenance as well as rhizosphere food webs or mycorrhizae, and so provides the majority of soil detrital C. Plants allocate C to minimize water and nutrient limitations (Friedlingstein et al. 1999), shifting allocation belowground to support root growth and resource acquisition at the expense of aboveground growth as soil resources become limiting (Haynes and Gower 1995; Albaugh et al. 1998). In general, as site GPP increases, the proportion of GPP allocated belowground decreases (Litton et al 2007). Because 20–80% of GPP is allocated belowground (Litton et al. 2007), small changes in this percentage in response to climate change may greatly alter ecosystem productivity and soil C storage.

The C use efficiency (CUE) or the ratio of NPP to GPP reflects the ecosystems C storage potential. CUE is relatively conservative at the individual plant level because physiologically respiration is linked tightly to photosynthesis (Amthor 1994). Gifford (2003) suggested that a constant CUE also operates at the ecosystem scale. Waring et al. (1998) found that CUE was relatively constant (≈0.47) across a range of forest types and environmental conditions, suggesting that the ecophysiological controls on NPP and GPP are the same in most forest ecosystems. A constant CUE greatly simplifies modeling NPP and GPP because it obviates the need to measure component R_A, a notoriously difficult process to measure (Landsberg and Sands 2011). However, the constant CUE reported in Waring et al. (1998) may be an artifact of estimating both GPP and R_A from NPP (Medlyn and Dewar 1999). DeLucia et al. (2007) analyzed a large number of studies where GPP was estimated independent of NPP, and found that CUE varied widely between forest types (0.2–0.8) and that it decreased with stand age. A better understanding of the ecophysiology R_A is needed to reliably predict how climate change effects forest CUE (DeLucia et al. 2007).

Heterotrophic Respiration and Net Ecosystem Productivity

As the products of NPP die, dead organic matter or detritus accumulates on the forest floor and in the soil. Decomposition of dead organic matter releases C to the atmosphere through heterotrophic respiration (R_H) and is a major component of ecosystem metabolism (Figure 8.2). Generally, factors that favor high NPP also contribute to high R_H (Lambers et al. 2006). Detrital C not used in R_H accumulates in the soil and is retained in chemically bound organomineral aggregates resistant to decomposition, which can persist for hundreds to thousands of years. Soil C is an important C pool and accounts for most of the ecosystem C in some forest ecosystems.

Net ecosystem productivity (NEP) is the rate that C accumulates in living biomass, detritus, and soil, and defines the ecosystems ability to sequester C. In most forest ecosystems, NEP is the difference between GPP and ecosystem respiration (R_E), where R_E is the sum of R_A and R_H (Figure 8.2). As defined here, NEP does not include C movement through leaching or lateral transfer of dissolved inorganic and organic C or loss through emissions of volatile organic compounds, methane, or CO, which may represent important C fluxes over large spatial or temporal scales or in certain ecosystems (e.g., forest wetlands) (Chapin et al. 2006). NEP is relatively small compared to its component fluxes of GPP and R_E. If GPP exceeds R_E, then NEP is positive and the system is capturing C (i.e., C sink); however, if R_E exceeds GPP, then NEP is negative and the system is losing C (i.e., C source). NEP can vary from year to year owing to different environmental responses of GPP, R_A, and R_H. Changes in temperature, precipitation, length of growing season, or CO_2 will directly affect GPP, resulting in either increased or decreased NEP. However, the interannual variability of R_E is more complicated because of the disparate response of R_A and R_H to temperature, moisture (Falge et al. 2002), and substrate availability (Johnsen et al. 2007).

Quantifying NEP, GPP, and R_E is a formidable task even for a single site. The ecological "bottom-up" approach uses biometric measurements of NPP and chamber-based measurements of R_A and R_H to estimate C flux of component C pools and then sums these values over space and time to estimate NEP (Figure 8.2). This approach allows for determining how ecosystem components contribute to NEP and how these components respond to the environment. The meteorological "top-down" approach utilizes eddy-covariance measurements of energy and mass exchange between the forest canopy and the atmosphere to estimate net ecosystem CO_2 exchange (NEE) (Landsberg and Sands 2011). It differs from the ecological approach in that it integrates NEE over large areas and thus provides direct ecosystem level estimates of GPP and R_E. Conceptually, NEP estimated from ecological analysis and NEE is the same because both comprise the difference between GPP and R_E; however, each method measures these components at different spatial and temporal scales and thus do not always agree (Baldocchi 2003). NEE also differs from NEP in that it does not account for nongaseous transfer of C into and out of the system via leaching, lateral transfer, herbivory, and harvest (Chapin et al. 2006). In addition, annual estimates of NEE and NEP can diverge because of a lag between C fixation and biomass growth; however, estimates between the two methods converge when averaged over multiple years (Baldocchi 2003; Gough et al. 2008). In the scientific literature, NEE and NEP are often expressed in opposite sign. This is because atmospheric scientists define NEE as the net C flux from the ecosystem to the atmosphere, whereas ecologists define NEP as the net C flux from the atmosphere into the ecosystem (Chapin et al. 2006). While the eddy-covariance technique has become a standard method for estimating NEE, in most studies, it is combined with the ecological method to estimate NEP and component processes (Baldocchi 2003). This complementary approach has provided robust estimates of NEP and component processes for a number of southern forest ecosystems (Ehman et al. 2002; Curtis et al. 2002; Lai et al. 2002; Clark et al. 2004; Noormets et al. 2010; Goulden et al. 2011).

Disturbance and Net Ecosystem Productivity

Forests in the Southern United States are characterized by frequent disturbances from natural (e.g., fire, wind and ice storms, drought, insects, and disease) and human-induced (e.g., harvesting) causes (Dale et al. 2001). The type of disturbance (consumptive or nonconsumptive) and management intensity will determine the amount and type of detrital C and the trajectory or recovery of NEP over time (Sprugel 1985). Nonconsumptive disturbances such as windstorms or disease transfer carbon directly from living biomass to forest floor and soil detrital pools, leaving almost all biomass on site. Consumptive disturbances (e.g., fire), remove large amounts of carbon in live biomass, forest floor detritus, and potentially soil C, transferring it directly to the atmosphere. Similarly, forest harvesting, a consumptive disturbance, removes large amounts of biomass; however, in contrast to fire, the forest floor and soil are usually left intact. The C dynamics following disturbance have been intensively studied in southern pine plantations (Gholtz and Fisher 1982; Maier et al. 2004; Noormets et al. 2010; McCarthy et al. 2010; Maier et al. 2012); however, less is known about natural pine ecosystems (Powell et al. 2008) or mixed deciduous forests (Ehman et al. 2002; Curtis et al. 2002).

Changes in NEP following a disturbance can be described in four phases (Figure 8.3). For example, tree harvesting of a southern pine plantation transfers some live residual biomass C to detrital C

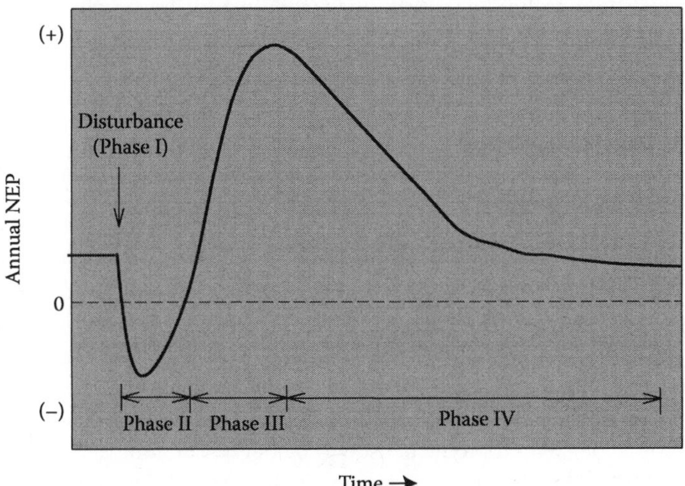

FIGURE 8.3 Conceptual pattern of annual net ecosystem productivity (NEP, Mg C ha^{-1} year^{-1}) following a stand disturbance.

pools (phase I). Residual forest floor biomass following harvest of pine plantations can range from 15 to 50 Mg C ha^{-1} (Eisenbies et al. 2009; Maier et al. 2012), depending on whether the site was stem-only or whole-tree harvested. The second phase is a period of negative NEP, where R_E exceeds GPP. The magnitude and duration of this phase is strongly influenced by disturbance intensity, soil characteristics, and NPP of the regenerating stand (Maier et al. 2004). During this phase, increased R_H from accelerated decomposition of forest floor and soil organic matter dominates R_E. The post disturbance pulse of R_H can be quite high (>20 Mg C ha^{-1} year^{-1}) and is influenced heavily by the type of site preparation (e.g., burning, disking, bedding, weed control) (Gough et al. 2005). The recovery period or the point of time to the transition between negative to positive NEP can take <3 years on highly productive sites (Gholtz and Fisher 1982; Clark et al. 2004) and as long as 15 years on poor sites (Thornton et al. 2002; Sampson et al. 2006). In phase III, NEP transitions to an extended period of rapid C accumulation as tree growth moves into the exponential growth phase and reaches a peak near crown closure as canopy biomass and leaf area stabilize. In young, rapidly growing pine plantations, NEP can reach 5–7.5 Mg C ha^{-1} year^{-1} (Clark et al. 1999; Hamilton et al. 2002; Maier et al. 2004). During this phase, NEP is strongly linked to NPP and can be manipulated by management. For example, 4 years of fertilization in a loblolly pine plantation growing on a poor site shifted NEP at age 12 from C neutral (nonfertilized; NEP ≈ 0 Mg C ha^{-1} year^{-1}) to a strong C sink (fertilized: NEP = 6.4 Mg C ha^{-1} year^{-1}) (Maier et al 2004) (Figure 8.2). Fertilization doubled GPP, but only increased R_E by 48%. Increased ecosystem C storage occurred mainly in perennial woody biomass. Phase IV is a period characterized by a gradual decline in NEP (Chapin et al. 2002). The ecophysiological mechanisms responsible for reduced NEP in phase IV are unclear, but are probably a function of reduced NPP following canopy closure (Gower et al. 1996). Age-related decline in NPP has been linked to increased respiration costs as the wood:foliage biomass ratio increases (Hunt et al. 1999; Goulden et al. 2011), decreasing GPP due to nutrient and/or water limitations (Ryan et al. 1997; Hubbard et al. 1999), or some combination of the two. Ryan et al. (2008) concluded that the decline in aboveground NPP was due primarily to reduced GPP and secondarily to a shift in partitioning of GPP to foliage respiration and belowground allocation. Alternatively, reduced NPP could be due to changes in stand structure where competition-induced mortality and lower individual tree resource-use efficiencies reduce stand growth (Binkley et al. 2002; Ryan et al. 2008). Regardless of the mechanisms for declining NEP, it is likely that management designed to increase NPP will result in increased NEP. NEP of intensively managed pine plantations can range between 5 and 7 Mg C ha^{-1} year^{-1}, much higher than naturally regenerated stands (<2 Mg C ha^{-1} year^{-1}, Powell et al. 2008). A more complete understanding of the ecophysiological mechanisms controlling NEP and its component processes will aid in the development of strategies for actively managing carbon.

EX SITU CARBON SEQUESTRATION

Carbon is stored in forest products that are currently in use and those in landfills. The effectiveness of *ex situ* C sequestration is dependent on the fate of the wood fiber; that is, whether it is converted into products where the C is confined from the atmosphere for relatively long or short durations of time (Skog and Nicholson 1998; Skog 2008). The amount of C per unit of wood from southern forests has been estimated to be 270.7 and 317.5 kg m^{-3} (Skog and Nicholson 1998) for softwood and hardwood, respectively—the highest values estimated for the United States. The half-life (time until it ends its initial use) of wood used to build single-family homes before 1939 was 78 years and has increased since then (Skog and Nicholson 2008). In contrast, paper has an estimated half-life of only 2.5 years (Skog 2008). Wood products used to be burned in dumps. Now, wood products are disposed of in landfills, where due to low oxygen they decompose slowly at rates ranging from 3% for solid wood up to 38% for office paper, over a 50-year period (Micales and Skog 1997). It should be noted, however, that landfills do produce methane, which is 25 times more effective than CO_2 as a greenhouse gas (Ryan et al. 2010). Lastly, waste wood (i.e., sawdust) is used as a fuel in wood manufacturing plants, pulp mills, and paper processing plants and so acts as a direct replacement for fossil fuels, making the production of these products more energy efficient. Case Study 3 on the impact of silviculture on C sequestration provides an example of an analysis that takes both C *in situ* and *ex situ* sequestration pools into account.

PULSE AND PRESS DISTURBANCES, FOREST PRODUCTIVITY, AND CARBON SEQUESTRATION

It is likely the effects of climate change will occur as exogenous or endogenous disturbance events of as-yet unknown duration, frequency, or intensity in southern forest stands and landscapes. These disturbance events will act as either "press" or "pulse" disturbances (Chapter 2). Press, or persistent, disturbances include fundamental changes associated with climate change itself, such as regional and localized temperature increases, altered precipitation patterns and increased atmospheric CO_2. Changes in structure and function associated with press events may not be immediately evident, but manifest over time and include changes in species distribution, concomitant shifts in natural ranges of species, and long-term changes in forest growth and productivity. Unlike press events, pulse events occur as discrete and distinct disturbance events in time and space. Pulse disturbances associated with climate change are expected to increase in frequency and/or intensity, and include an increase in the frequency and intensity of wind- or storm-related events, wildfire, insect and disease outbreaks, and flooding. Although some pulse events have only a short-term effect on structure and function, other events may perturb the system in such a way that long-term forest productivity is altered. These climate-change related pulse and press disturbance events will likely have both positive and negative consequences on the production and sustainability of the Southern United States timber and fiber supply, as well as the C storage potential of affected forest stands.

EXAMPLES OF PRESS DISTURBANCES

Increased Temperature and/or Decreased Precipitation on Productivity

The control over individual tree, stand-, and forest-level growth and productivity is defined by complex interactions among climate, edaphic conditions, genetics, endogenous and exogenous disturbance patterns, and competition for water, nutrients, light, and other resources. The immediate effects of climate change on tree growth will be caused by changes in regional and localized patterns of precipitation and temperature, which directly affect the phenology (e.g., Vitasse et al. 2009) and physiological processes that control C uptake and assimilation (Kozlowski et al. 1991).

Of the abiotic factors that control photosynthesis (e.g., water, nutrients, light, temperature, CO_2), it is temperature and water availability that most limit photosynthetic activity (Salisbury and Ross 1992) and, hence, tree growth and productivity. A recent study by Way and Oren (2010) suggests that temperate tree species are currently growing at temperatures below their maximum thresholds, and that the increased temperatures associated with climate change may stimulate tree growth. While this generalization regarding the positive relationship between temperature and tree growth is supported by provenance trials for a variety of temperate tree species (Schmidtling 1994; Carter 1996; McLane et al. 2011) and retrospective dendrochronology studies (Friend and Hafley 1989; Pichler and Oberhuber 2007; White et al. 2011), the amount by which growth is increased appears to be dependent upon genetics (McKeand et al. 1997; Sonesson and Eriksson 2000), species (Nedlo et al. 2009), and functional group, with deciduous tree species generally experiencing a greater response to increased temperature than evergreen tree species (Way and Oren 2010).

In addition to possible temperature-related increases in growth due to increased photosynthetic activity (Way and Oren 2010), changes in growing season length and associated changes in phenology have the potential to substantially impact not only tree growth and C assimilation (White et al. 1999; Rötzer et al. 2004), but also the timing of flowering and reproductive success of many species

(Walkovsky 1998; Beaubien and Freeland 2000; Peñuelas and Filella 2001). Despite the inherent temporal and spatial variation in growing season length (White et al. 1999), a correlation between temperature and length of growing season has been documented. In the Eastern United States, for example, an increase in growing season length of five days for every 1°C increase in average annual temperature has been observed (White et al. 1999). This increase in growing season length has been corroborated by other studies across both North America (Schwartz and Reiter 2000) and Europe (Menzel 2000; Chmielewski and Rötzer 2001). In North America, the onset of biological spring was advanced by an average of 5 to 6 days between 1900 and 1997 (Schwartz and Reiter 2000), while in Europe, the average annual growing season has lengthened between 8 (Chmielewski and Rötzer 2001) and 11 days since the 1960s (Menzel 2000). The effect of increased growing season length on tree growth and productivity is less understood than the effect of temperature on growing season length itself. Models that incorporate the timing of bud burst into growth simulations suggest that the timing of bud burst and, consequently, length of growing season, have a positive effect on tree growth (Menzel and Fabian 1999), although the responsiveness of bud burst and leaf flushing to temperature is species specific (Vitasse et al. 2009).

Although increased temperatures are predicted to stimulate tree growth, patterns of precipitation in a changing climate may greatly alter or even negate the theoretical increase in tree growth expected with increased temperatures (Way and Oren 2010). Holding edaphoclimatic factors constant, increased temperatures increase evapotranspiration rates. In forest stands with low soil water-holding capacity in particular, this increase in evapotranspiration and soil moisture deficits eventually leads to decreased C assimilation due to stomatal closure (Kozlowski et al. 1991), altered tree allometry and biomass partitioning (Callaway et al. 1994; McDowell et al. 2006; Landsberg and Sands 2011), and, ultimately, decreased cell expansion and tree growth (Kozlowski and Pallardy 1997; Henderson and Grissino-Mayer 2009; White et al. 2011). Provided that the increase in temperature predicted to occur over the next 100 years is concomitant with a decrease in precipitation, substantial reductions in tree growth and productivity could occur despite the positive effects that increased temperatures may have on tree growth and productivity.

Current models suggest that average annual precipitation in some of the sub-regions in the South may decrease as much as 24% over the next 40 years (McNulty et al. 2010), and at a regional scale the frequency and severity of episodic drought events may increase (Dale et al. 2001). If, as the data suggest, precipitation is a strong determinant of annual rates of tree growth for many tree species in the Southern United States (e.g., Orwig and Abrams 1997; Pan et al. 1997; Henderson and Grissino-Mayer 2009; Speer et al. 2009), tree growth and productivity should be expected to decrease as temperature increases and precipitation, and subsequent soil moisture availability, decrease. However, dendrochronology studies have demonstrated that the temporal distribution of precipitation throughout any given year is as important, if not more important, than total annual precipitation. For example, studies have shown that oak species (*Quercus*) are most prone to reduced growth when soil water balance is reduced early in the current growing season (Tardif et al. 2006; Speer et al. 2009) or when the water balance of a stand is low late in the growing season of the year prior to ring formation (i.e., preconditioning) (Jacobi and Tainter 1988; White et al. 2011). In contrast, loblolly (*Pinus taeda*) and longleaf pine (*Pinus palustris*) do not demonstrate any preconditioning requirements and generally experience a decrease in annual growth when precipitation during the spring and summer months is reduced (Jordon and Lockaby 1990; Henderson and Grisso-Mayer 2009).

Retrospective dendrochronology studies have proven useful in quantifying climate–growth relations. However, the response of tree growth to future climate will vary among and within tree species, making broad-scale generalizations regarding the future impact of climate change on tree growth based on dendrochronology studies difficult. The sensitivity of annual growth to future climate will depend greatly upon species and edaphic conditions as well as local temperature and precipitation patterns, including the amount and temporal distribution of precipitation within and among years. Neither precipitation nor temperature alone determines growth dynamics. Rather, tree growth is heavily influenced by a given site's water balance (Kozlowski and Pallardy 1997; Littell et al. 2008), which is influenced by temperature, precipitation, and edaphic conditions (e.g., soil moisture-holding capacity). Relatively few dendrochronology studies have quantified the variation in climate–growth relationships across environmental gradients, with most concluding that climate–growth relationships for any given species are not uniform across the landscape (Tardiff and Bergeron 1997; Orwig and Abrams 1997; Case and Peterson 2005; Griesbauer and Green 2010; White et al. 2011). For example, the decrease in growth that an individual tree growing on a low-quality site (e.g., a site with low water-holding capacity) experiences during periods of high evapotranspirative demand (i.e., drought) is generally greater than that of an individual tree of similar size and species found on a highly productive site (e.g., a site with high water-holding capacity) (Orwig and Abrams 1997; Case and Peterson 2005; White et al. 2011), suggesting an individual tree's resistance to reduced water availability is greater on high- versus low-quality sites. However, recovery to predrought growth levels (i.e., resilience to drought) may occur more quickly on the lower quality sites due to an individual's acclimation to local growing conditions (Orwig and Abrams 1997) suggesting greater resilience in sites of low versus high productivity. This differential response of tree growth to a climate across ecological gradients emphasizes the variation in resistance and resilience of species to changing climate across a diverse and complex landscape.

Forest Soil C and Climate Change

Over 2700 Gt of C are stored in soils globally, more than double the combined amount of C contained in the atmosphere (780 Gt) and stored in biomass (575 Gt) (Lal 2008a). The vast pool size and slow turnover of soil C make it central to global C cycling, but also make it difficult to study or manipulate experimentally (Hungate et al. 2009). Across an array of biomes, land–air temperature is rising and the rate is expected to increase in the coming decades (IPCC 2007a). In the scientific community there is a general consensus that this warming (currently ~0.3 C per decade) is due to greenhouse gas forcing by anthropogenic increases of greenhouse gases (e.g., CO_2, CH_4) in the atmosphere and it is very unlikely that this warming could be produced by natural causes (IPCC 2007a). Attention is being focused on the role forests play in sequestering some of the anthropogenic C inputs to the atmosphere in biomass and soils, while conserving existing C stocks through informed resource management. Predicted warming of 3–6°C over the next century (IPCC 2007a) may cause changes in forest productivity, nutrient availability, additions to belowground C pools from autotrophs and accelerated decomposition of soil C. There remains a great deal of uncertainty in forecasting whether future forests will be more productive in a warmer world and store more C in soil or release more C into the atmosphere via higher rates of soil organic matter decomposition accelerating warming, that is, "positive feedback" (Luo 2007). The effect of gradual warming on long-term soil C storage will be determined by the balance of autotrophic inputs and losses of stored C soil. A universal theory on the temperature sensitivity of soil C decomposition is desirable, but the myriad of differences in climate, existing soil C stocks, nutrient availability, vegetation, soil mineralogy, soil microorganisms, and land-use practices add complexity and uncertainty requiring regional assessments.

Conversion of forest and other natural systems to agriculture leads to declines in soil organic C (SOC); in temperate regions it is estimated that 30–50% of SOC is lost in 50–100 years after conversion (Lal 2008b). In much of the Southern United States, soil C and soil fertility have already been depleted through past agricultural practices (Giddens and Garman 1941; Jackson et al. 2005). Conventional wisdom in managed southern forests has been that limiting erosion, maintaining canopy cover, and good silvicultural practices would lead to C accumulation in soil organic matter and eventually in the mineral fraction. Given the level of degradation, it seemed there was little additional soil C that could be lost on upland sites. In wet or seasonally inundated forests, where SOC is protected from oxidation by anoxic conditions and has accumulated for centuries, maintaining hydrology is key to continued C retention. Atmospheric C is fixed via photosynthesis and enters the soil via decomposition of biomass or other more ephemeral photosynthate products such as root exudates. Relatively labile C in biomass is then retained via humification (breakdown by microbes into more resistant forms), aggregation (formation organo-mineral complexes, especially with clay), and formation of biochar (Lal 2008b). Biochar or black C (BC) formed by incomplete combustion of biomass is resistant to decay and has the potential to persist in soils for thousands of years (Schmidt and Noack 2000). The dynamics of BC in soil are still poorly understood, but BC could be an important sink in fire-adapted ecosystems. Frequent prescribed fires (2- to 5-year return) are being used to restore longleaf pine ecosystems in the Coastal Plain and Piedmont regions. The rate of BC accumulation and the importance of this pool are unknown in these systems, but it is the focus of active research. Incorporation of logging debris directly into the soil after harvest has been explored as a means to accelerate humification and long-term C storage in managed forests (Buford and Stokes 2000). However, 8 years after establishment, a forest slash incorporation study on both mineral and organic soils did not significantly SOC or soil chemistry (Sanchez et al. 2009).

Seasonal variation in field measurements of soil respiration is frequently described as a response to changes in soil temperature and moisture. When moisture is not limiting, the response of soil respiration to soil temperature is exponential (e.g., Palmroth et al. 2005); this widely observed phenomenon likely contributed to speculation that runaway C releases from soil were possible in a warmer climate. To avoid some of the experimental artifacts of laboratory incubations with disturbed soils, a variety of soil-warming experiments were established in forest (Rustad and Fernandez 1998; Melillo et al. 2002), prairie (Luo et al. 2001), and agricultural systems to evaluate the fate of soil C and plant productivity.

The soil warming experiment at the Harvard Forest in Massachusetts was established in 1991, using buried heating cables to create a 5°C elevation in soil temperature in a mature hardwood forest. Soil respiration and soil organic matter decomposition were higher in heated plots for a few years, but the response declined until there was no difference between the treatments (Melillo et al. 2002). Warming increased nitrogen (N) mineralization, and Melillo et al. (2002) speculated that this could stimulate C storage in biomass in N-limited environments. The results of this study call into question projections of large long-term releases of C from soils with limited labile soil C. An earlier study in a low elevation spruce-fir forest showed that soil respiration rates had not acclimated to heated soils after 3 years and were 25–40% higher than controls (Rustad and Fernandez 1998). It is unclear whether the 3-year experiment duration was long enough to deplete labile C reserves. There is evidence that soil C decay acclimates to higher soil temperatures after several years of warming (Luo et al. 2001; Melillo et al. 2002; Bradford et al. 2008), and that decomposition of organic C in mineral soil does not vary with temperature (Giardina and Ryan 2000), which would weaken any potential positive feedback between soil C decay and the climate system. However, the debate is far from settled with others asserting that nonlabile SOC is more sensitive to temperature than labile SOC and the duration of current warming studies is not long enough to detect these changes (Knorr et al. 2005). In this scenario, the positive feedback between SOC decay and climate would be greater than predicted. In the eroded soils of the southeast, soil C stocks are generally low on upland sites and soil respiration and SOC decay would be expected to follow the acclimation scenario described by Melillo et al. (2002) and Lou et al. (2001). However, the effects of warming in forests on nutrient-poor clay soils in the southeastern United States, subtropics, and tropics are understudied. Newly installed warming studies in oak forests in the Piedmont region of Georgia and North Carolina will provide experimental results in the coming years (Machmuller et al. 2011).

Lacking a grand, manipulative warming experiment in the Southern United States to guide our understanding, it is necessary to make inferences from the literature on the effect of forest management and disturbance on soil C. In general, harvesting forests has little or no effect on soil C or N, especially when whole tree harvesting is avoided (Johnson and Curtis 2001). In a meta-analysis by Johnson and Curtis (2001), on average, harvesting conifers led to an increase of soil C and N by approximately 25% and harvesting hardwood stands led to small declines in soil C (−8%), and small increases in soil N. In the course of a rotation, trees accumulate biomass and contribute leaf litter to the forest floor, the forest floor is shaded and a small quantity of C may be stabilized in the mineral soil. After harvest, the forest floor, no longer shaded, heats to temperatures commensurate to the increase predicted a century from now; for example, soil in a clear cut was +3°C warmer than an adjacent 27-year-old slash pine stand (Castro et al. 2000). Over the course of a rotation, the soil is exposed to sudden warming followed by gradual cooling as trees shade the forest floor and soil C still accumulates, more for conifers, less for hardwoods (Johnson and Curtis 2001).

Protecting existing soil C is important in the Southern United States, but the more relevant question is how management practices and climate change will affect additional C inputs. Worldwide, forests occupy approximately 50% of the land surface area, but tend to be located in temperature- and light-limited areas. In general, productivity of forests will be positively affected by increasing atmospheric CO_2, temperature, and precipitation, positively affected by N deposition (at least in the short term, where not at toxic levels), and negatively affected by pollutants, for example, O_3 under current climate change predictions (Boisvenue and Running 2006; Hungate et al. 2009). Atmospheric deposition of N varies annually based on weather patterns and anthropogenic emissions; in the Southern United States, it currently is less than one-third of deposition rates in the Midwest or Northeast (National Atmospheric Deposition Program, N deposition maps, nadp.sws.uiuc.edu). Forest productivity and soil C accumulation under various climate change scenarios can only be maintained if there is a minimum level of available N (Hungate et al. 2009), and this is a concern in southern forests where N limitations are common. Based on available information and models, southern forest soil C stocks will likely be retained and slowly accumulate as long as forest productivity is maintained or increases, and forestry practices that enhance productivity and soil fertility are used.

Elevated Atmospheric CO_2

Owing primarily to fossil fuel burning and deforestation, atmospheric CO_2 levels have increased nearly 35% since preindustrial times, from ~280 ppm to ~380 ppm (IPCC 2007b). Depending on the growth and emissions scenario used, atmospheric CO_2 may rise as high as 550–850 ppm by 2100 (IPCC 2007b). While CO_2 is the primary driver of anthropogenic climate change, it is also the basis of plant photosynthesis. Given that plant photosynthesis is not saturated at current CO_2 levels, anthropogenic increases in CO_2 will almost certainly lead to higher photosynthesis. However, greater photosynthesis may not translate to significantly greater forest productivity and plant C storage, and gains in productivity may not be sustainable over the long term. There are a number of questions that must be answered in order to assess the likely effect of elevated CO_2 on future forest productivity. These questions include: How much can elevated CO_2 increase forest productivity? How variable is the response across and within forests? How much of the CO_2 enhanced productivity will be allocated to woody biomass versus other tissues such as foliage and fine roots? Finally, will elevated CO_2 actually increase forest productivity in the long term, or will responses be constrained by the availability of other growth resources?

Effects of elevated CO_2 on plant growth have been studied extensively for more than 30 years. These investigations began with greenhouse or potted plant seedling studies, which often demonstrated large responses of photosynthesis, leaf area, and growth. However, the use of young plants in the exponential growth phase made it difficult to separate acceleration of development from direct CO_2 effects (see reviews of Drake and González-Meler 1997; Curtis and Wang 1998). Subsequently, open top chamber experiments allowed for the study of larger vegetation in field conditions. However, it was still difficult to predict the effects of increased CO_2 on entire ecosystems, especially those in intermediate or mature stages where canopy closure has occurred and plants are competing for resources. With the development of free air CO_2 enrichment (FACE) technology, it became possible to study the response of intact forest ecosystems to elevated CO_2 (Hendrey et al. 1999). These experiments (described in detail below) form the basis for our current understanding of the effect of increased CO_2 levels on forest productivity and C storage.

FACE technology allows for elevation of CO_2 concentrations in unenclosed patches of forest, with minimal alteration of microclimate and other stand conditions (Hendrey et al. 1999). Four FACE experiments have been conducted in established or establishing forest stands. These FACE sites contain 6–12 circular plots (22–30 m in diameter), where half of the plots are exposed to "ambient" CO_2 concentrations and half are exposed to "elevated" CO_2 concentrations. In "elevated" plots, pure CO_2 is mixed with ambient air to create air with a CO_2 concentration of ~550 ppm (~200 ppm greater than the current level), which is then released from vertical pipes surrounding the circular plot (Hendrey et al. 1999). "Ambient" CO_2 plots have a similar infrastructure, but receive air with current CO_2 concentrations (~376 ppm across time and sites; Norby et al. 2005). Two FACE experiments are located in southern forests, one at Duke Forest (Chapel Hill, NC) in a planted loblolly pine forest, and the other in a planted sweetgum forest in Oak Ridge, TN. In these forests, the CO_2 elevation experiments were conducted when the forests were 13–27 (Duke FACE) and 10–21 (ORNL FACE) years old. Two other forest FACE experiments were conducted on newly planted forests: a mixed aspen/birch/maple forest in Rhinelander, WI (operated for 12 years), and two rotations of poplar species in Viterbo, Italy (6 years total; i.e., the stand grew for 3 years, was coppiced and monitored for an additional 3 years). Detailed descriptions of the experimental setups are provided for Duke FACE (Hendrey et al. 1999), ORNL FACE (Norby et al. 2001), AspenFACE (Karnosky et al. 1999, 2005), and POP-EUROFACE (Miglietta et al. 2001).

The average enhancement of NPP observed in forest FACE experiments (after 2–6 years of CO_2 exposure) was surprisingly consistent at 23% (Norby et al. 2005). In forests with low native LAI, such as the coniferous Duke FACE, the increase in productivity was driven almost entirely by increased light capture due to higher LAI (e.g., ~16% higher LAI at Duke FACE; McCarthy et al. 2007) under elevated CO_2 (Norby et al. 2005; McCarthy et al. 2006). In contrast, for forests with high native LAI (such as the broadleaf ORNL site), elevated CO_2 did not increase LAI and light interception (Gielen et al. 2003; Norby et al. 2003, 2005), and the bulk of the enhancement of productivity resulted from increased photosynthetic efficiency (Norby et al. 2005; McCarthy et al. 2006).

Despite the appearance of a predictable, constant enhancement of NPP with elevated CO_2, analysis of within-site data reveals a great deal of variability in the growth response of forests to increased CO_2. Much attention has been focused in the Duke FACE experiment on quantifying the interaction of elevated CO_2 enhancement with other growth resources, particularly nitrogen availability and water (Oren et al. 2001; Finzi et al. 2002; McCarthy et al. 2006, 2010). At Duke FACE, NPP enhancement has been closely correlated with soil N availability. The greater the available N, the larger the NPP enhancement with elevated CO_2 (McCarthy et al. 2010). Conversely, the NPP response to elevated CO_2 decreased with decreasing soil N availability (McCarthy et al. 2010), corresponding with studies in which forests with very low availability of nutrients (e.g., in *Pinus taeda* on sandy soils and *Picea abies* on sandy glacial till), had no detectable response to elevated CO_2 (Oren et al. 2001; Ward et al. 2008). In both of these studies, significant CO_2 responses were observed in fertilized trees, demonstrating that the lack of response was related to low N availability. Additionally, water availability also influences productivity response to elevated CO_2. Interannual variability in both basal area (Moore et al. 2006) and NPP (Finzi et al. 2006; McCarthy et al. 2010) at Duke FACE were correlated with water availability (assessed as growing season precipitation or growing season precipitation minus potential evapotranspiration; P-PET), with NPP enhancement under elevated CO_2 greatest when P-PET was highest (McCarthy et al. 2010). However, N availability was a stronger driver of variability in CO_2 response than water availability, as NPP under elevated CO_2 was ~130 g C m^{-2} greater at the highest versus lowest N availability, and only ~30 g C m^{-2} greater at the highest versus lowest P-PET (where both factors spanned the full range of possible values; McCarthy et al. 2010). Furthermore, results from AspenFACE, where half of the ambient and elevated CO_2 plots were also exposed to elevated (1.5 times ambient) ozone (O_3), suggest that O_3 pollution may completely offset (i.e., negate) CO_2 induced growth enhancements (Karnosky et al. 2003; King et al. 2005). Overall, these findings demonstrate that there can be great spatial variability in how forest productivity is affected by elevated CO_2, and that some forests may exhibit no productivity enhancement at all.

To fully understand the impact of elevated CO_2 on future forests it is necessary to understand not only how elevated CO_2 may affect overall stand productivity, but to account for how the additional NPP may be partitioned to different tree biomass components. The implications of elevated CO_2 for timber and fiber production or C storage will be much different if the majority of additional NPP gained under elevated CO_2 concentrations is invested in nonwoody biomass such as foliage and fine roots versus wood biomass. In the extreme case, stimulation of NPP by elevated CO_2 could do little to increase C storage or wood production. Although allocation is frequently poorly understood, even under current environmental conditions (Litton et al. 2007), the traditional view of allocation is that plants allocate their resources (C and nutrients) in order to optimize their gain of further resources (including water, e.g., Thornley 1972; Dewar 1993; McConnaughay and Coleman 1999). Therefore, an optimal allocation strategy for trees growing under elevated atmospheric CO_2 would be to allocate proportionally more C to root formation, in order to more fully exploit soil resources (i.e., increase water and nutrient uptake). The magnitude of this shift should be driven by soil resource availability, with more nutrient- or water-limited systems showing a greater increase in fine root allocation (Palmroth et al. 2006; Litton et al. 2007). In practice, results from forest FACE sites have been mixed regarding whether elevated CO_2 causes shifts in the proportion of C allocated to different plant pools. The sweetgum plantation at ORNL FACE showed a dramatic shift in C partitioning, with up to 80% of the extra CO_2-enhanced NPP being partitioned to wood during the first 2 years of the experiment, but only 25% to wood after 3 years (Norby et al. 2002, 2004). The remainder of NPP was allocated largely to short-lived, nonwoody biomass, doing little to increase standing biomass (Norby et al. 2004) but ultimately increasing soil C storage (Iversen et al. 2012). To a lesser degree, at the end of the first rotation in the POP-EUROFACE experiment on *Populus* species, root pools were increased relatively more under elevated CO_2 than aboveground woody components (Gielen et al. 2005). However, root-to-shoot ratio was unchanged, and the fraction of NPP allocated to woody aboveground biomass was high, ranging among species from 53% to 67% (Calfapietra et al. 2003; Gielen et al. 2005). Unlike the first rotation in which relative accumulation of biomass in stems and roots did not change, during the second (coppice) rotation of *Populus* species at POP-EUROFACE, elevated CO_2 resulted in greater C accumulation in branches and lesser accumulation stems as compared to ambient CO_2 trees; the ratio of above- and belowground biomass remained the same (Liberloo et al. 2006). On the other hand, being exposed to elevated CO_2 since planting, *Populus tremuloides* exhibited no changes to the fraction of standing biomass in various pools, nor the partitioning of NPP (King et al. 2005). These differing outcomes suggest that elevated CO_2 does not have a uniform effect on biomass allocation, and should be considered together with other site factors, for example, LAI (Palmroth et al. 2006).

In the context of forests managed for timber, an important distinction is whether forests under elevated CO_2 will accumulate more tree biomass in the long term or whether elevated CO_2 will simply accelerate the process of stand development, allowing canopies to close and trees to reach their maximum sizes more quickly (Körner 2006). Results from natural CO_2 springs, showing decline in the enhancement of stem growth with age, suggest that the elevated CO_2 may not increase steady-state stem biomass pools (Hättenschwiler et al. 1997). However, little information is available to address this issue as elevated CO_2 experiments have been shorter than the life of a forest stand. At the Duke FACE site, after 14 years of CO_2, mortality under ambient and elevated CO_2 was similar (~2.5% for pines and ~1% for understory hardwoods), and analysis of average tree biomass versus stand density does not suggest that elevated CO_2 modified the expected relationship of tree size and density (H. McCarthy, unpublished data). Thus far, there is little evidence that elevated CO_2 drastically increases site carrying capacity. However, more rapid accumulation of biomass could allow for shorter rotation lengths and more rapid timber and fiber production.

Finally, it is necessary to consider whether productivity gains resulting from elevated CO_2 are sustainable over the long term. Elevated CO_2 experiments, by necessity, induce a step-change in CO_2

concentrations. Thus, there is the possibility that productivity changes observed in such experiments represent only temporary responses, resulting from disequilibrium of C and N pools in the ecosystem. Higher production under elevated CO_2 must be supported by some combination of increased uptake of N and increased efficiency of N use. Given that most forest ecosystems are nitrogen limited, many models and long-term simulations that directly evaluate CO_2 effects (unlike most global scale models, which do not link CO_2 stimulation with N availability) predict that the growth enhancement observed in relatively short (2–15 years) FACE experiments cannot be sustained over the long term (Luo et al. 2004). The leading hypothesis that predicts a decline in CO_2 stimulation is progressive N limitation (PNL). In this scenario, NPP initially increases, litter production increases while litter quality changes, decomposition decreases, and N mineralization and thus availability is reduced, ultimately feeding back to a lower NPP (Luo et al. 2004). In short, much of the ecosystem N gets tied up in plant biomass, and there is reduced N availability for future biomass production. This process is commonly observed in developing stands (Richter et al. 2000), but elevated CO_2 may further accelerate this process. Many studies have examined N cycling in forest FACE sites, in order to look for evidence of the onset of PNL, and to answer questions regarding how forests are able to acquire the extra N necessary to support CO_2 enhanced growth. Across the four forest FACE sites, three sites have been found to support increased NPP through increases in N uptake, despite these forests being considered N limited (Finzi et al. 2007). Only the N unlimited (due to previous agricultural land use) POP-EUROFACE demonstrated an increase in nutrient use efficiency. Thus far, there have been conflicting findings regarding PNL in different forest types. At the Duke FACE, initial results from the FACE prototype plot (a plot established prior to the main experiment to test the FACE approach) showed a loss of NPP stimulation after 3 years of elevated CO_2 (Oren et al. 2001). This was attributed to N limitation, as fertilized plots showed no such reduction in CO_2 enhancement (Oren et al. 2001). However, over the longer term in the Duke FACE experiment, the average NPP enhancement has been maintained with no decline over time (McCarthy et al. 2010; H. McCarthy, unpublished data). The longevity of this NPP stimulation has been attributed to greater allocation of C belowground (to roots, mycorrhizal symbionts, and labile C exudates), which in turn makes N more available and accessible (Drake et al. 2011). In contrast, PNL has clearly been observed in the sweetgum forest at ORNL FACE (Norby et al. 2010). Six years into the experiment, NPP (of both ambient and elevated plots) began declining, and NPP in elevated CO_2 plots was no longer significantly greater than NPP in ambient plots (Norby et al. 2010). This loss of CO_2 enhancement was correlated with declining N availability, where N availability declined more rapidly with elevated CO_2 (Norby et al. 2010).

Based on current knowledge, elevated CO_2 may increase short-term productivity and C storage in southern forests. However, the response is likely to be quite variable, depending on the availability of N and water in different regions and sites, with some forests exhibiting little or no CO_2 response due to very low resource availability. Furthermore, increases in productivity may not translate into increases in wood production or C storage, as some forests may allocate much of their extra CO_2-induced biomass into nonwoody biomass. Finally, there is evidence to suggest that productivity gains may not be sustainable in the long term, and that NPP under future elevated CO_2 concentrations may not be much greater than current levels, due to the inability of forests to acquire the N necessary to support increased growth.

EXAMPLE OF A PULSE DISTURBANCE

HURRICANES

Hurricanes (i.e., tropical cyclones with sustained winds ≥119 km/h^{-1}) can cause substantial economic damage to forests. In 2005; Hurricane Katrina resulted in massive damage to forests along the Louisiana and Mississippi gulf coasts (Chambers et al. 2007; Kupfer et al. 2007; Stanturf et al. 2007). McNulty (2002) estimated that a single hurricane can obviate the equivalent of 10% of the annual C sequestered in the United States. Owing to its size, intensity and trajectory, Hurricane Katrina may have had 6–14 times that impact (Chambers et al. 2007). In 2005, winds from Hurricane Katrina damaged 22 million m^3 of timber estimated to be valued between $1.4 and $2.4 billion. Impacts are not limited to loss of wood volume and quality; ecosystem services provided by these forests can also be impaired. Subsequent decomposition of dead biomass has been estimated to be reducing C sequestration capacity of Gulf Coast forests by an amount equal to the total U.S. net annual forest C sink (Chambers et al. 2007).

Although not necessarily linked to climate change, hurricane activity has increased since the mid-1990s and this higher activity has been projected to last for the next 10–40 years (Goldenberg et al. 2001). Four main factors are related to the extent and intensity of wind damage on forests: climate, soils, topography, and stand conditions (Wilson 2004). Hurricanes obviously represent an extreme climatic event. Sites with soil conditions that restrict root growth and depth are consistently more prone to uprooting. Variation in windthrow along topographical gradients is more complicated and often confused with species and soil variation. There are many stand attributes that help determine the susceptibly of stands to windthrow. These include height-to-diameter ratios, height, spacing, recent thinning, and impacts of previous disturbance on creating exposed edges that contain trees more vulnerable to windthrow. Species composition may also impact the degree of damage from hurricanes and represents a stand attribute that can be manipulated by forest managers.

Some evidence suggests that longleaf pine might be more tolerant of high winds than either slash pine or loblolly pine. In a study of the Hobcaw Forest in coastal South Carolina after Hurricane Hugo, Gresham et al. (1991) reported that longleaf pine suffered less damage than loblolly pine. It was noted that species native to the coastal plain are possibility better adapted to the disturbance regimes found there; for example, longleaf pine, baldcypress (*Taxodium distichum*), and live oak (*Quercus virginiana*) suffered less damage than forest species with broader distribution ranges.

Johnsen et al. (2009) studied wind damage of these pine species in a common garden experiment in southeast Mississippi following Hurricane Katrina, which directly impacted the stand in August 2005. The experiment, a factorial arrangement of silvicultural treatments established in 1960, included one hundred twenty 100-tree plots covering about 22 ha. Following the hurricane, diameter at breast height (DBH) was measured on all trees and each tree was rated with respect to mortality from wind damage. Longleaf pine suffered lower mortality (7%) than loblolly pine (26%) (Figure 8.4). Longleaf pine lost significantly fewer stems ha^{-1} and less basal area than loblolly pine. Differences in mortality between species were not a function of mean plot tree height or plot density.

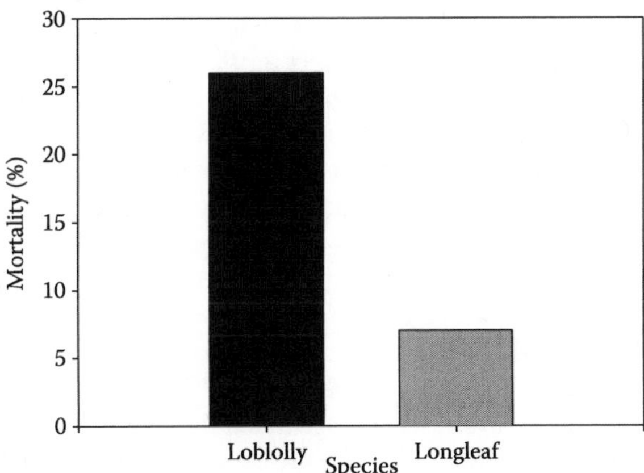

FIGURE 8.4 Mortality of loblolly and longleaf pine from a Mississippi experiment hit directly from Hurricane Katrina in 2005. Species effects were statistically significant at $\alpha = 0.05$. (Adapted from Johnsen, K.H. et al. 2009. *Southern Journal of Applied Forestry* 33:178–181.)

It is not possible to unequivocally state that longleaf pine has adapted to be more tolerant to wind damage than loblolly pine. Wind damage increases with tree size, but the frequency and severity varies with species, site, wind parameters, and stand characteristics (i.e., canopy evenness and age distribution), making blanket statements regarding species fitness an oversimplification (Gresham et al. 1991; Peterson 2007). As per the risk map shown in Stanturf et al. (2007), the southern coastal plain of the United States (the center of the historical range of longleaf pine) is highly prone to hurricane events. Intense hurricanes occur two out of every 3 years across the Eastern United States (McNulty 2002). Similar to historical natural fire regimes, the selection pressure of frequent high-velocity winds appears to have been high. This, and the results above, support the supposition that longleaf pine has evolved to have higher resistance to wind damage than loblolly pine.

REGIONAL ESTIMATES OF FOREST PRODUCTIVITY AND RESPONSES TO CLIMATE CHANGE

The effect of and responses to future climate change on forest productivity at regional scales can only be estimated via forest productivity models. We used the WaSSI-C model (Sun et al. 2011a) to examine the annual gross ecosystem productivity (GEP), ecosystem respiration (R_E), and net ecosystem exchange (NEE) of the five ecoregions of the Southern United States. The WaSSI-C model is a water-centric model that simulates the monthly water balances (precipitation, evapotranspiration [ET], water yield, and soil moisture storage) of 2103 basins (or called 8-digit Hydrologic Unit Code; HUC) across the lower 48 states including the 674 watersheds in the 13 southern states. The average size of the basin with mixed land use and land cover is 3662 km^2, ranging from 184 km^2 to 20,515 km^2. The key component of watershed water balance is ET, which accounts for as much as 85% of precipitation input at the annual scale in the Southern United States. Monthly ET of a watershed with mixed land-covers is modeled as a function of potential ET (calculated based on a temperature-driven model), precipitation, and LAI, that is, $ET = f(PET, P, LAI)$ (Sun et al., 2011b). Ecosystem C fluxes of each land cover within a watershed were modeled as a series of linear functions of ET. Mathematically, $GEP = a * ET$, $R_E = b + c * GEP$, and $NEE = R_E - GEP$. The parameters, a, b, c, are derived from global eddy flux data (Sun et al. 2001a). An ecosystem with a negative NEE is considered a C sink; otherwise, the ecosystem is a C source. At the watershed scale, all modeled variables (ET, GEP, R_E, NEE) were calculated by land-cover type, and then averaged by land-cover proportion to generate the mean for each watershed. The WaSSI-C model has been validated with both USGS runoff and MODIS ET and GEP products at the continental scale (Sun et al. 2011a). Four cornerstone climate change scenarios (CSIROA1B, CSIROB2, HADB2, and MIROCA1B) were applied to simulate monthly and annual ET, GEP, and NEE. Historic LAI data were derived from MODIS remote sensing products for the period of 2000–2006. Given the complexity and uncertainty of climate–vegetation–biogeochemical interactions at the regional scale, we focused on impacts of the physical climate on water availability and its influences on C balances. We assumed LAI for each land-cover type does not change over the simulation time frame (2002–2060) and CO_2 fertilization effects on water use efficiency and plant growth are not considered (see Elevate Atmospheric CO_2 section above for rationale). WaSSI-C modeling results for the first three future climate scenarios were averaged to represent the mean response to climate change among these scenarios. We did not average the climate, but rather averaged the simulated response variables. It appears that the MIROCB2 scenario represents extreme hot and dry conditions (precipitation decreased 8% and PET increased 20% nationally) and differs significantly from other three scenarios. Thus, for this analysis, we amalgamated the first three scenarios to represent the "mean" conditions, and used the MIROCB2 as the "worst-case" scenario. We define "2060 Results" as the mean between 2051 and 2060, while the baseline was considered the years 2002–2010. The "change" was estimated as the difference between results projected for 2060 and 2010.

As a whole, under the amalgamated ensemble "mean" condition, the Southern United States was projected to increase in ecosystem productivity and C sequestration strength due to a warming climate that will result in increased potential ET (PET), actual water loss (ET), photosynthesis (GEP), and ecosystem respiration (R_E) (Figure 8.5). At the southern regional scale, GEP, R_E, and NEE were projected to increase by 2.6%, 2.1%, and 3.8% respectively, in 2060, however, there was large spatial variability. Dry regions were expected to show a decrease in ecosystem productivity, presumably due to water stress caused by reduced precipitation and increased PET. The causal effects between productivity and water stress are more pronounced under the MIROCA1B scenario. Under this scenario, because of reduced precipitation and elevated PET, large water stress was predicted to occur. Ecosystem productivity (GEP) and C sequestration (NEE) were projected to decrease greatly under the MIROCA1B scenario due to reduced ET and water stress (Figure 8.6). At the regional scale, GEP, R_E, and NEE were expected to decrease by 6.7%, 5.4%, and 9.8% in 2060. Similarly, the dry arid Mid-South had the highest negative impacts from this climate change scenario (Figures 8.6 and 8.7).

Simulation results show that the Mid-South had the lowest GEP and NEE due to low water availability under a dry climate and the impacts on this subregion were most pronounced under all climate change scenarios (Figures 8.6 and 8.7). The Coastal Plain and Mississippi Alluvial Valley subregions had the highest GEP and NEE due to a warm and wet climate (Figures 8.6 and 8.7). These scenario modeling exercises demonstrate that future regional forest productivity patterns are generally controlled by the combination of changes in air temperature and precipitation. Precipitation is a key factor that should be examined carefully, especially in the traditionally climate transition zones where current precipitation levels just barely support forests. Our analyses also indicate that because the existing climate models do not agree on the future trends of precipitation, the projections of future change in forest productivity have large uncertainty.

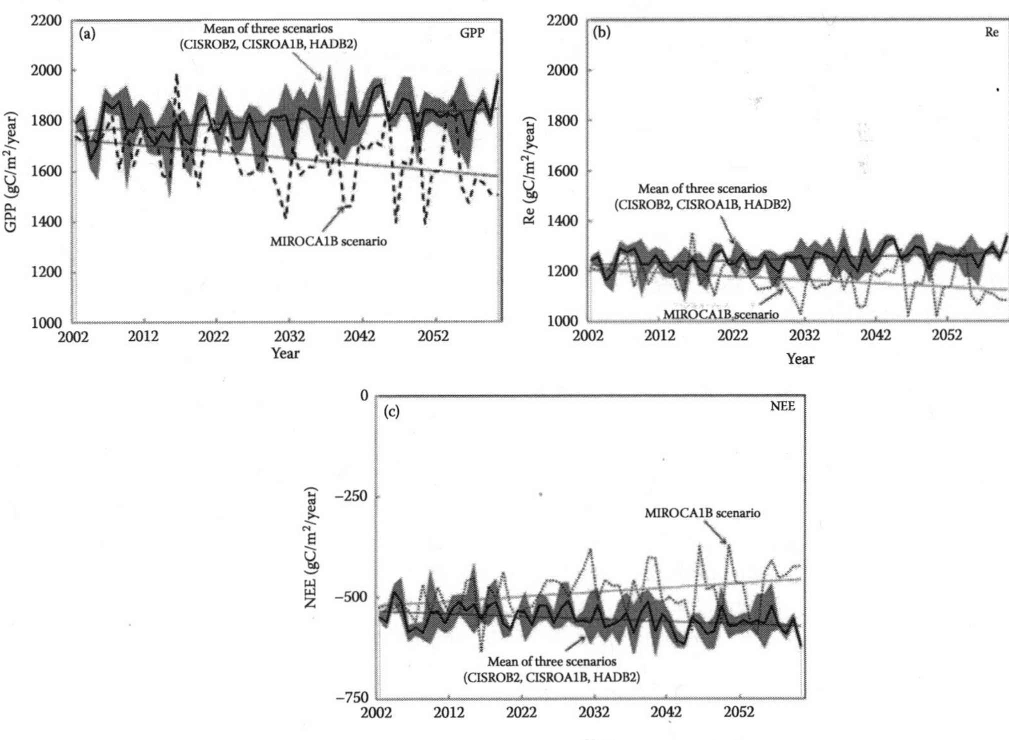

FIGURE 8.5 Predicted mean annual (a) Gross Ecosystem Productivity (GEP), (b) Ecosystem Respiration (Re), and (c) Net Ecosystem Exchange (NEE) across 674 watersheds in the Southern United States under four climate change scenarios.

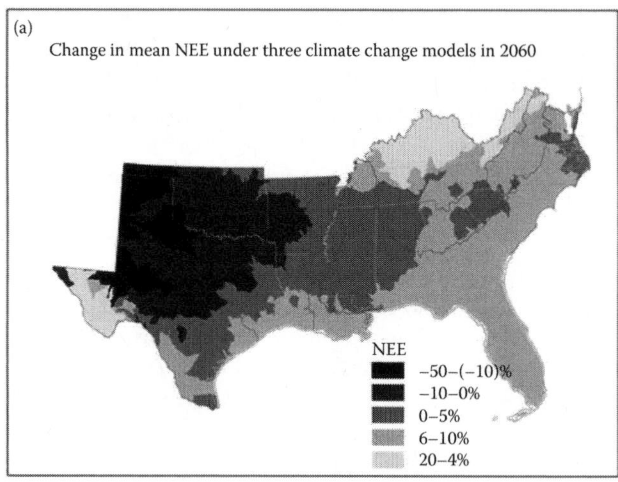

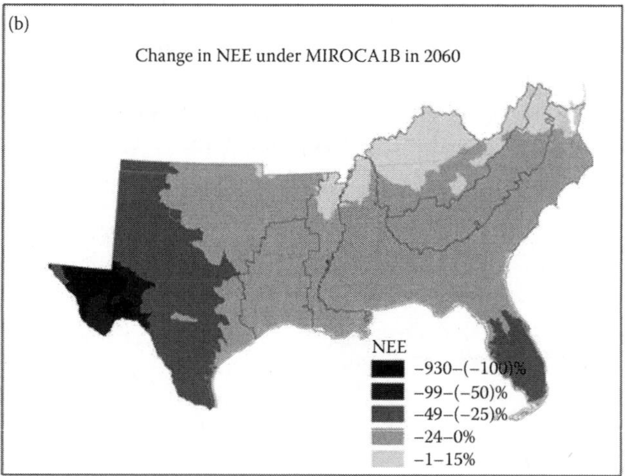

FIGURE 8.6 Predicted mean response of annual net ecosystem exchange (NEE) across 674 watersheds in the Southern United States to: (a) three climate change scenarios (CSIROA1B, CSIROB2, HADB2), and (b) MIROCA1B climate change scenario.

CASE STUDIES

The following case studies are presented to provide tangible examples of ways forest management can impact forest productivity and C sequestration. The first case study is a retrospective analysis of growth and aboveground *in situ* C sequestration comparing loblolly and longleaf pine planted in Mississippi in 1961. Using a combination of long-term plot data and modeling, the second case study examines the impact of thinning and future climate scenarios on aboveground productivity of yellow poplar. The third case study presents a modeling analysis on the impacts of rotation length, thinning, and planting density on *in situ* and *ex situ* C sequestration in loblolly and slash pine plantations.

Case Study 1—Relative Growth, Stocking, and Carbon Accumulation of Loblolly and Longleaf Pines in the Mississippi Gulf Coast

In the United States Gulf Coast region, climate change is anticipated to alter not only temperature and precipitation, but also the frequency and severity of tropical storms and hurricanes. Comprehensive

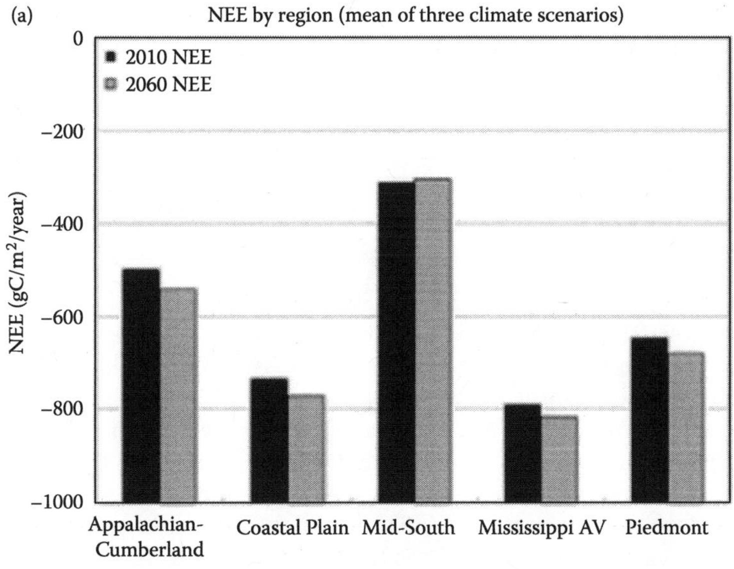

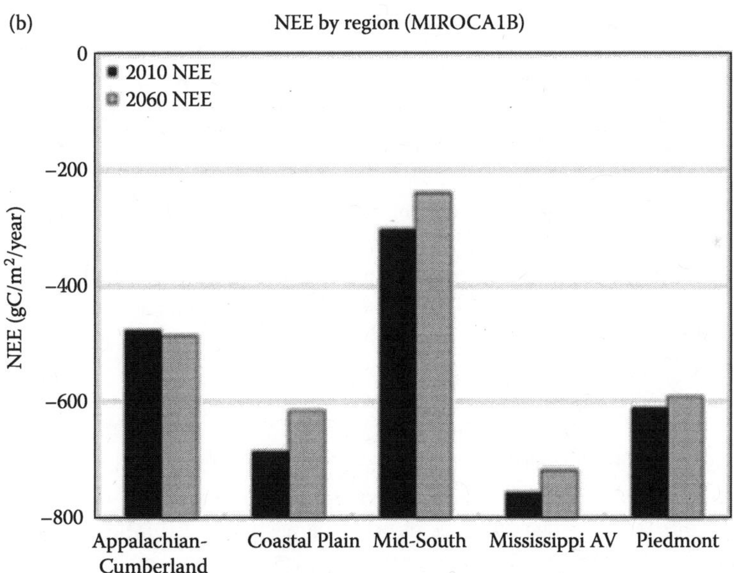

FIGURE 8.7 Summary of change in net ecosystem exchange (NEE by) subregion (a) under three climate scenarios (CSIROA1B, CSIROB2, HADB2), and (b) under MIROCA1B.

management of forests for wood products, fuel load reduction, C sequestration, and ability to withstand frequent disturbance is highly desirable, but the outcomes of prior management decisions are difficult to assess decades after they are made. Long-term studies on experimental forests can be quite valuable to objectively assess the impacts of management as forests mature. In this case history of the Species by Management Intensity study at the Harrison Experimental Forest (Smith and Schmidtling 1970), we highlight the impact of pine species, site preparation, and fertilization on aboveground productivity. While this study was originally intended to examine pine species' response to management and the heritability and expression of desirable wood properties in 1960, by inventorying the experiment over the years it has proven valuable for assessing species-specific growth and yield, susceptibility to wind damage, and the cumulative impact of frequent storms on stocking and C accumulation. In this case history, loblolly pine (*Pinus taeda*), a fast growing, easy to propagate pine that is very responsive to intensive forest management is compared with longleaf pine (*Pinus palustris*), which was historically more dominant in the Gulf Coast, but was passed over due to difficulty with propagation and slow early growth in favor of other pine species. This comparison will show the importance of selecting the most appropriate genetic material (in this case, species) suited to the current and future site conditions and microclimates to meet management goals.

Materials and Methods

Site and experimental design: The site (30.65N, 89.04W, elevation 50 m) is located 32 km north of Gulfport, Mississippi. The soils are variable, but best described by the Poarch series (coarse-loamy, siliceous, semiactive, thermic Plinthic Paleudults) and the Saucier-Susquehanna complex with well-drained upland, fine sandy loams and slopes from 1% to 4%. This case history uses a subset of the treatments and species from an experimental planting established in 1960 after a second rotation longleaf pine stand was clear cut, as originally described by Schmidtling (1973). Using a split-plot design with four blocks, the whole plots were randomly assigned one of two species (loblolly and longleaf pine) and the split plots were cultural treatments of varying intensity. The three split plots were: (1) no cultivation or fertilization (CON); (2) cultivated with no fertilization (CULT); (3) cultivated with a single application of 112, 224, or 448 kg ha^{-1} of NPK fertilizer (10-5-5) (CULT + F). Cultivated plots were cleared of all stumps and slash, plowed, and then disked prior to planting. They were then disked three times each season for 3 years to reduce woody competition and then mowed in years 4 and 5. Fertilizer was applied 1 year after planting.

Tree growth and aboveground C determination: In February and March 1961, 100 1-year-old bare root seedlings were bar-planted with 3.05 m spacing in each square measurement plot enclosed by an additional two rows of buffer trees. Growth through 25 years (fall 1984) has been reported by Smith and Schmidtling (1970), Schmidtling (1973), and Schmidtling (1987). This case history uses height data collected at ages 2, 3, 4, 5, 8, 9, 25, and 39, and diameter at breast height at ages 8, 9, 25, and 39. Longleaf pine aboveground biomass (AGB) was determined by fitting an equation that estimates ln(AGB) as a function of $\ln(D^2)$ using data from Garbett (1977). Loblolly pine AGB was estimated from equations reported by Jokela and Martin (2000). Aboveground biomass was converted to aboveground C by multiplying by 0.5.

Results

Height and diameter: In both loblolly and longleaf pines, mean tree height increased with intensity of cultural treatments in the first 15 years. Trees were smaller in the CON and increased from CULT to CULT + F (Figure 8.8). By age 39, trees of both species were still taller in the CULT + F treatment, but mean tree height in CON was larger than in CULT. Despite earlier gains in height by loblolly pine, by age 25, longleaf pine was taller than loblolly pine within the same cultural treatment (Figure 8.8). At age 9, DBH increased with cultural intensity and loblolly was greater than longleaf (Figure 8.9). Through age 39, loblolly in the highest intensity of management (CULT + F) maintained the largest diameter, closely followed by longleaf pine. By this age, diameters of unfertilized loblolly and cultivated longleaf were 27% lower than fertilized loblolly and longleaf. Diameters of longleaf in CON were only 9% lower than fertilized treatments by the end of the study, though it is likely that early mortality led to fewer larger trees, confounding direct comparisons.

Stocking and aboveground carbon accumulation: While mean tree statistics describe the individual trees in a plot, differences in survivorship between species and treatments led to differences in stocking, affecting mean tree parameters. Throughout the experiment loblolly pine stocking was very consistent and there were no substantive differences between treatments (Figure 8.10). Longleaf pine suffered mortality in the first several years of the experiment. Survivorship in CON was particularly poor and stocking remained much lower than the other treatments for the rest of the experiment (Figure 8.10). Without fire, cultivation, or herbicide for weed control, longleaf pine can

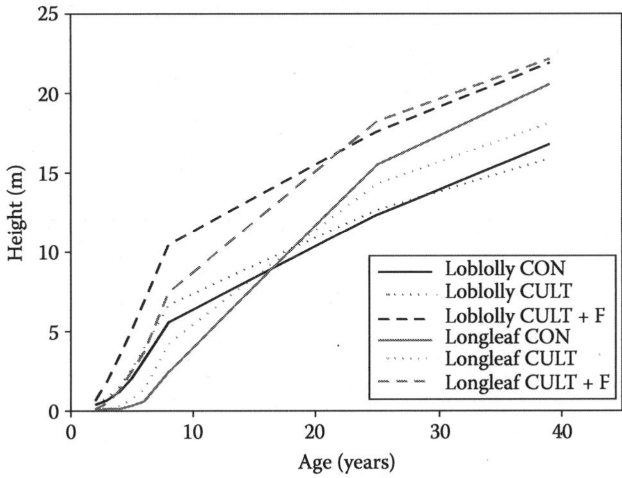

FIGURE 8.8 Height of loblolly and longleaf pines at the Species by Management Intensity study at the Harrison Experimental Forest, located in Saucier, Mississippi, through age 39. Abbreviations in legend are as in text.

be quickly outcompeted. By age 25, longleaf pine stocking in CULT and CULT + F was higher than any of the loblolly treatments. Total aboveground biomass converted to Mg C ha^{-1} is a useful metric that combines individual tree parameters with stocking data, giving the sum total C accumulated in an experimental plot. Loblolly pine in CULT + F accumulated 36% less C than longleaf CULT + F, while loblolly CULT was 45% less than longleaf CULT; there was virtually no difference between the species in CON (Figure 8.11). Clearly, if stocking can be maintained through management activities in early years, longleaf pine will accumulate more C than loblolly pine on this Mississippi Gulf Coast site, albeit given the genetic sources used in this study.

Effects of management intensity and species on soil C and N: In addition, despite large differences in aboveground C, soil C and N were not influenced by species selection at age 49 (Butnor et al. 2011). The treatments did result in long-term differences in soil C and N. In the upper 10-cm of soil, CULT had 24% less C than the average C content of CON and CULT + F, which were not

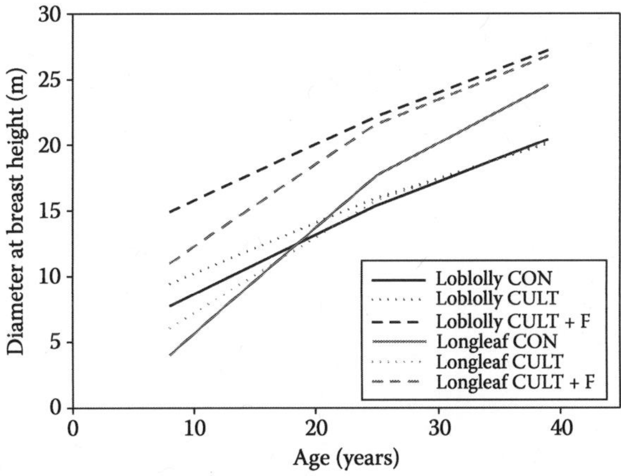

FIGURE 8.9 Diameter at breast height of loblolly and longleaf pines at the Species by Management Intensity study located at the Harrison Experimental Forest, located in Saucier, Mississippi, through age 39. Abbreviations in legend are as in text.

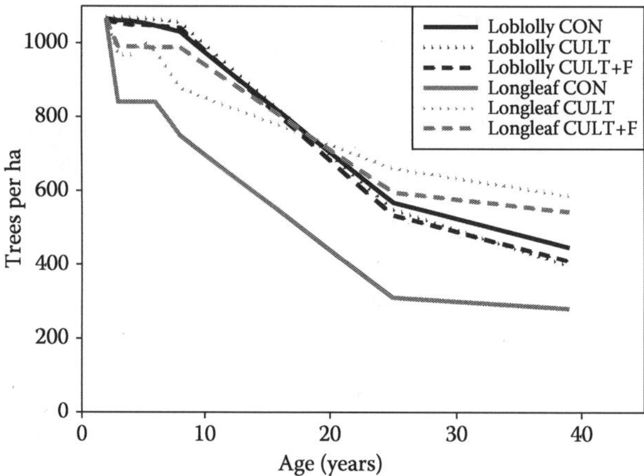

FIGURE 8.10 Stocking of loblolly and longleaf pines at the Species by Management Intensity study located at the Harrison Experimental Forest, located in Saucier, Mississippi, through age 39. Abbreviations in legend are as in text.

significantly different; in the 10- to 20-cm depth interval, that difference grew to 32%. Soil N content in CULT was 38% lower than the average C content of CON and CULT + F, which were not significantly different (Butnor et al. 2011).

Discussion

This case history comparing loblolly and longleaf pine demonstrates the importance of selecting appropriate genetic material (in this case species) suited to the current and future site conditions and microclimates to meet management goals. On productive soils or with the addition of fertilizer, loblolly pine will have greater productivity than longleaf pine in most of its range. This is especially true if longleaf pine seedling survival is poor. Over the long term, in disturbance prone environments, the advantage shifts to longleaf pine. Being more resistant to wind-related mortality loblolly pine (Johnsen et al. 2009), it is better able to survive in regions with frequent storm intervals. Given the rapid height growth of loblolly early in the rotation (Figure 8.8), there has been widespread

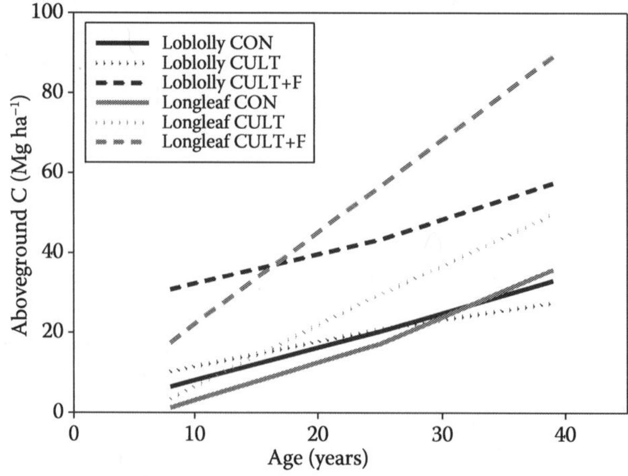

FIGURE 8.11 Cumulative aboveground C accumulation in loblolly and longleaf pines through age 39. Abbreviations in legend are as in text.

planting of loblolly in the range historically dominated by longleaf. If storm frequency and intensity increase as predicted in coming decades, deciding to plant loblolly pine stands will entail a level of risk. The Harrison Experimental Forest is impacted by damaging winds every 5–10 years, resulting in gradual declines in stocking of planted pines. The cumulative effects are more pronounced in loblolly than longleaf as stands age. There is always the risk of catastrophic losses from direct impacts of hurricanes, but if storm frequency increases in regions where they did not occur before, growth and yield are likely to suffer. Industrial forestland owners and managers may be in a position to potentially mitigate these risks by using intensive management and short rotations, but managers of public-owned land and small private landowners may find that deploying better suited, more resilient genetic material to be most feasible.

The results of this experiment show that in the conditions found at the Harrison Experimental forest, longleaf pine will accumulate more aboveground C than loblolly if early survival is maintained. Much less is known about the relative differences in belowground C storage and the recalcitrance of roots and decaying materials between these species. After 49 years, there was no significant effect of pine species on soil C, but soil C accumulates slowly and it may require multiple rotations for the effects to become evident, if at all. Stump removal caused reductions in soil C and soil N in CULT, which was mitigated by a one-time fertilizer application in CULT + F. Thus, the importance of residual stumps for soil C sequestration and site fertility was pronounced.

Case Study 2—Influence of Past and Future Climate on the Growth of Yellow Poplar (*Liriodendron tulipifera* L.) in the Southern Appalachian Mountains

Yellow poplar (*Liriodendron tulipifera* L.) is a mesophytic species that grows in highly productive forest stands in the eastern portion of the Central Hardwood Region. According to FIA estimates (Thompson 1998; Schweitzer 2000; Brown 2004; Rose 2007), yellow poplar constitutes ~15% of the total live-tree volume in the southern Appalachians. This shade-intolerant tree species is an aggressive competitor on mid- to high-quality sites throughout the region. Although a highly productive timber species, yellow poplar is sensitive to drought, with reduced growth during periods of reduced precipitation (Beck 1985; Kolb 1990; Elliott and Swank 1994; Orwig and Abrams 1997; Klos et al. 2009).

Although the role of climate in regulating tree growth is well established, relatively few studies have addressed the role active management may have in altering climate–growth relationships. In Finland, Mäkinen et al. (2002) observed that the growth of Norway spruce was negatively correlated with temperature and positively correlated with precipitation, and the variation in growth was similar in stands thinned to different densities. In contrast, Misson et al. (2003) found the growth of Norway spruce in heavily thinned stands was less affected by drought than trees in lightly thinned or unthinned stands.

In the southern Appalachians, minor changes in topography have a significant influence on climatic patterns, species composition, and site quality, all of which influence productivity. Yellow poplar is one of the most productive forest types in the southeastern United States. Information regarding climate–growth relationships under past climatic conditions as well as under varying management histories and site qualities may be used to inform practitioners of management activities that may increase resilience of a dominant tree species to climate change. Results presented here will help identify stands most susceptible to climate-related reductions in productivity, and provide guidelines on management activities that best offset those negative effects.

Methods

Study area: This study uses data collected as part of a long-term study examining the growth and yield of yellow poplar throughout the southern Appalachians. Between 1961 and 1964, 1410.1 ha growth and yield plots were established in yellow poplar stands throughout the Blue Ridge and northern Ridge and Valley Provinces of the southern Appalachian Mountains. Plots were located in northern Georgia, western North Carolina, and southern Virginia. All plots were established in

TABLE 8.1
Pre-Thinning (1961) and Post-Thinning (1966) Stand Attributes ($n = 104$)

Stand Attribute	Mean	Min.	Max.	Standard Deviation
Yellow poplar SI (m)	32.3	22.9	40.2	3.3
Age				
Pre-thinning	49	18	76	15
Post-thinning	54	23	81	15
Trees/ha				
Pre-thinning	245	80	700	134
Post-thinning	194	70	510	101
Dq (cm)				
Pre-thinning	34.4	15.9	51.4	8.3
Post-thinning	38.3	17.3	54.5	8.2
Basal area (m^2/ha)				
Pre-thinning	19.6	8.9	35.0	6.8
Post-thinning	19.9	9.2	34.7	6.7

even-aged stands dominated by yellow poplar across a range of initial ages, SI, and structures (Table 8.1).

Data collection: At the time of plot establishment and prior to the thinning treatment, all trees >11.4 cm diameter at breast height (DBH; 1.37 m above ground line) within each plot were tagged. For all trees, species, DBH, and total height were recorded. Following the initial inventory, plots received a low thinning to a residual basal area (BA; m^2/ha) at least one 6 m^2/ha BA class less than the pre-thinning BA. After the second inventory cycle (1966–1969), 128 of the 141 permanent plots were thinned for a second time to the originally assigned residual BA. No subsequent thinnings followed. Re-measurement of all plots occurred every 5 years through 1991. During each inventory cycle, the status of all tagged trees was assessed and DBH was recorded on all live trees.

Statistical analysis: Statistical analysis was performed using only those plots located in the Blue Ridge Physiographic Province in Georgia and North Carolina. An additional 10 plots were removed from analysis due to harvesting and incomplete data, making 104 of the original 141 plots available for use in data analysis. Regression analysis was used to model the effects of age, stand structure (i.e., BA), SI, and climate on BA at 5-year intervals. Because weather data obtained from stationary weather stations do not correspond to precipitation observed at a given plot (Beck 1985), average

TABLE 8.2
Five-Year Average Growing Season Precipitation (May through September), Maximum Growing Season Temperature, June and July Precipitation, and the 5-Year Average Ratio of June to July Precipitation (q) Used in the Model Building Process

Variable	Mean	Min.	Max.	Standard Deviation
Precipitation (mm)	725	446	1025	141
Maximum temperature (°C)	25.9	23.1	27.8	0.9
Q	1.16	0.74	2.29	0.26

Source: Daly, C. et al. 2004. Up-to-date monthly climate maps for the conterminous United States. Proceedings of 14th AMS Conference on Applied Climatology, 84th AMS Annual Meeting Combined Preprints. American Meteorological Society: Seattle, WA, January 13–16, 2004, Paper P5.1, http://ams.confex.com/ams/pdfpapers/71444.pdf.

Note: Climate data were obtained for each of the 104 plots using the climate interpolation program, PRISM.

monthly climate data between 1961 and 1999 for each plot (Table 8.2) were obtained using the PRISM climate model, which is specifically designed to interpolate climate data for mountainous terrain (Daly et al. 1994). Climate variables tested included average 5-year total growing season precipitation (mm), with growing season defined as the months of May through September, average 5-year maximum growing season temperature (°C), and the 5-year average ratio of June to July precipitation (q). This q ratio was investigated because work by Beck (1985) suggests that the amount of precipitation received during the month of July in addition to the ratio of precipitation in June versus July, as opposed to total growing season precipitation, best explains annual diameter growth in yellow poplar stands.

Models were fitted using stand and site attributes SI, age, and BA along with the climate variables. Interactions between climate variables and stand and site attributes were also examined. The best model was chosen based on the model resulting in the lowest Akaike's information criterion (AIC) values. The covariance structure used to account for the autocorrelation among measurements that occurred on each independent plot was modeled using first-order autoregressive structure that allowed for heterogeneous variances. Basal area was $\log_e$-transformed to achieve normality and homoskedasticity. The model was fitted using Proc Mixed in SAS/STAT® software, version 9.3 (SAS Institute, Inc.) with a significance level of $\alpha = 0.05$.

Effects of future climate on the productivity of yellow poplar: To compare the productivity in terms of BA development of the 104 stands located on the Blue Ridge Providence of the Appalachian Ecoregion, the climate-sensitive BA model developed above was used to project BA at two different time intervals. The first interval was the time period between 1960 and 2010, and the second interval was for the time period between 2010 and 2060. In the first interval, PRISM data provided the requisite climate data, while stand attributes (e.g., age, SI, BA) observed following the second thinning (Table 8.1) were used as starting points in the modeling process. In the second projection interval, future climate data downscaled to the county level provided the requisite climate data (Coulson et al. 2010), while again, stand attributes observed after the second thinning were used as starting points in the modeling process. Future climate data were obtained from two IPCC climate scenarios coupled with two GCM combinations: (1) A1B/MIROC scenario, which predicts extreme future climatic conditions, and, therefore, represents the upper bound of the productivity predictions; and (2) B2/HAD scenario, which predicts a more moderated climate, and, therefore, represents the lower bound of the productivity predictions. Basal area for each plot under the 1960–2010 and two future climatic conditions were compared and contrasted at each time step (i.e., 5-year interval).

Results

The climate data obtained from PRISM for the 1961–1999 time period coinciding with the periodic inventories of the 104 plots was variable (Table 8.2). The 5-year average ratio of June to July precipitation (q) coinciding with the inventory cycles was the most significant predictor of BA over time relative to the other climate variables examined. When q was included in the model, the 5-year average rainfall received in July, as suggested by Beck (1985), was not significant and, therefore, not included in the model. The addition of a temperature variable increased AIC, and was therefore not included. The final climate-sensitive BA model for yellow poplar predicted BA as a function of SI, age, BA, q, along with the interactions between q and age and q and BA:

$$\ln \text{BA}_2 = b_0 - b_1\,(1/\text{SI}) + b_2\,(1/A_2) + b_3\,((A_1/A_2)\ln \text{BA}_1) - b_4\,(q) - b_5\,(q \times A_2) - b_6\,(q \times \ln \text{BA}_1)$$

where SI is site index, A_1 is stand age at the previous inventory, A_2 is stand age at the current inventory, BA_1 is stand BA at the previous inventory, q is the 5-year average ratio of June to July precipitation, and b_0, b_1, b_2, b_3, b_4, b_5, and b_6 are estimated model parameters (Table 8.3).

Climate under the A1B/MIROC and B2/HAD future scenarios varied substantially from the climate normals used in the baseline scenario (Table 8.4). Although average q for the baseline and A1B/MIROC scenario were similar, the A1B/MIROC scenario had the greatest variability

TABLE 8.3
Estimated Model Parameters (Standard Error) and Associated AIC and Approximate R^2 Values for the Final Climate-Sensitive Basal Area (BA) Projection Model

b_0 (intercept)	b_1 (1/SI)	b_2 ($1/A_2$)	b_3 ((A_1/A_2) * ln BA_1)	b_4^* (q)	b_5 ($q * A_2$)	b_6 (q * ln BA_1)	AIC	R^2
0.2813 (0.0671)	2.6528 (0.7496)	15.0290 (0.4148)	0.9565 (0.0181)	0.0443 (0.0385)	0.0007 (0.0001)	0.0345 (0.0118)	−2386.9	0.98

Note: The p-values for all parameters were <0.05 except for those noted.
* Parameter estimate not significant at $\alpha = 0.05$.

in climate over the 50-year projection period. In contrast, average q under the B2/HAD scenario was 21% lower than under the baseline, with variability surrounding q lower than in both the A1B/MIROC and baseline scenarios.

The effects of climate, precipitation in particular, were significant in the final model. However, on average there was little difference in BA of yellow poplar stands over time under the climate scenarios examined in this study. This was likely due to the high degree of variability, as opposed to a strict increase or decrease in precipitation, during the 50-year projection period in both the future and baseline scenarios. At the end of the 50-year projection period (2060), the average difference between predicted BA under the baseline scenario and the future A1B/MIROC scenario was negligible at 0.07 m^2/ha while the average difference in total BA between the baseline scenario and the future B2/HAD scenario was 1.12 m^2/ha. The increase in BA predicted to occur in the 104 plots between 2010 and 2060 varied between 29% and 270% for the baseline scenario, 28% and 278% for the A1B/MIROC scenario, and 27% and 253% for the B2/HAD scenario.

TABLE 8.4
Five-Year Average Growing Season Precipitation (May through September), 5-Year Average Maximum Growing Season Temperature, and Average 5-Year Ratio of June to July Precipitation (q) Corresponding to the Baseline (1961–2006), A1B/MYROC (2011–2060), and B2/HAD (2011–2060) Scenarios

	Mean	Min.	Max.	Standard Deviation
	Baseline Scenario with Climate Normals (1961–2006)			
Precipitation (mm)	665	455	1049	110
Maximum temperature (°C)	25.8	22.6	27.6	1.2
Q	1.16	0.69	1.75	0.27
	A1B/MYROC (2010–2060)			
Precipitation (mm)	537	353	832	102
Maximum temperature (°C)	29.0	24.8	32.9	1.7
Q	1.12	0.57	2.09	0.41
	B2/HAD (2010–2060)			
Precipitation (mm)	712	502	1036	112
Maximum temperature (°C)	28.1	23.9	30.9	1.6
Q	0.87	0.59	1.21	0.17

Note: Climate normals used in the baseline scenario were only available from 1961 to 2006.

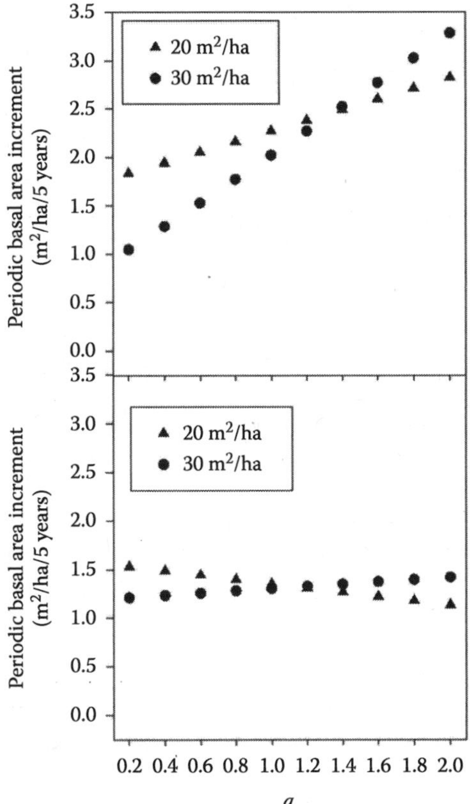

FIGURE 8.12 Predicted 5-year periodic basal area increment (BAI; m^2/ha/5 years) for a 50-year-old stand (a) and 100-year-old stand (b) in response to density (20 m^2/ha versus 30 m^2/ha) and *q* values (as described in text).

Although the average BA for all 104 plots under the two future climate scenarios after 50 years was relatively similar to the average BA projected using past climate normals, there was substantial variability in the 5-year periodic projections. Regardless of age, lower density stands experience an increase in BA increment relative to higher density stands at low *q* values. This trend is reversed when *q* increases beyond a given value. The *q* value at which BA increment in higher density stands exceeds that in lower density stands is dependent upon stand age (Figure 8.12). The reversal in the trend in BA increment occurs at slightly higher *q* values for younger rather than older stands. Although the patterns in BA increment in young versus old stands can be described by normal stand development/production patterns, trends between BA increment and climate are visible. Young stands experience an increase in BA increment with increasing *q* values. In older stands, BA increment appears to increase with increasing *q* values in higher density stands, while BA increment decreases with increasing *q* values in higher density stands (Figure 8.12).

Discussion

Based on this, as well as other published studies (Pan et al. 1997; Fekedulegn et al. 2008; Klos et al. 2009), the growth of yellow poplar can be considered to be sensitive to past and future fluctuations in precipitation. Contrary to the findings of Pan et al. (1997), temperature was not found to significantly influence the accumulation of stand-level BA over time. Using BA projections based on past climate normals as a reference, the effects of altered precipitation predicted to occur under the A1B/MYROC and B2/HAD scenarios on BA accumulation were minor. Although the overall average BA accumulation differed little among the climate scenarios, substantial variation among 104 plots was observed. Much of this variation was due to differences in the interaction between *q* and age and *q* and initial stocking among the 104 plots.

The fact that stand structure and age, two factors that can be controlled through active management, significantly interact with climate to control growth and productivity suggests there are management strategies that can be utilized to increase resiliency of yellow poplar stands to a changing climate. According to the model, as stocking increases, the sensitivity to decreased precipitation increases. Although the sensitivity of the accumulation of BA to *q* decreases as stands age, the general decline in stand productivity associated with stand age (Ryan et al. 1997) is evident. Aging of forest stands in the southern Appalachians coupled with extensive forest management may leave highly productive yellow poplar stands susceptible to decreased productivity over the next 50 years. Thinning may make yellow poplar less vulnerable to decreased BA production under future climate scenarios during summers when June rainfall is low relative to July rainfall (i.e., low *q* values). The decision to thin these stands, for a relatively small increase in production may, however, conflict with the economics surrounding thinning in what are generally extensively managed systems.

The model presented here does not include loss of production due to mortality. Consequently, any decrease in stand-level production due to drought-induced mortality is not accounted for in the model. Given that most of the yellow poplar mortality that occurred in the dataset used in this study was due to stem breakage and/or wind-related (e.g., uprooted) events, it does not appear that drought is a significant or common cause of yellow poplar mortality (Klos et al. 2009). Under past climatic conditions, precipitation throughout the Blue Ridge was equally distributed throughout the year. Diameter growth begins in June and is complete by the end of August (Beck 1985). Changes in the length of growing season, in particular earlier leaf-out due to increased spring temperatures, could alter when diameter growth commences and terminates. Consequently, the model presented here, where June and July rainfall are the only climate variables influencing BA growth, may not adequately describe future conditions.

Although yellow poplar is sensitive to altered precipitation, the effect of climate on BA accumulation is dwarfed by the influence of stand age, site quality, and stand structure. The nonexistent role of absolute temperature or precipitation in the model is likely a function of two factors: (1) the fact the model was built using 5-year average climate and stand data as opposed to annual climate and stand data; and (2) the position on the landscape where yellow poplar is predominant. Even under the most extreme future climate scenario, A1B/MIROC, average growing season precipitation approximates the minimum total annual precipitation that occurs throughout the current range of yellow poplar (Beck 1990). In the southern Appalachians, yellow poplar grows best on moist, well-drained soils with high available soil moisture common to stream bottoms, coves, and moist slopes (Beck and Della-Bianca 1981). The plots used to develop the climate-sensitive growth model are all highly productive stands where available moisture exceeds potential evapotranspiration even during periods where precipitation may be below normal (Beck 1985). Consequently, this model may not accurately describe the growth response of yellow poplar outside high-quality sites in the Blue Ridge Physiographic Province (e.g., ridge-tops, Piedmont region, southern extent of its range, etc.).

Case Study 3—Effects of Silviculture on Carbon Balance of Loblolly Pine and Slash Pine Plantations in Southeastern United States

An important advantage of forest management approaches to CO_2 mitigation is that silvicultural technologies are well developed, in place, and inexpensive to apply. In managed forests, C stocks can be divided into two major pools: *in situ* C in standing biomass (above- and belowground) and soil organic matter, and *ex situ* C sequestered in products created from harvested wood (Marland and Marland 1992). Sustainable forest management has the potential to greatly influence both *in situ* and *ex situ* C pools (Johnsen et al. 2001a). Loblolly pine and slash pine play an important role in mitigation of CO_2 emissions due to their high productivity and extensive planting throughout the southeastern United States. Accurate determinations of C stocks and understanding of factors controlling C dynamics in both species are essential for C offset projects and the development of sustainable management systems.

In this case study, we assessed the effects of silvicultural management on forest C stocks per unit area in loblolly pine and slash pine plantations established in the southeastern U.S. Coastal Plain. We used the models reported by Gonzalez-Benecke et al. (2010 and 2011a), to account for net C stock dynamics, and to address the following questions: (1) To what extent do extended rotations increase C sequestration in southern pine forests? (2) Is initial planting density an important factor for C storage? (3) What is the effect of thinning on net C storage for each species? (4) Under similar site quality and silvicultural treatments, which species accumulates more C? The models used in this case study were validated against published data of net ecosystem production, forest floor mass, and above- and belowground living pine biomass.

Materials and Methods

Models: Allometric and biometric equations were combined with growth and yield models to estimate C stocks and dynamics for loblolly pine and slash plantations in the southeastern United States (Gonzalez-Benecke et al. 2010 and 2011a). *In situ* C stock (C stored in living pine trees + understory + forest floor + coarse woody debris + standing dead trees) was determined using growth and yield models for loblolly pine (Harrrison and Borders 1996; Logan et al. 2002) and slash pine (Pienaar et al. 1996; Martin et al. 1997; Yin et al. 1998; Bailey et al. 1980), and combined with allometric and biometric equations reported for both species (for details of equations see Gonzalez-Benecke et al. 2010, 2011a). The models use quadratic mean diameter (QMD) and number of trees per hectare, estimated by the growth and yield models, as inputs for biomass equations to determine above- and belowground C stock. Projected LAI and litterfall were estimated from the model reported by Gonzalez-Benecke et al. (2012), where LAI was set to be proportional to SI and stand density index, and current year needlefall was set to be correlated with previous year mean annual projected LAI. The model also determined litterfall from needlefall using age-related needlefall-to-litterfall ratios for loblolly and slash pine. Forest floor biomass accumulation was determined as the sum of yearly litterfall inputs corrected for decay loss using the equation to estimate decay rate of the forest floor as reported by Radtke et al. (2009). Understory biomass accumulation was estimated from the equation reported by Gonzalez-Benecke et al. (2010), which predicts understory biomass as a function of stand LAI. At the time of thinning, reductions in pine LAI were set to be proportional to reductions in BA due to thinning and, therefore, needlefall, litterfall, and forest floor and understory biomass were affected due to their LAI-dependence. Standing dead trees, estimated from mortality equations of the growth and yield models, were also incorporated into the dead component of total biomass. At thinning and final harvest (clear-cut), logging slash (root and crown biomass plus stem residues) from harvested trees were also included into flux calculations and incorporated into the dead biomass pool. Stem residues were obtained by assuming a harvest efficiency of 88% and 87% of stand volume, for loblolly pine and slash pine, respectively (Bentley and Johnson 2008a, 2008b, 2009). Initial C accumulated from the previous rotation in coarse root debris, forest floor, and aboveground coarse woody debris was assumed to be ~55.4 Mg C ha^{-1} (Pehl et al. 1984; Van Lear and Kapeluck 1995; Clark et al. 2004; Miller et al. 2006; Eisenbies 2009).

To estimate *ex situ* C pool (C stored in wood products), harvested groundwood estimated from thinnings or clear cuts with the growth and yield models was assigned to three main product classes, sawtimber (ST), chip-and-saw (CNS), and pulpwood (PW), using the models proposed by Harrison and Borders (1996) and Pienaar et al. (1996). Industrial conversion efficiencies of 65%, 65%, and 58% were assigned to ST, CNS, and PW, respectively (Spelter and Alderman 2005; Smith et al. 2006). In addition, all the product types were divided into four life span categories (Liski et al. 2001; Gundimeda 2001) and adapted to loblolly pine and slash pine utilization patterns in the southeastern United States (Row and Phelps 1991; Skog and Nicholson 1998).

The model also accounts for C costs of silvicultural operations. Emissions of C by silvicultural activities were determined from Markewitz (2006) and White et al. (2005). C release in transportation of raw material from the forest to the mill was estimated according to White et al. (2005), assuming an average distance of 100 km from forest to mill, load per logging truck of 24 m^3, and fuel economy of diesel logging truck of 2.6 km L^{-1}.

Net C stock (Mg C ha^{-1}) was defined as follows: Net C stock = Total C *in situ* (C stored in living pine trees + understory + forest floor + coarse woody debris + standing dead trees) + Total C *ex situ* (C stored in wood products ST + CNS + PW) – Total C cost (silvicultural activities, including transportation of logs to mills). Carbon mass (Mg C ha^{-1}) was calculated by using an average C content of 50% for pine and understory biomass components (Clark et al. 1999; Johnsen et al. 2004). The models did not include changes in soil C. It was assumed that C storage in soil was not affected by forest management in southern pines plantations (Gholtz and Fisher 1982; Harding and Jokela 1994; Han et al. 2007; Johnson and Curtis 2001).

Silvicultural management scenarios: The effects of silvicultural treatments (planting density, thinning, and rotation length) on C sequestration were analyzed by simulating C flux under different scenarios of loblolly pine and slash pine plantations established in the southeastern U.S. Lower Coastal Plain. The initial stand parameters used for simulations (set equal for both species) were: base SI = 22 m, bedding, weed control at planting and at age 1, and NP fertilization at ages 5 (135 kg ha^{-1} N + 28 kg ha^{-1} P) and 10 (225 kg ha^{-1} N + 28 kg ha^{-1} P). After site preparation, fertilization, and weed control treatments, the model estimated that base SI was increased to effective SI of 25.3 and 25.8 m, for loblolly pine and slash pine, respectively. The effect of initial stand density was evaluated by running the model under planting densities of 500, 1500, and 2500 trees ha^{-1}. Rotation length effects on C sequestration were assessed by evaluating the model for 15, 30, and 45 years harvesting age. Both planting density and rotation length analysis were carried out under unthinned conditions. The effects of thinning (as a percentage of living trees removed) were assessed by evaluating the model under different combinations of thinning age (8, 12, and 16 years) and removal intensity (20%, 40%, and 60% of living trees removal), for stands with planting density of 1500 trees ha^{-1} and 25 years rotation length. For all simulations, estimates of average C stock were reported as the average of all yearly values for a simulation length of 175–180 years, depending on rotation age used. In the case of scenarios with rotation lengths of 15, 25, 30, and 45 years, the number of rotation cycles used to calculate net C stock were 12, 7, 6, and 4, respectively.

Results

Silvicultural management effects on C sequestration: Under conditions used in the simulations for unthinned loblolly pine stands, the average net C stock for a 180-years simulation length increased from 131 to 173 Mg C ha^{-1}, when rotation length was increased from 15 to 30 years (Table 8.5). If the rotation length was extended to 45 years, the average net C stock increased to 181 Mg C ha^{-1}, respectively. On unthinned slash pine stands with rotation lengths of 15, 30, and 45 years, the average net C stock was 111, 175, and 183 Mg C ha^{-1}, respectively. For both species, across rotation lengths, *in situ* C stock accounted for between 71% and 76% of the gross C sequestration (not including silvicultural C costs). The magnitude of emissions associated with silvicultural activities (including transportation) was between 1.6% and 1.9% of the gross C stock, with larger proportion on stands with shorter rotations (Table 8.5). The relative impact of the different woody products varied depending on rotation length scenario. For example, for unthinned loblolly pine stands, ST accounted for 1%, 11%, and 15% of gross C sequestration, for rotation lengths of 15, 30, and 45 years, respectively. In contrast, CNS followed an opposite trend, accounting for 27%, 15%, and 7%, for the same rotation length scenarios, respectively. In slash pine stands the proportion of ST was lower, accounting for 1%, 9%, and 10%, while CNS accounted for 25%, 17%, and 8% of gross C sequestration for the same rotation length scenarios, respectively (Table 8.5). Across different rotation lengths, the forest floor + dead trees + coarse woody debris (FF_D) components averaged 42 and 38 Mg C ha^{-1}, for loblolly pine and slash pine stands, respectively (between 31% and 34% of

TABLE 8.5
Average Carbon Stock for Loblolly Pine (LOB) and Slash Pine (SLA) Plantations for 175–180 Years Simulation Length under Different Rotation Length Scenarios

	15 Years		30 Years		45 Years	
C Pool	LOB	SLA	LOB	SLA	LOB	SLA
In situ	93.4	80.7	130.3	131.3	144.4	150.9
Living pine	50.7	44.2	87.3	90.0	102.4	110.1
Understory	1.1	1.7	1.0	1.8	0.9	1.8
Forest floor + dead trees	41.6	34.8	42.0	39.5	41.1	39.1
Ex situ	40.0	32.1	45.3	47.0	40.1	34.6
Sawtimber	1.5	0.4	18.9	16.3	27.2	18.6
CNS	35.8	28.3	25.6	29.6	12.6	15.5
Pulpwood	2.7	3.4	0.8	1.1	0.3	0.5
Silviculture emissions	−2.5	−2.1	−3.1	−2.8	−3.3	−3.0
Net C stock	130.9	110.6	172.5	175.4	181.3	182.5

Note: All units are average carbon stock (Mg C ha^{-1}) for first five rotations. All scenarios included base site index = 22 m, planting density = 1500 trees ha^{-1}, bedding, weed control at planting and at age 1, and NP fertilization at age 5 and 10 years.

total *in situ* C stock) and the understory averaged 1.0 and 1.7 Mg C ha^{-1}, for loblolly pine and slash pine stands, respectively (less than 2% of total *in situ* C stock). Between species, for 15-year rotation length, average living tree biomass for the first five rotations of loblolly pine stands was 6.5 Mg C ha^{-1} larger than that of slash pine, but as rotation length was extended to 45 years, slash pine living tree biomass catches up and exceeds loblolly pine's, averaging 7.6 Mg C ha^{-1}more.

The effect of planting density on average net C stock was similar for both species: planting more trees increased stand C sequestration. Loblolly pine and slash pine stands planted with 2500 trees ha^{-1} can store 43 and 36 Mg C ha^{-1} more than stands planted with 500 trees ha^{-1}, respectively (Table 8.6). For both species, the effect of planting density on average net C stock was largely reflected

TABLE 8.6
Average Carbon Stock for Loblolly Pine (LOB) and Slash Pine (SLA) Plantations for 180 Years Simulation Length under Three Different Planting Density Scenarios

	500 Trees ha^{-1}		1500 Trees ha^{-1}		2500 Trees ha^{-1}	
C Pool	LOB	SLA	LOB	SLA	LOB	SLA
In situ	94.7	92.8	122.2	119.2	138.6	135.9
Living pine	61.9	58.9	78.9	78.2	88.8	91.7
Understory	1.1	1.8	1.0	1.7	1.0	1.7
Forest floor + dead trees	31.7	32.2	42.4	39.2	48.9	42.5
Ex situ	45.8	49.8	46.2	47.4	45.7	43.5
Sawtimber	31.8	40.5	13.9	11.5	7.1	1.3
CNS	13.7	9.1	31.2	34.4	36.9	38.5
Pulpwood	0.3	0.2	1.1	1.5	1.7	3.7
Silviculture emissions	−2.9	−2.5	−3.3	−3.0	−3.4	−3.4
Net C Stock	137.5	140.1	165.2	163.5	180.9	176.1

Note: All units are average carbon stock (Mg C ha^{-1}) for first five rotations. All scenarios included base site index = 22 m, rotation length = 25 years, bedding, weed control at planting and at age 1, and NP fertilization at age 5 and 10 years.

in *in situ* rather than *ex situ* C pools. Across species, living pine tree biomass and FF_D were largely increased from ~60 and 32 Mg C ha^{-1} to ~90 and 46 Mg C ha^{-1}, respectively, as planting density increased from 500 to 2500 trees ha^{-1} (Table 8.6). The C storage in woody products was reduced by 0.1 and 6.2 Mg C ha^{-1}, for loblolly pine and slash pine stands, respectively, as planting density increased from 500 to 2500 trees ha^{-1}. Even though C storage in woody products was affected little by planting density, the C stored in different woody products varied depending on planting density scenario. For instance, ST decreased from 32 and 41 Mg C ha^{-1} to 7 and 1.3 Mg C ha^{-1}, for loblolly pine and slash pine stands, respectively, when planting density was increased from 500 to 2500 trees ha^{-1}. In contrast, CNS followed an opposite trend, increasing from 14 and 9 Mg C ha^{-1} to 37 and 39 Mg C ha^{-1}, for loblolly pine and slash pine stands, respectively, for the same planting density increment (Table 8.6). From a total silviculture C cost perspective, fertilization accounted for 0.57 Mg C ha^{-1} (two fertilizations), representing between 18% and 27% of the total silvicultural C emissions. Harvest and transportation of woody products accounted for more than 50% of the total silvicultural C emissions.

In general, after 200–250 years, C flux in the woody products converged to stable values, reaching quasi-equilibrium minimum and maximum values. At rotation ages of 15 and 45 years, *in situ* C stocks were 141 and 178 Mg C ha^{-1} for loblolly pine stands (Figure 8.13a and b), and 136 and 193 Mg C ha^{-1} for slash pine stands, respectively (Figure 8.13c and d). For loblolly pine stands, total woody products C stock increased each rotation from 57 and 102 Mg C ha^{-1} during the first rotation, up to 91 and 153 Mg C ha^{-1} at the end of the fifth rotation, for rotation ages of 15 and 45 years, respectively (Figure 8.13a and b). In the case of slash pine stands, *ex situ* C stock increased from 50 and 105 Mg C ha^{-1} during the first rotation, up to 77 and 152 Mg C ha^{-1} at the end of the fifth rotation, for rotation ages of 15 and 45 years, respectively (Figure 8.13c and d). Differences in tree size (diameter and height) and number of trees remaining due to different rotation age scenarios created different woody products pools that had different life spans. While for 15 years rotation length, PW represented 12% and 15% of *ex situ* C stock for loblolly pine and slash pine stands, respectively, ST accounted for 12% and 4% of *ex situ* C stock for loblolly pine and slash pine stands, respectively.

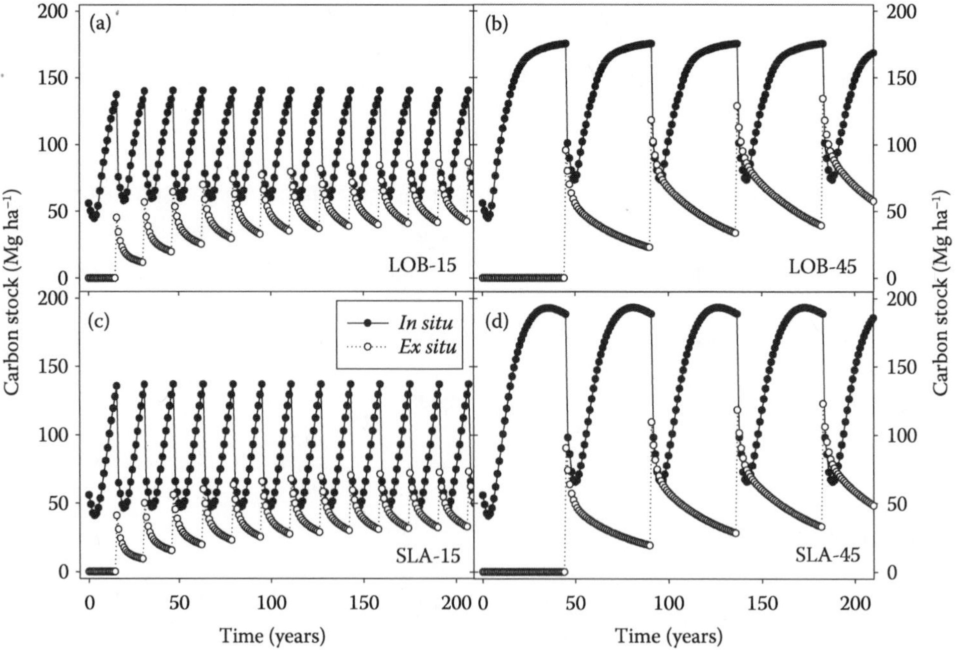

FIGURE 8.13 *In situ* and *ex situ* C stock for unthinned loblolly pine (LOB) and slash pine (SLA) plantations under different rotation length for a 200-year simulation period.

On the other hand, for 45-years rotation length, PW accounted for 2% of total *ex situ* C stock for both species, and ST accounted for 69% and 56% of total *ex situ* C stock for loblolly pine and slash pine stands, respectively (Figure 8.13a–d). In general, C stored in products derived from PW (e.g., paper, packing material, office supplies, etc.) had a negligible effect on net C sequestration; between harvest events (thinning or clear cutting), the amount of C stored diminished toward zero, while C stored in CNS and ST increased between harvests (data not shown).

In general, for a stand planted on land with a SI = 22 m with 1500 trees ha^{-1} and managed under a 25-year rotation, thinning had a small effect on net C stock. Even though the effect was minimal for loblolly pine (Figure 8.14e), there was a 4.9% overall reduction in net C stock in slash pine (Figure 8.14f). For both species, for any given thinning intensity, there was a small effect due to the age of thinning on *in situ*, *ex situ,* and net C stocks (Figure 8.14). Increments in thinning intensity produced a quasi-constant decline in *in situ* C stock, independent of thinning age. *Ex situ* C storage had an opposite response to thinning: the more intensive the thinning regime, the more gain in woody products C storage, and this effect was only slightly affected by thinning age (Figure 8.14c and d). The reduction of *in situ* C stock due to thinning was counteracted by increasing the *ex situ* C stock, producing a null effect on net C stock. However, for loblolly pine plantations with thinning intensity of 60%, there was a small (<2%) increment on net C stock. In the case of slash pine plantations, on the other hand, thinning reduced net C stock by 3.6%. Across thinning age and intensity, there was an average change in net C stock of 1.0 and –5.9 Mg C ha^{-1}, for loblolly pine and slash pine stands, respectively (Figure 8.14e and f).

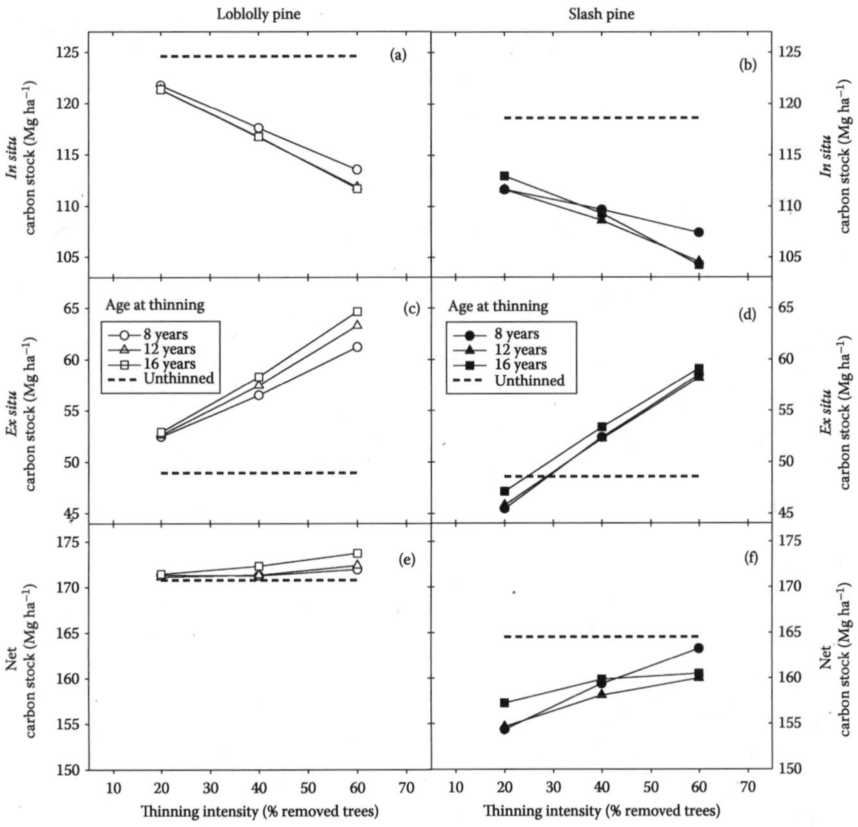

FIGURE 8.14 Effects of stand age at thinning and thinning intensity on average *in situ*, *ex situ*, and net C stock plantations for first five rotations for loblolly pine (open symbols, a, c, e) and slash pine (black symbols, b, d, f) plantations.

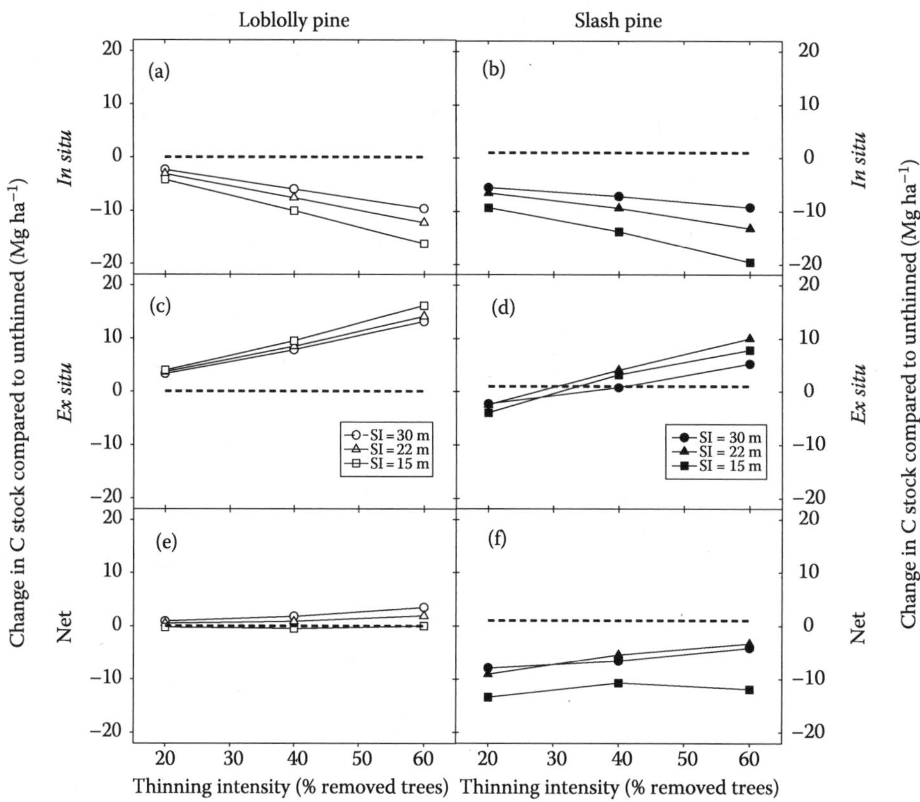

FIGURE 8.15 Effects of site index and thinning intensity on average *in situ*, *ex situ*, and net C stock for first five rotations for loblolly pine (open symbols, a, c, e) and slash pine (black symbols, b, d, f) plantations.

The combined effect of thinning intensity and SI was expressed as the change in C stock compared to unthinned stands. Overall, independent of stand SI, thinning has a small effect on average net C stock (Figure 8.15a–f). However, *in situ* C stock was more reduced in stand with base SI = 15 m (Figure 8.15a and b). The effect of low thinning intensity (i.e., 20%) on *in situ* C stock was similar across SI. For example, the reduction in *in situ* C stock was between 2.3 and 4.2 Mg C ha^{-1}, for loblolly pine stands, and between 5.5 and 9.3 Mg C ha^{-1}, for slash pine stands, both thinned with 20% removal intensity. When thinning intensity was increased to 60% removal of living trees, the reduction in *in situ* C stock was between 9.6 and 16.2 Mg C ha^{-1}, and 9.3 and 19.6 Mg C ha^{-1}, for loblolly pine and slash pine stands, respectively.

Discussion

When stand C density was compared between loblolly and slash pine under similar levels of site quality (base SI = 22 m) and silvicultural inputs over a 25-year rotation length, living pine C stocks of loblolly and slash pine were generally similar. Nevertheless, there was a trend that under shorter rotation (i.e., 15 years), net C stock of loblolly pine was 18.3% larger than slash pine, mostly due to faster initial growing rates of loblolly pine trees. This general trend has been documented elsewhere. For example, Colbert et al. (1990) reported, for 4-year-old trees, that loblolly pine stands had 63% more needle biomass and 36% more aboveground biomass than slash pine stands growing under similar site and silvicultural characteristics. At age 6, at the same site, Dalla-Tea and Jokela (1991) pointed out that loblolly pine stands had 40% more LAI and 68% more aboveground biomass than slash pine stands. Burkes et al. (2003) also found for 4-year-old trees, that leaf and fine root biomass of slash pine stands were significantly smaller than in the loblolly pine stands. However, on the same stands measured by Colbert et al. (1990) and Dalla-Tea and Jokela (1991), Vogel et al. (2011) reported that at age 26 years slash pine had 4.7% more *in situ* C stock than loblolly pine. The latter authors reported that under unfertilized conditions, even with sustained elimination of understory vegetation, living tree C stocks of slash pine were larger than loblolly pine. When nutritional limitations were eliminated through fertilizer additions, living tree stand biomass of loblolly pine was larger than of slash pine. As nutritional demands and the responses to fertilization for loblolly pine tend to be larger than slash pine (Jokela et al. 2010), differences in nutrient requirements and nutrient use efficiency between the two species should be taken in account when developing sustainable and ecological forestry regimes. In our analysis, the fertilization regime used included two applications, which may not be sufficient to support the demands of loblolly pine, especially under longer rotations scenarios.

Increasing initial planting density in the range tested in this study had a positive effect on net C storage, and the effects of planting density on C storage were most apparent in the *in situ* C pool, affecting both living tree biomass and FF_D biomass accumulation. Even though raising the planting density increased the proportion of fixed C used in stem production in loblolly pine (Burkes et al. 2003), this effect was not reflected in the *ex situ* C pool. As planting densities increased, there was a tendency to decrease sawtimber products yields, affecting the average *ex situ* C pools; however, the increase in forest floor, coarse woody debris, and total living tree C storage largely counteracted that negative effect.

Increasing the rotation length increased C stock in both species. Reports for other conifer species (Liski et al. 2001; Harmon and Marks 2002) have indicated similar effects of rotation length on C storage; that is, extended rotations increased C sequestration in conifer forest plantations. As Canadell and Raupach (2008) pointed out, longer harvesting cycles represent one of the major management strategies used to increase forest C density. Nevertheless, the inclusion of biomass harvest for fossil fuel offset might change our conclusions, especially when shorter rotations include provisions for improved technology at the end of each rotation. Further research is needed in this area, and this model is a tool to address these types of questions.

Due to compensating effects of *in situ* and *ex situ* C storage, net C stocks were only minimally affected by thinning. Most of the studies that have addressed the impacts of thinning on C budgets in pine ecosystems have only reported the responses of living pine biomass (Balboa-Murrias et al. 2006; Skovsgaard et al. 2006; Chiang et al. 2008; Finkral and Evans 2008), and only a few studies have reported the impacts of thinning on total *in situ* C (Boerner et al. 2008; Jiménez et al. 2011) or NEP fluxes (Campbell et al. 2009; Dore et al. 2010). All the previously cited studies concluded that there was a reduction in pine C sequestration after thinning. Garcia-Gonzalo (2007), in a similar analysis that included *ex situ* C pools for mixed coniferous stands in Finland, reported a net reduction between 25 and 33 Mg C ha^{-1} in living trees biomass and a net increase between 30 and 45 Mg C ha^{-1} in harvested timber. Even though the wood extracted in thinning was primarily pulpwood that had an impact on *ex situ* C sequestration, increased growth of residual trees due to thinning promoted the production of larger tree size classes at final harvest. These long-lived products increased the *ex situ* C pool, compensating for the reduction in *in situ* C associated with thinning. When *ex situ* pools were considered, the possible economic benefits of thinning were not in opposition to maintaining or increasing net C stock.

Reporting of C stock in harvested wood is not mandatory under the Intergovernmental Panel on Climate Change (IPCC 2006), but the enhancement of that C pool could provide important GHG emission offsets. The *ex situ* C pool could be influenced by both the final utilization of particular products, and also by substituting wood for more C-intensive materials. If waste wood and forest biomass residues were used as substitutes for fossil fuels (Galik 2009), or if long-lasting wood products take the place of more C-intensive materials like concrete or steel (Perez-Garcia 2005), then the mitigation impacts of *ex situ* C stocks could be even larger.

Using published models to evaluate the effects of planting density, thinning, and rotation length on C sequestration for loblolly pine and slash pine plantations in the southeastern United States, we conclude that: (1) shorter rotations were not as effective for C sequestration as extended rotations that increased average net C stock; (2) increasing initial planting density had a positive effect on net C storage; (3) if woody products, which accounted for ~30% of the net C stock, were incorporated into the C inventory, thinning will have a tendency to be C neutral because of the counteracting effects of *in situ* and *ex situ* C stocks; and (4) emissions due to silvicultural and harvest activities were small compared to the magnitude of the total stand C stock.

MANAGEMENT OPTIONS TO DEAL WITH CLIMATE CHANGE IMPACTS

THE POTENTIAL FOR AND APPLICATION OF BIOENERGY IN SOUTHERN FORESTS

In 2009, total energy usage in the United States was estimated at approximately 94.5 quadrillion BTUs (Quads). Sources of energy in decreasing order of importance were petroleum (37%), natural gas (25%), coal (21%), nuclear (9%), biomass (4%), and other sources such as hydroelectric, wind, geothermal, and solar (4% in total) (US DOE 2011). This dependence upon many nonrenewable and high C emissions sources places the United States in a challenging economic, political, and ecological position. In recent years, increasing emphasis has been placed on developing increased bioenergy production capacity in the United States. Excellent reviews of the early technological developments and pathways forward have been provided by Tolbert and Wright (1998) and Tuskan (1998). Later summaries of further developments and outstanding challenges have been provided by Mayfield et al. (2007), Buford and Neary (2010), and Dale et al. (2011). There are several driving forces behind this push for bioenergy, including potential reductions in greenhouse gas emissions from use of nonrenewable C sources, desire to maintain or increase direct C sequestration, the desire to decrease United States' dependence upon foreign energy supplies, and the desire to increase economic diversity and potential economic developments of many depressed rural economies. These diverse goals can be addressed through the increased production and utilization of renewable bioenergy crops in the United States. Increased bioenergy production goals were formalized in the Energy Independence and Security Act of 2007 (EISA). The U.S. Renewable Fuels Standards (RFS2) is an integral part of this Act and became effective in 2010. As part of this act, the United States has committed to producing and using 136 billion liters of renewable transportation fuel per year by 2022. A significant component of this target is currently being met through corn-based ethanol. Other assessments have also addressed potential bioenergy production scenarios in the United States and while not yet formally incorporated into governmental laws, they view the bioenergy production potentials in a longer-term and more aggressive way. In 2005 the Department of Energy (DOE) and Oak Ridge National Laboratory developed a report titled "Biomass as a feedstock for a bioenergy and bioproducts industry: the technical feasibility of a billion-ton annual supply," which was updated in 2011 (US DOE 2011). Total available biomass at any given time period is a function of the rate of technological advancements made with potential bioenergy crops and the prices paid by biomass consumers (US DOE 2011). Total biomass available in dry Mg is projected to be approximately 430 million Mg in 2012 based on current growth projections. This increases to 1.0–1.5 billion Mg under low and high rates of technological advancements, with a base cost of $54 per Mg of biomass delivered (US DOE 2011). Under these modeling approaches, two important changes occur that dramatically increase the supply of woody biomass. At higher prices, a portion of woody biomass in the southern states normally harvested for pulpwood production is redirected toward bioenergy production. In addition, dedicated energy crop production increases dramatically. By 2030, energy crop production has been projected to increase to approximately 635 million Mg under some production scenarios. This is approximately 45% of 1.5 billion Mg.

In order for these aggressive deployment projections to occur, continued technology developments are needed. For the southeastern United States, potential candidate lists of dedicated energy crops were first presented by Wright (1994). This list of candidate species has been updated and for the Southeast now includes switchgrass (*Panicum virgatum* L.), *Populus* species or hybrids, various tropical grasses, loblolly (*Pinus taeda* L.) or slash (*Pinus elliottii* Engelm) pine, sweetgum (*Liquidambar styraciflua* L.) and sorghum [*Sorghum bicolor* (L.) Moench] (Dale et al. 2011). In the more southern sections of the southeastern zone, eucalyptus species and hybrids are being investigated as a potential woody feedstock. Kline and Coleman (2010) summarized the potential advantages and disadvantages of the woody crop species based on previous experiences growing these species across the South. For *Populus*, the primary challenges to deployment included narrow site adaptabilities, variable productivity rates, intensive management and input requirements, and a general lack of drought tolerance. Potential advantages include high growth capacity under ideal conditions, extensive ongoing genetic research including genomic mapping, the existence of commercial stands, and the potential for further improvements in the medium term (Kline and Coleman 2010). For sweetgum, its advantages include being the most adaptable hardwood species across the region, being native to the region, having existing baseline research to support further development, and having the potential for improvement in the medium term. Disadvantages for sweetgum include more moderate productivity levels and more limited commercial deployment (Kline and Coleman 2010). Eucalypts possess the primary advantage of having the greatest potential growth rates among the woody species being considered. They are also generally adaptable to marginal site conditions, and have the potential for rapid improvement. Their primary limitation is susceptibility to cold damage, limiting potential deployment range. There are also potential environmental concerns with the deployment of an exotic species. In addition, their water use patterns merit additional research (Kline and Coleman 2010).

Among these potential species, there are wide variations in the relative stage of development of both the genetic resources of each species and the development of associated cropping regimes needed to support high yields needed to make production of these crops viable. Loblolly pine and slash pine are sharp contrasts to the species previously described. Pine plantation production systems are well established and widely deployed across the South and southeast. Current pine plantation regeneration and management techniques reflect over 60 years of concentrated research to optimize genetic sources, plant production techniques, site preparation procedures, plantation establishment densities and configurations and optimized competition control and nutrient management techniques. These sustained efforts have led to dramatic yield improvements and widespread adoption of pine plantation management across the South (Fox et al. 2007). Pine plantation acreage dramatically increased from less than a million hectares in 1952 to more than 13 million hectares in 1999 (Weir and Greis 2002). This represents a large and potentially available resource pool that could serve a key role in bioenergy production in the South.

Ultimately, the bioenergy crops that are developed and deployed across the South may be dependent upon current and future developments of conversion technologies. If direct burning of woody biomass or production of biofuels through gasification become the primary production processes, then pine may become the primary feedstock for these processes (Hinchee et al. 2009). Wood pellet production for heating and fuel purposes represents one significant and rapidly expanding market for pine biomass in the Southern United States. Global production expanded from approximately 8 million metric tons in 2007 to over 13 million tons in 2009. North American production constituted over 50% of worldwide wood pellet production in 2009 (Pirraglia et al. 2010). Consumption rates in Europe alone are projected to exceed 50 million tons per year in 2020 (Pirraglia et al. 2010). These growth rates are driven primarily by European Union agreements that require at least 20% of energy consumption to be provided by renewable energy sources by 2020. Rapid expansion in pellet production facilities is occurring across the United States with greatest production capacity expansions occurring across the South. Many of these facilities utilize abundant pine wood resources to meet growing demands and it seems that production and utilization of this resource will become increasingly important in the future.

In contrast to ethanol produced from corn, there is no commercial-scale cellulosic ethanol production in the United States currently and significant obstacles to commercial development exist.

These limitations include: (1) lignocellulosic feedstock costs and availability, (2) high pretreatment costs associated with recalcitrant cellulosic feedstocks, (3) enzyme costs, (4) excessive capital investment requirements, and (5) low ethanol prices (Gonzalez et al. 2011). These inherent limitations are further compounded when suitability of pine wood characteristics are considered. Pine wood is generally much more difficult to process than many alternate hardwood fibers and is currently much more suitable for bioenergy processes that utilize gasification or direct-firing technologies (Hinchee et al., 2009). It is possible to alter wood chemical or physical characteristics through either traditional breeding or genetic engineering approaches and these approaches are currently being used to reduce or modify lignin types and increase cellulose concentrations in woody crops. If successful, these approaches could facilitate cellulosic ethanol production efficiencies. These modifications, however, can have unintended, undesirable secondary effects including reductions in photosynthetic rates and increased cavitation potential (Nehra et al. 2005; Hinchee et al. 2009). These approaches can be further complicated since feedstock optimization for utilization by one bioenergy process actually make that feedstock less suitable for utilization by other processes. As an example, the caloric value of lignin is approximately 40% greater than the caloric value of cellulose (Hinchee et al. 2009). Decreasing lignin concentrations could make a feedstock more suitable for cellulosic ethanol production, but could actually make this material less desirable for pyrolysis-based technologies. Advancements in this area are advancing rapidly and this suggests production and processing changes may interact to influence production technology.

At the current time, southern pine including loblolly and slash pine represent a large and potentially accessible resource for bioenergy production in the Southern United States. Southern pines have been widely studied and their primary productivity limitations are fairly well understood. Among the characteristics that have led the widespread deployment of pine plantations across the South are the relatively broad site adaptability of pine and its relatively low sensitivity to soil moisture stress. Changing these inherent soil characteristics can be difficult or expensive and other potential bioenergy species having similar adaptations to these factors as pine would be desirable. Irrigation for woody crop bioenergy production would be cost prohibitive for all but the most unique niche applications. Selection of crop species that exhibit high productivity under only optimal soil moisture conditions would limit widespread deployment. Loblolly pine has demonstrated site adaptability and water use characteristics that are desirable. Munsell and Fox (2010) estimated that dedicated biomass production management regimes utilizing loblolly pine could produce approximately greater than 5 dry Mg C ha^{-1} $year^{-1}$ across wide areas of the South using currently available technologies, but that these regimes would be unlikely to maximize returns to landowners. Under current scenarios, integrated bioenergy and solid wood production systems are likely to maximize economic returns to landowners and it is assumed that the pine biomass produced in these integrated production systems could be utilized for either solid wood products or bioenergy wood depending upon market.

Jokela et al. (2004) summarized growth responses of loblolly pine in seven integrated silvicultural studies across the South, and these studies found growth responses ranging from 2- to 3.5-fold in the best treatments within each individual study. Stem biomass production in these studies ranged from approximately 32 to 83 Mg C ha^{-1} $year^{-1}$ at age 15 (annual production of approximately 2–5.7 Mg C ha^{-1} $year^{-1}$). Farnum et al. (1983) estimated that if the exhibited trends continued, the biological growth potential of loblolly pine could approach approximately 7 Mg C ha $year^{-1}$ if other resource limitations were alleviated. Sampson and Allen (1999) calibrated the BIOMASS model and estimated that loblolly pine annual stem growth could approach 8 Mg C ha^{-1} $year^{-1}$ with treatments that ameliorated resource limitations. Our most complete understanding of the role of water and nutrient limitations of stand productivity for southern pines is provided by studies that allow complete experimental control of water and nutrient availability. The Southeast Tree Research and Education Site (SETRES) study was established in the Sand Hills of North Carolina to understand these relationships. In this study, loblolly pine plantations are grown in a factorial experiment with two nutrient levels and two water levels (Albaugh et al. 1998, 2004) available. Results from this study supported the conclusion that nutrient limitations rather than water limitations were the primary limit to productivity in young to mid-rotation loblolly pine stands in the southeast and dramatic growth increases are possible with optimized stand nutrition.

In contrast to pine, hardwood silvicultural and tree improvement efforts in the South date back to the 1960s, but until recently hardwood fiber demand and the prices paid for hardwood sources were insufficient to sustain continuous, aggressive research efforts. Interest in and support for hardwood research programs have varied widely over the last 50 years and sustained progress has not been made. To date, progress with any of these species has limited their operational deployment to a mere fraction of the acreage dedicated to pine plantations. Dvorak and Hodge (1998) estimated that there were approximately 80,000 hectares of hardwood plantations across the South. Technological or economic limitations currently restrict wider deployment. Of the potential woody perennial bioenergy species listed for possible deployment across the South, long-term integrated research is far behind the current state of knowledge for pine research. This creates technical hurdles that limit successful implementation of a large-scale hardwood bioenergy production program (Kline and Coleman 2010).

Development of future bioenergy feedstocks will be closely linked to the productivity levels that can be achieved and the resulting economic benefits the landowner can accrue (US DOE 2011). Deployment of rapidly growing eucalyptus plantations has, in select geographic regions, the potential to produce large quantities of biomass on short rotations. Eucalyptus has been a genera of interest since the 1960s because of the remarkable productivity that can be obtained under ideal circumstances, that is, Brazil, but deployment in the southern United States has been limited by winter cold damage. Estimates of eucalyptus productivity in Brazil can range from approximately 4 Mg C ha^{-1} $year^{-1}$ for low intensity silviculture to yields exceeding 15 Mg C ha^{-1} $year^{-1}$ using intensive management regimes and elite clones (Stape et al. 2010). In the United States, eucalyptus research has focused on developing plants with sufficient cold tolerance to survive and grow well in the Southern United States. In the 1970s through mid-1980s, hundreds of eucalyptus species and individual seed sources were tested across the South and initial results were promising (Hunt and Zobel 1978), but a series of cold winters in the late 1970s and early 1980s, which caused catastrophic damage to test plantings, tempered enthusiasm for further research efforts (Jahromi 1982). In recent years, new research has been initiated to identify additional cold-tolerant eucalyptus genotypes by both private companies and by the Forest Productivity Cooperative at North Carolina State University and Virginia Tech. Current research indicates significant progress toward identifying species with good growth and cold tolerance. Several promising cold-tolerant eucalyptus species have been identified including *Eucalyptus benthamii, Eucalyptus macarthurii,* and *Eucalyptus viminalis.* In preliminary testing, these plants have demonstrated cold tolerance through USDA hardiness zones 8A with low winter temperatures of −12.2°C to −9.5°C.

There are significant differences between the bioenergy production scenarios currently being examined for the Southern United States and traditional forestry practices developed and practiced in the region. Management of bioenergy plantations is likely to be substantially more intensive than traditional forestry practices. Integrated cropping regimes will be utilized that combine the most appropriate genetics sources with more intensive site preparation, competition control practices, and nutrient management regimes. During peak plantation production ages, bioenergy plantations will carry more leaf area than traditional plantations. This will have the potential to create greater nutrient demands and the potential for greater water use to sustain high productivity. These changes will place greater demands on site resources to sustain productivity. Understanding and managing these potential changes will be critical to effectively manage these production systems. Process models may be the best available tool to integrate responses from a wide diversity of existing studies (Johnsen et al. 2001b). Various models have been developed for loblolly pine and many of these models focus on the relationships between nutrient availability, nutrient uptake, leaf area production, and subsequent plantation growth responses (Landsberg et al. 2001; Ducey and Allen 2001; Comerford et al. 2006; Sampson et al. 2006). Use of these existing or development of new models directly tailored to bioenergy production scenarios may be the most efficient way to evaluate effects of bioenergy production in the Southern United States.

Previous studies have indicated that biomass removals associated with harvest do have the ability to reduce nutrient capital on sites, but it is more difficult to demonstrate that plantation productivity is consistently and negatively impacted by increasing biomass removals (Johnson and Todd 1998; Powers et al. 2005). Concerns for excessive nutrient removals may be more valid after several subsequent rotations and when hardwood species rather than pine are the crop species of interest. Cation removals in particular can be higher for hardwoods compared to pine and these removals increase with increasing harvest utilization, that is, whole-tree harvests versus stem-only harvests (Johnson et al. 1982). Whole tree versus stem only harvests had N, P, K, and Ca removals 2.9, 3.1, 3.3, and 2.6 times greater, respectively, than removals for bole only harvests in Tennessee. The total removal of nutrients during harvesting was closely related to total biomass removals in harvesting, which were 2.6 times greater in the whole-tree harvest. Fifteen years following harvest in the same study there were no differences in tree growth, soil C or N concentrations or soil bulk density. There were however differences in soil and plant K, Ca, and Mg between the treatments (Johnson and Todd, 1998). These results suggest that greater removal of biomass does have the potential to alter soil nutrient status for some nutrients but that these changes could be relatively small and difficult to detect especially over shorter sampling intervals.

Harvesting impacts including varying levels of biomass removals and soil compaction effects have been investigated in other studies. In the late 1980s and 1990s, the U.S. Forest Service installed a series of studies called the Long-Term Soil Productivity Studies (LTSP) across the United States. These studies include differing levels of organic matter removals and soil compaction as core treatments. In general, whole-tree harvest did not negatively impact plant growth relative to stem-only harvest treatments. Assessments after 10 seasons across a range of 26 sites indicated that plant productivity was unaffected by biomass removals during harvest. Soil compaction treatments again yielded variable responses, with plant production increasing on sandy soils in response to compaction, and decreasing on compacted clay soils (Powers et al. 2005). Harvesting impacts were also examined by Eisenbies et al. (2006, 2007). In this study, loblolly pine growth and stand level productivity were not adversely affected by organic matter removal or soil physical disturbance if these factors were ameliorated by bedding treatments. The complex and often subtle responses to organic matter removal or soil physical disturbance suggest a degree of resiliency in many systems to short-term changes. Over longer time periods the full effects are unknown. Given the projected yields of short rotation woody crop species and short harvest intervals, it is likely that potential nutrient removals will need to be balanced by subsequent nutrient amendments to increase or even sustain productivity (Heilman and Norby 1998). There could be negative long-term impacts, but the relative responsiveness in many systems to silvicultural treatments suggests that these potential limitations could be corrected.

The final factor influencing bioenergy crop production, which may in turn be influenced by bioenergy cropping regimes, is water availability. The inherent assumption in current bioenergy developmental plans is that bioenergy crops will be developed on marginal farmlands, pastures, or other underutilized lands. Conversion of crop lands to perennial woody crop systems has the potential to increase water use. This occurs through two factors: generally greater canopy interception and associated water losses and greater annual evapotranspiration. Annual transpiration will be greater for woody perennial crops due to longer active growth periods. Merely converting from idle cropland to any highly productive bioenergy crop would be expected to increase water use. The magnitude of these water use increases can be substantial. Amatya et al. (2002) calculated that loblolly pine stands in North Carolina could have annual evapotranspiration rates approximately 30% greater than adjacent croplands.

Other Management Options

Pine plantation management provides a clear opportunity for managing forests for productivity and C sequestration in light of climate change (Groninger et al 1999; Johnsen et al. 2001a, 2004). Over the past 50 years, the productivity of planted pine has tripled (Fox et al. 2007a). In the Southern United States, N availability typically limits pine productivity (Fox et al. 2007b). Over 400,000 ha year^{-1} of pine plantations are now fertilized with N (Albaugh et al. 2007). Nitrogen fertilization increases productivity directly but it may also act synergistically to increase productivity by eliminating nutrition as a limiting factor (see above section on elevated CO_2). Fertilization can also directly decrease C losses via soil respiration (Butnor et al. 2003) under high and low atmospheric CO_2 concentrations, likely by decreasing heterotrophic respiration (Gough and Seiler 2004). Intensive pine management often includes site preparation; on wetter sites this often includes bedding. In an extensive review of the literature, Jandl et al. (2007) indicate that soil disturbance is the clearest means of decreasing soil C and abruptly increasing soil respiration (Gough et al. 2005). Other management tools that directly impact C sequestration include species selection (see case study 1), modifying initial planting density and rotation length (see case study 3), and thinning (see case studies 2 and 3).

Genetics

Rapid warming and/or changes in precipitation will likely result in local genetic populations being less well adapted. In the Western United States many species have steep altitudinal and latitudinal clines, and such within-species maladaptation could be substantial (St. Clair and Howe 2007). In the east, clinal variation of important conifers with extensive ranges appears less distinct but is still important, and such information on geographic variation is used to deploy seed sources. Deployment guidelines (Schmidtling 2001) were partly based on data from the Southwide Southern Pine Seed Source Study (SSPSSS). For example, the SSPSS for loblolly pine was conducted using 15 wild seed sources at many different locations across the species range. Data were collected between the 1950s and 1980s, and therefore under historical weather conditions. Schmidtling (1994) used these data to create regression models to relate growth to temperature variables. He predicted that "local" loblolly pine sources will experience a 10% decrease in height if mean annual temperature increases by 4°C. Longleaf pine and shortleaf pine were estimated to decrease by 12% and 5%, respectively. Such predictions indicate that seed transfer or breeding zones will have to be modified in the future (Spittlehouse and Stewart 2004). Schmidtling's (1994) approach only predicted responses to temperature but there is also variation in disease, insect, and drought resistance (Wells and Wakeley 1966, van Buijtenen et al. 1976; Wells 1983; Gonzalez-Beneke and Martin 2010). Also, provenances of loblolly pine west of the Mississippi River appear more drought tolerant than Atlantic Coastal Plain sources. However, because they grow faster, Atlantic Coastal Plain sources are commonly planted in the Mid-South subregion (Lambeth et al. 2005). Such a deployment strategy may need to be reconsidered as the climate in the Mid-South becomes hotter and drier.

Nearly all planted loblolly pine are the result of at least one cycle of genetic improvement (McKeand et al. 2006). Family variation in drought tolerance has been documented in other species (Johnsen et al. 1999). Breeding for ideotypes (Martin et al. 2001; Nelson and Johnsen 2008), genotypes with a specific combination of desirable traits, has been proposed as a way to match genotype to site. Maier et al. (2012) identified two highly productive loblolly pine genotypes that carry disparate amounts of leaf area, suggesting they also differ in resource water and/or nutrient efficiency. Ideotype development and deployment, using relatively short rotation lengths, might provide a means to respond to rapidly changing environmental conditions. Additionally, more efficient nutrient and water use might result in such genotypes taking better advantage of increased atmospheric CO_2 concentrations.

In the Southern United States, such extensive provenance and progeny test data, as discussed above, are mostly available for the major southern pines. For less intensively managed species whose ranges are predicted to shift dramatically, contract, or disappear, diversity will need to conserved via *in situ* and *ex situ* genetic conservation (Ledig 1986). For many species, decisions on genetic deployment or conservation in light of climate must be made based on evaluations of life history and ecological traits (Aitken et al. 2008). Such an approach is being taken by the

National Forest (Region 8) in the Southern United States. Potter and Crane (2010) have developed the "Forest Tree Genetic Risk Assessment System: A Tool for Conservation Decision-Making in Changing Times." Although not solely addressing climate change impacts, it accounts for threats to species or species populations as they interact with climate change. The system deals with intrinsic characteristics (population structure, fecundity, and seed dispersal) and external threats (climate change, insects, and diseases). Twelve species and species groups have been identified for conservation (Crane 2011): Atlantic white cedar, red spruce, seven threatened or endangered oaks, several ashes, butternut, longleaf pine, pitch pine, Fraser fir, September elm, yellow buckeye, and Ozark chinquapin. One method proposed to engender *ex situ* and *in situ* conservation of such species is to establish restoration seed reserves (RSR) (Echt et al. 2011). RSRs will be made up of progeny from trees selected across the species range planted *ex situ* and culled to produce seed production areas high in genetic variation. Populations of seedlings produced from seed from RSRs will presumably genetically buffer forests from climate change effects and other biotic threats, creating more resilient forests.

REFERENCES

Aitken, S.N., S. Yeaman, J.A. Holliday, T. Wang, and S. Curtis-McLane. 2008. Adaptation, migration or extirpation: Climate change outcomes for tree populations. *Evolutionary Applications* 1:95–111.

Albaugh, T.J., H.L. Allen, P.M. Dougherty, L.W. Kress, and J.S. King. 1998. Leaf-area and above- and below-ground growth responses of loblolly pine to nutrient and water additions. *Forest Science* 44:317–328.

Albaugh, T.J., H.L. Allen, P.M. Dougherty, and K.H. Johnsen. 2004. Long term growth responses of loblolly pine to optimal nutrient and water resource availability. *Forest Ecology and Management* 192:3–19.

Albaugh, T.A., H.L. Allen, and T.R. Fox. 2007. Historical patterns of forest fertilization in the southern United States from 1969 to 2004. *Southern Journal of Applied Forestry* 31:129–137.

Amatya, D.M., G.M. Chescheir, R.W. Skaggs, and G.P. Fernandez. 2002. Hydrology of poorly drained coastal watersheds in Eastern North Carolina. In *ASAE Annual International Meeting/CIGR XVth World Congress*. Online, https://elibrary.asabe.org/azdez.asp?JID=5&AID=9852&CID=cil2002&T=2

Amthor, J.S. 1994. Plant respiratory responses to the environment and their effects on the carbon balance. In: *Plant—Environment Interactions*. 501–554. M. Dekker, New York, NY, USA.

Bailey, R.L., N.C. Abernethy, and E.P. Jones. 1980. Diameter distributions models for repeatedly thinned slash pine plantations. In *Proceedings of the 1st Biennial Southern Silvicultural Research Conference*, ed. J.P. Barnett, New Orleans, LA: U.S. 115–126. Department of Agriculture, Forest Service, Southern Forest Experiment Station G.T.R. SO-34. New Orleans, LA, USA.

Balboa-Murrias, M.A., R. Rodriguéz-Soalleiro, A. Merino, and J.G. Alvarez-González. 2006. Temporal variations distribution of carbon stocks in aboveground biomass of radiata pine and maritime pine pure stands under different silvicultural alternatives. *Forest Ecology and Management* 237:29–38.

Baldocchi, D.D. 2003. Assessing the eddy covariance technique for evaluating carbon dioxide exchange rates of ecosystems: Past, present and future. *Global Change Biology* 9:479–492.

Beaubien, E.G. and H.J. Freeland. 2000. Spring phenology trends in Alberta, Canada: Links to ocean temperature. *International Journal of Biometeorology* 44:53–59.

Beck, D.E. 1985. Is precipitation a useful variable in modeling diameter growth of yellow poplar? In: *Proceedings of the Third Biennial Southern Silvicultural Research Conference,* ed. E. Shoulders. Forest Service Southern Forest Experiment Station G.T.R. S0–54. 555–55.

Beck, D.E. 1990. *Liriodendron tulipifera* L.: Yellow poplar. In: *Silvics of North America*: *2. Hardwoods,* eds. R.M. Burns and B.H. Honkala, 406–416. U.S. Department of Agriculture Agricultural Handbook 654. Washington DC, USA.

Beck, D.E. and L. Della-Bianca. 1981. *YellowPoplar: Characteristics and Management.* U.S.D.A. Agricultural Handbook 583. Washington DC, USA.

Bentley, J.W. and T.G. Johnson. 2008a. *South Carolina Harvest and Utilization Study, 2007.* Forest Service Southern Research Station Resource Bulletin RS-140. Asheville, NC, USA.

Bentley, J.W. and T.G. Johnson. 2008b. *Alabama Harvest and Utilization Study, 2008.* Forest Service Southern Research Station Resource Bulletin RS-141. Asheville, NC, USA.

Bentley, J.W. and T.G. Johnson. 2009. *Florida Harvest and Utilization Study, 2008.* Forest Service Southern Research Station Resource Bulletin RS-162. Asheville, NC, USA.

Binkley, D., J.L. Stape, M.G. Ryan, H.R. Barnard, and J. Fownes. 2002. Age-related decline in forest ecosystem growth: An individual-tree, stand structure hypothesis. *Ecosystems* 5:58–67.

Boerner, R.E.J., H. Huang, and S.C. Hart. 2008. Fire, thinning, and the carbon economy: Effects of fire surrogate and treatments on estimated carbon storage and sequestration rate. *Forest Ecology and Management* 255:3081–3097.

Boisvenue, C. and S.W. Running. 2006. Impacts of climate change on natural forest productivity—Evidence since the middle of the 20th century. *Global Change Biology* 12:862–882.

Bradford, M.A., C.A. Davies, S.D. Frey et al. 2008. Thermal adaptation of soil microbial respiration to elevated temperature. *Ecology Letters* 11:1316–1327.

Brown, M.J. 2004. Forest statistics for the mountains of North Carolina, 2002. In: *Resources Bulletin, SRS-87*. Forest Service Southern Research Station. Asheville, NC, USA.

Buford, M.A. and B.J. Stokes. 2000. Incorporation of biomass into forest soils for enhanced productivity, restoration, and biostorage: A modelling study to evaluate research needs. *New Zealand Journal of Forest Research* 30:130–137.

Buford, M.A. and D.G. Neary. 2010. Sustainable biofuels from forests: Meeting the challenge. In: *Ecological Society of America Biofuels Sustainability Reports*. Washington, DC. Online, http://www.esa.org/biofuelsreports

Burkes, E.C., R.E. Will, G.A. Barron-Gafford, R.O. Teskey, and B. Shiveret. 2003. Biomass partitioning and growth efficiency of intensively managed *Pinus taeda* and *Pinus elliottii* stands of different planting densities. *Forest Science* 49:224–234.

Butnor, J.R., K.H. Johnsen, R. Oren, and G.G. Katul. 2003. Reduction of forest floor respiration by fertilization on both carbon dioxide-enriched and reference 17-year-old loblolly pine stands. *Global Change Biology* 9:849–861.

Butnor, J.R., K.H. Johnsen, F.G. Sanchez, and C.D. Nelson. 2011. Impacts of pine species, cultivation and fertilization on soil properties half a century after planting. *Canadian Journal of Forest Research* 42:675–685.

Calfapietra, C., B. Gielen, A.N.J. Galema et al. 2003. Free-air CO_2 enrichment (FACE) enhances biomass production in a short-rotation poplar plantation. *Tree Physiology* 23:805–814.

Callaway, R.M., E.H. DeLucia, and W.H. Schlesinger. 1994. Biomass allocation of montane desert ponderosa pine: An analog for response to climate change. *Ecology* 75:1474–1481.

Campbell, J., G. Alberti, J. Martin, and B.E. Law. 2009. Carbon dynamics of a ponderosa pine plantation following a thinning treatment in the northern Sierra Nevada. *Forest Ecology and Management* 257:453–463.

Canadell, J.G. and M.R. Raupach. 2008. Managing forests for climate change mitigation. *Science* 320: 1456–1457.

Carter, K.K. 1996. Provenance tests as indicators of growth response to climate change in 10 north temperate tree species. *Canadian Journal of Forest Research* 26:1089–1095.

Case, M.J. and D.L. Peterson. 2005. Fine-scale variability in growth-climate relationships of Douglas-fir, North Cascade Range, Washington. *Canadian Journal of Forest Research* 35:2743–2755.

Castro, M.S., H.L. Gholz, K.L. Clark, and P.A. Steudler. 2000. Effects of forest harvesting on soil methane fluxes in Florida slash pine plantations. *Canadian Journal of Forest Research* 30:1534–1542.

Chambers, J.Q., J.I. Fisher, H. Zeng, E.L. Chapman, D.B. Bake, and G.C. Hurtt. 2007. Hurricane Katrina's carbon footprint on U.S gulf coast forests. *Science* 318:1107.

Chapin, F.S., P.A. Matson, and H.A. Mooney. 2002. *Principles of Terrestrial Ecosystem Ecology*. Springer-Verlag, New York, USA.

Chapin, F.S., G.M. Woodwell, J.T. Randerson et al. 2006. Reconciling carbon-cycle concepts, terminology, and methods. *Ecosystems* 9:1041–1050.

Chiang, J.M., R.W. McEwan, D.A. Yaussy, and K.J. Brown. 2008. The effects of prescribed fire and silvicultural thinning on the aboveground carbon stocks and net primary production of overstory trees in an oak-hickory ecosystem in southern Ohio. *Forest Ecology and Management* 255:1584–1594.

Chmielewski, F.M. and T. Rötzer. 2001. Response of tree phenology to climate change across Europe. *Agricultural Forestry Meterology* 108:101–112.

Churkina, G. and S.W. Running. 1998. Contrasting climatic controls on the estimated productivity of global terrestrial biomes. *Ecosystems* 1:206–215.

Clark, K.L., H.L. Gholz, J.B. Moncrieff, F. Cropley, and H.W. Loescher. 1999. Environmental controls over net exchanges of carbon dioxide from contrasting Florida ecosystems. *Ecological Applications* 9:936–947.

Clark, K.L., H.L. Gholz, and M.S. Castro. 2004. Carbon dynamics along a chronosequence of slash pine plantations in North Florida. *Ecological Applications* 14:1154–1171.

Colbert, S.R., E.J. Jokela, and D.G. Neary. 1990. Effects of annual fertilization sustained weed control on dry matter partitioning, leaf area, and growth efficiency of juvenile loblolly and slash pine. *Forest Science* 36:995–1014.

Comerford, N.B., W.P. Cropper Jr., L. Hua et al. 2006. Soil supply and nutrient demand (SSAND): A general nutrient uptake model and an example of its application to forest management. *Canadian Journal of Soil Science* 86:665–673.

Coulson, D.P., L.A. Joyce, D.T. Price et al. 2010. *Climate Scenarios for the conterminous United States at the county spatial scale using SRES scenarios A1B and A2 and PRISM climatology*. Fort Collins, CO: U.S. Forest Service, Rocky Mountain Research Station. Online, http://www.fs.fed.us/rm/data_archive/data-acess/US_ClimateScenarios_county_A1B_A2_PRISM.shtml.

Crane, B.S. 2011. *CC projects summary R8 Southern Region Genetic Resource Management Program*. Online, ftp://ftp2.fs.fed.us/incoming/r6/ro/genetics/cc_wksp/R8/.

Curtis, P.S. and X. Wang. 1998. A meta-analysis of elevated CO_2 effects on woody plant mass, form, and physiology. *Oecologia* 2:299–313.

Curtis, P.S., P.J. Hanson, P. Bolstad et al. 2002. Biometric eddy-covariance based estimates of annual carbon storage in five eastern North American deciduous forests. *Agricultural Forestry Meterology* 113:3–19.

Dale, V.H., L.A. Joyce, S. Mcnulty et al. 2001. Climate change forest disturbances. *BioScience* 51:723–734.

Dale, V.H., K.L. Kline, L.L. Wright, R.D. Perlack, M. Downing, and R.L. Graham. 2011. Interactions among bioenergy feedstock choices, landscape dynamics, and land use. *Ecological Applications* 21:1039–1054.

Dalla-Tea, F. and E.J. Jokela. 1991. Needlefall, canopy and light interception productivity of young intensively managed slash and loblolly pine stands. *Forest Science* 37:1298–1313.

Daly, C., R.P. Neilson, and D.L. Phillips. 1994. A statistical-topographic model for mapping climatologically precipitation over mountainous terrain. *Journal of Applied Meterology* 33:140–158.

Daly, C., W.P. Gibson, M. Doggett, J. Smith, and G. Taylor. 2004. Up-to-date monthly climate maps for the conterminous United States. Proceedings of 14th AMS Conference on Applied Climatology, 84th AMS Annual Meeting Combined Preprints. American Meteorological Society: Seattle, WA, January 13–16, 2004, Paper P5.1, http://ams.confex.com/ams/pdfpapers/71444.pdf.

DeLucia, E.H., J.E. Drake, R.B. Thomas, and M. Gonzalez-Melers. 2007. Forest carbon use efficiency: Is respiration a constant fraction of gross primary production?. *Global Change Biology* 13:1–11.

Dewar, R.C. 1993. A root shoot partitioning model-based on carbon-nitrogen water interactions and Munch phloem flow. *Functional Ecology* 7:356–368.

Dore, S., T.E. Kolb, M. Montes-Helu et al. 2010. Carbon and water fluxes from ponderosa pine forests disturbed by wildfire and thinning. *Ecological Applications* 20:663–683.

Drake, B.G. and M.A. González-Meler. 1997. More efficient plants: A consequence of rising atmospheric CO_2? *Physiology and Plant Molecular Biology* 48:609–639.

Drake, J.E., A. Gallet-Budynek, K.S. Hofmocke et al. 2011. Increases in the flux of carbon below ground stimulate nitrogen uptake and sustain the long-term enhancement of forest productivity under elevated CO_2. *Ecology Letters* 14:349–357.

Ducey, M.J. and H.L. Allen. 2001. Nutrient supply fertilization efficiency in midrotation loblolly pine plantations: A modeling analysis. *Forest Science* 47:96–102.

Dvorak, W.S. and G.R. Hodge. 1998. Wood supply strategies in countries with fast-growing plantations. *Paper Age* March: 25–27.

Echt, C.S., B.S. Crane, and C.D. Nelson. 2011. Establishing restoration seed reserves in national forest system seed orchards. In: *Proceedings of the 31st Southern Tree Improvement Conference*, eds. C.D. Nelson, R.J. Rousseau, and C. Yuceer, pp. 46–49, http://www.sftic.org.

Ehman, J.L, H.P. Schmid, C.S.B. Grimmond et al. 2002. An initial intercomparison of micrometerological ecological inventory estimates of carbon exchange in a mid-latitude deciduous forest. *Global Change Biology* 8:575–589.

Eisenbies, M.H., J.A. Burger, W.M. Aust, S.C. Patterson, and T.R. Fox. 2006. Assessing change in soil-site productivity on intensively managed loblolly pine plantations. *Soil Science Society of America Journal* 70:130–140.

Eisenbies, M.H., J.A. Burger, W.M. Aust, and S.C. Patterson. 2007. Changes in site productivity recovery of soil properties following wet- dry-weather harvesting disturbance in the Atlantic Coastal Plain for a stand of age 10 years. *Canadian Journal of Forestry* 37:1336–1348.

Eisenbies, M.H., E.D. Vance, W.M. Aust, and J.R. Seiler. 2009. Intensive utilization of harvest residues in southern pine plantations: Quantities available and implications for nutrient budgets and sustainable site productivity. *Bioenergy Research* 2:90–98.

Elliott, K.J. and W.T. Swank. 1994. Impact of drought on tree mortality growth in a mixed hardwood forest. *Journal of Vegetation Science* 5:229–236.

Falge, E., D.D. Baldocchi, J. Tenhunen et al. 2002. Seasonality of ecosystem respiration gross primary production as derived from FLUXNET measurements. *Agricultural Forestry Meterology* 113:53–74.

Farnum, P., R. Timmis, and J.L. Kulp. 1983. Biotechnology of forest yield. *Science* 219:694–702.

Fekedulegn, D., R.R. Hicks Jr., and J.J. Colbert. 2008. Influence of topographic aspect, precipitation drought on radial growth of four major tree species in an Appalachian watershed. *Forest Ecology and Management* 177:409–425.

Finkral, A.J. and A.M. Evans. 2008. The effects of a thinning treatment on carbon stocks in a northern Arizona ponderosa pine forest. *Forest Ecology and Management* 255:2743–2750.

Finzi, A.C., E.H. DeLucia, J.G. Hamilton, D.D. Richter, and W.H. Schlesinger. 2002. The nitrogen budget of a pine forest under free air CO_2 enrichment. *Oecologia* 132:567–578.

Finzi, A.C., D.J.P. Moore, E.H. DeLucia et al. 2006. Progressive N limitation of ecosystem processes under elevated CO_2 in a warm-temperature forest. *Ecology* 87:15–25.

Finzi, A.C., R.J. Norby, C. Calfapietra et al. 2007. Increases in nitrogen uptake rather than nitrogen-use efficiency support higher rates of temperate forest productivity under elevated CO_2. *Proceedings of the National Academy of Sciences* 104:14014–14019.

Fox, T.R., E.J. Jokela, and H.L. Allen. 2007a. The development of pine plantation silviculture in the southern United States. *Journal of Forestry* 105:337–347.

Fox, T.R., H.L. Allen, T.J. Albaugh, R. Rubilar, and C.A. Carlson. 2007b. Tree nutrition forest fertilization of pine plantations in the Southern United States. *Southern Journal of Applied Forestry* 31:5–11.

Friedlingstein, P., G. Joel, C.B. Field, and I.Y. Fung. 1999. Toward an allocation scheme for global terrestrial carbon models. *Global Change Biology* 5:755–770.

Friend, A.L. and W.L. Hafley. 1989. Climatic limitations to growth in loblolly shortleaf pine (*Pinus taeda* and *P. echinata*): A dendroclimatological approach. *Forest Ecology and Management* 26:113–122.

Galik, C.S., R.C. Abt, and Y. Wu. 2009. Forest biomass supply in the southeastern United States-implications for industrial roundwood bioenergy production. *Journal of Forestry* 107:69–77.

Garbett, W.S. 1977. Aboveground biomass and nutrient content of a mixed slash-longleaf pine stand in Florida. In M.S. thesis. University of Florida, Gainesville, FL, USA.

Garcia-Gonzalo, J., H. Peltola, E. Briceño-Elizondo, and S. Kellomäki. 2007. Changed thinning regimes may increase carbon stock under climate change: A case study from a Finnish boreal forest. *Climatic Change* 81:431–454.

Gholtz, H.L. and R.F. Fisher. 1982. Organic matter production distribution in slash pine (*Pinus elliottii*) plantations. *Ecology* 63:1827–1839.

Giardina, C.P. and M.G. Ryan. 2000. Evidence that decomposition rates of organic carbon in mineral soil do not vary with temperature. *Nature* 404:858–861.

Giddens, J.E. and W.H. Garman. 1941. Some effects of cultivation on the Piedmont soils of Georgia. *Proceedings of the Soil Science Society of America* 6:439–446.

Gielen, B., C. Calfapietra, M. Lukac et al. 2005. Net carbon storage in a poplar plantation (POPFACE) after three years of free-air CO_2 enrichment. *Tree Physiology* 25:1399–1408.

Gielen, B., M. Liberloo, J. Bogaert et al. 2003. Three years of free-air CO_2 enrichment (POPFACE) only slightly affect profiles of light and leaf characteristics in closed canopies of *Populus*. *Global Change Biology* 9:1022–1037.

Gifford, R.M. 2003. Plant respiration in productivity models: Conceptualization, representation issues for global terrestial carbon-cycle research. *Functional Plant Biology* 30:171–186.

Goldenberg, S.B., C.W. Landsea, A.M. Mestas-Nunez, and W.M. Gray. 2001. The recent increase in Atlantic hurricane activity: Causes and implications. *Science* 293:474–479.

Gonzalez, R., T. Treasure, R. Phillips, H. Jameel, and D. Saloni. 2011. Economics of cellulosic ethanol production: Green liquor pretreatment for softwood and hardwood, greenfield repurpose scenarios. *Bioresources* 6: 2551–2567.

Gonzalez-Benecke, C.A. and T.A. Martin. 2010. Water availability genetic effects on water relations of an 11 year-old loblolly pine (*Pinus taeda*) plantation. *Tree Physiology* 30:376–392.

Gonzalez-Benecke, C.A., T.A. Martin, W.P. Cropper Jr., and R. Bracho. 2010. Forest management effects on *in situ* ex situ slash pine forest carbon balance. *Forest Ecology and Management* 260:795–805.

Gonzalez-Benecke, C.A., T.A. Martin, and E. Jokela. 2011a. A flexible hybrid model of life cycle carbon balance of loblolly pine (*Pinus taeda* L.) management systems. *Forests* 2:749–776.

Gonzalez-Benecke, C.A., E.J. Jokela, and T.A. Martin. 2012. Modeling the effects of stand development, site quality, silviculture on leaf area index, litterfall, forest floor accumulations in loblolly slash pine plantations. *Forest Science* 58:457–471.

Gough, C.M. and J.R. Seiler. 2004. Belowground carbon dynamics in loblolly pine (*Pinus taeda*) immediately following diammonium phosphate fertilization. *Tree Physiology* 24:845–851.

Gough, C.M., J.R. Seiler, P.E. Wiseman, and C.A. Maier. 2005. Soil CO_2 efflux in loblolly pine (*Pinus taeda* L.) plantations on the Virginia Piedmont South Carolina Coastal plain over a rotation-length chronosequence. *Biogeochemistry* 73:127–147.

Gough, C.M., C.S. Vogel, H.P. Schmid, and P.S. Curtis. 2008. Controls on annual forest carbon storage: Lessons from the past predictions for the future. *BioScience* 58:609–622.

Goulden, M.L., A.M.S. McMillan, G.C. Winston et al. 2011. Patterns of NPP, GPP, respiration, NEP during boreal forest succession. *Global Change Biology* 17:855–871.

Gower, S.T., R.E. McMurtrie, and D. Murty. 1996. Aboveground net primary production decline with stand age: Potential causes. *Trends in Ecology Evolution* 11:378–382.

Gresham, C.A., T.M. Williams, and D.J. Lipscomb. 1991. Hurricane Hugo wind damage to southeastern U.S. coastal forest tree species. *Biotropica* 223:420–426.

Griesbauer, H.P. and D.S. Green. 2010. Regional ecological patters in interior Douglas-fir climate–growth relationships in British Columbia, Canada. *Canadian Journal of Forest Research* 40:308–321.

Groninger, J.W., K.H. Johnsen, J.R. Seiler, R.E. Will, D.S. Ellsworth, and C.A. Maier. 1999. Elevated carbon dioxide in the atmosphere: What might it mean for loblolly pine plantation forestry? *Journal of Forestry* 97:4–10.

Gundimeda, H. 2001. A framework for assessing carbon flow in Indian wood products. *Environment, Development and Sustainability* 3:229–251.

Hamilton, J.G., E.H. DeLucia, K. George, S.L. Naidu, A.C. Finzi, and W.H. Schlesinger. 2002. Forest carbon balance under elevated CO_2. *Oecologia* 131:250–260.

Han, F.X., M.J. Plodinec, Y. Su, D.L. Monts, and Z. Li. 2007. Terrestrial carbon pools in southeast south-central United States. *Climatic Change* 84:1919–202.

Harding, R.B. and E.J. Jokela. 1994. Long-term effects of forest fertilization on site organic matter nutrients. *Soil Science Society of America Journal* 58:216–221.

Harmon, M.E. and B. Marks. 2002. Effects of silvicultural practices on carbon stores in Douglas-fir–western hemlock forests in the Pacific Northwest, U.S.A.: Results from a simulation model. *Canadian Journal of Forest Research* 32:863–877.

Harrison, W.M. and B.E. Borders. 1996. Yield prediction growth projection for site-prepared loblolly pine plantations in the Carolinas, Georgia, Alabama and Florida. PMRC Technical Report 1996-1. Athens, GA, USA.

Hättenschwiler, S., F. Miglietta, A. Raschi, and C.H. Körner. 1997. Thirty years of *in situ* tree growth under elevated CO_2: A model for future forest responses? *Global Change Biology* 3:463–471.

Haynes, B.E. and S.T. Gower. 1995. Belowground carbon allocation in unfertilized fertilized red pine plantations in northern Wisconsin. *Tree Physiology* 15:317–325.

Heilman, P. and R.J. Norby. 1998. Nutrient cycling fertility management in temperate short rotation forest systems. *Biomass Bioenergy* 14:361–370.

Henderson, J.P. and H.D. Grissino-Mayer. 2009. Climate-tree growth relationships of longleaf pine (*Pinus palustris* Mill.) in the Southeastern Coastal Plain, USA. *Dendrochronologia* 27:31–43.

Hendrey, G.R., D.S. Ellsworth, K.F. Lewin, and J. Nagy. 1999. A free-air enrichment system for exposing tall vegetation to elevated atmospheric CO_2. *Global Change Biology* 5:293–309.

Hicke, J.A., G.P. Asner, J.T. Randerson et al. 2002. Trends in North American net primary productivity derived from satellite observations, 1982–1998. *Global Biogeochemical Cycles* 16:1–11.

Hinchee, M., W. Rottmann, L. Mullinax et al. 2009. Short-rotation woody crops for bioenergy biofuels application. *In vitro Cellular Developmental Biology—Plant* 45:619–629.

Hubbard, R.M., B.J. Bond, and M.G. Ryan. 1999. Evidence that hydraulic conductance limits photosynthesis in old *Pinus ponderosa* trees. *Tree Physiology* 19:165–172.

Huggett, R., D.N. Wear et al. [In press]. Forest forecasts. In: Wear, D.N., Greis, J.G., eds. The Southern Forest Futures Project: Technical Report.

Hungate, B.A., K.J. van Groenigen, J. Six et al. 2009. Assessing the effect of elevated carbon dioxide on soil carbon: A comparison of four meta-analyses. *Global Change Biology* 15:2020–2034.

Hunt, E.R., M.B. Lavigne, and S.E. Franklin. 1999. Factors controlling the decline of net primary production with stand age for balsam fir in Newfoundland assessed using an ecosystem simulation model. *Ecological Modeling* 122:151–164.

Hunt, R. and B. Zobel. 1978. Frost-hardy Eucalyptus grow well in the Southeast. *Southern Journal of Applied Forestry* 1:6–10.

Intergovernmental Panel on Climate Change (IPCC). 2006. IPCC Guidelines for National Greenhouse Gas Inventories. In: *Agriculture, Forestry and Other Land Use,* Vol. 4. eds. H.S. Eggleston, L. Buendia, K. Miwa, T. Ngara and K. Tanabe K. IGES, Japan.

Intergovernmental Panel on Climate Change (IPCC). 2007a. Climate Change 2007: Synthesis Report. In: *Contribution of Working Groups I, II and III to the Fourth Assessment Report of the Intergovernmental Panel on Climate Change*. eds. R.K. Pachauri and A. Reisinger, A. IPCC, Geneva, Switzerland.

Intergovernmental Panel on Climate Change (IPCC). 2007b. Summary for Policymakers. In: *Climate Change 2007: The Physical Science Basis*. Contribution of Working Group I to the Fourth Assessment Report of the Intergovernmental Panel on Climate Change. eds. S. Solomon, D. Qin, M. Manning, Z. Chen, M. Marquis, K.B. Averyt, M.Tignor, and H.L. Miller. Cambridge University Press, Cambridge, United Kingdom and New York, NY, USA.

Iversen, C.M., J.K. Keller, C.T. Garten, and R.J. Norby. 2012. Soil carbon and nitrogen cycling and storage throughout the soil profile in a sweetgum plantation after 11 years of CO_2 enrichment. *Global Change Biology* 18:1684–1697.

Jackson, C.R., J.K. Martin, D.S. Leigh, and L.T. West. 2005. A southeastern piedmont watershed sediment budget; evidence for a multi-millennial agricultural legacy. *J. Soil Water Cons*. 60:298–310.

Jacobi, J.C. and F.H. Tainter. 1988. Dendroclimatic examination of white oak along an environmental gradient in the Piedmont of South Carolina. *Castanea* 53:252–262.

Jahromi, S,T. 1982. Variation in cold resistance growth in *Eucalyptus viminalis*. *Southern Journal of Applied Forestry* 4:221–225.

Jandl, R., M. Lindner, L. Vesterdal et al. 2007. How strongly can forest management influence soil carbon sequestration?. *Geoderma* 137:253–268.

Jiménez, E., J.A. Vega, C. Fernández, and T. Fonturbel. 2011. Is pre-commercial thinning compatible with carbon sequestration? A case study in a maritime pine stand in northwestern Spain. *Forestry* 84:149–157.

Johnsen, K.H., L.B. Flanagan, D.A. Huber, and J.E. Major. 1999. Genetic variation in growth and carbon isotope discrimination in Picea mariana: Analyses from a half-diallel mating design using field grown trees. *Canadian Journal of Forest Research* 29:1727–1735.

Johnsen, K.H., L. Kress, and C.E. Maier. 2005. Quantifying root lateral distribution and turnover using pine trees with a distinct stable carbon isotope signature. Functional Ecology 19:81–87.

Johnsen, K.H., D.N. Wear, R. Oren et al. 2001a. Carbon sequestration via southern pine forestry. *Journal of Forestry* 99:14–21.

Johnsen, K.H., L. Samuelson, R. Teskey, S. McNulty, and T. Fox. 2001b. Process models as tools in forestry research management. *Forest Science* 47:2–8.

Johnsen, K.H., R.O. Teskey, L. Samuelson et al. 2004. Carbon sequestration in loblolly pine plantations: methods, limitations, and research needs for estimating storage pools. 373–381. In *Southern Forest Science: Past, Present, and Future*. Forest Service Southern Research Station G.T.R. SRS-75. Asheville, NC, USA.

Johnsen, K.H., C. Maier, F. Sanchez et al. 2007. Physiological girdling of pine trees via phloem chilling: Proof of concept. *Plant, Cell Environment* 30:128–134.

Johnsen, K.H., J.R. Butnor, J.S. Kush, R.C. Schmidtling, and C.D. Nelson. 2009. Longleaf pine displays less wind damage than loblolly pine. *Southern Journal of Applied Forestry* 33:178–181.

Johnson, D.W. and D.E. Todd. 1998. The effects of harvesting on long-term changes in nutrient pools in a mixed oak forest. *Soil Science Society of America Journal* 62:1725–1735.

Johnson, D.W. and P.S. Curtis. 2001. Effects of forest management on soil C N storage: Meta analysis. *Forest Ecology and Management* 140:227–238.

Johnson, D.W., D.C. West, D.E. Todd, and L.K. Mann. 1982. Effects of sawlog vs whole-tree harvesting on the nitrogen, phosphorus, potassium, calcium budgets of an upland mixed oak forest. *Soil Science Society of America Journal* 46:1304–1309.

Jokela, E.J., P.M. Dougherty, and T.A. Martin. 2004. Production dynamics of intensively managed loblolly pine stands in the southern United States: A synthesis of seven long-term experiments. *Forest Ecology and Management* 192:117–130.

Jokela, E.J. and T.A. Martin. 2000. Effects of ontogeny soil nutrient supply on production, allocation leaf area efficiency in loblolly slash pine stands. Canadian *Journal of Forest Research* 30:1511–1524.

Jokela, E.J., T.A. Martin, and J.G. Vogel. 2010. Twenty-five years of intensive forest management with southern pines: Important lessons learned. *Journal of Forestry* 108:338–347.

Jordon, D.N. and B.G. Lockaby. 1990. Time series modeling of relationships between climate long-term radial growth of loblolly pine. *Canadian Journal of Forest Research* 20:738–742.

Karl, T.R., J.M. Melillo, and T.C. Peterson. 2009. *Global Climate Change Impacts in the United States*. Cambridge University Press, Cambridge, UK.

Karnosky, D.F., B. Mankovska, K. Percy et al. 1999. Effects of tropospheric O_3 on trembling aspen: Interaction with CO_2: Results from an O_3-gradient and a FACE experiment. *Water, Air Soil Pollution* 116:311–322.

Karnosky, D.F., K.S. Pregitzer, D.E. Zak et al. 2005. Scaling ozone responses of forest treess to the ecosystem level in a changing climate. *Plant, Cell Environment* 28:965–981.

Karnosky, D.F., D.E. Zak, K.S. Pregitzer et al. 2003. Tropospheric O_3 moderates responses of temperate hardwood forests to elevated CO_2: A synthesis of molecular to ecosystem results from the Aspen FACE project. *Functional Ecology* 17:289–304.

King, J.S., M.E. Kubiske, K.S. Pregitzer et al. 2005. Tropospheric O_3 compromises net primary production in young stands of trembling aspen, paper birch and sugar maple in response to elevated atmospheric CO_2. *New Phytologist* 168:623–636.

Kline, K.L. and M.D. Coleman. 2010. Woody energy crops in the southeastern United States: Two centuries of practitioner experience. *Biomass Bioenergy* 34:1655–1666.

Klos, F.J., G.G. Wang, W.L. Bauerle, and J.R. Rieck. 2009. Drought impact on forest growth mortality in the southeast USA: An analysis using Forest Health Monitoring data. *Ecological Applications* 19:699–708.

Knorr, W., I.C. Prentice, J.I. House, and E.A. Holland. 2005. Long-term sensitivity of soil carbon turnover to warming. *Nature* 433:298–301.

Kolb, T.W., K.C. Steiner, L.H. McCormic, and T.W. Bowersox. 1990. Growth response of northern red-oak yellow poplar seedlings to light, soil moisture, nutrients in relation to ecological strategy. *Forest Ecology and Management* 38:65–78.

Körner, C. 2006. Plant CO_2 responses: An issue of definition, time and resource supply. *New Phytologist* 172:393–411.

Kozlowski, T.T. and S.G. Pallardy. 1997. *Physiology of Woody Plants*. Second Edition. Academic Press, San Diego, USA.

Kozlowski, T.T., P.J. Kramer, and S.G. Pallardy. 1991. *The Physiological Ecology of Woody Plants*. Academic Press, San Diego, USA.

Kupfer, J.A., A.T. Myers, S.E. McLane, and G.N. Melton. 2007. Patterns of forest damage in a southern Mississippi landscape caused by Hurricane Katrina. *Ecosystems* 11:45–60.

Lai, C.-T., G. Katul, J.R. Butnor et al. 2002. Modelling the limits on the response of net carbon exchange to fertilization in a south-eastern pine forest. *Plant, Cell Environment* 25:1095–1119.

Lal, R. 2008a. Sequestration of atmospheric CO_2. *Energy Environmental Science* 1:86–100.

Lal, R. 2008b. Soil carbon stocks under present future climate with specific reference to European ecoregions. *Nutrient Cycling in Agroecosystems* 81:113–127.

Lambers, H., F.S. Chapin, and T.L. Pons. 2006. *Plant Physiological Ecology*. Springer, New York, USA.

Lambeth, C.S., S. McKeand, R. Rousseau, and R. Schmidtling. 2005. Planting nonlocal seed sources of loblolly pine: Managing benefits and risks. *Southern Journal of Applied Forestry* 29:96–104.

Landsberg, J.J. and P. Sands. 2011. *Physiological Ecology of Forest Production: Principles, Processes, and Models*. Vol. 4. Academic Press, San Diego, USA.

Landsberg, J.J. and R.H. Waring. 1997. A generalized model of forest productivity using simplified concepts of radiation-use efficiency, carbon balance and partitioning. *Forest Ecology and Management* 95:209–228.

Landsberg, J.J., K.H. Johnsen, T.J. Albaugh, H.L. Allen, and S.E. McKeand. 2001. Applying 3-PG, a simple process-based model designed to produce practical results, to data from loblolly pine experiments. *Forest Science* 47:743–751.

Ledig, F.T, J.J. Vargas Hernández, and K.H. Johnsen. 1986. The conservation of forest genetic resources: Case histories from Canada, Mexico, and the United States. *Forest Ecology and Management* 14:77–90.

Liberloo, M., C. Calfapietra, M. Lukac et al. 2006. Woody biomass production during the second rotation of a bio-energy Populus plantation increases in a future high CO_2. *Global Change Biology* 12:1094–1106.

Liski, J., A. Pussinen, K. Pingoud, R. Mäkipää, and T. Karjalainen. 2001. Which rotation length is favorable to carbon sequestration? *Canadian Journal of Forest Research* 31:2004–2013.

Littell, J.S., D.L. Peterson, and M. Tjoelker. 2008. Douglas-fir growth in mountain ecosystems: Water limits tree growth from stand to region. *Ecological Monographs* 78:349–368.

Litton, C.M., J.W. Raich, and M.G. Ryan. 2007. Review: Carbon allocation in forest ecosystems. *Global Change Biology* 13:2089–2109.

Logan, S.R., B.D. Shiver, and W.M. Harrison. 2002. Polymorphism of southern pine site index curves resulting from different cultural treatments. In *PMRC Technical Report 2002-1*. Athens, GA, USA.

Luo, Y. 2007. Terrestrial carbon-cycle feedback to climate warming. *Annual Review of Ecology, Evolution Systematics* 38:683–712.

Luo, Y., S. Wan, D. Hui, and L.L. Wallace. 2001. Acclimatization of soil respiration to warming in a tall grass prairie. *Nature* 413:622–625.

Luo, Y., B. Su, W.S. Currie et al. 2004. Progressive nitrogen limitation of ecosystem responses to rising atmospheric carbon dioxide. *BioScience* 54:731–739.

Machmuller, M.B., J. Mohan, J.S. Clark, and J.M. Melillo. 2011. The effect of experimental warming on soil respiration in two mixed deciduous forests. In *Proceedings of the 96th Ecological Society of America Annual Meeting*. Austin, Texas, August 7–12, 2011.

Maier, C.A. and L.W. Kress. 2000. Soil CO_2 evolution and root respiration in 11 year-old loblolly pine (*Pinus taeda*) plantations as affected by moisture and nutrient availability. *Canadian Journal of Forest Research* 30:347–359.

Maier, C.A., K.H. Johnsen, P. Dougherty, D. McInnis, P. Anderson, and S. Patterson. 2012. Effect of harvest residue management on tree productivity and carbon pools during early stand development in a loblolly pine plantation. *Forest Science* 58:430–445.

Maier, C.A., T.J. Albaugh, H.L. Allen, and P.M. Dougherty. 2004. Respiratory carbon use carbon storage in mid-rotation loblolly pine (*Pinus taeda* L.) plantations: The effect of site resources on the stand carbon balance. *Global Change Biology* 10:1335–1350.

Mäkinen, H., P. Nöjd, and A. Isomäki. 2002. Radial, height volume increment variation in *Picea abies* (L.) Karst stands with varying thinning intensities. *Scandinavian Journal of Forest Research* 17:304–316.

Markewitz, D. 2006. Fossil fuel carbon emissions from silviculture: Impacts on net carbon sequestration in forests. *Forest Ecology and Management* 236:153–161.

Marland, G. and S. Marland. 1992. Should we store carbon in trees? *Water, Air and Soil Pollution* 64:181–195.

Martin, T.A., K.H. Johnsen, and T.L. White. 2001. Ideotype development in southern pines: Rationale strategies for overcoming scale-related obstacles. *Forest Science* 47:21–28.

Martin, S.W., R.L. Bailey, and E.J. Jokela. 1997. Models for unfertilized fertilized slash pine plantations: In *CRIFF B 400 B500 Series*. In *PMRC Technical Report, 1997-3*. Athens, GA, USA.

Mayfield, C.A., C.D. Foster, C.T. Smith, J. Gan, and S. Fox. 2007. Opportunities, barriers, and strategies for forest bioenergy bio-based product development in the southern United States. *Biomass Bioenergy* 31:631–637.

McCarthy, H.R., R. Oren, A.C. Finzi et al. 2007. Temporal dynamics and spatial variability in the enhancement of canopy leaf area under elevated atmospheric CO_2. *Global Change Biology* 13:2479–2497.

McCarthy, H.R., R. Oren, A.C. Finzi, and K.H. Johnsen. 2006. Canopy leaf area constrains [CO]-induced enhancement of productivity and partitioning among aboveground carbon pools. *Proceedings of the National Academy of Sciences* 103:16356–19361.

McCarthy, H.R., R. Oren, K.H. Johnsen et al. 2010. Reassessment of plant carbon dynamics at the Duke free air CO_2 enrichment site: Interactions of atmospheric [CO2] with nitrogen water availability over stand development. *New Phytologist* 185:502–513.

McConnaughay, K.D.M. and J.S. Coleman. 1999. Biomass allocation in plants: Ontogeny or optimality? A test along three resource gradients. *Ecology* 80:2581–2593.

McDowell, N.G., H.D. Adams, J.D. Bailey, M. Hess, and T.E. Kolb. 2006. Homeostatic maintenance of ponderosa pine gas exchange in response to stand density changes. *Ecological Applications* 16:1164–1182.

McKeand, S.E., R.P. Crook, and H.L. Allen. 1997. Genotypic stability effects on predicted family responses to silvicultural treatments in loblolly pine. *Southern Journal of Applied Forestry* 21:84–89.

McKeand, S.E., E.J. Jokela, D.A. Huber et al. 2006. Performance of improved genotypes of loblolly pine across different soils, climates and silvicultural inputs. *Forest Ecology and Management* 227:178–184.

McKinley, D.C., M.G. Ryan, R.A. Birdsey et al. 2011. A synthesis of current knowledge on forests carbon storage in the United States. *Ecological Applications* 21:1902–1924.

McLane, S.C., L.D. Daniels, and S.N. Aitken. 2011. Climate impacts on lodgepole pine (*Pinus contorta*) radial growth in a provenance experiment. *Forest Ecology and Management* 262:115–123.

McNulty, S.G. 2002. Hurricane impacts on U.S. forest carbon sequestration. *Environmental Pollution* 116: S17–S24.

McNulty, S., J. Moore Meyers, P. Caldwell, and G. Sun. 2010. Climate Change. In: *Southern Forest Futures Assessment*, eds. D.N. Wear and J.G. Greis. 153–173. Forest Service Southern Research Station G.T.R. SRS-53, Asheville, NC, USA.

McNulty, S., J.M. Myers, P. Caldwell, and G. Sun. In press. Climate change. In: *The Southern Forest Futures Project: Technical Report*, eds. Wear, D.N. and J.G. Greis. Forest Service Southern Research Station G.T.R.

Medlyn, B.E. and R.C. Dewar. 1999. Comment on the article by R.H. Waring, J.J. Landsberg M. Williams relating net primary production to gross primary production. *Tree Physiology* 19:137–138.

Medlyn, B.E., R.A. Duursma, and M.J.B. Zeppel. 2011. Forest productivity under climate change: A checklist for evaluating model studies. *Wiley Interdisciplinary Reviews-Climate Change* 2:332–355.

Melillo, J.M., P.A. Steudler, J.D. Aber et al. 2002. Soil warming carbon-cycle feedbacks to the climate system. *Science* 298:2173–2176.

Menzel, A. and P. Fabian. 1999. Growing season extended in Europe. *Nature* 397:659.

Menzel, A. 2000. Trends in phonological phases in Europe between. *International Journal of Biometeorology* 44:76–81.

Micales, J.D. and K.E. Skog, 1997. The decomposition of forest products in landfills. *International Biodeterioration & Biodegradation*. 39(2):145–158.

Mickler, R.A., J.E. Smith, and L.S. Heath. 2004. Forest carbon trends in the southern United States. In: *Southern Forest Science: Past, Present, and Future,* eds. H.M. Rausher and K.H. Johnsen, 383–394. G.T.R. SRS-075, Asheville, NC, USA.

Miglietta, F., A. Peressotti, F.P. Vaccari, A. Zaldei, P. DeAngelis, and G. Scarascia-Mugnozza. 2001. Free-air CO_2 enrichment (FACE) of a poplar plantation: The POPFACE fumigation system. *New Phytologist* 150: 465–476.

Miller, A.T., H.L. Allen, and C.A. Maier. 2006. Quantifying the coarse-root biomass of intensively managed loblolly pine plantations. *Canadian Journal of Forest Research* 36:12–22.

Misson, L., C. Vincke, and F. Devillez. 2003. Frequency responses of radial growth series after different thinning intensities in Norway spruce (*Picea abies* (L.) Karst.) stands. *Forest Ecology and Management* 177:51–63.

Moore, D.J.P., S. Aref, R.M. Ho, J.S. Pippen, J.G. Hamilton, and E.H. DeLucia. 2006. Annual basal area increment growth duration of *Pinus taeda* in response to eight years of free-air carbon dioxide enrichment. *Global Change Biology* 12:1367–1377.

Munsell, J.F. and T.R. Fox. 2010. An analysis of the feasibility for increasing woody biomass production from pine plantations in the southern United States. *Biomass Bioenergy* 34:1631–1642.

Nedlo, J.E., T.A. Martin, J.M. Vose, and R.O. Teskey. 2009. Growing season temperatures limit growth of loblolly pine (*Pinus taeda* L.) seedlings across a wide geographic transect. *Trees* 23:751–759.

Nehra, N.S., M.R. Becwar, W.H. Rottmann et al. 2005. Forest biotechnology: Innovative methods, emerging opportunities. *In Vitro Cellular and Developmental Biology-Plant* 41:701–717.

Nelson, C.D. and K.H. Johnsen. 2008. Genomic and physiological approaches to advancing forest tree improvement. *Tree Physiology* 28:1135–1143.

Noormets, A., M.J. Gavazzi, S.G. McNulty et al. 2010. Response of carbon fluxes to drought in a coastal plain loblolly pine forest. *Global Change Biology* 16:272–287.

Norby, R.J., E.H. DeLucia, B. Gielen et al. 2005. Forest response to elevated CO_2 is conserved across a broad range of productivity. *Proceedings of the National Academy of Sciences* 102:18052–18056.

Norby, R.J., P.J. Hanson, E.G. O'Neill et al. 2002. Net primary productivity of a CO_2-enriched deciduous forest: The implications for carbon storage. *Ecological Applications* 12:1261–1266.

Norby, R.J., J. Ledford, C.D. Reilly, N.E. Miller, and E.G. O'Neill. 2004. Fine-root production dominates response of a deciduous forest to atmospheric CO_2 enrichment. *Proceedings of the National Academy of Sciences* 101:9689–9693.

Norby, R.J., J.D. Sholtis, C.A. Gunderson, and S.S. Jawdy. 2003. Leaf dynamics of a deciduous forest canopy: No response to elevated CO_2. *Oecologia* 136:574–584.

Norby, R.J., D.E. Todd, J. Fults, and D.W. Johnson. 2001. Allometric determination of tree growth in a CO_2-enriched sweetgum stand. *New Phytologist* 150:477–487.

Norby, R.J., J.M. Warren, C.M. Iversen, B.E. Medlyn, and R.E. McMurtrie. 2010. CO_2 enhancement of forest productivity constrained by limited nitrogen availability. *Proceedings of the National Academy of Sciences* 107:19368–19373.

Oren, R., D.S. Ellsworth, K.H. Johnsen et al. 2001. Soil fertility limits carbon sequestration by forest ecosystems in a CO_2-enriched atmosphere. *Nature* 411:469–472.

Orwig, D.A. and M.D. Abrams. 1997. Variation in radial growth responses to drought among species, site, and canopy strata. *Trees* 11:474–484.

Palmroth, S., C. Maier, H. McCarthy et al. 2005. Contrasting responses to drought of forest floor CO_2 efflux in a loblolly pine plantation and a nearby oak-hickory forest. *Global Change Biology* 11:421–434.

Palmroth, S., R. Oren, H.R. McCarthy et al. 2006. Aboveground sink strength in forests controls the allocation of carbon belowground and its CO_2-induced enhancement. *Proceedings of the National Academy of Sciences* 103:19362–19367.

Pan, C., S.J. Tajchman, and J.N. Kochenderfer. 1997. Dendroclimatological analysis of major forest species of the central Appalachians. *Forest Ecology and Management* 98:77–87.

Pehl, C.E., C.L. Tuttle, J.N. Houser, and D.M. Moehring. 1984. Total Biomass nutrients of 25-year-old loblolly pines (*Pinus taeda* L.). *Forest Ecology and Management* 9:155–160.

Peñuelas, J. and I. Filella. 2001. Responses to a warming world. *Science* 294:793–795.

Perez-Garcia, J., B. Lippke, J. Comnick, and C. Manriquez. 2005. An assessment of carbon pools, storage, wood products market substitution using life-cycle analysis results. *Wood Fiber Science* 37:140–148.

Peterson, C.J. 2007. Consistent influence of tree diameter species on damage in nine eastern North America tornado blowdowns. *Forest Ecology and Management* 250:96–108.

Pichler, P. and W. Oberhuber. 2007. Radial growth response of coniferous forest trees in an inner Apline environment to a heat-wave in 2003. *Forest Ecology and Management* 242:688–699.

Pienaar, L.V., B.D. Shiver, and J.W. Rheney. 1996. Yield prediction for mechanically site-prepared slash pine plantations in the southeastern coastal plain. PMRC Technical Report 1996-3.

Pirraglia, A., R. Gonzalez, and D. Saloni. 2010. Wood pellets: An expanding market opportunity. *Biomass Magazine*. June 2010. Online, http://biomassmagazine.com/articles/3853/wood-pellets-an-expanding-market-opportunity.

Potter, K. and B.S. Crane. 2010. *Forest Tree Genetic Risk Assessment System: A tool for Conservation Decision-Making in Changing Times*. Online, http://www.forestthreats.org/current-projects/project-summaries/genetic-risk-assessment-system-description-120610.pdf.

Powell, T.L., H.L. Gholz, K.L. Clark, G.S. Starr, W.P. Cropper, and T.A. Martin. 2008. Carbon exchange of a mature, naturally regenerated pine forest in north Florida. *Global Change Biology* 14:2523–2538.

Powers, R.F., D.A. Scott, F.G. Sanchez et al. 2005. The North American long-term soil productivity experiment: Findings from the first decade of research. *Forest Ecology and Management* 220:31–50.

Prestemon, J.P. and R.C. Abt. 2002. Timber products supply demand. In*: Southern Forest Resource Assessment.* eds. D.N. Wear and J.G. Greis, Forest Service Southern Research Station G.T.R. SRS-53. Asheville, NC, USA.

Radtke, P.J., R.L. Amateis, S.P. Prisley et al. 2009. Modeling production decay of coarse woody debris in loblolly pine plantations. *Forest Ecology and Management* 257:790–799.

Richter, D.D., D. Markewitz, P.R. Heine et al. 2000. Legacies of agriculture forest regrowth in the nitrogen of old-field soils. *Forest Ecology and Management* 138:233–248.

Rose, A.K. 2007. Virginia's forests, 2001. *Forest Service Southern Research Station Resource Bulletin SRS-120*. Asheville, NC, USA.

Rötzer, T., R. Grote, and H. Pretzsch. 2004. The timing of bud burst its effect on tree growth. *International Journal of Biometeorology* 48:109–118.

Row, C. and R.B. Phelps. 1991. Carbon cycle impacts of future forest products utilization and recycling trends. In *Agriculture in a World of Change,* Proceedings of USDA Outlook 1991: U.S. Department of Agriculture. Washington, DC, USA.

Running, S.W., R.R. Nemani, F.A. Henisch, M. Zhao, M. Reeves, and H. Hashimoto. 2004. A continuous satellite-derived measure of global terrestrial primary production. *BioScience* 54:547–560.

Rustad, L.E. and I.J. Fernandez. 1998. Experimental soil warming effects on CO_2 and CH_4 flux from a low elevation spruce–fir forest soil in Maine. *Global Change Biology* 4:597–605.

Ryan, M.G., D. Binkley, and J.H. Fownes. 1997. Age-related decline in forest productivity: Pattern process. *Advances in Ecological Research* 27:213–262.

Ryan, M.G., D. Binkley, and J.L. Stape. 2008. Why don't our stands grow even faster? Control of production carbon cycling in eucalypt plantations. *Southern Forests* 70:99–104.

Ryan, M.G., M.E. Harmon, R.A. Birdsey et al. 2010. A synthesis of the science on forests carbon for U.S. forests. *Issues in Ecology* 13:1–16.

Ryan, M.G., J.L. Stape, D. Binkley et al. 2010. Factors controlling Eucalyptus productivity: How water availability stand structure alter production carbon allocation. *Forest Ecology and Management* 259:1695–1703.

Salisbury, F.B. and C.W. Ross. 1992. *Plant Physiology*, Fourth Edition. Wadsworth Publishing Company, Belmont, CA, USA.

Sampson, D.A. and H.L. Allen. 1999. Regional influences of soil available water-holding capacity climate, leaf area index on simulated loblolly pine productivity. *Forest Ecology and Management* 124:1–12.

Sampson, D.A., R.H. Waring, C.A. Maier, D.M. Gough, M.J. Ducey, and K.H. Johnsen. 2006. Fertilization effects on forest carbon storage exchange net primary production a new hybrid process model for stand management. *Forest Ecology and Management* 221:91–109.

Sanchez, F.G., E.A. Carter, and Z.H. Leggett. 2009. Loblolly pine growth soil nutrient stocks eight years after forest slash incorporation. *Forest Ecology and Management* 257:1413–1419.

Schmidt, M.W.I. and A.G. Noack. 2000. Black carbon in soils sediments: Analysis, distribution, implications, current challenges. *Global Biogeochemical Cycles* 14:777–793.

Schmidtling, R.C. 1973. Intensive culture increases growth without affecting wood quality of young southern pines. *Canadian Journal of Forest Research* 3:565–573.

Schmidtling, R.C. 1987. Relative performance of longleaf compared to loblolly slash pines under different levels of intensive culture. In *Proceedings of the Fourth Biennial Southern Silvicultural Research Conference*, ed. D. Philips, Forest Service Southeastern Forest Experiment Station. G.T.R. SE-42, Asheville, NC, USA. 395–400.

Schmidtling, R.C. 1994. Use of provenance tests to predict response to climate change: Loblolly pine and Norway spruce. *Tree Physiology* 14:805–817.

Schmidtling, R.C. 2001. Southern pine seed sources. Forest Service Southern Research Station G.T.R. SRS-44. Asheville, NC, USA.

Schwartz, M.D. and B.E. Reiter. 2000. Changes in North American spring. *International Journal of Climatology* 20:929–932.

Schweitzer, C.J. 2000. *Forest statistics for East Tennessee Resources Bulletin*. Forest service Southern Research Station SRS-51. Asheville, NC, USA.

Skog, K.E. 2008. Sequestration of carbon in harvested wood products for the United States. *Forest Products Journal* 58:56–72.

Skog, K.E. and G.A. Nicholson. 1998. Carbon cycling through wood products: The role of wood paper products in carbon sequestration. *Forest Products Journal* 48:75–83.

Skovsgaard, K.E., I. Stupak, and L. Vesterdal. 2006. Distribution of biomass carbon in even-aged stands of Norway spruce (*Picea abies* (L.) Karst.): A case study on spacing thinning effects in northern Denmark. *Scandinavian Journal of Forest Research* 21:470–488.

Smith, J.E., L.S. Heath, K.E. Skog, and R.A. Birdsey. 2006. Methods for calculating forest ecosystem harvested carbon with standard estimates for forest types of the United States. Forest Service Northeastern Research Station G.T.R. NE-343. Newtown Square, PA, USA.

Smith, B.W., P.D. Miles, C.H. Perry, and S.A. Pugh. 2007. *Forest Resources of the United States, 2007*. Forest Service G.T.R. WO-78. Washington, DC, USA.

Smith, L.F. and R.C. Schmidtling. 1970. Cultivation fertilization speed early growth of planted southern pines. *Tree Planters Notes* 21:1–3.

Sonesson, J. and G. Eriksson. 2000. Genotypic stability genetic parameters for growth biomass traits in a water x temperature factorial experiment with *Pinus sylvestris* L. seedlings. *Forest Science* 46:487–495.

Speer, J.H., H.D. Grissino-Mayer, K.H. Orvis, and C.H. Greenberg. 2009. Climate response of five oak species in the eastern deciduous forest of the southern Appalachian Mountains, USA. *Canadian Journal of Forest Research* 39:507–518.

Spelter, H. and M. Alderman. 2005. Softwood sawmills in the United States Canada. Forest Products Laboratory Research Paper FPL-630. Madison, WI, USA.

Spittlehouse, D.L. and R.B. Stewart. 2004. Adaptation to climate change in forest management. *British Columbia Journal of Ecosystems Management* 4:1–11.

Sprugel, D.G. 1985. Natural disturbance and ecosystem energetics. In: *The Ecology of Natural Disturbance and Patch Dynamics*, eds. S.T.A. Pickett and P.S. White, 335–352. Academic Press, New York, N.Y., USA.

Stanturf, J.A., S.L. Goodrick, and K.W. Outcalt. 2007. Disturbance coastal forests: A strategic approach to forest management in hurricane impact zones. *Forest Ecology and Management* 250:119–135.

Stape, J.L., D. Binkley, M.G. Ryan et al. 2010. The Brazil eucalyptus potential productivity project: Influence of water, nutrients stand uniformity on wood production. *Forest Ecology and Management* 259:1684–1694.

St. Clair, J.B. and G.T. Howe. 2007. Genetic maladaptation of coastal Douglas-fir seedlings to future climates. *Global Change Biology* 13:1441–1454.

Sun, G., P. Caldwell, A. Noormets et al. 2011a. Upscaling key ecosystem functions across the conterminous United States by a water-centric ecosystem model. *Journal of Geophysical Research* 116:1–16.

Sun, G, K. Alstad, J. Chen et al. 2011b. A general predictive model for estimating monthly ecosystem evapotranspiration. *Ecohydrology* 4:245–255.

Tardif, J. and Y. Bergeron. 1997. Comparative dendroclimatological analysis of two black ash two white cedar populations from contrasting sites in the Lake Duparquet region, northwestern Quebec. *Canadian Journal of Forest Research* 27:108–116.

Tardif, J.C., F. Conciatori, P. Nantel, and D. Gagnon. 2006. Radial growth climate responses of white oak (*Quercus alba*) northern red oak (*Quercus rubra*) at the northern distribution limit of white oak in Quebec. *Journal of Biogeography* 33:1657–1669.

Thompson, M.T. 1998. Forest statistics for north Georgia. *Forest Service Southern Research Station Resources Bulletin*, SRS-34. Asheville, NC, USA.

Thornley, J.H.M. 1972. A balanced quantitative model for root: Shoot ratios in vegetative plants. *Annals of Botany* 36:431–441.

Thornton, P.E., B.E. Law, H.L. Gholz et al. 2002. Modeling and measuring the effects of disturbance history climate on carbon water budgets in evergreen needleleaf forests. *Agricultural Forest Meteorology* 113:185–222.

Tolbert, V.R. and L.L. Wright. 1998. Environmental enhancement of U.S. biomass crop technologies: Research results to date. *Biomass Bioenergy* 15:93–100.

Turner, D.P., G.J. Koerper, M.E. Harmon, and J.J. Lee. 1995. A carbon budget for forests of the conterminous United States. *Ecological Applications* 5:421–436.

Tuskan, G.A. 1998. Short-rotation woody crops supply systems in the United States: What do we know what do we need to know? *Biomass Bioenergy* 14:307–315.

van Buijtenen, J.P., M.V. Bilan, and R.H. Zimmerman. 1976. Morphophysiological characteristics related to drought resistance in *Pinus taeda*. In *Tree Physiology Yield Improvement,* eds. M.G.R. Cannel and F.T. Last, 349–360. Academic Press, London, England.

Van Lear, D.H. and P.R. Kapeluck. 1995. Above- and below-stump biomass nutrient content of a mature loblolly pine plantation. *Canadian Journal of Forest Research* 25:361–367.

Vitasse, Y., A.J. Porté, A. Kremer, R. Michalet, and S. Delzon. 2009. Responses of canopy duration to temperature changes in four temperate tree species: Relative contributions of spring autumn leaf phenology. *Oecologia* 161:187–198.

Vogel, J.G., L. Suau, T.A. Martin, and E.J. Jokela. 2011. Long term effects of weed control fertilization on the carbon nitrogen pools of a slash loblolly pine forest in north central Florida. *Canadian Journal of Forest Research* 41:552–567.

Walkovsky, A. 1998. Changes in phenology of the locust tree (*Robinia pseudoacacia* L.) in Hungary. *International Journal of Biometeorology* 41:155–160.

Ward, E.J., R. Oren, B.J. Sigurdsson, P.G. Jarvis, and S. Linder. 2008. Fertilization effects on mean stomatal conductance are mediated through changes in the hydraulic attributes of mature Norway spruce trees. *Tree Physiology* 28:579–596.

Waring, R.H., J.J. Landsberg, and M. Williams. 1998. Net primary production of forests: A constant fraction of gross primary production? *Tree Physiology* 18:129–134.

Way, D.A. and R. Oren. 2010. Differential responses to changes in growth temperature between trees from different functional groups biomes: A review synthesis of the data. *Tree Physiology* 30:669–688.

Wear, D.N. 2002. Land use. In: *Southern Forest Resource Assessment*. eds. D.N. Wear and J.G. Greis, Forest Service Southern Research Station SRS-53. Asheville, NC, USA.

Wear, D.N. and J.G. Greis. 2002. Southern Forest Resource Assessment. Forest Service Southern Research Station G.T.R. SRS-53, Asheville, NC, USA.

Wear, D.N. and J.G. Greis. 2012. The Southern Forest Futures Project: Summary Report. Forest Service Southern Research Station G.T.R. SRS-168. Asheville, NC, USA.

Wear, D.N., and J.G. Greis, eds. In press. The Southern Forest Futures Project: Technical Report. U.S. Department of Agriculture Forest Service, Southern Research Station G.T.R., Asheville, NC, USA.

Wells, O.O. 1983. Southwide pine seed source study—Loblolly pine at 25 years. *Southern Journal of Applied Forestry* 25:763–771.

Wells, O.O. and P.C. Wakeley. 1966. Geographic variation in survival, growth, fusiform rust infection of planted loblolly pine. *Forest Science Monographs* 11:40 p.

White, M.A., S.W. Running, and P.E. Thornton. 1999. The impact of growing season length variability on carbon assimilation evapotranspiration over. *International Journal of Biometeorology* 42:139–145.

White, M.K., S.T. Gower, and D.E. Ahl. 2005. Life-cycle inventories of roundwood production in Wisconsin-inputs into an industrial forest carbon budget. *Forest Ecology and Management* 219:13–28.

White, P.B., S.L. Van De Gevel, H.D. Grissino-Mayer, L.B. LaForest, and G.G. Deweese. 2011. Climatic response of oak species across an environmental gradient in the southern Appalachian Mountains, USA. *Tree-Ring Research* 67:27–37.

Wilson, J. 2004. Vulnerability to wind damage in managed landscapes of the coastal Pacific Northwest. *Forest Ecology and Management* 191:341–351.

Wright, L.L. 1994. Production technology status of woody herbaceous crops. *Biomass and Bioenergy* 6:191–209.

Yin, R., L.V. Pienaar, and M.E. Aronow. 1998. The productivity profitability of fiber farming. *Journal of Forestry* 96:13–18.

9 Managing Forest Water Quantity and Quality under Climate Change

Daniel A. Marion, Ge Sun, Peter V. Caldwell, Chelcy F. Miniat, Ying Ouyang, Devendra M. Amatya, Barton D. Clinton, Paul A. Conrads, Shelby Gull Laird, Zhaohua Dai, J. Alan Clingenpeel, Yonqiang Liu, Edwin A. Roehl Jr., Jennifer A. Moore Meyers, and Carl Trettin

CONTENTS

Water is a critical resource of the Southern United States and is intimately linked to other ecosystem and societal values. The South is known for its warm climate, rich water resources (Figure 9.1), and large acreage of forest lands that provide an ideal place for people to live. Indeed, water availability is central to sustaining an economy that relies on irrigation agriculture, forestry, recreation,

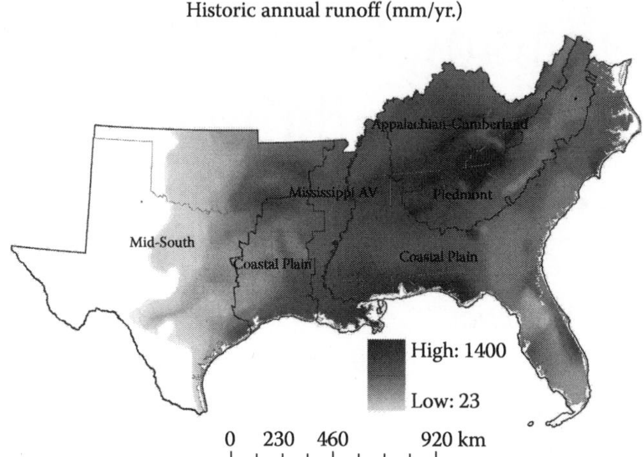

FIGURE 9.1 Mean annual runoff (1960–1980) distribution across the subregions of the South.

industry, power generation, transportation, and most importantly, the long-term future of natural ecosystems and human society (Hossian et al. 2011).

As in many other regions in the United States, the quality and quantity of water resources in the South are at risk from climate change, land conversion from forests to urban uses, increasing water demand from a growing population, and sea level rise (Lockaby et al. in press). A recent national climate change assessment (Karl et al. 2009) suggests that droughts, floods, and water quality problems are all likely to be amplified by climate change in most regions in the United States.

Understanding the potential impacts of climate change and associated stresses on water resources is the key to developing overall adaptation responses to minimize negative consequences at the local level. Today's forest managers face new challenges to maintain forest ecosystem services that are threatened by multiple stressors (Pacala and Sokolow 2004; Ryan et al. 2010). Adapting management practices to climate change by creating more resilient ecosystems is an essential step toward sustainability (Baron et al. 2009). But how much leverage can we expect using forest management in efforts to adapt to the more extreme hydrologic regimes caused by climate change (Ford et al. 2011b)? And what specific options and barriers will affect implementation of management strategies at the stand and landscape scales? Those are the kind of questions that scientists and managers must address if they expect to respond to climate change (National Research Council of the National Academies 2008).

The direct impacts of climate change on water resources will depend on how climate change alters the amount, form (snow vs. rain), and timing of precipitation and how this subsequently affects baseflow, stormflow, groundwater recharge, and flooding (Karl et al. 2009). Climate change poses profound threats to water quantity and quality across the South. The diverse climate, geology, and topography (Sun et al. 2004) and resulting physiographic differences create complex seasonal and spatial hydrologic patterns and water resource distributions (Figure 9.1). Approximately 8000 km of southern coastline are highly vulnerable to sea level rise, adding another level of complexity to water resource management. The large contrast of hydrometeorological and socioeconomic characteristics across the five broad subregions of the South (Coastal Plain, Piedmont, Appalachian-Cumberland Highland, Mississippi Alluvial Valley, and Mid-South) creates a wide range of water quantity and quality responses to climate change, adaptation, and mitigation management options (Table 9.1).

GOALS AND APPROACHES

The goals of this chapter are to provide land managers and policy makers a state-of-the-science summary of potential water resource responses to anticipated climate change, and to identify relevant

TABLE 9.1
Topographic and Hydroclimatic Characteristics of the Five Subregions of the South and Key Water Quantity and Quality Concerns Related to Climate Change

Ecoregion	Topography	Climate and Hydrology[a]	Key Concerns of Climate Change
Appalachian-Cumberland	Rugged topography with steep hillslopes and high elevations (300–2000 m)	High precipitation (1100–2000 mm/year), high runoff (400–1200 mm/year) and runoff ratio (>50%); low evapotranspiration caused by low air temperature	Increased flooding; droughts; erosion and sedimentation; and water temperature
Coastal Plain	Poorly to well drained flat lands with elevation <100 m above sea level	High precipitation (>1200 mm/year) and warm temperature, generally exceeding potential evapotranspiration leaving excess moisture, high evapotranspiration, variable runoff amount (300–800 mm/year), low runoff ratio (<30 percent most area) and slow-moving streams	Changes in wetland hydroperiod and water quality; hurricane impacts; inundation and saltwater intrusion due to sea level rise
Mid-South	Highly variable from flat coastal areas and Great Plains to rugged topography in the Ouachita-Ozark Mountains (0–2500 m)	A large climatic gradient; highly variable precipitation (400–1500 mm/year), runoff (50–550 mm/year) and runoff ratio (5–45%)	Increased flooding in the mountains; droughts in the plains; increased sedimentation due to higher rainfall intensity
Mississippi Alluvial Valley	Flat lands, low elevation (<50 m)	High precipitation, 1200–1500 mm/year, with a decreasing south–north gradient; low runoff amount (450–550 mm/year) and runoff ratio (<35%)	Change in wetland hydrology; inflows from upstreams; water quality from agricultural runoff and river transportation
Piedmont	Moderate slopes, elevation 100–300 m	High precipitation (1200–1400 mm/year), moderate runoff amount (200 to 800 mm/year), and ratio 20–60%	Increasing drought; water use demand; reactivation of sediment

[a] Runoff ratio is the ratio of mean annual runoff to mean annual precipitation, the balance of precipitation is lost to the atmosphere as evapotranspiration.

mitigation and adaptation practices to better manage forest watershed resources in the South. Our specific objectives are to (1) document and project the consequences of climate change and variability in altering the quantity, quality, and timing of water supplies at multiple scales; (2) present case studies to demonstrate how management can mitigate and adapt to climate change in different subregions; and (3) discuss the vulnerability of southern forest watersheds to water resource impacts resulting from climate change and other important stresses over the next 50 years.

This chapter provides an examination of historical and projected future climate changes and the impacts of rising sea levels on key hydrologic processes, water supply and demand related to human use, and water quality in southern forest watersheds. We examined the interactions of climate and water using long-term monitoring data collected at numerous sites (both U.S. Geological Survey and U.S. Forest Service stations) throughout the South. Statistical and process-based simulation models were used to project potential changes in water resources under multiple climate scenarios developed by combining four commonly used general circulation models (MIROC3.2, CSIROMK2, CSIROMK3.5, and HadCM3) with two emissions storylines (Intergovernmental Panel on Climate Change 2007): A1B representing low population/high economic growth and high-energy use, and

TABLE 9.2
Expected Response (Increase, Decrease, or Both) by Water Quantity and Quality Parameters to Climate Change and Other Key Stressors in the Southern United States

	Climate Change and Other Stressors			
Water Parameter	**Higher Air Temperature**	**Precipitation Change**	**Rising Sea Level**	**Land Use Changes (Population Growth, Urbanization)**
Water supply	Decrease	Both	Decrease	Decrease
Evapotranspiration	Increase	Both	Increase	Decrease
Peakflow, stormflow	Decrease	Both	Increase	Increase
Low flow	Decrease	Both	Increase	Both
Wetland hydroperiod	Decrease	Both	Increase	Decrease
Consumptive water use	Increase	Increase	Increase	Increase
Water temperature	Increase	Increase	Decease	Increase
Soil erosion, sedimentation	Both	Both	Increase	Increase
Chemical loading	Both	Both	Increase	Increase

B2 representing moderate growth and low-energy use. Most of the sections that follow utilize one or more of the following climate scenarios: CSIROMK3.5 A1B, CSIROMK2 A1B, HadCM3 B2, and MIROC3.2 A1B.

The chapter is organized in two main parts. First, we evaluate the vulnerability of water resources to future changes in climate and associated stresses at two different spatial scales: region-wide analyses and subregion case studies. The region-wide sections examine how key water resource processes are expected to respond to climate change across the entire South. We then propose adaptation methods based on our current understanding of how hydrologic processes are likely to respond to climate change over the next 50 years. Second, case studies provide more detailed investigations of particular processes within a given subregion, A response matrix (Table 9.2) that relates stressors and water quantity and quality parameters is also provided to summarize how key hydrologic parameters are expected to respond to future climate and development stressors.

WHAT WE ALREADY KNOW AND DO NOT KNOW

Climate change in the South—both observed and projected—is spatially complex. Across the region, mean air temperature increased 0.9°C from 1970 to 2008 (Karl et al. 2009). Although the amount of very heavy rainfall increased 15–20% from 1958 to 2007, the change in annual precipitation varied from a decrease of up to 15% in the Carolinas to an increase of 15% in areas of the Mid-South. Consequently, drought frequency has increased in this traditionally wet region (Karl et al. 2009). Mean annual temperatures in the region are expected to increase by an additional 2.5–3.5°C over the next 50 years (McNulty et al. in press), and water resources will likely become increasingly stressed (Lockaby et al. in press).

Climate change can impact hydrologic processes and water resources directly by altering precipitation, evapotranspiration, groundwater table, soil moisture, or streamflow; and indirectly by degrading water quality or reducing the water available for irrigation. Long-term streamflow data suggest that over the past century mean annual streamflow has increased in the Southern United States; and this increase is associated with increased precipitation (Intergovernmental Panel on Climate Change 2007; Karl and Knight 1998; Lins and Slack 1999). However, the general circulation models do not agree on the predicted change in direction (increases or decreases in total precipitation) of future precipitation events for the Eastern United States (Intergovernmental Panel on Climate Change 2007). A more consistently predicted aspect of future precipitation is that the frequency of the extremes will increase. Most general circulation models predict that as the climate warms, the frequency of extreme precipitation will increase across the globe (O'Gorman and Schneider 2009). Indeed, many regions of the United States have experienced an increased frequency of precipitation extremes over the last 50 years (Easterling et al. 2000a; Huntington 2006; Intergovernmental Panel on Climate Change 2007). However, unlike frequency, projections of the timing and spatial distribution of extreme precipitation are among the most uncertain aspects of future climate predictions (Allen and Ingram 2002; Karl et al. 1995). Despite these uncertainties, recent experience with droughts and low flows in many areas of the United States indicate that even small changes in drought severity and frequency have a major impact on agricultural productions and ecosystem services, including drinking water supplies (Easterling et al. 2000b; Luce and Holden 2009).

Although the properties that define water quantity and quality presented in Table 9.2 are intimately coupled, we discuss these two major response categories separately. For water quantity, we focus on key hydrologic fluxes including evapotranspiration, annual water yield, and low flows. For water quality, we examine water temperature and rainfall erosivity, the two parameters that are directly linked to climate change. We also quantify and map the locations with water supply stress to humans by linking water availability and human water use at a basin scale.

WATER QUANTITY

EVAPOTRANSPIRATION

Evapotranspiration is the combination of evaporation of water from plant and ground surfaces and transpiration. The proportion of rainfall that does not get consumed for evapotranspiration is available for streamflow and soil/groundwater recharge. Thus, evapotranspiration is a major regulator of streamflow, soil moisture, and groundwater recharge. In the South, a large proportion of annual precipitation (50–85%) is returned back to the atmosphere as evapotranspiration, leaving less than half to produce runoff (Lu et al. 2009). Most of the factors that control evapotranspiration—solar radiation, air temperature, precipitation, atmospheric carbon dioxide (CO_2) concentration, and vegetation characteristics—are expected to change under climate change (Sun et al. 2011b). Therefore, the effects of climate change on water resources and ecosystem function can be partially explained by alterations in evapotranspiration (Sun et al. 2011a).

Effects of CO_2 increase from climate change: The effects of increasing CO_2 concentration on evapotranspiration are complex, varying with scale, stand age, and other factors. At the plant's leaf surface, a variety of stomatal responses generally cause transpiration rates to decrease in response to long-term CO_2 increases (Beerling 1996; Franks and Beerling 2009; Prentice and Harrison 2009; Warren et al. 2011). However, at the stand level, the effects of CO_2 increase are highly dependent on the age of the stand. For example, in observational and modeling studies, evapotranspiration decreased by 11% in older stands, but increased in younger stands as a result of leaf area increases (Leuzinger and Korner 2007; Warren et al. 2011). Other modeling studies also suggest that the physiological effect of CO_2 on forest stand transpiration or evapotranspiration may vary between increases and decreases or may not change at all, depending on forest type and climate (Hanson et al. 2005; Ollinger et al. 2008; Tague et al. 2009). However, the effect of increasing CO_2 might be modest when the responses of leaf phenology and growth rate adjustment are compared to changes in other climatic variables such as precipitation, temperature, and humidity (Cech et al. 2003; Leuzinger and Korner 2010).

Forest evapotranspiration rates and the sensitivity of evapotranspiration to climate change are affected by tree species and forest type. For example, Ford et al. (2011a) showed that for a given tree diameter, yellow-poplar (*Liriodendron tulipifera*) transpiration rates were nearly twofold larger than hickory (*Carya* spp.) and fourfold larger than oaks (*Quercus* spp.). Yellow-poplar transpiration rates were also much more responsive to climatic variation compared to oaks and hickories (Ford et al.

2011a). In general, pine (*Pinus* spp.) forests are much more responsive to climatic variation than deciduous forests (Ford et al. 2011a; Stoy et al. 2006).

Changes in forest species composition in response to climate change may also affect evapotranspiration. Using climate change scenarios, Iverson et al. (2008) estimated that of 134 tree species examined, about 66 species would gain and 54 species would lose at least 10% of their suitable habitat, with most of the species habitat moving generally northeast. As species distributions change in response to climate change, especially if large areas of forests experience catastrophic mortality from droughts (Breshears et al. 2005), pest outbreaks, or other factors, evapotranspiration will also likely change.

Canopy and litter interception: Canopy and litter surface area can intercept precipitation and reduce the amount reaching the soil surface. Referred to as interception, this is an important component of forest evapotranspiration in the Eastern United States (Helvey 1967; Helvey and Patric 1965; Savenije 2004). Past work in western North Carolina (Helvey 1967), Georgia (Bryant et al. 2005), and coastal South Carolina (Amatya et al. 1996; Harder et al. 2007) found that interception losses varied from 12% to 22% of total precipitation, depending on forest type and elevation. Canopy and forest litter interception rates are affected by storm characteristics (wind speed, rainfall duration, intensity and drop size, and number of events), and vegetation and litter properties (canopy structure, leaf area index, and litter quality and amount); thus, climate change will likely alter canopy and litter interception losses (Crockford and Richardson 2000).

Evidence of climate change effects: Evidence of evapotranspiration response to climate change largely comes from observational studies that use proxies for evapotranspiration, such as estimates of evapotranspiration from the balance of precipitation and river discharge records, herbarium samples of leaves collected over time, agriculturally important climate indices, and remotely sensed forest greenup and senescence images. However, whether evapotranspiration is increasing or decreasing at continental and global scales is debatable, based on existing evidence. For example, Walter et al. (2004) concluded that evapotranspiration has been increasing across most of the United States, in six major basins, at a rate of 10.4 mm per decade. The increase of evapotranspiration was likely due to increase of irrigation, groundwater withdrawal, and increase precipitation (Walter et al. 2004), or expansion of forest areas in the southeast (Trimble et al. 1987). In contrast, Labat et al. (2004) have shown that global river discharge has been increasing at a rate of 4% for each 1°C increase in global temperature, suggesting a reduction in evapotranspiration. This increase in discharge has been directly attributed to the physiological effect of CO_2 decreasing evapotranspiration and not to the effect of changing land use (Gedney et al. 2006). A study of leaf samples collected from 1890 to 2010 showed a significant decline in stomatal density and maximum stomatal conductance (Lammertsma et al. 2011), supporting the notion of a downward trend in plant transpiration. Therefore, the combined effects of physical and chemical climate change on evapotranspiration are uncertain for certain ecosystems or regions.

Effects of growing-season length: The potential growing season for deciduous forests is increasing over time, which could increase annual evapotranspiration and thereby offset any possible decrease brought about by the effects of CO_2 on stomatal conductance. For example, Kunkel et al. (2004) show that the frost-free season across the United States has lengthened by about two weeks, but this phenomenon is spatially variable with more of an increase in the West than the East. The impacts of warmer winters and springs on deciduous forest greenup and senescence are difficult to generalize (Hänninen and Tanino 2011). For several boreal and temperate tree species, growth cessation in the autumn might come earlier with increasing temperatures. For others, spring bud burst might be delayed by warmer temperatures. For example, Zhang et al. (2007) show that North American lower-latitude forests have a delayed greenup over time, presumably because they do not receive the requisite chilling hours (Schwartz and Hanes 2010); in higher latitudes where chilling requirements are still being met, greenup is occurring sooner. This means that evapotranspiration in the lower latitudes should be delayed over time while evapotranspiration in the higher latitudes should be advancing over time.

Along with being affected by regional climate change, forest phenology can also be affected by local climate, specifically the heat-island effect and other influences from urban centers (Elmore et al. 2012). Lastly, important interactions between climate and abiotic conditions can also constrain evapotranspiration. For example, the potential increase in evapotranspiration resulting from a lengthened growing season can be constrained by reduced water availability and droughts that often arise late in the growing season (Jung et al. 2010; Zhao and Running 2010).

Low Flows

Low flows refer to streamflow during prolonged dry weather, a seasonal phenomenon and an integral component of the flow regime of any river (Smakhtin 2001). Low flows are normally derived from groundwater discharge and usually occur in the same season every year. Low flow is affected by climate, topography, geology, soils, and human activities (Smakhtin 2001). Effects of climate change on low flows have important consequences for water supply to reservoirs, transportation, power generation, aquatic habitats, and water quality (dissolved oxygen concentration, water temperature, salinity, and nutrient levels).

Previous studies suggest that the low flow characteristics have been changing in the South. For example, Lins and Slack (1999, 2005) reported significantly upward trends in annual minimum flows and 10th-percentile flows from 1940 to 1999 at most sites in the Appalachian-Cumberland Highland, Mississippi Alluvial Valley, and Mid-South; whereas many sites in the Coastal Plain and Piedmont exhibited significantly downward trends. McCabe and Wolock (2002) corroborated this result and found that the upward trends in streamflow were a result of a steep increase in precipitation beginning in the early 1970s.

Modeling study at watershed and regional scales: We used two different methods, the 7Q10 method (Telis 1992) and the minimum flows and levels method (Neubauer et al. 2008), to determine low flows in this study. The 7Q10 method, defined as the lowest mean flow that occurs for a consecutive 7-day period with a recurrence interval of 10 years, has been used for identifying extremely low flows (Reilly and Kroll 2003; Telis 1992). Conversely, the minimum flow and levels method is defined as the minimum water flows and/or levels required to prevent significant harm to water resources that are subject to water withdrawals (St. Johns River Water Management District 2010). This second method calculates how often and for how long the high, mean, and low water levels and/or flows need to occur to prevent significant harm.

For detecting historical low flow trends, we selected three forest-dominated headwater watersheds: Cache River at Forman, IL; Big Creek at Pollock, LA; and St. Francis River at Wappapello, MO—all located in the Mississippi Alluvial Valley. These watersheds have gauging stations maintained by the U.S. Geological Survey and have rather long (60–90 years) streamflow record periods.

To predict future low flow conditions, we applied an improved monthly scale Water Supply Stress Index model (Caldwell et al. 2012; Sun et al. 2008). The model uses a monthly water balance procedure that is sensitive to land cover and climate. It accounts for evapotranspiration, infiltration, soil storage, snow accumulation and melt, surface runoff, and baseflow processes within each basin, and conservatively routes discharge through the stream network from upstream to downstream watersheds. The model does not account for flow regulation by dams or for water diversion projects such as interbasin transfers. It has been evaluated against measured streamflow and evapotranspiration data at watershed to regional scales (Sun et al. 2011a). Sun et al. (2008) used this model to assess the impact of future climate change on monthly flows across some 2100 U.S. watersheds defined at the 8-digit Hydrologic Unit Code (HUC) scale. We focused on the 677 watersheds that encompass the South, some of which receive water from outside the region.

Climate data were derived using the CSIROMK3.5 A1B, CSIROMK2 A1B, HadCM3 B2, and MIROC3.2 A1B climate scenarios and downscaled to the 8-digit HUCs (Coulson et al. 2010a,b). Model results are presented as the mean low flow response under these four climate scenarios.

Using the 80% exceedance probability in their flow duration curves, we found that the low flow rates were 0.16 m^3/s at the Cache, 0.48 m^3/s at the Big Creek, and 5.01 m^3/s at the St. Francis gauging stations. The exceedance probabilities for the 60- and 90-day low flow durations at the three stations have increased during the study periods (Table 9.3). It appears that the low flows are occurring more frequently over time as the watersheds have become drier.

Water supply modeling shows that low monthly mean flows are projected to decrease 6.1% per decade across the South into the first half of the twenty-first century (Figure 9.2), with the largest decreases occurring in the Appalachian-Cumberland Highland, Mississippi Alluvial Valley, Mid-South, and western Coastal Plain (Table 9.4 and Figure 9.3). The large decrease in the Mississippi Alluvial Valley is partially the result of decreasing flows from streams outside of the southern region. Portions of the Coastal Plain and Mid-South were projected to have upward trends in low flows; however, these increases were not statistically significant (Figure 9.2). In all subregions, general low flows amounts were projected to decrease from 2010 to 2060 (Figure 9.3).

Total Water Yield

Water yield is the sum of surface runoff and groundwater discharge from a watershed. Changes in water yield from streams can greatly affect the health of aquatic ecosystems (Postel and Richter 2003) as well as public water supply capabilities (Postel and Carpenter 1997). To assess how climate change and increased water demand from population growth may affect future water supply stress in the South, we used an improved water supply stress model (Sun et al. 2008) to predict future water yields and water supply stresses.

Current trends: The effect of climate change on water yield has been analyzed using historical flow records at long-term gauging sites in watersheds with minimal human alteration. Lins and Slack (1999, 2005) reported significantly upward trends from 1940 to 1999 in all percentiles of streamflow—except maximum flows—at most sites in the Appalachian-Cumberland Highland, Mississippi Alluvial Valley, and Mid-South, but only at a few sites in the Coastal Plain and Piedmont. Groisman et al. (2003) and McCabe and Wolock (2002) corroborated this result and found that the upward trends in streamflow occurred as a result of a steep increase in precipitation beginning in the early 1970s. Conversely, Krakauer and Fung (2008) argued that increased evapotranspiration from climate change will ultimately lead to decreasing streamflows.

Future projections: Using the mean water yield response under the CSIROMK3.5 A1B, CSIROMK2 A1B, HadCM3 B2, and MIROC3.2 A1B climate scenarios, we found that annual water yield would decrease Southwide by approximately 10 mm per decade (3.7% of the 2001–2010 levels), or 50 mm (18% of the 2001–2010 levels) by 2060 (Figure 9.4 and Table 9.5). Although we found considerable seasonal variability in the projected water yield, the general trend was a statistically significant decrease ($p < 0.05$). Likewise, although we found considerable variability in the magnitude of water yield changes among the four climate scenarios, all four scenarios result in downward trends.

Mean water yield varied considerably across the South as well (Table 9.5 and Figure 9.5), with most of the Appalachian-Cumberland Highland and Mississippi Alluvial Valley and parts of the Mid-South and Coastal Plain, exhibiting statistically significant decreases of more than 2.5% per decade. Although portions of the Coastal Plain and Mid-South are projected to have upward trends in water yield, these increases were not statistically significant at the 0.05 level. The projected downward trends in water yield are not constant through the entire 2010–2060 period. In all subregions, water yield is projected to decrease from 2010 to 2025, level off from 2025 to 2045, and decrease again after 2045 (Figure 9.6).

Impacts on water stress: Decreases in water supply and increases in water demand as a result of population growth will likely combine to increase water supply stress to humans and ecosystems. We define water supply stress as the ratio of human-related water demand/withdrawal by all economic sectors to the total amount of water available (Sun et al. 2008). Water demand includes

TABLE 9.3
Exceedance Probability and Recurrence Interval for the 60- and 90-day Low Flows over Consecutive 10- or 20-year Periods at Three U.S. Geological Survey Stations Draining Forest Lands of the Mississippi Alluvial Valley

Site Name	Station Number	Low Flow (m^3/s)	Probability (%)	RI[a] (years)	Probability (%)	RI[a] (years)	Probability (%)	RI[a] (years)	Probability (%)	RI[a] (years)
Cache River at Forman, IL	3612000	0.16	20 years (1922–1942)		20 years (1942–1962)		20 years (1962–1982)		20 years (1982–2002)	
			60-day flow duration							
			0.54	1.9	0.57	1.75	0.62	1.62	0.67	1.5
			90-day flow duration							
			0.4	2.5	0.33	3	0.43	2.33	0.52	1.9
Big Creek at Pollock, LA	7373000	0.48	20 years (1942–1962)		20 years (1962–1982)		20 years (1982–2002)		10 years (2002–2011)	
			60-day flow duration							
			0.43	2.3	0.62	1.6	0.28	3.5	0.8	1.25
			90-day flow duration							
			0.38	2.6	0.57	1.75	0.29	3.4	0.7	1.4
St. Francis River at Wappapello, MO	7039500	5.01	20 years (1940–1960)		20 years (1960–1980)		20 years (1980–2000)		10 years (2000–2011)	
			60-day flow duration							
			0.61	1.62	0.62	1.61	0.65	1.58	0.58	1.7
			90-day flow duration							
			0.5	2	0.52	1.9	0.6	1.7	0.3	3.2

[a] Recurrence interval.

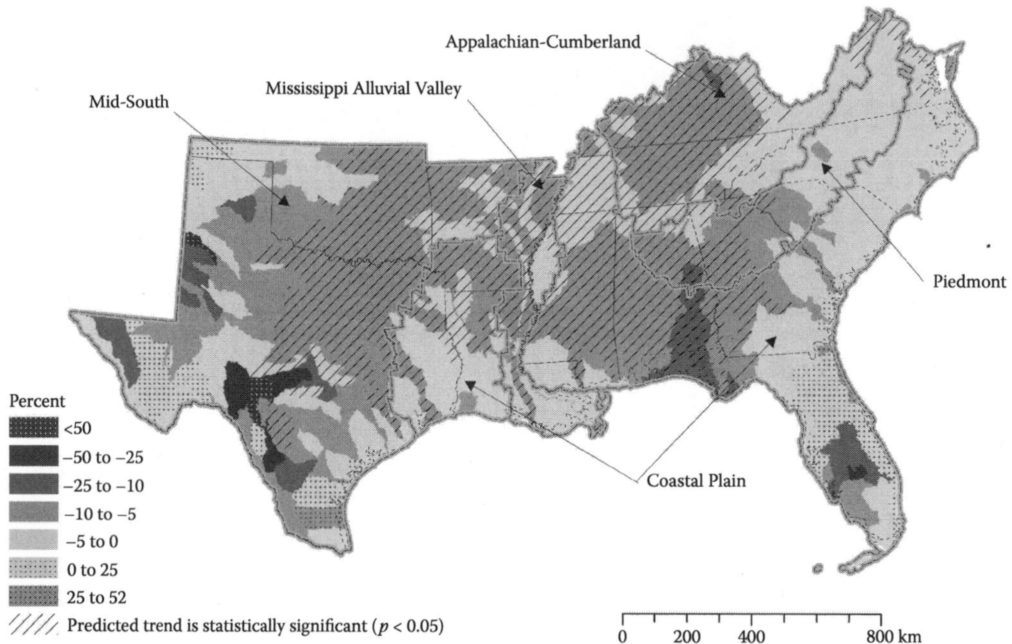

FIGURE 9.2 Predicted trend in annual minimum monthly streamflow for 2010–2060, normalized using the mean of the 2001–2010 annual minimum monthly flow.

water withdrawals by seven water use sectors: commercial, domestic, industrial, irrigation, livestock, mining, and thermoelectric. For future projections, we consider population growth impacts on water demand as well as land use and climate change impacts on water supply.

The impact of declining water yield and increasing population is projected to increase water supply stress over much of the South by 2060, particularly in developing watersheds (Figure 9.7). For example, the Upper Neuse River watershed, providing water supply for the Raleigh-Durham metropolitan area, is projected to experience a climate-induced, 14% decrease in water supply and a growth-induced 21% increase in water demand. This results in an increase in water supply stress from 0.30 in the first decade (2001–2010) to 0.44 in the last (2051–2060). The 0.40 value has been established as a general threshold at which a watershed begins to experience water supply stress (Alcamo et al. 2000; Vörösmarty et al. 2000), although stress may occur at lower or higher values depending on local water infrastructure and management protocols.

TABLE 9.4

Ensemble Mean Predicted Trends in Regional Annual Minimum Monthly Flows Compared to a Baseline Period (2001–2010)

	Baseline	Predicted Trend	
Region/Subregion	**(1 million m³/month)**	**(1 million m³/month/decade)**	**(percent of 2001 to 2010 mean/decade)**
All South	197	−12	−6.1
Appalachian-Cumberland	427	−20	−4.7
Coastal Plain	154	−9.7	−6.3
Mid-South	48	−3.0	−6.3
Mississippi Alluvial Valley	1261	−86	−6.8
Piedmont	41	−2.1	−5.1

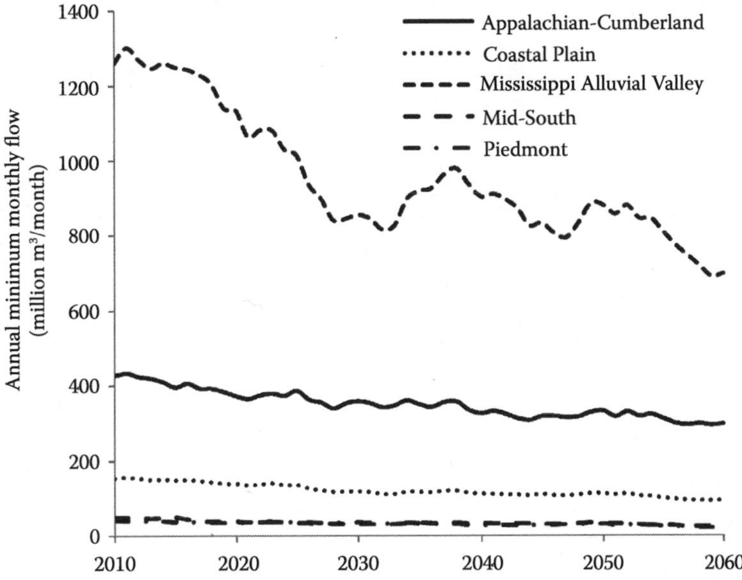

FIGURE 9.3 The annual minimum monthly streamflow by subregion in the Southern United States predicted for 2010–2060 using the mean value of four climate scenarios and displayed using a 10-year moving mean. The MIROC3.2-A1B, CSIROMK2-B2, CSIROMK3.5-A1B, and HadCM3-B2 climate scenarios for the Southern United States (McNulty et al. in press) each combine a general circulation model (MIROC3.2, CSIROMK2, CSIROMK3.5, HadCM3) with an emissions storyline (A1B storyline representing moderate population growth and high-energy use, and B2 representing lower population growth and energy use).

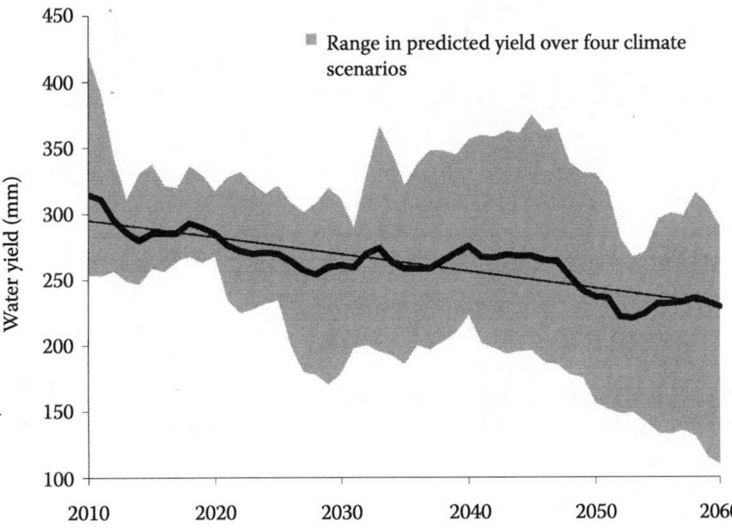

FIGURE 9.4 Predicted southwide annual water yield for 2010–2060. The shaded area represents the range in predicted water yield over the four climate scenarios; the heavy line is the 10-year moving mean; and the thin line is the trend.

TABLE 9.5
Mean of Predicted Trends in Regional Annual Water Yield Based on Four Climate Projections Compared to a Baseline Period (2001–2010)

Region/Subregion	Baseline (mm)	Predicted Trend[a] (mm/decade)	Predicted Trend[a] (% of 2001–2010 mean/decade)
All South	322	−10.2[b]	−3.2[b]
Appalachian-Cumberland	588	−16.2[b]	−2.8[b]
Coastal Plain	392	−11.5[b]	−2.9[b]
Mississippi Alluvial Valley	556	−25.1[b]	−4.5[b]
Mid-South	131	−5.4	−4.1
Piedmont	410	−8.7	−2.1

[a] The MIROC3.2 A1B, CSIROMK2 B2, CSIROMK3.5 A1B, and HadCM3 B2 climate scenarios for the Southern United States each combine a general circulation model (MIROC3.2, CSIROMK2, CSIROMK3.5, HadCM3) with an emissions storyline (A1B storyline representing moderate population growth and high-energy use, and B2 representing lower population growth and energy use).

[b] Significant at the 0.05 level.

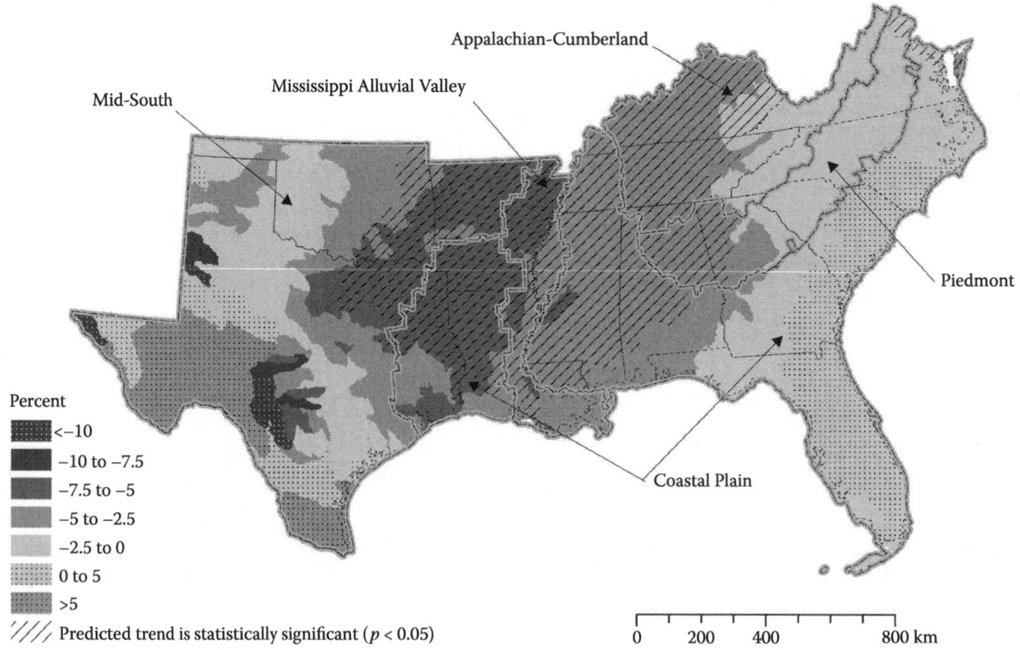

FIGURE 9.5 Predicted trend in mean annual water yield for 2010–2060, normalized using the mean of the 2001–2010 annual water yield.

WATER QUALITY

Climate change not only affects water quantity but also affects water quality (Cruise et al. 1999; Murdoch and Baron 2000; Whitehead et al. 2009). A warming climate may elevate water temperature sufficiently to harm aquatic life (Mohseni and Stefan 1999; Webb et al. 2008). Changes in precipitation amount or storm intensity can affect soil erosion potential by changing the runoff

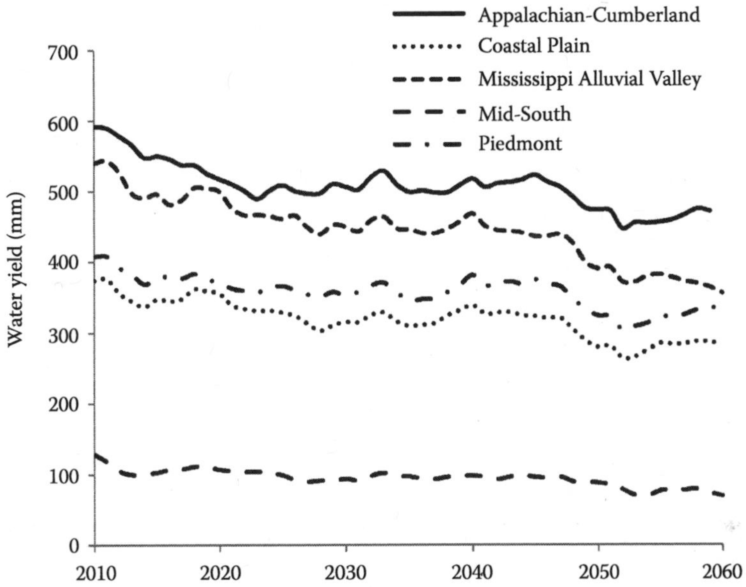

FIGURE 9.6 Predicted trend in annual water yield by subregion in the Southern United States, based on four climate scenarios and using a 10-year moving mean. The MIROC3.2 A1B, CSIROMK2 B2, CSIROMK3.5 A1B, and HadCM3 B2 climate scenarios for the Southern United States (McNulty et al. in press) each combine a general circulation model (MIROC3.2, CSIROMK2, CSIROMK3.5, HadCM3) with an emissions storyline (A1B storyline representing moderate population growth and high energy use, and B2 representing lower population growth and energy use).

amount, the kinetic energy of rainfall, or the vegetation cover that resists erosion. Increased erosion results in increased sediment delivery to streams and lakes. Moreover, water temperatures and sediment concentrations may increase in combination with decreased flow rates and velocities, thereby magnifying their individual impacts to aquatic life.

Water Temperature

Climate change is of particular concern for coldwater fish in the Southern Appalachian Mountains. For example, eastern brook trout (*Salvelinus fontinalis*) need dissolved oxygen levels in excess of 8 mg/L, striped bass (*Morone saxatilis*) prefer dissolved oxygen levels above 5 mg/L, and most warm-water fish need dissolved oxygen in excess of 2 mg/L. Increases in water temperature as a result of climate change would decrease instream dissolved oxygen concentrations. Streams in the Southern Appalachians provide crucial habitat for eastern brook trout and other coldwater species such as rainbow trout (*Oncorhynchus mykiss*) and brown trout (*Salmo trutta*). The lethal temperature limit for such species is approximately 25°C (Matthews and Berg 1997; Meisner 1990).

Several natural factors influence the extent to which changes in air temperature impact stream temperature; these include total streamflow, relative groundwater contribution to flow (Bogan et al. 2003; Matthews and Berg 1997; Poole and Berman 2001; Sullivan et al. 1990; Webb et al. 2008), and canopy cover over the stream. In addition, human-related factors that influence the air/water temperature relationship include runoff from impervious surfaces (Nelson and Palmer 2007), thermal discharges (Webb and Nobilis 2007), and reservoir releases (Webb and Walling 1993).

Several studies have assessed climate change impacts on U.S. stream temperatures using a variety of scales and resolutions, either by trend analysis of historical stream temperatures (Kaushal et al. 2010) or by developing air/water temperature models and projecting into the future (Eaton and Scheller 1996; Mohseni et al. 1998; O'Neal 2002, van Vliet et al. 2011). We adopted the latter

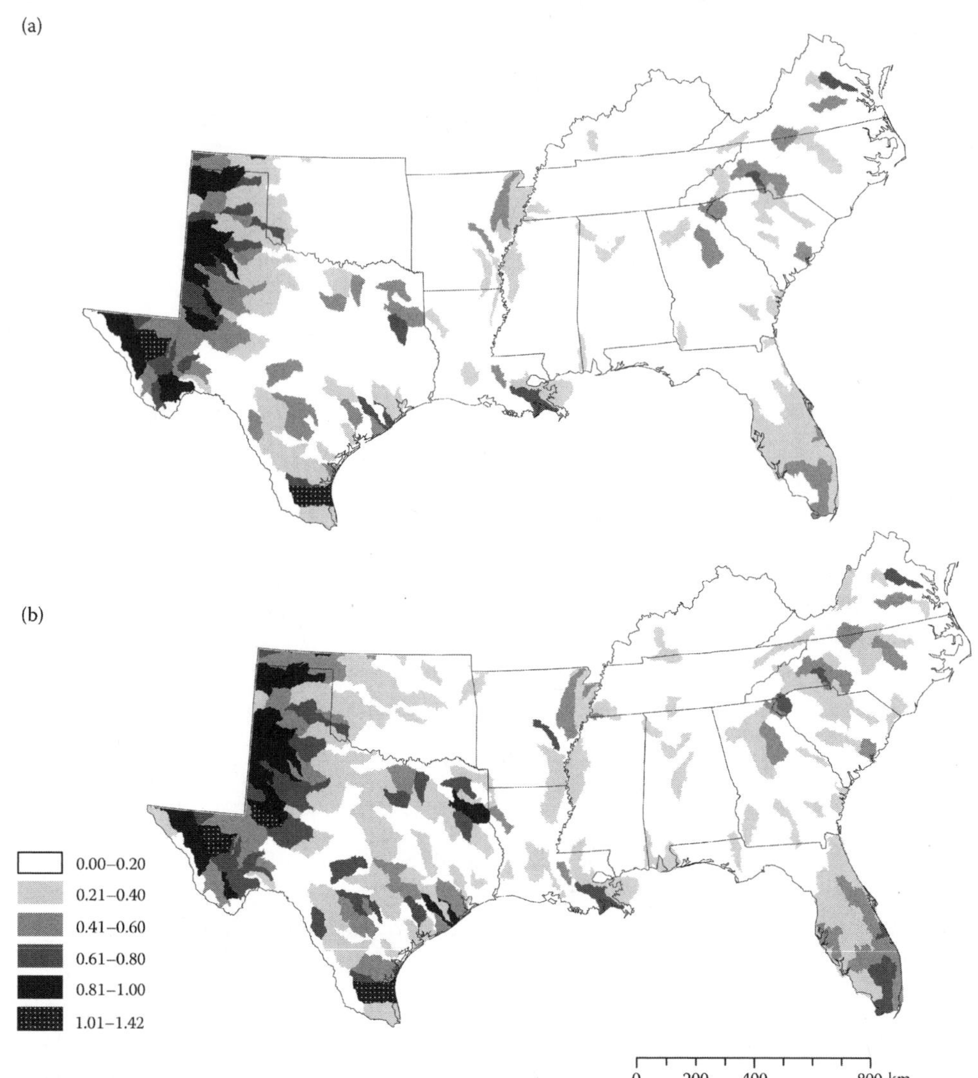

FIGURE 9.7 Mean annual Water Supply Stress Index (a ratio of water demand/water supply) based on four climate scenarios for (a) 2001 to 2010 (baseline), and (b) 2051 to 2060 (predicted). The MIROC3.2 A1B, CSIROMK2 B2, CSIROMK3.5 A1B, and HadCM3 B2 climate scenarios for the Southern United States (McNulty et al. in press) each combine a general circulation model (MIROC3.2, CSIROMK2, CSIROMK3.5, HadCM3) with an emissions storyline (A1B storyline representing moderate population growth and high energy use, and B2 representing lower population growth and energy use).

approach, developing monthly air/water temperature models using long-term historical data for 91 relatively undisturbed watersheds in the South, and then predicting how stream temperature changes under future climate scenarios.

Datasets: The stream temperature observations used in this study came from two sources. Seventy sites were selected from the Hydro-Climatic Data Network (Slack et al. 1993), which are all in watersheds with limited hydrologic alteration by humans. An additional 21 sites are located on smaller Forest Service experimental watersheds in the Appalachian-Cumberland Highland and Piedmont subregions of North Carolina. For both datasets, monthly mean stream temperature values were computed from mean daily values. The resulting database includes stream temperature observations collected during some portion of the 1960 to 2011 period in streams with drainage areas that range from 0.1 to 101,033 km^2 (Figure 9.8).

Air temperatures representing the recent time period came from 4-km by 4-km resolution monthly weather data, available from the PRISM Climate Group (Gibson et al. 2002). Data from the PRISM climate grid cells containing the stream temperature measurement sites were used to represent the air temperature at those sites. For future mean monthly air temperature, we used data from the CSIROMK2 B2, CSIROMK3.5 A1B, HadCM3 B2, and MIROC3.2 A1B climate scenarios (Coulson et al. 2010a,b).

Model development: S-curve monthly air/stream temperature models (Mohseni et al. 1998) were developed for each site. A monthly time step was selected because this is the resolution at which most future climate predictions are available. Models were developed using the most recent two years of stream temperature data at each site, leaving the remaining data to validate the models' accuracy in predicting stream temperature outside the periods used for model development.

Once the air/stream temperature models were generated, stream temperature values were computed for the historical time period (1960–2007) using the PRISM air temperature data, and the

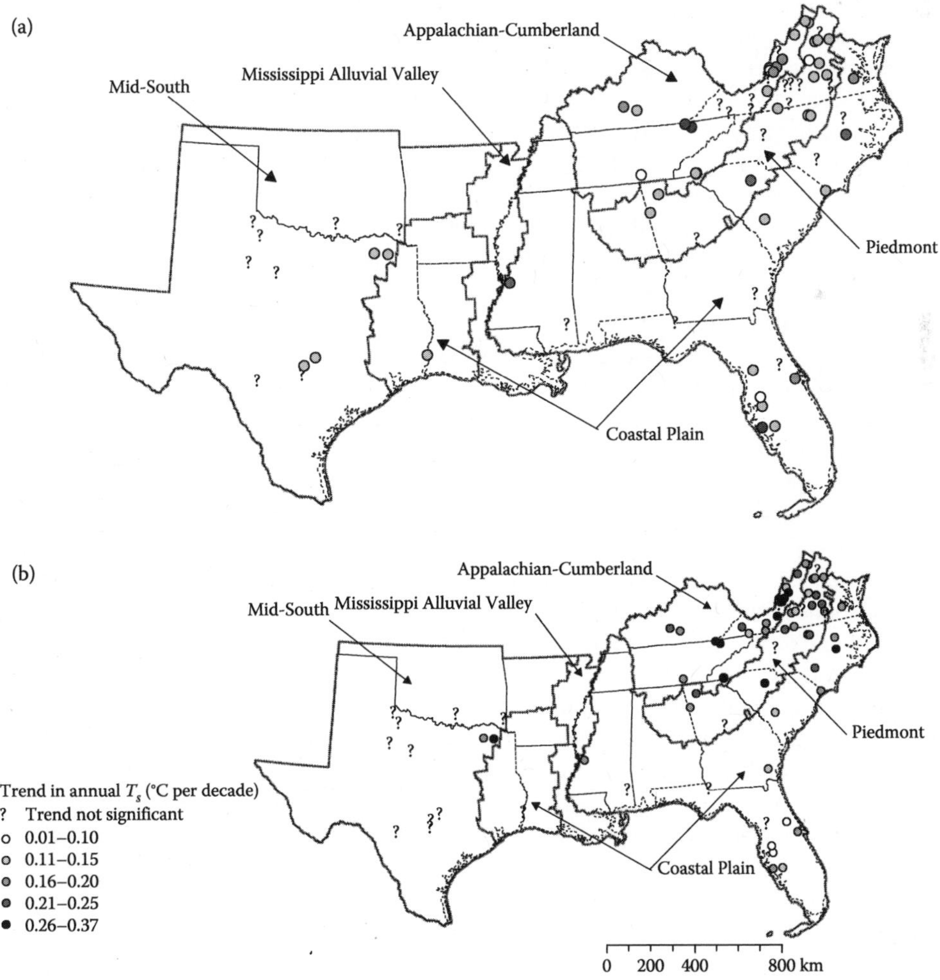

FIGURE 9.8 Estimated trend (1960–2007) in (a) mean annual and (b) annual maximum monthly stream water temperature across the Southern United States (°C/decade).

future (2011–2060) using each of the four climate scenarios to generate monthly time series of historical and future stream temperature at each site. For future predictions, the mean stream temperature over the four climate scenarios was computed to represent the central future condition, and the range in stream temperature among the future climate predictions was computed to represent the uncertainty in these predictions.

Stream temperature was regressed against time during both the historical and future time periods, with the slope of the regression representing the change in stream temperature over time. If the change for either period was significantly different from zero ($\alpha = 0.05$), then we reasoned that the stream temperature had changed in response to air temperature during that period.

Estimated historical trends: Sixty-two of the 91 sites showed significant changes, where mean annual stream temperature increased from 1960 to 2007 (Figure 9.8). The mean temperature increase across the 62 sites was 0.14°C per decade, with a range of 0.08–0.29°C per decade. For the sites with significant changes, the mean changes among subregions did not differ, although five of the largest increases were for sites in the Appalachian-Cumberland Highland.

More relevant to aquatic ecosystems than mean annual stream temperature are the extreme temperature conditions, such as the annual maximum monthly stream temperature. Seventy-one of the 91 sites showed significant changes in annual maximum monthly stream temperature from 1960 to 2007 (Figure 9.8). The mean increase in annual maximum monthly stream temperature for the 71 sites was 0.20°C per decade, with a range of 0.04–0.37°C per decade. The mean for sites in the Appalachian-Cumberland Highland (0.23°C per decade) was significantly different from that for sites in the Coastal Plain (0.15°C per decade) and Piedmont (0.15°C per decade). Significant changes in annual maximum monthly stream temperature were generally larger than the changes in mean annual stream temperature, suggesting that historical changes in climate have had more impact on the extremes in stream temperature than on mean stream temperature.

Predicted future trends: All 91 sites exhibited significant upward trends in mean annual stream temperature from 2011 to 2060 under all four of the future climate scenarios. The mean upward trend over all sites and climate scenarios was 0.26°C per decade, with means ranging from 0.21°C per decade under the HadCM3 B2 scenario (Figure 9.9) to 0.35°C per decade under the MIROC3.2 A1B scenario. The changes between subregions did not differ when compared using the mean of all climate scenarios.

The projected 2011–2060 change in annual maximum stream temperature increased at all 91 sites, and was significant at 27 sites under CSIROMK3.5 A1B, 74 sites under CSIROMK2 B2, 76 sites under HadCM3 B2, and all 91 sites under MIROC3.2 A1B (Figure 9.9). The mean upward trend among all climate scenarios for all sites with significant changes was 0.25°C per decade, ranging from 0.21°C per decade under CSIROMK2 B2 to 0.30°C per decade under MIROC3.2 A1B. Under HadCM3 B2, the mean change for sites in the Appalachian-Cumberland Highland (0.29°C per decade) and Piedmont (0.27°C per decade) was significantly different from the mean change in the Coastal Plain (0.18°C per decade).

Our modeling indicates that stream temperature values in the historical period have likely already increased in much of the South as a result of increasing air temperature, and suggests that they will continue to do so at an accelerated rate in the near future. In particular, sites in the Appalachian-Cumberland Highland and Piedmont have the highest upward trends in mean annual and especially annual maximum monthly stream temperature. Sensitivity of stream temperature to air temperature at a given site can be controlled or influenced by management practices, which will be discussed later in this chapter.

Soil Erosion and Sediment

Sediment is one of the primary threats to water quality in the South (West 2002), where surface erosion is the primary process by which sediment is created. Surface erosion is the detachment and removal of soil or mineral grains from the ground surface. Rainfall and surface runoff are

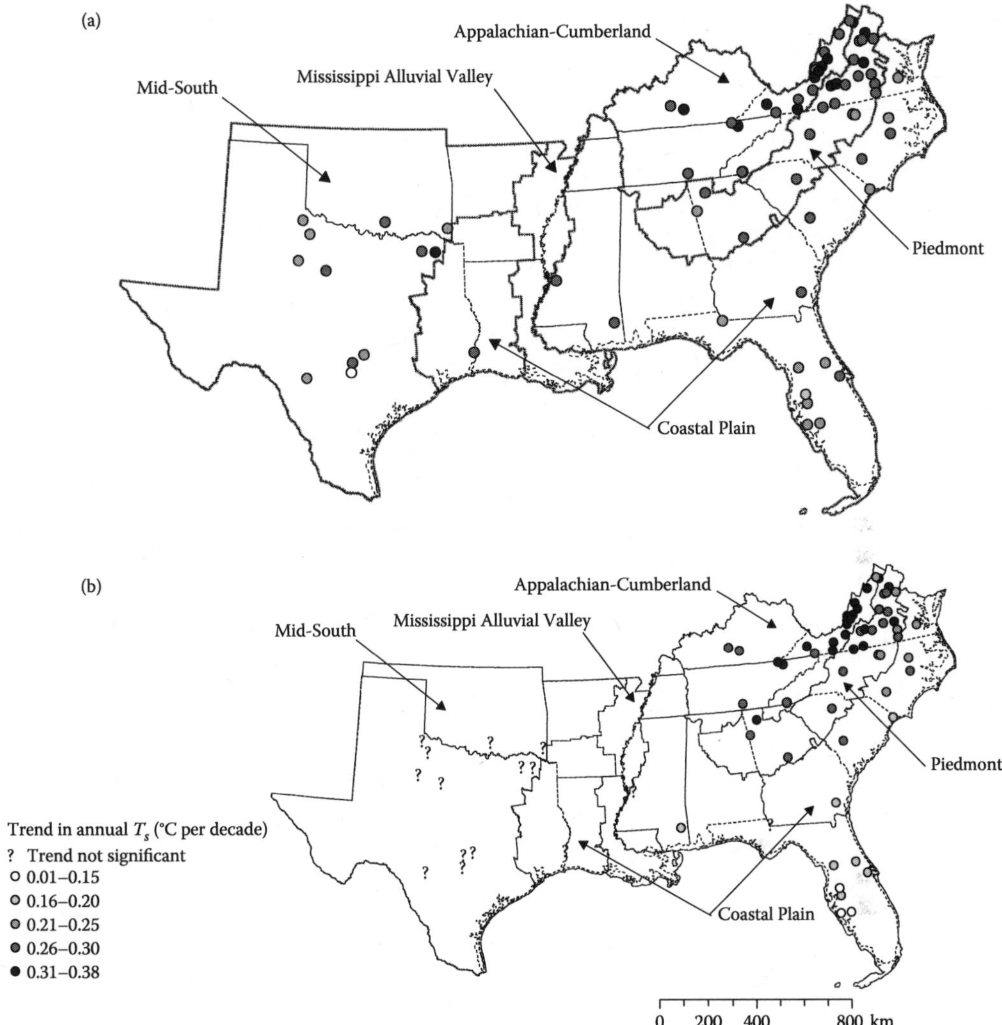

FIGURE 9.9 Predicted trend (2011–2060) in (a) mean annual and (b) annual maximum monthly stream water temperature across the Southern United States (°C/decade) under the HadCM3 B2 climate scenario. The HadCM3 B2 climate scenario for the Southern United States (McNulty et al. in press) combines the HadCM3 general circulation model with the B2 emissions storyline (representing lower population growth and energy use).

the primary forcing agents by which surface erosion occurs and sediment is transported to water bodies.

The Universal Soil Loss equation (Wischmeier and Smith 1978) and its successor, the Revised Universal Soil Loss equation (Renard et al. 1997), have long been used to estimate the amount of surface erosion associated with different environmental conditions and land use activities in the South and elsewhere. The Universal Soil Loss equation was originally developed for estimating surface erosion from cultivated lands, with the Revised Universal Soil Loss equation extending its applicability to rangelands (Renard et al. 1997). Dissmeyer and Foster (1984) showed how the Universal Soil Loss equation could be used in forest lands, and Dissmeyer and Stump (1978) provided specific data for doing so in the South.

The factors used within both equations represent the major parameters affecting surface erosion. Two of these factors are climate related: the rainfall–runoff erosivity factor (R) and the cover-management factor (C). The C-factor is indirectly affected by climate in that temperature and precipitation have long-term influences over the type and density of vegetation that can be sustained in a given area without irrigation or other human interventions. The R-factor provides an index of the intensity and amount of rainfall occurring at a given location over a long period of time, and as such is directly affected by climate.

The R-factor provides a useful surrogate for assessing potential changes in future surface erosion related to climate change. Past research has demonstrated that the amount of surface erosion increases with increased R-factor values (Pruski and Nearing 2002a,b; Renard et al. 1997). R-factor values can be estimated using monthly or annual precipitation data (Renard and Freimund 1994). Therefore, the change in R-factor values is used here to evaluate how future surface erosion may be affected by projected climate changes.

Past evaluations: Changes in R-factor values resulting from climate change have not been demonstrated in the South by directly comparing R-factors computed from recent precipitation data to those originally computed by Wischmeier and Smith (1965) from data for the mid-20th century. Possibly this did not happen because R-factor values were significantly revised by Renard et al. (1997) using additional climate stations and more recent climate data. The more recent R-factor map (Renard et al. 1997) is generally similar to the earlier map in the South, with some marked differences along the Gulf of Mexico in Louisiana and Florida and within the Appalachian-Cumberland Highland. However, it is not clear whether these differences are the result of actual climate change or the inclusion of new stations within these areas.

Predicted precipitation based on general circulation models and emissions storylines has been used to assess how future climate change may affect R-factor values within the conterminous 48 states. In general, the R-factor value changes determined by Phillips et al. (1993) showed little consistency in the South among the four climate scenarios they considered. The one exception was in the Appalachian-Cumberland Highland where all scenarios indicated increased R-factor values. Given their methods, Phillips et al. (1993) warned against overemphasizing the details of their results for individual regions such as the South.

Results from Nearing (2001) also showed little geographic agreement in R-factor predictions for the South under the two climate scenarios considered. This analysis was also at the national scale, but used more recent general circulation models, more sophisticated emissions storylines, finer scale climate projections, and statistical models for predicting R-factor values.

Methods for predicting future trends: Our examination used a somewhat more conservative emissions storyline than past research, and a finer scale climate projection (HadCM3 B2) than past studies. We focused our analysis on the South and used similar-sized drainage basins to display the results. R-factor values are computed using the statistical models of Renard and Freimund (1994) based on mean annual precipitation. Renard and Freimund (1994) also derived models for predicting the R-factor based on the modified Fournier index (Arnoldus 1977), which is calculated using both mean monthly and mean annual precipitation. However, the models based on mean annual precipitation alone predicted R-factor values that more closely matched the values computed by Renard et al. (1997) for the South. Like Nearing (2001), the mean annual precipitation values we used to compute R-factor values were derived from continuous 20-year periods, which in our case were 2011 to 2030, 2031 to 2050, and 2051 to 2070.

To evaluate how much R-factor values would change in the future, we defined 1961 to 1990 as the baseline or "historical" period. R-factor values for this period were computed using precipitation data generated by rescaling 4-km × 4-km PRISM data (Daly et al. 1994) to the 677 southern watersheds whose boundaries are defined by the U.S. Geological Survey 8-digit HUC naming system. The mean watershed area was 3660 km^2. HUC boundaries often extend beyond state boundaries along the interior perimeter of the South to include entire drainage basins. Future precipitation projections were derived from a county-level dataset (Coulson et al. 2010a,b) that was overlaid onto the HUC layer and used to compute area-weighted values for each of the 677 watersheds analyzed.

Determining threshold of concern: Because recommended limits for soil loss caused by surface erosion have not been established for most forest lands in the South, the R-factor change required to exceed a given soil loss limit cannot be determined. Instead, the "threshold of concern" approach (Sassaman 1981) is used here to define the R-factor change limits beyond which undesirable increases in erosion might occur. Although our concern is surface erosion, we used the change in R-factor values to identify when and where mitigation or adaptation actions to future climate change may be most likely.

Soils that have similar physical characteristics exhibit similar surface erosion responses. "Major land resource areas" are U.S. geographic areas composed of similar soils that have been mapped by the U.S. Department of Agriculture (2006). Soils within these areas have developed over millennia in response to a range of rainfall erosivity that is characteristic of the climate in which the soils occur. It seems reasonable that soils developed under higher rainfall erosivity conditions should be able to tolerate a higher relative change in R-factor values than soils developed under lower erosivity conditions. For this analysis we assumed that the threshold for R-factor change is ±10% of the maximum R-factor value identified by Dissmeyer and Stump (1978) for each major land resource area in the South (Figure 9.10).

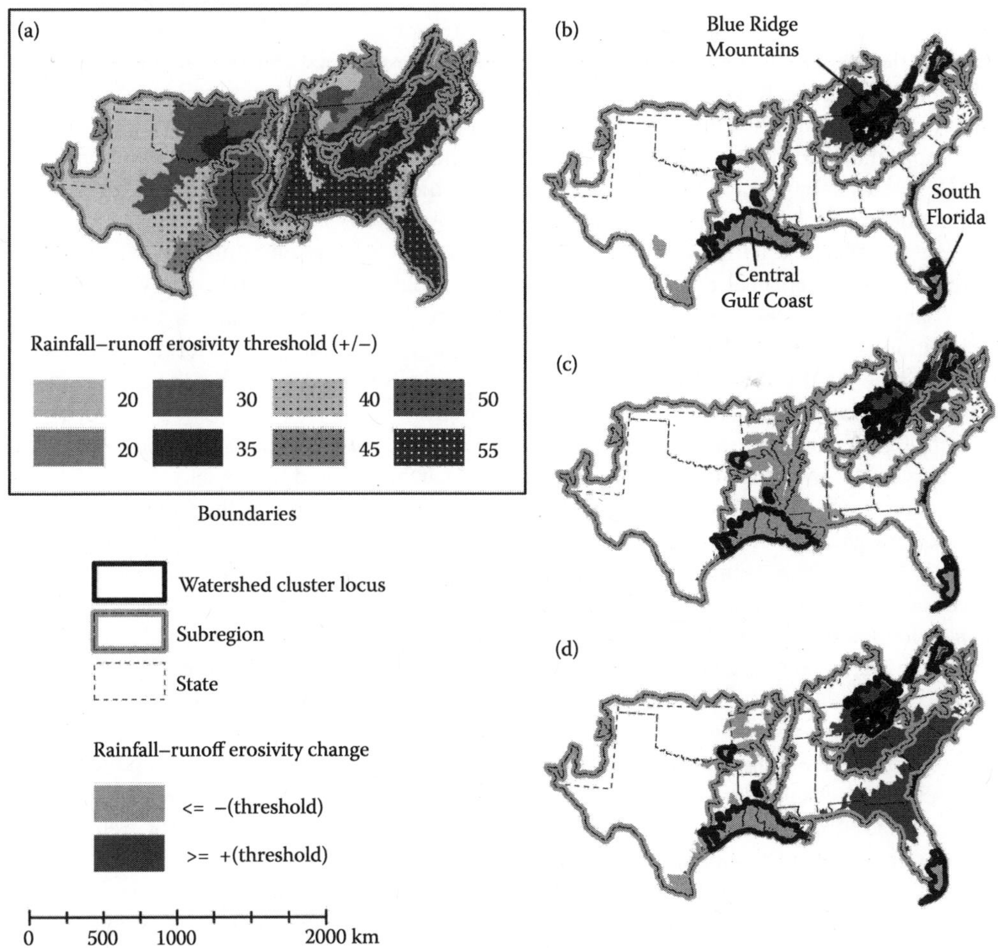

FIGURE 9.10 Changes in future rainfall/runoff erosivity factor for Southern U.S. watersheds based on the HadCM3 B2 climate change scenario, showing defined change thresholds for rainfall–runoff erosivity (a) and watersheds predicted to exceed the thresholds during the (b) 2011–2030, (c) 2031–2050, and (d) 2051–2070 periods. The HadCM3 B2 climate scenario for the Southern United States (McNulty et al. in press) combines the HadCM3 general circulation model with the B2 emissions storyline (representing lower population growth and energy use).

Thresholds for both positive and negative R-factor changes are used. Using a positive change threshold seems obvious because future surface erosion should increase if R-factor values increase. However, a decrease in R-factor values may also result in increased erosion. Using simulation modeling, Pruski and Nearing (2002b) found that erosion could increase when R-factor values either increased or decreased. This occurred under certain conditions because the reduced rainfall associated with decreased R-factor values also caused biomass to decrease and thereby reduced the erosion resisting characteristics of the vegetation.

Projected changes in R-factor: The R-factor changes from the baseline period (1961–1990) for each of the three future periods are discussed below. Hereinafter, only those watersheds that exceed either the positive or negative thresholds are considered.

The future change in R-factor values shows several important trends. Watersheds where the R-factor change is projected to exceed either the positive or negative threshold tend to cluster together (Figure 9.10). These clusters are geographically distinct with each having a locus exhibiting changes in all three future time periods, and a zone of adjoining watersheds where change is apparent in two of the three time periods. For discussion, these three clusters are referred to as the Central Gulf Coast, Blue Ridge Mountains, and South Florida. Each of the clusters shows a consistent but different change behavior. The Central Gulf Coast watersheds all exhibit notable decreases in R-factors, the Blue Ridge Mountain watersheds exhibit increases, and in South Florida both response types occur. The clusters differ in their expansion over time. The Blue Ridge Mountains cluster appears to steadily increase in geographic extent over time, while the extent of the Central Gulf Coast cluster appears to fluctuate somewhat over the study period. The South Florida cluster remains stable with little size change over time.

The geographic and temporal trends in R-factor changes are caused by the combined effect of future precipitation changes (Figure 9.11) and the thresholds defined for the soil areas (see Figure 9.10). Clearly, the clusters noted above correspond closely to those areas showing relatively large precipitation changes. The fact that both the Central Gulf Coast and South Florida clusters occur in areas where the thresholds are among the highest in the South suggests that these areas may be particularly susceptible to surface erosion in the future.

Central Gulf Coast cluster: Historically, the watersheds within or adjoining the Central Gulf Coast cluster all have relatively high R-factor values as demonstrated by the correspondingly high thresholds (30–55) defined for them (Figure 9.10). The primary locus of the Central Gulf Coast cluster is in southeastern Texas and southern Louisiana, where 27 contiguous watersheds exhibit notable R-factor decreases in all three time periods. An additional three watersheds (one in northern Louisiana and two in southeastern Oklahoma) are disconnected from the cluster, but also exhibit notable changes in all three periods. For the period ending in 2030, 46 watersheds have notable R-factor decreases, with all but a few occurring along the Gulf coast or slightly inland. By 2050, the affected area has extended up the Mississippi River and eastward along the Gulf of Mexico to include 102 watersheds. Almost all of Louisiana and most of Arkansas are involved in this period, but 8 watersheds along the coast and in interior southern Texas that previously exhibited notable decreases no longer do so. By 2070, the area affected contracts to only 56 watersheds with the Louisiana and Texas coastal areas continuing to be involved, as are a number of watersheds in Arkansas and eastern Oklahoma.

Over the entire 2011–2070 study period, the change in R-factor for the Central Gulf Coast cluster is most consistent in the Coastal Plain and Mid-South watersheds (Figure 9.12). Mean decreases in R-factors are consistently around 70 for Coastal Plain watersheds and just below 50 for watersheds in the Mid-South. The watersheds in the Mississippi Alluvial Valley are more variable with mean decreases between those of the Coastal Plain and Mid-South watersheds, but with a few actually showing decreases larger than those in Coastal Plain watersheds.

Blue Ridge Mountain cluster: The Blue Ridge Mountain cluster has two groupings of watersheds that exhibit consistent increases in R-factor for all three time periods. The larger locus of 15 watersheds occupies the eastern half of the Appalachian-Cumberland Highland where Virginia,

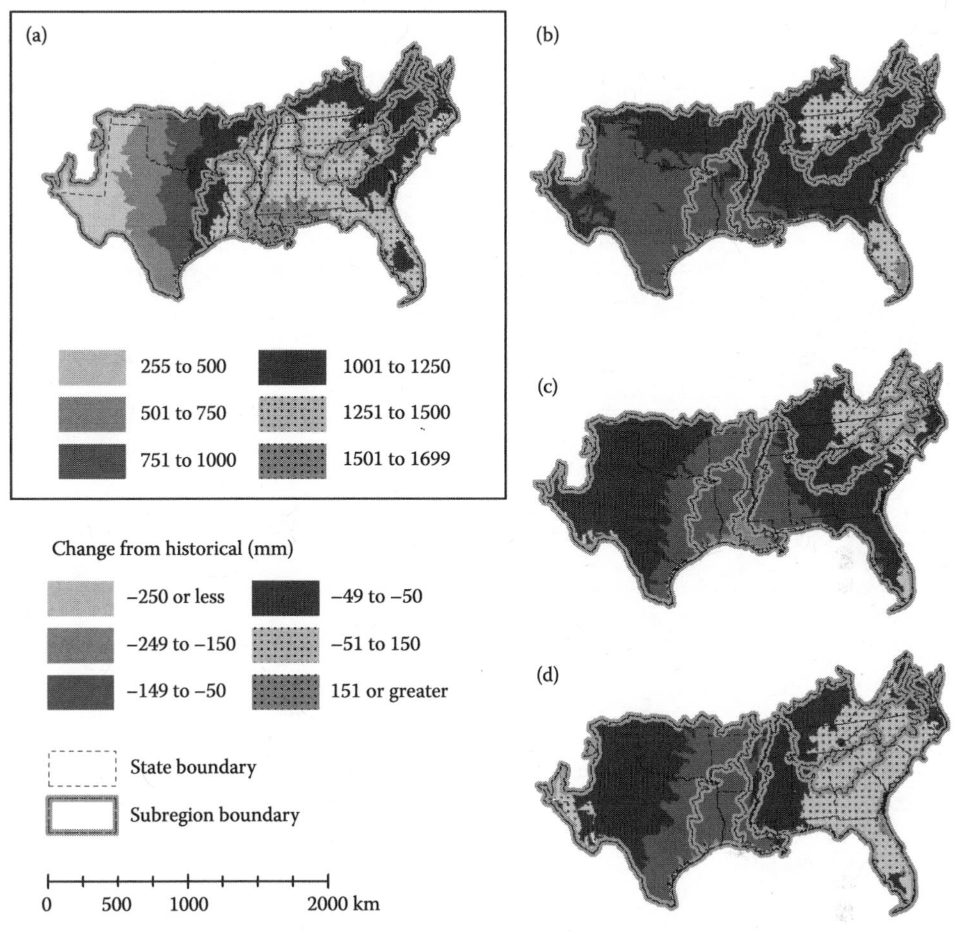

FIGURE 9.11 Future change in mean annual precipitation for Southern U.S. watersheds as shown by (a) the mean for the historical period (based on PRISM modeling), and predicted increase (+) or decrease (−) from the historical mean for the (b) 2011–2030, (c) 2031–2050, and (d) 2051–2070 periods based on the HadCM3 B2 climate change scenario. The HadCM3 B2 climate scenario for the Southern United States (McNulty et al. in press) combines the HadCM3 general circulation model with the B2 emissions storyline (representing lower population growth and energy use).

Kentucky, Tennessee, and North Carolina come together. The smaller locus consists of five watersheds that occur mostly within the northern Piedmont in Virginia (see Figure 9.10). Historically, watersheds in this core area have lower relative R-factor values (200–350), but projected increases in annual precipitation suggest a marked increase in future rainfall erosivity.

The number of affected watersheds increases steadily over time. By 2030, 51 watersheds show notable R-factor increases. By 2050, the area affected shifts somewhat eastward into the Piedmont and the total number increases to 63. By 2070, the affected area has spread southeast into the Coastal Plain and the number of affected watersheds doubles (126).

R-factor changes in affected watersheds in the Appalachian-Cumberland Highland and Piedmont are fairly consistent over time, with means between 35 and 45 (Figure 9.12). In contrast, R-factor changes are more variable over time for affected watersheds in the Coastal Plain, reflecting their shifting location within the Coastal Plain. In 2030, the single watershed that exceeds the threshold occurs on the coast of the Florida panhandle (although not discernible in Figure 9.10) where

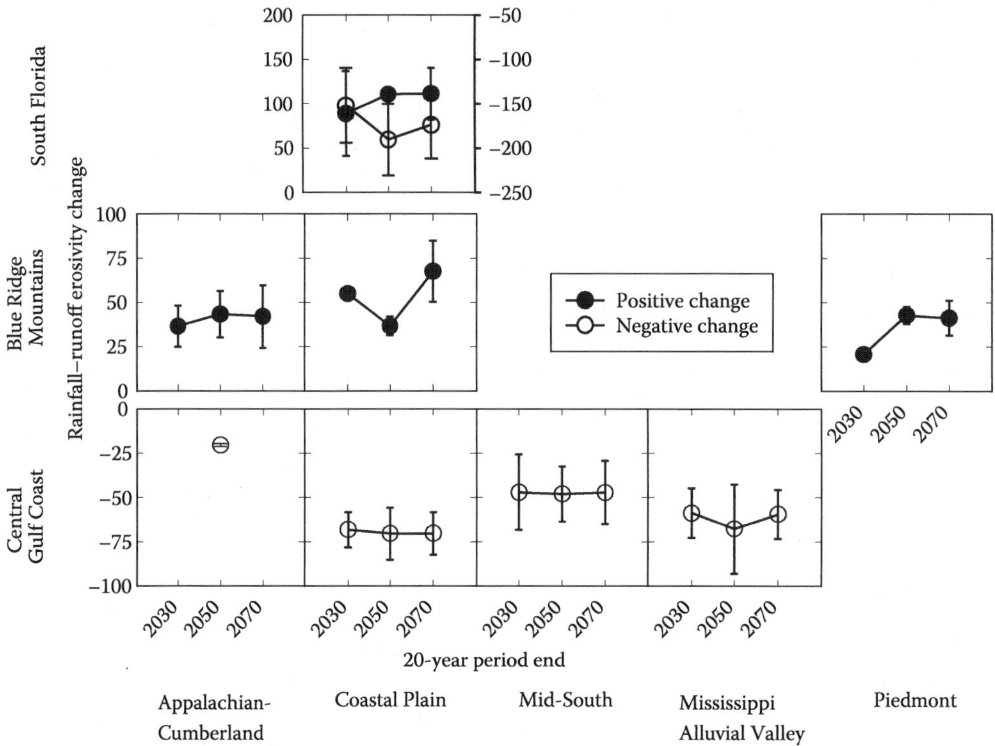

FIGURE 9.12 The mean change in rainfall–runoff erosivity factor for Southern U.S. watersheds by "geographic change cluster" (Central Gulf Coast, Blue Ridge Mountains, and South Florida) and subregion for three 20-year study periods.

historical R-factors and thresholds are relatively high. By 2050, the affected Coastal Plain watersheds are all located along the Chesapeake Bay, where historical R-factors are lower, and the mean R-factor change is now 37 (Figure 9.12). By 2070 the mean change has increased to 68 because the affected watersheds have shifted southward and R-factor changes are larger. The effect of R-factor increases on surface erosion within this cluster may be amplified by the area's steeper terrain, particularly in the Appalachian-Cumberland Highland.

South Florida cluster: The South Florida change cluster is unusual in that its watersheds exhibit both notable increases and decreases over a relatively small geographic area. The locus of the cluster consists of three watersheds on the eastern side of the southern Florida peninsula and illustrates the extreme contrasts projected to occur in the future (Figure 9.10). Two of the watersheds exhibit consistent R-factor decreases ranging from −122 to −219, whereas the adjoining watershed to the south (encompassing the Florida Keys) exhibits increases of 111–177. If not attributable to model error, these remarkable differences indicate tremendous climatic variability over short distances.

All three time periods reflect this marked geographic variability. By 2030, seven watersheds show notable changes with two decreasing and five increasing. By 2050, only the watersheds within the cluster locus are affected; by 2070 one additional watershed joins the increase total.

Throughout the study period, R-factor changes in the South are largest in South Florida. Watersheds exhibiting notable increases have mean values between 89 and 111 (Figure 9.12). Those showing notable decreases have means of −152 to −190.

Other considerations: Past research has noted that the response of surface erosion to R-factor changes may not be a one-to-one ratio. Erosion simulation studies using agricultural conditions and the Revised Universal Soil Loss equation showed that soil loss did not respond in equal proportion to changes in the R-factor; rather, the percent change in soil loss is somewhat less than the change in R-factor value (Pruski and Nearing 2002a; Renard and Ferreira 1993). Pruski and Nearing (2002a) attribute this to increased biomass production under wetter conditions that thereby increases the ability of vegetation to resist erosion. Renard and Ferreira (1993) found that temperature changes actually affected soil loss amounts more than rainfall because of the effect that temperature has on residue decomposition rates.

The way rainfall occurs can also affect erosion response. Pruski and Nearing (2002b) found that varying the number of days with rain, the rainfall intensity, and the season when rainfall occurred all affected erosion response, and sometimes resulted in a percent change in soil loss exceeding that of the R-factor. It should also be noted that the steeper slopes that often occur in forest lands could compound the effect of R-factor changes.

In summary, our modeling results show that climate change has the potential to produce marked changes in R-factor values in the South, that R-factor changes would be largely restricted to three geographic clusters (the Central Gulf Coast, Blue Ridge Mountains, and South Florida), and that R-factor changes are likely to persist out to 2070. Land managers working within areas covered by these clusters may well need to consider mitigation and adaptation measures in planning future land disturbing activities.

Salinity Intrusion

Saltwater intrusion into freshwater aquifers and drainage basins can threaten the biodiversity of freshwater tidal marshes and contamination of municipal, industrial, and agricultural water supplies (Bear et al. 1999). The balance between the hydrologic flow conditions within a coastal drainage basin and fluctuation in sea level governs the magnitude, duration, and frequency of salinity intrusion into coastal rivers. Future climate change is likely to alter these hydrologic balances on the East Coast of the United States by increasing air temperatures, changing regional precipitation regimes, and causing sea levels to rise.

We examined saltwater intrusion at two municipal intakes—on the Atlantic Intracoastal Waterway and the Waccamaw River along the South Carolina Grand Strand near Myrtle Beach (Figure 9.13)—to illustrate how future sea level rise and a reduction in streamflows can potentially affect salinity intrusion, threatening municipal water supplies and the biodiversity of freshwater tidal marshes (Furlow et al. 2002).

Modeling methods and databases: An updated salinity intrusion model of the Pee Dee River Basin (Conrads and Roehl 2007) was used to evaluate the potential effects of climate change on salinity intrusion. The model was developed using data-mining techniques, including empirical modeling using multilayer perceptron (Rosenblatt 1958) artificial neural network (Jensen 1994) models.

The U.S. Geological Survey maintains a real-time stream-gauging network of recorders for streamflow (>50 years), water-level, and specific conductance (<25 years) in the Pee Dee and Waccamaw River Basins (Figure 9.13). During the past 25 years of data collection, the estuarine system has experienced various extreme conditions including large 24-h rainfalls, the passing of tropical systems and major offshore hurricanes, and record droughts.

The artificial neural network models that were developed used a subset of the Pee Dee/Waccamaw network. Model outputs are the specific conductance at seven coastal gauging stations using data from five upland streamflow recorders, tidal water levels at the northern end of the Intracoastal Waterway, and wind speed and direction from a coastal meteorological station. The models were validated using historical measurements of specific conductance at selected coastal stream gauging stations.

To simulate the effects of sea level rise, the mean coastal water levels were incremented by 0.5-foot segments to simulate sea level rise up to 3 feet. To simulate the effect of reduced streamflow to the coast, daily historical streamflows were reduced by increments of 5% up to 25%. Daily specific

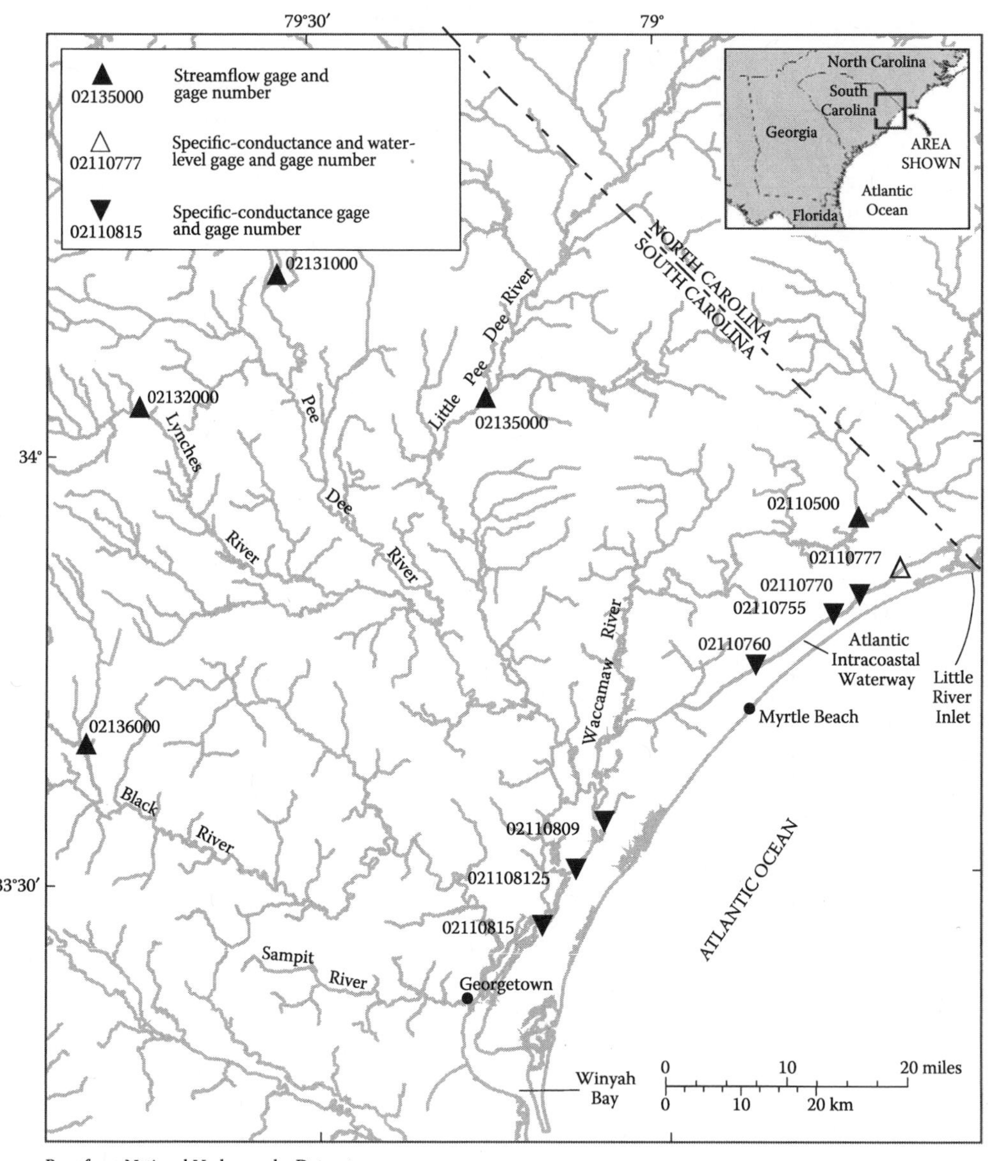

Base from National Hydrography Dataset;
UTM projection, zone 17, datum NAD83

FIGURE 9.13 Location of study monitoring stations within the Pee Dee and Waccamaw River Basins and Atlantic Intracoastal Waterway in South Carolina.

conductance values were simulated for each incremental rise in sea level and each incremental reduction in streamflow from July 1995 to August 2009.

Findings: Results for the Pawleys Island stream gauge (station 021108125), just downstream from a municipal freshwater intake, were selected for this analysis (Figure 9.13). The model satisfactorily simulates the specific conductance in the 2000-μS/cm range and accurately simulates the high intrusion events in the fall of 2002 and 2008 that exceeded 10,000 μS/cm.

Operations at municipal water treatment plants become more difficult if specific conductance values for source water exceed 1000 to 2000 μS/cm. Figure 9.14 shows the number of days that specific conductance values exceeded thresholds of 1000, 2000, and 3000 μS/cm for the 14-year

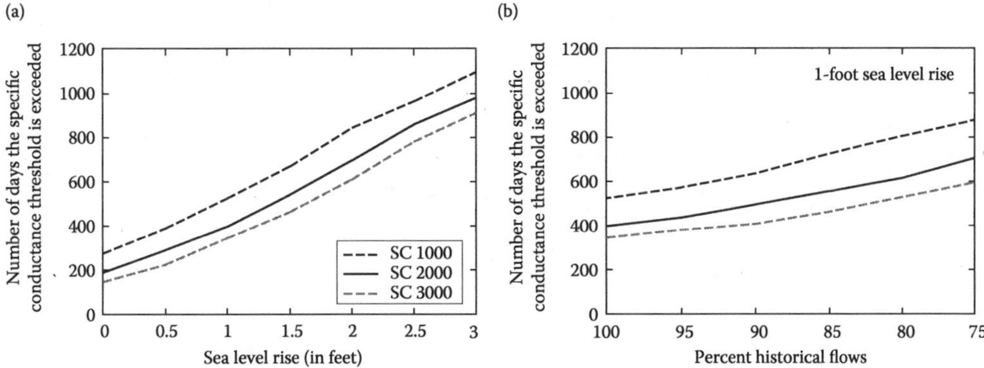

FIGURE 9.14 Number of days that specific conductance (SC) thresholds would be exceeded at Pawleys Island Station (gauge 021108125) in South Carolina for (a) sea level rise up to to 3 feet, and (b) with decreased historical flows concurrent with a 1-foot sea level rise.

simulation period. For example, daily specific conductance >2000 μS/cm historically occurred for almost 200 days (Figure 9.14), during which the municipal intake was unavailable. A one-foot sea level rise would double the number of days the municipal intake is unavailable to 400 days and a two-foot rise increases the unavailability to nearly two years (700 days).

Changes in precipitation patterns that are caused by changes in the climate have the potential to decrease streamflow to the coast. Salinity intrusion occurs during low streamflow periods; a decrease in streamflow combined with a sea level rise could increase the duration of salinity intrusion. For a specific conductance threshold of 2000 μS/cm, a one-foot sea level rise combined with a 10% decrease in historical streamflow would increase the days the municipal intake is unavailable by 25%, or an additional 100 days (Figure 9.14). A 25% reduction of low streamflow levels increases the incidences of intake unavailability to >700 days.

MANAGEMENT OPTIONS FOR MITIGATION AND ADAPTATION

Our analyses clearly show that climate change and its impacts on water quantity and quality have occurred in several subregions of the South. Although climate scenarios for the next several decades do not agree about the magnitude or direction of the expected changes for some variables (especially precipitation), they all point toward a climatic regime that the region has not experienced before (Milly et al. 2008). This would require a reevaluation of forest water resource management methods, even for those practices that have been successful in the past (Galloway 2011), because they were developed for previous climate characteristics. Existing historical records from individual forest monitoring sites may be too short to be useful in accurately forecasting the impacts of climate change, but the accumulated data from many sites provide a foundation for developing mitigation and adaptation methods in forest environments of the future. Computer simulation models provide valuable tools for identifying the potential risks and consequences of climate change and for helping land managers design response actions that minimize adverse impacts.

Various risks to natural ecosystems (Carlisle et al. 2010) and society can result from climate change impacts on water resources (Table 9.6). Innovative adaptation options are needed to reduce or adapt to the severe consequences of climate changes, such as water supply shortages, habitat loss, and increased wildfires. Best management practices (BMPs) that have been found to be most effective in reducing nonpoint source pollution will likely need to be enhanced and revisited so that they best reflect future hydrologic and management conditions. Integrated watershed management that enhances ecosystem resilience to climate disturbances and maintains ecosystem services—including climate moderation and mitigation—would improve adaptation efforts. Below, we discuss

TABLE 9.6
Potential Adaptation Options to Managing Hydrologic Impact and Risks from Climate Change

Hydrologic Impacts	Risks to Ecosystems and Society	Adaptation Options
Water supply stress increase	Water shortage; drying up of drinking wells; consequences to aquatic ecosystems, socioeconomics, and businesses	Maintain watershed health, thinning forests; reduce groundwater and surface water use for irrigation of croplands and lawns; enhance water conservation
Evapotranspiration increase	Hydrologic droughts; wildfires; insect, disease outbreaks	Use native tree species; reduce tree stocking
Increase of peakflow, stormflow volume, floods	Flooding; increased soil erosion and sedimentation	Reduce impervious areas; increase stormwater retention ponds; increase evapotranspiration by increasing forest coverage
Low flow decrease	Water quality degradation; fish habitat loss; reduced transportation capacity	Reduce off stream water withdrawal; adjust water outflow from reservoirs
Wetland hydroperiod change	Wildlife habitat loss; carbon dioxide emission	Plug ditches; adjust water outflow from reservoirs
Stream water temperature increase	Water quality degradation; loss of cold fish habitat	Maintain riparian buffers and shading
Soil erosion, sedimentation increase	Water quality degradation; siltation of reservoirs; increase of cost of water treatment	Enhance best management practices (BMPs) for forest roads; redesign riparian buffers; minimize direct discharge of runoff from roads to streams
Chemical loading increase	Water quality degradation; higher cost of water treatment	Maintain streamflow quantity; forest best management practices (BMPs)

management options and offer some recommendations to address specific water resource concerns related to climate change.

Gunn et al. (2009) recommend a toolbox approach with the three strategies of resistance, resilience, and response. Resistance strategy is a set of short-term approaches to address immediate threats and focuses on minimizing the impacts of disturbance that are exacerbated by climate change. Resilience strategy is both a short- and long-term approach to address the capacity of a forest stand or community to recover from a disturbance and return to a reference state. Response—the most costly strategy to facilitate the movement of species over time—requires acceptance of a level of uncertainty that will be uncomfortable for many forest managers and relies on close collaboration with forest ecologists, silviculturists, and other specialists.

Sea Level Rise and Hurricanes

Close proximity to the ocean makes some Coastal Plain ecosystems extremely vulnerable to hurricanes and sea level rise, both of which are linked to climate change and variability. Louisiana, Mississippi, Alabama, and Florida are projected to see significant increases in salinity levels associated with saltwater intrusion, as well as a degraded quality of the inflows. Forests in low-lying areas such as mangroves in Florida and cypress (David et al. 2010) are particularly vulnerable to sea level rise and increased air temperature. Other concerns are more wildfires and more areas where pine species are apt to be favored. More frequent, high-intensity summer rainfall may lead to downstream flooding in certain areas. Higher water temperatures and changes in freshwater delivery will likely alter estuarine stratification, residence time, and eutrophication. Occurrence of many stressors—such as pollution, harvesting, habitat destruction, invasive species, land and resource use, and extreme natural events—if concurrent would intensify hydrologic responses.

A first step might be to redesign forest road cross-drainage structures using projected high-intensity storms. Coastal areas are expected to see more land development and urbanization with continued population growth. Handling increased stormwater runoff from intense rainfall on flat lands and poorly drained soils requires special designs of urban BMPs. Protection of natural wetlands would maintain their hydrologic functions and buffer disturbances at the landscape level. Resilience and vulnerability evaluation of artificially drained managed forests will be more accurate if they incorporate projected future hydrology.

Although increases of sea level rise and decreases in streamflows show substantial effects that would have operational consequences for municipal water-treatment plants in coastal areas, the climate scenarios that were used for our projections would help water resource managers plan for mitigation efforts to minimize the effect of increased salinity of source water. Mitigation efforts may include timing of withdrawals during outgoing tides, increased storage of untreated runoff, timing larger releases of regulated flows so that the saltwater–freshwater interface moves downstream, and blending higher conductance surface water with lower conductance water from an alternative source such as groundwater.

Extreme Precipitation Events and Low Flows

Our results suggest that forest management activities that fall short of changing forest composition or stand structure after a clear-cut will not substantially alter streamflow responses to extreme precipitation events. Although often statistically significant, differences among annual streamflow responses were small under mean climate conditions versus observed or projected extremes in annual drought or high annual precipitation, and would result in changes that range from 7% increases to 5% decreases. These results suggest a limited capacity to create watershed conditions more resilient to extreme annual precipitation than native hardwood forests managed using traditional forestry practices that rely on natural regeneration. In contrast, management activities that converted deciduous hardwoods to pine monocultures substantially lowered streamflow in times of extreme annual precipitation; such actions may reduce flood risk but also exacerbate drought. This argues for careful consideration of the tradeoff between managing forests for opposite extremes during contingency land use planning.

Landowners and policy makers both look to forests as a means to offset climate change effects (Pacala and Sokolow 2004) and to forest managers to create ecosystems that are more resilient to extremes and changing climate (Baron et al. 2009). We have shown that in Appalachian-Cumberland Highland watersheds, changing forest cover through species conversions affects streamflow and thus downstream water supply in ways other than what would be expected from unmanaged forests; however, forest cover change also affects many other ecosystem services, including carbon sequestration (Liao et al. 2010). Forests cannot be managed solely for water resources without affecting carbon sequestration, and vice versa (Jackson et al. 2005). Whether increasing forest cover or converting deciduous to pine forest cover will mitigate climate change is uncertain (Jackson et al. 2008), but our study shows the potential of forest management to mitigate against the hydrologic impacts of extreme annual precipitation associated with climate change.

In light of predicted decreases in low flows, water resource conservation and hydrological system protection will become vitally important for keeping the minimum streamflow required to prevent significant harm to water resources. Management practices that help to reduce water use, reclaim wastewater, and enhance infiltration and groundwater discharge to streams may become important tools in mitigating climate change impacts on low flows.

Water Quality

Climate change has the potential to degrade water quality. Because our research indicates that climate change will likely increase stream temperatures, the most effective mitigation options would be those that focus on decreasing stream temperature through management, particularly in the summer months (Swift 1973, 1982; Wooldridge and Stern 1979). Maintaining or increasing shading from solar radiation through riparian buffer retention, conservation, or restoration has been shown to decrease stream temperatures (Burton and Likens 1973; Kaushal et al. 2010; Peterson and Kwak 1999; Swift 1973, 1982; Swift and Messer 1971).

Wide BMP usage has helped control nonpoint source pollution, including sediment loading to water bodies. To mitigate climate change impacts, BMP adjustments may be necessary to address anticipated increases in storm flow, soil erosion, and sediment. If climate change alters rainfall patterns as predicted, the size of large runoff events (such as the 20-year storm) will likely increase; to put it another way, the current 20-year storm will occur more frequently. Storm intensity in general is predicted to increase, requiring larger capacity in forest road drainage structures and BMP improvements to reduce the effects of increased storm water discharge. Possibilities for BMP innovations to handle future climate conditions in forested watersheds are:

Broad-based dips: Current standards are based on road grade. Compensating for increased runoff will require adjustments to the relationship between road grade and distance from one dip to the next; for a given road grade, the distance from one broad-based dip to the next would need to be shorter with increased rainfall amount/intensity. The alternative would be to modify design criteria of the dip itself to allow increased volumes of storm flow.

Culvert and other cross-drainage structure size: Increasing the size of culverts, causeways for forest roads and drop structures, and gully check dams for a given drainage area may be required. An alternative would be to install hooded inlets that increase culvert capacity. In situations where road elevation is limiting, hooded inlets would decrease the need for additional fill at channel crossings.

Riparian buffers: Increasing the use of riparian buffers would be most helpful. In addition, increased buffer widths may be needed to improve the capacity of undisturbed stream side zones to absorb potential overland flow resulting from more frequent, extreme storm events. The Effective Functional Width is a tool for illustrating how increased overland flow during intense storms can decrease the capacity of a riparian buffer of a given width to protect water quality. Figure 9.15 depicts

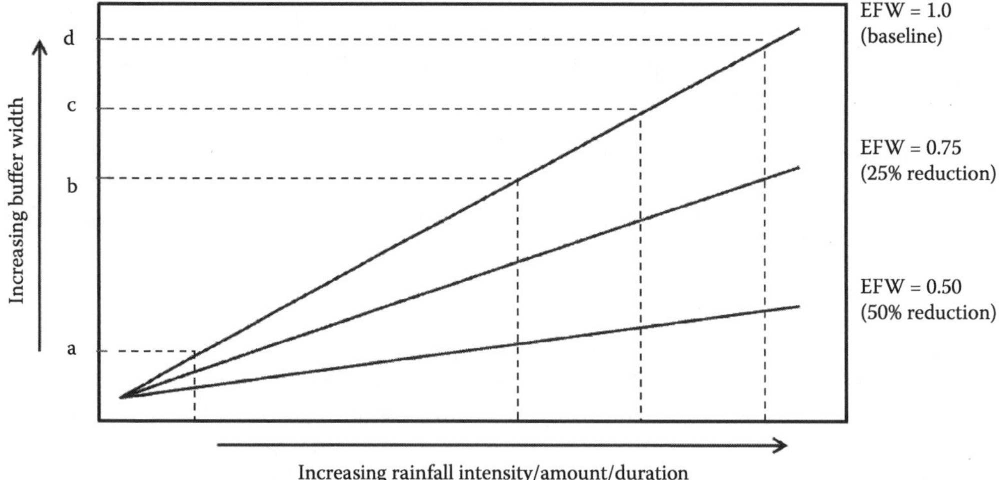

FIGURE 9.15 Theoretical model for maintaining the effectiveness of a stream buffer for reducing surface water quality degradation; EFW is the effective buffer width.

the decision space for buffer width determination. Under undisturbed conditions a given parameter has a value of 1.0, indicating an inherent capacity of the riparian zone to absorb materials being transported downslope as a result of prevailing levels of erosion and decomposition. After disturbance (such as a timber harvest), the condition of the parameter deteriorates by some factor depending on its sensitivity to the effects of the land-disturbing activity. In order to maintain an acceptable level or condition of a given parameter, riparian buffer width adjustments would be needed. If a 25% or even 50% reduction in the condition of the parameter is tolerable, and is itself the most limiting or most sensitive parameter in that system, then narrower buffers may be assigned. Further, buffer width is a function of some measure of rainfall intensity so that for a given maximum expected intensity/amount/duration, buffer width may be adjusted to maintain an acceptable level of parameter condition.

Storm water management: Often, the management of forested watersheds is intended to regulate water yield, a paradigm that may need reorienting toward management of peak flows to face the challenges of climate change. Road design and construction practices may be needed that more effectively disconnect the road system from the stream network; this would slow flow routing during extreme events and increase belowground storage. Storm water management that maximizes opportunities for infiltration would reduce the impact of extreme storm events on peak flows. Further, predicted increases in peak storm water discharge would place increased demand on water treatment facilities, often resulting in discharge of untreated effluent. Improvements such as water gardens, porous parking lots, sediment basins, and sustainable urban drainage systems that incorporate urban forestry practices would encourage infiltration, reduce needed treatment facility capacity, and ultimately reduce the risk of degraded surface water quality.

Other human influences not related to forest management may help mitigate the impact of climate change on stream water quantity and quality. For example, deep-water releases from reservoirs can have a cooling effect similar to groundwater, and municipal wastewater from a deep underground pipe can cool streams in the summer and increase baseflow (Bogan et al. 2003). Wastewater discharge as a means to reduce stream temperatures should be used with caution, as it may also have a warming effect in winter months. Re-using treated wastewater helps to reduce higher-temperature effluent volume entering streams (Kinouchi et al. 2007), and decreasing water withdrawals from streams through water conservation may help maintain a more stable temperature (Webb and Nobilis 1995). Increasing shaded stormwater retention wetlands, increasing urban tree canopy over any runoff surface, and reducing impervious surface area (Peterson and Kwak 1999) can all help decrease stream temperatures, particularly in urban areas or areas where the riparian vegetation has been degraded.

Moderating Climate and Mitigating Impacts through Forest Restoration and Afforestation

Natural and managed forests provide the best water quality among land uses (Lockaby et al. in press). In addition, forests can also modulate regional climate by controlling energy and water transfers between the atmosphere and forested land-surface (Chen et al. 2012; Liu 2011; Liu et al. 2008; Sun et al. 2010). Past agriculture-to-forest land conversion in the South may have led to significant changes in land-surface energy and water processes, including an increase in solar radiation absorbed by the land surface caused by the lower light reflection of forests (Liu 2011). An increase in absorbed radiation provides more energy for heat transfer from the land surface to the atmosphere through sensible and latent heat fluxes and increasing evapotranspiration, which ultimately reduce runoff and water yield. These changes in land-surface processes can further change atmospheric dynamics and hydrologic processes, including precipitation and soil moisture. Some simulation studies have indicated that precipitation is likely to increase with large-scale afforestation (Liu 2011).

Forest restoration and afforestation are expected to play an important role in mitigating the impacts of climate change. A plan to plant over 7 million ha of new trees by 2020 to replace pasture and farming lands in the South, as well as in the Great Lake states and the Corn Belt states (Watson 2009) is even larger than the one carried out by the Civilian Conservation Corps during the Great

Depression, which planted 3 billion trees from 1933 to 1942. Although its primary purpose would be to mitigate the greenhouse effect by increasing the capacity of forests to remove atmospheric CO_2, this effort would also be useful in mitigating the impacts of the greenhouse effect on forest water conditions.

Mitigating the hydrologic impacts of climate change through forest restoration and afforestation is a complex issue. Achieving the mitigation goal requires consideration of many factors when making restoration or afforestation plans. First, approaches for changing forest hydrologic conditions in an afforested area can be different or even opposite of those for surrounding landscapes. Differences can even exist within the afforested area, depending on local atmospheric patterns and physiography. Second, vegetation types are expected to change in the future as a result of climate change. For example, some deciduous forests in the region's upper latitudes are projected to change to savanna communities, and some coastal mixed forests to conifer woodlands (Neilson et al. 2005). Ideally, future afforestation would be done in the places where forests are projected to be most degraded by climate change. Finally, one mitigation tool could have mixed outcomes. For example, afforestation may increase soil water storage by increasing precipitation and decreasing runoff, but may also reduce it by increasing evapotranspiration.

SUMMARY OF REGION-WIDE RESPONSES

Climate changes (increases in rainfall variability and air temperature) are happening across the South and are predicted to continue in the near future.

- Climate warming would result in increased water loss through evapotranspiration through increased evaporative potential and plant species shift, and thus, under a similar precipitation regime, climate warming could cause decreased total streamflow, low flow rates, and regional water supply.
- Decreased precipitation in the western Coastal Plain and Mid-South would have serious impacts on water availability and aquatic habitat.
- Water supply stress is projected to increase significantly by 2050, the result of hydrologic alterations caused by climate change and increased water use by key economic sectors such as domestic water supply, irrigation agriculture, and power plants; water stress will likely be more severe in the summer season.
- Runoff and soil erosion are projected to increase in some areas, the result of changes in rainfall that either increase rainfall erosivity or decrease vegetative cover protection.
- Water temperature is projected to increase with air temperature rise, resulting in possible impacts on coldwater fish habitat in the Appalachian-Cumberland Highland.
- Salinity intrusion in coastal fresh water systems is likely to increase in response to sea level rise and decreased fresh-water inputs from uplands—both consequences of climate change.
- Forest management has the potential to mitigate or reduce damage from hydrologic extremes and degraded water quality caused by climate change, but practices must be applied judiciously to avoid unintended consequences.
- Existing BMPs will need adjustments and enhancement to increase watershed resilience to the likely adverse impacts of climate change on water quantity and quality.

CASE STUDIES

The South has diverse physiographic and socioeconomic characteristics, and thus water resource distribution, water use patterns, and response to climate change vary across the region. This section uses four case studies to provide a closer look at some of the water resource issues facing the Coastal Plain, Mississippi Alluvial Valley, Mid-South, and Appalachian-Cumberland Highland.

Booming coastal development and climate change are two major environmental concerns in the Coastal Plain (Williams et al. 2012). Much of the South's lower Coastal Plain (along the Gulf of Mexico and the southeastern coast of the Atlantic Ocean) consists of natural and managed forests, depressional wetlands, pine flatwoods, riparian buffers, and bottomland hardwoods on brackish waters (Lockaby et al. in press). Research from Santee Experimental Forest in South Carolina and two sites in North Carolina is used to examine how climate change is affecting lowland forests in the Coastal Plain.

The rich soils and relatively flat landscape of the Mississippi Alluvial Valley have long supported an economy dominated by agriculture and forestry. However, in recent years concerns are growing over future water supplies as water demands by agriculture increase. Using data from the Yazoo River basin in Mississippi, we examine how climate change may affect the amount and timing of water supplies in this subregion.

The Appalachian-Cumberland Highland has the highest concentrations of interior forest in the South (Wear and Gries 2002) and the most rugged terrain. Most climate scenarios disagree on whether precipitation amounts will increase or decrease for this subregion, but agree that an increase in frequency of precipitation extremes will occur. Research from the Coweeta Hydrologic Laboratory in North Carolina is used to investigate how typical forest management systems might affect peak streamflow amounts under climate change scenarios in forests of the Appalachian-Cumberland Highland.

Some of the largest and most continuous forest areas in the Mid-South occur in the Ouachita Mountains. An important concern of forest managers here and elsewhere in the Mid-South is how climate change may affect surface erosion and sediment delivery to streams and water bodies. Using simulation modeling, we examine how sediment risks would change in the Ouachita National Forest of Arkansas and Oklahoma under future climate conditions.

Case Study 1: Coastal Plain

The forested landscapes of the lower Coastal Plain are characterized by poorly drained soils with near-surface water tables and low topographic relief. When managed properly, these lands are highly productive for agriculture and commercial forestry (Amatya and Skaggs 2001). Because of their close proximity to sensitive estuarine ecosystems, these landscapes are vulnerable to large freshwater runoff from upstream, tropical storms, tidal surges, and saltwater intrusion from rising sea levels.

Flow generation within forested wetland watersheds may reflect a mechanism wherein saturation of the upper soil layer and forest floor produces surface runoff and rapid lateral transfers within this highly conductive layer (La Torre Torres et al. 2011; Sun et al. 2000b). Rapid rise of shallow water tables during rainfall events is common in the Coastal Plain (Amatya et al. 1996; Sun et al. 2000b; Williams 1978; Young and Klawitter 1968), and unlike upland watersheds, usually dominates hydrologic responses. Total flow depends largely on the dynamics of the water table (hydroperiod), which are driven by rainfall and evapotranspiration (Amatya et al. 1996; Amatya and Skaggs 2011; La Torre Torres et al. 2011; Lu et al. 2009; Pyzoha et al. 2007). Because of high evapotranspiration in the lower Coastal Plain, even third-order streams can lose all surface flow in normal dry periods (Amatya and Radecki-Pawlik 2007; Amatya et al. 2009; Dai et al. 2010a; Williams et al. 2012). In contrast, flooding problems during the winter wet periods are not uncommon within the Coastal Plain (Amatya et al. 2006a; Sheridan 2002; Young and Klawitter 1968). Summer droughts in the Coastal Plain are also on the rise (Karl et al. 2009).

Several general studies have been conducted to examine impacts of potential climate change and variability on stream and drainage outflows, evapotranspiration, and water table depths on the coasts of the lower Coastal Plain (Amatya et al. 2006b; Cruise et al. 1999; Dai et al. 2010b, 2011a,b; Hatch et al. 1999; Lu et al. 2009; Sun et al. 2000b), with findings similar to those noted in earlier sections. We use data from three research areas to determine if climate change impacts have already occurred in the lower Coastal Plain or are likely to occur in the future. They show that (1) the lower

Coastal Plain has seen an increase in air temperature with minimum temperatures increasing more than the maximum; however, despite changes in temperature, changes in annual streamflows and water table depths in one watershed were not observed; (2) increased frequency of large storms is likely to impact forest plant communities and increase flood occurrence; and (3) increased spring and summer droughts will likely make forest vegetation vulnerable to stresses caused by high evapotranspiration demands.

Santee Experimental Forest: The U.S. Forest Service established the Santee Experimental Forest within the Francis Marion National Forest (about 55 km northwest of Charleston, South Carolina) to conduct scientific research on forests and water in a Coastal Plain setting (Amatya and Trettin 2007). This site is characteristic of the subtropical area of the Atlantic coast with long and hot summers followed by short, warm, and humid winters. The mean annual temperature (1946–2007) is 18.5°C, and the mean annual precipitation is 1370 mm. In the past 60 years, extreme temperatures have exceeded 40°C in summers and fallen below –14°C in winters.

Dai et al. (2011b) reported that global warming caused a rapid (0.19°C per decade) increase in surface air temperature during the last 63 years at the Santee (Figure 9.16). The increase in mean annual temperature was slow or within normal variability from 1946 to 1969 but accelerated after 1969. The increase rate of the mean daily minimum temperature (0.26°C per decade) was twice as much as that of the mean daily maximum temperature (0.13°C per decade). The annual minimum temperature increased at a rate of 0.38°C per decade in the last 63 years ($p < 0.05$), accompanied by a large fluctuation. However, there was a downward trend in the annual maximum of about 0.19°C per decade ($p < 0.05$). These results indicate that the current warming trend has had a larger effect on lower temperatures than on higher ones in this area (Dai et al. 2011b).

The temperature increase rate at the Santee (0.19°C per decade) in the last 63-year period is substantially higher than the global mean increase rate of 0.07°C per decade during the twentieth century (Intergovernmental Panel on Climate Change 2001), about 0.06°C per decade higher than the global rate for land in the same time period (Hansen et al. 2010), and approximately the same as the global mean of 0.2°C per decade since 1976 (Hansen et al. 2006; Trenberth et al. 2007). The Santee results indicate that warming in this area became apparent six years prior to it becoming discernible globally (Hansen et al. 2006).

The year-to-year variability of precipitation in the Santee was large over the last 63 years. Annual precipitation ranged from 835 to 2026 mm, with a mean of 1370 mm (Dai et al. 2011b). Although the variations in annual precipitation exhibited an upward trend from 1946 to 2007, the increase

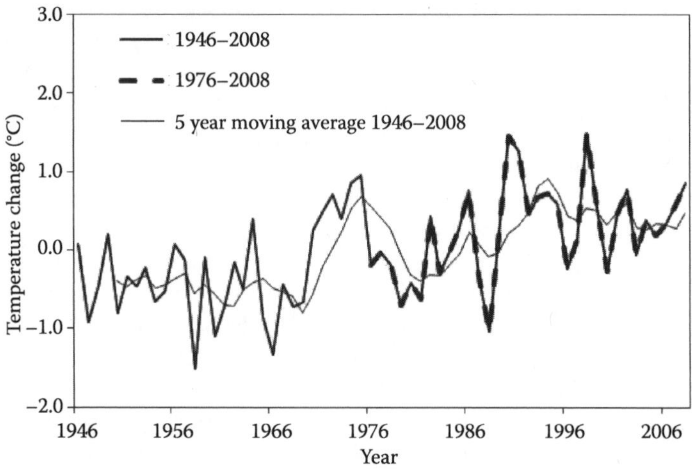

FIGURE 9.16 Temperature change from 1946 to 2008 at the Santee Experimental Forest on the South Carolina coast.

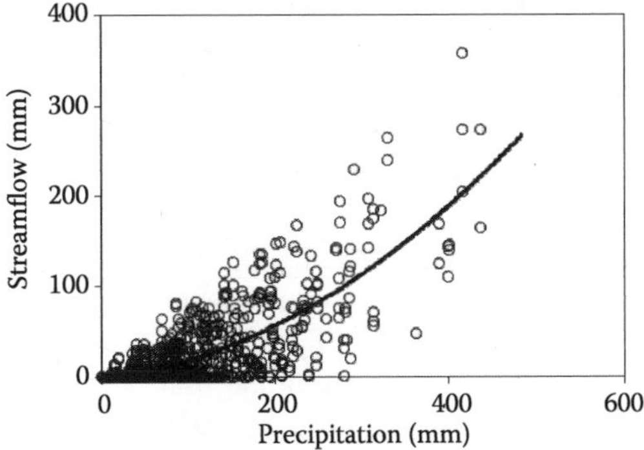

FIGURE 9.17 Relationship of monthly streamflow to precipitation from 1950 to 2000 at the Santee Experimental Forest on the South Carolina coast.

rate was not statically significant ($p > 0.1$). This upward trend was probably a result of a three-year drought period in the 1950s or from a multi-decade precipitation fluctuation. The mean number of large storm event (>50 mm) increased from 4.4 times per year from 1946 to 1981 to 5.7 times per year from 1982 to 2008. However, there was no significant upward trend in the last 63 years as a whole. This suggests that an increase in storm frequency may still be developing in this area.

Streamflow within the low-gradient forest watersheds largely depends on precipitation and evapotranspiration. As a result, first-order streams can be without flow during summer with high evapotranspiration demand and may flood during wet winters or summers that see large events like tropical storms. A scatter plot of monthly streamflows from all first- and second-order watersheds with monthly rainfall at Santee shows a significant ($p < 0.01$) relationship (Figure 9.17) showing the strong dependence of streamflow on precipitation. If storm intensity increases because of climate change, the frequency of high streamflows that produce flooding is likely to increase as well.

Vegetation damage from Hurricane Hugo caused streamflow impacts that continued for several years following the disastrous 1989 storm (Figure 9.18). The observed streamflow within the 14-year period after the hurricane is above the trend line, although mean annual precipitation was about 45 mm lower than the mean in the observation period (1965 to 2007). The high streamflow

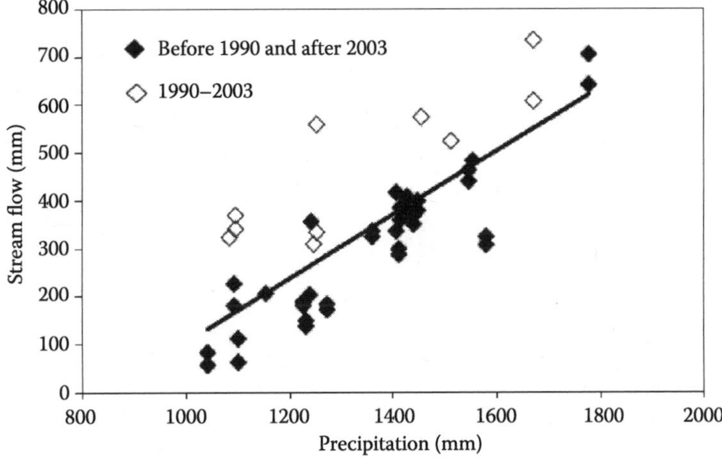

FIGURE 9.18 Impact of annual precipitation on annual streamflow on watershed 77 and watershed 80 from 1990 and 2003 at the Santee Experimental Forest on the South Carolina coast. The blank diamonds represent the relationship between precipitation and streamflow between 1990 and 2003 after Hurricane Hugo in 1989.

rate in the years following the hurricane was caused by destruction of over 80% of the canopy at this site (Hook et al. 1991), which undoubtedly caused a substantial decrease in the evapotranspiration demands of plants. These results are consistent with those reported by Shelby et al. (2005) for sites in the North Carolina Coastal Plain. However, the flow recovered about 10 years after the hurricane as the vegetation regenerated on the North Carolina watersheds (Williams et al. 2012).

A significant decrease ($p < 0.02$) in water table depth occurred from 1964 to 1993 because of annual precipitation increases (Figure 9.19). High water table level in this area in summers and autumns is mostly related to high precipitation (Amatya and Skaggs 2011; Amatya et al. 2009; Sun et al. 2000a).

The water table level was high throughout most of the winter–spring (December–February) seasons (Dai et al. 2011b) consistent with other studies by Amatya and Skaggs (2011). However, precipitation in these seasons was much lower than that in summers. The high water table level in winters or early springs is primarily caused by a low demand in evapotranspiration, indicating that evapotranspiration is one of the key factors that influence groundwater depth in these first-order watersheds (Amatya and Skaggs 2011; Amatya et al. 2009). These results indicate that if evapotranspiration increases because of climate change, water table depths may also increase in the Santee watersheds.

Williams et al. (2012) examined how Hurricane Hugo affected streamflow using historical streamflow data from a control (watershed 80, a first-order, forested watershed within the Santee, not harvested) and a treatment (watershed 77, salvage harvest after damage) watershed. They found that the watershed 80 had lower flow than watershed 77 during the pre-Hugo period, a relationship that reversed in 1993. The reversal continued for nearly 10 years, after which the streamflow of watershed 77 returned to near reference level. One explanation may be that increased evapotranspiration by the regenerating vegetation caused the lower streamflow (Amatya et al. 2006a; Williams et al. 2012).

To predict likely hydrologic response to future climate change, we applied a physically based distributed wetland hydrological model, MIKE SHE (Danish Hydraulic Institute 2005; Lu et al. 2009), to watershed 80 (Dai et al. 2010a). The MIKE SHE model links the hydrology with forest vegetation through the leaf-area index, rooting depth, and canopy storage capacity. Modeling results (Figure 9.20) suggest that the annual mean streamflow would increase or decrease by 2.4% with a 1% increase or decrease in precipitation, and decrease with an increase in air temperature. A quadratic polynomial relationship (Dai et al. 2011a) described the relationship between changes in groundwater table depth and precipitation (Figure 9.19). Both the mean annual water table depth

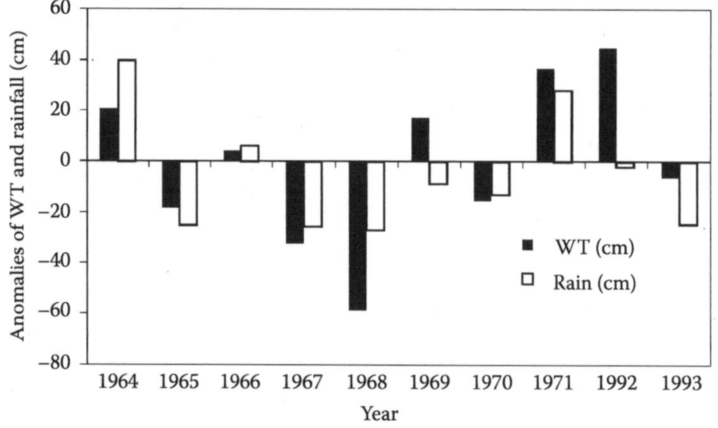

FIGURE 9.19 Impact of annual precipitation on annual mean water table on watershed 77 at the Santee Experimental Forest on the South Carolina coast (Dai et al. 2011b). WT is annual mean water table depth; rain is annual precipitation.

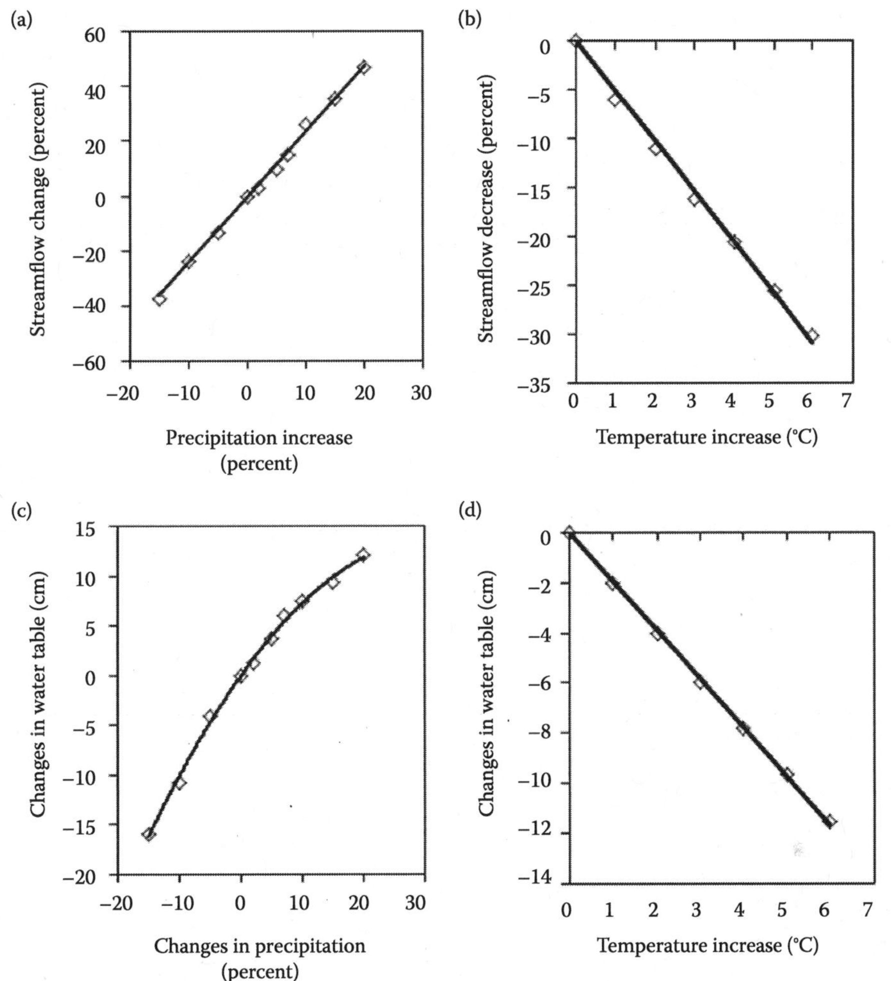

FIGURE 9.20 Simulated sensitivity of hydrologic responses to potential climate change at watershed 80 at the Santee Experimental Forest on the South Carolina coast: (a) streamflow response to precipitation, (b) water table response to precipitation, (c) streamflow response to air temperature, and (d) water table response to air temperature.

and mean annual streamflow decreased with an increase in temperature within the range of 0–6°C (Figure 9.20).

Carteret site: The Carteret site is on a typical Coastal Plain landscape in Carteret County in North Carolina and is owned by the Weyerhaeuser Company. The research site, established in late 1987, consists of three artificially drained experimental watersheds (D1, D2, and D3), each about 25 ha, covered with loblolly pine plantations. Details of all the hydrometeorological measurements and forest hydrologic studies at the site have been documented elsewhere (Amatya and Skaggs 2011; Amatya et al. 1996, 2000, 2003, 2006a, 2006b; Tian et al. 2011).

Annual rainfall ranged from 852 mm in 2001 to 2308 mm in 2003, with the 21-year mean of 1517 mm, which was 8% higher than the 50-year (1951–2000) long-term mean of 1390 mm observed at the nearby Morehead City weather station. Similar to the Santee, no statistically significant change ($\alpha = 0.05$) was apparent in the 21-year rainfall series at the Carteret site. January showed the largest decrease followed by October and December. Hurricanes occurred in 7 of the 21 years, mostly in September. Consequently, the mean rainfall of 552 mm from July to

September was significantly higher than other seasons, accounting for a third of the mean annual rainfall.

Although the 0.44°C rise in observed mean annual temperature during the entire 21-year period was not statistically significant ($\alpha = 0.05$), the 0.85°C rise from 2000 to 2008 was highly significant (Figure 9.21) and is consistent with other recent studies (Chapman and Davis 2010). The mean annual increase in net radiation (in evaporation equivalent energy units) of 430 mm from 1988 to 2008 was also significant. This radiation increase produced an estimated 270 mm increase in the annual Penman–Monteith potential evapotranspiration for a grass reference during the same period (Figure 9.21). The annual runoff coefficient, defined as the percent of rainfall occurring as streamflow, varied from 5% in 2001 to as large as 56% in 2003, the wettest year. The mean annual runoff coefficient of 32% was somewhat higher than the published data for similar conditions. Annual streamflow did not change during the 21-year period. Similarly, the 21-year annual mean water table

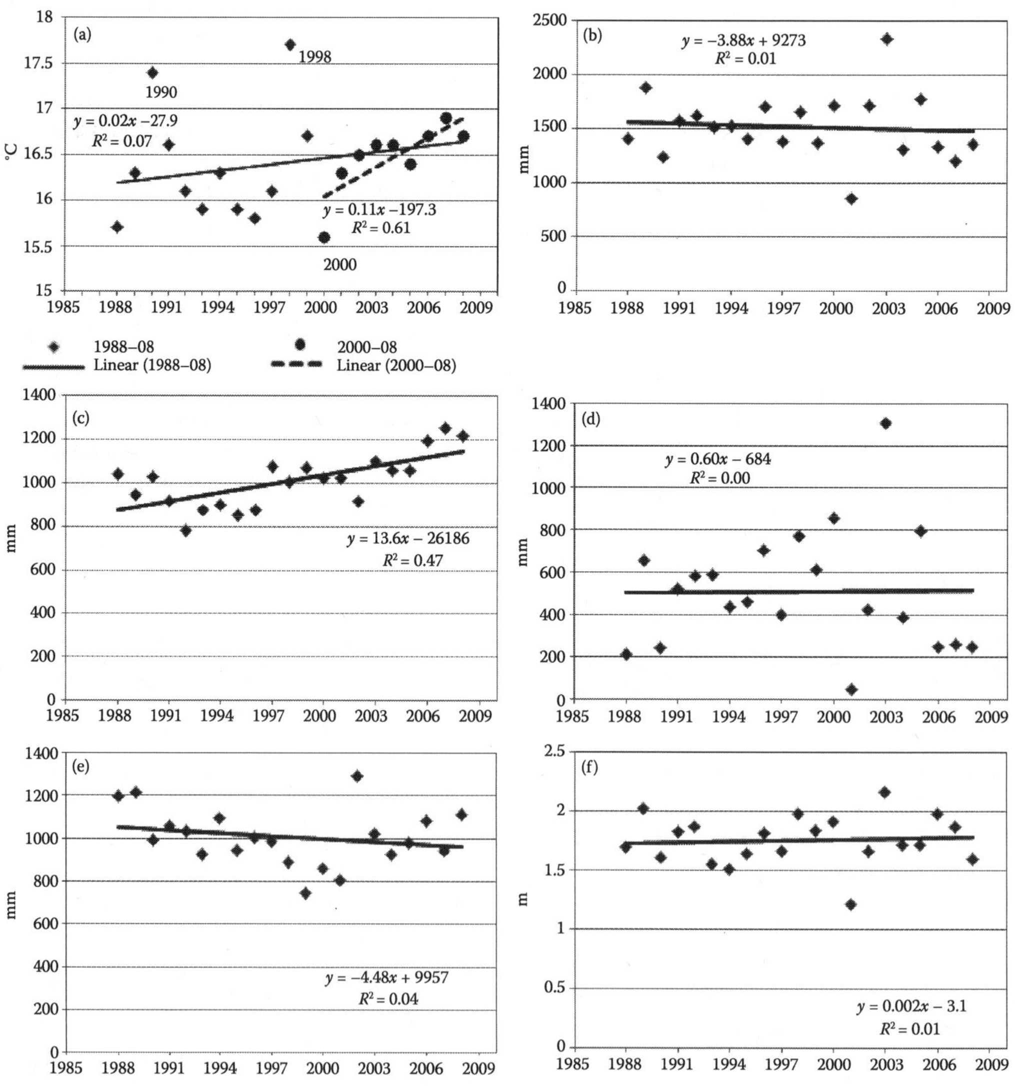

FIGURE 9.21 Annual trends in 21-year hydrologic variables at the Carteret study site in North Carolina: (a) temperature, (b) potential evapotranspiration, (c) evapotranspiration, (d) rainfall, (e) outflow, and (f) water table elevation. (Adapted from Dai, Z. et al. 2011a. *Atmosphere* 2: 330–357, doi:10.3390/atmos2030330.)

depth (96 cm) did not change despite the pine trees growing from 14 to 35 years stand age during that period and increases in temperature and potential evapotranspiration (Figure 9.21). Field data suggest the site is not water-limited except in extreme dry years (Amatya and Skaggs 2011; Sun et al. 2002). The combination of precipitation and air temperature rise appears to dictate future hydrologic change for this study site.

Parker Tract site: The Parker Track study site near Plymouth, North Carolina was established in late 1995 by Weyerhaeuser Company and North Carolina State University (Amatya et al. 2003). The landscape has a flat topography and poorly drained mineral or organic soils. Vegetation ranges from second-growth mixed hardwood and pine (*Pinus* spp.) forest to loblolly pine (*P. taeda*) plantations of various stand ages. Intensive hydrometeorological measurements at this site were conducted from 1996 to 2004 to support a variety of watershed studies (Amatya et al. 2003, 2004, 2006a, 2006b; Appelboom et al. 2008; Fernandez et al. 2006, 2007). Since 2005 studies have been shifted to understand how the coupling of management and climate disturbance affects forest ecosystem carbon and water balances (Noormets et al. 2010; Sun et al. 2010).

A simulation study was conducted to evaluate the potential effects of climate change on the hydrology of a 2950-ha managed pine forest at the Parker Tract site (Amatya et al. 2006b). The DRAINWAT model (Amatya et al. 1997) was first validated with a 5-year (1996–2000) data series from the study site (Amatya et al. 2004) and then run for the 1951-to-2000 period using historical weather data from Plymouth to determine the long-term hydrology. Separate simulations were conducted with 2001-to-2025 climate datasets projected by two General Circulation Models: CGC1 and the HadCM2 (Kittel et al. 1997). Simulation results (Figure 9.22) indicated that the CGC1 model yielded a significantly ($p < 0.0001$) lower mean total outflow (167 mm) than historical climate data or the HadCM2 model (380 mm). The decrease in outflow resulted from the drier conditions predicted by the CGC1 model relative to those predicted using the historical data. The 5% increase in rainfall using the HadCM2 model appeared to have no effect on runoff but may increase the evapotranspiration of this pine forest. The drier climate scenario (CGC1) did increase forest evapotranspiration compared to the historical data and was similar to predictions by the wetter (HadCM2) climate scenario (Figure 9.22). The water table depths based on the CGC1 model were deeper compared to those based on the HadCM2 model. However, the deeper annual water table depths predicted using both climate scenarios did not reduce the water table depth to the extent that evapotranspiration was limited by soil water depletion in the root zone. We concluded that both climate scenarios had very little or no impact on evapotranspiration, indicating the temperature increase has less of an effect on the soil moisture than does increased rainfall.

Case Study 2: Mississippi Alluvial Valley

Located within the Mississippi Alluvial Valley, the Yazoo is the largest river basin in Mississippi, with a total drainage area of 34,600 km^2. The goal of this study was to estimate the potential impact of future climate change on hydrologic characteristics in the Yazoo River Basin using an integrated hydrologic simulation model. This study developed a site-specific model for the Yazoo based on watershed, meteorological, and hydrological conditions. The calibrated and validated model was applied to predict the potential impact of future climate changes upon streamflow under multiple climate change scenarios.

Study watershed and data acquisition: The Yazoo basin is separated equally into two distinct topographic areas: the Bluff Hills and the Mississippi Alluvial Delta (Guedon and Thomas 2004; Mississippi Department of Environmental Quality 2008; Shields et al. 2008). The Bluff Hills area is a dissected, hilly, upland area where streams originate from a mixture of pine (*Pinus* spp.), pine-hardwood, and oak-hickory (*Quercus* spp., *Carya* spp.) forests, and pastures. The residual and alluvial soils in this area are originally derived from loess and are highly erodible. This area supports a variety of land uses with forestry and small-scale agriculture predominating. The Delta area is a flat lowland with a highly productive agricultural economy (Mississippi Department of Environmental

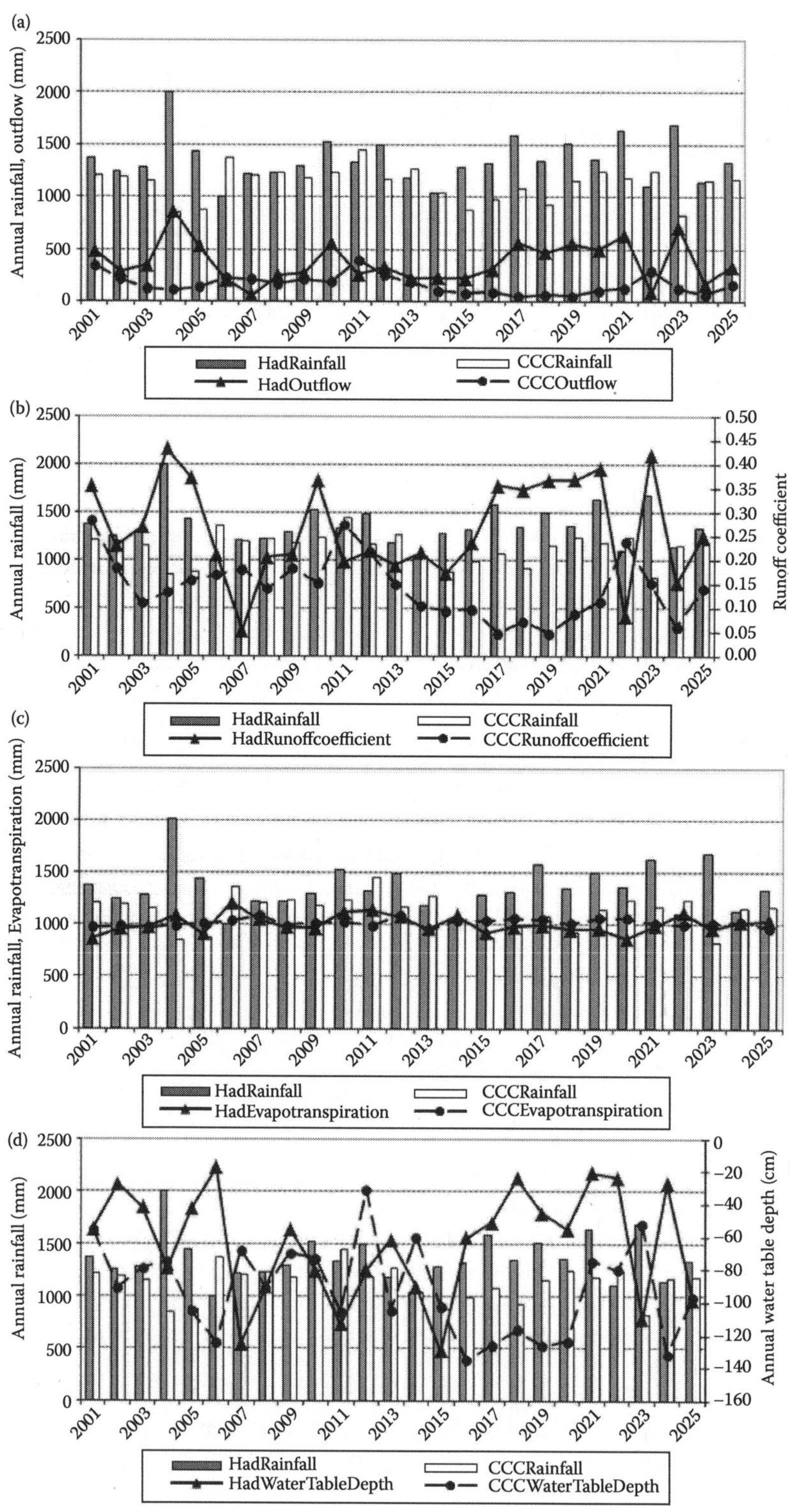

FIGURE 9.22 Annual rainfall at the Parker Tract site predicted by two general circulation models (HadCM2 and CGCI) for (a) drainage outflows, (b) runoff coefficients, (c) evapotranspiration, and (d) mean water table depths, as predicted by DRAINWAT using a 25-year climate scenario data series.

Quality 2008). The fine-textured, alluvial soils in this area are derived from past deposition by the ancestral Mississippi and Ohio Rivers (Guedon and Thomas 2004). Overall, the Yazoo area considered in this study is comprised of 31.8% (196 km^2) cropland, 4.4% (27.3 km^2) wetland, 0.3% (1.8 km^2) barren, 1.8% (11.5 km^2) urban, and 61.7% (381.2 km^2) forest uses. Environmental, climatic, and hydrologic data for the Yazoo (watershed 08030208) were obtained from a variety of public sources. Past climate data were obtained from weather stations or streamflow gauging stations within the Yazoo. These data were adjusted to represent the entire Yazoo basin. Potential evapotranspiration data were computed based on air temperature. These past climate data were used only for model calibration and validation purposes.

Four future climate scenarios were used in this study: HadCM B2, CSIROMK2 B2, CSIROMK3.5 A1B, and MIROC3.2 A1B. The data used are monthly air temperature and precipitation for a period from 2000 to 2050, which were computed for each U.S. Geological Survey 8-digit Hydrologic Unit Code (HUC) watershed within the Yazoo. These four climate scenarios were used to assess the impact of future climate changes on water discharge, evaporative loss, and water yield.

Model development: Two modeling systems are used in this study: the Better Assessment of Science by Integrating Point and Nonpoint Sources (BASINS) model (U.S. Environmental Protection Agency 2010) and the Hydrological Simulation Program-FORTRAN (HSPF) model (Bicknell et al. 2001). The BASINS system integrates a geographic information system, standardized environmental databases, and state-of-the-art modeling tools into one convenient package (U.S. Environmental Protection Agency 2010). The hydrological model is a comprehensive model for simulating many hydrologic processes within watersheds of almost any size and complexity (Bicknell et al. 2001).

Model calibration was done using a 5-year period from January 1, 2000 to December 31, 2004. To reduce uncertainty in the calibration process, only the six hydrologic parameters most sensitive to the hydrological model predictions (Donogian et al. 1984) were adjusted. The differences between the observed and predicted annual water outflow volumes for the calibration and validation period were within an acceptable range (Bicknell et al. 2001).

Effects of climate change on streamflow: Comparison of mean annual water-yield for the past 10 years (2001–2011) versus 40 years in the future (2011–2050) under the four climate scenarios indicates a continuing decrease (Table 9.7). The percent change in mean annual water yield varied from 29.47% for the CSIROMK3.5 A1B scenario to 18.51% for the MIROC3.2 A1B scenario, with all four climate scenarios indicating continuing decreases out to 2050. The same downward trends were observed for maximum annual water yields. The decreases in mean and maximum annual water yields were primarily the result of the projected precipitation decrease. Mixed results were found for the mean annual evaporative loss, with the CSIROMK2 B2 scenario indicating a long-term increase and the other three scenarios indicating a long-term decrease. Further research is thus necessary to better determine how evaporative losses will respond in the future.

Figure 9.23 shows the minimum, mean, and maximum values for monthly discharge rate and water yield (measured at the outlet of the watershed) under the four climate scenarios during the 40-year simulation period (2011–2050). In general, the MIROC3.2 A1B scenario produced the highest monthly minimum, mean, and maximum discharge rate and flow volume in most years because it forecasted the highest annual precipitation (Table 9.7). Overall, the mean and maximum annual discharge rates were projected to decrease in the Yazoo because of a decrease in precipitation during the next 40 years. Projected precipitation changes had profound impacts on flow discharge rate and annual water yield in this Mississippi Alluvial Valley watershed.

Case Study 3: Appalachian-Cumberland Highland

Extreme events often present greater challenges to water resource managers than those presented by average conditions. In this case study, we asked whether climate impacts within the Appalachian-Cumberland Highland may be either mitigated or exacerbated by forest management practices that alter land cover. We used a retrospective analysis of long-term climate and streamflow data

TABLE 9.7
Comparison of the Sum and Mean Values for Precipitation, Evaporative Loss, and Water Outflow between the Past 10 Years and Future 40 Years in the Yazoo River Basin

	Precipitation			Evaporative Loss			Water Yield		
Climate Scenario[a]	2001–2010 (mm/year)	2011–2050 (mm/year)	Change (%)	2001–2010 (mm/year)	2011–2050 (mm/year)	Change (%)	2001–2050 (mm/year)	2011–2050 (mm/year)	Change (%)
Mean Annual									
HadCM3 B2	1477	1341	−10.1	1102	1002	−10.0	237	192	−23.7
MIROC3.2 A1B	1397	1269	−10.1	1186	1032	−14.9	154	134	−15.1
CSIROMK3.5 A1B	1646	1462	−12.6	1483	1199	−23.7	127	98	−29.6
CSIROMK2 B2	1387	1388	0.1	1143	1243	8.0	124	117	−6.3
Maximum Annual									
HadCM3 B2	1831	1958	6.5	1419	1317	−7.8	348	384	9.3
MIROC3.2 A1B	1605	2517	36.2	1380	2073	33.4	188	326	42.2
CSIROMK3.5 A1B	2504	2174	−15.2	1896	1748	−8.5	352	385	8.6
CSIROMK2 B2	1615	2095	22.9	1295	1892	31.5	111	139	20.2

[a] The MIROC3.2 A1B, CSIROMK2 B2, CSIROMK3.5 A1B, and HadCM3 B2 climate scenarios for the Southern United States each combine a general circulation model (MIROC3.2, CSIROMK2, CSIROMK3.5, HadCM3) with an emissions storyline (A1B storyline representing moderate population growth and high-energy use, and B2 representing lower population growth and energy use).

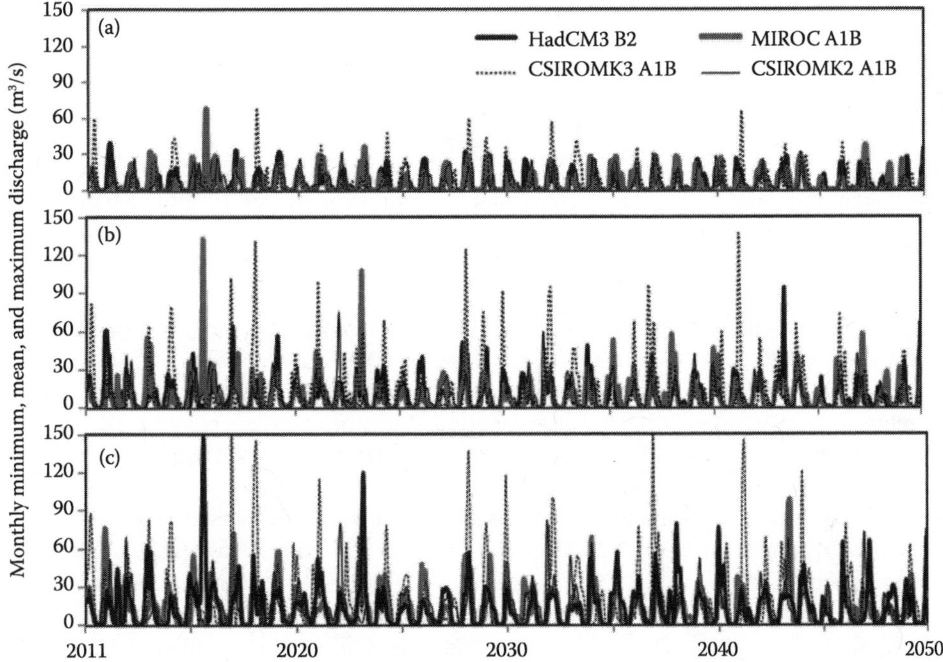

FIGURE 9.23 Simulated (a) monthly minimum, (b) mean, and (c) maximum discharge for the Yazoo River Basin from 2011 to 2050, based on four climate change scenarios. The MIROC3.2 A1B, CSIROMK2 B2, CSIROMK3.5 A1B, and HadCM3 B2 climate scenarios for the Southern United States (McNulty et al. in press) each combine a general circulation model (MIROC3.2, CSIROMK2, CSIROMK3.5, HadCM3) with an emissions storyline (A1B storyline representing moderate population growth and high energy use, and B2 representing lower population growth and energy use).

collected at the Coweeta Hydrologic Laboratory to determine whether streamflow from managed watersheds responds differently from unmanaged reference watersheds when subjected to variation in air temperature and extremes in annual precipitation.

The Coweeta Hydrologic Laboratory, located in western North Carolina, is a 2185-ha basin wherein climate and hydrologic monitoring and forest watershed experimentation began in the early 1930s. The basin has a marine, humid temperate climate with frequent, small, low-intensity rainfall events and annual precipitation averaging 1800 to 2300 mm, depending on elevation. Trend analysis suggests that air temperature at Coweeta has increased significantly since the early 1980s (Laseter et al. 2012). For this study, extreme precipitation years, both wet and dry, were identified according to Guttman (1999).

Forest management treatments: We used the long-term streamflow records from six watersheds, each with a different management and land-use history. Two watersheds (watershed 1 and watershed 17) were species conversion experiments from deciduous forests to evergreen, eastern white pine (*Pinus strobus*) plantations at 1.8-m × 1.8-m spacing. Two watersheds at high and low elevations (watershed 37 and watershed 7, respectively) were clear-cut using a range of techniques. Another watershed (watershed 13) was subjected to successive clear-cutting separated by a 23-year interval; the result was a multispecies coppice stand, in which the vegetation recovered via stump sprouting from existing, well-established root stock (Leopold et al. 1985). The remaining treatment watershed (watershed 6) underwent old-field succession following conversion of a mixed hardwood forest to a grass cover. All these management activities are still widely practiced on both publicly and privately managed forests in the Eastern United States, and thus represent prevalent land management strategies.

Modeling the interactions of management and precipitation on annual streamflow: The effect of management treatments on annual water yield was determined using the paired-watershed approach. Our goal was to predict the streamflow response to management knowing streamflow from an unmanaged watershed, a model of watershed response, and a model of the interaction of the watershed response and precipitation. Our approach was conceptually similar to the classic paired watershed regression approach detailed fully in Ford et al. (2011b). We found that interactions between watershed vegetation recovery and temperature were never significant and therefore air temperature is not included in the final model.

We used statistically downscaled mean annual precipitation from two climate scenarios, MIROC3.2 A1B and CSIROMK3.5 A1B, representing low and high forecasts of mean annual precipitation, respectively, to project streamflow from 2010 to 2050. Along with forecasted streamflow from the reference watershed, we modeled watershed treatment responses from the end of the observed series (2010) out to the year 2050. We assumed that forest communities on the treated catchments were comparable to the reference watershed in 2010, and that the original treatments were applied to the same catchments in 2011 (meaning that forest age was reset to 0 in 2011). We assumed that species composition and structure (such as stocking and leaf area index) recovered over the ~40-year posttreatment period comparable (both spatially and temporally) to what was observed in the actual treated watersheds.

Watershed recovery, management, and climate effects on streamflow: In all scenarios, land management altered the expected level of streamflow (Figure 9.24). Initial increases in streamflow compared to what would be expected had the catchment remained unmanaged ranged from 21.4 cm/year (19%) to 38.1 cm/year (70%) larger. Increases in streamflow immediately after treatment persisted generally only in the species conversions stands (3.5–5 years), and were associated with controlling competition so the new forest could establish. In general, streamflow recovered to pretreatment levels in 6–7 years. The rate of watershed recovery depended on management and

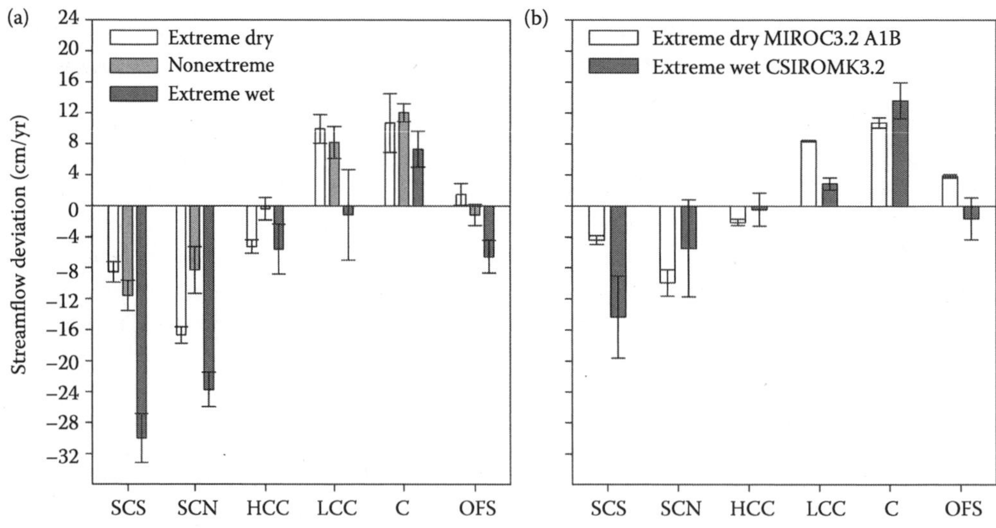

FIGURE 9.24 At the Coweeta Hydrologic Laboratory in the Appalachian-Cumberland Highland, (a) observed streamflow responses to six management treatments during extreme wet years, extreme dry years, and nonextreme years from the beginning of recordkeeping to 2010 as compared to (b) mean of extreme wet and dry years modeled repeating entire management cycle in all watersheds—reset stand age to 0 in 2011, forecast out to 2050—for both CSIROMK3.5A1B and MIROC3.2 A1B climate change scenarios. The MIROC3.2 A1B and CSIROMK3.5 A1B climate scenarios for the Southern United States (McNulty et al. in press) each combine a general circulation model (MIROC3.2 and CSIROMK3.5) with the A1B emissions storyline (representing moderate population growth and high-energy use). SCS, SCN = south- and north-facing species conversion; HCC, LCC = high- and low-elevation clearcut; C = copice; OFS = old field conversion.

location. Watershed recovery was faster for several low-elevation treatments compared to the high-elevation treatment, and also faster for south-facing stands compared to north-facing stands. The second cutting in the coppice stand experienced the fastest watershed recovery rate among all management treatments compared to the first cutting, which was among the slowest to recover.

At any forest age, the physical and biological components controlling hydrologic processes in the treatment watersheds responded differently than the reference watersheds to variation in precipitation (not temperature). The only exception to this was the coppice-managed stand. In all other watersheds, as precipitation increased, the treatment effect increased. This effect was most pronounced in the old-field succession, the south-facing species conversion, and the low-elevation clear-cut. In addition, these three land uses differed significantly from one another in the magnitude of the effect on streamflow deficit. The largest land use and climate interaction was seen in the old-field succession, followed by the two species conversions and the high-elevation clear-cut, then by the low-elevation clear-cut. In general, the drier the year, the more managed watersheds responded like the reference watersheds. Likewise, the wetter the year, the larger the differences were between control and treated watersheds.

Management simulated under climate change scenarios: Forecasted temperature and precipitation out to year 2050 were significantly different from the mean of observed conditions, depending on the climate scenario used. The MIROC3.2 A1B forecasted warmer and drier conditions for the Southern Appalachian Mountains, with drought frequency increasing appreciably: in a 10-year period, 6 years would be defined as drought years, compared to 1 year in 10 for the observed record. Extreme wet years were completely eliminated. Conversely, CSIROMK3.5 A1B forecasted wetter conditions for the Southern Appalachians, with frequency of extreme wet years increasing to 3 years in 10, compared to the 1 year in 10 for the observed record; droughts were completely eliminated.

We simulated watershed responses to future climate scenarios by assuming that the same set of management treatments were repeated in 2011 and we modeled responses to 2050 using forecasted climate. The most notable differences were observed for the south-facing species conversion and the coppice treatments (Figure 9.24). Streamflow responses for both management treatments were different from the observed record, primarily because the simulation period was relatively short (~40 years); thus the proportionally longer time period that the vegetation was in a younger, less-mature state than the reference watershed.

Potential for mitigation or exacerbation by forest management practices: Management affected the resulting vegetation structure and function, and the vegetation response to climate was different from the reference watersheds. Whether the effects of extreme wet or dry precipitation years exacerbated or mitigated the streamflow response depended on the management treatments. For example, managing a catchment with a species conversion treatment reduced annual streamflow during both extreme wet and dry event years, which may potentially exacerbate low flows and drought, but it also may mitigate high flows and flood risk. This conversion could decrease the apparent frequency of observed extreme wet event years on average by a factor of three (Figure 9.25). For example, in 2010 the annual streamflow generated from 2500 mm of precipitation would have been 130 cm without management; whereas with management, it was only 102 cm. The precipitation amount required to produce 102 cm on the watershed had it not been treated is only 2391 mm. The probabilities associated with 2500 mm compared to 2391 mm differ by a factor of three (0.0006 vs 0.002). Thus, the apparent frequency of extreme precipitation events decreased by threefold. This management treatment moves the observed streamflow distribution toward lower flows and could help mitigate high flows under a wetter future climate.

Pine forests in this temperate area intercept precipitation year-round with their evergreen canopies, as well as transpire year-round in the area's relatively mild winter temperatures. Both processes divert precipitation from streamflow in ways that differ from hardwood stands. Higher evapotranspiration from pine forests means that soils have a larger capacity to store excess water during wet years. This may be a good option under a future climate with increased precipitation, but a poor choice for a climate projected to be drier because the higher evapotranspiration also means that less soil water is available during drought conditions (Farley et al. 2005).

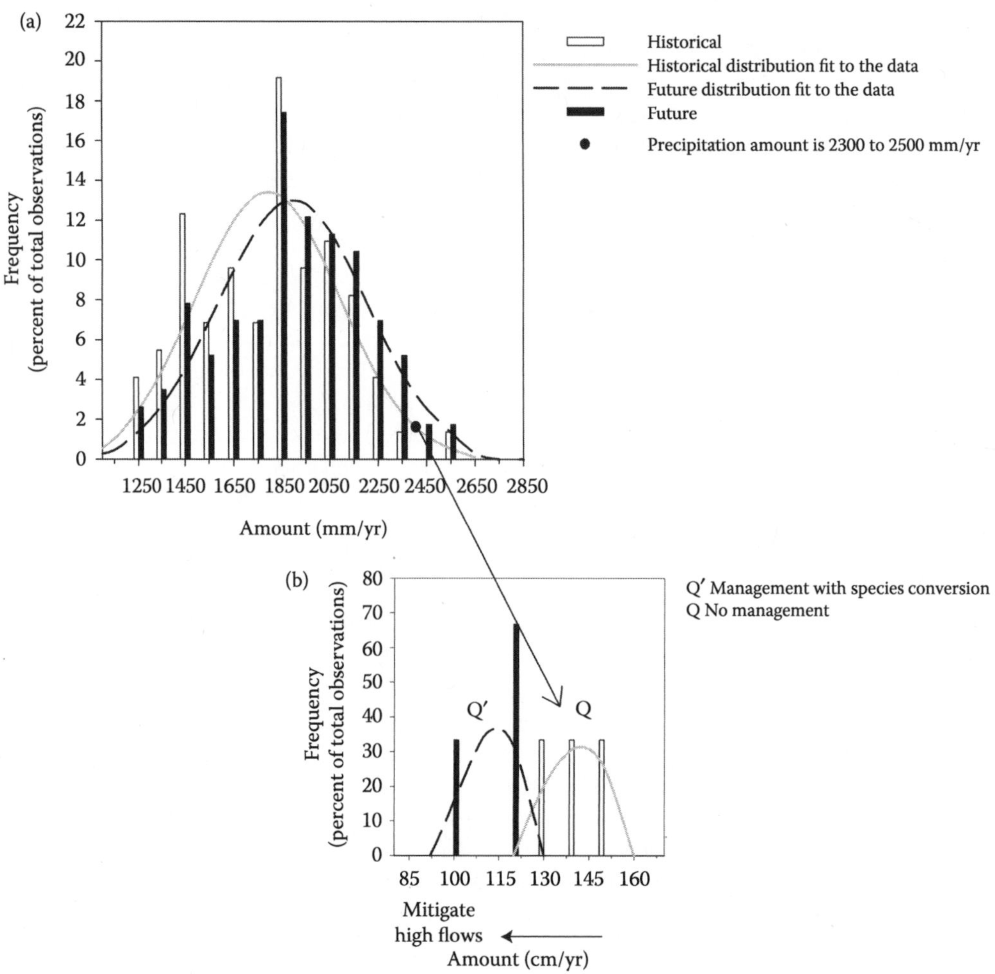

FIGURE 9.25 At the Coweeta Hydrological Laboratory in the Appalachian-Cumberland Highland, (a) frequency distribution of future annual precipitation—based on the CSIROMK3.5 A1B climate change scenario—and historical annual precipitation, and (b) observed streamflow frequency distribution resulting from management with a species conversion or no management for the given precipitation amount. The CSIROMK3.5 A1B climate scenario for the Southern United States (McNulty et al. in press) combines the CSIROMK3.5 general circulation model with the A1B emissions storyline (representing moderate population growth and high energy use).

Impacts on stream temperature in a headwater watershed: Watershed 32 (0.4-km^2 drainage area) at Coweeta is a control watershed with mixed hardwood stands that have remained undisturbed since 1927. This site had a significantly upward trend in estimated mean annual stream temperature of 0.12°C per decade from 1960 to 2007 (Figure 9.26). The mean change in stream temperature was predicted to increase to 0.19°C per decade from 2011 to 2060 under the HadCM3 B2 scenario. There is some variability in the estimated change of mean annual stream temperature among climate scenarios. For example, the change ranged from 0.16°C per decade under the CSIROMK3.5 A1B and CSIROMK2 B2 scenarios to 0.27°C per decade under the MIROC3.2 A1B scenario. Despite the variability in the change in future stream temperature, the trend was increasing and significant ($p < 0.05$) regardless of the climate scenario selected. The conservative HadCM3 B2 scenario predicted upward trends in annual monthly minimum (0.23°C per decade) and maximum stream temperature (0.24°C per decade) at this site. Thus, the

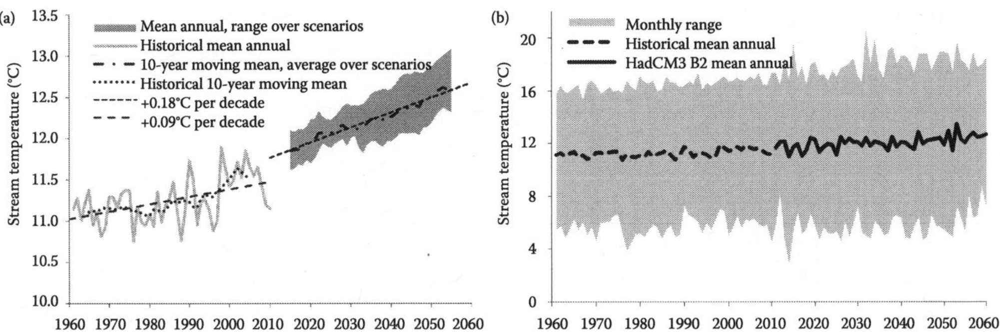

FIGURE 9.26 For watershed 32 at the Coweeta Hydrologic Laboratory in the Appalachian-Cumberland Highland, (a) 10-year moving mean of mean annual stream temperature and (b) mean annual stream temperature and annual range in monthly stream temperature under historic and HadCM3 B2 future climate conditions. The HadCM3 B2 climate scenario for the Southern United States (McNulty et al. in press) combines the HadCM3 general circulation model with the B2 emissions storyline (representing lower population growth and energy use).

mean annual maximum stream temperature was predicted to increase from 11.9°C to 12.5°C by the decade ending in 2060.

Case Study 4: Mid-South

Located in the northeast corner of the Mid-South, the Ouachita National Forest covers over 6800 km^2 of the Ouachita Mountains in western-central Arkansas and eastern Oklahoma (Figure 9.27). Surface erosion and sediment delivery to streams and water bodies is a major ongoing concern of

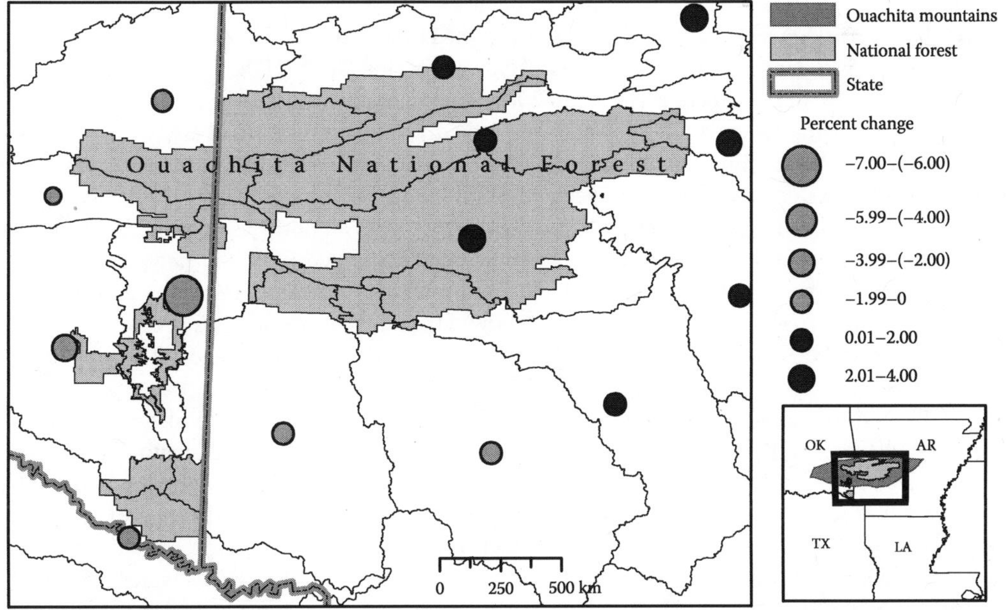

FIGURE 9.27 Percent change in mean annual precipitation from historical for the 2041–2060 period in watersheds draining the Ouachita National Forest based on an ensemble general circulation model with the A1B emissions storyline (representing moderate population growth and high-energy use). (Adapted from Mote, T.L. and J.M. Shepherd. 2011. Monthly ensemble downscaled climate model data for the Southeast U.S. http://shrmc.ggy.uga.edu/downscale.php. Accessed: November 07, 2011.)

the Forest (U.S. Department of Agriculture, Forest Service 2005) and of forest managers throughout the subregion. Within the forest boundaries, the predominant sediment-generating process is surface erosion related to timber harvesting, off-highway vehicle trails, and forest roads. A limited amount of private agricultural land occurs within or adjacent to the Ouachita, and these areas can also be important sediment sources. Concern about sediment impacts to stream habitats and water quality prompted the development of the Aquatic Cumulative Effects (ACE) model (Clingenpeel and Crump 2005) to assess sediment risks within the Ouachita.

Anticipated changes in the climate affecting the Ouachita suggest that future surface erosion risks may change. Predicted temperature increases can affect vegetation in ways that reduce ground cover and root binding of the soil (Pruski and Nearing 2002b). Changes in rainfall amount and intensity can directly affect erosion drivers like runoff amount and raindrop impact energy, as well as indirectly affect the amount and types of vegetation that can be sustained. Rainfall changes are the primary concern as snowfall represents only a very small portion of annual precipitation and its effect on sediment production is negligible.

To assess the overall sediment risk of climate change on the Ouachita, we conducted a series of simulations to determine how plausible changes in rainfall amount, rainfall intensity, and vegetation cover affect sediment risk under a future climate. Future precipitation and temperatures were projected using an ensemble general circulation model and the A1B emissions storyline representing low population/high economic growth and high-energy use (Intergovernmental Panel on Climate Change 2001) and hereafter referred to as the "ensemble" scenario. Based on future temperature and rainfall, different levels for each of the factors that affect sediment production were determined. The ACE model (Clingenpeel and Crump 2005) was used to assess how changes in these factors might affect sediment risks across the Ouachita. Finally, the effect of increasing road maintenance on future sediment risks was also evaluated.

Study methods: The period from 2041 to 2060 was selected as the future period for which sediment risks would be simulated. The climate characteristics computed were annual precipitation, monthly precipitation, and mean monthly temperature for each year within the 20-year period. These characteristics were computed for the watersheds that drain the majority of the Ouachita lands using the U.S. Geological Survey 8-digit Hydrologic Unit Code (HUC) scale.

The ACE model (Clingenpeel and Crump 2005) computed the estimated sediment yield within a12-digit HUC (hereafter called a "subwatershed"). Sediment yield from road areas was estimated using the WEPP model (Elliot 2004); for all other areas it was estimated using a version of the Universal Soil Loss equation (Wischmeier and Smith 1965) with model factor values developed for forestry and related management practices (Dissmeyer and Stump 1978). For the 2041 to 2060 period, the distribution of different land-use practices was assumed to be the same as that which currently occurs. The combined sediment yield from road and unroaded areas was used to determine the sediment risk class (high, moderate, or low) for each subwatershed. A review of the ACE model is given in Marion and Clingenpeel (2012). For this analysis, we computed the sediment risk for all 190 subwatersheds within the Ouachita.

The levels of three factors were varied to explore how Ouachita sediment risks might respond to climate change: the R-factor, the C-factor, and the number of wet days per month. The R-factor is a measure of precipitation erosivity (Wischmeier and Smith 1965) and is used in the ACE model for estimating sediment losses in unroaded areas. Three different R-factor values were considered. The first is the R-factor predicted using the mean of the projected annual precipitations for the 2041–2060 period (Renard and Freimund 1994). The other two R-factor values were ±15% of the predicted R-factor for each watershed. R-factors were computed for each watershed and then used to compute mean values for each Ouachita subsection (see Clingenpeel and Crump 2005 for an explanation of how subsections are used in the ACE model). The subsection means were then assigned to all subwatersheds within each subsection.

The C-factor is a measure of the relative effect of the ground cover on sediment loss. The higher the C-factor value, the more sediment is lost through erosion. Two C-factor levels were considered:

no change from the current C-factor value, and a 15% increase. The 15% increase is based on Pruski and Nearing (2002b), who found that under certain soil and climate conditions, a decrease in precipitation could result in an increase in sediment loss caused by a decrease in biomass. The simulations of Pruski and Nearing (2002b) only considered agricultural conditions, but the locations that produced this response are those that are geographically closest to the Ouachita. Lacking other guidance for how C-factors might change under forest conditions, we assumed that the same response is possible on the Ouachita. C-factors were assigned to subwatersheds by subsection.

The number of wet days per month is a variable in the WEPP-Road (Elliot 2004) routine used by the ACE model to estimate sediment yield from road areas. This variable was used here to simulate increases in rainfall intensity by decreasing the number of wet days per month. Two factor levels were analyzed: a 20% decrease and at 50% decrease. A "no change" level was not considered because precipitation intensity has already been shown to be increasing in the Mid-South (White 2011).

Three additional factors were used to capture other climate change effects on road sediment losses in the 2041–2060 period: monthly maximum temperature, monthly minimum temperature, and mean monthly precipitation. These factors are all used by the WEPP-Road routine. They were taken directly from the climate scenarios and held constant during all simulations. The minimum and maximum from each 20-year mean monthly temperature series were used to estimate the respective minimum and maximum monthly temperatures, and the mean of each series was used to estimate the mean monthly precipitation.

Lastly, we also examined how increased maintenance on roads might reduce the sediment risks. All factor combinations were computed assuming the current maintenance level will be used from 2041 to 2060, and then again assuming an increased maintenance level.

Findings: The ensemble scenario indicates that by 2060 mean annual air temperature will increase 2.6–3.4°C relative to the current period in the Ouachita watersheds. Mean annual precipitation in the watersheds is projected to decrease somewhat (0.9–7.0%) on the western side of the Ouachita (Figure 9.27); but increase somewhat (0.4–2.6%) on the eastern side. The change in R-factor values between the current period and the 2041-to-2060 period follows the precipitation change pattern, with R-factor values decreasing 1.8–13.7% on the western side and increasing 0.6–5.3% on the eastern side.

The combined effects of the different changes in total rainfall amount, rainfall intensity, and vegetation cover on the sediment risk ratings are shown in Figure 9.28. The predicted changes in R-factors have little effect on the risk classifications of the Ouachita subwatersheds unless a concomitant change in C-factors occurs. Using the predicted R-factors, the numbers of subwatersheds in each risk class do not change appreciably from their current number. If R-factors are 15% less (low R condition), noticeable improvement occurs, with 25 subwatersheds moving from a higher to a lower risk class. The sediment risk worsens when R-factors are 15% larger than the predicted values (high R condition), with 15 subwatersheds shifting up to a higher risk class. These results indicate that future conditions would have to become wetter than is projected by the ensemble model before erosivity changes alone would worsen the current sediment risk.

However, if C-factors change concurrently with the R-factors, the effect on sediment risk is magnified. A 15% increase in the C-factor (reduced vegetation cover) causes the number of high-risk subwatersheds to increase well above the current number under both the predicted and high R levels, and to be equivalent to the current number under the low R level. Curiously, the number of low risk subwatersheds at the low R level actually increases above that under current conditions when the C-factor is increased 15%. This result may be important because if the future climate is drier than predicted by the ensemble climate scenario, the chance of C-factors increasing would be relatively higher than under the wetter conditions represented by the predicted and high R levels.

Surprisingly, increasing the rainfall intensity by decreasing the number of wet days from −20% to −50% had almost no effect on the sediment risk classifications. This suggests that the possible increase in road sediment generated by increasing the rainfall intensity is not sufficient by itself to

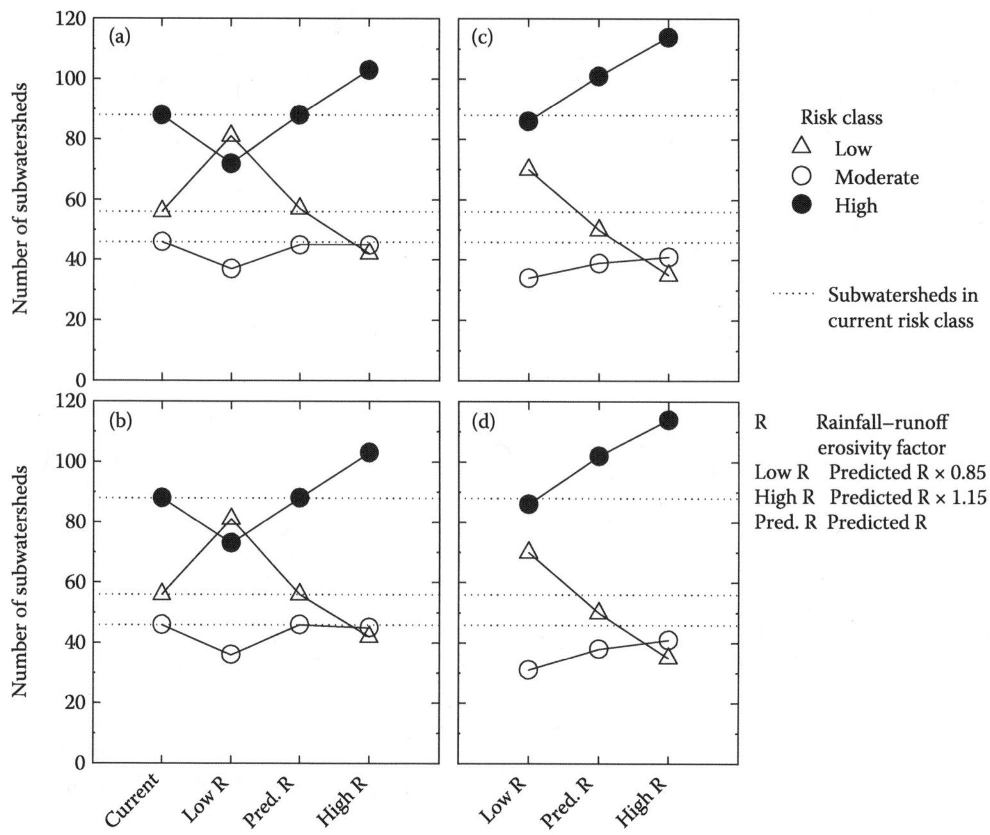

FIGURE 9.28 Effects of current and predicted (2041–2060) changes in total rainfall amount, rainfall intensity, and vegetation cover factor on the sediment risk rankings for Ouachita National Forest subwatersheds: (a) 20% decrease in wet days with no change in vegetation cover factor, (b) 50% decrease in wet days with no change in vegetation cover factor, (c) 20% decrease in wet days with 15% increase in vegetation cover factor, and (d) 50% decrease in wet days with 15% increase in vegetation cover factor. The rainfall–runoff erosivity and vegetation cover factors (Wischmeier and Smith 1965), and the number of wet days per month (Elliot 2004) are used within the Aquatic Cumulative Effects model (Clingenpeel and Crump 2005) to predict sediment risk ranking. Rainfall–runoff erosivity and number of wet days are based on precipitation predicted for the 2041–2060 period using an ensemble general circulation model and the A1B emissions storyline (representing moderate population growth and high-energy use) (Mote and Shepherd 2011).

alter the effects caused in unroaded areas by changes in the R- and C-factors. However, increasing rainfall intensity does have an effect when road maintenance is considered.

Increasing road maintenance could reduce possible increases in sediment risk resulting from the climate changes discussed so far. The effect of increasing road maintenance above current levels on sediment risk classification is shown in Figure 9.29. Increasing road maintenance always decreases road erosion in the ACE model; therefore, it is not surprising that the number of improved subwatersheds (those moving to a lower risk class) increased for all factor-level combinations. The number of improved subwatersheds always increases when the R-factor level decreases (moving down on the *y*-axis) or the C-factor level decreases (moving from left to right). The general lack of effect from increasing rainfall intensity is also evident when values for the two intensity levels are compared within a given combination of R- and C-factor levels. However, a small effect is apparent when the C-factors are unchanged and the R-factors are either at the predicted or at the low level. For predicted R-factors, the number of improved subwatersheds with increased road maintenance

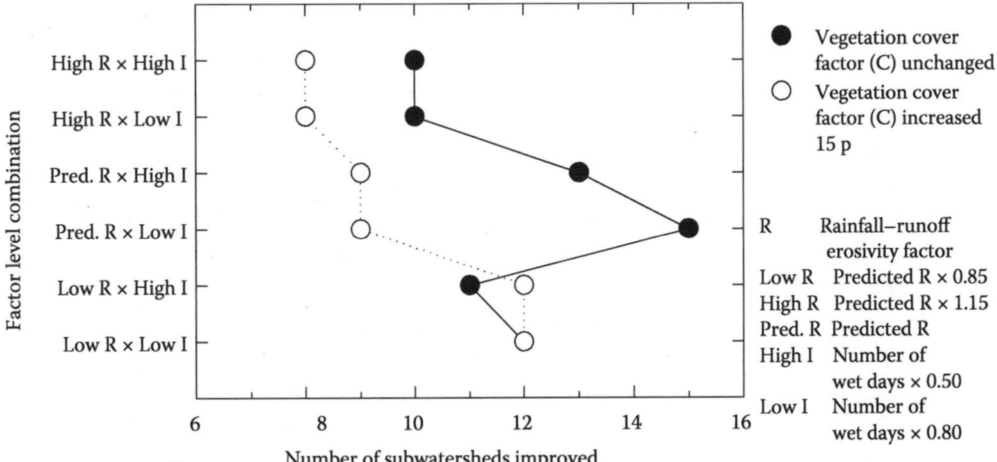

FIGURE 9.29 Effect of increased road maintenance on the number of Ouachita National Forest subwatersheds with improved sediment risk rankings—higher to lower risk class—under varying future rainfall and forest conditions. The rainfall–runoff erosivity and vegetation cover factors (Wischmeier and Smith 1965), number of wet days per month (Elliot 2004), and road maintenance levels are used within the Aquatic Cumulative Effects model (Clingenpeel and Crump 2005) to predict sediment risk ranking. Rainfall–runoff erosivity and number of wet days are based on precipitation predicted for the 2041–2060 period using an ensemble general circulation model and the A1B emissions storyline (representing moderate population growth and high-energy use) (Mote and Shepherd 2011).

decreases by two when rainfall intensity increases to the high level, but it decreases by one for the low R condition.

In summary, results from these simulations indicate that changes in mean annual precipitation predicted from a climate projection for the Ouachita from 2041 to 2060 and an assumed increase in rainfall intensity based on 20% fewer wet days are not sufficient by themselves to worsen the current sediment risk ratings. However, if these changes happen concurrently with decreased vegetative cover, then such a situation is possible. Decreased vegetative cover is a distinct possibility given the increased air temperatures predicted for this period. To a substantial degree, potential increases in sediment risk could be mitigated by increasing road maintenance within the forest.

SUMMARY

Forest managers and policy makers in the South will face growing challenges in mitigating or adapting to the effects of climate change on water resources in the near future. This chapter identified the key water resource characteristics related to forestlands that are mostly likely to be affected and how they will respond based on past research and new analyses. Future climate conditions were projected using climate scenarios derived from commonly used general circulation models and emission storylines. The region-wide sections show that in the future, important increases in evapotranspiration, water temperature, soil erosion, and coastal salinity intrusion are projected to occur either across broad areas of the South or within specific localities. The discussion of mitigation and adaptation options notes which BMPs are likely to be effected by changing climate conditions and how they may need to be modified. The case studies provide more detailed assessments of impacts to particular water resources within each of the subregions of the South. The information in this chapter should provide forest managers and policy makers a sound knowledge base from which to identify the potential issues they will need to consider in their future, location-specific plans, as well as a rich set of examples of analysis approaches they might use in accomplishing their own assessments.

REFERENCES

Alcamo, J., T. Henrichs, and T. Rosch. 2000. World water in 2025: Global modeling and scenario analysis for the World Commission on Water for the 21st century. Kassel World Water Series, Kassel, Germany. Report No 2.

Allen, M.R. and W.J. Ingram. 2002. Constraints on future changes in climate and the hydrologic cycle. *Nature* 419: 224–232.

Amatya, D.M., T.J. Callahan, C.C. Trettin et al. 2009. Hydrologic and water quality monitoring on Turkey Creek Watershed, Francis Marion National Forest, SC. *Annual ASABE International Meeting*, Reno, Nevada. Paper 09–5999.

Amatya, D.M., G.M. Chescheir, G.P. Fernandez et al. 2003. Lumped parameter models for predicting nitrogen transport in lower Coastal Plain watersheds. Water Resources Research Institute, Raleigh, North Carolina. Report 347.

Amatya, D.M., G.M. Chescheir, G.P. Fernandez et al. 2004. DRAINWAT-based methods for estimating nitrogen transport on poorly drained watersheds. *Transactions of the ASAE* 43(3): 677–687.

Amatya, D.M., J.D. Gregory, and R.W. Skaggs. 2000. Effects of controlled drainage on storm event hydrology in a loblolly pine plantation. *Journal of the American Water Resources Association* 36(1): 175–190.

Amatya, D.M., M. Miwa, C.A. Harrison et al. 2006a. Hydrology and water quality of two first order watersheds in coastal South Carolina. *American Society of Agricultural Engineers*. St. Joseph, Michigan. Paper 062182.

Amatya, D.M. and A. Radecki-Pawlik. 2007. Flow dynamics of three forested watersheds in coastal South Carolina, U.S.A. *Acta Scient. Polonorum—Formatio Circumiectus* 6(2): 3–17.

Amatya, D.M. and R.W. Skaggs. 2011. Long-term hydrology and water quality of a drained pine plantation in North Carolina, U.S.A. *Transactions of the ASABE* 54(6): 2087–2098 (special issue "Advances in Forest Hydrology").

Amatya, D.M., R.W. Skaggs, and J.D. Gregory. 1996. Effects of controlled drainage on the hydrology of a drained pine plantation in the North Carolina coastal plains. *Journal of Hydrology* 181: 211–232.

Amatya, D.M., R.W. Skaggs, and J.D. Gregory. 1997. Evaluation of a watershed scale forest hydrologic model. *Journal of Agricultural Water Management* 32(1997): 239–258.

Amatya, D.M., G. Sun, R.W. Skaggs et al. 2006b. Hydrologic effects of global climate change on a large drained pine forest. *In*: Williams and J.E. Nettles (eds). *Proceedings of the International Conference on Hydrology and Management of Forested Wetlands*. pp. 383–394. ASABE, St. Joseph, Michigan.

Amatya, D.M. and C.C. Trettin. 2007. Development of watershed hydrologic studies at Santee Experimental Forest, South Carolina. *In:* M.J. Furniss et al. (eds). *Proceedings of the Forest Service National Earth Sciences Conference*, Volume I. pp. 180–190. US Department of Agriculture, Forest Service, Pacific Northwest Research Station, Portland, Oregon. PNW-GTR-689.

Amatya, D.M. and R.W. Skaggs. 2001. Hydrologic modeling of pine plantations on poorly drained soils. *Forest Science* 47(1): 103–114.

Appelboom, T.W., G.M. Chescheir, R.W. Skaggs et al. 2008. Nitrogen balance for a plantation forest drainage canal on the North Carolina coastal plain. *Transactions of the ASABE* 51(4): 1215–1233.

Arnoldus, H.M.J. 1977. Methodology used to determine the maximum potential average annual soil loss due to sheet and rill erosion in Morocco. *Food and Agriculture Organization Soils Bulletin* 34: 39–51.

Baron, J.S., L. Gunderson, C.D. Allen et al. 2009. Options for National Parks and Reserves for adapting to climate change. *Environmental Management* 44: 1033–1042, doi:10.1007/s00267-009-9296-6.

Bear, J., A.H.D. Cheng, S. Sorek et al., eds. 1999. *Seawater Intrusion in Coastal Aquifers—Concepts, Methods and Practices*. Kluwer Academic Publishers, Dordrecht, the Netherlands. 640 p.

Beerling, D.J. 1996. Ecophysiological responses of woody plants to past CO_2 concentrations. *Tree Physiology* 16: 389–396.

Bicknell, B.R., J.C. Imhoff, J.L. Kittle Jr. et al. 2001. *Hydrological Simulation Program—Fortran, HSPF User's Manual, Version 12*. US EPA, Office of Research and Development, National Exposure Research Laboratory, Athens, Georgia.

Bogan, T., O. Mohseni, and H.G. Stefan. 2003. Stream temperature—equilibrium temperature relationship. *Water Resources Research* 39(9): 1245.

Breshears, D.D., N.S. Cobb, P.M. Rich et al. 2005. Regional vegetation die-off in response to global-change-type drought. *Proceedings of the National Academy of Sciences* 102(42): 15144–15148.

Bryant, M.L., S. Bhata, and J.M. Jacobs. 2005. Measurements and modeling of throughfall variability for five forest communities in the Southeastern US. *Journal of Hydrology* 312: 95–108.

Burton, T.M. and G.E. Likens. 1973. The effect of strip-cutting on stream temperature in the Hubbard Brook Experimental Forest, New Hampshire. *BioScience* 23(7): 433–435.

Caldwell, P.V., G. Sun, S.G. McNulty et al. 2012. Impacts of impervious cover, water withdrawals, and climate change on river flows in the conterminous U.S. *Hydrology and Earth System Sciences* 16: 2839–2857.

Carlisle, D.M., D.M. Wolock, and M.R. Meador. 2010. Alteration of streamflow magnitudes, and potential ecological consequences: A multiregional assessment. *Frontiers in Ecology and the Environment* 2010, doi:10.1890/100053.

Cech, P.G., S. Pepin, and C. Körner. 2003. Elevated CO_2 reduces sap flux in mature deciduous forest trees. *Oecologia* 137: 258–268.

Chapman, D.S. and M.G. Davis. 2010. Climate change: Past, present, and future. *EOS* 91(37): 325–326.

Chen, G., M. Notaro, Z. Liu et al. 2012. Simulated local and remote biophysical effects of afforestation over the southeast United States in boreal summer. *Journal of Climate* 25: 4511–4522.

Clingenpeel, J.A. and M.A. Crump. 2005. *A manual for the Aquatic Cumulative Effects model.* Unpublished paper. Ouachita National Forest, Supervisors Office, Hot Springs. Arkansas.

Conrads, P.A. and E.A. Roehl Jr. 2007. Analysis of salinity intrusion in the Waccamaw River and the Atlantic Intracoastal Waterway near Myrtle Beach, South Carolina, 1995–2002. US Geological Survey. Scientific Investigations Report 2007–5110.

Coulson, D.P., L.A. Joyce, D.T. Price et al. 2010a. Climate scenarios for the conterminous United States at the county spatial scale using SRES scenarios B2 and PRISM climatology. US Forest Service, Rocky Mountain Research Station. http://www.fs.fed.us/rm/data_archive/dataaccess. Accessed: November 1, 2012.

Coulson, D.P., L.A. Joyce, D.T. Price et al. 2010b. Climate scenarios for the conterminous United States at the county spatial scale using SRES scenarios A1B and A2 and PRISM climatology. US Forest Service, Rocky Mountain Research Station. http://www.fs.fed.us/rm/pubs_other/rmrs_2010_coulson_d005.pdf. Accessed: November 1, 2012.

Crockford, R.H. and D.P. Richardson. 2000. Partitioning of rainfall into throughfall, stemflow and interception: Effect of forest type, ground cover and climate. *Hydrological Processes* 14: 2903–2920.

Cruise, J.F., A.S. Limaye, and N. Al-Abed. 1999. Assessment of impacts of climate change on water quality in the southeastern United States. *Journal of the American Water Resources Association* 35(6): 1539–1550.

Dai, Z., D. Amatya, G. Sun et al. 2011a. Climate variability and its impact on forest hydrology on South Carolina Coastal Plain of USA. *Atmosphere* 2: 330–357, doi:10.3390/atmos2030330.

Dai, Z., C. Li, C. Trettin et al. 2010a. Bi-criteria evaluation of MIKE SHE model for a forested watershed on South Carolina coastal plain. *Hydrology and Earth Systems Science*. 14: 179–219, doi:10.5194/hess-14-1033-2010.

Dai, Z., C.C. Trettin, and D.M. Amatya. 2011b. Impacts of climate variability on forest hydrology and carbon sequestration at Santee Experimental Forest in coastal South Carolina. U.S. Department of Agriculture, Forest Service, Southern Research Station. General Technical Report (in review).

Dai, Z., C. Trettin, C. Li et al. 2010b. Sensitivity of stream flow and water table depth to potential climatic variability in a coastal forested watershed. *Journal of the American Water Resources Association* 46(5): 1036–1048, doi:10.1111/j.1752-1688. 2010.

Daly, C., R.P. Neilson, and D.L. Phillips. 1994. A statistical-topographic model for mapping climatological precipitation over mountainous terrain. *Journal of Applied Meteorology* 33: 140–158.

Danish Hydraulic Institute. 2005. MIKE SHE Technical Reference, Version 2005. DHI Water and Environment. Danish Hydraulic Institute, Denmark.

David, J.L., B.A. Middleton, and R.O. Anyah. 2010. Modeled climate change impacts on the production of bald cypress swamps across a latitudinal gradient. *Proc. Society of Wetland Scientists Conference*, Salt Lake City, Utah.

Dissmeyer, G.E. and G.R. Foster. 1984. A guide for predicting sheet and rill erosion on forestland. U.S. Department of Agriculture, Forest Service. Gen. Tech. Publ. R8-TP 6. 40 p.

Dissmeyer, G.E. and R.F. Stump. 1978. Predicted erosion rates for forest management activities and conditions sampled in the Southeast. Administrative report. U.S. Department of Agriculture Forest Service, State and Private Forestry, Southern Region, Atlanta, Georgia. 26 p.

Donogian, A.S., J.C. Imhoff, B.R. Bicknell et al. 1984. Application guide for hydrological simulation program-FORTRAN (HSPF). EPA, Athens, Georgia. EPA-600/3-84-065.

Easterling, D.R., J.L. Evans, P.Y. Groisman et al. 2000a. Observed variability and trends in extreme climate events: A brief review. *Bulletin of the American Meteorological Society* 81: 417–425.

Easterling, D.R., G.A. Meehl, C. Parmesan et al. 2000b. Climate extremes: Observations, modeling, and impacts. *Science* 289: 2068–2074, doi:10.1126/science.289.5487.2068.

Eaton, G.J. and R.M. Scheller. 1996. Effects of climate warming on fish thermal habitat in streams of the United States. *Limnology & Oceanography* 41(5): 1109–1115.

Elliot, W.J. 2004. WEPP internet interfaces for forest erosion prediction. *Journal of the American Water Resources Association* 40(2): 299–309.

Elmore, A.J, S.M. Guinn, B.J. Minsley et al. 2012. Landscape controls on the timing of spring, autumn, and growing season length in mid-Atlantic forests. *Global Change Biology* 18: 656–674. doi: 10.1111/j.1365-2486.2011.02521.x.

Farley, K., E. Jobbagy, and R.B. Jackson. 2005. Effects of afforestation on water yield: A global synthesis with implications for policy. *Global Change Biology* 11:1565–1576.

Fernandez, G.P., G.M. Chescheir, R.W. Skaggs et al. 2006. DRAINMOD-GIS: A lumped parameter watershed scale drainage and water quality model. *Agricultural Water Management* 81(2006): 77–97.

Fernandez, G.P., G.M. Chescheir, R.W. Skaggs et al. 2007. Application of DRAINMOD-GIS to a lower coastal plain watershed. *Transactions of the ASABE* 50(2): 439–447.

Ford, C.R., R.M. Hubbard, and J.M. Vose. 2011a. Quantifying structural and physiological controls on variation in canopy transpiration among planted pine and hardwood species in the southern Appalachians. *Ecohydrology* 4: 183–195.

Ford, C.R., S.H. Laseter, W.T. Swank et al. 2011b. Can forest management be used to sustain water-based ecosystem services in the face of climate change? *Ecological Applications* 21: 2049–2067.

Franks, P.J. and D.J. Beerling. 2009. Maximum leaf conductance driven by CO_2 effects on stomatal size and density over geologic time. *Proceedings of the National Academy of Sciences* 106(25): 10343–10347.

Furlow, J., J.D. Scheraga, R. Freed et al. 2002. The vulnerability of public water systems to sea level rise. *In:* J.R. Lesnik, (ed). *Proceedings of the Coastal Water Resource Conference*. pp. 31–36. American Water Resources Association, Middleburg, Virginia.

Galloway, G.E. 2011. If stationarity is dead, what do we do now? *Journal of the American Water Resources Association* 47(3): 563–570, doi: 10.1111/j.1752-1688.2011.00550.x.

Gedney, N., P.M. Cox, R.A. Betts et al. 2006. Detection of a direct carbon dioxide effect in continental river runoff records. *Nature* 439: 835–838.

Gibson, W.P., C. Daly, T. Kittel et al. 2002. Development of a 103-year high-resolution climate data set for the conterminous United States. *In: Proceedings, 13th AMS Conference on Applied Climatology*. pp. 181–183. American Meteorological Society, Portland, Oregon.

Groisman, P.Y., R.W. Knight, T.R. Karl et al. 2003. Contemporary changes of the hydrological cycle over the contiguous United States: Trends derived from *in situ* observations. *Journal of Hydrometeorology* 5(1): 64–85.

Guedon, N.B. and J.V. Thomas. 2004. State of Mississippi Water Quality Assessment Section 305(b) Report Appendum. Mississippi Department of Environmental Quality, Jackson, Mississippi.

Gunn, J.S., J.M. Hagan, and A.A. Whitman. 2009. Forestry adaptation and mitigation in a changing climate. A forest resource manager's guide for the northeastern United States. Natural Capital Initiative at Manomet. Ecosystem Services Resources Center Report NCI-2009-1.

Guttman, N.B. 1999. Accepting the standardized precipitation index: A calculation algorithm. *Journal of the American Water Resources Association* 35: 311–322.

Hänninen, H. and K. Tanino. 2011. Tree seasonality in a warming climate. *Trends in Plant Science* 16: 412–416.

Hansen, J., M. Sato, R. Ruedy et al. 2006: Global temperature change. *Proceedings of the National Academy of Sciences* 103: 14288–14293.

Hansen, J.E., R. Ruedy, M. Sato et al. 2010. Global surface temperature change. *Reviews of Geophysics* 48: RG4004, doi:10.1029/2010RG000345.

Hanson, P.J., S.D. Wullschleger, R.J. Norby et al. 2005. Importance of changing CO_2, temperature, precipitation, and ozone on carbon and water cycles of an upland oak forest: Incorporating experimental results into model simulations. *Global Change Biology* 11: 1402–1423.

Harder, S.V., D.M. Amatya, T.J. Callahan et al. 2007. A hydrologic budget of a first order forested watershed, South Carolina. *Journal of the American Water Resources Association* 43(3): 1–13.

Hatch, U., S. Jagtap, J. Jones et al. 1999. Potential effects of climate change on agricultural water use in the southeast US. *Journal of the American Water Resources Association* 35(6): 1551–1561.

Helvey, J.D. 1967. Interception by eastern white pine. *Water Resources Research* 3(3): 723–729.

Helvey, J.D. and J.H. Patric. 1965. Canopy and litter interception of rainfall by hardwoods of eastern United States. *Water Resources Research* 1: 193–206.

Hook, D.D., M.A. Buford, and T.M. Williams. 1991. The impact of Hurricane Hugo on the South Carolina coastal plain forest. *Journal of Coastal Research*, special issue 8: 291–300.

Hossian, F., D. Niyogi, J. Adegoke et al. 2011. Making sense of the water resources that will be available for future use. *EOS* 92(17): 144–145.

Huntington, T.G. 2006. Evidence for intensification of the global water cycle: Review and synthesis. *Journal of Hydrology* 319: 83–95.

Intergovernmental Panel on Climate Change. 2001. *Climate Change 2001: The Scientific Basis. Contribution of Working Group I to the Third Assessment Report of the Intergovernmental Panel on Climate Change.* J.T. Houghton, Y. Ding, D.J. Griggs et al. (eds). Cambridge University Press, Cambridge, United Kingdom and New York, New York.

Intergovernmental Panel on Climate Change. 2007. Contribution of Working Groups I, II and III to the Fourth Assessment Report of the Intergovernmental Panel on Climate Change. *In*: Core Writing Team, R.K. Pachauri, and A. Reisinger (eds). *Climate Change 2007: Synthesis Report.* Geneva, Switzerland.

Iverson, L.R., A.M. Prasad, S.N. Matthews et al. 2008. Estimating potential habitat for 134 eastern US tree species under six climate scenarios. *Forest Ecology and Management* 254: 390–406.

Jackson, R.B., E.G. Jobbagy, R. Avissar et al. 2005. Trading water for carbon with biological carbon sequestration. *Science* 310: 1944–1947, doi:10.1126/science.1119282.

Jackson, R.B., J.T. Randerson, J.G. Canadell et al. 2008. Protecting climate with forests. *Environmental Research Letters* 3: 044006, doi:10.1088/1748-9326/3/4/044006.

Jensen, B.A. 1994. *Expert Systems—Neural Networks, Instrument Engineers' Handbook*, Third Edition. Chilton, Radnor, Pennsylvania.

Jung, M., M. Reichstein, P. Ciais et al. 2010. Recent deceleration of global land evapotranspiration due to moisture supply limitation. *Nature* 467: 951–954.

Karl, T.R. and R.W. Knight. 1998. Secular trends of precipitation amount, frequency, and intensity in the USA. *Bulletin of the American Meteorology Society* 79: 231–241.

Karl, T.R., R.W. Knight, and N. Plummer. 1995. Trends in high-frequency climate variability in the twentieth century. *Nature* 377: 217–220.

Karl, T.R., J.M. Mellillo, and T.C. Peterson (eds). 2009. *Global Climate Change Impacts in the United States.* Cambridge University Press, Cambridge, United Kingdom.

Kaushal, S.S., G.E. Likens, N.A. Jaworski et al. 2010. Rising stream and river temperatures in the United States. *Frontiers in Ecology and the Environment* 8: 461–466.

Kinouchi, T., H. Yaki, and M. Miyamoto. 2007. Increase in stream temperature related to anthropogenic heat input from urban wastewater. *Journal of Hydrology* 335: 78–88.

Kittel, T.G.F., J.A. Royle, C. Daly et al. 1997. A gridded historical (1895–1993) bioclimate dataset for the conterminous United States. *In: Proceedings of the 10th Conference on Applied Climatology*, American Meteorological Society, Boston, Massachusetts. pp. 219–222.

Krakauer, N.Y. and I. Fung. 2008. Mapping and attribution of change in streamflow in the conterminous United States. *Hydrology and Earth Systems Science* 12: 1111–1120.

Kunkel, K.E., D.R. Easterling, K. Hubbard et al. 2004. Temporal variations in frost-free season in the United States: 1895–2000. *Geophysical Research Letters* 31: L03201.

Labat, D., Y. Godderis, J.L. Probst et al. 2004. Evidence for global runoff increase related to climate warming. *Advances in Water Resources* 27(6): 631–642.

Lammertsma, E.I., H. Jan de Boer, S.C. Dekker et al. 2011. Global CO_2 rise leads to reduced maximum stomatal conductance in Florida vegetation. *Proceedings of the National Academy of Sciences* 1100371108v1–201100371.

Laseter, S.H., C.R. Ford, J.M. Vose et al. 2012. Long-term temperature and precipitation trends at the Coweeta Hydrologic Laboratory, Otto, North Carolina, USA. *Hydrology Research,* 43: 890–901.

La Torre Torres, I., D.M. Amatya, T.J. Callahan et al. 2011. Seasonal rainfall-runoff relationships in a lowland forested watershed in the southeastern U.S.A. *Hydrological Processes* 25: 2032–2045.

Leopold, D.J., G.R. Parker, and W.T. Swank. 1985. Forest development after successive clearcuts in the Southern Appalachians. *Forest Ecology and Management* 13: 83–120.

Leuzinger, S. and C. Korner. 2007. Water savings in mature deciduous forest trees under elevated CO_2. *Global Change Biology* 13: 2498–2508.

Leuzinger, S. and C. Korner. 2010. Rainfall distribution is the main driver of runoff under future CO_2 concentration in a temperate deciduous forest. *Global Change Biology* 16: 246–254.

Liao, C., Y. Luo, C. Fang et al. 2010. Ecosystem carbon stock influenced by plantation practice: Implications for planting forests as a measure of climate change mitigation. *PLoS ONE* 5: e10867, doi:10.1371/journal.pone.0010867.t001.

Lins, H.F. and J.R. Slack. 1999. Streamflow trends in the United States. *Geophysical Research Letters* 26: 227–230.

Lins, H.F. and J.R. Slack. 2005. Seasonal and regional characteristics of US streamflow trends in the United States from 1940–1999. *Physical Geography* 26: 489–501.

Liu, Y.-Q. 2011. Hydrological impacts of forest restoration in the southern United States. *Ecohydrology* 4: 299–314, doi:10.1002/eco.178.

Liu, Y.-Q., J. Stanturf, and H. Lu. 2008. Modeling the potential of the northern China forest shelterbelt in improving hydroclimate conditions. *Journal of the American Water Resources Association* 44(5): 1176–1192.

Lockaby, G., C. Nagy, J.M. Vose et al. (in press). Water and Forests. *In:* D.N. Wear and J.G. Greis (eds). The Southern Forest Futures Project: Technical Report. US Department of Agriculture, Forest Service, Southern Research Station, Asheville, North Carolina. General Technical Report.

Lu, J., G. Sun, S.G. McNulty et al. 2009. Sensitivity of pine flatwoods hydrology to climate change and forest management in Florida, USA. *Wetlands* 29: 826–836.

Luce, C.H. and Z.A. Holden. 2009. Declining annual streamflow distributions in the Pacific Northwest United States, 1948–2006. *Geophysical Research Letters* 36: L16401, doi:10.1029/2009GL039407.

Marion, D.A. and J.A. Clingenpeel. 2012. Methods used for analyzing the cumulative watershed effects of fuel management on sediment in the eastern United States. *In*: LaFayette, R., M.T. Brooks, J.P. Potyondy, L. Audin, S.L. Krieger, and C.C. Trettin (eds). *Cumulative Watershed Effects of Fuel Management in the Eastern United States. Forest Service*, Southern Research Station, Asheville, North Carolina. General Technical Report SRS-161: 308–326.

Matthews, K.R. and N.H. Berg. 1997. Rainbow trout responses to water temperature and dissolved oxygen stress in two southern California stream pools. *Journal of Fish Biology* 50(1): 50–67, doi: 10.1111/j.1095-8649.1997.tb01339.x.

McCabe, G.J. and D.M. Wolock. 2002. A step increase in streamflow in the conterminous United States. *Geophysical Research Letters* 29: 2185.

McNulty, S.G., J.A. Moore Myers, P. Caldwell et al. (in press). Climate Change. *In*: D.N. Wear and J.G. Greis (eds). *The Southern Forest Futures Project: Technical Report.* US Department of Agriculture, Forest Service, Southern Research Station, Asheville, North Carolina. General Technical Report.

Meisner, J.D. 1990. Effect of climatic warming on the southern margins of the native range of brook trout, *Salvelinus fontinalis*. *Canadian Journal of Fisheries and Aquatic Science* 47(6): 1065–1070.

Milly, P.C.D., J. Betancourt, M. Falkenmark et al. 2008. Stationarity is dead: Whither water management. *Science* 319: 573–574.

Mississippi Department of Environmental Quality. 2008. *Sediment TMDL for the Yalobusha River Yazoo River Basin*. Mississippi Department of Environmental Quality, Jackson, Mississippi.

Mohseni, O. and H.G. Stefan. 1999. Stream temperature air temperature relationship: A physical interpretation. *Journal of Hydrology* 218(3–4): 128–141.

Mohseni, O., H.G. Stefan, and T.R. Erickson. 1998. A nonlinear regression model for weekly stream temperatures. *Water Resources Research* 34(10): 2685–2692.

Mote, T.L. and J.M. Shepherd. 2011. Monthly ensemble downscaled climate model data for the Southeast U.S. http://shrmc.ggy.uga.edu/downscale.php. Accessed: November 07, 2011.

Murdoch, P.S. and J.S. Baron. 2000. Potential effects of climate change on surface-water quality in North America. *Journal of the American Water Resources Association* 36: 347–366.

National Research Council of the National Academies. 2008. *Hydrologic Effects of a Changing Forest Landscape*. National Research Council of the National Academies, The National Academies Press, Washington, District of Columbia.

Nearing, M.A. 2001. Potential changes in rainfall erosivity in the U.S. with climate change during the 21st century. *Journal of Soil and Water Conservation* 56(3): 229–232.

Neilson, R.P., L.F. Pitelka, A. Solomon et al. 2005. Forecasting regional to global plant migration in response to climate change: Challenges and directions. *BioScience* 55: 749–759.

Nelson, K.C. and M.A. Palmer. 2007. Stream temperature surges under urbanization and climate change: Data, models, and responses. *Journal of the American Water Resources Association* 43(2): 440–452.

Neubauer, C.P., G.B. Hall, E.F. Lowe et al. 2008. Minimum flows and levels method of the St. Johns River Water Management District, Florida. *Environmental Management* 42(6): 1101–1114.

Noormets, A., M. Gavazzi, S.G. McNulty et al. 2010. Response of carbon fluxes to drought in a coastal plain loblolly pine forest. *Global Change Biology* 16: 272–287.

O'Gorman, P.A. and T. Schneider. 2009. The physical basis for increases in precipitation extremes in simulations of 21st-century climate change. *Proceedings of the National Academy of Sciences* 106: 14773–14777, doi:10.1073/pnas.0907610106.

Ollinger, S.V., C.L. Goodale, K. Hayhoe et al. 2008. Potential effects of climate change and rising CO_2 on ecosystem processes in northeastern US forests. *Mitigation and Adaptation Strategies for Global Change* 13: 467–485.

O'Neal, K. 2002. *Effects of Global Warming on Trout and Salmon in US Streams*. Defenders of Wildlife, Washington, District of Columbia.

Pacala, S. and R. Sokolow. 2004. Stabilization wedges: Solving the climate problem for the next 50 years with current technologies. *Science* 305: 968–972, doi:10.1126/science.1100103.

Peterson, J.T. and T.J. Kwak. 1999. Modeling the effects of land use and climate change on riverine smallmouth bass. *Ecological Applications* 9(4): 1391–1402.

Phillips, D.L., D. White, and C.B. Johnson. 1993. Implications of climate change scenarios for soil erosion potential in the United States. *Land Degradation and Rehabilitation* 4: 61–72.

Poole, G.C. and C.H. Berman. 2001. An ecological perspective on in-stream temperature: Natural heat dynamics and mechanisms of human-caused thermal degradation. *Environmental Management* 27(6): 787–802.

Postel, S. and S. Carpenter. 1997. Freshwater ecosystem services. *In:* G.C. Daily (ed). *Nature's Services: Societal Dependence on Natural Ecosystems*. pp. 195–214. Island Press, Washington, District of Columbia.

Postel, S. and B. Richter. 2003. *Rivers for Life: Managing Water for People and Nature*. Island Press, Washington, District of Columbia. 220 p.

Prentice, I.C. and S.P. Harrison. 2009. Ecosystem effects of CO_2 concentration: Evidence from past climates. *Climate of the Past* 5: 297–307.

Pruski, F.F. and M.A. Nearing. 2002a. Climate-induced changes in erosion during the 21st century for eight U.S. locations. *Water Resources Research* 38(12): 1298, doi:10.1029/2001WR000493.

Pruski, F.F. and M.A. Nearing. 2002b. Runoff and soil loss responses to changes in precipitation: A computer simulation study. *Journal of Soil and Water Conservation* 57(1): 7–16.

Pyzoha, J.E., T.J. Callahan, G. Sun et al. 2007. A conceptual hydrologic model for a forested Carolina bay depressional wetland on the Coastal Plain of South Carolina, USA. *Hydrologic Processes* 22(14): 2689–2698, doi:10.1002/hyp.68.

Reilly, C.F. and C.N. Kroll. 2003. Estimation of 7-day, 10-year low-streamflow statistics using baseflow correlation. *Water Resources Research* 39: 1236.

Renard, K.G. and V.A. Ferreira. 1993. RUSLE model description and database sensitivity. *Journal of Environmental Quality* 22: 458–466.

Renard, K.G. and J.R. Freimund. 1994. Using monthly precipitation data to estimate the R factor in the revised USLE. *Journal of Hydrology* 157: 287–306.

Renard, K.G., G.R. Foster, G.A. Weesies et al. 1997. *Predicting Soil Erosion by Water: A Guide To Conservation Planning with the Revised Universal Soil Loss Equation (RUSLE). Agriculture Handbook* Number 703. US Department of Agriculture, Washington, District of Columbia.

Rosenblatt, F. 1958. The perceptron: A probabilistic model for information storage and organization in the brain. *Psychological Review* 65: 386–408.

Ryan, M.G., M.E. Harmon, R.A. Birdsey et al. 2010. A synthesis of the science on forests and carbon for US forests. *Issues in Ecology* 13: 1–16.

Sassaman, R.W. 1981. Threshold of concern: A technique for evaluating environmental impacts and amenity values. *Journal of Forestry* 79: 84–86.

Savenije, H.G. 2004. The importance of interception and why we should delete the term evapotranspiration from our vocabulary. *Hydrological Processes* 18: 1507–1511.

Schwartz, M.D. and J.M. Hanes. 2010. Continental-scale phenology: Warming and chilling. *International Journal of Climatology* 30: 1595–1598.

Shelby, J.D., G.M. Chescheir, R.W. Skaggs et al. 2005. Hydrologic and water quality response of forested and agricultural lands during the 1999 extreme weather conditions in eastern North Carolina. *Transactions of the ASAE* 48: 2179–2188.

Sheridan, J.M. 2002. Peak flow estimates for coastal plain watersheds. *Transactions of the ASAE* 45(5): 1319–1326.

Shields, F.D., Jr., C.M. Cooper, S. Testa III et al. 2008. Nutrient transport in the Yazoo River Basin. Research Report 60. US Department of Agriculture, Agricultural Research Service, National Sedimentation Laboratory, Oxford, Mississippi.

Slack, J.R., A.M. Lumb, and J.M. Landwehr. 1993. Hydro-Climatic Data Network (HCDN) streamflow data set, 1874–1988. US Geological Survey, Water Resources Investment Report 93–4076.

Smakhtin, V.U. 2001. Low flow hydrology: A review. *Journal of Hydrology* 240: 147–186.

St. Johns River Water Management District. 2010. St. Johns River Water Management District, Chapter 40c-8, F.A.C. Minimum Flows and Levels. St. Johns River Water Management District, Palatka, Florida.

Stoy, P.C., G.G. Katul, M.B.S. Siqueira et al. 2006. Separating the effects of climate and vegetation on evapotranspiration along a successional chronosequence in the southeastern US. *Global Change Biology* 12: 2115–2135.

Sullivan, K., J. Tooley, K. Doughty et al. 1990. *Evaluation of Prediction Models and Characterization of Stream Temperature Regimes in Washington.* Timber/Fish/Wildlife Rep. No. TFW-WQ3-90-006. Washington Department of Natural Resources, Olympia, Washington. 224 pp.

Sun, G., K. Alstad, J. Chen et al. 2011a. A general predictive model for estimating monthly ecosystem evapotranspiration. *Ecohydrology* 4: 245–255, doi:10.1002/eco.194.

Sun, G., D.M. Amatya, S.G. McNulty et al. 2000a. Climate change impacts on the hydrology and productivity of a pine plantation. *Journal of the American Water Resources Association* 36(2): 367–374.

Sun, G., P. Caldwell, A. Noormets et al. 2011b. Upscaling key ecosystem functions across the conterminous United States by a water-centric ecosystem model. *Journal of Geophysical Research* 116: G00J05, doi:10.1029/2010JG001573.

Sun, G., S.G. McNulty, D.M. Amatya et al. 2002. A comparison of the watershed hydrology of coastal forested wetlands and the mountainous uplands in the southern US. *Journal of Hydrology* 63: 92–104.

Sun, G., S.G. McNulty, J.A. Moore Myers et al. 2008. Impacts of multiple stresses on water demand and supply across the southeastern United States. *Journal of the American Water Resources Association* 44: 1441–1457.

Sun, G., A. Noormets, M. Gavazzi et al. 2010. Energy and water balances of two contrasting loblolly pine plantations on the lower coastal plain of North Carolina, USA. *Forest Ecology and Management* 259: 1299–1310.

Sun, G., H. Riekerk, and L.V. Kornhak. 2000b. Ground-water table rise after forest harvesting on cypress-pine flatwoods in Florida. *Wetlands* 20(1): 101–112.

Sun, G., M. Riedel, R. Jackson et al. 2004. Influences of management of southern forests on water quantity and quality. *In*: H.M. Rauscher and K. Johnsen (eds). *Southern Forest Sciences: Past, Current, and Future.* pp. 195–226. US Department of Agriculture, Forest Service, Southern Research Station, Asheville, North Carolina. Gen. Tech. Rep SRS-75.

Swift, L.W. 1982. *Duration of Stream Temperature Increases Following Forest Cutting in the Southern Appalachian Mountains.* International Symposium on Hydrometeorology, Denver, Colorado.

Swift, L.W., Jr. 1973. Lower water temperatures within a streamside buffer strip. US Department of Agriculture, Forest Service, Southeastern Forest Experiment Station, Asheville, North Carolina. Res. Note SE-193.

Swift, L.W., Jr. and J.B. Messer. 1971. Forest cuttings raise temperatures of small streams in the Southern Appalachians. *Journal of Soil and Water Conservation* 26(3): 111–116.

Tague, C., L. Seaby, and A. Hope. 2009. Modeling the eco-hydrologic response of a Mediterranean type ecosystem to the combined impacts of projected climate change and altered fire frequencies. *Climatic Change* 93: 137–155.

Telis, P.A. 1992. Techniques for estimating 7-day 10-year low flow characteristics for ungauged sites on streams in Mississippi. US Geologic Survey. Water-Resources Investigations Report 91–4130.

Tian, S., M.A. Youssef, R.W. Skaggs et al. 2011. Modeling water, carbon and nitrogen dynamics for two drained pine plantations under intensive management practices. *Forest Ecology and Management* 264(2012): 20–36.

Trenberth, K.E., P.D. Jones, P. Ambenje et al. 2007. Observations: Surface and atmospheric climate change. Chapter 3 *In:* S. Solomon, D. Qin, M. Manning et al. (eds). *Climate Change 2007: The Physical Science Basis. Contribution of Working Group I to the Fourth Assessment Report of the Intergovernmental Panel on Climate Change.* Cambridge University Press, Cambridge, United Kingdom and New York, New York. http://www.ipcc.ch/publications_and_data/ar4/wg1/en/ch3.html. Accessed: November 1, 2012.

Trimble, S.W., F.H. Weirich, and B.L. Hoag. 1987. Reforestation and the reduction of water yield on the Southern Piedmont since circa 1940. *Water Resources Research* 23(3): 425–437, doi:10.1029/WR023i003p00425.

US Department of Agriculture. 2006. *Land Resource Regions and Major Land Resource Areas of the United States, the Caribbean, and the Pacific Basin.* US Department of Agriculture, Natural Resource Conservation Service, Washington, District of Columbia. Agriculture Handbook 296.

US Department of Agriculture, Forest Service. 2005. Draft environmental impact statement for the proposed revised land and resource management plan: Ouachita National Forest. US Department of Agriculture. Management Bulletin R8-MB 124 C.

US EPA. 2010. BASINS 4.0 (Better Assessment Science Integrating Point & Non-Point Sources) Description. http://water.epa.gov/scitech/datait/models/basins/BASINS4 _index.cfm. Accessed: November 1, 2012.

van Vliet, M.T.H., F. Ludwig, J.J.G. Zwolsman et al. 2011. Global river temperatures and sensitivity to atmospheric warming and changes in river flow. *Water Resources Research* 47: W02544, doi:10.1029/2010WR009198.

Vörösmarty, C.J., P. Green, J. Salisbury et al. 2000. Global water resources: Vulnerability from climate change and population growth. *Science* 289: 284–288.

Walter, M.T., D.S. Wilks, J.-Y. Parlange et al. 2004. Increasing evapotranspiration from the conterminous United States. *Journal of Hydrometeorology* 5: 405–408.

Warren, J.M., E. Pötzelsberger, S.D. Wullschleger et al. 2011. Ecohydrologic impact of reduced stomatal conductance on forests exposed to elevated CO_2. *Ecohydrology* 4: 196–210.

Watson, T. 2009. Climate plan calls for forest expansion, USA Today. www.usatoday.com/news/nation/environment/2009–08-19-forest_N.htm. Accessed: November 1, 2012.

Wear, D.N. and J.G. Greis. 2002. Southern forest resource assessment: Summary report. Gen. Tech. Rep. SRS 53. US Department of Agriculture, Forest Service, Southern Research Station, Asheville, North Carolina. 103 p.

Webb, B.W. and F. Nobilis. 1995. Long term water temperature trends in Austrian rivers. *Hydrological Sciences Journal* 40: 83–96.

Webb, B.W. and F. Nobilis. 2007. Long-term changes in river temperature and the influence of climatic and hydrological factors. *Hydrological Sciences Journal* 52(1): 74–85.

Webb, B.W. and D.E. Walling. 1993. Longer-term water temperature behavior in an upland stream. *Hydrological Processes* 7(1): 19–32.

Webb, B.W., D.M. Hannah, R.D. Moore et al. 2008. Recent advances in stream and river temperature research. *Hydrological Processes* 22(7): 902–918, doi: 10.1002/hyp.6994.

West, B. 2002. Water quality in the South. *In*: D.N. Wear and J.G. Greis (eds). *Southern Forest Resource Assessment*. pp. 455–476. US Department of Agriculture, Forest Service, Southern Research Station, Asheville, North Carolina. General Technical Report SRS-54.

White, E. 2011. Trends in heavy precipitation over the southern region. *Southern Climate Monitor* 1(7): 2–7.

Whitehead, P.G., R.L. Wilby, R.W. Battarbee et al. 2009. A review of the potential impacts of climate change on surface water quality. *Hydrological Sciences Journal* 54: 101–123.

Williams, T., D.M. Amatya, A. Jayakaran et al. 2012. Runoff generation from shallow water table southeastern forests: Unusual behavior of paired watersheds following a major disturbance. ASABE Paper # 121337045, St. Joseph, Michigan: ASABE. http://elibrary.asabe.org/azdez.asp?JID=5&AID=42110&CID=dall2012&T=2. Accessed: June 26, 2013.

Williams, T.M. 1978. Response of shallow water tables to rainfall. *In*: W.E. Balmer (ed). *Proceedings: Soil Moisture, Site Productivity Symposium*. pp. 363–370. US Department of Agriculture, Forest Service, Southeastern Area, State and Private Forestry.

Wischmeier, W.H. and D.D. Smith. 1965. *Predicting Rainfall Erosion Losses from Cropland East of the Rocky Mountains: Guide for Selection of Practices for Soil and Water Conservation*. US Department of Agriculture, Washington, District of Columbia. Agriculture Handbook No. 282.

Wischmeier, W.H. and D.D. Smith. 1978. *Predicting Rainfall Erosion Losses—A Guide to Conservation Planning*. US Department of Agriculture, Washington, District of Columbia. Agriculture Handbook No. 537.

Wooldridge, D.D. and D. Stern. 1979. Relationships of silvicultural activities and thermally sensitive forest streams. University of Washington, College of Forest Resources, Seattle, Washington. Rep. DOE 79-5a-5.

Young, C.E. and R.A. Klawitter. 1968. Hydrology of wetland forest watersheds. *In:* F.C. Alley, K. Lehotsky, and J.T. Ligon (eds). *Proceedings of Hydrology in Water Resources Management*. pp. 29–38. Water Resources Research Institute Report #4. Clemson University, Clemson, South Carolina.

Zhang, X., D. Tarpley, and J.T. Sullivan. 2007. Diverse responses of vegetation phenology to a warming climate. *Geophysical Research Letters* 34: L19405.

Zhao, M. and S.W. Running. 2010. Drought-induced reduction in global terrestrial net primary production from 2000 through 2009. *Science* 329: 940–943.

10 Climate-Induced Migration of Native Tree Populations and Consequences for Forest Composition

W. Henry McNab, Martin A. Spetich, Roger W. Perry, James D. Haywood, Shelby Gull Laird, Stacy L. Clark, Justin L. Hart, Scott J. Torreano, and Megan L. Buchanan

CONTENTS

The climate of the 13 Southern United States is generally thought to be changing in response to global and continental scale influences; and by 2060, average annual temperature is predicted to be higher and precipitation lower than for the year 2000, the date defined as current for the purposes of this analysis (Figure 10.1). Some southern forest species and communities may be highly vulnerable to the effects of changing climate, possibly resulting in conversion of woodland to savanna or grassland (Bosworth et al. 2008). In addition to the effects of temperature and precipitation on regeneration and growth, future forests will be affected by other factors contributing to climate change—such as carbon dioxide (CO_2) emissions resulting from economic and population growth—and to length of growing season, insect pollinators, plant demography, and other environmental influences not addressed here.

In this chapter "climate" is viewed as average annual temperature and precipitation, and "change" is the increase or decrease of those two variables from 2000 to 2060. In response to a changing climate in the Southern Region, the three objectives of this chapter are to: (1) assess vulnerability of major tree species; (2) evaluate risk to forest communities consisting of groups of species; and (3) briefly review silvicultural adaptation options available to managers and landowners. Our method used graphical displays of temperature and precipitation limits for individual tree species in the Southern Regions (Coastal Plain, Piedmont, Appalachian-Cumberland Highland, Mississippi

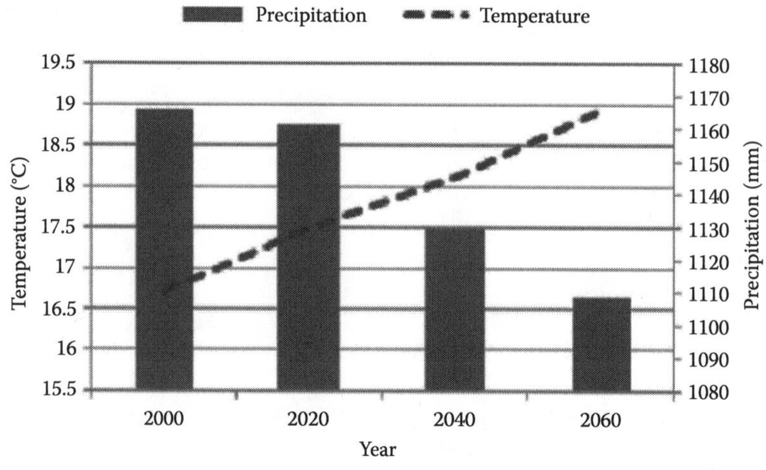

FIGURE 10.1 Average annual temperature and precipitation for the Southern Region from 2000 (current) and through 2060 (predicted).

Alluvial Valley, and Mid-South) (Burns and Honkala 1990), but we rely on spatial analysis of climatic data (Wear and Greis 2012).

This chapter, like others in this book, uses much of the framework described in Chapter 1 and outcomes of the Southern Forest Futures Project (SFFP, Wear and Greis 2012; Chapter 2). For example, climate models and production of CO_2 emissions based on economic projections are taken from the SFFP (Chapter 2). This chapter is organized around three major components:

- *Species vulnerability assessment*: Defined as the threat presented by future climate scenarios to the current distribution of selected upland tree species. Results are based almost entirely on future temperature and precipitation predictions generated by the SFFP (Chapter 2). Extensive graphic displays of current and possible future climates are presented for subdivisions of the major physiological provinces of the Southern Region.
- *Ecosystem risk analysis*: Defined as determining the probability that the climate scenarios will affect tree diversity (either negatively or positively) of ecosystems over time. Biodiversity will be quantified as species richness and defined as the number of tree species in an area of interest. This definition reduces the influence of individual species in ecosystems and considers only the total number reacting to climate change, which could include nonnative species. Biodiversity, or species richness, is largely a function of climate, soil, and disturbance, and has been used as an expression of ecosystem health (Kimmins 1997). Analysis of the effects of climate change on selected high-risk communities are presented in detailed case studies at the end of this chapter, which draw on either new research or available knowledge in the published literature combined with data from one or more climate scenarios.
- *Adaptation options*: Management techniques available to reduce vulnerability and risk for selected species and ecosystems. Mitigation is defined as the actions that address causes of climate change (namely, increasing CO_2 and other hydrocarbon gases such as methane), which is viewed primarily as the function of governments, achieved through policy and regulation, and therefore are beyond the scope of this chapter. This part of the vegetation chapter will reference and draw heavily on material presented in other chapters of the book.

Response of vegetation to climate change cannot be associated only with the stresses of changing temperature and precipitation. Other influences on vegetation include disturbances from insects, disease, and fire (Dale et al. 2001). Also, vegetation changes will affect wildlife habitat, water yields, recreational experiences, and demand for certain forest products; and will in turn be influenced by other environmental stresses and disturbances. The contents of this chapter, therefore, provide only a portion of the likely effects of climate change and should be considered with other chapters for a clearer evaluation of the overall future environment likely to be present in the ecologically diverse Southern Region.

Our approach to evaluating risk and vulnerability of vegetation to climate change will be based primarily on spatial analysis of climatic data across a range of scales. The broadest scale will be subregions of the Southern Region, which are large, ecologically similar areas that extend across several states. The smallest scale will be the individual counties within a state, which will be treated as sample units. A middle scale is sections, which are ecologically uniform areas of a dozen or more counties that nest within subregions.

This chapter can be used both as a stand-alone document and a source of information referenced by other chapters. As a stand-alone document, it presents the information that supports the assessment and analysis of climate change on vegetation and thereby facilitates presentation of information in a compact format without requiring the reader to consult other chapters. Vegetation information should be combined with that from other chapters, however, to provide a complete view of the effects of climate change on forests of the Southern Region that may occur over the next half century. The scope of information in this chapter is limited to trees because of their economic significance, their importance as a major component of wildlife habitat, and because more is known about the effects of climate on tree distribution than for shrubs and grasses. Also, this chapter should not be viewed as providing an exhaustive and detailed review of the literature on what is known about climate change on vegetation in the South; much of that type of information is provided in the SFFP (Wear and Greis 2012) and elsewhere in this book. Rather, this chapter uses a simple approach to assess vulnerability and risks to southern forests based on data from several climate scenarios that encompass a range of potential future environments for the Southern Region.

SOUTHERN FORESTS TODAY

Forests are the dominant vegetation on about 128 million ha in the 13 southern states, and comprise nearly 59% of a region where climate is particularly well suited for the occurrence of a large number of conifer and hardwood tree species (Conner and Hartsell 2002). The climate of much of the Southern Region is classified as humid subtropical, where a zone of hot, humid summers and mild-to-cool winters is controlled largely by maritime influences from the Atlantic Ocean and Gulf of Mexico (Bailey 1995). The average annual temperature throughout the Southern Region decreases from south to north and has a relatively narrow range, from 23.9°C in the Florida keys to 12.8°C in western Oklahoma. Annual precipitation averages about 127 cm throughout most of the region from Arkansas and Louisiana eastward, the exceptions being areas of mountainous topography, such as the escarpment of the Southern Appalachian Mountains where precipitation is higher. Annual precipitation decreases sharply from eastern Texas and Oklahoma westward, resulting in a near arid climate farthest west. In contrast, the Everglades of southern Florida represent a subtropical, hydric environment.

The Southern Region has diverse climates and vegetation. Except for drier parts of central and western Texas, most of the Southern Region consists of humid temperate climate dominated by forests of tall broadleaf deciduous hardwood and evergreen conifers (Bailey 1995), which are the focus of this chapter. Barbour and Billings (2000) list a number of hardwood and conifer forest types that have boundaries largely defined by climate.

COMMON TREE SPECIES

From eastern Texas and Oklahoma to the Atlantic Ocean, forests are the potential natural vegetation cover of about 80% of the Southern Region. Currently, however, only 40% of this region is classified as forest land, which ranges from 10% in Texas to 70% in Alabama (Smith et al. 2004). Deciduous forests of the Southern Region are dominated mainly by upland hardwoods, particularly the oak–hickory (*Quercus* spp–*Carya* spp) type (Table 10.1). Except for small areas of pinyon–juniper (*Pinus edulis–Juniperus* spp) woodlands in central and western Texas and high-elevation spruce–fir (*Picea rubens–Abies fraseri*) in the Southern Appalachians, the conifer forest types consist almost entirely of southern pines: shortleaf (*Pinus echinata*), longleaf (*Pinus palustris*), loblolly (*Pinus taeda*), and slash (*Pinus elliottii*). Hardwood forests, both upland and lowland, are dominated by one or more species of oak and hickory, with minor amounts of other species including some conifers, such as cypress (*Taxodium* spp) and tupelo (*Nyssa* spp). Uncommon tree communities are associated with small areas of topography and soils that combine to produce unusual environmental conditions; for example, the sand pine (*Pinus clausa*) forests on the deep, excessively drained soils of central Florida. The sand pine forest is one example of an ecosystem [others include cedar glades of central Tennessee (Barbour and Billings 2000) and granite outcrops of Georgia] within the generalized southern pine forest cover type that may be vulnerable to a changing climate but were not included in our assessment. Possible effects of climate change on spruce–fir and other selected tree-dominated ecosystems and some key understory components—such as wiregrass (*Aristida stricta*)—will be described in more detail in case studies. Overall, however, southern forests are largely characterized by high species diversity and high productivity as a result of favorable current

TABLE 10.1
Extent of Major Forest Types in the Southern United States

Forest Type[a]	Area (million ha)	Percent
Oak–hickory	32.9	37.8
Loblolly–shortleaf pine	21.0	24.2
Oak–gum–cypress	12.4	14.3
Oak–pine	12.2	14.0
Longleaf–slash pine	5.6	6.5
Elm–ash–cottonwood	1.1	1.3
Other[b]	0.7	0.8
Maple–beech–birch	0.4	0.5
White–red–jack pine	0.3	0.4
Nonstocked	0.2	0.2
Spruce–fir	<0.1	<0.1
Total	86.9	100.0

Source: Adapted from Smith, W.B. et al. 2004. Forest resources of the United States: 2002. Gen. Tech. Rep. NC-241. St. Paul, MN: U.S. Department of Agriculture, Forest Service, North Central Research Station. 137pp.

[a] Excludes 40,000 ha of nonforested lands, mostly in central and western Texas and Oklahoma.

[b] Includes pinyon–juniper woodlands in Texas and Oklahoma.

climatic conditions, but also due to fertility resulting from underlying geologic formations and various types of disturbance regimes. Ownership of forest land in the Southern Region is primarily by individuals (71%), followed by forest industry (17%), and government (12%) (reclassified from Smith et al. 2004).

Threats to Forest Health

As elsewhere in the United States, southern forests are facing threats from many sources, mainly from population growth, economic development, invasive species, diseases, and climate change (Wear and Greis 2012). Excluding environmental factors, the area covered by southern forests is projected to decline by 1.2 million ha between the years 1997 and 2050, largely in the vicinity of metropolitan areas in response to population growth and associated economic activities (Alig and Butler 2004). Climate change and environmental stresses can trigger interrelated health problems, such as nonnative plant invasions, increased wildfire risk, reduced prescribed burning, and insect and disease outbreaks—all of which directly result in rapid and large-scale changes in ecosystem function, structure, and productivity (Boisvenue and Running 2006).

The distribution of vegetation across a range of scales is strongly influenced by environmental factors. Climate clearly determines forest types at the upper scales, from continental to regional. At the lower scales of stand and site, moisture availability during the growing season affects species composition as well as productivity. Disturbance by fire is equally important for maintaining some vegetation types, such as the southern pines in general and longleaf pine in particular.

Temperature versus Precipitation Effects

The distribution of tree species across a range of geographic scales is limited primarily by temperature and precipitation and the integrated influences of relative humidity, potential evapotranspiration, and moisture deficits (Leininger 1998; Stephenson 1998; Whittaker 1975; Woodward 1987). Minimum annual average temperature typically limits the northern range of distribution of many species by affecting the growing degree days necessary for physiological processes, such as spring budburst, flowering, pollen production, and seed production (Karlsson 2002; Linkosalo et al. 2000). Recent evidence suggests species are rapidly expanding their northern ranges in response to warming climates (Chen et al. 2011; Woodall et al. 2009). In the Southern Region, the ranges of tree species are limited to the south by the Gulf of Mexico and to the west by availability of water during the growing season (Schmidtling 2001; Stephenson 1998). Favorable temperatures and availability of water during the growing season are critical to optimum tree growth (Leininger 1998). Detrimental physiological impacts can occur when the limits of temperature and precipitation are reached for a given tree species (Kozlowski et al. 1991). For instance, drought can result in disruption and blockage of water movement through xylem vessels (Sperry and Sullivan 1992). Water availability and temperature are probably the most important environmental factor affecting the geographic distribution and structure of vegetation (Chen et al. 2011; Woodall et al. 2009; Woodward 1987). Trees have lower limits of annual precipitation that determine their regional distribution, and these limits have been reported for many important U.S. species (Burnes and Honkala 1990).

Stress to trees under a changing climate can produce changes in survival and distribution. For instance, in work reviewed by Hansen et al. (2001) climate change resulted in a 32% reduction in loblolly-shortleaf pine habitat in the United States. In that same study it was predicted that shortleaf pine distribution would shift north and west while being replaced in its current zone by oak–pine habitat. Also, it was predicted that the area of oak–hickory forests would increase by 34% in the Eastern United States.

Drought during the frost-free season, whether induced by climate change or other mechanisms, is considered a major contributor to forest decline (Leininger 1998; Manion 1981). For example, a drought- induced oak decline event in Arkansas and Missouri from 2000 to 2005 affected up to 120,000 ha in the Ozark National Forest of Arkansas alone (Starkey et al. 2004). VanMantgem et al. (2009) attributed widespread increase of tree mortality in the Western United States to water deficit and regional warming. Many other recent drought-related tree mortality occurrences have taken place across the globe (Allen et al. 2010).

Other Associated Studies

Some studies have used a habitat approach to assessing the effects of climate change on vegetation, where altitude, aspect, soil characteristics, and other factors are considered in addition to climate. Schwartz et al. (2001) used habitat variables to predict the future distribution of tree species in Ohio under differing future climate scenarios. In an extensive study using multiple environmental variables that affect the range of many eastern tree species, Iverson et al. (2008) found that annual or seasonal precipitation and temperature are important factors, in combination with habitat variables (e.g., elevation, soil) associated with the current range of 134 tree species in the Eastern United States. Similar approaches to assess climate change effects on trees based on habitat modeling have been done in southern Africa (O'Brien 1988, 1993). Prediction equations with both habitat and climate can be used to approximate the future ranges of species, but the importance of climatic variables cannot be separated from the effects of soil and topography, or their interaction.

METHODS, MODELS, AND DATA SOURCES

As described in other chapters of this book and in the SFFP (Wear and Greis 2012), changes in the species composition of forests in the Southern Region are influenced primarily by components of climate, such as temperature, precipitation, and potential evapotranspiration. Other important factors include insects and disease, and subtle influences such as presence of insect pollinators and soil microfauna. Climate, however, is the principal factor addressed in this chapter.

Climate Predictions

We used three downscaled general circulation models described by the SFFP and summarized in detail in Chapter 2 (Wear and Greis 2012):

- *MIROC3.2*: Model for Interdisciplinary Research on Climate developed by the Center for Climate System Research at the University of Tokyo; includes essential components of ocean temperature models and dynamic thermodynamic sea-ice sheet models.
- *CSIROMK2 and MK3.5*: Models developed by the Commonwealth Scientific and Industrial Research Organization, the Australian national science agency.
- *HadCM3*: Model developed by the Hadley Centre for Climate Change, an organization established in England in 1990 to coordinate all climate change research.

Although more than 20 general circulation models are currently being evaluated as appropriate for predicting future global meteorological conditions (Jun et al. 2008), the rationale for selecting these three is presented by Wear and Greis (2012).

Two future socioeconomic scenarios affecting CO_2 emissions were used (Intergovernmental Panel on Climate Change 2007):

- *A1B*: representing low population/high economic growth and high-energy use.
- *B2*: representing moderate growth and low-energy use.

Land-use change is an important driver of future climates, particularly use resulting from food production, and is implicitly considered in the two storylines (Wear and Greis 2012).

The three general circulation models and two emissions scenarios were grouped into four outcomes (Table 10.2) used in this chapter to assess effects of climate change on vegetation in the Southern Region. The combinations of the three general circulation models and two economic possibilities for the year 2060 were selected to encompass the range of likely outcomes of climate change. For the 13 southern states, the four scenarios are variable in their predictions for future temperature and precipitation, but in general suggest a somewhat warmer (Figure 10.2) and slightly dryer climate (Figure 10.3). The four scenarios suggest greater variability in temperature than precipitation for the entire Southern Region. Evaluation of climate change based on the Southern Region as a whole is too broad a scale as evidenced by the existing variation of temperature and precipitation. Smaller ecological units are required to more fully evaluate the effects of climate change.

TABLE 10.2
The Principal Cornerstone Future Forecasts of General Climate Models Utilized by Southern Forest Futures Project[a]

	Population and Income Growth Rate[b]	
Timber Prices	**High**	**Low**
High	Cornerstone A (MIROC + A1B)	Cornerstone C (CSIROMK2 + B2)
Low	Cornestone B (CSIROMK3.5 + A1B)	Cornerstone D (HadCM3 + B2)

[a] Not shown are two cornerstones that are based on rates of replanting harvested lands.
[b] The three general circulation models (MIROC, CSIRO, Hadley) used in this subchapter are representative of many developed to predict possible changes in temperature and precipitation over time (Jun et al. 2008). Iverson et al. (2008), for example, based their species range projections for the Hadley and two other GCMs (Jun et al. 2008).

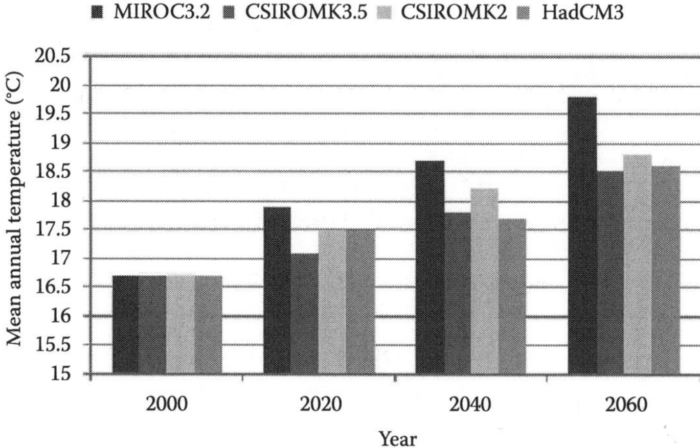

FIGURE 10.2 Average annual temperature 2000 (current) through 2060 (predicted) for four climate scenarios in the Southern United States (Wear et al. in press); scenarios were developed from three general circulation models (MIROC3.2, CSIROMK2 and 3.5, and HadCM3) and two emissions storylines (A1B represents low population growth/high-energy use and B2 represents moderate population growth/low-energy use).

Climate change in the Southern Region was assessed in greater detail by downscaling to five ecological subregions (Figure 10.4). The largest subregion is the Mid-South, which includes much of Texas, Oklahoma, and Arkansas and accounts for about a third of the region, followed by the Coastal Plain and Piedmont, each of which occupies about a quarter of the region. The subregions, each of which represents a multi-state area of relatively uniform climate and vegetation, are based on ecoregionalization work by Cleland et al. (2007). For example, the Appalachian-Cumberland Highland is generally dominated by hardwoods, whereas the forests of the Coastal Plain are largely pines. The strongest connection of subregions with vegetation and climate change is in the Mid-South, where diminishing annual precipitation results in transition from closed canopy pine forests currently found in Arkansas and eastern Texas, to sparse oak woodlands and shrublands of central Texas and Oklahoma, and to grasslands farther west (Allred and Mitchell 1955).

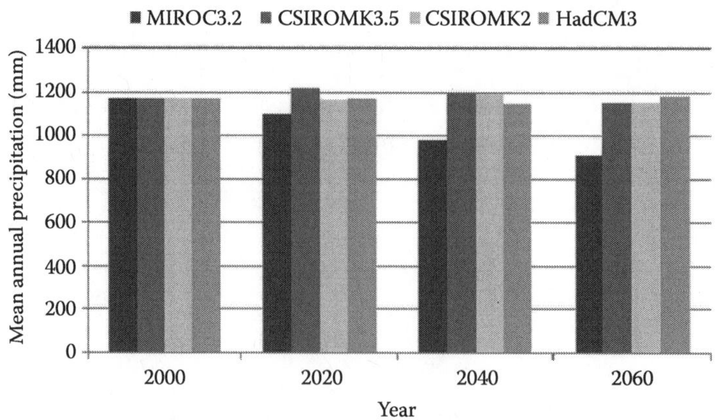

FIGURE 10.3 Average annual precipitation 2000 (current) through 2060 (predicted) for four climate scenarios for the Southern United States (Wear et al. in press); scenarios were developed from three general circulation models (MIROC3.2, CSIROMK2 and 3.5, and HadCM3) and two emissions storylines (A1B represents low population growth/high-energy use and B2 represents moderate population growth/low-energy use).

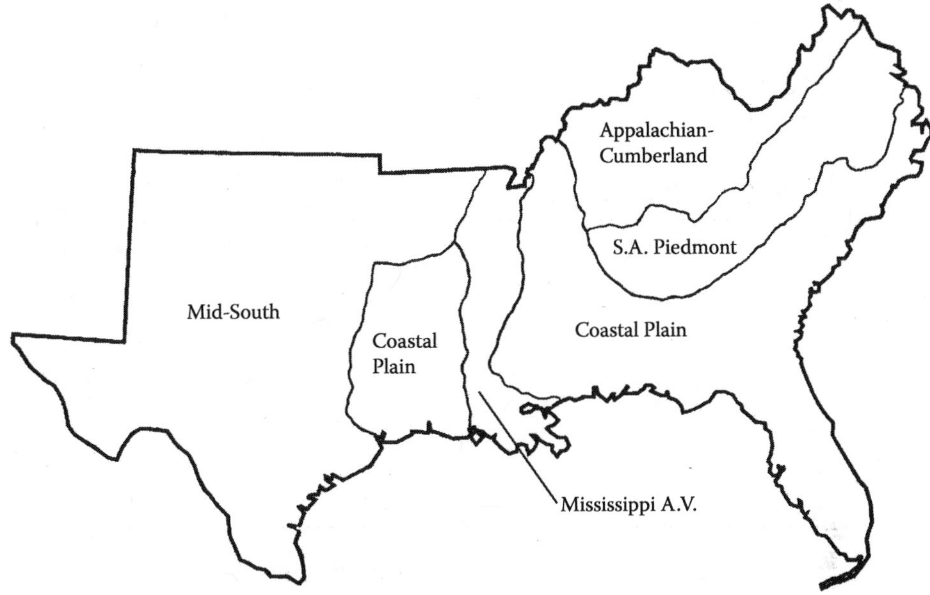

FIGURE 10.4 Subregions of the Southern United States delineated by the Southern Forest Futures Project. (Wear and Greis 2012) that represent areas of ecological similarity.

Temperature and precipitation predictions: Trends of climate change, manifested by variation of annual temperature and precipitation, are expected to be relatively uniform within subregions of the Southern Region when predictions of the four climate scenarios are averaged (Figures 10.5 and 10.6). From 2000 to 2060, temperature is predicted to increase by about 3°C and precipitation will decrease by an annual average of 100 mm. The most change in average annual precipitation is predicted to occur in the Mississippi Alluvial Valley and parts of the Mid-South; least change is expected in the Piedmont and Appalachian-Cumberland Highland.

Variations among models: Predictions by the individual scenarios suggest the potential for greater variability of temperature (Figure 10.7) and precipitation (Figure 10.8) across the Southern Region. The MIROC3.2 A1B scenario, for example, predicts the largest increase of temperature in the Mississippi Alluvial Valley by 2060 and smaller increases in the Coastal Plain and Mid-South.

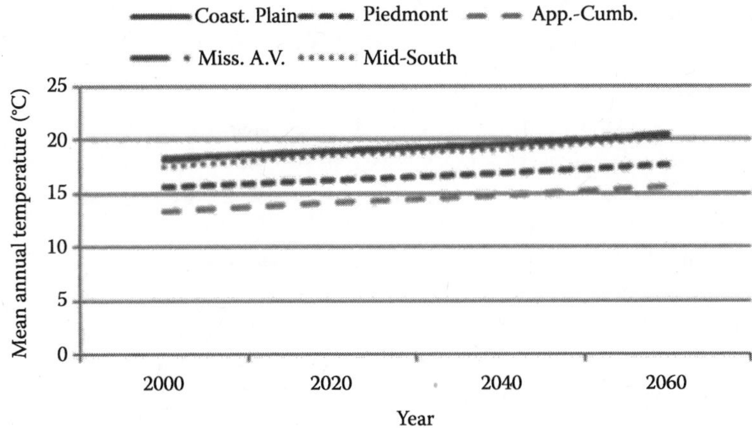

FIGURE 10.5 Average annual temperature from 2000 (current) to 2060 (predicted) for the five subregions of the Southern United States.

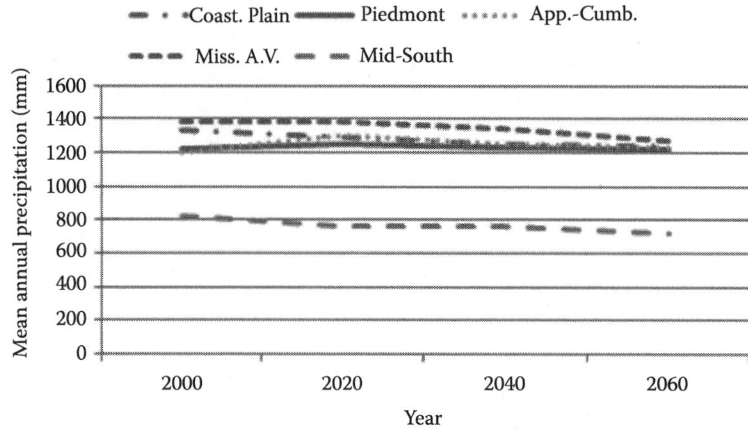

FIGURE 10.6 Average annual precipitation from 2000 (current) to 2060 (predicted) for the five subregions in the Southern United States.

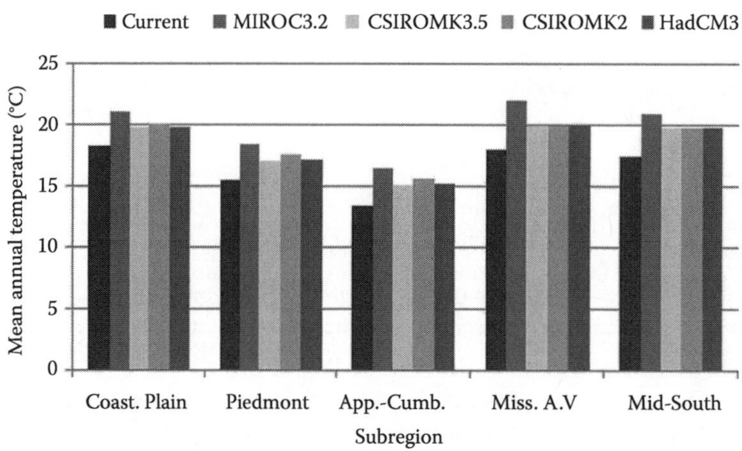

FIGURE 10.7 Average annual temperature currently (2000) and predicted through 2060 by the four climate scenarios for the five subregions of the Southern United States (Wear et al. in press); scenarios were developed from three general circulation models (MIROC3.2, CSIROMK2 and 3.5, and HadCM3) and two emissions storylines (A1B represents low population growth/high energy use and B2 represents moderate population growth/low energy use).

Future precipitation predicted by the MIROC3.2 A1B scenario for these same three subregions, however, indicates similar decreases for the Mississippi Alluvial Valley and the Coastal Plain, with a smaller decrease for the Mid-South. Climate change forecasts by the other scenarios show similar patterns of variability among subregions, but with smaller differences in magnitude. Consistency of predictions among the four scenarios is problematic when used to assess the effects of climate changes with some degree of confidence because both the moisture and temperature budgets are important (and sometimes counteracting) influences on the composition and productivity of forests (Whittaker 1975; Woodward 1987).

Tree Species Distribution Effects

Predicting the effects of future temperature and precipitation on forest vegetation is only an approximation; models are lacking that directly associate the magnitude of environmental variables with

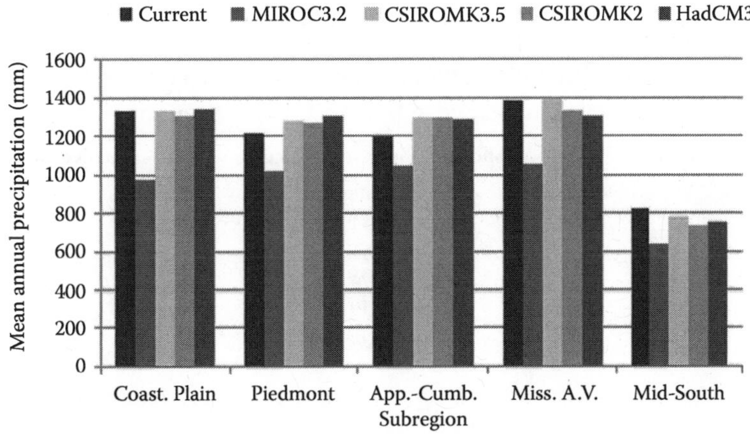

FIGURE 10.8 Average annual precipitation currently (2000) and predicted through 2060 by the four climate scenarios for the five subregions of the Southern United States (Wear et al. in press); scenarios were developed from three general circulation models (MIROC3.2, CSIROMK2 and 3.5, and HadCM3) and two emissions storylines (A1B represents low population growth/high energy use and B2 represents moderate population growth/low energy use).

the reproduction and growth of plants. Although considerable knowledge is available on the species and community relationships with associated gradients of temperature and moisture at the stand level (Whittaker 1975) and for selected commercial species such loblolly pine (Nedlo et al. 2009), relatively few prediction models provide information that is relevant to most of the common forest tree species and that can be applied uniformly throughout the Southern Region. Simple correlation models that are available to evaluate the effects of climate change on vegetation are likely the most appropriate given the lack of agreement of future conditions forecasted by the scenarios (Agren et al. 1991; Hijmans and Graham 2006).

We evaluated vulnerability based on the range of distribution of tree species commonly occurring in each subregion of the Southern Region. Projecting the effects of future climate on plant distribution was based on the premise proposed by Woodward (1987) that the annual water budget (precipitation minus evapotranspiration) is a critical environmental climatic variable. We reduced complexity of the water budget approach by limiting our analysis to temperature and precipitation only; evapotranspiration was excluded because estimates of soil moisture storage and losses would be required, which was beyond the scope of our study. The northern limit of shortleaf pine, for example, is associated with a minimum annual temperature of about 10°C; its western range coincides with minimum annual precipitation of about 1020 mm (Burns and Honkala 1990). In a similar manner, they reported the minimum temperature and precipitation requirements for many common forest species in the Southern Region. We used temperature and precipitation recorded for 2000 and predicted for 2060 by each climate scenario to compare the current and potential future range of common tree species using the 1342 counties of the 13 southern states as unbiased sample units.

Our method of analysis is illustrated by examining the current occurrence of shortleaf pine in the Southern Region and its predicted occurrence for the four climate scenarios (Table 10.3). Shortleaf pine currently (2000) occurs in 1084, or 81%, of the 1342 counties in the 13 southern states. Under the four climate scenarios, the potential suitable environment for shortleaf pine ranges from 451 to 1070 counties. The most severe change could occur with the MIROC3.2 A1B scenario in the western part of the region where precipitation is predicted to be reduced, thereby increasing stress and likely resulting in a reduction in the range of shortleaf pine. Temperature, however, is forecasted to increase, which would allow a potential expansion of the northern limits of the species. Considering both climate factors simultaneously, however, suggests unlikely northern expansion of

TABLE 10.3
Current (2000) and Predicted (2060) Range of Shortleaf Pine in the Southern United States in Response to Four Climate Scenarios[a]

Climate Scenario	Shortleaf Pine Occurrence	Relative Occurrence
	Counties[b]	%
Current (2000)	1084	81
MIROC3.2 (2060)	451	34
CSIROMK3.5 (2060)	1070	80
CSIROMK2 (2060)	1042	78
HadCM3 (2060)	1045	78

Source: Adapted from Wear, D.N., R. Huggett, and J.G. Greis. In press. Constructing alternative futures. *Southern Forest Futures Project Technical Report.*

[a] Scenarios were developed from three general circulation models (MIROC3.2, CSIROMK2 and 3.5, and HadCM3) and two emissions storylines (A1B represents low population growth/high-energy use and B2 represents moderate population growth/low-energy use).

[b] The 13 southern states are subdivided into 1342 counties.

its range with favorable temperature if precipitation is limited. The predicted effect on the range of shortleaf pine for the other three scenarios would be similar but much less severe, particularly for the CSIROMK3.5 A1B scenario.

Scope of assessment: A more complete assessment of potential climate change on forests of the Southern Region would need to include other tree species in the temperature and precipitation analysis, in addition to shortleaf pine. Burns and Honkala (1990) present information on the range of 63 native southern tree species that are associated with annual temperature and precipitation (Table 10.4), as well as several introduced species, such as tree of heaven (*Ailanthus altissima*). Some species—such as sourwood (*Oxydendrum arboreum*), downy serviceberry (*Amelanchier arborea*), and Chinese tallow tree (*Sapium sebiferum*)—are excluded from our analysis because generally recognized climatic limits associated with their ranges are not available.

Our method also excluded subtropical red mangrove (*Rhizophora mangle*) and other coastal species associated with normally hydric to subhydric habitats as well as species and communities of freshwater estuaries and floodplains of major river systems, such as bottomland hardwoods, bald cypress (*Taxodium distichum*), and tupelo swamps. These species and communities are expected to adjust their local distribution primarily in relation to changing sea levels and tidal influences and not resulting from changes in temperature and precipitation (see Chapter 13, Wear and Greis 2012).

Our vulnerability assessment, which uses a simplified measure of species' presence, was based on the assumption that a species will be present if the minimum precipitation threshold is met. As recognized earlier, this simplistic approach to assessing vulnerability of species to climate change has shortcomings, but appears appropriate considering the scope and information available for this assessment.

We recognize that other environmental components also influence the range of tree species, such as geologic parent material, disturbance, and soils (Fletcher and McDermott 1957), but temperature and moisture have the most important effect on distribution in the northern and western limits of tree ranges (Schmidtling 2001). Although we included both temperature and precipitation in our evaluations, we placed greater emphasis on the latter variable because its effects would likely be most apparent on survival, growth, and regeneration during the relatively short period of 60 years of our assessment. An unaccounted for source of variation in our simplified model of tree distribution is the effect of increased temperature on transpiration and evaporation (i.e., evapotranspiration) on availability of soil moisture, which could be more important than precipitation (McNulty et al.

TABLE 10.4
Average Annual Temperature (°C) and Precipitation (mm) Associated with the Northern and Western Range Limits of Tree Species Occurring Throughout Each Subregion[a]

			CP	P	AC	MAV	MS			
Species[b]	**Annual Temperature (°C)[c]**	**Annual Precipitation (mm)[d]**	**All Sections**	**All Sections**	**All Sections**	**All Sections**	**Ozark-Ouachita**	**Cross Timbers**	**High Plains**	**West Texas**
Gray oak	13	250								*
Lacey oak	15	250							*	
Mexican pinyon	4	250								*
Oneseed juniper	11	250							*	*
Plains cottonwood	0	250							*	
Tree of heaven[e]	0	360	*	*	*	*				
Eastern redcedar	4	380	*	*	*		*	*		
American elm	0	380	*	*	*	*				
Eastern cottonwood	0	380	*	*		*				
Bur oak	0	380					*	*		
Boxelder	0	380		*	*	*				
Green ash	0	380				*				
Blackjack oak	10	420	*	*	*		*	*		
Black willow	4	460	*	*	*	*				
Cedar elm	14	460						*		
Chinkapin oak	4	500		*	*		*	*		
Sugarberry	12	510	*	*		*				
Sugar maple	0	510			*		*			
Eastern redbud	7	510						*		
Silver maple	4	510		*	*	*				
Honeylocust	7	510		*	*	*				
Slippery elm	4	540		*	*	*				
American basswood	0	540			*		*			
Post oak	10	560	*	*	*		*	*		
Red maple	0	640	*	*	*	*	*			
Bitternut hickory	4	640	*		*		*	*		
Black walnut	7	640		*	*		*			
Black oak	7	760	*	*	*		*	*		
Flowering dogwood	7	760	*	*	*		*			
Sassafras	7	760	*	*	*	*	*			
White oak	4	760	*	*	*		*			
Pignut hickory	7	760	*	*	*		*			
American sycamore	7	760	*	*	*	*				
Yellow poplar	7	760	*	*	*					
Persimmon	10	760	*	*	*	*				
White ash	4	760		*	*		*			
American beech	4	760					*			
Northern red oak	4	760			*		*			
Shagbark hickory	4	760			*		*			
Scarlet oak	10	760		*	*					
Chestnut oak	10	810			*					
Mockernut hickory	10	880	*	*	*		*			
Cucumber tree	7	880			*		*			

continued

TABLE 10.4 (continued)
Average Annual Temperature (°C) and Precipitation (mm) Associated with the Northern and Western Range Limits of Tree Species Occurring Throughout Each Subregion[a]

			CP	P	AC	MAV	MS			
Species[b]	Annual Temperature (°C)[c]	Annual Precipitation (mm)[d]	All Sections	All Sections	All Sections	All Sections	Ozark-Ouachita	Cross Timbers	High Plains	West Texas
Virginia pine	10	880		*	*					
Black cherry	4	960	*	*	*		*			
Red mulberry	7	1010	*	*	*	*				
Winged elm	13	1010		*	*		*	*		
Sweetgum	10	1020	*	*	*	*	*			
Southern red oak	12	1020	*	*	*	*	*			
Loblolly pine	13	1020	*	*			*			
American holly	13	1020	*	*	*	*	*			
Water hickory	15	1020	*			*				
Black gum	7	1020	*	*	*	*		*		
Paulownia[e]	7	1020	*		*	*				
Red bay	16	1020	*							
Umbrella magnolia	12	1020					*			
Shortleaf pine	10	1020		*	*		*			
Black locust	10	1020			*					
Longleaf pine	16	1090	*							
Shumard's oak	10	1140	*	*	*			*		
Overcup oak	13	1140	*	*		*				
Sweet birch	7	1140			*					
Sweetbay	16	1220	*							
Laurel oak	16	1250	*							
Slash pine	17	1270	*							
Southern magnolia	18	1270	*							
Water oak	16	1270	*	*		*				

[a] Coastal plain (CP), Piedmont (P), Appalachian-Cumberland Highland (AC), Mississippi Alluvial Valley (MAV)—and sections within the Mid-South subregion (MS) of the Southern United States.

[b] Asterisk indicates that species within subregions and sections were evaluated for effects of climate change.

[c] Approximated for some species by comparing maps of their distribution with average annual temperature isolines.

[d] Species regional precipitation limit values are mostly from Burns and Honkala (1990) except for some species, such as blackjack oak and black gum, for which the precipitation limits were taken from the Natural Resource Conservation Service plants database at plants.usda.gov/characteristics.html.

[e] Nonnative species.

1998). Other limitations include competition by other species, soil fertility, mycorrhizal relationships, and interactions with the wildlife necessary for seed dispersal (Schwartz 1993). Confers and many hardwood species are wind pollinated but for some species, such as American basswood (*Tilia americana* var. *heterophylla*) and yellow poplar (*Liriodendron tulipifera*), climate change could affect occurrence of the necessary insect pollinators.

Despite our simple approach, we believe our methods are appropriate for several reasons: (1) lack of consensus on the accuracy of the GCMs for predicting future environmental conditions (especially for precipitation at finer spatial scales), (2) lack of available information on physiological response of forest species to variation of temperature and precipitation, and (3) our goal of making a broad scale assessment across a large region of how various tree species might be affected only by a changing climate. Although the four scenarios generally agree in forecasting warmer temperatures, which would allow for potential expansions of species ranges, there were considerable differences for predicted precipitation in some subregions, which would result in altered ranges for many species. An example of using regression to model the effects of temperature, precipitation, and potential evapotranspiration on the current and future ranges of a tree species is presented in a case study in this chapter.

Smaller, more detailed southern ecological units (subregions and sections) will be used to identify specific areas of potentially highest vulnerability and risk to trees that could result from a changing climate (Figure 10.9). With the exception of one subregion (Mid-South), a single group of endemic tree species occurring throughout was used for assessment of vulnerability to climate change within all of the included sections. In the large and climatically variable Mid-South subregion, where temperature and precipitation decrease markedly with increasing northerly latitude and westerly longitude, we used a separate group of species for each section to better account for the effects of climate change.

Other data sources: Our assessment also included additional plant communities or species for which we used different analyses. We included several uncommon or rare plant communities because these species are more vulnerable and inhabit the southernmost edge of their ranges or are isolated; an example is the island-like spruce–fir communities that are limited to a few peaks of the highest altitudes of the Southern Appalachians. Another example is *Eucalyptus grandis*, an introduced species from temperate to subtropical areas of Australia, which is being grown commercially for a range of products.

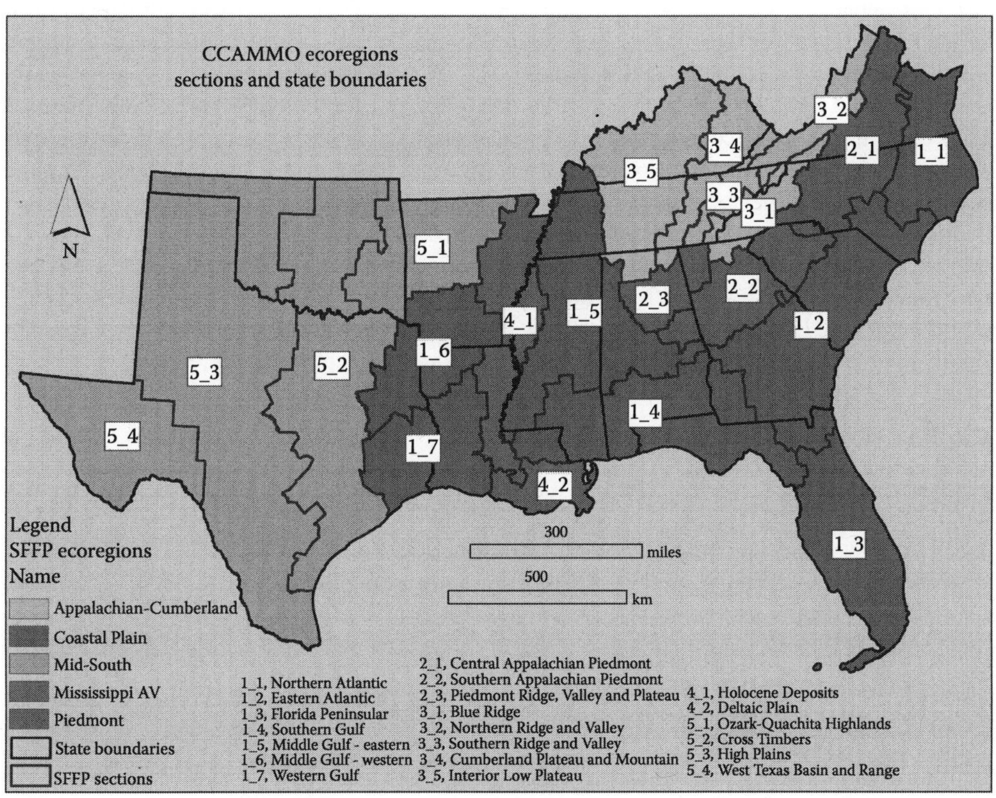

FIGURE 10.9 U.S. subregions (Cleland et al. 2007) and sections (Wear and Greis 2012) developed for the Southern Forest Futures Project (SFFP).

Our assessment of vulnerability of vegetation to climate change is based entirely on prediction of future temperature and precipitation by the four climate scenarios. Except for the study of climate on the regional productivity and distribution of an occasional commercial species (McNulty et al. 1998) and the habitat approach of Iverson et al. (2008), we found no references pertaining to both major and minor tree species throughout the Southern Region that included variables (temperature, precipitation) for use in assessment of the effects of climate. Using the four scenario predictions of future climate with simple spreadsheet-type models based on temperature and precipitation limits allowed us to analyze the Southern Region as a whole rather than drawing inferences from a patchwork of unrelated small studies with unknown ranges of applicability. Finally, our systematic, quantitative process of vulnerability assessment on individual species provided data for statistical analysis of risk to forest composition.

Forest Diversity Effects

Assessment of climate change on individual tree species may be beneficial for some forestry activities such as future conditions that might affect stand establishment and growth, environmental stress, insect problems, and other pine plantation management concerns. However, a more complete assessment of climate change on forests can be obtained by examining the group of species predicted to be present on a particular landscape. To accomplish this, we determined species richness for each climate scenario. Richness, defined as the number of species occurring per sample unit (county), is a measure of biological diversity represented by overlapping ranges of species. The value ranges from zero to the number of tree species that was used in the precipitation analysis. Richness included both native and nonnative species.

Species richness was analyzed to provide a basis for comparison of the four climate scenarios. The hypothesis tested was that predictions of future climate by each climate scenario result in no real difference in tree diversity between 2000 and 2060. We used the nonparametric paired Wilcoxon signed-rank test to account for likely violation of normality (Zar 1996). We selected a systematic sample of 50% of the total counties from the population within a subregion or section and applied a finite population correction (Zar 1996). Tests of significant differences were made at the $p < 0.05$ level.

Our assessment of vulnerability proceeds in geographic sequence of decreasing current temperature and precipitation, beginning with the humid eastern Coastal Plain, westward through the Piedmont, Appalachian-Cumberland Highland, Mississippi Alluvial Valley, and finally the xeric environment of the Mid-South of Texas and Oklahoma. Vulnerability analysis for two sections of the Mid-South, the High Plains and West Texas Basin and Range, was more limited than for subregions in other sections because only a few tree species are present throughout.

COASTAL PLAIN

The Coastal Plain occupies about 37% (801,081 km^2) of the southern region. Conifers are the characteristic forest type of the subregion with upland forests dominated by one or more species of southern yellow pines, particularly loblolly and formerly longleaf. Extensive areas of hardwoods are present along the floodplains of the major rivers. This subregion has been subdivided into seven sections: Northern Atlantic (9.1%), Eastern Atlantic (20.4%), Florida Peninsular (13.0%), Southern Gulf (18.8%), Middle Gulf-East (18.3%), Middle Gulf-West (10.6%), and Western Gulf (9.7%).

Climate Scenario Predictions

Temperatures in the Coastal Plain are predicted to increase from 2000 to 2060 based on an average of forecasts across the four climate scenarios (Figure 10.10). The rate of increase is approximately 0.5°C per decade for each of the seven sections. Annual temperature is highest in the Florida Peninsular section and lowest for the North Atlantic section. With the exception of the Middle

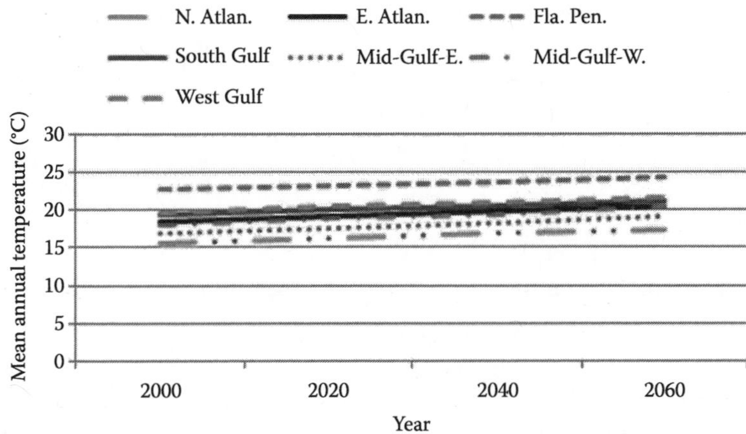

FIGURE 10.10 Average annual temperature from 2000 (current) to 2060 (predicted) for sections of the southern Coastal Plain; bidecadal temperatures for each section are the average of predictions from four climate scenarios (Wear et al. in press); that were developed from three general circulation models (MIROC3.2, CSIROMK2 and 3.5, and HadCM3) and two emissions storylines (A1B represents low population growth/high-energy use and B2 represents moderate population growth/low-energy use).

Gulf-East, the sections are quite similar. For precipitation, the average of predictions from the climate scenarios indicates a long-term trend of decreasing amounts (Figure 10.11); again, the trend for the Middle Gulf-East section differs from the other sections—in this instance, an increase from 2000 to 2020, then a decrease to the 2000 level. An average of the four climate scenarios suggests a total temperature increase of about 3°C and a precipitation decrease of about 100 mm across all sections, although amounts vary for each climate scenario.

Three of the four scenarios forecast only slight changes in average annual precipitation from 2000 to 2060 for most sections of the Coastal Plain; only the MIROC3.2 A1B scenario predicts a large decrease across all sections. Future precipitation predictions by the other scenarios are

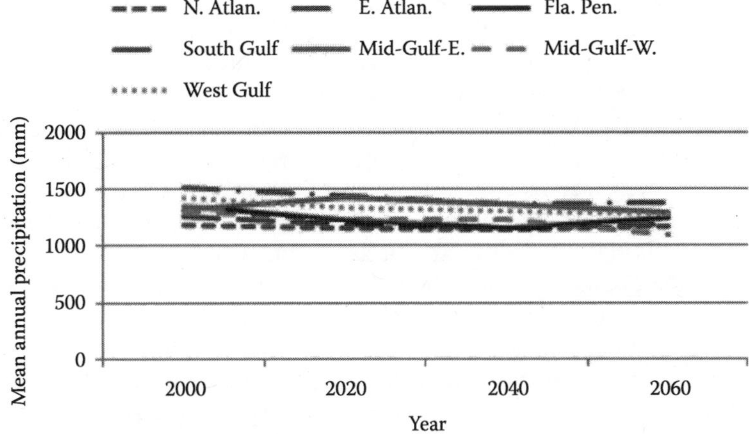

FIGURE 10.11 Average annual precipitation from 2000 (current) to 2060 (predicted) for sections of the southern Coastal Plain; bidecadal precipitation for each section represents the average of four climate scenarios (Wear et al. in press); that were developed from three general circulation models (MIROC3.2, CSIROMK2 and 3.5, and HadCM3) and two emissions storylines (A1B represents low population growth/high-energy use and B2 represents moderate population growth/low-energy use).

variable among sections, ranging from similar amounts for the Middle Gulf-East section to somewhat variable amounts for the Western Gulf section. Comparison of predictions indicates the largest increase in precipitation for the CSIROMK3.5 A1B scenario and largest decrease in precipitation for the MIROC3.2 A1B scenario. The CSIROMK3.5 A1B scenario predicts a slight precipitation increase, by about 50 mm, in the Florida Peninsular and Middle Gulf-East and a decrease in the Southern Gulf and Middle Gulf-West sections. Overall, the climate of the Coastal Plain is predicted to become warmer and dryer by the year 2060.

Effects on Tree Species Distribution in Coastal Plain Sections

Thirty-seven tree species were used for assessment of the effects of precipitation on forests in the Coastal Plain (Table 10.4). Minimum limits for temperature and precipitationrequirements for native species ranged from 0°C and 380 mm for American elm (*Ulmus americana*) to 18°C and 1270 mm for southern magnolia (*Magnolia grandiflora*). Temperature and precipitation predicted by three of the climate scenarios is expected to be greater than the minimum requirements for all species. For the fourth, MIROC3.2 A1B, precipitation is expected to be slightly less than adequate for 10 species and much less for eight. Tree species with annual precipitation requirements greater than 950 mm, such as slash pine and water oak (*Q. nigra*), are predicted to decrease in areas of occurrence.

Effects of climate change on the occurrence of selected tree species in the Coastal Plain are likely to be highest in the Florida Peninsular section. In Lee and Lakeland Counties, for example, precipitation is predicted to be 400 mm less than the current level of 1700 mm. Temperature for those counties is predicted to decrease by an average of 1°C, which, when considered with reduced precipitation, will possibly affect several of the tree species that have the largest annual water requirements, such as sweetbay and water oak.

Northern Atlantic: Vulnerability of tree species to climate change in the Northern Atlantic section of the Coastal Plain is predicted to be highest under the MIROC3.2 A1B scenario, which would result in an average 23% decrease in the range of the 37 tree species (Table 10.5). An average 47% decline is predicted for 18 species, with the greatest decline (80%) for longleaf pine under the MIROC3.2 A1B scenario. An increase in range of eight species is predicted for two scenarios (CSIROMK2 B2 and HadCM3 B2). Little or no change of area of occurrence is predicted for 19 species. Overall, effects of climate change in the Northern Atlantic section of the Coastal Plain are predicted to be minimal because of minimal changes in annual temperature and precipitation.

Eastern Atlantic: Vulnerability of selected tree species to climate change in the Eastern Atlantic section of the Coastal Plain is predicted to be highest under the MIROC3.2 A1B scenario, which would result in an average 42% decrease in the range of the 37 tree species (Table 10.6). An average 81% reduction in area of occurrence is predicted for 19 species, including red mulberry (*Morus rubra*), southern red oak (*Q. falcata*), and loblolly pine, under the MIROC3.2 A1B scenario. An increase in range of seven species is predicted by the HadCM3 B2 scenario. Little or no change of area of occurrence is predicted for 18 species, including post oak (*Q. stellata*), red maple (*Acer rubrum*), and white oak (*Q. alba*). Overall, effects of climate change on species in the Eastern Atlantic section of the Coastal Plain are predicted to be moderate because of small-to-moderate changes in annual temperature and precipitation.

Florida Peninsular: Vulnerability of tree species to climate change in the Florida Peninsular section of the Coastal Plain is predicted to be highest under the MIROC3.2 A1B scenario, which would result in an average 66% decrease in the range of the 37 tree species (Table 10.7). An average 90% reduction in area of occurrence is predicted for 17 species, including sweetgum (*Liquidambar styraciflua*), loblolly pine, and water hickory (*Carya aquatica*), under the MIROC3.2 A1B scenario. A small increase in range of four species (laurel oak, slash pine, southern magnolia, and water oak) is predicted by the CSIROMK3.5 A1B scenario based on changes of annual temperature and precipitation. Little or no change of area of occurrence is predicted for 10 species including American elm, blackjack oak (*Q. marilandica*), and red maple. Overall, effects of climate change

TABLE 10.5
Predicted Change (%) in Area of Distribution of Selected Tree Species in the North Atlantic Section of the Coastal Plains from 2000 to 2060 in Response to Four Climate Scenarios[a]

	Species Area Change (%)			
Species[b]	**MIROC3.2**	**CSIROMK3.5**	**CSIROMK2**	**HadCM3**
Black cherry	−7.1	0.0	0.0	0.0
Red mulberry	−55.7	0.0	0.0	0.0
Sweetgum	−61.4	−1.4	0.0	0.0
Southern red oak	−61.4	−1.4	0.0	0.0
Loblolly pine	−61.4	−1.4	0.0	0.0
American holly	−61.4	−1.4	0.0	0.0
Water hickory	−61.4	−1.4	0.0	0.0
Black tupelo	−61.4	−1.4	0.0	0.0
Paulownia	−61.4	−1.4	0.0	0.0
Red bay	−61.4	−1.4	0.0	0.0
Longleaf pine	−80.0	4.3	7.1	8.6
Shumard's oak	−65.7	−14.3	22.9	25.7
Overcup oak	−65.7	−14.3	22.9	25.7
Sweetbay	−31.4	−11.4	14.3	21.4
Laurel oak	−15.7	0.0	24.3	25.7
Slash pine	−11.4	0.0	18.6	28.6
Southern magnolia	−11.4	0.0	18.6	28.6
Water oak	−11.4	0.0	18.6	28.6
Average[c]	−22.9	−1.3	4.0	5.2

Source: Adapted from Wear, D.N., R. Huggett, and J.G. Greis. In press. Constructing alternative futures. *Southern Forest Futures Project Technical Report.*

[a] Scenarios were developed from three general circulation models (MIROC3.2, CSIROMK2 and 3.5, and HadCM3) and two emissions storylines (A1B represents low population growth/high-energy use and B2 represents moderate population growth/low-energy use).

[b] Species with no change in each of the four scenarios are not listed.

[c] Species with no change are included.

in the Florida Peninsular section are predicted to be severe because of increased temperature and reduction of annual precipitation.

Southern Gulf: Vulnerability of tree species to climate change in the Southern Gulf section of the Coastal Plain is predicted to be highest under the MIROC3.2 A1B scenario, which would result in an average 31% decrease in the range of the 37 tree species (Table 10.8). An average 58% reduction in area of occurrence is predicted for 19 species, with five species, including sweetbay, slash pine, and southern magnolia (*Magnolia grandiflora*) being extirpated under the MIROC3.2 A1B scenario. None of the scenarios predicts an increase in range of any species. Little or no change of area of occurrence is predicted for 18 species including eastern redcedar (*Juniperus virginiana*), sugarberry (*Celtis laevigata*), and black oak (*Q. velutina*). Overall, effects of climate change on tree species in the Southern Gulf section are predicted to be moderate because of small-to-medium reductions in annual temperature and precipitation.

Middle Gulf-East: Vulnerability of tree species to climate change in the Middle Gulf-East section of the Coastal Plain is predicted to be highest under the MIROC3.2 A1B scenario, which would result in an average 19% decrease in the range of the 37 tree species (Table 10.9). An average 41% reduction in area of occurrence is predicted for 17 species including overcup oak (*Q. lyrata*), Shumard's oak,

TABLE 10.6
Predicted Change (%) in Area of Distribution of Selected Tree Species in the Eastern Atlantic Section of the Coastal Plains from 2000 to 2060 in Response to Four Climate Scenarios[a]

	Species Area Change (%)			
Species[b]	MIROC3.2	CSIROMK3.5	CSIROMK2	HadCM3
Mockernut hickory	−55.4	0.0	0.0	0.0
Black cherry	−86.0	0.0	0.0	0.0
Red mulberry	−96.7	0.0	0.0	0.0
Sweetgum	−96.7	0.0	0.0	0.0
Southern red oak	−96.7	0.0	0.0	0.0
Loblolly pine	−96.7	0.0	0.0	0.0
American holly	−96.7	0.0	0.0	0.0
Water hickory	−96.7	0.0	0.0	0.0
Black tupelo	−96.7	0.0	0.0	0.0
Paulownia	−96.7	0.0	0.0	0.0
Red bay	−96.7	0.0	0.0	0.0
Longleaf pine	−97.5	0.0	0.0	0.0
Shumard's oak	−96.7	1.7	0.8	3.3
Overcup oak	−96.7	1.7	0.8	3.3
Sweetbay	−66.1	−2.5	−4.1	31.4
Laurel oak	−46.3	−2.5	1.7	34.7
Slash pine	−41.3	−4.1	−5.0	29.8
Southern magnolia	−41.3	−4.1	−5.0	29.8
Water oak	−41.3	−4.1	−5.0	29.8
Average[c]	−41.6	−0.4	−0.4	4.4

Source: Adapted from Wear, D.N., R. Huggett, and J.G. Greis. In press. Constructing alternative futures. *Southern Forest Futures Project Technical Report.*

[a] Scenarios were developed from three general circulation models (MIROC3.2, CSIROMK2 and 3.5, and HadCM3) and two emissions storylines (A1B represents low population growth/high-energy use and B2 represents moderate population growth/low-energy use).

[b] Species with no change in each of the four scenarios are omitted.

[c] Species with no change are included.

and laurel oak (*Q. laurifolia*) under the MIROC3.2 A1B scenario. An increase in range of six species is predicted by the CSIROMK3.5 A1B and CSIROMK2 B2 scenarios. Little or no change of area of occurrence is predicted for 20 species including red maple, sweetbay (*Persea borbonia*), yellow poplar, and black oak. Overall, effects of climate change in the Middle Gulf-East section are predicted to be moderate primarily because of moderate reductions in annual precipitation.

Middle Gulf-West: Vulnerability of tree species to climate change in the Middle Gulf-West section of the Coastal Plain is predicted to be largest under the MIROC3.2 A1B scenario, which would result in an average 42% decrease in the range of the 37 tree species (Table 10.10). An average 82% reduction in area of occurrence is predicted for 19 species, including overcup oak, longleaf pine, Shumard's oak, sweetbay, and laurel oak under the MIROC3.2 A1B scenario. None of the climate scenarios predicts an increase in the range of tree species. Little or no change in area of occurrence is predicted for 18 species under any scenario, including sassafras (*Sassafras* spp), white oak, and persimmon (*Diospyros* spp). Overall, effects of climate change in the Middle Gulf-West section are predicted to be moderate-to-severe because of moderate reductions in annual precipitation.

TABLE 10.7
Predicted Change (%) in Area of Distribution of Selected Tree Species in the Florida Peninsular Section of the Coastal Plains from 2000 to 2060 in Response to Four Climate Scenarios[a]

	Species Area Change (%)			
Species[b]	MIROC3.2	A1B CSIRO	CSIROMK2	HadCM3
Black oak	–69.4	0.0	0.0	0.0
Flowering dogwood	–69.4	0.0	0.0	0.0
Sassafras	–69.4	0.0	0.0	0.0
White oak	–69.4	0.0	0.0	0.0
Pignut hickory	–69.4	0.0	0.0	0.0
American sycamore	–69.4	0.0	0.0	0.0
Yellow poplar	–69.4	0.0	0.0	0.0
Persimmon	–69.4	0.0	0.0	0.0
Mockernut hickory	–97.2	0.0	0.0	0.0
Black cherry	–100.0	0.0	0.0	0.0
Red mulberry	–100.0	0.0	0.0	0.0
Sweetgum	–100.0	0.0	0.0	0.0
Southern red oak	–100.0	0.0	0.0	0.0
Loblolly pine	–100.0	0.0	0.0	0.0
American holly	–100.0	0.0	0.0	0.0
Water hickory	–100.0	0.0	0.0	0.0
Black tupelo	–100.0	0.0	0.0	0.0
Paulownia	–100.0	0.0	0.0	0.0
Red bay	–100.0	0.0	0.0	0.0
Longleaf pine	–100.0	0.0	0.0	0.0
Shumard's oak	–100.0	0.0	0.0	0.0
Overcup oak	–100.0	0.0	0.0	0.0
Sweetbay	–100.0	0.0	0.0	0.0
Laurel oak	–97.2	2.8	–8.3	2.8
Slash pine	–94.4	5.6	–5.6	0.0
Southern magnolia	–94.4	5.6	–5.6	0.0
Water oak	–94.4	5.6	–5.6	0.0
Average[c]	–65.8	0.5	–0.7	0.1

Source: Adapted from Wear, D.N., R. Huggett, and J.G. Greis. In press. Constructing alternative futures. *Southern Forest Futures Project Technical Report.*

[a] Scenarios were developed from three general circulation models (MIROC3.2, CSIROMK2 and 3.5, and HadCM3) and two emissions storylines (A1B represents low population growth/high-energy use and B2 represents moderate population growth/low-energy use).

[b] Species with no change in each of the four scenarios are not listed.

[c] Species with no change are included.

Western Gulf: Vulnerability of tree species to climate change in the Western Gulf section of the Coastal Plain is predicted to be highest under the MIROC3.2 A1B scenario, which would result in an average 32% decrease in the range of the 37 tree species (Table 10.11). An average 52% reduction in area of occurrence is predicted for 19 species including laurel oak, southern magnolia, and longleaf pine under the MIROC3.2 A1B scenario. None of the scenarios predicts an increase in range of the selected tree species. Little or no change in area of occurrence is predicted for 18 species including black willow (*Salix nigra*), white oak, Shumard's oak, overcup oak, and flowering dogwood

TABLE 10.8
Predicted Change (%) in Area of Distribution of Selected Tree Species in the Southern Gulf Section of the Coastal Plains from 2000 to 2060 in Response to Four Climate Scenarios[a]

	Species Area Change (%)			
Species[b]	MIROC3.2	CSIROMK3.5	CSIROMK2	HadCM3
Mockernut hickory	−5.9	0.0	0.0	0.0
Black cherry	−30.6	0.0	0.0	0.0
Red mulberry	−41.2	0.0	0.0	0.0
Sweetgum	−44.7	0.0	0.0	0.0
Southern red oak	−44.7	0.0	0.0	0.0
Loblolly pine	−44.7	0.0	0.0	0.0
American holly	−44.7	0.0	0.0	0.0
Water hickory	−44.7	0.0	0.0	0.0
Black tupelo	−44.7	0.0	0.0	0.0
Paulownia	−44.7	0.0	0.0	0.0
Red bay	−44.7	0.0	0.0	0.0
Longleaf pine	−58.8	0.0	0.0	0.0
Shumard's oak	−77.6	0.0	0.0	0.0
Overcup oak	−77.6	0.0	0.0	0.0
Sweetbay	−100.0	−1.2	0.0	0.0
Laurel oak	−100.0	−3.5	−1.2	0.0
Slash pine	−100.0	−5.9	−4.7	0.0
Southern magnolia	−100.0	−5.9	−4.7	0.0
Water oak	−100.0	−5.9	−4.7	0.0
Average[c]	−31.1	−0.6	−0.4	0.0

Source: Adapted from Wear, D.N., R. Huggett, and J.G. Greis. In press. Constructing alternative futures. *Southern Forest Futures Project Technical Report.*

[a] Scenarios were developed from three general circulation models (MIROC3.2, CSIROMK2 and 3.5, and HadCM3) and two emissions storylines (A1B represents low population growth/high-energy use and B2 represents moderate population growth/low-energy use).

[b] Species with no change in each of the four scenarios are not listed.

[c] Species with no change are included.

(*Cornus florida*) under any scenario. Overall, effects of climate change in the Western Gulf section are predicted to be moderate because of moderate reductions in annual precipitation.

Summary: Overall effects of future climate change will likely be variable among the seven sections of the Coastal Plain. The highest threats to forest tree species are predicted to occur in the Florida Peninsular, Southern Gulf, and Middle Gulf-West sections. The lowest threats are likely to occur in the Northern Atlantic and Middle Gulf-East sections. Tree species most vulnerable to climate change in the Coastal Plain could include slash pine, southern red oak, Shumard's oak, and water oak. Among climate scenarios, the MIROC3.2 A1B scenario forecasts the highest threats to plants resulting from possible climate change by 2060; the lowest threats are associated with the CSIROMK2 B2 and HadCM3 B2 scenarios.

Forest Diversity Effects

Future diversity of forest tree species in the Coastal Plain is predicted to significantly decrease according to the MIROC3.2 A1B climate scenario (down from an average of 35.5 species currently

TABLE 10.9
Predicted Change (%) in Area of Distribution of Selected Tree Species in the Middle Gulf-East Section of the Coastal Plains Subregion from 2000 to 2060 in Response to Four Climate Scenarios[a]

	Species Area Change (%)			
Species[b]	**MIROC3.2**	**CSIROMK3.5**	**CSIROMK2**	**HadCM3**
Red mulberry	−2.0	0.0	0.0	0.0
Sweetgum	−5.9	0.0	0.0	0.0
Southern red oak	−5.9	0.0	0.0	0.0
Loblolly pine	−5.9	0.0	0.0	0.0
American holly	−5.9	0.0	0.0	0.0
Water hickory	−5.9	0.0	0.0	0.0
Black tupelo	−5.9	0.0	0.0	0.0
Paulownia	−5.9	0.0	0.0	0.0
Red bay	−5.9	0.0	0.0	0.0
Longleaf pine	−71.6	0.0	0.0	0.0
Shumard's oak	−71.6	0.0	0.0	0.0
Overcup oak	−96.1	1.0	1.0	1.0
Sweetbay	−88.2	9.8	11.8	3.9
Laurel oak	−85.3	11.8	14.7	−2.0
Slash pine	−76.5	16.7	16.7	0.0
Southern magnolia	−76.5	16.7	16.7	0.0
Water oak	−76.5	16.7	16.7	0.0
Average[c]	−18.7	2.0	2.1	0.1

Source: Adapted from Wear, D.N., R. Huggett, and J.G. Greis. In press. Constructing alternative futures. *Southern Forest Futures Project Technical Report.*

[a] Scenarios were developed from three general circulation models (MIROC3.2, CSIROMK2 and 3.5, and HadCM3) and two emissions storylines (A1B represents low population growth/high-energy use and B2 represents moderate population growth/low-energy use).

[b] Species with no change in each of the four scenarios are not listed.

[c] Species with no change are included.

to 22.7 in the year 2060), and to fluctuate by section according to the other three scenarios (Table 10.12). The CSIROMK3.5 A1B and CSIROMK2 B2 scenarios predicted a significant decrease in tree species richness, but the B2HadCM3 B2 scenario predicted no change.

Coastal Plain section scale: Among the seven sections of the Coastal Plain, predicted richness is most variable for the MIROC3.2 A1B scenario, ranging from 11.3 to 28.7 species in 2060 (Table 10.12). Predictions of diversity by the other three scenarios were generally consistent and may increase slightly in some areas of the Coastal Plain, particularly in the Northern Atlantic and Middle Gulf-East sections. The highest risks to diversity were in the Middle Gulf-West section where all scenarios predicted significant declines from the current level of 35.7 species to between 20.5 and 30.1 species in the year 2060. In contrast, the other three scenarios predicted little or no overall changes in diversity.

County scale: Changes in tree species diversity at the county level in the Coastal Plain indicate that the largest declines would occur in the Middle Gulf-West section. In that section, average diversity across all climate scenarios could decrease by 10 species for Morris and Smith Counties in Texas. Large decreases could also occur for Osceola County and Glades County in Florida, the Florida Peninsular section, and San Jacinto County and Montgomery County in the Texas portion of the Western Gulf section. Diversity was predicted to increase in small areas of the Middle

TABLE 10.10
Predicted Change (%) in Area of Distribution of Selected Tree Species in the Middle Gulf-West Section of the Coastal Plains from 2000 to 2060 in Response to Four Climate Scenarios[a]

	Species Area Change (%)			
Species[b]	**MIROC3.2**	**CSIROMK3.5**	**CSIROMK2**	**HadCM3**
Mockernut hickory	−24.4	0.0	0.0	0.0
Black cherry	−55.6	0.0	0.0	0.0
Red mulberry	−80.0	0.0	−8.9	−6.7
Sweetgum	−88.9	0.0	−11.1	−6.7
Southern red oak	−88.9	0.0	−11.1	−6.7
Loblolly pine	−88.9	0.0	−11.1	−6.7
American holly	−88.9	0.0	−11.1	−6.7
Water hickory	−88.9	0.0	−11.1	−6.7
Black tupelo	−88.9	0.0	−11.1	−6.7
Paulownia	−88.9	0.0	−11.1	−6.7
Red bay	−88.9	0.0	−11.1	−6.7
Longleaf pine	−100.0	−11.1	−37.8	−20.0
Shumard's oak	−97.8	−31.1	−46.7	−37.8
Overcup oak	−97.8	−31.1	−46.7	−37.8
Sweetbay	−93.3	−60.0	−80.0	−68.9
Laurel oak	−86.7	−66.7	−84.4	−73.3
Slash pine	−71.1	−62.2	−71.1	−64.4
Southern magnolia	−71.1	−62.2	−71.1	−64.4
Water oak	−71.1	−62.2	−71.1	−64.4
Average[c]	−42.2	−10.5	−16.4	−13.3

Source: Adapted from Wear, D.N., R. Huggett, and J.G. Greis. In press. Constructing alternative futures. *Southern Forest Futures Project Technical Report.*

[a] Scenarios were developed from three general circulation models (MIROC3.2, CSIROMK2 and 3.5, and HadCM3) and two emissions storylines (A1B represents low population growth/high-energy use and B2 represents moderate population growth/low-energy use).

[b] Species with no change in each of the four scenarios are not listed.

[c] Species with no change are included.

Gulf-East section, where species richness would be greater by 2-to-4 species for Marshall County and Calloway County in Kentucky. Diversity will also likely increase slightly, by several species, for Green County in North Carolina and Lexington County in South Carolina.

Overall, the MIROC3.2 A1B scenario consistently predicts decreased tree species diversity in 2060, but predictions by the other climate scenarios are variable, ranging from little or no change to small increases. Future diversity averaged across the four climate scenarios indicated a decrease throughout most of the Coastal Plain (Figure 10.12).

PIEDMONT

The Piedmont is a transition zone between the Coastal Plain and Appalachian-Cumberland Highland that occupies about 9% (197,155 km^2) of the Southern Region. This subregion was heavily disturbed by settlement during the 1800s, when much of the forest land was cleared for cotton production and subsistence agriculture. Forests are a mixture of loblolly pine and hardwoods on uplands, and hardwoods on floodplains of major rivers. The Piedmont is subdivided into three sections: Central

TABLE 10.11
Predicted Change (%) in Area of Distribution of Selected Tree Species in the Western Gulf Section of the Coastal Plains from 2000 to 2060 in Response to Four Climate Scenarios[a]

	Species Area Change (%)			
Species[b]	**MIROC3.2**	**CSIROMK3.5**	**CSIROMK2**	**HadCM3**
Mockernut hickory	−11.4	0.0	0.0	0.0
Black cherry	−31.4	0.0	0.0	0.0
Red mulberry	−42.9	0.0	0.0	0.0
Sweetgum	−45.7	0.0	0.0	0.0
Southern red oak	−45.7	0.0	0.0	0.0
Loblolly pine	−45.7	0.0	0.0	0.0
American holly	−45.7	0.0	0.0	0.0
Water hickory	−45.7	0.0	0.0	0.0
Black tupelo	−45.7	0.0	0.0	0.0
Paulownia	−45.7	0.0	0.0	0.0
Red bay	−45.7	0.0	0.0	0.0
Longleaf pine	−71.4	0.0	−2.9	−5.7
Shumard's oak	−97.1	0.0	−8.6	−20.0
Overcup oak	−97.1	−2.9	−22.9	−31.4
Sweetbay	−94.3	0.0	−22.9	−34.3
Laurel oak	−94.3	0.0	−22.9	−34.3
Slash pine	−94.3	−5.7	−25.7	−34.3
Southern magnolia	−94.3	−5.7	−25.7	−34.3
Water oak	−94.3	−5.7	−25.7	−34.3
Average[c]	−32.1	−0.5	−4.2	−6.2

Source: Adapted from Wear, D.N., R. Huggett, and J.G. Greis. In press. Constructing alternative futures. *Southern Forest Futures Project Technical Report.*

[a] Scenarios were developed from three general circulation models (MIROC3.2, CSIROMK2 and 3.5, and HadCM3) and two emissions storylines (A1B represents low population growth/high-energy use and B2 represents moderate population growth/low-energy use).

[b] Species with no change in each of the four scenarios are not listed.

[c] Species with no change are included.

Appalachian Piedmont (45.4%), Southern Appalachian Piedmont (38.5%), and Piedmont Ridge, Valley, and Plateau (16.2%).

Climate Scenario Predictions

The long-term trend of temperature in the Piedmont is predicted to increase from 2000 to 2060 based on averaged forecasts across the four climate scenarios (Figure 10.13). The rate of increase is approximately 0.2°C per decade for each of the three sections. Annual predicted change in temperature is highest in the Central Appalachian Piedmont section and lowest in the Central Appalachian Piedmont section. With the exception of the Central Appalachian Piedmont, changes in temperature among the sections are quite similar. For precipitation, the average prediction from the climate scenarios indicates a long-term trend of almost uniform amounts over time (Figure 10.14). The trend for the Piedmont Ridge, Valley, and Plateau section differs from the others in that precipitation there increases from the year 2000 to 2020 then returns to the 2000 level. An average of the four climate scenarios suggests a total temperature increase of about 3°C and precipitation remaining

TABLE 10.12
Forest Diversity, Expressed as Average Tree Species Richness (Standard Deviation) for Coastal Plain Sections from 2000 to 2060 Estimated by a Model Based on Predictions of Annual Precipitation from Four Climate Scenarios[a]

Section	Current Diversity	Future (2060) Diversity Predicted by Each Scenario			
		MIROC3.2	CSIROMK3.5	CSIROMK2	HadCM3
		N Tree Species (SD)			
Northern Atlantic	32.1 (2.2)	23.5 (4.7)[b]	31.8 (2.3)	33.5 (2.5)[b]	34.1 (2.5)[b]
Eastern Atlantic	34.2 (2.3)	19.1 (2.5)[b]	33.9 (2.2)	33.8 (2.2)	35.9 (1.8)[b]
Florida Peninsula	37.0 (0)	11.3 (3.1)[b]	37.0 (0)	36.6 (1.3)	36.8 (0.7)
Southern Gulf	37.0 (0)	25.2 (5.8)[b]	36.7 (1.1)	36.8 (0.9)	37.0 (0)
Mid Gulf—East	36.0 (1.9)	28.7 (2.5)[b]	36.8 (0.7)[b]	36.8 (0.7)[b]	36.0 (1.7)
Mid Gulf—West	35.7 (2.1)	20.5 (3.6)[b]	32.1 (2.3)[b]	29.6 (4.2)[b]	30.8 (4.0)[b]
Western Gulf	37.0 (0)	24.1 (5.7)[b]	36.6 (1.0)	35.0 (2.6)[b]	34.2 (3.2)[b]
Coastal Plains subregion	35.2 (2.4)	22.7 (6.1)[b]	34.9 (2.6)[b]	34.8 (3.0)[b]	35.3 (2.7)

Source: Adapted from Wear, D.N., R. Huggett, and J.G. Greis. In press. Constructing alternative futures. *Southern Forest Futures Project Technical Report.*

[a] Scenarios were developed from three general circulation models (MIROC3.2, CSIROMK2 and 3.5, and HadCM3) and two emissions storylines (A1B represents low population growth/high-energy use and B2 represents moderate population growth/low-energy use).

[b] Indicates significant difference between current (2000) and future (2060) species richness at the $p = 0.05$ level of probability.

almost constant across all sections, although amounts vary among the climate scenarios. Based on warmer temperatures alone, the ranges of several species could expand, but a warmer temperature will increase moisture stress from potential evapotranspiration.

Three of the climate scenarios forecast only small changes in average annual precipitation from 2000 to 2060 for most sections of the Piedmont; only the MIROC3.2 A1B scenario predicts a large decrease across all sections. The trend of future precipitation by the other three scenarios is similar among the sections, with a small increase of about 50 mm above current amounts. The scenarios predict the smallest increase of future precipitation will occur in the Central Appalachian Piedmont section and the largest increase will be in the Piedmont Ridge and Valley section. Comparison of precipitation predictions among climate change models indicates the largest increase for the HadCM3 B2 scenario and largest decrease in precipitation for the MIROC3.2 A1B scenario. Overall, climate of the Appalachian Piedmont subregion in 2060 is predicted to be warmer and drier as a result of little change in precipitation and increased evapotranspiration.

Effects on Tree Species Distribution in the Piedmont Sections

Thirty-nine selected tree species were used for assessment of the effects of temperature and precipitation in 2060 on forests in the Piedmont (Table 10.4). The forecasted future temperatures exceeded the minimum limits for all species evaluated. Minimum precipitation requirements for native species ranged from 380 mm for boxelder (*Acer negundo*) to 1270 mm for water oak. Precipitation predicted for three of the scenarios is expected to be greater than the minimum requirements for all species except for water oak. For the fourth, MIROC3.2 A1B, precipitation is expected to be slightly less than adequate for eight species and much less for three. Tree species with annual precipitation requirements greater than 1020 mm, such as Shumard's oak and water oak, are regarded as vulnerable and their area of occurrence could decrease.

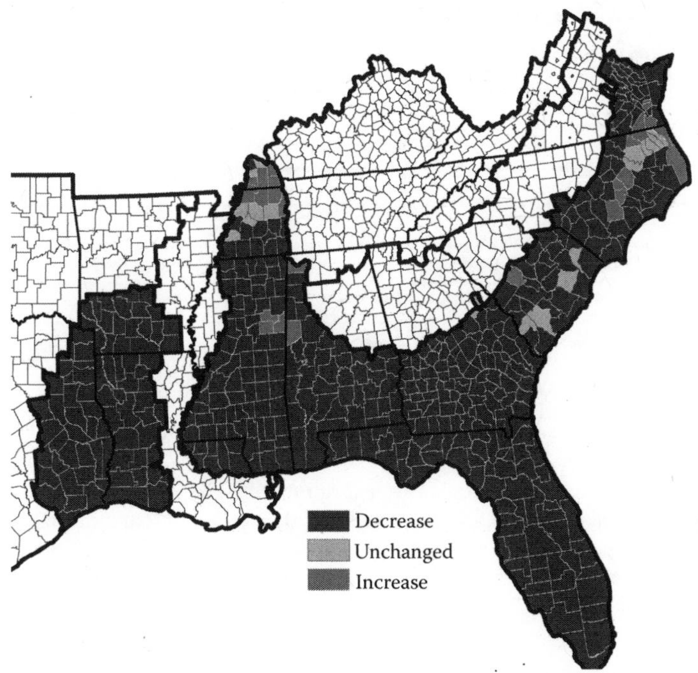

FIGURE 10.12 Future (2060) diversity (species richness) expressed as a percent of current (2000) diversity averaged across four climate scenarios for the southern Coastal Plain (Wear et al. in press) that were developed from three general circulation models (MIROC3.2, CSIROMK2 and 3.5, and HadCM3) and two emissions storylines (A1B represents low population growth/high-energy use and B2 represents moderate population growth/low-energy use). Diversity is predicted to decrease in counties shaded dark gray (<97.5% of current), remain generally unchanged in counties shaded light gray (97.5 to 102.5% of current) and increase in counties shaded medium gray (>102.5% of current).

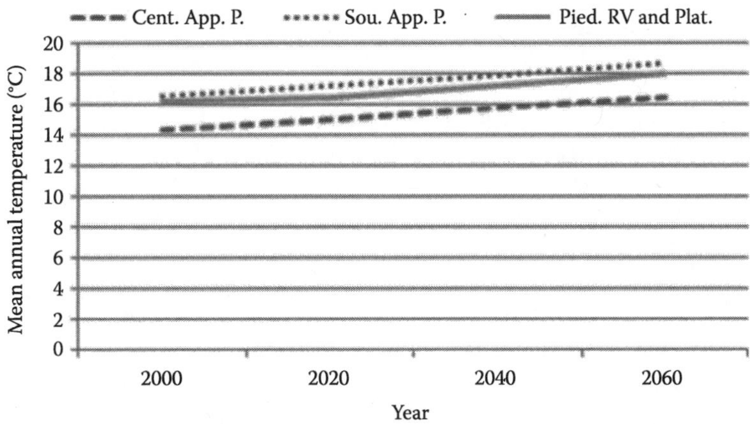

FIGURE 10.13 Average annual temperature from 2000 (current) to 2060 (predicted for sections of the southern Piedmont; temperature for each section represents the average of four climate scenarios (Wear et al. in press); that were developed from three general circulation models (MIROC3.2, CSIROMK2 and 3.5, and HadCM3) and two emissions storylines (A1B represents low population growth/high-energy use and B2 represents moderate population growth/low-energy use).

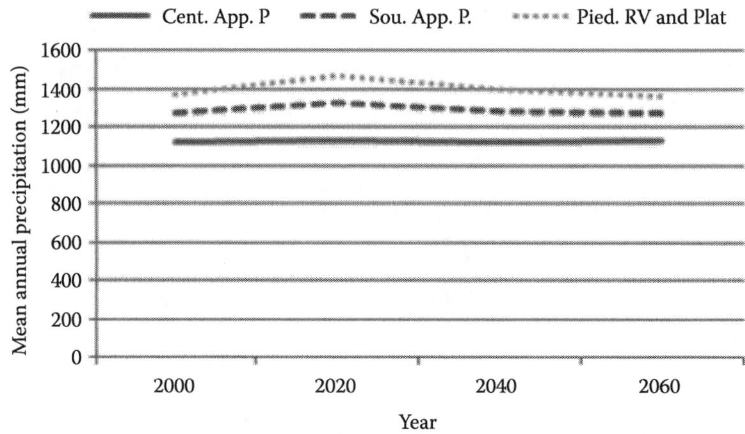

FIGURE 10.14 Average annual precipitation from 2000 (current) to 2060 (predicted) for sections of the southern Piedmont; precipitation for each section represents the average of four climate scenarios (Wear et al. in press); that were developed from three general circulation models (MIROC3.2, CSIROMK2 and 3.5, and HadCM3) and two emissions storylines (A1B represents low population growth/high-energy use and B2 represents moderate population growth/low-energy use).

TABLE 10.13
Predicted Change (%) in Area of Distribution of Selected Tree Species in the Central Appalachian Piedmont Section of the Piedmont from 2000 to 2060 in Response to Four Climate Scenarios[a]

	Species Area Change (%)			
Species[b]	**MIROC3.2**	**CSIROMK3.5**	**CSIROMK2**	**HadCM3**
Black cherry	−37.5	0.0	0.0	0.0
Red mulberry	−83.0	2.3	0.0	0.0
Sweetgum	−83.0	3.4	3.4	3.4
Southern red oak	−83.0	3.4	3.4	3.4
Loblolly pine	−83.0	3.4	3.4	3.4
American holly	−83.0	3.4	3.4	3.4
Blackgum	−83.0	3.4	3.4	3.4
Shortleaf pine	−83.0	3.4	3.4	3.4
Winged elm	−83.0	3.4	3.4	3.4
Shumard's oak	−33.0	27.3	28.4	46.6
Overcup oak	−33.0	27.3	28.4	46.6
Water oak	−3.4	9.1	6.8	14.8
Average[c]	−19.8	2.3	2.2	3.4

Source: Adapted from Wear, D.N., R. Huggett, and J.G. Greis. In press. Constructing alternative futures. *Southern Forest Futures Project Technical Report.*

[a] Scenarios were developed from three general circulation models (MIROC3.2, CSIROMK2 and 3.5, and HadCM3) and two emissions storylines (A1B represents low population growth/high-energy use and B2 represents moderate population growth/low-energy use).

[b] Species with no change in each of the four scenarios are not listed.

[c] Species with no change are included.

Central Appalachian Piedmont: Vulnerability of tree species to climate change in the Central Appalachian Piedmont section is predicted to be highest under the MIROC3.2 A1B scenario, which would result in an average 20% decrease in the range of the 39 tree species (Table 10.13). An average 64% reduction in area of occurrence is predicted for 12 species including sweetgum, southern red oak, and laurel oak under the MIROC3.2 A1B scenario. Three of the scenarios predict an increase in the occurrence of 10–11 species including southern red oak, black gum, and Shumard's oak. Overall, effects of climate change in the Central Appalachian Piedmont section of the Piedmont are predicted to be minor to moderate because of small increases in annual temperature and minimal reductions in annual precipitation.

Southern Appalachian Piedmont: Vulnerability of tree species to climate change in the Southern Appalachian Piedmont section is predicted to be highest under the MIROC3.2 A1B scenario, which would result in an average 17% decrease in the range of the 39 tree species (Table 10.14). An average 55% reduction in area of occurrence is predicted for 12 species including red mulberry, Shumard's oak, loblolly pine, and overcup oak under the MIROC3.2 A1B scenario. The other three climate scenarios predict an increase in the range of Shumard's oak, overcup oak, and water oak. Little or no change is predicted for the ranges of 27 species, including chinkapin oak (*A. Muehlenbergii*), red maple, and sassafras. Overall, the effects of climate change in the Southern Appalachian Piedmont section are predicted to be minor because of little or no changes in annual precipitation.

Piedmont Ridge, Valley, and Plateau: Vulnerability of tree species to climate change in the Piedmont Ridge, Valley, and Plateau section is predicted to be highest under the MIROC3.2 A1B scenario, which would result in an average 6% decrease in the range of the 39 tree species (Table 10.15).

TABLE 10.14
Predicted Change (%) in Area of Distribution of Selected Tree Species in the Southern Appalachian Piedmont Section of the Piedmont from 2000 to 2060 in Response to Four Climate Scenarios[a]

	Species Area Change (%)			
Species[b]	**MIROC3.2**	**CSIROMK3.5**	**CSIROMK2**	**HadCM3**
Black cherry	−27.3	0.0	0.0	0.0
Red mulberry	−50.6	0.0	0.0	0.0
Sweetgum	−54.5	0.0	0.0	0.0
Southern red oak	−54.5	0.0	0.0	0.0
Loblolly pine	−54.5	0.0	0.0	0.0
American holly	−54.5	0.0	0.0	0.0
Black tupelo	−54.5	0.0	0.0	0.0
Shortleaf pine	−54.5	0.0	0.0	0.0
Winged elm	−54.5	0.0	0.0	0.0
Shumard's oak	−79.2	6.5	6.5	6.5
Overcup oak	−79.2	6.5	6.5	6.5
Water oak	−41.6	15.6	11.7	44.2
Average[c]	−16.9	0.7	0.6	1.5

Source: Adapted from Wear, D.N., R. Huggett, and J.G. Greis. In press. Constructing alternative futures. *Southern Forest Futures Project Technical Report.*

[a] Scenarios were developed from three general circulation models (MIROC3.2, CSIROMK2 and 3.5, and HadCM3) and two emissions storylines (A1B represents low population growth/high-energy use and B2 represents moderate population growth/low-energy use).

[b] Species with no change in each of the four scenarios are not listed.

[c] Species with no change are included.

TABLE 10.15
Predicted Change (%) in Area of Distribution of Selected Tree Species in the Northern Ridge and Valley Section of the Piedmont from 2000 to 2060 in Response to Four Climate Scenarios[a]

	Species Area Change (%)			
Species[b]	**MIROC3.2**	**CSIROMK3.5**	**CSIROMK2**	**HadCM3**
Shumard's oak	−70.8	0.0	0.0	0.0
Overcup oak	−70.8	0.0	0.0	0.0
Water oak	−100.0	0.0	0.0	0.0
Average[c]	−6.2	0.0	0.0	0.0

Source: Adapted from Wear, D.N., R. Huggett, and J.G. Greis. In press. Constructing alternative futures. *Southern Forest Futures Project Technical Report.*

[a] Scenarios were developed from three general circulation models (MIROC3.2, CSIROMK2 and 3.5, and HadCM3) and two emissions storylines (A1B represents low population growth/high-energy use and B2 represents moderate population growth/low-energy use).

[b] Species with no change in each of the four scenarios are not listed.

[c] Species with no change are included.

An average 81% reduction in area of occurrence is predicted for three species, Shumard's oak, water oak, and overcup oak under the MIROC3.2 A1B scenario. None of the climate scenarios predicts an increase in range of the selected tree species. Little or no change is predicted for the ranges of 36 species, including silver maple (*Acer saccharinum*), flowering dogwood, and white ash (*Fraxinus americana*). Overall, the effects of climate change in the Piedmont Ridge, Valley, and Plateau section are predicted to be minor because of minimal reductions in annual precipitation.

Summary: Overall effects of future climate change will likely be relatively consistent among the three sections of the Piedmont. The most serious threat to forest tree species is predicted to occur along the southern edge of the Southern Appalachian section bordering the Eastern Atlantic section of the Coastal Plain subregion. The lowest threat may occur in the Piedmont Ridge, Valley, and Plateau section and in the northern part of the Central Appalachian Piedmont section, where species richness could increase in several counties. Tree species most vulnerable to climate change in the Coastal Plain will probably include Shumard's oak and water oak. Among climate scenarios, the MIROC3.2 A1B scenario forecasts the highest threats to plants resulting from possible climate change by 2060; the lowest threats are associated with the CSIROMK2 B2 and HadCM3 B2 scenarios.

Forest Diversity Effects in the Piedmont Region

Future diversity of forest tree species throughout the Piedmont could significantly change according to predictions by all climate scenarios, although the direction (increase or decrease) is unclear (Table 10.16). The MIROC3.2 A1B climate scenario predicted a large decrease in diversity (averaging about six species), compared to predictions of small increases by the other three scenarios.

Piedmont section scale: Predicted future tree diversity is most variable among the three sections of the Piedmont for the MIROC3.2 A1B scenario, where species richness could decrease from an average of 7.4 to 2.5 species in the year 2060 (Table 10.16). Predictions among the other three climate scenarios were less variable and some areas of the Piedmont, particularly in the Central and Southern Appalachian Piedmont sections, could actually experience increases in species richness. The highest risks to changes of future diversity are associated with the Northern Ridge and Valley

TABLE 10.16
Forest Diversity, Expressed as Average Tree Species Richness (Standard Deviation) for Piedmont Sections from 2000 to 2060 Estimated by a Model Based on Predictions of Annual Precipitation from Four Climate Scenarios[a]

Section	Current Diversity	Future (2060) Diversity Predicted by Each Scenario			
		MIROC3.2	CSIROMK3.5	CSIROMK2	HadCM3
		N Tree Species (SD)			
Central Piedmont	36.5 (2.1)	29.1 (3.4)[b]	37.4 (1.7)[b]	37.2 (1.8)[b]	37.6 (1.8)[b]
Southern Piedmont	38.4 (0.6)	32.1 (4.6)[b]	38.6 (0.5)[b]	38.6 (0.5)[b]	38.9 (0.2)[b]
Pied. Ridg. and Valley	39.0 (0.9)	36.5 (0.9)[b]	39.0 (0.0)	39.0 (0.0)	39.0 (0.0)
Piedmont Subregion	37.2 (1.8)	31.3 (4.5)[b]	38.1 (1.4)[b]	38.0 (1.5)[b]	38.3 (1.4)[b]

Source: Adapted from Wear, D.N., R. Huggett, and J.G. Greis. In press. Constructing alternative futures. *Southern Forest Futures Project Technical Report.*

[a] Scenarios were developed from three general circulation models (MIROC3.2, CSIROMK2 and 3.5, and HadCM3) and two emissions storylines (A1B represents low population growth/high-energy use and B2 represents moderate population growth/low-energy use).

[b] Indicates significant difference between current (2000) and future (2060) species richness at the $p = 0.05$ level of probability.

section, where the four scenarios predicted either a significant decline or continuation at the current level of 39.0 species. If the large decreases predicted by the MIROC3.2 A1B scenario were excluded, the other sections of the Piedmont could be expected to experience a small but significant decrease in diversity.

County scale: Changes in tree species diversity at the county level in the Piedmont indicate that the largest declines would occur in the Central Appalachian Piedmont section, where diversity averaged across all climate scenarios could decrease by up to four species for South Boston County in Virginia. Areas that could experience decreases in diversity of two to three species are McCormick County in South Carolina and Meriwether County in Georgia. Increased diversity was predicted in small areas of the Piedmont section, where richness will be greater by four to five species for Fredericksburg County and Stafford County in Virginia. Three models predict that diversity of trees will increase slightly, by several species, for Gwinnett County and Madison County in Georgia. The most consistent effects of climate change on diversity will occur in the lower Piedmont of Georgia and South Carolina and in most of Virginia, where numbers of tree species are predicted to decrease slightly in all counties.

Overall, the MIROC3.2 A1B scenario consistently predicted the likely loss of two to three species in most counties, but predictions by the other climate scenarios were variable, ranging from little or no change to small increases for some counties. Future diversity averaged across the four climate scenarios indicated little or no change in about half of the counties of the Piedmont and a decrease of more than three species in the others (Figure 10.15).

APPALACHIAN-CUMBERLAND HIGHLAND

Forests of this hilly to mountainous subregion cover about 11% (242,596 km^2) of the South; upland forests consist mainly of deciduous hardwoods dominated by one or more species of oak and hickory. This subregion was subdivided into five sections: Blue Ridge (13.7%), Northern Ridge and Valley (14.0%), Southern Ridge and Valley (7.2%), Cumberland Plateau and Mountain (18.9%), and Interior Low Plateau (46.1%).

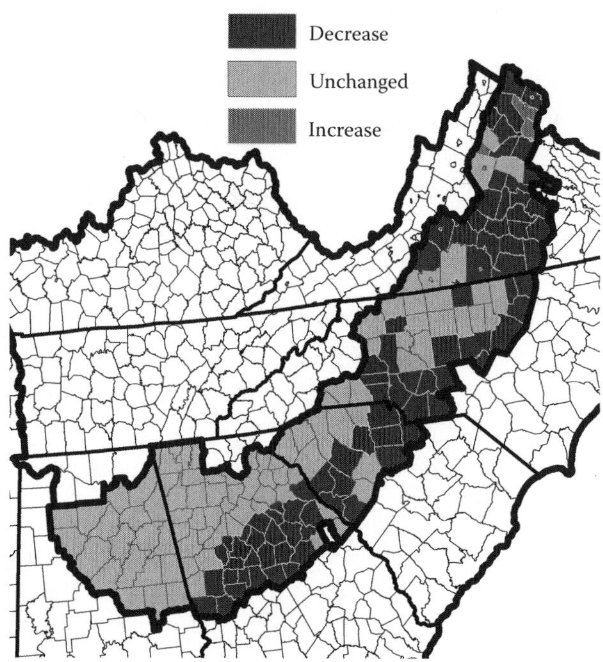

FIGURE 10.15 Future (2060) diversity (species richness) expressed as a percent of current (2000) diversity averaged across four climate scenarios for the Piedmont (Wear et al. in press); that were developed from three general circulation models (MIROC3.2, CSIROMK2 and 3.5, and HadCM3) and two emissions storylines (A1B represents low population growth/high-energy use and B2 represents moderate population growth/low-energy use). Diversity is predicted to decrease in counties shaded dark gray (<97.5% of current), remain generally unchanged in counties shaded light gray (97.5 to 102.5% of current) and increase in counties shaded medium gray (>102.5% of current).

Climate Scenario Predictions

The long-term trend of temperature in the Appalachian-Cumberland Highland is predicted to increase from the year 2000 to 2060 based on averaged forecasts across the four climate scenarios (Figure 10.16). The rate of increase is approximately 0.3°C per decade for each of the five sections. Predicted changes in annual temperature are greatest in the Interior Low Plateau section and least in the Northern Ridge and Valley section. With the exception of the Blue Ridge and Northern Ridge and Valley, predicted changes in temperature among the sections are quite similar. For precipitation, the average of predictions from the climate scenarios indicates a long-term trend of nearly constant amounts (Figure 10.17); again. The trend for the Northern Ridge and Valley section differs from the other sections—in this instance, a smaller increase from 2000 to 2020 and then little change through 2060. An average of the four climate scenarios suggests a total temperature increase of about 3°C and a precipitation decrease of about 100 mm across all sections, although amounts vary for each climate scenario.

Three of the four climate scenarios forecast moderate changes in average annual precipitation from the year 2000 to 2060 for most sections of the Appalachian-Cumberland Highland; only the MIROC3.2 A1B scenario predicts a large decrease across all sections. Future precipitation predictions by the other scenarios vary among sections, ranging from similar amounts for the Cumberland Plateau and Mountain section to somewhat variable amounts for the Interior Low Plateau section. Comparison of predictions indicates the largest increase of precipitation for the CSIROMK3.5 A1B and HadCM3 B2 scenarios and largest decrease in precipitation for the MIROC3.2 A1B scenario. For all climate scenarios except the MIROC3.2 A1, precipitation is predicted to increase slightly, by

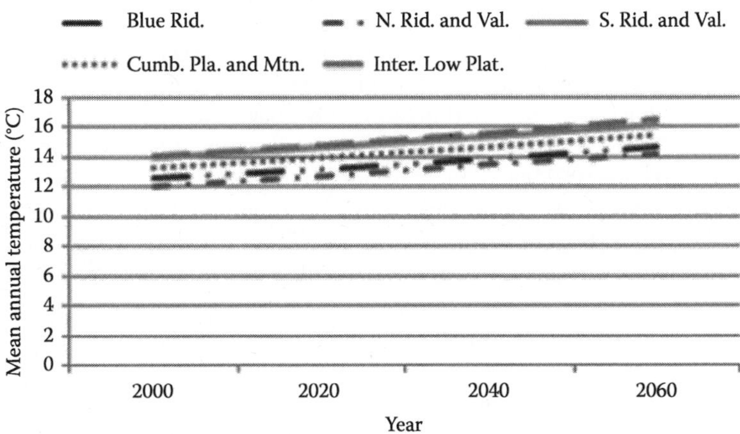

FIGURE 10.16 Average annual temperature from 2000 (current) to 2060 (predicted) for sections of the Appalachian-Cumberland Highland; temperature for each section represents the average of four climate scenarios (Wear et al. in press); that were developed from three general circulation models (MIROC3.2, CSIROMK2 and 3.5, and HadCM3) and two emissions storylines (A1B represents low population growth/high-energy use and B2 represents moderate population growth/low-energy use).

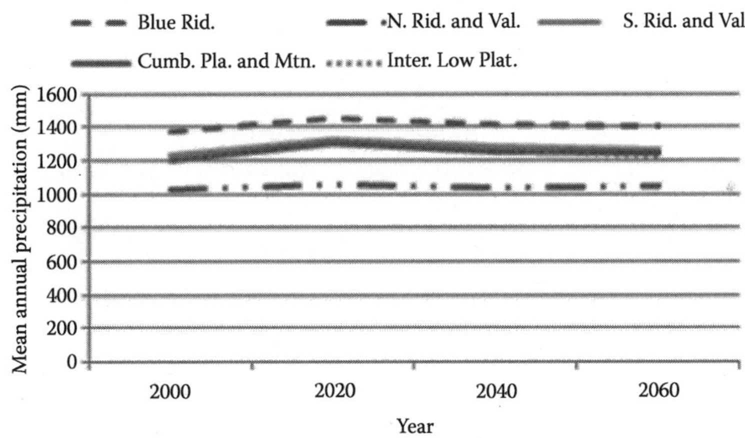

FIGURE 10.17 Average annual precipitation from 2000 (current) to 2060 (predicted) for sections of the Appalachian-Cumberland Highland; precipitation for each section represents the average of four climate scenarios (Wear et al. in press); that were developed from three general circulation models (MIROC3.2, CSIROMK2 and 3.5, and HadCM3) and two emissions storylines (A1B represents low population growth/high-energy use and B2 represents moderate population growth/low-energy use).

about 60 mm, in the Blue Ridge and Southern Ridge and Valley sections and remain above-current levels in all sections. Overall, climate of the Appalachian-Cumberland Highland is predicted to be warmer and wetter in 2060.

Effects on Tree Species Distribution in Appalachian-Cumberland Highland Sections

Forty-four selected tree species were used for assessment of the effects of temperature and precipitation on forests in the Appalachian-Cumberland Highland (Table 10.4). Minimum annual temperature limits on the distribution of evaluated species ranged from below 0°C for boxelder to 13°C for winged elm (*Ulmus alata*). Precipitation requirements for native species ranged from 380 mm for American elm to 1260 mm for Shumard's oak. Temperature and precipitation predicted by three of the scenarios is expected to be greater than the minimum requirements for all species. For the fourth, MIROC3.2 A1B, precipitation is expected to be slightly greater than adequate for nine species and slightly less for two. Tree species with annual precipitation requirements greater than 1050 mm, such as sweet birch (*Betula lenta*) and Shumard's oak, are predicted to decrease in area of occurrence.

Blue Ridge: Vulnerability of tree species to climate change in the Blue Ridge section of the Appalachian-Cumberland Highland is predicted to be highest under the MIROC3.2 A1B scenario, which would result in an average 6% decrease in the range of the 44 tree species (Table 10.17). A 21% average reduction in area of occurrence is predicted for 12 species including sweetgum, shortleaf pine, and sweet birch under the MIROC3.2 A1B scenario, whereas an increase in the range of 10 species is predicted for the other three climate scenarios. Little or no change is predicted for the ranges of 32 species, including boxelder, honeylocust (*Gleditsia triacanthos*), and red maple. Overall, the effects of climate change in the Blue Ridge section are predicted to be minor because of minimal reductions in annual precipitation.

Northern Ridge and Valley: Vulnerability of tree species to climate change in the Northern Ridge and Valley section of the Appalachian-Cumberland Highland is predicted to be greatest under the MIROC3.2 A1B scenario, which would result in an average 15% decrease in the range of the 44 tree species (Table 10.18). An average 41% reduction in area of occurrence is predicted for 16 species, including sweetgum, southern red oak, and shortleaf pine under the MIROC3.2 A1B scenario, whereas an increase in range is predicted for 15 species under the other three scenarios.

TABLE 10.17
Predicted Change (%) in Area of Distribution of Selected Tree Species in the Blue Ridge Section of the Appalachian-Cumberland Highland from 2000 to 2060 in Response to Four Climate Scenarios[a]

	Species Area Change (%)			
Species[b]	**MIROC3.2**	**CSIROMK3.5**	**CSIROMK2**	**HadCM3**
Black cherry	−5	0	0	0
Red mulberry	−20	0	0	0
Sweetgum	−20	2.5	2.5	2.5
Southern red oak	−20	2.5	2.5	2.5
American holly	−20	2.5	2.5	2.5
Black tupelo	−20	2.5	2.5	2.5
Paulownia	−20	2.5	2.5	2.5
Shortleaf pine	−20	2.5	2.5	2.5
Black locust	−20	2.5	2.5	2.5
Winged elm	−20	2.5	2.5	2.5
Shumard's oak	−35	10	10	10
Sweet birch	−35	10	10	10
Average[c]	−5.8	0.9	0.9	0.9

Source: Adapted from Wear, D.N., R. Huggett, and J.G. Greis. In press. Constructing alternative futures. *Southern Forest Futures Project Technical Report.*

[a] Scenarios were developed from three general circulation models (MIROC3.2, CSIROMK2 and 3.5, and HadCM3) and two emissions storylines (A1B represents low population growth/high-energy use and B2 represents moderate population growth/low-energy use).

[b] Species with no change in each of the four scenarios are not listed.

[c] Species with no change are included.

TABLE 10.18
Predicted Change (%) in Area of Distribution of Selected Tree Species in the Northern Ridge and Valley Section of the Appalachian-Cumberland Highland from 2000 to 2060 in Response to Four Climate Scenarios[a]

	Species Area Change (%)			
Species[b]	**MIROC3.2**	**CSIROMK3.5**	**CSIROMK2**	**HadCM3**
Chestnut oak	−5.4	0.0	0.0	0.0
Mockernut hickory	−29.7	5.4	5.4	5.4
Cucumber tree	−29.7	5.4	5.4	5.4
Virginia pine	−29.7	5.4	5.4	5.4
Black cherry	−70.3	10.8	8.1	10.8
Red mulberry	−54.1	37.8	21.6	24.3
Sweetgum	−54.1	32.4	18.9	24.3
Southern red oak	−54.1	32.4	18.9	24.3
American holly	−54.1	32.4	18.9	24.3
Black tupelo	−54.1	32.4	18.9	24.3
Paulownia	−54.1	32.4	18.9	24.3
Shortleaf pine	−54.1	32.4	18.9	24.3
Black locust	−54.1	32.4	18.9	24.3
Winged elm	−54.1	32.4	18.9	24.3
Shumard's oak	−2.7	35.1	18.9	35.1
Sweet birch	−2.7	35.1	18.9	35.1
Average[c]	−14.9	9.0	5.3	7.2

Source: Adapted from Wear, D.N., R. Huggett, and J.G. Greis. In press. Constructing alternative futures. *Southern Forest Futures Project Technical Report.*

[a] Scenarios were developed from three general circulation models (MIROC3.2, CSIROMK2 and 3.5, and HadCM3) and two emissions storylines (A1B represents low population growth/high-energy use and B2 represents moderate population growth/low-energy use).

[b] Species with no change in each of the four scenarios are not listed.

[c] Species with no change are included.

Little or no change is predicted for 28 of the species, including blackjack oak, slippery elm (*Ulmus rubra*), and black oak. Overall, the effects of climate change in the Northern Ridge and Valley section are predicted to be minor because of minimal reductions in annual precipitation.

Southern Ridge and Valley: Vulnerability of tree species to climate change in the Southern Ridge and Valley section of the Appalachian-Cumberland Highland is predicted to be greatest under the MIROC3.2 A1B scenario, which would result in an average 10% decrease in the range of the 44 tree species (Table 10.19). An average 36% reduction in area of occurrence is predicted for 12 species, including sweetgum, shortleaf pine, and black locust (*Robinia pseudoacacia*) under the MIROC3.2 A1B scenario, whereas an increase in range of two species is predicted for the other three scenarios. Little or no change is predicted for the ranges of 32 species, including black willow, post oak, and northern red oak (*Q. rubra*). Overall, the effects of climate change in the Southern Ridge and Valley section are predicted to be minor because of minimal reductions in annual precipitation.

Cumberland Plateau and Mountain: Vulnerability of tree species to climate change in the Cumberland Plateau and Mountain section of the Appalachian-Cumberland Highland is predicted to be greatest under the MIROC3.2 A1B scenario, which would result in an average 11% decrease in the range of the 44 tree species of trees (Table 10.20). An average 42% reduction in area of occurrence is predicted for 12 species including sweetgum, southern red oak, and American holly (*Ilex*

TABLE 10.19
Predicted Change (%) in Area of Distribution of Selected Tree Species in the Southern Ridge and Valley Section of the Appalachian-Cumberland Highland from 2000 to 2060 in Response to Four Climate Scenarios[a]

	Species Area Change (%)			
Species[b]	**MIROC3.2**	**CSIROMK3.5**	**CSIROMK2**	**HadCM3**
Black cherry	−5.9	0.0	0.0	0.0
Red mulberry	−29.4	0.0	0.0	0.0
Sweetgum	−35.3	0.0	0.0	0.0
Southern red oak	−35.3	0.0	0.0	0.0
American holly	−35.3	0.0	0.0	0.0
Black tupelo	−35.3	0.0	0.0	0.0
Paulownia	−35.3	0.0	0.0	0.0
Shortleaf pine	−35.3	0.0	0.0	0.0
Black locust	−35.3	0.0	0.0	0.0
Winged elm	−35.3	0.0	0.0	0.0
Shumard's oak	−58.8	23.5	23.5	23.5
Sweet birch	−58.8	23.5	6.3	6.3
Average[c]	−9.9	1.1	0.7	0.7

Source: Adapted from Wear, D.N., R. Huggett, and J.G. Greis. In press. Constructing alternative futures. *Southern Forest Futures Project Technical Report.*

[a] Scenarios were developed from three general circulation models (MIROC3.2, CSIROMK2 and 3.5, and HadCM3) and two emissions storylines (A1B represents low population growth/high-energy use and B2 represents moderate population growth/low-energy use).

[b] Species with no change in each of the four scenarios are not listed.

[c] Species with no change are included.

opaca) under the MIROC3.2 A1B scenario, whereas a small increase in the range of two species is predicted for the other three scenarios. Little or no change is predicted for the ranges of 32 species, including American basswood, red maple, and American sycamore (*Platanus occidentalis*). Overall, the effects of climate change in the Cumberland Plateau and Mountain section are predicted to be minor because of minimal reductions in annual precipitation.

Interior Low Plateau: Vulnerability of tree species to climate change in the Interior Low Plateau section of the Appalachian-Cumberland Highland is predicted to be highest under the MIROC3.2 A1B scenario, which would result in an average 11% decrease in the range of the 44 tree species (Table 10.21). An average 41% reduction in area of occurrence is predicted for 12 species including black cherry (*Prunus serotina*), shortleaf pine, Shumards's oak, and winged elm under the MIROC3.2 A1B scenario. An increase in the ranges of two species is predicted by the other three. Little or no change is predicted for the ranges of 32 species, including boxelder, post oak, and flowering dogwood. Overall, the effects of climate change in the Interior Low Plateau section are predicted to be minor because of minimal reductions in annual precipitation.

Summary: Overall effects of future climate change will likely be relatively consistent among the five sections of the Appalachian-Cumberland Highland. The highest threat to forest tree species could occur in the Northern Ridge and Valley section and the lowest threat may occur in the Blue Ridge section. Tree species most vulnerable to climate change in this subregion will probably include sweet birch and black locust. Among climate scenarios, the MIROC3.2 A1B scenario forecasts the highest threats to trees by 2060; the lowest threats are associated with the CSIROMK3.5 A1B and HadCM3 B2 scenarios.

TABLE 10.20
Predicted Change (%) in Area of Distribution of Selected Tree Species in the Cumberland Plateau and Mountain Section of the Appalachian-Cumberland Highland from 2000 to 2060 in Response to Four Climate Scenarios[a]

	Species Area Change (%)			
Species[b]	MIROC3.2	CSIROMK3.5	CSIROMK2	HadCM3
Black cherry	−23.1	0.0	0.0	0.0
Red mulberry	−36.5	0.0	0.0	0.0
Sweetgum	−44.2	0.0	0.0	0.0
Southern red oak	−44.2	0.0	0.0	0.0
American holly	−44.2	0.0	0.0	0.0
Black tupelo	−44.2	0.0	0.0	0.0
Paulownia	−44.2	0.0	0.0	0.0
Shortleaf pine	−44.2	0.0	0.0	0.0
Black locust	−44.2	0.0	0.0	0.0
Winged elm	−44.2	0.0	0.0	0.0
Shumard's oak	−44.2	17.3	21.2	21.2
Sweet birch	−44.2	17.3	21.2	21.2
Average[c]	−11.4	0.8	1.0	1.0

Source: Adapted from Wear, D.N., R. Huggett, and J.G. Greis. In press. Constructing alternative futures. *Southern Forest Futures Project Technical Report.*

[a] Scenarios were developed from three general circulation models (MIROC3.2, CSIROMK2 and 3.5, and HadCM3) and two emissions storylines (A1B represents low population growth/high-energy use and B2 represents moderate population growth/low-energy use).

[b] Species with no change in each of the four scenarios are not listed.

[c] Species with no change are included.

Forest Diversity Effects

Future diversity in the Appalachian-Cumberland Highland could significantly decrease according to predictions by the MIROC3.2 A1B climate scenario, where average species richness could drop from 42.8 currently to 37.9 by 2060 (Table 10.22). Predictions from the other three scenarios suggest possible benefits from climate change resulting from small but significant increases in diversity of major tree species.

Appalachian-Cumberland Highland section scale: Among the five sections of the Appalachian-Cumberland Highland, predicted diversity is most variable for the MIROC3.2 A1B scenario, where future species richness could range from 31.2 to 40.8 species in 2060 (Table 10.22). Predictions of diversity by the other three climate scenarios differed, but generally indicated no decline in species richness and increases in richness for some areas, particularly in the Northern Ridge and Valley and Cumberland Plateau and Mountain section, where the changes are significant at the $p < 0.05$ level of probability. The highest risks to diversity are associated with climate change in the Blue Ridge and Southern Ridge and Valley sections, where, on average, the climate scenarios predicted either a significant decline or no significant difference between the current and future species richness.

County scale: Changes in tree species diversity at the county level indicate that the largest declines would occur in the Interior Low Plateau section—specifically northern Alabama, central Tennessee, and western Kentucky—and that average diversity across all climate scenarios could decrease by an average of almost four species for Carroll County in Kentucky. Decreases in diversity of about three species could occur also in the Blue Ridge section (Carroll County and Floyd County in Virginia),

TABLE 10.21
Predicted Change (%) in Area of Distribution of Selected Tree Species in the Interior Low Plateau Section of the Appalachian-Cumberland Highland from 2000 to 2060 in Response to Four Climate Scenarios[a]

	Species Area Change (%)			
Species[b]	MIROC3.2	CSIROMK3.5	CSIROMK2	HadCM3
Black cherry	−12.6	0.0	0.0	0.0
Red mulberry	−33.6	0.0	0.0	0.0
Sweetgum	−40.3	0.0	0.0	0.0
Southern red oak	−40.3	0.0	0.0	0.0
American holly	−40.3	0.0	0.0	0.0
Black tupelo	−40.3	0.0	0.0	0.0
Paulownia	−40.3	0.0	0.0	0.0
Shortleaf pine	−40.3	0.0	0.0	0.0
Black locust	−40.3	0.0	0.0	0.0
Winged elm	−40.3	0.0	0.0	0.0
Shumard's oak	−63.0	16.0	26.1	16.8
Sweet birch	−63.0	16.0	26.1	16.8
Average[c]	−11.2	0.7	1.2	0.8

Source: Adapted from Wear, D.N., R. Huggett, and J.G. Greis. In press. Constructing alternative futures. *Southern Forest Futures Project Technical Report.*

[a] Scenarios were developed from three general circulation models (MIROC3.2, CSIROMK2 and 3.5, and HadCM3) and two emissions storylines (A1B represents low population growth/high-energy use and B2 represents moderate population growth/low-energy use).

[b] Species with no change in each of the four scenarios are not listed.

[c] Species with no change are included.

the Southern Ridge and Valley section (Grainger County in Tennessee), and the Cumberland Plateau and Mountain section (Perry County and Knott County in Kentucky). Diversity of trees is predicted to increase by six to seven species in Waynesboro County and Augusta County in Virginia, and gain five species in Buena Vista County, Virginia. The most consistent effects of climate change on diversity will likely occur in Scott County, Tennessee, and other counties in the Cumberland Plateau and Mountain section, where the four models predicted little or no change of diversity.

Overall, the effects of climate change on diversity at the county scale are expected to be variable, probably resulting from the Appalachian-Cumberland Highland's diverse topography effects on precipitation, with diversity generally declining by several species throughout, but increasing by two to five species in several counties of most sections. The MIROC3.2 A1B scenario consistently predicted the likely loss of two to three species in most counties, but predictions by the other climate scenarios were variable, ranging from little or no change to small increases in some counties. Future diversity averaged across the four climate scenarios indicated little or no change will likely occur in most counties in the Appalachian-Cumberland Highland, but a decline in diversity could occur in many of the northern counties of Kentucky and an increase is possible in the northern Virginia counties of the Northern Ridge and Valley section (Figure 10.18).

MISSISSIPPI ALLUVIAL VALLEY

The extensive floodplain of the Mississippi River is the smallest of the five subregions of the Southern Region, occupying about 5% (114,702 km^2) of the total area. Except for the wetlands near

TABLE 10.22
Forest Diversity, Expressed as Mean Tree Species Richness (Standard Deviation) for Appalachian-Cumberland Sections from 2000 to 2060 Estimated by a Model Based on Predictions of Annual Precipitation from Four Change Scenarios[a]

Section	Current Diversity	Future (2060) Diversity Predicted by Each Scenario			
		MIROC3.2	CSIROMK3.5	CSIROMK2	HadCM3
		N Tree Species (SD)			
Blue Ridge	43.8 (0.6)	40.8 (4.7)[b]	44.0 (0.0)	44.0 (0.0)	44.0 (0.0)
Northern Ridge and Valley	37.3 (5.2)	31.2 (1.7)[b]	41.5 (3.4)[b]	39.7 (4.6)[b]	41.1 (4.8)[b]
S. Ridge and Valley	43.5 (0.9)	37.9 (5.0)[b]	44.0 (0.0)	44.0 (0.0)	44.0 (0.0)
Cumb. Plat. and Mtn.	43.8 (0.8)	38.3 (5.1)[b]	43.9 (0.4)[b]	43.9 (0.4)[b]	43.9 (0.4)[b]
Interior Low Plat.	43.6 (0.8)	38.5 (4.7)[b]	43.6 (0.6)	44.0 (0.3)[b]	43.8 (0.6)
App.-Cumb. Subregion	42.8 (3.0)	37.9 (5.2)[b]	43.5 (1.5)[b]	43.4 (2.2)[b]	43.4 (2.2)[b]

Source: Adapted from Wear, D.N., R. Huggett, and J.G. Greis. In press. Constructing alternative futures. *Southern Forest Futures Project Technical Report.*

[a] Scenarios were developed from three general circulation models (MIROC3.2, CSIROMK2 and 3.5, and HadCM3) and two emissions storylines (A1B represents low population growth/high-energy use and B2 represents moderate population growth/low-energy use).

[b] Indicates significant difference between current (2000) and future (2060) species richness at the $p = 0.05$ level of probability.

the confluence of the Mississippi River with the Gulf of Mexico, much of this subregion of largely alluvial soils has been cleared for agriculture. Historically, the region was dominated by bottomland hardwood forests, consisting mainly of oaks, blackgum, sweetgum, and bald cypress. The Mississippi Alluvial Valley is subdivided into two sections: Holocene Deposits (74.8%) and Deltaic Plain (25.2%).

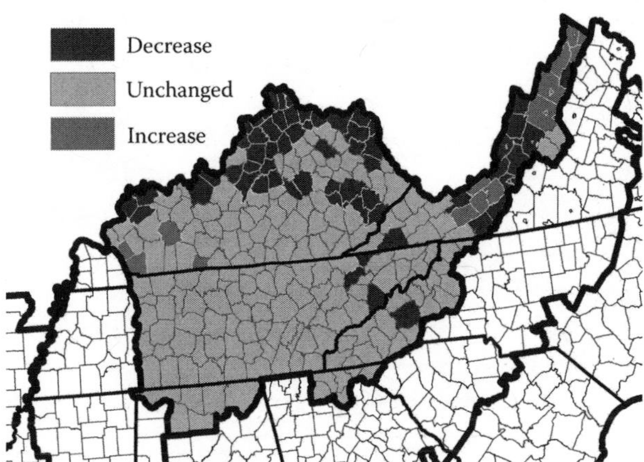

FIGURE 10.18 Future (2060) diversity (species richness) expressed as a percent of current (2000) diversity averaged across four climate scenarios for the Appalachian-Cumberland Highland (Wear et al. in press); that were developed from three general circulation models (MIROC3.2, CSIROMK2 and 3.5, and HadCM3) and two emissions storylines (A1B represents low population growth/high-energy use and B2 represents moderate population growth/low-energy use). Diversity is predicted to decrease in counties shaded dark gray (<97.5% of current), remain generally unchanged in counties shaded light gray (97.5 to 102.5% of current) and increase in counties shaded medium gray (>102.5% of current).

Climate Scenario Predictions

The long-term temperature trend in the Mississippi Alluvial Valley is predicted to increase from 2000 to 2060 when averaged across the four climate scenarios (Figure 10.19). The rate of increase is approximately 0.3°C per decade for each of the two sections. The difference in average temperature (~3°C) between the two sections remains relatively constant during the 60-year forecast period. For precipitation, the average of predictions from the climate scenarios indicates a long-term trend of decreasing amounts (Figure 10.20). The trend for the Deltaic Plain section

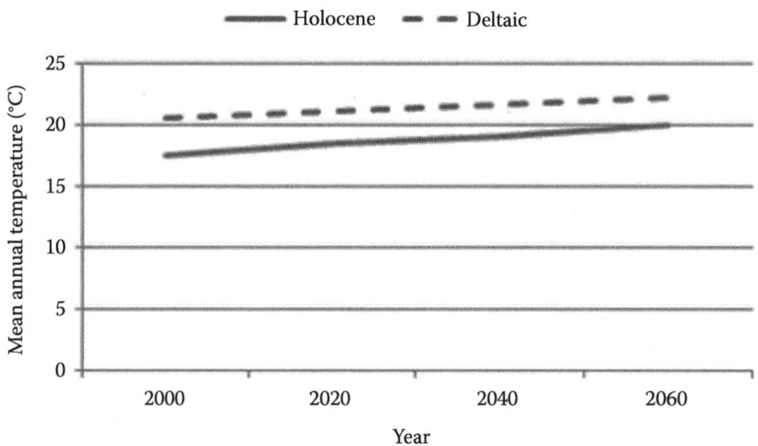

FIGURE 10.19 Average annual temperature from 2000 (current) through 2060 (predicted) for sections of the Mississippi Alluvial Valley; temperature for each section represents the average of four climate scenarios (Wear et al. in press); that were developed from three general circulation models (MIROC3.2, CSIROMK2 and 3.5, and HadCM3) and two emissions storylines (A1B represents low population growth/high-energy use and B2 represents moderate population growth/low-energy use).

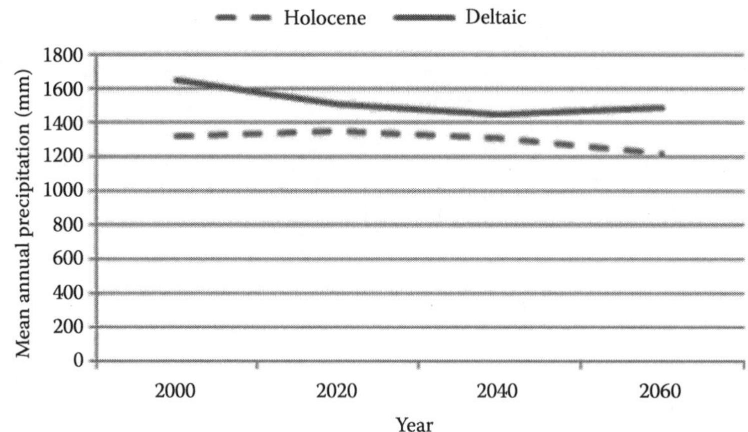

FIGURE 10.20 Average annual precipitation from 2000 (current) through 2060 (predicted) for sections of the Mississippi Alluvial Valley; precipitation for each section represents the average of four climate scenarios (Wear et al. in press); that were developed from three general circulation models (MIROC3.2, CSIROMK2 and 3.5, and HadCM3) and two emissions storylines (A1B represents low population growth/high-energy use and B2 represents moderate population growth/low-energy use).

decreases from 2040 to 2060, compared to a slight increase for the Holocene Deposits section. An average of the four climate scenarios suggests a total temperature increase of about 5°C and a precipitation decrease of about 150 mm across the two sections, although amounts vary for each climate scenario.

Three of the four climate scenarios forecast only slight changes in average annual precipitation from 2000 to 2060 for the two sections of the Mississippi Alluvial Valley. Only the MIROC3.2 A1B scenario predicts a large decrease across both sections. Future precipitation predictions by the other climate scenarios vary between the sections, forecasting similar amounts for the Holocene Deposits section but relatively different amounts for the Deltaic Plain section. Comparison of predictions indicates equal or increased precipitation for the CSIROMK3.5 A1B scenario and the largest decrease in precipitation for the MIROC3.2 A1B scenario. The HadCM3 B2 scenario predicts that precipitation will decrease slightly, by about 50 mm, in the Holocene Deposits section and will decrease by almost 100 mm in the Deltaic Plain section. Overall, climate of the Mississippi Alluvial Valley is predicted to be warmer and dryer in 2060.

Effects on Tree Species Distribution in Mississippi Alluvial Valley Sections

Twenty-three selected tree species were used for assessment of the effects of precipitation on forests in the Mississippi Alluvial Valley (Table 10.4). Minimum temperature limits on the occurrence of species ranged from 0°C for American elm to 16°C for water oak. Minimum precipitation requirements for native species ranged from 380 mm for eastern cottonwood (*Populus deltoides*) and American elm to 1270 mm for water oak. Mean annual temperatures predicted by the four climate scenarios are likely to be higher than the minimums associated with the northern limits of the distributions for all species. Precipitation predicted by three of the climate scenarios is expected to be greater than the minimum requirements for all species. For the MIROC3.2 A1B scenario, precipitation is expected to be slightly more than adequate for seven species but less than adequate for water oak and overcup oak. Tree species with annual precipitation requirements greater than 1050 mm, such as water oak and overcup oak, could decrease in area of occurrence.

Holocene Deposits: Vulnerability of tree species to climate change in the Holocene Deposits section of the Mississippi Alluvial Valley is predicted to be highest under the MIROC3.2 A1B scenario, which would result in an average 21% decrease in the range of the 23 tree species (Table 10.23). An averaged 54% reduction in area of occurrence is predicted for nine species including southern red oak, water oak, and overcup oak under the MIROC3.2 A1B scenario. None of the climate scenarios predicts an increase in the distribution of species. Little or no change is predicted for the ranges of 14 species, including green ash (*Fraxinus pennsylvanica*), red maple, water oak, and black willow. Overall, the effects of climate change on trees in the Holocene Deposits section are predicted to be small because of minor influence of slightly higher temperature and minimal reductions in annual precipitation.

Deltaic Plain: Vulnerability of tree species to climate change in the Deltaic Plain section of the Mississippi Alluvial Valley is predicted to be highest under the MIROC3.2 A1B scenario, which would result in an average 5% decrease in the range of the 23 tree species (Table 10.24). An average 63% reduction in area of occurrence is predicted for only two species: overcup oak and water oak. None of the climate scenarios predicts an increase in the distribution of species. Little or no change is predicted for the ranges of 21 of the 23 species studied, including sugarberry, slippery elm, and red maple. Overall, the effects of climate change on trees in the Deltaic Plain section could be minor because of minimal reductions in annual precipitation and probably little immediate influence of slightly higher temperatures.

Summary: Overall effects of future climate change will likely vary between the two sections of the Mississippi Alluvial Valley as a result of precipitation rather than temperature. The ranges of tree species will probably not be affected by slightly higher temperatures. A higher threat to forest

TABLE 10.23
Predicted Change (%) in Area of Distribution of Selected Tree Species in the Holocene Deposits Section of the Mississippi River Alluvial Valley from 2000 to 2060 in Response to Four Climate Scenarios[a]

	Species Area Change (%)			
Species[b]	**MIROC3.2**	**CSIROMK3.5**	**CSIROMK2**	**HadCM3**
Red mulberry	−42.3	0.0	0.0	0.0
Sweetgum	−50.0	0.0	0.0	0.0
Southern red oak	−50.0	0.0	0.0	0.0
American holly	−50.0	0.0	0.0	0.0
Water hickory	−50.0	0.0	0.0	0.0
Black tupelo	−50.0	0.0	0.0	0.0
Paulownia	−50.0	0.0	0.0	0.0
Overcup oak	−94.2	0.0	−1.9	−15.4
Water oak	−53.8	0.0	−13.5	−9.6
Average[c]	−21.3	0.0	−0.7	−1.1

Source: Adapted from Wear, D.N., R. Huggett, and J.G. Greis. In press. Constructing alternative futures. *Southern Forest Futures Project Technical Report.*

[a] Scenarios were developed from three general circulation models (MIROC3.2, CSIROMK2 and 3.5, and HadCM3) and two emissions storylines (A1B represents low population growth/high-energy use and B2 represents moderate population growth/low-energy use).

[b] Species with no change in each of the four scenarios are not listed.

[c] Species with no change are included.

tree species could occur in the Holocene Deposits section, where future annual precipitation could decrease. The tree species most vulnerable to climate change in this subregion will probably include overcup oak and water oak. Among climate scenarios, the MIROC3.2 A1B scenario forecasts the highest threat to plants resulting from possible climate change by 2060; the lowest threats are associated with the CSIROMK3.5 A1B and HadCM3 B2 scenarios.

TABLE 10.24
Predicted Change (%) in Area of Distribution of Selected Tree Species in the Deltaic Plain Section of the Mississippi River Alluvial Valley from 2000 to 2060 in Response to Four Climate Scenarios[a]

	Species Area Change (%)			
Species[b]	**MIROC3.2**	**CSIROMK3.5**	**CSIROMK2**	**HadCM3**
Overcup oak	−25.0	0.0	0.0	0.0
Water oak	−100.0	0.0	0.0	0.0
Average[b]	−5.4	0.0	0.0	0.0

Source: Adapted from Wear, D.N., R. Huggett, and J.G. Greis. In press. Constructing alternative futures. *Southern Forest Futures Project Technical Report.*

[a] Scenarios were developed from three general circulation models (MIROC3.2, CSIROMK2 and 3.5, and HadCM3) and two emissions storylines (A1B represents low population growth/high-energy use and B2 represents moderate population growth/low-energy use).

[b] Species with no change in each of the four scenarios are not listed.

[c] Species with no change are included.

Forest Diversity Effects

Tree diversity in the Mississippi Alluvial Valley is likely to significantly decrease according to predictions of future climate change by the MIROC3.2 A1B scenario, or to remain near current levels according to forecasts by the other three scenarios (Table 10.25). The largest change in diversity resulted from predictions by the MIROC3.2 A1B scenario, where the mean tree species richness could decrease from 22.6 currently to 18.8 in 2060. For the subregion as a whole, the data analysis suggests little change to tree diversity from the year 2000 to 2060 if the MIROC3.2 A1B scenario is considered unlikely to occur and is disregarded.

Mississippi Alluvial Valley section scale: For both sections of the Mississippi Alluvial Valley, the predicted future diversity was most variable for the MIROC3.2 A1B scenario, where species richness could decrease from the current average of 22.6 species in both regions to 18.1 species for the Holocene Deposits section and 21.8 species for the Deltaic Plain section in 2060 (Table 10.25). Predictions of diversity by the other scenarios were generally consistent and indicated no significant differences between current and future diversity. Overall risk to future diversity resulting from climate change is predicted to be low.

County scale: Changes in tree species diversity at the county level in the Mississippi Alluvial Valley indicate that the greatest declines would occur in the Holocene Deposits section, which extends in a broad band along the Mississippi River from central Louisiana to northern Arkansas. In that section, diversity averaged across all climate scenarios could decrease by an average of about three species for Bolivar County in Mississippi and Chicot County in Arkansas. Diversity would remain at current levels for Coahoma County and Tunica County in Mississippi, but would decrease slightly, by an average of one to two species, in other counties. Decreases in diversity of about one species could occur uniformly throughout much of the Deltaic Plain section in southern Louisiana, with no particular hotspot of obvious decline of tree species.

Overall, the effects of climate change on diversity at the county scale in the Mississippi Alluvial Valley could be a more or less uniform decline of several species throughout. The MIROC3.2 A1B scenario consistently predicted the likely loss of two to three species in most counties, but predictions by the other scenarios were variable, ranging from little or no change to small decreases

TABLE 10.25
Forest Diversity, Expressed as Mean Tree Species Richness (Standard Deviation) for Mississippi Alluvial Valley Sections from 2000 to 2060 Estimated by a Model Based on Predictions of Annual Precipitation from Four Climate Scenarios[a]

Section	Current Diversity	Future (2060) Diversity Predicted by Each Scenario			
		MIROC3.2	CSIROMK3.5	CSIROMK2	HadCM3
		N Tree Species (SD)			
Holocene Deposits	22.5 (0.6)	18.1 (3.6)[b]	22.5 (0.6)	22.3 (0.6)	22.3 (0.7)
Deltaic Plain	23.0 (0.4)	21.8 (0.4)[b]	23.0 (0.0)	23.0 (0.0)	23.0 (0.0)
Miss. Valley Subregion	22.6 (0.5)	18.8 (3.5)[b]	22.6 (0.5)	22.5 (0.6)	22.4 (0.7)[b]

Source: Adapted from Wear, D.N., R. Huggett, and J.G. Greis. In press. Constructing alternative futures. *Southern Forest Futures Project Technical Report.*

[a] Scenarios were developed from three general circulation models (MIROC3.2, CSIROMK2 and 3.5, and HadCM3) and two emissions storylines (A1B represents low population growth/high-energy use and B2 represents moderate population growth/low-energy use).

[b] Indicates significant difference between current (2000) and future (2060) species richness at the $p = 0.05$ level of probability.

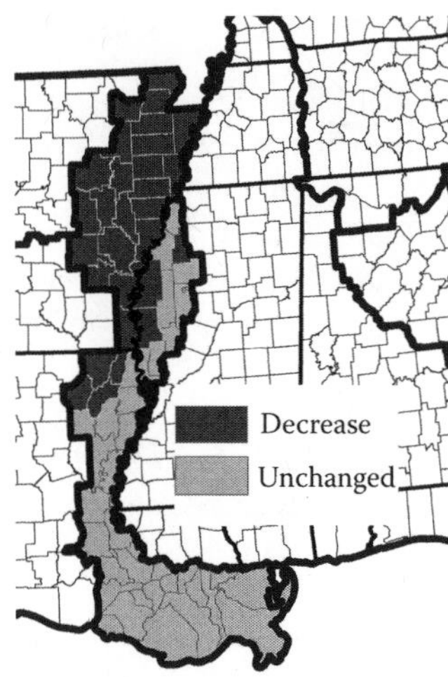

FIGURE 10.21 Future (2060) diversity expressed as a percent of current (2000) diversity averaged across four climate scenarios for the Mississippi Alluvial Valley (Wear et al. in press); that were developed from three general circulation models (MIROC3.2, CSIROMK2 and 3.5, and HadCM3) and two emissions storylines (A1B represents low population growth/high-energy use and B2 represents moderate population growth/low-energy use). Diversity is predicted to decrease in counties shaded dark gray (<97.5% of current), remain generally unchanged in counties shaded light gray (97.5 to 102.5% of current) and increase in counties shaded medium gray (>102.5% of current).

for some counties. Future diversity averaged across the four scenarios indicated a decrease in the northern part of the subregion and little or no change likely in southern part (Figure 10.21).

MID-SOUTH

Vegetation of the Mid-South, which occupies about 37% (806,944 km^2) of the Southern Region, is variable, ranging from oak and pine forests in the east, through a transition woodlands zone of Cross Timbers, to the grassland plains of central Texas, and finally to the scrublands of near desert conditions in western Texas. The four sections of this subregion are the Ozark-Ouachita Highlands (11.1%), Cross Timbers (27.2%), High Plains (48.7%), and the West Texas Basin and Range (13.0%). From east to west, they define zones of progressively decreasing precipitation.

Ecotones are likely to be particularly sensitive to changes in precipitation and temperature since tree species that exist there are at their climatic range limits (Risser 1995; Stahle and Hehr 1984). Post oak in the Cross Timbers section ecotone is one example. A study conducted in Arkansas, Oklahoma, and Texas by Stahle and Hehr (1984) showed increased climate sensitivity of post oak with declining rainfall. Most of the sites in their study were in the Ozark-Ouachita Highlands and Cross Timbers sections that we examine below.

Climate Scenario Predictions

The long-term trend of temperature in the Mid-South is predicted to increase from the year 2000 to 2060 based on forecasts for all sections averaged over the four climate scenarios (Figure 10.22). The rate of increase is approximately 0.5°C per decade for each of the four sections. An almost constant difference of about 2°C separates the Ozark-Ouachita Highlands section from the other three for the 60-year forecast period. For precipitation, the averaged prediction from the climate scenarios indicates a long-term trend of decreasing amounts (Figure 10.23). The precipitation trend for the West Texas Basin and Range section remains about constant from the year 2000 to 2060, compared

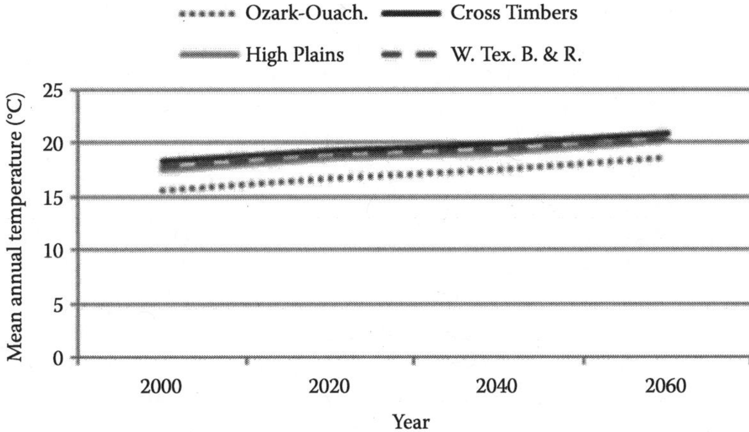

FIGURE 10.22 Average annual temperature from 2000 (current) to 2060 (predicted) for sections of the Mid-South; temperature for each section represents the average of four climate scenarios (Wear et al. in press); that were developed from three general circulation models (MIROC3.2, CSIROMK2 and 3.5, and HadCM3) and two emissions storylines (A1B represents low population growth/high-energy use and B2 represents moderate population growth/low-energy use).

to a decrease for the other sections (especially the Ozark-Ouachita Highlands and Cross Timbers sections). An average of the four climate scenarios suggests a total temperature increase of about 5°C and a precipitation decrease of about 100 mm across the four sections, although amounts vary for each climate scenario.

Three of the four climate scenarios forecast only slight changes in average annual precipitation from the year 2000 to 2060 for the four sections of the Mid-South; only the MIROC3.2 A1B scenario predicts a large decrease across all sections. Future precipitation predictions by the other climate scenarios differ among sections, ranging from similar amounts for the High Plains and West Texas Ranges and Basins sections to differing amounts for the Ozark-Ouachita Highlands section. The CSIROMK3.5 A1B scenario predicts precipitation to decrease by about 150 mm in

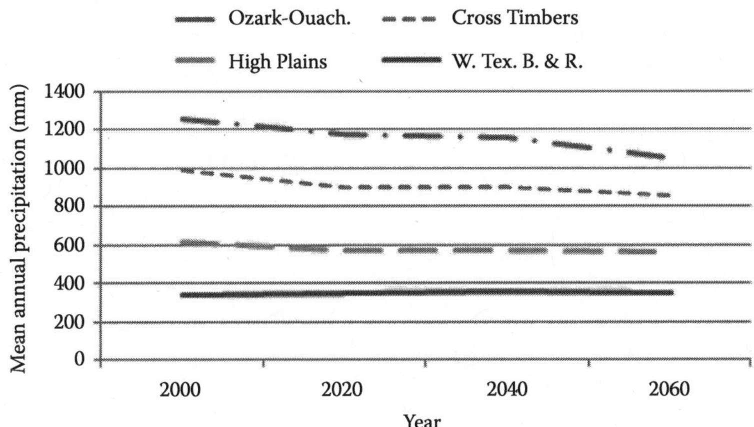

FIGURE 10.23 Average annual precipitation from 2000 (current) to 2060 (predicted) for sections of the Mid-South; precipitation for each section represents the average of four climate scenarios (Wear et al. in press); that were developed from three general circulation models (MIROC3.2, CSIROMK2 and 3.5, and HadCM3) and two emissions storylines (A1B represents low population growth/high-energy use and B2 represents moderate population growth/low-energy use).

the Ozark-Ouachita Highlands section and almost 100 mm in the Cross Timbers section. With the exception of the MIROC3.2 A1B scenario, all climate scenarios forecast future precipitation to remain constant or increase slightly in the High Plains and West Texas sections. Overall, climate of the Mid-South is predicted to be warmer and dryer in 2060.

Under all four climate scenarios, temperature is predicted to increase while precipitation is predicted to decrease over the 60-year period from 2000 to 2060 in the Ozark-Ouachita Highlands and Cross Timbers regions. Of the four climate scenarios, MIROC3.2 A1B predicts the largest reduction in average annual precipitation and largest increase in temperature: a 301-mm precipitation decrease and 4.2°C temperature increase in the Ozark-Ouachita Highlands; a 231-mm precipitation decrease and 4.0°C temperature increase in the Cross Timbers section of Oklahoma; and a 50-mm precipitation decrease and 3.1°C temperature increase in the Texas portion of the Cross Timbers section. Although both sections are predicted to follow a generally decreasing trend in annual precipitation, the Cross Timbers section is predicted to receive at least 200 mm less than the Ozark-Ouachita Highlands over the 60-year period.

The Mid-South was subdivided in two zones (eastern and western) to facilitate evaluation of vulnerability. Tree cover in eastern zone, consisting of the Ozark Plateau and Cross Timbers sections, includes over a dozen species. In the drier western zone (High Plains and West Texas Ranges and Basins sections), shrubs predominate and trees are limited to two or three species that are associated with increased moisture in ravines or higher mountains.

Eastern zone: The Ozark-Ouachita Highlands, also known as the Interior Highlands, encompasses both the Ozark Mountains and the Ouachita Mountains in Arkansas and Oklahoma. The Ozark area is mainly forested, consisting primarily of upland hardwoods with pine as a secondary but economically important component, whereas the Ouachita Mountains are dominated by shortleaf pine with a hardwood component. The Cross Timbers section runs from central Texas through central Oklahoma and into the southeastern corner of Kansas. This section forms an ecotone between the treeless Great Plains to the west and the forests to the east. Dominant tree species are blackjack and post oak with eastern redcedar being a major invader where fire has been suppressed.

Leininger (1998) reported on a study that examined the impact of elevated temperature and drought on tree seedlings. He predicted that trees over a broad area could be susceptible to decline that is induced by a prolonged reduction in annual precipitation. Additionally, the combination of increased temperature and declining precipitation could significantly change both the frequency and severity of southern pine beetle (*Dendroctonus frontalis*) outbreaks in shortleaf pine stands between now and 2060 (McNulty et al. 1998). Gan (2004) predicted that under future climate change the southern pine beetle could kill from 4 to 7.5 times the 2004 value of trees killed by the beetle. Changes in winter climatic conditions could also affect populations of pine engraver beetles (*Ips* spp), which are also serious pests of southern pines (Lombardero et al. 2000).

Western zone: The High Plains section is an area of mostly hilly landscapes that extends in a broad zone from the panhandle of western Oklahoma southward through central Texas to Mexico, with vegetation that ranges from grasslands to shrub lands. The northern part of this zone borders on the grasslands of the Great Plains where predominant vegetation consists of short prairie grasses with some juniper woodlands. The middle part of the zone, in central Texas, occupies an area known as the Edwards Plateau, where the dissected, hilly topography is underlain by limestone and granite formations and vegetation ranges from oak woodlands to juniper in eroded areas of rock outcrops and steep bluffs. In the southern part of the High Plains section, bordering the Rio Grande River, topography becomes more plains-like with vegetation ranging from low trees on woodlands to thorny brush on shrublands.

The West Texas Basin and Range section, which occupies the Big Bend part of Texas, is northern part of the Mexican Chihuahua Desert and includes the most mountainous area of the Mid-South subregion. Desert grass and scrub lands separate the six to eight relatively short mountain ranges where tree vegetation is a scattered mixture of low scrub oaks and junipers. Elevations range from about 900 to 1200 m on valley floors, to 1500 to 2650 m on mountain peaks. Precipitation ranges from less than 20 cm in the low desert valleys to about 50 cm on the highest mountains.

Effects on Tree Species Distribution in Eastern Zone Sections of the Mid-South

Cross Timbers: Eleven tree species were selected to assess the effects of temperature and precipitation on forests in the Cross Timbers section of the Mid-South (Table 10.4). All four climate scenarios predicted higher temperature and lower precipitation. Vulnerability of tree species to climate change is predicted to be highest under the MIROC3.2 A1B scenario, which would result in an average 12% decrease in the range of the 11 species (Table 10.26). An averaged 5% reduction in area of occurrence is predicted under the CSIROMK3.5 A1B, CSIROMK2 B2, and HadCM3 B2 scenarios for black oak, winged elm, and Shumard's oak by 2060. Much of the possible change in area of predicted occurrence could result from reduced precipitation; the predicted, slightly higher temperatures could have little immediate effects on the distribution of tree species.

The average annual precipitation predicted by the MIROC3.2 A1B scenario in the year 2060 (736 mm) would drop below the average minimum precipitation requirement for Shumard's oak, winged elm, and black oak. However, in the westernmost counties of Oklahoma, the scenario predicts average annual precipitation to be 675 mm. This decreased level of precipitation would likely impact black oak more severely but not add species to the precipitation deficit list. In the rest of the area, average annual precipitation is 789 mm—above minimum annual precipitation for black oak but still below that of winged elm and Shumard's oak. Temperature would increase by 4°C in Oklahoma and 3.1°C in Texas. The other three climate scenarios predicted temperature increases of 3°C in Oklahoma and from 2.0 (CSIROMK3.5 A1B and CSIROMK2 B2) to 2.3°C (HadCM3 B2) in Texas.

Ozark-Ouachita Highlands: Thirty-one tree species were selected to assess the effects of temperature and precipitation on forests in the Ozark-Ouachita Highlands of Arkansas and Oklahoma (Table 10.4). All of the scenarios predicted temperatures above the minimum associated with the current range of the evaluated species. Only one of the four climate scenarios (MIROC3.2 A1B) predicts average precipitation to drop below the average minimum needed by any of the 31 species. Eleven species are predicted to be in a precipitation deficit situation, including nine hardwoods and two conifers. The two most important pines in the area, shortleaf and loblolly, are predicted to be affected. Vulnerability of tree species is predicted to be highest under the MIROC3.2 A1B scenario, which would result in an average 26% decrease of the 31 species (Table 10.27). By 2060, precipitation in the Ozark-Ouachita Highlands section (1050 mm) is forecasted to begin approaching the

TABLE 10.26
Predicted Change (%) in Area of Distribution of Selected Tree Species in the Cross-Timbers Section of the Mid-South from 2000 to 2060 in Response to Four Climate Scenarios[a]

	Species Area Change (%)			
Species[b]	MIROC3.2	CSIROMK3.5	CSIROMK2	HadCM3
Black oak	−51.4	−1.8	−18.9	−5.4
Winged elm	−44.1	−22.5	−29.7	−34.2
Shumard's oak	−17.1	−14.4	−14.4	−17.1
Average[c]	−11.9	−3.5	−5.9	−5.2

Source: Adapted from Wear, D.N., R. Huggett, and J.G. Greis. In press. Constructing alternative futures. *Southern Forest Futures Project Technical Report.*

[a] Scenarios were developed from three general circulation models (MIROC3.2, CSIROMK2 and 3.5, and HadCM3) and two emissions storylines (A1B represents low population growth/high-energy use and B2 represents moderate population growth/low-energy use).

[b] Species with no change in each of the four scenarios are not listed.

[c] Species with no change are included.

TABLE 10.27
Predicted Change (%) in Area of Distribution of Selected Tree Species in the Ozark-Ouachita Highlands Section of the Mid-South from 2000 to 2060 in Response to Four Climate Scenarios[a]

	Species Area Change (%)			
Species[b]	MIROC3.2	CSIROMK3.5	CSIROMK2	HadCM3
Mockernut hickory	−6.1	0.0	0.0	0.0
Cucumber tree	−6.1	0.0	0.0	0.0
Black cherry	−57.1	0.0	−20.4	0.0
Winged elm	−87.8	−4.1	−32.7	−12.2
Sweetgum	−91.8	−6.1	−32.7	−22.4
Southern red oak	−91.8	−6.1	−32.7	−22.4
Loblolly pine	−91.8	−6.1	−32.7	−22.4
American holly	−91.8	−6.1	−32.7	−22.4
Black gum	−91.8	−6.1	−32.7	−22.4
Umbrella magnolia	−91.8	−6.1	−32.7	−22.4
Shortleaf pine	−91.8	−6.1	−32.7	−22.4
Average[c]	−25.8	−1.5	−9.1	−5.5

Source: Adapted from Wear, D.N., R. Huggett, and J.G. Greis. In press. Constructing alternative futures. *Southern Forest Futures Project Technical Report.*

[a] Scenarios were developed from three general circulation models (MIROC3.2, CSIROMK2 and 3.5, and HadCM3) and two emissions storylines (A1B represents low population growth/high-energy use and B2 represents moderate population growth/low-energy use).

[b] Species with no change in each of the four scenarios are not listed.

[c] Species with no change are included.

2000 level in the Cross Timbers section (991 mm), a difference of only 59 mm. If the trend continues, the Ozark-Ouachita Highlands section could begin to emulate the vegetation and habitat characteristics of the Cross Timbers section.

Several nonnative invasive species that already threaten forests in the region could gain a competitive advantage as some competing species die and others become stressed at 2060 precipitation levels. The following species, listed with their lower precipitation requirements, could be particularly competitive: Russian olive (*Elaegnus angustifolia*)—305 mm; tree of heaven—356 mm; and Siberian elm (*Ulmus pumila*)—406 mm. Although Russian olive is not currently listed as a nonnative of concern in Arkansas, it has been reported in nearby Oklahoma, Texas, Missouri, and Tennessee (Invasive.org).

Summary: The trends of all four climate scenarios indicate declining precipitation and increasing temperature in the Cross Timbers and Ozark-Ouachita Highlands sections. Although predictions do not indicate some species to be in a deficit situation, trees may become stressed if the predicted decreases in annual precipitation continue to drop close to their limits.

Given these possibilities, in management over the next 15 years it would be wise to focus on species diversity to maintain all species components. Beyond the next 15 years, greater emphasis could be given to the maintenance of those species most likely to survive through 2060. However, periodic reassessment will be needed as part of an adaptive management strategy.

Effects on Tree Species Distribution in Western Zone Sections of the Mid-South

High Plains: Three selected tree species—plains cottonwood (*Populus deltoides* spp), Lacey oak (*Q. laceyi*), and oneseed juniper (*Juniperus monosperma*)—were used for assessment of the effects of temperature and precipitation in 2060 on forests in the High Plains section of the Mid-South (Table 10.4). The minimum temperatures associated with their current northern and western distributions ranged from 0°C for plains cottonwood to 15°C for Lacey oak. Minimum precipitation requirement was 250 mm for all species. Temperature and precipitation predicted for 2060 by the four climate scenarios is expected to be greater than the minimum requirements for all species.

Vulnerability of tree species to climate change in the High Plains section is predicted to be similar under all climate scenarios and should have little effect on the current range of the three tree species (Table 10.28). Overall, effects of climate change on tree species in the High Plains section of the Mid-South are predicted to be minor because of minimal changes in annual temperature and precipitation.

West Texas Basin and Range: Three selected tree species—gray oak (*Q. grisea*), oneseed juniper, and Mexican pinyon (*Pinus cembroides*)—were used for assessment of the effects of temperature and precipitation in the year 2060 on forests in the West Texas Basin and Range section of the Mid-South (Table 10.4). The minimum temperature associated with their current northern and western distributions ranged from 4°C for Mexican pinyon to 13°C for Gray oak. The minimum precipitation requirement is 250 mm for all species. Precipitation predicted for 2060 by the four climate scenarios is expected to be greater than the minimum requirements for all species.

Vulnerability of tree species to climate change in the West Texas Basin and Range section is predicted to be similar under all climate scenarios and should have little effect on the current range of the three selected tree species (Table 10.29). Overall, effects of climate change on tree species in this section of the Mid-South are predicted to be minor because of little likely immediate effects of somewhat higher temperature and minimal changes in annual precipitation.

Summary: Overall effects of future climate change on vegetation will likely be relatively small for both the High Plains and West Texas Basin and Range sections of the Mid-South. Warmer temperatures will likely affect potential evapotranspiration, but will be above the minimums that limit the northern and western distribution of tree species. A slightly greater threat to forest tree species could occur in the High Plains section, where future annual precipitation could decrease; precipitation in the West Texas Range and Basins section is predicted to either remain similar to current amounts or increase slightly. Tree species are about equal in vulnerability to reduced precipitation and their future distributions should show little changes. Among climate scenarios, the MIROC3.2 A1B scenario forecasts the highest threat to trees, followed by the CSIROMK2 B2 scenario.

TABLE 10.28
Predicted Change (%) in Area of Distribution of Selected Tree Species in the High Plains Section of the Mid-South from 2000 to 2060 in Response to Four Climate Scenarios[a]

	Species Area Change (%)			
Species	**MIROC3.2**	**CSIROMK3.5**	**CSIROMK2**	**HadCM3**
Oneseed juniper	0.0	0.0	0.0	0.0
Plains cottonwood	0.0	0.0	0.0	0.0
Lacey oak	0.0	0.0	0.0	0.0
Average	0.0	0.0	0.0	0.0

Source: Adapted from Wear, D.N., R. Huggett, and J.G. Greis. In press. Constructing alternative futures. *Southern Forest Futures Project Technical Report.*

[a] Scenarios were developed from three general circulation models (MIROC3.2, CSIROMK2 and 3.5, and HadCM3) and two emissions storylines (A1B represents low population growth/high-energy use and B2 represents moderate population growth/low-energy use).

TABLE 10.29
Predicted Change (%) in Area of Distribution of Selected Tree Species in the West Texas Basin and Range Section of the Mid-South from 2000 to 2060 in Response to Four Climate Scenarios[a]

	Species Area Change (%)			
Species	**MIROC3.2**	**CSIROMK3.5**	**CSIROMK2**	**HadCM3**
Oneseed juniper	−6.7	0.0	0.0	6.7
Mexican pinyon	−6.7	0.0	0.0	6.7
Gray oak	−6.7	0.0	0.0	6.7
Average	−6.7	0.0	0.0	6.7

Source: Adapted from Wear, D.N., R. Huggett, and J.G. Greis. In press. Constructing alternative futures. *Southern Forest Futures Project Technical Report.*

[a] Scenarios were developed from three general circulation models (MIROC3.2, CSIROMK2 and 3.5, and HadCM3) and two emissions storylines (A1B represents low population growth/high-energy use and B2 represents moderate population growth/low-energy use).

Forest Diversity Effects

Future tree diversity will likely decrease in the eastern zone of the Mid-South and remain unchanged in the dry western zone (Table 10.30). In the eastern zone, differences between the current and future species richness are largest for the MIROC3.2 A1B climate scenario (3.6 species) compared to an average decrease of about 1.1 species for the other climate scenarios. All scenarios predict no change in diversity for the western zone.

TABLE 10.30
Forest Diversity, Expressed as Mean Tree Species Richness (Standard Deviation) for Mid-South Sections from 2000 to 2060 Estimated by a Model Based on Predictions of Annual Precipitation from Four Climate Scenarios[a]

Section and Zone	**Current Diversity**	**Future (2060) Diversity Predicted by Each Scenario**			
		MIROC3.2	**CSIROMK3.5**	**CSIROMK2**	**HadCM3**
			***N* Tree Species (SD)**		
Ozark-Ouachita	31.0 (0.0)	22.6 (2.0)[b]	30.3 (2.2)[b]	27.8 (4.3)[b]	29.1 (3.4)[b]
Cross Timbers	09.7 (0.9)	08.3 (0.7)[b]	09.2 (0.5)[b]	08.9 (0.7)[b]	09.0 (0.4)[b]
Eastern zone	16.2 (9.9)	12.6 (6.7)[b]	15.6 (9.9)[b]	14.6 (9.1)[b]	15.1 (9.5)[b]
High Plains	03.0 (0.0)	03.0 (0.0)	03.0 (0.0)	03.0 (0.0)	03.0 (0.0)
West Texas R&B	03.0 (0.0)	03.0 (0.0)	03.0 (0.0)	03.0 (0.0)	03.0 (0.0)
Western zone	03.0 (0.0)	03.0 (0.0)	03.0 (0.0)	03.0 (0.0)	03.0 (0.0)

Source: Adapted from Wear, D.N., R. Huggett, and J.G. Greis. In press. Constructing alternative futures. *Southern Forest Futures Project Technical Report.*

[a] Scenarios were developed from three general circulation models (MIROC3.2, CSIROMK2 and 3.5, and HadCM3) and two emissions storylines (A1B represents low population growth/high-energy use and B2 represents moderate population growth/low-energy use).

[b] Indicates significant difference between current (2000) and future (2060) species richness at the $p = 0.05$ level of probability.

Mid-South section scale: Predicted changes in future diversity were greatest for the Ozark-Ouachita Highlands section, where diversity in the year 2060 could decrease by an average of 0.7–8.4 species (Table 10.30). Prediction of diversity by the year 2060 in the Cross Timbers section varied less, with estimates of change ranging from 0.5 to 1.4 species. Risk of tree diversity being affected by climate change is likely to be moderate to high in the eastern sections. In the western two sections, risk of climate change affecting diversity is likely to be low.

County scale: Changes in tree species diversity at the county level in the Mid-South indicate that the largest declines would occur in the eastern zone, which includes the Ozark-Ouachita Highlands section (northwestern Arkansas and eastern Oklahoma) and the Cross Timbers section (central Oklahoma and eastern central Texas). For many counties in the eastern zone, including Polk and Saline in Arkansas, diversity of tree species would remain at current levels. Climate change would be most apparent in two Oklahoma hotspots, with richness declining by nine species in McIntosh County and seven species in Mayes County. Diversity changes in the adjacent Cross Timbers section to the west would be similar but not as severe. Average diversity was predicted to increase for Nueces County and Calhoun County near the Gulf of Mexico. An average of two tree species would be lost by Rogers County and Okfuskee County in Oklahoma, but little or no measurable change in diversity will occur in counties throughout much of this section.

The least change in tree-species richness at the county level in the Southern Region would occur in the western zone of the Mid-South subregion, where the current dry and hot environment is suitable for few species of trees. Throughout the 170 counties in this zone, diversity is predicted to decline only for Loving County in Texas (near New Mexico) and only by one tree species. The MIROC3.2 A1B scenario consistently predicted the likely loss of two to three species in most counties in the Cross Timbers section, but predictions by the other climate scenarios were slight, ranging from little or no change to small decreases for some counties. Future diversity averaged across the four climate scenarios indicated decreased diversity in the Cross Timbers section and little or no change likely in the High Plains and West Texas Basin and Range sections, where most species present are tolerant of drought (Figure 10.24).

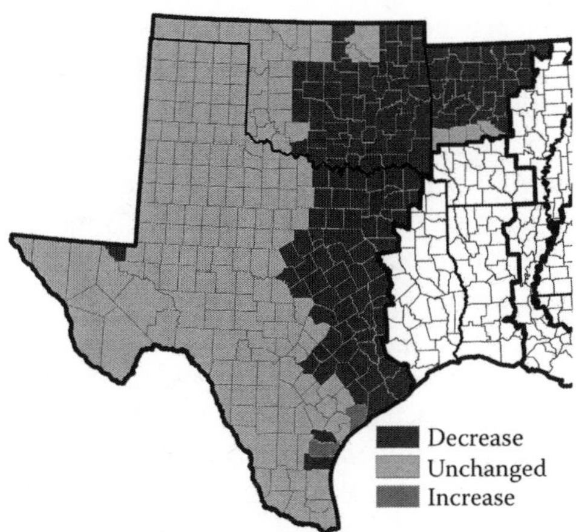

FIGURE 10.24 Future (2060) diversity expressed as a percent of current (2000) diversity averaged across four climate scenarios for the Mid-South (Wear et al. in press); that were developed from three general circulation models (MIROC3.2, CSIROMK2 and 3.5, and HadCM3) and two emissions storylines (A1B represents low population growth/high energy use and B2 represents moderate population growth/low energy use). Diversity is predicted to decrease in counties shaded dark gray (<97.5% of current), remain generally unchanged in counties shaded light gray (97.5 to 102.5% of current) and increase in counties shaded medium gray (>102.5% of current).

CASE STUDIES

Case Study 1: Understory Vegetation in Longleaf Pine Stands of the Coastal Plain's Western Gulf

If climate change results in more intensive droughts in the Coastal Plain's Western Gulf, understory vegetation would be affected as well as overstory trees (U.S. Global Change Research Program 2009). To examine drought patterns, Palmer Drought Severity Index values from the National Climatic Data Center (2011) were applied for the four climate scenarios over the historical range of longleaf pine (*Pinus palustris*)—central Louisiana, western central Louisiana, northern central Louisiana, and eastern Texas. From 1895 through 2009, droughts occurred 36% of the time in the longleaf pine range based on drought severity classifications from Hayes (2010), which were severe to extreme 6% of the time. In the last 50 years, droughts occurred 33% of the time with severe-to-extreme droughts occurring 5% of the time. This suggests that drought conditions are moderating, with the exception of 2000–2009, when they again occurred 36% of the time and were severe-to-extreme 6% of the time. Year-to-year climate is highly variable, but clearly, forests in the Western Gulf have been subjected to recurring droughts that become severe to extreme only occasionally.

However, should the pattern of drought intensify, as predicted by some climate scenarios (Global Change Research Program 2009), then forests that have adapted to mild-to-moderate droughts 36% of the time would become more stressed and at risk for wildfires (Outcalt and Wade 2004). This would favor the dominant bluestem grasses (*Andropogon* spp and *Schizachyrium* spp) and other common herbaceous plants (Haywood and Harris 2000; Wade et al. 2000) that are indicative of longleaf pine understories that Turner et al. (1999) described. Although favored by recurring fires, herbaceous plant yields are negatively affected by drought as expressed in the following relationship for longleaf pine stands (Wolters 1982):

$$\text{HP} = 2094.75 + 10.10\ [\text{April through October precipitation (cm)}] - 106.96 \times \text{BA},$$

where HP is herbaceous plant production in kg/ha and BA is basal area in m^2/ha.

Production of herbaceous plant material varied widely among the four climate scenarios between 2000 and 2060 (Figure 10.25). The MIROC3.2 model predicted a small increase of herbaceous production for 2020, followed by a sharp decline in 2040 and 2060. In contrast, the CSIROMK2 scenario predicts a decline for 2020, followed by recovery to current levels by 2040, and a large increase in 2060. The CSIROMK3.5 model predicts little or no change in production by 2060. The uncertainty of future herbaceous plant production likely arises from variation in estimates of growing season precipitation by the four scenarios because all other variables in the model were held constant, which demonstrates the challenges facing future resource management in consideration of a changing climate.

Changing climate has been found to be partly a response to rising CO_2 levels in the atmosphere (Intergovernmental Panel on Climate Change 2007). Three herbaceous plant species—wiregrass (*Aristida stricta*), rattlebox (*Crotalaria rotundifolia*), and butterfly milkweed (*Asclepias tuberosa*)—growing with longleaf pine in open-top chambers, had lower growth rates when exposed to elevated CO_2 concentrations compared to ambient CO_2 concentrations (Runion et al. 2006). However, longleaf pine had higher growth rates under the elevated CO_2 concentrations. Nevertheless, in field environments, higher temperatures increased respiration rates and droughts have been shown to reduce the amount of water available to plants and disrupt plant physiology (Sword Sayer and Haywood 2006), possibly leading to lower growth rates of trees as predicted in the Canadian climate scenario (U.S. Global Change Research Program 2009).

If the climate of the Southern United States becomes warmer with droughts occurring more frequently, wildfires will become more extensive and intense (U.S. Global Change Research Program

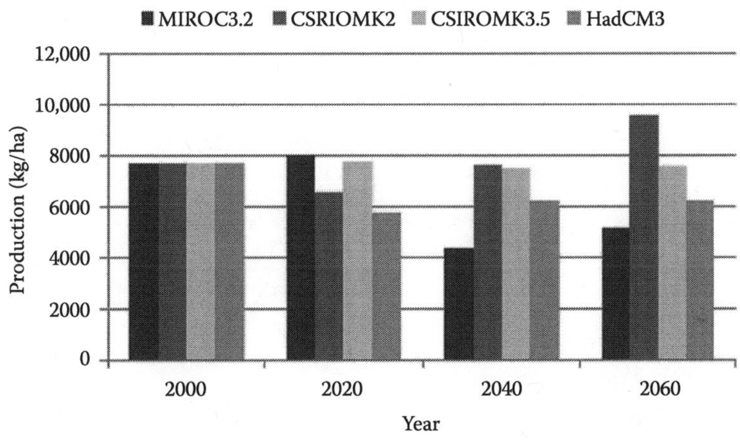

FIGURE 10.25 Herbaceous plant production in longleaf pine stands (25 m^2/ha) in the Western Gulf section of the southern Coastal Plain predicted by four climate scenarios (Wear et al. in press); scenarios were developed from three general circulation models (MIROC3.2, CSIROMK2 and 3.5, and HadCM3) and two emissions storylines (A1B represents low population growth/high-energy use and B2 represents moderate population growth/low-energy use).

2009; Chapter 5). Under normal climatic conditions, forest floor duff is moist and consumption of duff is minor, which protects tree roots during fires (Varner et al. 2007). However, as drought conditions lead to the drying out of the duff layer, virtually all of the forest floor can be consumed, and pine mortality can be severe (Outcalt and Wade 2004). Overstory and midstory pine and hardwood mortality can transform dense mixed pine and hardwood stands into open woodlands (Brockway et al. 2006; Haywood 2009; Haywood 2011; Wade et al. 2000), as occurs in national forests and other holdings where prescribed fire is continually reapplied, which is desirable to promote forest wildlife habitat.

Case Study 2: Effects of Climate Change on the Potential Range of *Eucalyptus grandis* in Florida

The genus Eucalyptus (*Eucalyptus*) consists of more than 700 species that are endemic primarily to Australia and Tasmania in regions of temperate conditions where winters are mild and the onset of freezing temperatures is gradual (Boland et al. 2006). The favorable tree form, wood properties, and growth rate of some species has resulted in extensive plantings of eucalypts in many countries beyond their native range, particularly Brazil, South Africa, Spain, and India (Booth and Pryor 1991). In the United States, commercial and trial plantings of eucalypts have been made in southern California and throughout much of Florida (Geary et al. 1983).

The potential for growing eucalypts in Florida has long been recognized (Meskimen and Francis 1990; Zon and Briscoe 1911). Eucalypts have been systematically evaluated for industrial wood products since the 1970s, initially as a source of hardwood pulp for fine papers and more recently for a range of other products, including fuelwood and mulch (Geary and others 1983; Meskimen and Franklin 1978; Rockwood 2012). Because the eucalypts vary in their sensitivity to frost, field trials have resulted in delineation of climatic zones based on severity of freeze damage to various species (Geary and others 1983). Most field research with eucalyptus has been done in southern Florida, where severe winter freezes (–5°C for 12 h) are uncommon (Meskimen et al. 1987). Within the southern climatic zone, planting sites vary primarily by their soil moisture regime, which ranges from excessively drained sandy ridges to poorly drained swampy prairies.

Considerable observations and data are available for *E. grandis*, a highly productive species that is suitable for a broad range of planting sites, except for the extreme driest or wettest soils (Geary

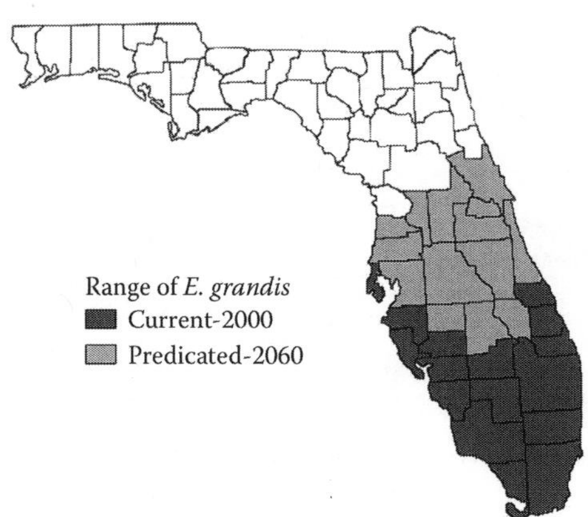

FIGURE 10.26 Current (2000) and predicted (2060) range of *Eucalyptus grandis* in Florida in response to the HadCM3 general circulation model and the B2 storyline (moderate population growth and low-energy use) scenario for a prediction model based on mean annual air temperature and mean annual potential evapotranspiration.

et al. 1983). *E. grandis* is sensitive to freezing temperatures and has long been recommended as suitable for planting in southern Florida, where the risk of frost damage is lowest (Figure 10.26). However, this species has been the subject of field trials and selection of clones for tolerance of freezing temperatures, which would allow it to be planted north of its currently recommended range (Keller et al. 2009; Meskimen et al. 1987; Rockwood 2012). The potential effects of a warmer climate on the potential area suitable for growing *E. grandis* has been reported for Australia (Hughes et al. 1996), but such information is not available for Florida.

The purpose of this case study was to investigate the probable effects of climate change on the area that might be suitable for management of *E. grandis* in Florida in 2060. A classification model was developed based on current values of commonly reported meteorological variables associated with the current range of *E. grandis* and the model was applied to estimate the potential future range of the species based on a future climate change scenario forecasted for Florida in 2060.

Methods: The source of current (2000) and predicted (2060) climate data for Florida was presented for four scenarios by Wear and others (in press). The four scenarios predict higher temperature and potential evapotranspiration for 2060 compared to 2000 (Table 10.31). Precipitation predictions will be variable, however, with the CSIROMK2 forecasting slightly less than current values and the MIROC3.2 scenario projecting much less. The HadCM3 scenario was selected because it forecasted moderate conditions that were generally between the extreme predictions of the other models.

E. grandis has been planted at a number of locations throughout Florida to evaluate survival, growth, and productivity (Geary et al. 1983). Evaluation of the field plantings revealed that the survival and growth of *E. grandis* in Florida appeared to be related to the magnitude and duration of subfreezing winter temperature (Meskimen and others 1987). The subpopulation of sites with satisfactory performance of *E. grandis* was used to identify counties currently suitable for its commercial production, thereby defining the current range of the species in Florida (Figure 10.1 in Geary et al. 1983). The current range of *E. grandis* passed through the interior of nine counties that represent a transition zone; transition counties were classified as either inside or outside the range based the greatest area of each category. Correlation analysis was used to evaluate the relationship of the climatic variables for all Florida counties with their geographic location (quantified by latitude and longitude). Analysis of variance was used to test for significant differences of individual climate variables for counties inside and outside of the current range for *E. grandis*. Results of both the correlation and variance analyses were used to interpret climate gradients for development of a predictive model that describes the current climate conditions associated with the range of *E. grandis* in Florida in 2000.

TABLE 10.31
Current (2000) and Predicted (2060) Levels of Climate Variables in Response to Four Climate Change Scenarios for Florida[a]

Climate Variable[b]	Current (2000)	Predicted Change (%) by Scenario for 2060			
		MIROC3.2	CSIROMK3.5	CSIROMK2	HadCM3
Temperature (°C)	21.5	110.4	107.2	107.2	106.1
Precipitation (mm)	1416.9	58.6	105.0	97.7	104.9
PET[c] (mm)	2588.8	118.1	111.0	111.1	108.5

Source: Adapted from Wear, D.N., R. Huggett, and J.G. Greis. In press. Constructing alternative futures. *Southern Forest Futures Project Technical Report.*

[a] Scenarios were developed from three general circulation models (MIROC3.2, CSIROMK2 and 3.5, and HadCM3) and two emissions storylines (A1B represents low population growth/high-energy use and B2 represents moderate population growth/low-energy use).

[b] Sample units are Florida counties ($n = 67$); climate variables were weighted by area and are expressed as average annual values.

[c] PET: potential evapotranspiration.

An incidence analysis of climate in the counties in which *E. grandis* currently occurs was performed to develop a function for predicting its potentially suitable range. Each county was assigned a binary code of 0 or 1, depending on whether it was outside or inside the current range. An incidence function was developed from three meteorological variables associated with current climate: average annual temperature, average annual precipitation, and average annual potential evapotranspiration. Maximum likelihood logistic regression was used to determine the interrelationships of climatic variables with the current occurrence of *E. grandis* (Crawley 2005). The model was

$$P\,(\text{presence} = 1) = \exp(\,b_iX_i)/1 + \exp(b_iX_i),$$

where P is the estimated probability of *E. grandis* occurrence, b_i is the vector of regression coefficients, X_i is the vector of explanatory variables, and exp is the base of the natural logarithms. Each of the three climatic variables was tested for its level of statistical significance in the regression analysis individually and in combination with the other variables. Effects of multicollinearity were reduced by excluding highly correlated variables from the final model. Predicted future occurrence of *E. grandis* in each county was classified as absent if the estimated probability (P) was ≤ 0.5 or as present if $P > 0.5$. The best model for prediction was determined based on the lowest Akaike Information Criterion (Menard 1995).

Results and discussion: The current annual averages of the three climatic variables were significantly correlated with latitude and longitude (Table 10.32). Current annual temperature decreased from southern to northern latitudes ($r = -0.97$) and from eastern to western longitudes ($r = -0.80$). Trends of precipitation were generally opposite those of temperature; it increased from south to north ($r = 0.63$) and from east to west ($r = 0.82$), particularly in the panhandle region of northern Florida. Potential evapotranspiration followed the trends of temperature, but with weaker geographical relationships.

The difference between average annual temperature and precipitation varied significantly inside compared to outside of the current range of *E. grandis* (Table 10.33). Current temperature was higher (23.2°C) inside compared to outside (20.9°C). Average annual precipitation, however, was 106.1 mm lower inside the range compared to outside. Potential evapotranspiration did not differ inside or outside the range.

TABLE 10.32
Pearson Correlation Coefficients among Climate Variables and Geographic Location of County Sample Units ($n = 67$) for Florida in 2000

Climate Variable[a]	Temperature	Precipitation	PET[b]	Latitude	Longitude
Temperature (°C)	1.00				
Precipitation (mm)	−0.75[c]	1.00			
PET[b] (mm)	0.44[c]	−0.70[c]	1.00		
Latitude (deg)	−0.97[c]	0.63[c]	−0.34[c]	1.00	
Longitude (deg)	−0.80[c]	0.82[c]	−0.54[c]	0.69[c]	1.00

[a] Sample units are Florida counties ($n = 67$); climate variables were weighted by area and are expressed as average annual values.
[b] PET: potential evapotranspiration.
[c] Correlation coefficient is significant at the $P < 0.05$ level of probability.

Seven formulations of logistic regression models were evaluated for predicting the current range of *E. grandis* (Table 10.34). Models formulated with a single climate variable, average annual precipitation, produced the highest level of significance ($p = 0.002$) followed by temperature ($p = 0.047$); potential evapotranspiration did not account for significant variation ($p = 0.480$) of *E. grandis* occurrence. AIC was lowest for the single-variable model based on temperature (13.192) than for either precipitation (64.137) or potential evapotranspiration (79.382). The prediction model based on both mean annual temperature and mean annual potential evapotranspiration produced the lowest AIC (12.503) among all formulations evaluated. I selected it to use for predicting the potential future range of *E. grandis* based on the climate forecasted by the HadCM3 scenario for 2060.

Model performance using the current climate data set resulted in correct classification of 97% of the counties. The two misclassified counties were in the transition zone and represented one case each of a false-positive and false-negative prediction. Application of the prediction model to the climate predicted by the HadCM3 scenario for 2060 resulted in correct classification of all 17 counties in the current range in southern Florida and expansion of the range of *E. grandis* into 14 additional counties in the central part of the state (Figure 10.26). The classification probability for the 31 counties predicted as inside the future range was high (>0.99) for all but one ($p = 0.77$); probabilities were near zero for the 36 counties predicted as outside the range of *E. grandis* in 2060. Findings

TABLE 10.33
Average Annual Climate Conditions (SD) of Florida Counties in Relation to the Current (2000) Range of *Eucalyptus grandis*

	County Location		
Climate Variable[a]	Inside Range ($n = 17$)	Outside Range ($n = 50$)	Difference
Temperature (°C)	23.23 (0.42)	20.92 (1.02)	2.31[b]
Precipitation (mm)	1337.68 (69.41)	1443.82 (111.13)	−106.14[c]
PET[d] (mm)	2596.88 (51.78)	2586.05 (56.00)	10.83

[a] Sample units are Florida counties ($n = 67$); climate variables were weighted by area and are expressed as average annual values.
[b] Difference is significant ($p < 0.00001$).
[c] Difference is significant ($p = 0.00046$).
[d] PET: potential evapotranspiration.

TABLE 10.34
Statistical Significance and Associated AIC for Full and Reduced Formulations of Logistic Regression Models Based on Combinations of Climate Variables to Predict the Occurrence of *Eucalyptus grandis* in Florida Counties in 2000

Climate Variables in Model Formulation	Climate Variable			
	Temperature	Precipitation	PET[b]	AIC[a]
Temperature (°C)	0.047	na[c]	na	13.192
Precipitation (mm)	na	0.002	na	64.137
PET (mm)	na	na	0.480	79.382
Temperature + precipitation	0.048	0.867	na	15.165
Temperature + PET	0.170	na	0.317	12.503
Precipitation + PET	na	0.001	0.034	61.033
Temperature + precipitation + PET	0.179	0.951	0.313	14.500

[a] Akaike Information Criterion.
[b] PET: potential evapotranspiration.
[c] na: not applicable.

from my study agree with those of others (Zon and Briscoe 1911; Geary et al. 1983; Rockwood 2012) that temperature is more important than precipitation for defining the area potentially suitable for planting *E. grandis* in Florida.

The area in central Florida potentially suitable for *E. grandis* with a moderate climate change is similar to the potential planting regions identified by Rockwood (2012) as suitable for current plantings of cold-tolerant cultivars. Rockwood (2012), however, includes 11 counties farther north than those classified as suitable for *E. grandis* based on 2060 climate forecasts by the HadCM3 scenario in my analysis. Because the HadCM3 scenario forecasted the smallest increase of temperature for 2060 (Table 10.32), a model based any of the other three scenarios (particularly the MIROC3.2) would have likely resulted in a proposed area suitable for *E. grandis* similar to that presented by Rockwood (2012).

In summary, predicting the results of a changing climate on the potential future range suitable for *E. grandis* will be more complex than illustrated by this study. For example, *E. grandis* is particularly sensitive to the abrupt onset of freezing temperatures in the normally temperate climate of south Florida, such as associated with strong fast-moving winter cold fronts from Arctic latitudes. Also, future survival of *E. grandis* on some sites could be influenced by moisture stresses associated with variation of precipitation resulting from climate change, which was not evaluated in my model. My model was based on *E. grandis* with average sensitivity to the winter climate of south Florida. Use of *E. grandis* planting stock from clones selected for increased tolerance of frost could provide an adaptive management strategy to increase the resilience of new plantings of the species to the effects of possible climate change.

Case Study 3: Effects of Climate Change on High-Elevation Southern Appalachian Forests

During the Pleistocene, boreal-type forests extended as far south as South Carolina 12,800 to 19,100 years ago; warming during the Holocene caused these forests to retreat north to their present locations (Delcourt and Delcourt 1998). Because of persistent cold climates found at high elevation sites in the Southern Appalachian Mountains, forests of red spruce (*Picea rubens*) and Fraser fir (*Abies fraseri*) remain on the highest peaks. These "sky islands" are some of the most uncommon forests of the South (White et al. 1993). They are widely dispersed and separated by a sea of lower elevation hardwood forest (Payne et al. 1989).

In Virginia, red spruce follows a 20°C July isotherm (Pielke 1981), and is associated with sites having >140 frost days, average annual temperatures ≤8°C and average precipitation ≥140 cm (Nowacki and Wendt 2010). Cogbill and White (1991) found that across the Appalachians the transition zone (ecotone) between spruce–fir and adjacent hardwood forest corresponds with a 17°C average annual July temperature and suggested that an elevation of 1610 m is the average lower boundary of continuous spruce–fir forest in the Southern Appalachians. However, pockets of this forest type may occur as much as 400 m lower in ravines, depressions, or other cold air pockets (Delcourt and Delcourt 1998). During the mid-Holocene period, temperatures may have been about 2°C higher than today (Delcourt and Delcourt 1998), and this short period of warming would have displaced the spruce–fir upward limits by 70–130 m, causing extinction on sites less than 1740 m high (Whittaker 1956) and explaining the rarity of these forests at lower elevations today.

To forecast persistence of these forests over the next 50 years, we modeled potential effects of climate change on high-elevation spruce–fir forests using best-case and worst-case scenarios, selected from four possible climate scenarios (Wear et al. in press).

Methods: The reduction in temperature with increasing elevation is termed the "lapse rate," which varies by season, climate, and topography. Cogbill and White (1991) suggested that spruce–fir forests ecotones correspond with a 17°C temperature in July. Thus, we used the average July lapse rate (–6.3°C/km) for the entire Appalachian Mountain range (Leffler 1981). Although steeper than the average annual lapse rate (–5.8°C/km), this lapse rate was similar to regional July rates modeled by Bolstad et al. (1998) for the Southern Appalachians (~6.5°C/km). Our selection of 1610 m as the current elevation for persistence of continuous spruce–fir forests in the Southern Appalachians was based on calculations by Cogbill and White (1991), and is congruent with modeling by Hayes et al. (2007) on combined average minimum elevation (1605 m) required for continuous spruce–fir forests on both previously logged (1698 m) and unlogged (1513 m) sites. Although spruce–fir forests generally do not occur on peaks lower than 1740 m in elevation, we included these peaks because our goal was to determine areas that could retain spruce–fir forests or may be suitable habitat for restoration of this forest type in the future.

In selecting the best-case scenario, we opted for the one that predicted wetter cooler summers (average temperature increase of 1.5°C in July; CSIROMK2 B2) over the one that predicted lower annual temperatures (1.5°C) but higher average temperatures in July (2.4°C; CSIROMK3.5 A1B). The best-case scenario (Figure 10.27) predicted an increase in July temperatures of 1.0–2.0°C and precipitation increases of 3–21 cm (9.2 cm average). Our worst-case scenario (MIROC3.2 A1B) predicted increased July temperatures of 2.5–3.3°C (2.9°C average) and decreases in precipitation of 11–24 cm (18.2 cm average) (Figure 10.28).

Results: Based on a July lapse rate of –6.3°C/km with an average annual increase in July temperatures of 1.5°C, the best-case scenario for climate change would produce a 238 m increase in the minimum elevation of continuous spruce–fir forests by the year 2060. An average annual increase of 2.9°C would increase the elevation of the spruce–fir ecotone by 460 m under the worst-case scenario. Based on these estimates, the elevation of the northern hardwood/spruce–fir ecotone (or elevations thermally suitable for restoration) would move upward from a minimum of 1610 m to 1848 m (an increase of 238 m) under the best-case scenario (Figure 10.29). Alternatively, the worst-case scenario predicts a raise in the lower limit of the spruce–fir ecotone to 2070 m (an increase of 460 m). Based on these estimates, approximately 25,785 ha are currently suitable for spruce–fir forests, but only around 1685 ha of suitable spruce–fir habitat would remain by 2060 under the best-case scenario, a reduction of 94%. Under the worst-case scenario, our estimated lower limit of spruce–fir forests (2,070 m) is higher than the highest elevation (2037 m, Mount Mitchell in North Carolina), which would mean extirpation of these forests from the Southern Appalachians. These results are in agreement with Delcourt and Delcourt (1998), who suggested that expiration would occur with a 3°C increase in average July temperature.

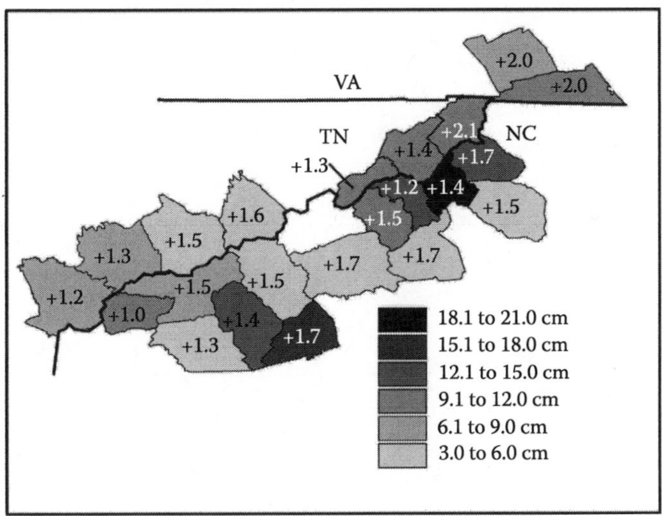

FIGURE 10.27 Projected changes from 2000 to 2060 in annual precipitation and projected rise in average July temperatures for high-elevation counties of the Southern Appalachian Mountains under a best-case scenario (CSIROMK2 B2), based on climate scenarios (Wear et al. in press); that were developed from three general circulation models (MIROC3.2, CSIROMK2 and 3.5, and HadCM3) and two emissions storylines (A1B represents low population growth/high-energy use and B2 represents moderate population growth/low-energy use); numbers within counties indicate expected rise in average temperature (°C). Overall average July temperature change is +1.5°C and average annual precipitation change is +9.2 cm for all counties combined.

Discussion: Four parameters complicate predictions of spruce–fir persistence. First, Cogbill and White (1991) suggested that the current lowest elevation where spruce–fir ecotones currently exist is 1610 m, which is the current basal elevation we used. However, Hayes et al. (2007) suggested the minimum elevation for spruce–fir in the Great Smoky Mountain National Park is 200 m higher than what was found before widespread logging in the past 100 years. Hayes et al. (2007) developed

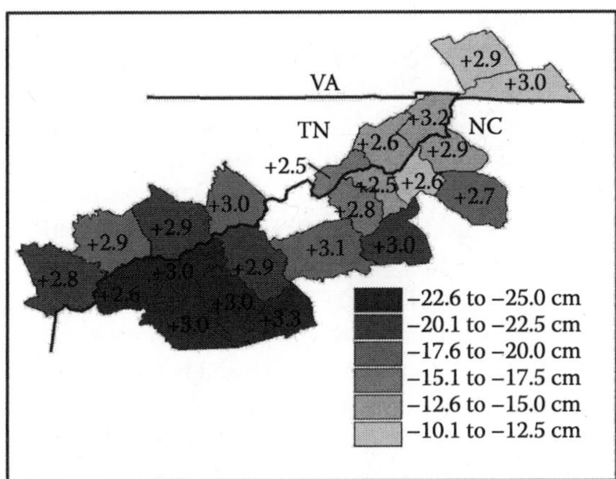

FIGURE 10.28 Projected changes from 2000 to 2060 in annual precipitation and projected rise in average July temperatures for high-elevation counties of the Southern Appalachian Mountains under a worst-case scenario (MIROC3.2 A1B), based on climate scenarios (Wear et al. in press); that were developed from three general circulation models (MIROC3.2, CSIROMK2 and 3.5, and HadCM3) and two emissions storylines (A1B represents low population growth/high-energy use and B2 represents moderate population growth/low-energy use); numbers within counties indicate expected rise in average temperatures (°C). Overall average July temperature change is +2.9°C and average annual precipitation change is –18.2 cm for all counties combined.

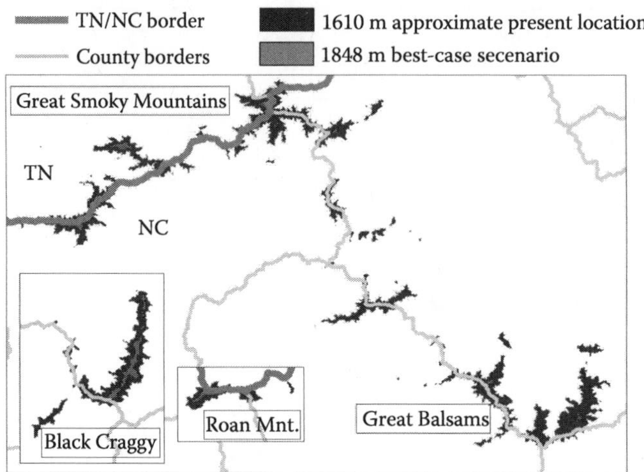

FIGURE 10.29 Estimated current location of continuous spruce–fir forests in the Southern Appalachian Mountains based on a minimum elevation of 1610 m above sea level, and potential location resulting from climate change by the year 2060 under a best-case scenario (MIROC3.2 A1B), based on climate scenarios (Wear et al. in press); that were developed from three general circulation models (MIROC3.2, CSIROMK2 and 3.5, and HadCM3) and two emissions storylines (A1B represents low population growth/high-energy use and B2 represents moderate population growth/low-energy use) (1848 m).

models that indicated spruce–fir forests typically occur >1513 m in areas not previously logged, compared to >1698 m at sites that were previously logged. Furthermore, minimum elevation on north and south slopes was similar in areas that were not previously logged, whereas minimum elevation was approximately 100 m higher on south slopes in areas that were previously logged. Thus, there is not a concrete consensus as to what minimum elevation spruce–fir forests can potentially inhabit given the effects of aspect and previous logging.

Second, estimating exact lapse rates at any one time or location is difficult because the lapse rates may vary by season, time of day, aspect, wind direction, humidity, topography, changes in elevation, presence of snow, atmospheric instability, cloud cover, and ground cover (Barry 1992). Lapse rates may also differ over mountain slopes compared to those in a free atmosphere, but average environmental lapse rates approximate –6°C/km (Barry 1992). Various average annual lapse rates, ranging from –4.9 to –5.8°C/km, have been used in studies to estimate temperature changes with elevation change in the Southern Appalachians (Cogbill et al. 1997; Delcourt and Delcourt 1998; Flebbe et al. 2006; Meisner 1990). Slight differences in the estimated lapse rate could lead to significantly different results, and variations in lapse rates resulting from topography could lead to greater or lesser than average changes in the minimum elevation associated with spruce–fir forests. Thus, constant or long-term average lapse rates should be used with caution (Bolstad et al. 1998).

Third, a retreat of spruce–fir forest could be exacerbated by reductions in precipitation predicted by the worst-case scenario. Spruce–fir forests are dependent on high atmospheric moisture, precipitation, and snowfall (Nowacki et al. 2010) and require moist conditions for germination and seedling survival (Hayes et al. 2007). Cloud cover also plays an important role (Hayes et al. 2007). The worst-case scenario predicted reduced atmospheric moisture, which could exacerbate the retreat and potentially reduce the possibility of remnant patches. Alternatively, the result of best-case scenario could lead to greater cloud cover, which might mitigate the effects of increased temperature. Thus, increases in the minimum elevation of spruce–fir forests from increased temperature could be lessened by increases in cloud cover.

Fourth, although large areas of continuous spruce–fir may be eliminated, some cold-pool areas and north slopes could potentially retain small patches of spruce–fir forests. These areas could retain spruce–fir forests, thus reducing the likelihood of complete elimination of spruce–fir under either scenario.

Although parameters for this modeling effort may be overly simplified, it illustrates the possible outcomes that could occur with increases in temperature. Given the uncertainty in precipitation changes among the various models and variable lapse rates, exact predictions would require substantially more effort but may still be tentative. An additional unknown is the interaction of climate change and the balsam woolly adelgid (*Adelges piceae*), which has killed large areas of mature forests of Fraser fir.

Case Study 4: Predicting Growth of Oak and Maple in an Old-Growth Forest in the Cumberland Plateau Tablelands

Old-growth forests are relatively rare in the Eastern United States, but some have been located in hardwood forests of the Cumberland Plateau, nearly all within plateau gorges (Clark et al. 2007; Martin 1975; Quarterman et al. 1972; Schmalzer et al. 1978). Only one large expanse of forest with old-growth potential (~600 ha) has been discovered on the tableland surface of the Cumberland Plateau (Haney and Lydic 1999). This forest is located in Tennessee's Grundy County on Savage Gulf Natural Area, which is listed as a National Natural Landmark by the U.S. Department of the Interior because of its biodiversity and unique geologic features (DeSelm and Clark 1975). The forest represents a unique opportunity to quantify long-term patterns of stand development, forest succession, and climate–tree growth relationships.

Using tree-ring and climate data to study old-growth forests can improve our understanding of the processes that influence forest communities (Henry and Swan 1974; Lorimer 1985; Oliver and Stephens 1977), eventually helping forest managers incorporate information about historical ranges of variability into silvicultural treatments that mimic past disturbance characteristics (Coates and Burton 1997; Swetnam et al. 1999; Webster and Lorimer 2005) and mitigating for the effects of climate change.

We examined climate–tree growth relationships of two dominant hardwoodgenera, oak (*Quercus* spp) and maple (*Acer* spp). Oaks have been continuously recruited at Savage Gulf for 300 years or more. Recently (~80 years) abundant, maple recruitment represents a species composition shift from the historically dominant oaks. This same species shift has been has been widely reported throughout the Central Hardwood Forest (Abrams 1998; McEwan et al. 2011; Nowacki and Abrams 2008). What is unknown is how these changing forests will respond to changes in climate.

Methods: We established 87, 0.04 ha fixed-radius plots throughout a 600-ha old-growth remnant, and we collected increment core samples from all oaks (*Q. alba*, *Q. prinus*, *Q. velutina*, *Q. coccinea*, and *Q. stellata*) and red maples (*A. rubrum*) ≥20 cm DBH. Once all rings were visually dated, we measured raw-ring width to the nearest 0.001 mm using a Velmex measuring stage interfaced with Measure J2X software; data were quality checked using the computer program COFECHA (Grissino-Mayer 2001; Holmes 1983). We created tree-ring chronologies using 98 oak series and 42 red maple series that had ≥0.5 species-specific interseries correlations to the composite chronology created by COFECHA. We used program ARSTAN (Cook, 1985) to produce the standard chronologies using negative exponential and negative linear detrending. We calculated predicted standardized tree-ring widths using instrumental climate variables from 1930 to 2009 with multiple regression. Our independent variables were 32 monthly climate variables (total precipitation and average daily temperature) from the year's previous July to current October. Instrumental climate data were downloaded for NOAA Climate Division 2 for Tennessee from the National Climate Data Center (U.S. Department of Commerce, National Oceanic and Atmospheric Administration, National Climatic Data Center 2011). We used a stepwise selection to limit results to the most important monthly variables ($p < 0.05$) that predicted tree-ring widths. We used the final regression models to predict tree-ring widths for 2001 to 2060 using monthly climate variables derived for an ensemble of three emission scenarios (Mote and Shepherd 2011; Chapter 2): A1B representing low population with high economic growth and energy use, A2 representing a divided world with local environmental sustainability, and B1 representing rapid change toward an service and information economy with a strong emphasis on clean and resource-efficient technologies (Intergovernmental Panel on Climate Change 2007).

Results: Patterns of annual tree growth in oak and red maple (Figure 10.30) could be significantly predicted by monthly variability in climate. Oak had the strongest relationship with temperature of the current year's June and this relationship was negative (Table 10.35). Oak tree-ring variability was also weakly and positively related to current year's June precipitation and the previous year's August precipitation. Unlike oak, red maple had the strongest relationships with previous year's October precipitation and weaker but significant relationships with current year's February precipitation and January temperature; and all three of these variables were positively related to red maple tree-ring growth. Red maple was weakly and positively related to current year's August precipitation and negatively related to current year's May temperature.

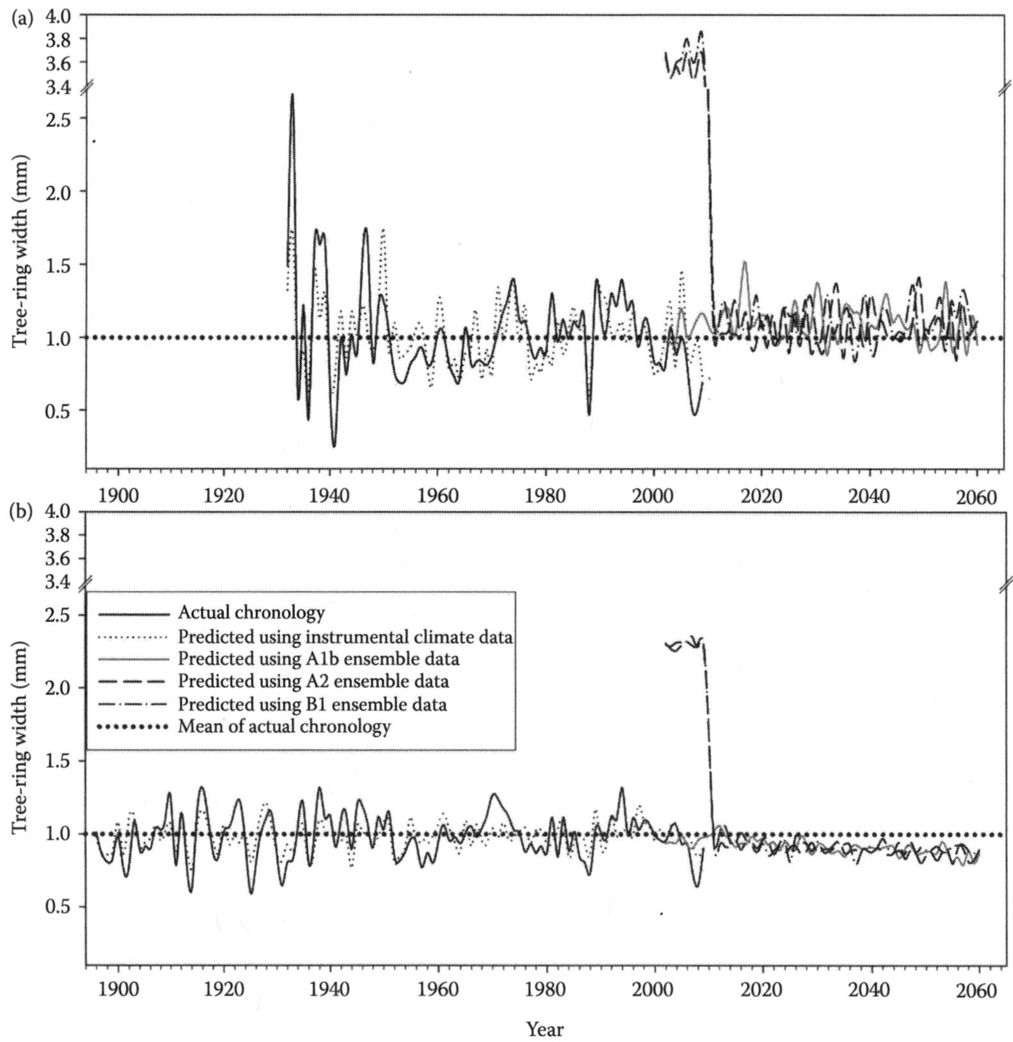

FIGURE 10.30 Actual and predicted tree-ring chronologies for (a) red maple and (b) oak species based on three emissions storylines (Wear et al. in press); scenarios were developed from three general circulation models (MIROC3.2, CSIROMK2 and 3.5, and HadCM3) and three emissions storylines (A1B represents low population growth/high-energy use, A2 represents a divided world with local environmental sustainability, and B2 represents moderate population growth/low-energy use).

TABLE 10.35

Regression Equations for Oak Species and Red Maple Derived from Instrumental Climate Data That Were Used to Predict Standardized Tree-Ring Widths from Ensemble Climate Model Data

Regression Equation	Independent Variable	Partial *R*-Square Value
	Oak	
$Y = 1.770 + (0.001 * \text{JunePrecip}) + (0.002 * \text{previousAugustPrecip}) + (-0.046 * \text{JuneTemp})$ ($R^2 = 0.37$, $p < 0.0001$)		
	Current June precipitation	0.18
	Current June temperature	0.10
	Previous August precipitation	0.09
	Red Maple	
$Y = 1.20530 + (0.00371 * \text{Octoberppt}) + (0.052025 * \text{JanuaryTemp}) + (0.00204 * \text{FebruaryPrecip}) + (-0.05741 * \text{MayTemp}) + (0.00173 * \text{AugustPrecip})$ ($R^2 = 0.47$, $p < 0.0001$)		
	Previous October precipitation	0.14
	Current January temperature	0.15
	Current February precipitation	0.11
	Current May temperature	0.05
	Current August precipitation	0.03

Source: Adapted from Wear, D.N., R. Huggett, and J.G. Greis. In press. Constructing alternative futures. *Southern Forest Futures Project Technical Report.*

The three predicted tree-ring chronologies are similar in terms of long-term trends within the two species. The only noticeable difference among the predicted chronologies is from 2000 to 2009: the A2 and B1 chronologies have unusually high rates of predicted growth, and they exceed the A1 chronologies by approximately 1.3 for oak and 2.6 mm for red maple. Compared to the average of the species actual chronologies (1.0 mm), the tree-ring chronologies predicted from the ensemble emissions storylines indicated that red maple would slightly increase in tree-ring growth from 2010 to 2060, but oak would show a decrease.

Discussion: Our data suggest that the effect of climate change on oak and red maple will depend largely on the season of drought. We predict that oak trees on the tablelands of the Savage Gulf Natural Area will be most productive during years with cool and wet springs preceded by a wet summer. The documented spring temperature signal likely corresponded to water stress of oak by the influence of temperature on evapotranspiration rates. Climate–growth investigations of various oak species in the Appalachian-Cumberland Highland have revealed similar relationships (Copenheaver et al. 2010; Speer et al. 2009; White et al. 2011). In contrast, a cold and dry winter preceded by a dry autumn would most negatively affect red maple. Moisture availability during the previous year supports the production and storage of carbohydrates and, when coupled with abundant moisture during the subsequent growing season, results in increased productivity for both species. Red maple was not negatively affected by summer temperature, and in fact, favored higher winter temperatures.

The ensemble emissions scenario for the Cumberland Plateau and for Grundy County predicts approximately no change in precipitation from 2010 to 2060 (<50 mm change), but show an approximate 1.0°C (B1 ensemble) to 1.5°C (A1 and A2 ensembles) increase in temperature for the same time period. The effect on red maple and the oak is speculative, but based on our data, we suggest that oak will be more negatively impacted by predicted temperature increases compared to red maple if precipitation stays the same or shows a marked decrease. We also predict that rising spring temperatures without a corresponding increase in spring and/or summer precipitation would negatively affect both species. Climate change could result in oak mortality at the Savage Gulf and elsewhere in the region if temperature increases are particularly high in the spring without a corresponding precipitation increase. While oaks are considered to be drought tolerant generally, drought has been documented to cause oak mortality in the Southern Appalachian Mountains (Clinton et al. 1993, 1994).

In contrast to oak, red maple's growth was favored by higher winter temperatures, suggesting temperature increases related to climate change could actually favor dominance of this species. However, red maple could be negatively affected if precipitation was lacking in winter and previous autumn months. Effect of drought on red maple is not well understood in long-term field studies, but short-term physiological tests show the species is less tolerant to drought than associated oak species (Abrams and Mostoller 1995).

The Savage Gulf offers the rare opportunity to study climate-tree growth relationships of a subxeric forest that contains some of the oldest hardwood and pine (*Pinus* spp) individuals recorded in the Eastern United States (Hart et al. in press). To maintain the historic oak component in this ecosystem in the face of predicted climate change, resource managers would likely need to reduce basal area and stem density of shade-tolerant competitors, particularly invading red maple, in all canopy strata. Klos et al. (2009) noted that drought-induced mortality was smallest on sites with low stem density and basal area, flat terrain, and high species richness. The tablelands of the Savage Gulf exhibit all four of these characteristics and may therefore, have low susceptibility to drought-induced oak mortality. This management protocol would help mitigate effects of predicted future temperature increases on oak, and would aid in the long-term sustainability (Abrams 1998; Nowacki and Abrams 2008). Because mechanical tree harvesting is not a management option in the Savage Gulf, the use of prescribed fire in conjunction with natural disturbances that create canopy gaps may be a viable option for maintaining and restoring conditions to the pre-invasion levels of the mid-1900s (Abrams 2005).

SUMMARY OF FINDINGS

Effects on Tree Species Distribution

Vulnerability of tree species to climate change in the Southern Region is potentially highest (measured by number of tree species predicted to decrease their range) in the Mid-Gulf-West of the Coastal Plain, the Central Appalachian Piedmont, and the Ozark-Ouachita Highlands section of the Mid-South. In contrast, the ranges of many tree species are predicted to expand in the Southern Appalachian Piedmont and in the Blue Ridge and Northern Ridge and Valley sections of the Appalachian-Cumberland Highland. For most areas of the Southern Region, however, the predicted range of tree species is projected to remain little changed.

Vulnerability assessments varied considerably among the four climate predictions, particularly when comparing the MIROC3.2 A1B scenario to the other three model and scenario combinations. This is not an unexpected result, as the MIROC3.2 A1B projected the most extreme changes in temperature and precipitation over the region (Chapter 2) and our simple models were driven by changes in those two climate variables. For the Coastal Plain, the MIROC3.2 A1B scenario predicted a decrease in range for half or more of the species across all sections. Vulnerability of species was particularly high in the Florida Peninsular section where 27 of the 37 species evaluated were predicted to reduce their ranges. Future climate and vulnerability in the Florida Peninsular section are uncertain because the other three scenarios predict an increase in the ranges of 1–4 species.

We also found conflicting vulnerability predictions for the Northern Ridge and Valley section of the Appalachian-Cumberland Highland, where the MIROC3.2 A1B scenario predicted that 16 of 44 species would potentially decrease their ranges. The other three climate scenarios, however, predicted that 15 of the 44 species would possibly increase their ranges. Vulnerability of species will likely be marginal in the Mississippi Alluvial Valley, where all scenarios except the MIROC3.2 A1B predicted few changes.

Decrease or increase in the ranges of single species offers a simplistic approach to assess vulnerability to climate change. With few exceptions (such as the slash pine and longleaf pine forests of central and northern Florida), most southern upland communities consist of multiple species. Mixtures of the species selected in each section for the vulnerability assessment generally may not occur naturally, but are used here as an index of diversity for comparison of predictions from the climate scenarios.

Risk to Future Forest Diversity

Tree species: The overall effect of climate change on diversity of southern trees was predicted to result in small reductions of species richness in most of the 21 ecological sections studied, although the number of species may remain constant or increase slightly in several sections by the year 2060. Future diversity (quantified by species richness) was predicted to range between 95% and 100% of current levels for over half of the region. The largest changes in diversity, where species richness could be less than 85% of current levels, occurred in three sections, all in the Coastal Plain. Adverse effects were predicted to be least in the Northern Ridge and Valley section of the Appalachian-Cumberland Highland, where species richness could increase to 105% of current levels. Although the magnitude of results varied somewhat among climate scenarios, the overall trends were generally consistent that tree diversity could decrease throughout the Southern Region by the year 2060, with the highest risks occurring in certain sections of the Coastal Plain, Piedmont, and eastern zone of the Mid-South.

The effects of climate change on diversity at the county scale are likely to occur in about half of the Southern Region. Potential loss of tree species will likely be largest throughout the Coastal Plain, the Piedmont, and the eastern zone of the Mid-South. Few or no changes in diversity are expected to occur in most of the Appalachian-Cumberland Highland, the western zone of the Mid-South, or the Mississippi Alluvial Valley. In the Deltaic Plain section of the Mississippi Alluvial Valley, however, some bottomland species excluded from our analysis rely on periodic flooding (e.g., baldcypress, *Taxodium distichum*) and could be affected by reduction in precipitation in other regions, such as the upper Midwest. However, because predictions of future climatic conditions—particularly annual precipitation—were variable among the four climate scenarios, determining which scenario is most relevant is currently largely conjecture; averaging predictions from an ensemble (Mote and Shepherd 2011) of the four models would be one way to account for climate changes in planning and reduce risk from making an incorrect choice.

Rare plant communities: Small areas of infrequently occurring plant communities have been identified in the Southern Region that may be particularly vulnerable to climate change (Grossman et al. 1994) (Table 10.36). Many of these communities are associated with unusual combinations of geology and topography that form unusually wet or dry environments occupied by characteristic species. For most of these communities, assessing vulnerability and risk from climate change was beyond the scope of this chapter, but consideration of special conservation measures is important as well as recognition that many species may be important where they occur.

MANAGEMENT OPTIONS TO REDUCE VULNERABILITY AND RISK

Because of the long-term nature of forestry, all landowners actively engaged in resource management will face challenges resulting from climate change. Landowners with holdings in areas of the Southern Region that face the highest threat to vulnerability for certain species (such as shortleaf pine in western Arkansas) and high risk to species diversity (such as in the Ozark-Ouachita Highlands section of the Mid-South) will be among those who need relevant information on resource management the earliest. Required information related to climate change is available for many species, especially loblolly pine (Huang et al. 2011; McNulty et al. 1998; Schmidtling 1994) and others (Ge et al. 2011) with high commercial value. However, information is lacking for other "nontimber" (or lower

TABLE 10.36
Examples of Currently Rare Plant Communities of the Southern United States (Grossman et al. 1994) That Have Increased Vulnerability to Climate Change Resulting from Increased Annual Temperature and Decreased Annual Precipitation

Species or Community	State
Florida Torreya	FL
Crowley's Ridge Forest	AR
Red Spruce–Fraser Fir Forest	NC, TN, VA
White Oak/*Vaccinium* spp Dwarf Forest	AR, OK
Sabal mexicana Wetland Forest	TX
Barrier Island Depression Forest	GA, NC, SC, VA
Coastal Plain Calcarious Mesic Forest	All
Gulf Coast Maritime Forest	AL, FL, MS
South Atlantic Island Maritime Forest	FL, GA, SC

commercial valued) species, particularly hardwoods that increase species richness by providing foraging and cover habitat for wildlife; examples are water oak, southern magnolia, and sweetgum. Detailed assessment of specific issues of resource management is beyond the scope of this chapter; however, the following paragraphs provide an overview of regeneration and other critical forest management issues that will be required to address the impacts of climate change on tree species in the Southern Region. In addition, other chapters address forest management issues that are associated with climate change, such as productivity and increased threat from wildfire and insects.

Changes to vegetation resulting from climate change will likely occur slowly in some subregions and may be inconspicuous in many areas of the Southern Region, particularly where temperature and precipitation remain similar to current levels. Because trees are relatively long lived, the earliest observed climate-related effects in forests may not be in species diversity, but changes of disturbances associated with drought, fire, insects, and storms (Dale et al. 2001). Previous widespread occurrences of both southern pine and engraver beetles have been associated with stress caused by lower than average rainfall during the growing season and warmer winters (Lombardero et al. 2000). The combination of increased temperature and declining precipitation could significantly change both the frequency and severity of bark beetle outbreaks in shortleaf pine stands (McNulty et al. 1998). Gan (2004) predicted that under future climate change, the southern pine beetle could kill from 4 to 7.5 times the 2004 value of trees killed annually by the beetle. As another example, the fungus that causes fusiform rust (*Cronartium quercuum* f. sp. *fusiforme*) is highly sensitive to temperature and humidity and further illustrates the effects of climate change on the threat of disease in southern pine (Chapter 6). This disease also affects hardwood species in upland oak hardwoods, particularly the northern red oaks, but also including water, willow, and laurel oaks. Additional information on effects of climate change on future insect and disease threats associated with the four specific climate scenarios may be found in Chapter 6.

Species composition changes are likely to occur in most subregions and many sections of the Southern Region. But those changes, driven by a failure to regenerate successfully (particularly for some species of oaks) will likely occur slowly and, therefore, could be less apparent than those associated with insects and diseases. Lack of oak regeneration success may result from reduced seed production by some species, such as many conifers and several hardwoods, such as yellow poplar and red maple. To maintain an historic oak component in some ecosystems in the likely event of climate change, land managers may need to consider reducing basal area and stem density of shade-tolerant competitors, such as red maple, in all canopy strata; or implementing management activities that favor more drought-tolerant oak species. In addition, even if seed production is adequate, regeneration failure may result from changing environmental conditions that do not support establishment and survival of newly established seedlings. Regeneration of stands using natural methods, such as shelterwood and single-tree selection (Chapter 7) are well suited for many upland hardwood types, but may not produce reliable results in some areas of the Southern Region, particularly where annual precipitation may be reduced and temperature may increase, thereby creating additional moisture stress. Planting has a number of advantages over natural methods as a means of controlling species composition in areas where climate change may adversely affect environmental conditions needed for successful regeneration. Schmidtling (1994) suggests that provenance tests offer an available source of information on performance of planted species under a changing climate. Although artificial methods may be more expensive to implement than natural regeneration, site preparation followed by planting will usually provide reliable results for certain species. In addition, planting allows the resource manager to select seedling sources that might be better suited for predicted future climatic conditions compared to that which currently exist on the site (Schmidtling 2001). Artificial regeneration of preferred tree species by planting could be a viable alternative to natural methods for resource managers concerned about short-term effects of climate change on species composition. Aitken et al. (2008) suggest that genetic selection from natural variability within a species offers a realistic method for resource managers to respond to climate change when trends of future temperature and precipitation become more apparent for an area. Additional information on using artificial regeneration for controlling species composition in response to climate change is presented in Chapter 7.

In conclusion, our simple analyses provide just a glimpse of how climate change may impact the distribution and diversity of some tree species in the Southern Region. Indeed, the rapid pace of climate change may preclude migration of species to new suitable habitats and the only viable approach to maintain species and communities will be through active and aggressive (e.g., facilitated migration) forest management. However, as noted in Chapter 7, it is unlikely that coordinated large-scale activities will be implemented across the vast array of ownerships in the South, at least in the short term. If some of the more extreme projections of climate change (such as MIROC3.2 A1B) are observed and disturbance regimes increase in frequency and severity as projected (Vose et al. 2012), then it is likely that forest composition and diversity will change in many areas in the South. These changes could have important ramifications for other resources such as wildlife habitat (Chapter 11) and water resources (Chapter 9).

REFERENCES

Abrams, M.D. 1998. The red maple paradox. *BioScience* 48: 355–364.

Abrams, M.D. 2005. Prescribing fire in eastern oak forests: Is time running out? *Northern Journal of Applied Forestry* 22: 190–196.

Abrams, M.D. and S.A. Mostoller. 1995. Gas exchange, leaf structure and nitrogen in contrasting successional tree species growing in open and understory sites during a drought. *Tree Physiology* 15: 361–370.

Agren, G.I., R.E. McMurtrie, W.J. Parton, J. Pastor, and H.H. Shugart. 1991. State-of-the-art of models of production–decomposition linkages in conifer and grassland ecosystems. *Ecological Applications* 1: 118–138.

Aitken, S.N., S. Yeaman, J.A. Holliday, T. Wang, and S. Curtis-McLane. 2008. Adaptation, migration or extirpation: Climate change outcomes for tree populations. *Evolutionary Applications* 1: 95–111,

Alig, R.J. and B.J. Butler. 2004. Area changes for forest cover types in the United States, 1952 to 1997, with projections to 2050. Gen. Tech. Rep. PNW-GTR-613. Portland, OR: U.S. Department of Agriculture, Forest Service, Pacific Northwest Research Station. 106 p.

Allen, C.D., A.K. Macalady, H. Chenchouni et al. 2010. A global overview of drought and heat-induced tree mortality reveals emerging climate change risks for forests. *Forest Ecology and Management* 259: 660–684.

Allred, B.W. and H.C. Mitchell. 1955. Major plant types of Arkansas, Louisiana, Oklahoma and Texas and their relation to climate and soils. *Texas Journal of Science* 7: 7–19.

Bailey, R.G. 1995. *Description of the Ecoregions of the United States*, 2nd ed. rev. and expanded (1st ed. 1980). Misc. Publ. No. 1391 (rev.), Washington, DC: United States Department of Agriculture, Forest Service, 108 p. with separate map at 1: 7,500,000.

Barbour, M.G. and W.D. Billings. 2000. *North American Terrestrial Vegetation*, 2nd ed.. Cambridge, UK: Cambridge University Press. 708pp.

Barry, R.G. 1992. *Mountain Weather and Climate*, 2nd ed. Routledge, Chapman & Hall, Inc. New York.

Boisvenue, C. and S.W. Running. 2006. Impacts of climate change on natural forest productivity—evidence since the middle of the 20th century. *Global Change Biology* 12: 862–882.

Boland, D.J., M.I.H. Brooker, G.M. Chippendale et al. 2006. *Forest Trees of Australia. Commonwealth Scientific and Industrial Research Organization Publishing*. Collingwood, Victoria, Australia. 768pp.

Bolstad, P.V., L. Swift, F. Collins, and J. Régnière. 1998. Measured and predicted air temperature at basin to regional scales in the southern Appalachian mountains. *Agricultural and Forest Meteorology* 91: 161–176.

Booth, T.H. and L.D. Pryor. 1991. Climatic requirements of some commercially important eucalypt species. *Forest Ecology and Management* 43: 47–60.

Bosworth, D., R. Birdsey, L. Joyce, and C. Millar. 2008. Climate change and the nation's forest: Challenges and opportunities. *Journal of Forestry* 106(4): 214–221.

Brockway, D.G., K.W. Outcalt, and W.D. Boyer. 2006. Longleaf pine regeneration ecology and methods. pp. 95–133. In: Jose, S., Jokela, E.J., and Miller, D.L. (eds.). *The Longleaf Pine Ecosystem Ecology*, Silviculture, and Restoration. Springer, New York. 438 p.

Burns, R.M. and B.H. Honkala (tech. coords). 1990. *Silvics of North America 1. Conifers; 2. Hardwoods*. Agriculture Handbook 654, Washington, DC: U.S. Department of Agriculture, Forest Service, 877 p.

Chen, I-C., J.K. Hill, R. Ohlemuller, D.B. Roy, and C.D. Thomas. 2011. Rapid range shifts of species associated with high levels of climate warming. *Science* 333: 1024–1026.

Clark, S.L., S.J. Torreano, D.L. Loftis, and L.D. Dimov. 2007. Twenty-two year changes in regeneration potential in an old-growth *Quercus* forest on the mid-Cumberland Plateau, Tennessee. In: Buckley, D.S., Clatterbuck, W.K. (eds.). *Proceedings 15th Central Hardwood Forest Conference*, Knoxville, TN, pp. 527–535, e-GTR-SRS-101.

Cleland, D.T., J.A. Freeouf, J.E. Keys, Jr. et al. 2007. *Ecological Subregions: Sections and Subsections of the Conterminous United States*. Gen. Tech. Rep. WO-76. Washington, DC: US Department of Agriculture, Forest Service. CD.

Clinton, B.D., L.R. Boring, and W.T. Swank. 1993. Canopy gap characteristics and drought influences in oak forests of the Coweeta Basin. *Ecology* 74: 1551–1558.

Clinton, B.D., L.R. Boring, and W.T. Swank. 1994. Regeneration patterns in canopy gaps of mixed-oak forests of the Southern Appalachians: Influences of topographic position and evergreen understory. *American Midland Naturalist*. 132: 308–319.

Coates, D.K. and P.J. Burton. 1997. A gap-based approach for development of silvicultural systems to address ecosystem management objectives. *Forest Ecology and Management*. 99: 339–356.

Cogbill, C.V. and P.S. White. 1991. The latitude–elevation relationship for spruce–fir forests and treeline along the Appalachian Mountain chain. *Vegetio* 94: 153–175.

Cogbill, C.V., P.S. White, and S.K. Wiser. 1997. Predicting treeline elevation in the southern Appalachians. *Castanea* 62: 137–146.

Conner, R.C. and A.J. Hartsell. 2002. Forest Area and Conditions, Chapter 16 in Southern Forest Resource Assessment (D.N. Wear and J.G. Greis eds.). Gen. Tech. Rep. SRS-53, Asheville, NC, U.S. Department of Agriculture, Forest Service, Southern Research Station, pp. 357–402.

Cook, E.R. 1985. A time-series analysis approach to tree-ring standardization. PhD dissertation, University of Arizona, Tucson.

Copenheaver, C.A., L.A. Hendrick, J.W. Houchins, and C.D. Pearce. 2010. Changes in growth and dendroclimatic response of trees growing along an artificial lake. *American Midland Naturalist* 163: 134–145.

Crawley, M.J. 2005. *Statistics—An Introduction Using R*. West Sussex, England: John Wiley & Sons, 327pp.

Dale, V.H., L.A. Joyce, S. McNulty et al. 2001. Climate change and forest disturbances. *BioScience* 51: 723–734.

Delcourt, P.A. and H.R. Delcourt. 1998. Paleoecological insights on conservation of biodiversity: A focus on species, ecosystems, and landscapes. *Ecological Applications* 8: 921–934.

DeSelm, H.R. and G.M. Clark. 1975. Potential national natural landmarks of the Appalachian Plateaus province of Alabama, Georgia, southern Kentucky, and Tennessee, Final Report. Prepared for the Appalachian Potential National Natural Landmark Program of West Virginia University, Morgantown.

Flebbe, P.A., L.D. Roghair, and J.L. Bruggink. 2006. Spatial modeling to project southern Appalachian trout distribution in a warmer climate. *Transactions of the American Fisheries Society* 135: 1371–1382.

Fletcher, P.W. and R.E. McDermott. 1957. *Influence of Geologic Parent Material and Climate on Distribution of Shortleaf Pine in Missouri*. Research Bulletin 625. Columbia, MO: Missouri Agricultural Experiment Station. 43 p.

Gan, J. 2004. Risk and damage of southern pine beetle outbreaks under global climate change. *Forest Ecology and Management*. 191: 61–71.

Ge, Z.M., S. Kellomaki, H. Peltola et al. 2011. Impacts of changing climate on the productivity of Norway spruce dominant stands with a mixture of Scots pine and birch in relation to water availability in southern and northern Finland. *Tree Physiology* 31: 323–338.

Geary, T.F., G.F. Meskimen, and E.C. Franklin. 1983. Growing eucalypts in Florida for industrial wood production. Gen. Tech. Rep. SE-23. Asheville, NC: U.S. Department of Agriculture, Forest Service, Southeastern Forest Experiment Station. 43 p.

Grissino-Mayer, H.D. 2001. Evaluating crossdating accuracy: A manual and tutorial for the computer program COFECHA. *Tree-Ring Research* 57: 205–221.

Grossman, D.H., K.L. Goodin, and C.L. Reuss (eds.). 1994. *Rare Plant Communities of the Conterminous United States, an Initial Survey*. Arlington, VA: The Nature Conservancy. 620pp.

Haney, J.C. and J. Lydic. 1999. Avifauna and vegetation structure in an old-growth oak–pine forest on the Cumberland Plateau, Tennessee (USA). *Natural Areas Journal* 19: 199–210.

Hansen, A.J., R.P. Neilson, V.H. Dale et al. 2001. Global change in forests: Responses of species, communities, and biomes. *BioScience* 51(9): 765–779.

Hart, J.C., S.L. Clark, S.J. Torreano, and M.L. Buchanan. In press. Composition, structure, and dendroecology of an old-growth *Quercus* forest on the tablelands of the Cumberland Plateau, USA. Forest Ecology and Management.

Hayes, M.A. Moody, P.S. White, and J.L. Costanza. 2007. The influence of logging and topography on the distribution of spruce–fir forests near their southern limits in the Great Smoky Mountains National Park, USA. *Plant Ecology* 189: 59–70.

Hayes, M.J. 2010. *What is Drought? National Drought Mitigation Center*, Univ. of Nebraska, Lincoln, NE. Available online at http://www.drought.unl.edu/whatis/indices.htm, accessed June 24, 2010.

Haywood, J.D. 2009. Eight years of seasonal burning and herbicidal brush control influence sapling longleaf pine growth, understory vegetation, and the outcome of an ensuing wildfire. *Forest Ecology and Management* 258: 295–305.

Haywood, J.D. 2011. Influence of herbicides and felling, fertilization, and prescribed fire on longleaf pine growth and understory vegetation through ten growing seasons and the outcome of an ensuing wildfire. *New Forests* 41: 55–73.

Haywood, J.D. and F.L. Harris. 2000. *Joint Fire Science Longleaf Pine Restoration Demonstration Sites*. The Joint Fire Science Program, Boise, Idaho, Grant No. USDI, BLM, 98-IA-189.

Henry, J.D. and J.M.A. Swan. 1974. Reconstructing forest history from live and dead plant material—An approach to the study of forest succession in southwest New Hampshire. *Ecology* 55: 772–783.

Hijmans, R.J. and C.H. Graham. 2006. The ability of climate envelope models to predict the effect of climate change on species distributions. *Global Change Biology* 12: 2272–2281.

Holmes, R.L. 1983. Computer assisted quality control in tree-ring dating and measurement. *Tree-Ring Bulletin* 43: 69–78.

Huang, J., B. Abt, G. Kinderman, and S. Ghosh. 2011. Empirical analysis of climate change impact on loblolly pine plantations in the southern United States. *Natural Resource Modeling* 24: 445–476.

Hughes, L., E.M. Cawsey, and M. Westoby. 1996. Climatic range sizes of Eucalyptus species in relation to future climate change. *Global Ecology and Biogeography Letters* 5: 23–29.

Intergovernmental Panel on Climate Change. 2007. IPCC fourth assessment report: Climate change 2007 (AR4). http://www.ipcc.ch/publications_and_data/publications_and_data_reports.htm. [Date accessed: July 7, 2010].

Iverson, L.R., A.M. Prasad, S.N. Matthews, and M. Peters. 2008. Estimating potential habitat for 134 eastern US tree species under six climate scenarios. *Forest Ecology and Management* 254: 390–406.

Jun, M., R. Knutti, and D.W. Nychka. 2008. Spatial analysis to quantify numerical model bias and dependence: How many climate models are there? *Journal of the American Statistical Association* 103: 934–947.

Karlsson, M.G. 2002. Flower formation in *Primula vulgaris* is affected by temperature, photoperiod and daily light integral. *Scientia Horticulturae* 95: 99–110.

Keller, G., T. Marchal, H. SanClemente et al. 2009. Development and functional annotation of an 11,303-EST collection from *Eucalyptus* for studies of cold tolerance. *Tree Genetics and Genomes* 5: 317–327.

Kimmins, J.P. 1997. Biodiversity and its relationship to ecosystem health and integrity. *The Forestry Chronicle* 73: 229–232.

Klos, R.J., G.G. Wang, W.L. Bauerle, and J.R. Rieck. 2009. Drought impact on forest growth and mortality in the southeast USA: An analysis using Forest Health and Monitoring data. *Ecological Applications* 19: 699–708.

Kozlowski, T.T., P.J. Kramer, and S.T. Pallardy. 1991. *The Physiological Ecology of Woody Plants*. Academic Press. St. Louis. 657pp.

Leffler, R.J. 1981. Estimating average temperatures on Appalachian summits. *Journal of Applied Meteorology* 20: 637–642.

Leininger, T.D. 1998. Effects of temperature and drought stress on physiological processes associated with oak decline. Chapter 35. In: Mickler, R.A. and Fox, S. (eds.). *The Productivity and Sustainability of Southern Forest Ecosystems in a Changing Environment*. Ecological Studies, Vol. 128. pp. 647–661. Springer-Verlag, Inc., New York. 865pp.

Linkosalo, T., T.R. Carter, R. Hakkinen, and P. Hari. 2000. Predicting spring phenology and frost damage of *Betula* spp. under climatic warming: A comparison of two models. *Tree Physiology* 20: 1175–1182.

Lombardero, M.J., M.P. Ayres, B.D. Ayres, and J.D. Reeve. 2000. Cold tolerance of four species of bark beetle (*Coleoptera: Scolytidae*) in North America. *Environmental Entomology* 29: 421–432.

Lorimer, C.G. 1985. Methodological considerations in the analysis of forest disturbance history. *Canadian Journal of Forest Research* 15: 200–213.

Manion, P.D. 1981. *Tree Disease Concepts*. Prentice-Hall, Inc. Englewood Cliffs, NJ. 399pp.

Martin, W.H. 1975. The Lilley Cornett Woods: A stable mixed mesophytic forest in Kentucky. *Botanical Gazette* 136: 171–183.

McEwan, R.W., J.M. Dyer, and N. Pederson. 2011. Multiple interacting ecosystem drivers: Toward an encompassing hypothesis of oak forest dynamics across eastern North America. *Ecography* 34: 234–256.

McNulty, S.G., J.M. Vose, and W.T. Swank. 1998. Predictions and projections of pine productivity and hydrology in response to climate change across the southern United States. Chapter 22. In: Mickler R.A. and Fox, S. (eds.). *The Productivity and Sustainability of Southern Forest Ecosystems in a Changing Environment. Ecological Studies*, Vol. 128. pp. 391–405. Springer-Verlag, Inc. New York. 865pp.

McNulty, S.G., P. L. Lorio, Jr., M.P. Ayres, and J.D. Reeve. 1998. Predictions of southern pine beetle populations using a forest ecosystem model. Chapter 33. In: Mickler, R.A. and Fox, S. (eds.). *The Productivity and Sustainability of Southern Forest Ecosystems in a Changing Environment. Ecological Studies*, Vol. 128. pp. 617–634. Springer-Verlag, Inc. New York. 865pp.

Meisner, J. D. 1990. Effects of climatic warming on the southern margins of the native range of the brook trout, *Salvelinus fontinalis*. *Canadian Journal of Fisheries and Aquatic Sciences* 47: 1065–1070.

Menard, S. 1995. *Applied Logistic Regression Analysis. Sage University Paper series on Quantitative Applications in the Social Sciences, 07-106*. Sage Publications Thousand Oaks, CA. 98pp.

Meskimen, G. and J.K. Francis. 1990. Rose gum *Eucalyptus*. In: Burns, R.M. and Honkala, B.H. (eds.). *Silvics of North America*. Vol. 2. pp. 505–512. Hardwoods. Agriculture Handbook 654. U.S. Department of Agriculture, Forest Service, Washington, DC. 877pp.

Meskimen, G. and E.C. Franklin. 1978. Spacing *Eucalyptus grandis* in southern Florida. *Southern Journal of Applied Forestry* 1: 3–5.

Meskimen, G.F., D.L. Rockwood, and K.V. Reddy. 1987. Development of Eucalyptus clones for a summer rainfall environment with periodic severe frosts. *New Forests* 3: 197–205.

Mote, T.L. and J.M. Shepherd. 2011. Monthly ensemble downscaled climate model data for the Southeast U.S. (data set accessed 2011-11-07 at http://shrmc.ggy.uga.edu/downscale.php.)

National Climatic Data Center 2011. *US Department of Commerce, National Oceanic and Atmospheric Administration*, Asheville, North Carolina. http://www.ncdc.noaa.gov.

Nedlo, J.E., T.A. Martin, J.M. Vose, and R.O. Teskey. 2009. Growing season temperatures limit growth of loblolly pine (*Pinus taeda* L.) seedlings across a wide geographic transect. *Trees—Structure and Function* 23: 751–759.

Nowacki, G., R. Carr, and M. Van Dyck. 2010. The current status of red spruce in the eastern United States: Distribution, population trends, and environmental drivers. In: Rentch, J.S. and T.M. Schuler (eds.). *Proceedings of the Conference on the Ecology and Management of High-Elevation Forests in the Central and Southern Appalachian Mountains*. Pp. 140–162. U.S. Department of Agriculture, Forest Service, Gen. Tech. Rep. NRS-P-64.

Nowacki, G. and D. Wendt. 2010. The current distribution, predictive modeling, and restoration potential of red spruce in West Virginia. In: Rentch, J.S. and T.M. Schuler (eds.). *Proceedings of the Conference on the Ecology and Management of High-Elevation Forests in the Central and Southern Appalachian Mountains*. Pp. 163–178. U.S. Department of Agriculture, Forest Service, Gen. Tech. Rep. NRS-P-64.

Nowacki, G.J. and M.D. Abrams. 2008. The demise of fire and "mesophication" of forests in the eastern United States. *BioScience* 58: 123–138.

O'Brien, E.M. 1988. Climatic correlates of species richness for woody 'edible' plants across southern Africa. *Monographs Systematic Botany Missouri Botanical Gardens* 25: 385–401.

O'Brien, E.M. 1993. Climatic gradients in wood plant species richness: Towards an explanation based on an analysis of southern Africa's woody flora. *Journal of Biogeography* 20: 181–198.

Oliver, C.D. and E.P. Stephens. 1977. Reconstruction of a mixed species forest in central New England. *Ecology* 58: 562–572.

Outcalt, K.W. and D.D. Wade. 2004. Fuels management reduces tree mortality from wildfires in southeastern United States. *Southern Journal of Applied Forestry* 28(1): 2834.

Payne, J.L., D.R. Young, and J.F. Pagels. 1989. Plant community characteristics associated with the endangered northern flying squirrel, *Glaucomys sabrinus*, in the southern Appalachians. *American Midland Naturalist* 121: 285–292.

Pielke, R.A. 1981. The distribution of spruce in west-central Virginia before lumbering. *Castanea* 46: 201–216.

Quarterman, E., B.H. Turner, and T.E. Hemmerly. 1972. Analysis of virgin mixed mesophytic forests in Savage Gulf, Tennessee. *Bulletin of the Torrey Botanical Club* 99: 228–232.

Risser, P.G. 1995. The status of the science examining ecotones. *Bioscience* 45: 318–325.

Rockwood, D.L. 2012. History and status of *Eucalyptus* improvement in Florida. *International Journal of Forestry Research* 2012, ID: 607879, 10pp.

Runion, G.B., M.A. Davis, S.G. Pritchard et al. 2006. Effects of elevated atmospheric carbon dioxide on biomass and carbon accumulation in a model regenerating longleaf pine community. *Journal of Environmental Quality* 35: 1478–1486.

Schmalzer, P.A., C.R. Hinkle, and H.R. DeSelm. 1978. Discriminant analysis of cove forests of the Cumberland Plateau of Tennessee. In: Pope, P.E. (ed.). *Proceedings of the 2nd Central Hardwoods Forest Conference*. pp. 62–86. West Lafayette, Indiana.

Schmidtling, R.C. 1994. Use of provenance tests to predict response to climate change: Loblolly pine and Norway spruce. *Tree Physiology* 14: 805–817.

Schmidtling, R.C. 2001. *Southern Pine Seed Sources*. Gen. Tech. Rep. SRS-44. Asheville, NC: U.S. Department of Agriculture, Forest Service, Southern Research Station. 25pp.

Schwartz, M.W. 1993. Modeling effects of habitat fragmentation on the ability of trees to respond to climatic warming. *Biodiversity and Conservation* 2: 51–61.

Schwartz, M.W., L.R. Iverson, and A.M. Prasad. 2001. Predicting the potential future distribution of four tree species in Ohio using current habitat availability and climatic forcing. *Ecosystems* 4: 568–581.

Smith, W.B., P.D. Miles, J.S. Vissage, and S.A. Pugh. 2004. Forest resources of the United States: 2002. Gen. Tech. Rep. NC-241. St. Paul, MN: U.S. Department of Agriculture, Forest Service, North Central Research Station. 137pp.

Speer, J.H., H.D. Grissino-Mayer, K.H. Orvis, and C.H. Greenberg. 2009. Climate response of five oak species in the eastern deciduous forest of the southern Appalachian Mountains, USA. *Canadian Journal of Forest Research* 39: 507–518.

Sperry, J.S. and J.E.M. Sullivan. 1992. Xylem embolism in response to freeze-thaw cycles and water stress in ring-porus, diffuse-porous, and conifer species. *Plant Physiology* 100: 605–613.

Stahle, D.W. and J.G. Hehr. 1984. Dendroclimatic relationships of post oak across a precipitation gradient in the south central United States. *Annals of the Association of American Geographers* 74(4): 561–573.

Starkey, D.A., F. Oliveria, A. Mangini, and M. Mielke. 2004. Oak decline and red oak borer in the Interior Highlands of Arkansas and Missouri: Natural phenomena, severe occurrences. In: Spetich, M. A. (ed). *Proceedings of the Upland Oak Ecology Symposium: History, Current Conditions and Sustainability*. pp. 217–222. 2002 October 7–10, Fayetteville, Arkansas. Gen. Tech. Rep. SRS-73. Asheville, NC: U.S. Department of Agriculture, Forest Service, Southern Research Station.

Stephenson, N.L. 1998. Actual evapotranspiration and deficit: Biologically meaningful correlates of vegetation distribution across spatial scales. *Journal of Biogeography* 25: 855–870.

Swetnam, T.W., C.D. Allen, and J.L. Betancourt. 1999. Applied historical ecology: Using the past to manage for the future. *Ecological Applications* 9: 1189–1206.

Sword-Sayer, M.A. and J.D. Haywood. 2006. Fine root production and carbohydrate concentrations of mature longleaf pine (*Pinus palustris* P. Mill.) as affected by season of prescribed fire and drought. *Trees* 20: 165–175.

Turner, R.L., J.E. Van Kley, L.S. Smith, and R.E. Evans. 1999. *Ecological Classification System for the National Forest and Adjacent Areas of the West Gulf Coastal Plain*. The Nature Conservancy, Nacogdoches, TX. 305pp.

US Global Change Research Program (USGCRP). 2009. US National Assessment of the Potential Consequences of Climate Variability and Change *Mega-Region: Southeast.* Available online at http://www.usgcrp.gov/usgcrp/Library/nationalassessment/6SE.pdf, accessed 21/06/2010.

Van Mantgem, P.J., N.L. Stephenson, J.C. Byrne et al. 2009. Widespread increase of tree mortality rates in the western United States. *Science* 323(23): 521–524.

Varner, J.M., III, J.K. Hiers, R.D. Ottmar et al. 2007. Overstory tree mortality resulting from reintroducing fire to long-unburned longleaf pine forests: The importance of duff moisture. *Canadian Journal of Forest Research* 37: 1349–1358.

Vose, J.M., D.L. Peterson, T. Patel-Weynand, (eds.). 2012. Effects of climatic variability and change on forest ecosystems: A comprehensive science synthesis for the U.S. forest sector. Gen. Tech. Rep. PNW-GTR-870. Portland, OR: U.S. Department of Agriculture, Forest Service, Pacific Northwest Research Station. 265pp.

Wade, D.D., B.L. Brock, P.H. Brose et al. 2000. Fire in eastern ecosystems. Chapter 4. pp. 61–104. In: Brown, J.K., and Smith, J.K. (eds.). *Wildland Fire in Ecosystems: Effects of Fire on Flora.* Gen. Tech. Rep. RMRS-GTR-42-vol 2, U.S. Department of Agriculture, Forest Service, Rocky Mountain Research Station, Ogden, UT. 257pp.

Wear, D.N. and J.G. Greis. 2012. The Southern Forest Futures Project: Summary Report. Gen. Tech. Rep. SRS-GTR-168. Asheville, NC: USDA-Forest Service, Southern Research Station. 54 p.

Wear, D.N., R. Huggett, and J.G. Greis. In press. Constructing alternative futures. In: Wear, D.N., Greis, J.G., (eds.). *Southern Forest Futures Project Technical Report.*

Webster, C.R. and C.G. Lorimer. 2005. Minimum opening sizes for canopy recruitment of midtolerant tree species: A retrospective approach. *Ecological Applications* 15: 1245–1262.

White, P.B., S.L. van de Gevel, H.D. Grissino-Mayer, L.B. LaForest, and G.G. DeWeese. 2011. Climatic response of oak species across an environmental gradient in the southern Appalachian Mountains, USA. *Tree-Ring Research* 67: 27–37.

White, P.S., E.R. Buckner, J.D. Pittillo, and C.V. Cogbill. 1993. High-elevation forests: Spruce–fir forests, northern hardwood forests, and associated communities. In: W.H. Martin, S.G. Boyce, and A.C. Echternacht (eds.). *Biodiversity of the Southeastern United States: Upland Terrestrial Communities.* Pp. 305–337. John Wiley and Sons, New York. 8: 921–934.

Whittaker, R.H. 1956. Vegetation of the great smoky mountains. *Ecological Monographs* 26: 1–80.

Whittaker, R.H. 1975. *Communities and Ecosystems.* Macmillan Publishing Co., Inc., New York. 385pp.

Wolters, G.L. 1982. Longleaf and slash pine decreases herbage production and alters herbage composition. *Journal of Range Management* 35(6): 761–763.

Woodall, C.W., C.M. Oswalt, J.A. Westfall et al. 2009. An indicator of tree migration in forests of the eastern United States. *Forest Ecology and Management* 257: 1434–1444.

Woodward, F.I. 1987. *Climate and Plant Distribution.* Cambridge University Press. New York. 174pp.

Zar, J.H. 1996. *Biostatistical Analysis.* Prentice-Hall, Inc. Upper Saddle River, NJ. 662pp.

Zon, R. and J.M. Briscoe. 1911. *Eucalypts in Florida. Bulletin 87.* Government Printing Office. U.S. Department of Agriculture, Forest Service. Washington, DC. 47pp.

11 Climate Change and Wildlife in the Southern United States

Potential Effects and Management Options

Cathryn H. Greenberg, Roger W. Perry, Kathleen E. Franzreb, Susan C. Loeb, Daniel Saenz, D. Craig Rudolph, Eric Winters, E.M. Fucik, M.A. Kwiatkowski, B.R. Parresol, J.D. Austin, and G.W. Tanner

CONTENTS

In the southeastern United States, climate models project a temperature increase of 2–10°C by 2100 (Intergovernmental Panel on Climate Change 2007). Climate change is already evident. Since the 1970s, average temperature has risen by about 1°C, with the greatest seasonal temperature increase during winter. Average precipitation during autumn has increased by 30% since 1901, but summer precipitation has decreased (U.S. Global Change Research Program 2012). Correspondingly, drought has affected a larger portion of the Southeast over the past three decades. The patterns and severity of storms are also changing, with more heavy downpours in many areas and the power of Atlantic hurricanes increasing (U.S. Global Change Research Program 2012).

Climate change is driven in part by activities associated with human population and economic growth that increase carbon dioxide (CO_2) emissions (Wear and Greis in press). Effects of climate change on wildlife are confounded by land-use changes associated with population and economic expansion that reduce and compromise the amount and continuity of habitat, and thus may limit their ability to respond. Strategic planning for wildlife conservation is hindered by uncertainty regarding levels of human population and economic growth, how and where climatic variables will change, and how wildlife species having widely differing life histories and habitat requirements are likely to respond. In this chapter, we explore potential impacts of climate change (2000–2060) on terrestrial vertebrates and butterflies within the five major subregions of the southeastern United States: the Mid-South, Coastal Plain, Mississippi Alluvial Valley, Piedmont, and Appalachian-Cumberland highlands (Wear and Greis in press), and discuss management options for mitigating impacts.

Studies of Pleistocene pollen and fossilized animals show correlations between glacial–interglacial fluctuations and latitudinal shifts in many species throughout much of the Southeast (Cooperative Holocene Mapping Project 1988; Hibbard et al. 1965; Root and Schneider 2002). During the transition from the last Ice Age to the present interglacial Holocene period, most plant species moved northward, but the rate of movement by individual species varied; this resulted in reassembly of forest communities different from those familiar to us today (Graham and Grimm 1990). In addition, habitat changes associated with warming and cooling periods likely forced the distribution of many animal species to shift in latitude or elevation, whereas others went extinct. Many species that are now extinct, including the American mastodon (*Mammut americanum*), Columbian mammoth (*Mammuthus columbi*), sabertooth cat (*Smilodon* spp), bison (*Bison antiquus*), giant ground sloth (*Eremotherium mirabile*), dire wolves (*Canis dirus*), and large salamanders lived in Florida and other southern states as recently as 10,000 years ago (Root and Schneider 2002; Webb 1974). Like plant communities, wildlife communities have reassembled as species responded differently to changing climate and associated changes in vegetation composition and structure (Graham and Grimm 1990).

As they did in the past, currently extant wildlife species will respond to predicted rapid climate change by: (1) going extinct, (2) exhibiting phenotypic or behavioral plasticity, or (3) undergoing evolutionary adaptive response (Austin et al. 2012; Holt 1990). Evolutionary adaptive response (natural selection) of a given species depends on rates of mutation, rates of gene flow, amount of genetic variation, level and consistency of selective pressure exerted on a particular trait, and generation time and age structure of the population (Austin et al. 2012). Because any of these factors can differ among populations and among locations, predicting and managing evolutionary change will be context-specific (Austin et al. 2012).

Today 1027 species of native terrestrial vertebrates occur in the Southeast, including 178 amphibians, 504 birds, 158 mammals, and 187 reptiles (Griep and Collins in press). The number of species and the species composition vary widely among subregions and ecosystems. Approximately 15% (152) of native terrestrial vertebrates are species of conservation concern, including 81 that are listed as endangered under the Endangered Species Act of 1973. Major factors contributing to population declines of these species include habitat destruction and fragmentation, isolation, small population size, low genetic diversity, diversion of waterways, introduction of nonnative invasive species, acid rain and other environmental pollutants, commercial development, human disturbance, and exploitation for trade. Climate change is an additional stress to wildlife that could be exacerbated by these factors (Griep and Collins in press).

Vulnerability of species to climate change depends on their exposure, sensitivity, and adaptive capacity (Glick et al. 2011; Intergovernmental Panel on Climate Change 2007). Species adapted to warmer or drier conditions could expand their current ranges. Conversely, the direct effects of altered temperature or rainfall may adversely affect animal species that are especially sensitive to air or water temperature, have specific moisture requirements, or rely on specific weather patterns for survival or successful breeding. Species that are sensitive to climate will not likely be affected by climate change if they occur in little-affected or unaffected geographic areas (exposure), or if they have the capacity to respond behaviorally (such as local movement among microhabitats for thermoregulation) or physiologically (such as becoming dormant during dry periods) (Glick et al. 2011). Generalist species may be less sensitive and adapt more easily to changes in climate than specialists. Similarly, wildlife species that occur in multiple ecosystems may be more resilient, as climate-driven alterations will likely differ among ecosystems. Even among species that are sensitive to climate, not all would be affected negatively. For example, the Florida scrub-jay (*Aphelocoma coerulescens*) and sand skink (*Neoseps reynoldsi*)—federally listed species of conservation concern that primarily inhabit young, recently disturbed stands of the Florida scrub ecosystem—could benefit from more frequent fires potentially associated with climate change (Chapter 5).

Differences in impacts across subregions and ecosystems are expected because predicted changes in temperature and precipitation are unlikely to be the same across the Southeast (Chapter 2), and because some ecosystems, rare habitats, and individual species are more vulnerable to direct and indirect effects of climate change than others. For example, in the Coastal Plain, changing fire regimes (Chapter 5) may affect inland ecosystems, whereas a rise in sea level could reduce or alter mangrove (*Avicennia* spp), coastal live oak (*Q. virginiana*), beach and dune, salt marsh, and freshwater coastal wetland habitats that are important to many plants and animals. Salinization of ground and soil water from rising sea levels may affect forests and freshwater wetlands in the Coastal Plain, along with their associated wildlife (Devall and Parresol 1998). In the Southern Appalachian Mountains, high elevation spruce–fir (*Picea* spp–*Abies* spp) forests and associated wildlife may decline in response to higher temperatures. Wildlife associated with vulnerable or rare habitats such as high-elevation bogs, may also be more susceptible to changes in temperature and precipitation than species that occur in more extensive habitat types. The Interior Highlands of Arkansas, Oklahoma, and Missouri and the Southern Appalachians may serve as important refugia for plants and animals in a changing climate, as they did during the Pleistocene (Devall and Parresol 1998; Dowling 1956).

Higher temperatures and altered weather patterns are likely to have major direct and indirect effects on biological diversity; these effects may differ substantially among animal populations or communities (at a broad scale) and species whose life histories and physiologies differ. For example, the ability of birds to move long distances may make them less vulnerable to climate change than reptiles and amphibians that generally have limited mobility (Root and Schneider 2002) and a narrow tolerance limit for temperature (particularly for amphibians) or moisture. Similarly, small mammals with a narrow range of habitats and small home-range sizes may be more affected by microclimate than larger carnivores or herbivores (Hibbard et al. 1965). Species that respond by moving longer distances may be able to survive, but range shifts could result in reassembly of wildlife communities in some locations and local extirpation of species in others. Thus, simple counts of species (richness) within subregions or landscapes may not change substantially in response to climate change because some species may be replaced.

Extreme weather events such as hurricanes, tornados, or storms, and associated changes in seasonal extremes of temperature and precipitation, may also significantly affect wildlife species. For example, in 1989 Hurrricane Hugo destroyed 80% of the nesting trees of the endangered red-cockaded woodpecker (*Picoides borealis*) in the Francis Marion National Forest, which had supported the largest known population of the woodpeckers (Watson et al. 1995). Increased frequency or intensity of hurricanes would likely have a major impact on the remnants of the longleaf pine (*Pinus palustris*) habitat that remain on the Coastal Plain. Prolonged droughts or altered hydroperiods (the length of time water is retained) in ephemeral wetlands could result in local extinctions of amphibian species that rely on these wetlands for breeding. Increased frequency or intensity of fires resulting from drought could alter habitat features required by some wildlife species, such as wiregrass (*Aristida stricta*) cover for Bachman's sparrows (*Peucaea aestivalis*) in the Coastal Plain or intact leaf litter required by ground nesting worm-eating warblers (*Helmitheros vermivorous*) in the central hardwood forest of the Cumberland-Appalachian highlands (Greenberg et al. 2007). Conversely, species such as eastern wood-pewees (*Contopus virens*) could benefit from more frequent or more intense fires (Greenberg et al. 2007).

Indirect and "cascading" impacts to wildlife are likely to be much greater than direct effects, but predictions are equally uncertain (Bagne et al. 2011). For example, shifts in vegetation and habitat structure caused by gradual climate change or altered natural disturbances (frequency, duration, or intensity of drought, wind, fire, or flooding) would likely have dramatic effects on many animal species. Changing amount, pattern, and composition of forests and other land uses both drive, and are driven by, climate change, and have a major impact on the ability of animals to disperse in response to climate change. Other indirect impacts could be changes in the amount or timing of food availability (caused by changes in synchrony between flowers and pollinators) and shifts in complex competitive or predator–prey interactions. Differences in the rate and direction of range shifts among species with symbiotic relationships, such as pollinators and host plants, can lead to extinctions (Schowalter et al. 1986).

The degree of exposure, sensitivity, and capacity to adapt to climate change are important in assessing vulnerability of species to climate change. Several methods for wildlife vulnerability assessment have been developed, such as the System for Assessing Vulnerability of Species, which uses a questionnaire based on habitat, physiology, phenology (timing of life-cycle events), and biotic interactions to develop vulnerability scores in the Western United States (Bagne et al. 2011). Criteria used to predict species vulnerability to climate change generally fall into four broad categories: habitat, physiology, phenology, and biotic interactions (Bagne et al. 2011; Glick et al. 2011). Another vulnerability assessment method, the NatureServe Climate Change Vulnerability Index (NatureServe 2012) uses available information about natural history, distribution, and landscape circumstances to help identify species that may be vulnerable to climate change.

Despite these criteria for identifying vulnerable species, vulnerability assessment for any given species is fraught with uncertainty. First, great uncertainty surrounds the accuracy of predicting changes in temperature, precipitation, or natural disturbance regimes at local levels (Vose et al. in press). Second, information on the basic natural history and requirements of many vertebrate species is incomplete. Thus, the accuracy of vulnerability assessments is limited by incomplete knowledge of (1) the level of exposure given the number of and the uncertainty surrounding climate change projections; (2) the basic natural history, physiology, and range of tolerance; (3) the behavioral or genetic adaptive capacity of most species; and (4) existing or potential competitive or predator–prey interactions. Generally, vulnerability to climate change may be increased for a species that is:

- At the edge of its range
- Restricted to high elevations
- Restricted to a specific vegetation association that could be altered by changes in temperature, precipitation, or natural disturbances
- A habitat specialist
- Dependent on another species to create specific habitats, such as burrows dug by gopher tortoise
- Narrow in its range of abiotic or biotic tolerance
- Low in genetic variation and ability to adapt evolutionarily
- Restricted in behavioral or phenotypic plasticity
- Wetland dependent, especially if dependent on a specific hydroperiod, water temperature, and water quality parameters
- Dependent on specific weather-related cues or "triggers" for breeding
- Unable to colonize rapidly due to low capacity for movement or dispersal
- Located in a habitat that is fragmented, an "island," or surrounded by inhospitable habitat
- Restricted to a specific temperature range for successful embryonic development, overwintering survival, or maintenance of its energy budget

- Dependent on specific timing of host plants or insects for important life history requirements
- Of conservation concern by virtue of low population levels, the threat of disease, or a shrinking habitat base

In this chapter, we combine literature synthesis, expert opinion, and data-based and expert-based case studies to address the vulnerability of species and possible responses to climate change in the southeastern United States using some or all of the four climate scenarios developed in the Southern Forest Futures Project (SFFP, Chapter 2; McNulty et al. in press). We focus on representative species or suites of species that have similar ecological requirements and thus may respond similarly to changes in climate or climate-driven changes to habitats. We include but do not focus solely on species of conservation concern, recognizing that some species of conservation concern may not be severely impacted, whereas other species that are not currently of conservation concern may be greatly impacted by climate change (Byers and Norris 2011). We organize our discussion broadly into taxonomic categories of vertebrates, including birds, amphibians, reptiles, and mammals. We also include a discussion on butterflies because of the important roles they serve in ecosystems as pollinators, prey, and herbivores. In addition, their visibility makes them easily monitored, early indicators of ecosystem health. Finally, we explore management options for wildlife conservation and mitigation management for a changing climate (Byers and Norris 2011).

Our case studies were selected to illustrate some of the many ways that climate change could affect wildlife in the southeastern United States through shifts in range, reproductive output, and critical habitat parameters (such as wetland hydroperiod); and by causing mismatched phenologies that disrupt biotic processes (such as the timing of anuran egg deposition and toxic leaf drop, or the timing of pollinators and flowering of host plants).

Subjects for case studies were in part selected to illustrate potential responses by broader groups of species that are likely to respond similarly to the same suite of environmental drivers, or to illustrate the very different ways that climate change might affect southeastern wildlife. For example, a case study of potential changes to hydroperiod in isolated, ephemeral ponds of the Coastal Plain that result from changes in amounts and seasonal patterns of precipitation, has implications for many pond-breeding amphibians in the Coastal Plain and other areas. Similarly, a case study on potential response to climate change by black-throated blue warblers (*Setophaga caerulescens*) in the Appalachian-Cumberland highlands illustrates potential effects on a suite of high-elevation bird species having similar requirements, such as the golden-crowned kinglet (*Regulus satrapa*), black-throated green warbler (*Setophaga virens*), and Canada warbler (*Wilsonia canadensis*). A case study that predicts Carolina northern flying squirrel (*Glaucomys sabrinus coloratus*) response to climate change scenarios illustrates the possible response of other species to potential reductions or elimination of high elevation, northern hardwood–spruce–fir forest.

MAMMALS

Because of their ability to regulate their body temperatures, mammals generally respond to climate indirectly through interactions with their food supply, predators, parasites, and habitat associations (Berteaux and Stenseth 2006). Most mammals are adapted to a particular vegetation community, the distribution of which is roughly determined by temperature and precipitation (Chapter 10). Significant changes in temperature and precipitation could alter habitats or reduce food resources and result in extirpation of some mammal species from their current range. For example, precipitation in areas that currently maintain forest could become highly variable or reduced to levels that support more drought-tolerant ecosystems such as shrublands or grasslands, resulting in a gradual shift away from forest- and woodland-associated mammals. Historically, North American forest–grassland ecotones (the boundary between different ecotypes) have retreated in a northeasterly direction and are expected to do so again if climate change continues its current trajectory (Frelich and Reich 2010).

Based on the processes of extinction and colonization and their ability to disperse, species are expected to shift their distributions or move to higher altitudes in response to a warming climate (Walther et al. 2002). Species that are isolated on small islands or island-type habitats with limited dispersal abilities, such as Carolina flying squirrels, and those in small, highly vulnerable habitats will be the most at risk. Although mammals are generally very mobile, human-modified landscapes may affect their movements to more favorable climates. Agriculture, urbanization, major highways, or other inhospitable environments may impede movements of terrestrial forest mammals to more favorable locations (Francl et al. 2010). Alternatively, human activity and modified landscapes could also contribute to faster colonization of new areas. For example, nine-banded armadillos (*Dasypus novemcinctus*) have expanded their range across the southern United States by approximately 7.8 km/year over the last 115 years as a result of both natural colonization and human-facilitated movements, such as using bridges to cross rivers and hitchhiking (Taulman and Robbins 1996). Responses by individual species to climate change may disrupt their interactions with other species (Walther et al. 2002), resulting in unforeseen consequences such as die-offs, disease spread, overpopulation, and becoming invasive when moving into a new area or when resident species vacate an area.

Studies on the effects of climate change on mammals are relatively rare (Barteaux and Stenseth 2006); thus, the potential effects of climate change on many southeastern mammals are unclear. However, studies on historical climate warming, such as the end of the Pleistocene, suggest that climate change can cause extinctions of mammals, changes in communities, and reductions in richness or diversity (Barnosky et al. 2003; Blois et al. 2010). Studies in mountainous areas suggest that mammals associated with low elevations may expand their ranges, whereas the ranges of high-elevation species may contract (Moritz et al. 2008).

Bats play an important role in forest ecosystems and suppress populations of night-flying insects, including forest and agricultural pests (Boyles et al. 2011; Kalka et al. 2008; Williams-Guillén et al. 2008). Eighteen species of bats inhabit the forested portions of the southeastern United States (Trani et al. 2007), and many of these species are federally endangered or considered species of concern. Bat species vary considerably in their habitat associations, roosting and foraging habits, and strategies for coping with harsh winter conditions such as cold temperatures and reduced food supply. Thus, responses to climate change will likely differ among species. Several characteristics of bats suggest that they may be quite sensitive to predicted changes in temperature and precipitation. Because reproduction, growth, and hibernation of temperate-zone bats are highly dependent on temperature, they are one group of mammals that may respond directly to changes in temperature regimes (Jones et al. 2009). Warm spring and summer temperatures can have a positive effect on reproductive success, resulting in a higher proportion of females that give birth, earlier fledging dates, and higher growth rates of young (Burles et al. 2009; Hoying and Kunz 1998). However, excessively high temperatures in roosts may result in increased energy expenditures, particularly when coupled with low humidity (Licht and Leitner 1967).

Climate change may result in changes in the distribution of bats during both summer and winter. For example, bat distributions in Europe are forecasted to change considerably over the next century based on various climate scenarios, with species in the Boreal Zone experiencing the greatest change and risk of extinction (Rebelo et al. 2010). One species may have already expanded its range within the United Kingdom (Lundy et al. 2010). In the Eastern United States, preferred hibernation temperatures of little brown bats (*Myotis lucifugus*) suggest that their winter distribution may show a pronounced northward movement (Humphries et al. 2002), whereas niche models suggest that both the little brown bat and the northern long-eared bat (*Myotis septentrionalis*) may expand farther southward in response to global climate change (Kalcounis-Ruepell et al. 2010). Brazilian free-tailed bats (*Tadarida brasiliensis*) in the Eastern United States have been expanding their range over the past several decades, which may be in response to a warming climate (Lee and Marsh 1978). Anecdotal evidence suggests that other species such as the Seminole bat (*Lasiurus seminolus*) may also be expanding their range northward (Bradley 2010; Wilhide et al. 1998). Climate change may also alter the timing of spring and autumn migrations of migratory bats (Newson et al. 2009). Autumn migratory activity is strongly affected by weather (Arnett et al. 2008; Baerwald and Barclay 2011); thus, extreme weather events could impact migratory patterns. Further, if changes in timing of insect emergence or activity are not synchronized with movement to winter and summer ranges, bats may arrive at summer habitats before food supplies are sufficient, a phenomenon that has been documented for birds (Inouye et al. 2000). Aside from studies by Cryan and Brown (2007), Perry et al. (2010), and Walters et al. (2006), little is known about the timing of bat migration in the southeastern United States or elsewhere, which complicates efforts to document changes in migratory timing that may result from climate change.

Small terrestrial mammals, including rodents and insectivores (shrews), typically comprise the largest group of mammals in most southeastern ecosystems (Trani et al. 2007). Of these, rodents are expected to experience most of the changes in mammal distributions resulting from climate change because they represent the most abundant group of mammal species found in many ecosystems. For example, a projected doubling in carbon dioxide (CO_2) may result in the loss of eight mammal species and a gain of 29 mammal species in the Great Smoky Mountains National Park, with the majority of this turnover being rodents (Burns et al. 2003). Small mammals play an important ecological role in forests. They are the primary prey for many raptors, snakes, and furbearers. They are also dispersers of mycorrhizal fungi spores (Johnson 1996; Trappe and Maser 1977) and seeds of important trees, such as oaks, hickories (*Carya* spp), and pines (Pank 1974; Smith and Aldous 1947; Steele et al. 1993; Vander Wall 1990). Predation of tree seeds by small mammals can affect forest regeneration and the distribution of forest ecosystems (Goheen and Swihart 2003). Underground tunneling by fossorial species may affect hydrological processes on forested watersheds (Ursic and Esher 1988), change soil properties (Huntly and Inouye 1988), and influence the composition of vegetation communities (Hobbs and Mooney 1985). Small mammals also consume the larvae and pupae of forest insect pests, which may reduce the severity of insect outbreaks (Hanski 1987). Consequently, changes in abundance of small mammals have important implications for forest ecosystems in the Southeast.

Rises in sea level from melting polar and glacial ice combined with increased intensity of hurricanes or a prolonged hurricane season associated with a warming climate (Haarsma et al. 1993) could negatively affect southeastern coastal ecosystems that support unique mammals (Michener et al. 1997). For example, along the northern Gulf of Mexico, five unique subspecies of beach mice (*Peromyscus polionotus*), four of which are federally endangered (Choctawhatchee, Alabama, St. Andrew, and Perdido Key), occur along a narrow strip of white sand dunes on the Alabama and Florida Gulf Coast (U.S. Department of the Interior, Fish and Wildlife Service 1987, 2010). Rising sea levels and beach erosion may lead to loss of habitat for these species, which currently occupy only about 80 km of coastal dunes (U.S. Department of the Interior, Fish and Wildlife Service 1987, 2010).

Past hurricanes have caused substantial mortality to a number of mammals across coastal areas of the Southeast. Muskrat (*Ondatra zibethicus*), raccoon (*Procyon lotor*), rabbit (*Sylvilagus* spp), and white-tailed deer (*Odocoileus virginianus*) were reduced by 60% in coastal Louisiana following Hurricane Audrey (Ensminger and Nichols 1958). Hurricane Hugo eliminated approximately 50% of the white-tailed deer, 65% of the squirrels, and most of the rabbits on Bulls Island in South Carolina (Cely 1991). Further, populations of island-dwelling species may be especially susceptible to reductions in habitat caused by rising sea levels. For example, a number of unique animals occur only in the Florida Keys where the maximum elevation is no more than about 6 m above sea level. These include federally endangered mammals such as the Key deer (*Odocoileus virginianus clavium*), Key Largo cotton mouse (*Peromyscus gossypinus allapaticola*), silver rice rat (*Oryzomys palustris natator*), Key Largo woodrat (*Neotoma floridana smalli*), and the Lower Keys marsh rabbit (*Sylvilagus palustris hefneri*). Models of climate-change induced rises in sea level for Big Key Island, Florida, indicate that a rise of 18 cm would inundate 34% of the island by the twenty-second century under a best-case scenario, whereas a rise of 1.4 m in sea level would inundate 96% of the island under a worst-case scenario (The Nature Conservancy 2011).

Carnivores, which are important top predators in many ecosystems, may be threatened by climate change as well. For example, rise in sea level could have significant negative impacts on large carnivores in peninsular Florida. Black bear (*Ursus americanus*) habitat would be reduced 9% by a 1-m sea level rise and 31% by a 5-m rise; and the endangered Florida panther (*Puma concolor coryi*) habitat would be reduced 29% by a 1-m rise and 90% by a 5-m rise (Whittle et al. 2008). Consequently, these and other species may need suitable habitats inland and farther north, along with forested corridors linking current and potential future ranges.

Changes in precipitation may present some of the most profound effects on mammals and ecosystems of the Southeast. Aside from the potential changes in vegetation that may result from long-term changes in precipitation (Chapter 12), increased variability or reductions in precipitation may affect disease outbreaks and mammal survival, physiology, and nutritional state. Decreased precipitation may directly affect bats. Because of their naked wings and exceptionally large lungs, bats have high rates of evaporative water loss (Neuweiler 2000); and insectivorous bats must rely on accessible water to maintain water balance (Kurta et al. 1989, 1990). For example, lactating fringed myotis (*Myotis thysanodes*) in Colorado drink 13 times as often as nonreproductive females (Adams and Hayes 2008) and reproduction in females of six bat species is adversely affected by unusually low precipitation in the Western United States (Adams 2010).

Changes in temperature or precipitation could affect food resources required by many groups of mammals. All bats of southeastern forests are insectivorous; therefore, climate change may indirectly affect bats by changing availability of their arthropod prey. Increased drought will likely influence abundance and diversity of prey for species, such as the little brown bat, that rely on emergent aquatic invertebrates (Frick et al. 2010; Rodenhouse et al. 2009). Drought may also affect bats that rely on insects associated with agriculture and forests as witnessed by daytime feeding of Brazilian free-tailed bats during the 2011 Texas drought (Mylea Bayless, personal communication, 2011, Conservation Programs Manager, Bat Conservation International, PO Box 162603, Austin, TX 78716). Although some bat species specialize on particular orders or families of insects (Agosta 2002; Lacki and Dodd 2011; Whitaker 2004), most show high spatial and temporal variation in food habits (Lee and McCracken 2005; Murray and Kurta 2002; Moosman et al. 2007). Thus, it is not clear how climate-induced changes in phenology or shifts in the distribution of insect prey would affect bats.

Changes in climate could lead to increased outbreaks of animal-transmitted diseases in humans (Harvell et al. 2002). Rodents carry or serve as reservoir hosts for many tick- and flea-borne diseases that infect humans, including Lyme disease, Rocky Mountain spotted tick fever, Tularemia, plague, and Ehrlichiosis (Gubler et al. 2001). Rodent-borne diseases frequently depend on rodent population levels, and disease outbreaks are often associated with small mammal population increases (Kuenzi et al. 2007; Mills et al. 1999), which in turn often depend on environmental conditions and food availability (Gubler et al. 2001). Climate variability, such as long wet periods followed by long dry periods, may create boom and bust cycles in rodent populations. For example, Hantavirus pulmonary syndrome is transmitted by rodents and often fatal to humans (Childs et al. 1994). Increased rainfall, which led to increased food abundance for rodents in the Four Corners Region of the Western United States, created a deadly Hantavirus outbreak in 1993 (Engelthaler et al. 1999). Hantavirus outbreaks in Europe are also associated with increased rodent populations that result from elevated average temperatures and increased rodent food (Klempa 2009; Piechotowski et al. 2008). Long periods of drought may also reduce rodent predators, such as hawks, owls, and snakes; thus, when rains follow, rodent populations may rebound faster than their predators can reproduce (Epstein 2001). Thus, increased variability in precipitation could produce similar boom and bust cycles in small mammal populations in the Southeast leading to more disease outbreaks.

It is clear that climate change has the potential to greatly affect mammal communities throughout the southeastern United States but predicting the magnitude and direction of those changes is difficult due to lack of targeted research in this area. Various modeling approaches are available that will allow managers to predict or forecast mammalian responses to climate change such as ecological niche models (e.g., Loehle 2012; Phillips et al. 2006; Wiens et al. 2009). The outcomes of these models can then be used to identify species that may be most at risk of extinction or are likely to be of management concern (e.g., become invasive).

At the end of this chapter, we present three examples that illustrate the potential effects of climate change on mammals in the southeastern United States. The first case study is for a relatively common and wide-ranging species, the eastern woodrat (*Neotoma floridana*); we examined the possible effects of climate change on the distribution of this species in the Cross Timbers section of the Mid-South, which is the western edge of its range. The second case study is on the Carolina northern flying squirrel, an endangered subspecies that depends on high elevation forests; we examined effects of climate change on persistence of the high-elevation habitats in the Blue Ridge section of the Appalachian-Cumberland highlands. In the third case study, we forecast changes in the summer maternity distribution of the Indiana bat (*Myotis sodalis*) across its entire range in response to four climate scenarios, and we estimate the relative importance of southeastern forests in the conservation and recovery of this endangered species under the four scenarios. The results of these models suggest that monitoring changes in mammal distributions, demography, and behavior in response to climate is critical for effective management.

BIRDS

Birds are especially appropriate for studying potential responses to climate change for two reasons: (1) long-term data sets on their abundance and distribution are available, and (2) their ranges are strongly associated with temperature (Root 1988b). Favorable weather conditions typically help to enhance avian reproductive success (Jarvinen 1982). For species that breed near their ecological or physiological limits, such as those near the boundaries of their geographical ranges, even small deviations from optimal climatic conditions can lead to considerable declines in survival or reproductive success (Martin 1987). The result could be declines in net recruitment (the process whereby surviving juveniles are added to a population), with potential profound effects on population size. Even small changes in net recruitment may have dramatic negative impacts on some populations of migrant landbirds that are presently declining (Askins et al. 1990; Robbins et al. 1989).

Climate modification also can indirectly affect birds by changing the amount, distribution, structure, and condition of habitats. Climate-driven changes in the amount or distribution of forest types, such as longleaf pine-wiregrass sandhills or spruce–fir forests, could affect species that are narrowly associated with those ecosystems such as red-cockaded woodpeckers and red crossbills (*Loxia curvirostra*), respectively. Similarly, increases or range extensions of exotic invasive plants due to changing climatic conditions could indirectly affect birds by changing plant food availability or vegetation structure. Climate-driven changes in fire frequency or severity may adversely affect some species while benefiting others. For example, frequent burning controls midstory hardwood vegetation, thus helping to maintain the open stand structure of pine woodlands, the primary habitat of endangered red-cockaded woodpeckers. This species is found primarily in the Florida Coastal Plain, the Piedmont, and the Mid-South. Frequent fire may also maintain young oak-scrub vegetation that is the optimal habitat for Florida scrub jays, a federally threatened species native to Florida. In contrast, more-frequent fires in the upland hardwood forests of the Appalachian-Cumberland highlands could reduce shrub cover, which is important habitat for species such as hooded warblers (*Wilsonia citrina*) (Greenberg et al. 2007). The eventual direct and indirect effects of climate change depend on a variety of factors, which make developing general predictions difficult and prone to uncertainty.

Bird responses to climate change could include adapting, changing the temporal or phenological aspects of their ecology, changing their spatial distribution, or moving to higher elevations. Many species may be capable of adapting to changes in abiotic conditions such as temperature and precipitation, and (or) biotic conditions such as vegetation (amount, type, and structure) and food resources. A certain amount of adaptation is predicted based on their natural plasticity. However, even highly adaptive species may reach their limits and would be forced to employ other measures to deal with climate change effects.

Temporal responses may appear as changes in migratory patterns, with birds arriving on the breeding grounds earlier than is typical. Phenology of bird migration is significantly affected by climatic conditions (Root et al. 2003; Sanz 2002; Sparks and Crick 1999), and recent shifts in migration dates are likely in response to recent climate change (Gordo 2007). Shifts in migration timing during the last few decades have typically trended toward earlier arrival when spring temperatures have increased (Gienapp et al. 2007; Root et al. 2005). These earlier occurrences of spring-like conditions may have negative impacts on nesting success if the trees, shrubs, or other vegetation have not leafed out or developed sufficiently to provide adequate concealment for eggs and hatchlings. Further, the earlier arrival may not be timed to match emergence of prey, which may impact adult as well as chick survival. Therefore, negative effects on nest success, breeding success, and population characteristics are possible.

Higher spring temperatures may also cause advances in the phenology of plants and insects (Schwartz et al. 2006). If birds arrive earlier on the breeding grounds, they will likely breed earlier. If the end of the breeding season is unchanged or if it is prolonged, more time would be available for reproduction, which could allow breeding pairs of some species to increase the number of clutches they produce. Thus, the annual reproductive rate (number of fledglings produced per pair) may actually increase. Further, there would be an extended period for nestlings to grow and for fledglings and adults to add body fat for the autumn migration and winter. These potential advantages of earlier arrival depend on possible changes in availability, distribution, and abundance of the food supply. However, whether the prey base will be sufficient to support birds and their young throughout an extended breeding period is unknown. Also unknown are additional consequences related to the presence and activities of animals that prey on birds. Weather events and climate change may impact departure dates of birds, progression and speed of migration, and stopover frequency and duration. Gordo (2007) suggested that additional research is needed to determine the actual relevance of each climatic variable on bird migratory phenology.

Some species may undergo distribution shifts, most likely by extending the northernmost boundary of their ranges. Hitch and Leberg (2007) found that the northern limits for many birds are significantly shifting northward, on average 2.35 km/year, and suggested this was a response to climate change. For some species, warming may have a beneficial effect if it results in a widening of the area where their climate requirements are met and if they can use this expanded area. The expanded area may contain food resources that differ from the current range; hence, the species must be capable of adjusting to new food resources and possibly different predators.

Some bird species may respond to climate change by shifting their elevation distributions. Because climate attributes—including temperature, windspeed, and precipitation—vary along elevational gradients, biotic variables such as vegetation composition and food resources also vary. In a study examining responses of birds along two elevation gradients in the French Alps over a 30-year period, Archaux (2004) found that 30% of bird species had significant shifts in elevation, with five species moving downward and three moving upward. Although there was a 2.3°C increase in spring temperature over the previous 25 years, most communities remained at their same elevation. He suggested that bird distributions by elevation have not yet been influenced by warming and that habitat suitability, interspecific interactions, or other site-specific factors likely accounted for the elevation shifts by some species. However, Berthold (1998) found an increase in elevation for several alpine species in Germany and Switzerland and concluded that these shifts were the result of temperature increases.

Waterfowl

A warming climate may bring some localized benefits to ducks, geese, and swans but the overall impact on waterfowl populations is likely to be negative (Glick 2005), largely because of changes to breeding and wintering habitats. Wetland habitats are particularly sensitive to changes in precipitation and temperature and may be degraded or eliminated if effects of climate change are not mitigated (North American Bird Conservation Initiative 2010). Expected temperature increases without accompanying increases in precipitation would affect these wetlands by reducing water inputs, reducing recharge capacity, changing the timing of wetland recharge, and increasing the frequency of droughts; and those wetlands that depend on snowmelt would be reduced or disappear. Moreover, the predicted increase in the intensity of storms and tornadoes would raise the likelihood of erosion from flash floods, reduce the length of time that water may be contained in wetlands, and compromise the predictability of changes in wetland water levels. These events could alter plant communities, prey abundance, and eventually the abundance of wetland species of birds.

In many areas of the United States, shallow wetlands known as potholes (shallow depressions that fill with water during the spring) and the wetland-dependent breeding birds that use them would be highly threatened by climate change (North American Bird Conservation Initiative 2010). This is particularly true of the prairie pothole region (spanning both sides of the U.S./Canadian border in the northern Great Plains), which is known for high productivity of waterfowl. With warming temperatures, many of these potholes are expected to disappear or be wet for shorter periods of time, making them unsuitable as breeding habitat (Glick 2005). Many of the waterfowl that travel through or overwinter in the Southeast breed in these wetlands; hence, reductions in waterfowl productivity in that area may affect populations in the five southeastern subregions.

Climate change is predicted to affect the timing and distance traveled during waterfowl migration. Wintering habitat for most species consists of freshwater lakes, river basins, deltas, coast marshes, and estuaries in the United States and Mexico. Many of the birds from the Atlantic flyway winter in locations along the mid-Atlantic coast, including the Chesapeake Bay and the Delaware Bay. Others fly further south into the Carolinas, Georgia, and Florida. Mississippi flyway and Central flyway species mostly winter in the Platte River Basin, the Mississippi Alluvial Valley, the lower Mississippi River Delta, the Playa Lakes Region, and in Coastal Plain marshes and flooded fields along the Gulf of Mexico. Thus, birds that breed within the Southeast or elsewhere, such as in the prairie pothole region, may be affected by conditions in several of the subregions in the Southeast.

Changes in sea level, precipitation patterns, and other climate conditions may influence the availability and quality of food and habitat for waterfowl. Glick (2005) suggested that a rise in sea level may reduce viable winter habitat in coastal wetlands by 17–43% in areas without structured protection, and by 20–45% in areas where sea walls and other protective structures are in place. She predicted an 8- to 86-cm rise in average sea level by 2100, which could eliminate up to 45% of coastal wetlands in the conterminous United States and threaten important wintering habitat including the shallow wetlands along the Gulf of Mexico and Atlantic coasts. Further, part of the Coastal Plain, Mississippi Alluvial Valley, and Mid-South would be particularly vulnerable (Glick 2005). Hence, climate change appears to have the potential for damaging breeding areas, migration patterns, and wintering areas of waterfowl, which spend at least part of their annual cycle in the Coastal Plain and Mississippi Alluvial Valley in the Southeast.

Shorebirds

Shorebirds likely are similar to waterfowl in terms of the potential effects of climate change on their productivity and habitat quality. Shorebirds rely on habitat found in low-lying coastal and intertidal areas during migration and wintering. According to the North American Bird Conservation Initiative (2010):

- The Gulf of Mexico and mid-Atlantic coasts (Mid-South and parts of the Coastal Plain) have seen the highest rates of relative rise in sea level and recent wetland loss in the United States.

- Rise in sea level is anticipated to inundate or fragment existing low-lying areas, including mudflats, barrier islands, and salt marshes—all habitats that shorebirds utilize.
- In areas that receive heavier rainfall, adverse effects of runoff with excess nutrients or changes in salinity are expected in coastal areas.

Galbraith et al. (2005) modeled changes in intertidal foraging habitat for shorebirds in response to sea level rise at five sites, one of which was in the Mid-South (Texas). They found that a conservative 2°C warming over the next 100 years would result in major intertidal habitat losses at four of the sites, with the Texas site experiencing an almost complete habitat loss by 2100 (which might be mitigated by construction of a new sea wall).

The quality of shorebird breeding habitat in far north latitudes also is likely to be affected by climate change. Adverse habitat changes already may be evident in arctic and subarctic areas of North America (Chapin et al. 1995). Increased temperatures and thawing of permafrost in arctic areas would deepen the active soil layer, allowing trees and shrubs to expand into areas that previously supported sedges, grasses, and dwarf shrubs. According to the North American Bird Conservation Initiative (2011), this change in habitat structure and plant species composition would have multiple effects. It would affect bird distributions and abundances, especially arctic and alpine shorebirds and waterfowl, by narrowing or eliminating their breeding habitats. However, an earlier onset of spring could initially increase productivity of nesting shorebirds if they migrate earlier to take advantage of earlier insect availability. Although precipitation would likely increase, warmer temperatures in the arctic could mean higher rates of evaporation, thus reducing soil moisture and the amount of tundra wetlands in the western and central arctic. Further, it could mean more frequent fires on the tundra. Thus, shorebirds that utilize the five subregions in the South for part of their life cycle may be adversely affected by climate change on their breeding habitat, migration routes, and wintering habitat.

Upland Game Birds

Extremes in climate are likely to impact game birds at all points of their life cycle, including the timing of breeding, selection of nest sites, and availability of food resources. Increased precipitation, which may result in unusually wet springs, could flood or wash away nests and drown chicks. Females of ground-nesting species may nest or re-nest above the high water, at sites of higher elevation than they would normally select. When the land dries, these areas may be more susceptible to predation (Wildlife Management Institute 2008). In contrast, increased drought frequency could also adversely affect upland game birds.

A warming climate may cause an accelerated breeding timetable, but how that would correspond to the emergence of insects and plants is unknown. For example, food resources may adapt to the evolving climate conditions at a different rate than game birds; thus, the presence of warm-season grasses and insects vital to chick growth may be lacking at the appropriate time. Extreme changes in climate would favor further invasion of nonnative plants and insect pests, which would degrade habitat quality.

Northern bobwhite (*Colinus virginianus*) and ruffed grouse (*Bonasa umbellus*)—two important, southeastern nonmigratory game birds with different habitat associations and range distributions—are predicted to differ in their response to climate change (Matthews et al. 2007). Bobwhite occur in all five southeastern subregions, but are least abundant in the Southern Appalachians (Matthews et al. 2007). Optimal bobwhite habitat in forested areas consists of early successional forests that are dominated by pines or hardwoods and that include both herbaceous and woody growth (Wildlife Management Institute 2008). Hot, dry summers could potentially reduce bobwhite recruitment by causing embryo mortality (Wildlife Management Institute 2008) or hyperthermia in young chicks (Evans 1997; Sumner 1935). However, increases in fire frequency associated with drier, hotter conditions (Chapter 5) will likely promote open or early successional habitat. Thus, reductions in annual productivity could be offset by higher quality habitat that could sustain more bobwhite breeding pairs (Wildlife Management Institute 2007).

The natural range of ruffed grouse in the Southeast is limited to the Appalachian-Cumberland highlands, which is the southern limit of their range, most of which is in Canada, Alaska, the upper Midwest, and the Northeast. In the East, ruffed grouse use deciduous forest and oak–savannah woodland and prefer young forest with abundant ground cover. Because they burrow in the snow to protect themselves from freezing temperatures, reduced snowfall or accelerated rates of melting snow associated with climate change could reduce their winter survival (Wildlife Management Institute 2008).

Matthews et al. (2007) modeled potential abundance and distribution of bobwhite and ruffed grouse in the Eastern United States under eight climate change scenarios. Each of their models for bobwhite (ranked as having high reliability) predicted an increase in range and (or) abundance in the Appalachian-Cumberland highlands and the Northern United States. In contrast, their models for ruffed grouse (ranked as having low reliability) predicted a range contraction and reduced abundance in the Southeast. Results of these models illustrate potential differences in climate change effects on different species with very different habitat associations, current ranges, and life history requirements.

Neotropical Migratory Birds

Climate change and associated habitat alterations are likely to cause significant losses in Neotropical migratory species in the conterminous United States (Price and Root 2001). For example, bioclimatic models developed by Rodenhouse et al. (2008) predicted that high-elevation bird species in the Northeast may experience a ≥50% reduction of suitable habitat with warming as little as 1°C. They also suggested that mid-elevation species are likely to experience declines in habitat quality that could ultimately affect their future viability. However, models predicting climate effects on the distribution of 13 Southern Appalachian forest-nesting birds indicated that naturally occurring climate oscillations may have a more immediate impact on species distributions, and ultimately population viability, than a slowly warming climate (Kim et al., unpublished report).

Climatic conditions in the Neotropic Ecozone (South and Central America, the Mexican lowlands, the Caribbean Islands, and southern Florida) likely influence the winter distribution and abundance of many Neotropical migratory birds (Root 1988a,b). Winter temperatures are predicted to increase 2–3°C, with average increases up to 156% over current levels (Smith and Tirpak 1989). These climate changes and potential associated changes in habitat quality, distribution, and availability on wintering grounds could affect populations of Neotropical migratory birds that breed in the southeastern United States.

Difficulties in predicting outcomes of climate change for Neotropical migratory birds is exacerbated by potential changes on their breeding grounds, wintering grounds (which for many migratory birds are primarily outside of the United States), and along migration routes. The distributions of some species may shift northward, move higher in elevation, or contract as warming occurs. Species that nest at high elevations in spruce–fir habitat are especially vulnerable, as that forest type is likely to diminish or disappear as temperatures warm (Chapter 11). Species that depend on specific vegetation associations may also be vulnerable, because most U.S. forests cannot rapidly respond to climate change (Root et al. 2003).

AMPHIBIANS

Shifts in climate can have negative effects on amphibian populations (Beebee 1995; Gibbs and Breisch 2001; Parmesan and Yohe 2003). Temperature and precipitation have major influences on the life cycle of amphibians, particularly their breeding activities (Busby and Brecheisen 1997; Donnelly and Crump 1998; Gosner and Black 1955; Saenz et al. 2006). Because most North

American amphibians lay eggs in water, the amount and timing of precipitation can affect their reproductive activities and yearly reproductive output (Conant and Collins 1998; Saenz et al. 2006). Water availability also affects adult amphibians because they are vulnerable to losing water from their skin and respiratory systems (Carey and Alexander 2003).

Warming climates influence the body condition of some amphibians (Corn 2005; Reading 1998; 2007; Walther et al. 2002). For example, Reading (2007) found that increased air temperatures are correlated with reductions in body condition indices of female European toads (*Bufo bufo*), rendering them less able to assimilate energy reserves during spring and summer and subjecting them to rapid depletion of stored energy reserves during mild winters compared to cold ones. Accelerated tadpole development caused by warmer water temperature can result in smaller body mass of young frogs at metamorphosis (Harkey and Semlitsch 1988).

Emissions and pollutants associated with climate change may also cause thinning of the stratospheric ozone, leading to increased atmospheric ultraviolet-B radiation (UV-B) that has been suggested as a cause of amphibian declines (Blaustein et al. 2003). Although experimental studies indicate that UV-B exposure may cause mortality or deformities in some amphibian species, no studies show a linkage between these findings and actual changes in abundance or distribution, or show that amphibians are being exposed to higher levels of UV-B in the wild (Corn 2005). Climate change also could alter the spread or virulence of emerging infectious diseases associated with amphibian population declines (Daszak et al. 2001), but responses would differ according to the ecology of the particular pathogen involved. For example, chytrid frog fungus (*Batrachochytrium dendrobatidis*), the cause of mass die-offs in some amphibian populations, thrives at moderate temperatures (23°C) but dies at 30°C; therefore, warmer, drier climates seem unlikely to promote the spread of this pathogenic fungus (Corn 2005).

Other consequences of a warming climate are changes in range and distribution of species and changes in interactions among species (Badeck et al. 2004; Blaustein et al. 2001; Collatz et al. 1998; Schierenbeck 2004; Walther et al. 2002). Parmesan and Yohe (2003) estimated that shifts in climate over the past 20–140 years caused more than half (59%) of the 1598 plant and animal species documented to experience measurable changes in phenology or distribution or both. Responses of amphibian phenology to climate change was double that of trees, birds, and butterflies and nearly eight times that of herbs, grasses, and shrubs (Parmesan 2007). In temperate areas, increases in temperature can trigger early amphibian emergence from hibernation and influence reproductive activities (Carey and Alexander 2003; Oseen and Wassersug 2002; Saenz et al. 2006).

It is well known that amphibian phenology in North America is influenced by weather. Gibbs and Breisch (2001) showed that breeding by wood frogs (*Lithobates sylvaticus*), spring peepers (*Pseudacris crucifer*), and bullfrogs (*Lithobates catesbeianus*) shifted to earlier dates during winter and spring when temperatures were higher than the previous year near Ithaca, New York. Saenz et al. (2006) found that several species in eastern Texas are sensitive to temperature and that temperature influences breeding activity in winter breeding species. Higher temperatures could alter the timing of breeding activity for several species, possibly increasing the risk of asynchronous timing with typical rainfall periods and availability of breeding sites or altering competitive interactions among tadpole species that would otherwise be temporally segregated.

Higher water temperatures also could affect the reproductive success of some amphibians, eventually altering their geographic distribution. Water temperature affects developmental rates of many amphibians, and larva of different species differ in their tolerance to minimum and maximum temperatures. For example, maturation was advanced and body mass increased for ornate chorus frog (*Pseudacris ornata*) tadpoles at day 32 with increasing temperature up to 25°C, only to reverse at 30°C; no tadpoles metamorphosed from water kept at 10°C after 111 days (Harkey and Semlitsch 1988). Geographic distribution and the timing of breeding and egg laying by some aquatic-breeding amphibians generally correspond with limits of larval temperature tolerance (Moore 1939). Compared to species that cannot tolerate very cold water temperatures, amphibians whose larvae tolerate low temperatures but are sensitive to higher temperatures are more northerly in their distributions or breed earlier in the spring or both (Moore 1939). For example, wood frogs are early breeders (March, in New York City) and range as far north as Canada, but also occur in the Southern Appalachians. Under experimental conditions, wood frog larvae successfully developed in waters ranging from 2°C to 24°C but died in waters about >25°C. In contrast, green frogs (*Lithobates clamitans*) are later breeders (June, in New York City), but do not range as far north as wood frogs and occur as far south as Florida. Under experimental conditions, green frog tadpoles successfully developed in waters ranging from 12°C to 33°C, but died at temperatures that were below (about 10°C) or above (about 36°C) this range (Moore 1939). These studies suggest that higher temperatures in the Southeast could cause some amphibians to shift their distributions northward because of tadpole temperature tolerance limits.

Impacts are expected to be stronger for species that occur farther north, where climate (particularly temperature) has changed more than at lower latitudes (Parmesan 2007). However, studies focused on southeastern amphibians suggest threats from climate change as well. Milanovich et al. (2010) predicted that increased temperatures projected in long-term weather models would cause a loss of salamander diversity in the Appalachians. Many mountain-top species may be near their thermal maxima and have limited dispersal ability. Increases in temperature would result in lost habitat for many species, and those with small geographic ranges will be at greater risk of extinction. McCallum (2010) predicted catastrophe for a subspecies of cricket frogs (*Acris crepitans blanchardi*) in Arkansas, based on long-term projections of a warming climate. He suggests that predicted climate change would significantly reduce the inclination of frogs to reproduce, which could induce population declines.

Change in precipitation, drought in particular, has been suggested as a threat to southeastern amphibians. Drier soil and leaf litter may create suboptimal conditions for terrestrial salamanders and the terrestrial stages of aquatic-breeding amphibians. Mole salamanders (*Ambystoma talpoideum*) are primarily burrowers (Ashton and Ashton 1988b) and likely are sensitive to soil moisture levels. Terrestrial salamanders (Plethodontidae) may decrease surface foraging in response to dry forest floor conditions; the resulting loss of energy could lead to poor body condition or delayed breeding (Petranka 1998). Similarly, some adult stream-breeding salamanders use upland habitats for foraging and overwintering or both (Ashton and Ashton 1978; Crawford and Semlitsch 2007). Studies have shown that abundance of adult stream salamanders, such as the Blue Ridge two-lined salamander (*Eurycea wilderae*), may be reduced in sites with reduced leaf litter depth, soil moisture, and overstory cover (Crawford and Semlitsch 2008; Moorman et al. 2011), suggesting that they are sensitive to moisture and temperature.

Amphibians that are primarily aquatic as adults may be adversely affected if changes in temperature or precipitation reduce the availability of permanent ponds or streams. Stream-dwelling salamanders rely on stable stream flows for larval development, which can take from several months to many years (Rodenhouse et al. 2009). Increased water temperatures may also be detrimental to larval salamanders in streams (Peterman and Semlitsch 2009; Semlitsch 2000). Long-term drying of wetlands could cause populations of species such as cricket frogs (*Acris* spp) or pig frogs (*Lithobates grylio*) to become more isolated and less widely dispersed across landscapes. Daszak et al. (2005) suggested that population declines of four species at the U.S. Department of Energy Savannah River Site in South Carolina are linked to a drying trend in the 1990s and shortened hydroperiods of breeding sites. In Arkansas, Trauth et al. (2006) suggested that drought and land leveling have caused population declines in Illinois chorus-frogs (*Pseudacris streckeri illinoensis*).

Changes in precipitation patterns could affect amphibian species that have distinct breeding seasons and depend on seasonal rainfall patterns and consequent hydroperiods in breeding ponds. For example, oak toads (*Bufo quercicus*) in Florida breed from May to September (Greenberg and Tanner 2005a); spring and summer droughts that result in dry breeding ponds could reduce juvenile recruitment. Widespread long-lasting drought could extend beyond the oak toad's expected lifespan of 2–4 years, producing dramatic or even catastrophic impacts to the species. Changing weather patterns also could alter amphibian-breeding cues (events that trigger breeding activity).

For example, explosive breeding by eastern spadefoot toads (*Scaphiopus holbrookii*) is triggered nearly exclusively by heavy rainfall events that fill previously dry ephemeral ponds (Greenberg and Tanner 2004, 2005b). In the absence of heavy rainfall events or in situations where ponds are already full during heavy rainfall, spadefoot toads would be unlikely to breed. Hydroperiod lengths are likely to affect amphibian populations differently because developmental rates of aquatic larvae differ among species. For example, spadefoot toad tadpoles develop and metamorphose in as little as two weeks, whereas most frog tadpoles require much longer, from three months for the gopher frog (*Lithobates capito*) to more than a year for bullfrogs (Ashton and Ashton 1988a).

Increases in hurricane activity in the Atlantic Ocean have been linked to higher sea surface temperatures in the North Atlantic (Goldenberg et al. 2001). More frequent storms are likely to result in higher storm surges and more wind damage in coastal areas. Schriever et al. (2009) found a drastic overall decrease in amphibian abundance in Louisiana following hurricanes Ivan, Katrina, and Rita. However, Gunzburger et al. (2010) found lasting changes in water chemistry in wetlands overwashed by storm surge during hurricanes in northwestern Florida, but they did not see a lasting impact on amphibian communities. They suggest that amphibian communities adjacent to marine habitats in the Southeast are resistant to the effects of storm surge overwash.

Climate-induced changes in the phenology of amphibian breeding and larval development could introduce unexpected interactions and results as well. For example, Fucik (2011) demonstrated that earlier breeding by winter-breeding frogs in Texas would increase the likelihood of adverse interactions with nonnative invasive Chinese tallow tree (*Triadica sebifera*) leaf litter, which can be lethal to developing tadpoles. Thus, a warming climate in the southeastern United States may lead to declines in some amphibian species as the result of multiple direct or indirect effects.

REPTILES

As a group, reptiles may be less vulnerable to climate change than amphibians because their scale-covered skin makes them less vulnerable to desiccation and better able to tolerate the drier, warmer conditions that are predicted (Pough et al. 2001). Their eggs are also protected from desiccation by calcareous shells. Further, their life cycles do not involve an aquatic egg or larval stage, which limits many amphibians to wetland habitats for reproduction. Many reptile species are highly mobile and capable of evading thermal stress; they also can travel long distances and have very large home ranges compared to amphibians (Brown 1993). Despite having several characteristics that should decrease vulnerability to climate change, reptiles could be affected by changes to primary habitats, temperature-driven energetic shortfalls, temperature-dependent sex determination, and changes in food availability.

Numerous reptile species are semi-aquatic, using wetlands or rivers as their primary habitat, and terrestrial habitats for egg-laying (Gibbons et al. 2000). In the Southeast, black swamp snakes (*Seminatrix pygaea*), water snakes (*Nerodia* spp), common mud turtles (*Kinosternon subrubrum*), American musk turtles (*Sternotherus* spp), and chicken turtles (*Deirochelys reticularia*), among others, are closely associated with ephemeral wetlands. Others, such as riverine map turtles (*Graptemys* spp), live in streams or rivers. Bog turtles (*Glyptemys muhlenbergii*) are restricted to bogs, seeps, and similar habitats. Increased frequency or duration of drought that affects these habitats could have dire effects on the reptile species that depend on them.

Reptile species with limited range distributions, such as the Louisiana pine snake (*Pituophis ruthveni*), may be particularly vulnerable to extinctions driven by climate change. Louisiana pine snakes are rare and declining throughout their range, which is limited to sandy soils in eastern Texas and west-central Louisiana (Rudolph et al. 2002). Baird's Pocket Gophers (*Geomys breviceps*), their primary prey (Rudolph et al. 2002), depend on a well-developed herbaceous layer maintained by frequent fires. The presence of major river barriers further impedes potential population migration of Louisiana pine snakes. Climate change or changes to habitat quality resulting from changes in fire frequency could potentially eliminate this species because of its limited and restricted distribution and barriers to movement.

The higher temperatures predicted over the next several decades could have major effects on the population characteristics and persistence of reptiles whose sex is determined by the temperature during incubation (Gibbons et al. 2000; Hawkes et al. 2007). This has been shown to occur in crocodilians (which include alligators and crocodiles) and in some turtles, with a greater proportion of females hatching with temperature increases ≥1°C (Janzen 1994). Hawkes et al. (2007) used sand temperatures and historical air temperatures at Bald Head Island in North Carolina to develop predictive models of climate change effects on hatchling production and gender of loggerhead sea turtles (*Caretta caretta*). They found that an increase of just 1°C tips the gender balance of hatchlings to female, which could lead to unfertilized clutches, lost cohorts, and eventual extinction, and that temperatures >3°C produce high mortality.

Higher temperatures associated with climate change could also have other implications for productivity in some reptile species. For example, Hawkes et al. (2007) found that earlier and prolonged nesting seasons for loggerhead sea turtles were correlated with warmer sea surface temperatures. In addition, several studies indicate that pond turtles exhibit increased juvenile growth rates and reached sexual maturity at an earlier age in warmer temperatures and longer growing seasons (Frazer et al. 1993). Gibbons et al. (1981) found that in a pond receiving thermal effluent at the Savannah River Site, male and female sliders (*Trachemys scripta*) grew faster than controls in nearby natural habitats; the males matured at an earlier age, but at the same size as the controls, while the females matured at the same age but at a larger size.

High temperatures that restrict foraging activity by reptiles could lead to energy shortfalls (Huey et al. 2010). Sinervo et al. (2010) correlated extinction probability of blue spiny lizard (*Sceloporus serrifer*) populations in Mexico with the number of hours per day that spring temperatures exceeded the body temperature at which the lizards are still capable of activity. They found that lizard populations persisted when temperatures acceptable for activity were over 3.85 h, but did not persist when the time available for foraging was shorter because net energy gain became insufficient for reproduction (Sinervo et al. 2010). Dunham's model for canyon lizards (*Sceloporus merriami*) in Texas corroborates the idea that restricted activity time may cause extinction, finding that even a 2°C increase in air temperature reduces activity time, causing a reduction in energy gain and population growth (see Gibbons et al. 2000). These studies demonstrate ecophysiological mechanisms that incorporate adaptive evasion (retreat to avoid high air temperatures) into evaluations of vulnerability to climate change and likelihood of extinction.

Reptile species with specialized diets also could be vulnerable to changes in climate that affect their food sources. For example, populations of eastern hog-nosed snakes (*Heterodon platirhinos*) and southern hog-nosed snakes (*Heterodon simus*) could decline in response to drought-induced population declines of frogs and toads, their primary prey (Ashton and Ashton 1988a).

BUTTERFLIES

Butterflies play a significant and critical role in ecosystem function as pollinators (Withgott 1999). Climate change, to the extent that it alters butterflies and other pollinator populations, would have cascading effects on entire ecosystems (Kremen and Ricketts 2000). As a consequence of their complex life cycles, diverse larval hosts, and frequent dependence on a particular suite of nectar resources, butterflies would respond to climate change in complex ways. Butterfly larvae are also important herbivores in terrestrial ecosystems and critical prey for birds and other predators. Therefore, the effects of climate change on butterfly populations would have consequences for both host plant and predator species (Schowalter et al. 1986).

Substantial literature documents geographical and phenological shifts in a wide range of butterfly species that are correlated with recent climate change. Briefly, during the past century, northward shifts in temperature isotherms by an average of 120 km in Europe are correlated with northward shifts in ranges of 35–240 km of 63% of nonmigratory butterfly species (Parmesan et al. 1999). Likewise, phenological variables of butterflies in Great Britain have shifted earlier by 2–3 days per decade (Roy and Sparks 2000). In the southeastern United States, many butterfly species are already experiencing similar patterns in relation to climate change, and other species are likely to join the ranks of the affected in the future. Other ecological processes (such as fire regimes) that will likely be altered by climate change would also affect butterfly populations.

The distribution of butterfly and skipper species in the southeastern United States is reasonably well documented and provides the basis for continent-wide examination of geographic patterns (Scott 1986), offering insights into the factors that control the overall distribution of these species. In general, butterflies are more diverse at lower latitudes and topographic complexity increases local diversity (Parmesan et al. 1999). In most of the southeastern United States, little topographic complexity leads to a rather homogeneous fauna as well as a weak diversity gradient from north to south. The most prominent exception to this pattern is a marked increase in species diversity in southern Texas and peninsular Florida, the result of a mild temperature regime, especially in the winter months. In these areas, numerous tropical species, which lack the physiological mechanisms to survive freezing temperatures, can persist as residents throughout the year.

The study of most southeastern butterfly species in relation to their environments is sufficiently advanced to provide a basis for examining the potential effects of climate change (Scott 1986). Host plant relationships, nectar resources, and individual species phenology will all interact in complicated ways as species respond to climate change (McLaughlin et al. 2002; Roy and Sparks 2000). To simplify our analysis, we divide the species into two categories based on their physiological ability to survive low temperatures. The first group consists of species that can survive subfreezing temperatures during some stage of the life cycle. Consequently, these species can maintain populations in both temperate and more extreme climates. They typically diapause (effectively hibernate) at specific stages of the life cycle, which differs among individual species. With few exceptions, these species are resident at any given locality.

The second group, wihich is more diverse, consists of species that lack the ability to diapause during periods of subfreezing temperatures. Not surprisingly, temperatures that are lethal to some species are not lethal to others, but 0°C is a reasonable approximation for most. A subset of these species—with primarily tropical affinities—is strictly resident, primarily in southern Texas and peninsular Florida. These species persist throughout the year without undergoing diapause. A second and highly variable subset includes species that are migratory (using migratory in the broadest possible sense). At one end of this spectrum are the intergenerational migrants—monarch (*Danaus plexippus*) and painted lady (*Vanessa cardui*)—that travel hundreds to thousands of kilometers to avoid low temperatures. At the other end are the species that reside permanently in southern latitudes but expand their ranges northward, often by hundreds of kilometers, during the warmer months.

These species groups can be expected to react to climate change in different ways. In relation to temperature patterns such as a northward shift in temperature isotherms, resident species with the ability to diapause will tend to adjust their ranges accordingly. The details of those adjustments will depend on concurrent shifts in host–plant distribution and the ability of individual species to move or disperse in an increasingly fragmented and rapidly changing landscape (Parmesan et al. 1999; Roy and Sparks 2000). Sedentary species that are resident in the southern-most sector, where the ability to diapause in response to low temperatures is generally not required, will tend to expand their ranges northward.

The ability of individual species to successfully shift their ranges in response to climate change will presumably be highly variable. Species that regularly migrate and recolonize the northern portions of their ranges each year will face few obstacles beyond adjustments to the geographical details and timing that may be required by changing climatic patterns. Issues related to habitat fragmentation will presumably be minimal as these species are already moving across a highly fragmented landscape every year. The presence of suitable host plants could be an issue; however, many of these species utilize a wide range of host plants over a wide geographical range.

Resident species, both southern species that do not diapause and more northerly species that do, may face more difficulties. Many of these species are thought to be quite sedentary, although the evidence against the ability of most species to migrate is not strong. Many species will have difficulty altering their ranges in fragmented landscapes, although this response is highly species dependent. Species that are adapted to disturbed or widespread habitats will be able to adapt to climate change relatively easily, but habitat specialists may not (Hoegh-Guldberg et al. 2008). An additional complication is the availability of suitable host plants. Many species, especially those with narrow host–plant requirements will be constrained by the response of host plants to climate change. Substantial lag times may result before host species colonize otherwise suitable habitats. Both factors, low capacity for movement and host–plant requirements, will undoubtedly result in many species being unable to find suitable habitat across a fragmented landscape impacted by climate change.

The response of other ecological processes to climate change could also alter butterfly community structure. Fire regimes, in particular, play a major role in structuring the abundance and species composition of butterfly communities in the southeastern United States (Rudolph and Ely 2000; Rudolph et al. 2006; Thill et al. 2004) by altering vegetation structure with profound implications for nectar resources and host plants. Climate change would potentially alter fire regimes through changes in fuel loads, precipitation patterns, and temperature regimes (Chapter 5). Prediction of fire regimes under climate change is fraught with difficulties, particularly because current regimes are primarily driven by human factors, especially prescribed fire. The fire regime before European settlement was typified by large-scale, low-intensity fires (Frost 1998). Fire return intervals were typically short, often in the range of two to five years in forested habitats, especially those dominated by pine and oak. Five centuries of escalating ecological change following European settlement have greatly altered fire regimes of the southeastern United States. Although predicting ecological responses due to changes in fire regimes is in its infancy, less fire typically results in fewer butterflies, primarily as a result of reduced nectar resources (Rudolph et al. 2006; Thill et al. 2004).

Based on predicted estimates of climate change, butterfly communities are likely to change dramatically. Predicting the responses of individual species is extremely difficult because most species are thought to respond to weather events in addition to long-term climate averages. However, changes over the last century strongly suggest that many species will expand their ranges substantially northward, contract their ranges less dramatically in the southern portions of their ranges, and adjust their distributions in relation to altitude. Other species that lack the ability to colonize new habitats rapidly will likely be locally or globally extirpated (McLaughlin et al. 2002), with resultant loss of diversity and critical ecosystem services.

MANAGEMENT OPTIONS FOR CLIMATE CHANGE

Historically, climate change has been an important driver in range shifts, extinction, colonization, and evolution through adaptation of species. Given the limitations imposed by the physiology and plasticity of each species, many of these outcomes were mediated by the availability of dispersal corridors in a landscape relatively free of modern human influence. However, the southeastern landscape is highly fragmented, with obstacles to dispersal imposed by urban and rural development, roads and highways, dams, degraded habitats, land ownership, and land use patterns. Using the past as a model, "preservation" of biological diversity—as if it were a static and unchanging entity that should remain as we know it today—is unrealistic in the face of unprecedented rates of climate change. However, conservation of biological diversity can be achieved if we recognize and allow the same processes of wildlife movement and reassembly of communities to occur by promoting landscape features that permit movement.

For some endemic island subspecies, such as the Florida Key deer, Key Largo woodrat, and Key Largo cotton mouse, their habitats could disappear, eliminating isolated areas that have maintained their genetic uniqueness. For these subspecies, establishing populations elsewhere may not be feasible because of the potential for interbreeding with mainland subspecies and "swamping" of their unique genetic traits.

Management for the conservation of biological and genetic diversity in a changing climate can involve multiple approaches that can be implemented by land use and highway planners, landowners (especially those with large land holdings), and state and federal land managers. Listed below are some critical steps toward mitigating adverse effects of climate change.

- Increase both amount and connectivity of wild lands and habitats through acquisitions and conservation programs. To allow wildlife to travel in response to climate change, give particular attention to interconnected habitats that run from north to south, and from higher to lower elevations (Root and Schneider 2002).
- Consider wildlife movement and road-kill mortality when planning or improving roads and highways by providing (as needed) elevated sections of highways, or by building wildlife underpasses to allow uninterrupted migration corridors (Root and Schneider 2002).
- Manage ecosystems to restore and maintain conditions that promote optimal habitat and larger populations, and thereby enhance the resilience of associated native wildlife species.
- Protect and conserve coastal habitats through strategic planning, zoning, and building codes. In areas of predicted rising sea levels, address inland migration of coastal wetlands in plans for new developments.
- Use caution when building barriers for flood control or during construction as they can result in elimination of existing wetlands. Establish practices that reduce the susceptibility of coastal habitats to sea level rise; examples include eliminating ditches to restore the hydrologic regime and limit saltwater intrusion, and assisting in the development of vegetation by planting salt-tolerant species and building oyster reefs to buffer shorelines from storm events and wave action.
- Restore riparian zones and watersheds to protect and maintain water quality, quantity, flow, hydrologic processes, and temperatures in wetlands and streams.
- Conserve species and special habitats, paying special attention to restoration and management of rare species or ecosystems.
- Aim for representation, resiliency, and redundancy by creating networks of intact habitats that represent the full range of species and ecosystems in a region, with multiple robust examples of each habitat.
- Reduce existing ecosystem stressors, such as habitat loss and alteration, pollution, ozone depletion, invasive species, and pathogens.
- Foster partnerships among agencies, organizations, scientists, and citizens to develop science-driven, landscape-scale strategies to maximize the use of scarce resources.
- Encourage policies that reduce the "carbon footprint," support wildlife and their habitats, and reduce climate change stresses.
- Mitigate climate change by reducing CO_2 emissions and manage forests to promote carbon sequestration where appropriate.
- Monitor, model, and implement adaptive management in response to unforeseen consequences of climate change such as trophic cascades (whereby the addition or removal of top predators impacts populations of their prey and the plants they eat).
- Increase management for species in areas where they are expected to advance, such as the northern limits of their ranges.
- Consider short-distance assistance to movements of species across artificial and natural impediments to migration, such as large rivers and areas dominated by intense agriculture.

Landscape Conservation Cooperatives and Climate Science Centers were established by the U.S. Department of the Interior to provide a philosophical foundation and organizational structure for improved coordination, cooperation, and partnerships in science and conservation across federal and state agencies, tribes, conservation organizations, and universities within geographically defined areas (Austen 2011). The goal of these cooperatives is to "develop landscape-level strategies for understanding and responding to climate change impact" by identifying and coordinating scientific research priorities, identifying conservation needs, using science to inform conservation actions on the ground, and providing a national network of resource managers, interested citizens, and private organizations (Austen 2011). This coordinated, integrated approach to climate change research and conservation provides an important benefit to the knowledge base and effective planning for climate change mitigation management.

Nonconventional management options could also be considered, but warrant serious examination from a philosophical perspective that includes land management and conservation ethical concerns. For species at risk of extinction, *ex situ* gene conservation options may be appropriate (U.S. Department of Agriculture, Forest Service 2011). These could include captive breeding programs, zoo-based population maintenance, and storing gametes for potential future use. Because translocation of species to geographic areas or habitats (assisted migration) where they do not naturally occur has been proposed as a management option for addressing climate change, Hoegh-Goldberg et al. (2008) developed a decision framework for determining the need for assisted migration that is based on a species' risk of extinction otherwise. However, the assumption that any and all means are acceptable to avoid extinction should be questioned. Assisted migration would effectively introduce nonnative species into intact wildlife communities, potentially causing unexpected repercussions for interspecific competition, genetics and natural selection, predator–prey interactions, and disease ecology (Ricciardi and Simberloff 2009). "Maverick" or unsupervised translocations and a "laissez-faire" scenario that allows extinctions to occur through inaction are the two opposite poles on a wide spectrum of approaches. What is needed is a comprehensive policy developed by conservationists, scientists, and land managers and adhered to by all (McLachlan et al. 2007).

CASE STUDIES

Case Study 1: Eastern Woodrat Distribution at the Western Edge of Range in Texas

Eastern woodrats (*Neotoma floridana*) are a common species associated with forests and woodlands throughout the Eastern United States, but can also be found in swamps and marshes as well as on the Great Plains, where they are associated with woody vegetation such as shelterbelts and fence lines (Beckmann et al. 2001; Wiley 1980). In Texas (Figure 11.1), they occupy the pineywoods, crosstimbers, and post oak savanna areas that occur across the eastern third of the state (Davis and Schmidly 2004). The range of eastern woodrats adjoins that of two similar woodrat species in central Texas; the three species likely diverged approximately 155,000 years ago (Edwards et al. 2001). White-throated woodrats (*Neotoma albigula*) are associated with deserts and semi-arid shrublands of the desert southwest (Davis and Schmidly 2004), whereas southern plains woodrats (*Neotoma micropus*) occur in brushlands of the semi-arid area between forests and the arid deserts to the west (Davis and Schmidly 2004). There is broad overlap in the ranges of these latter two species but in areas where they co-occur, habitat associations are likely the major factor maintaining their genetic integrity (Edwards et al. 2001). There is only slight overlap between the range of eastern woodrats and these other two species in Texas. The limited contact between the ranges of eastern woodrats and these other two species, along with differences in habitat requirements, likely play a major role in preventing hybridization (Edwards et al. 2001).

In the Mid-South, the Cross Timbers section of Oklahoma and Texas represents the historic western edge of continuous forests in the Southeast and the transition zone between eastern forests and the grasslands of the Great Plains (Chapter 10). Dominant ecotypes are woodlands and savannas of post oak (*Quercus stellata*) and blackjack oak (*Quercus marilandica*), intermixed with tallgrass prairies, all of which were historically maintained by frequent fire. Within this area, there is an east-west continuum; denser, moister forests with higher tree diversity occur in the east and drier forests with

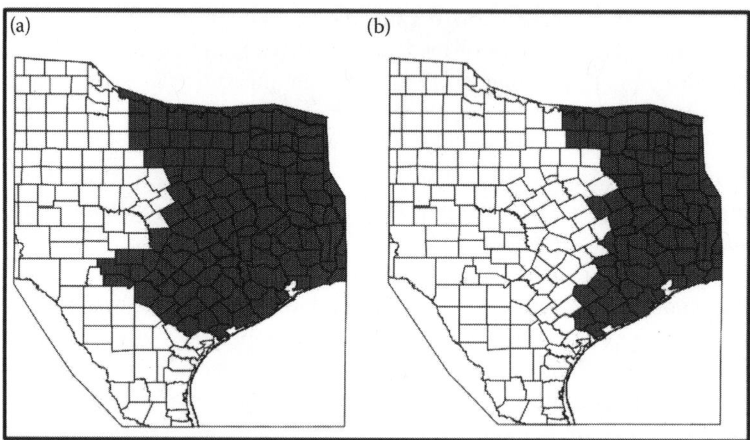

FIGURE 11.1 County-wide distribution of eastern woodrats in Texas: (a) currently, based on county records; and (b) by 2060, under the worst-case climate scenario—MIROC3.2 A1B—which predicts a mean annual reduction of 219-mm in precipitation. County records were derived from Davis and Schmidly (2004) with modifications from Birney (1976) and Edwards et al. (2001) the MIROC3.2 A1B climate scenario (McNulty et al. in press) combines the MIROC3.2 general circulation model with the A1B emissions storyline, representing moderate population growth and high-energy use.

lower tree density and lower diversity occur along the western border with the Great Plains (Rice and Penfound 1959). This east–west continuum parallels the decline in annual precipitation (Dyksterhuis 1957; Stahle and Hehr 1984), and suggests that precipitation determines the forest–grassland transition zone (Borchert 1950). Climate conditions likely become unfavorable for tree growth along the western frontier (Stahle and Hehr 1984). Reductions in precipitation and increases in temperature in forest–grassland ecotones such as the Cross Timbers could reduce dominance of forest species such as post oak in their westernmost areas. Changes in climate could affect various aspects of tree persistence, including survival, growth, and reproduction. For example, periods of drought correlate with the absence of post oak recruitment (Peppers 2004; Stahle and Cleaveland 1988). Furthermore, increased abundance of drought-tolerant eastern redcedar (*Juniperus virginiana*) or ash juniper (*Juniperus ashei*), along with altered fire regimes, may result in reductions in hard mast (acorns), which are a primary food for many forest small mammals, including eastern woodrats.

The Cross Timbers form the western boundary of the range of at least eight species of southeastern small mammals that rely on forests, including the golden mouse (*Ochrotomys nuttalli*), woodland vole (*Microtus pinetorum*), southern flying squirrel (*Glaucomys volans*), eastern chipmunk (*Tamias striatus*), cotton mouse (*Peromyscus gossypinus*), eastern gray squirrel (*Sciurus carolinensis*), southern short-tailed shrew (*Blarina carolinensis*), and eastern woodrat. Increases in temperatures, coupled with reductions in precipitation could result in reductions in forest cover and increases in shrublands in the western reaches of these species' ranges, resulting in reductions of forest-associated small mammals in Oklahoma and Texas.

Since the end of the Pleistocene, the distribution of eastern woodrats has retreated eastward across the Southern United States as the climate became warmer and dryer in the southwest (Graham et al. 1996). During the late Pleistocene, eastern woodrats occurred in forest communities of the Southwest, including the Mexican state of Chihuahua, but that area is now a desert (Harris 1984; Van Devender et al. 1987). Consequently, increased temperature and reduced precipitation could cause more arid conditions where woodrats currently reside and associated habitat changes could be a driving factor that forces the distribution of this species to retreat farther eastward.

We compared the current climate along the western edge of the eastern woodrat's range in Texas with model projections of climate change for the area. We assumed that vegetation associated with the current distribution of the eastern woodrat would respond to these climate changes and eastern woodrats would respond to changes in vegetation as they have done in the past. We also assumed that precipitation is the primary factor driving the vegetation associations (Chapter 10), and hence, the distribution of eastern woodrats.

We evaluated average annual temperature and precipitation for counties where the eastern woodrat occurs. We derived the current distribution of eastern woodrats from county records presented by Davis and Schmidly (2004), with modifications from Birney (1976) and Edwards et al. (2001). Using a logistic-regression model, we tested the accuracy of using temperature and precipitation to predict occurrence of eastern woodrats. Over a 10-year average (2000–2010) for precipitation and temperature, this model indicated that precipitation was highly informative for predicting distribution of eastern woodrats, and it explained 92% of the variation in the data. Including temperature in the model increased the percentage of variation explained by the model by only 2%. Therefore, to predict potential occurrence of eastern woodrats in the future based on changes in precipitation, we used a minimum 10-year average annual precipitation of 72.0 cm, which was based on the minimum annual precipitation of counties where the species has been documented. With this average annual precipitation value as the cutoff, 91% of the counties where the species is currently believed to occur were accurately classified but 11 counties along the western edge where eastern woodrats have not been documented were also included. Thus, this model overestimated the current distribution.

We predicted the future distribution of eastern woodrats under the best-case scenario (CSIROMK3.5 A1B) and the worst-case scenario (MIROC3.2 A1B) for precipitation from 2000 and 2060 (McNulty et al. in press). Based on model projections for the best-case scenario, average annual precipitation across the current Texas range of the eastern woodrat would decrease from 107.0 to 101.4 cm/year, an annual reduction of 5.6 cm of precipitation per year. Only two counties (Kerr and Gillespie) would fall below the minimum precipitation of 72.0 cm/year for potential retention of eastern woodrats under the best-case scenario.

Under the worst-case scenario, average precipitation across the current Texas range of the eastern woodrat would average 76.6 cm/year, a decrease of 30.4 cm (28%). By 2060, the range would retreat eastward approximately 160 km (Figure 11.1).

Case Study 2: Carolina Northern Flying Squirrel and High Elevation Spruce–Fir Forest in the Southern Appalachian Mountains

The Carolina northern flying squirrel (*Glaucomys sabrinus coloratus*) is a federally endangered subspecies of the northern flying squirrel that occurs only in high-elevation forests of the Southern Appalachian Mountains. The range of this subspecies is limited to 17 counties in the mountainous areas of western North Carolina, eastern Tennessee, and southern Virginia, with potential habitat occurring in five additional counties (Trani et al. 2007). Carolina northern flying squirrels inhabit the transition zone (ecotone) between northern hardwoods and the high-elevation spruce–fir forests that are found on the highest peaks (Payne et al. 1989; Weigl et al. 1999). Elevations where they occur are usually above 1540 m (Weigl et al. 1999). They occupy cool and moist areas with cold winters, in forests that have a well-developed canopy, substantial ground cover, abundant moist downed wood, and organic substrates (Weigl 2007). Although this subspecies may use hardwood forests, spruce–fir forests and mixed spruce-fir/hardwood forests support growth of hypogeous mycorrhizal fungi (truffles), which are important to its diet (Loeb et al. 2000; Weigl et al. 1999).

Southern flying squirrels (*Glaucomys volans*) are often considered a major competitor (Weigl 2007). However, differences in habitat preferences, diets, and climatic tolerances between the two species suggest only limited competition (Bowen 1992; Bowman et al. 2005), and there is little evidence that competition is a significant factor in the conservation of Carolina northern flying squirrel (Weigl 2007). Southern flying squirrels are likely more sensitive to the cold, and they rely on stored hardwood nuts and seeds for overwinter survival (Bowman et al. 2005; Weigl 2007), whereas Carolina northern flying squirrels are capable of surviving extremely cold and damp conditions that are lethal to southern flying squirrels and red squirrels (*Tamiasciurus hudsonicus*) (Weigl et al. 1999). However, the nematode parasite *Strongyloides robustus*, which is carried by southern flying squirrels and apparently causes few ill effects, can be detrimental to Carolina northern flying squirrels (Weigl 2007). Cold, high-elevation forests may only intermittently support *Strongyloides robustus* because of its sensitivity to cold (Weigl 2007). Therefore, effects of warming climate conditions that favor invasion of higher peaks by southern flying squirrels on the persistence of Carolina northern flying squirrels are unknown (Weigl 2007). Because of their relationship to high-elevation spruce–fir forests, we sought to determine the potential effects of rising temperatures on the persistence of Carolina northern flying squirrels in the Southern Appalachians.

Based on an analysis of the potential effects of climate change on persistence of spruce–fir forests in the Southern Appalachians, the best-case scenario (CSIROMK2 B2) predicted a 94% decrease in spruce–fir forests, and the worse-case scenario (MIROC3.2 A1B) predicted extirpation of these forests by the year 2060 (Chapter 10). Based on these estimates, the available habitat for Carolina northern flying squirrels will likely diminish significantly or may disappear by the year 2060.

Various unknowns affect estimates of spruce–fir persistence in the southern Appalachians; they include current estimates of minimal elevation where spruce–fir can persist, estimates of lapse rates (the rate of temperature change with increasing elevation), changes in precipitation, and pockets of cold air that may persist at significantly lower elevations (Chapter 10). Furthermore, the extent to which Carolina northern flying squirrels may use northern hardwoods forests is unclear. The closely related Virginia northern flying squirrel (*Glaucomys sabrinus fuscus*) occurs in northern hardwood forests with little or no spruce, but with eastern hemlock (*Tsuga canadensis*) present (Ford et al. 2007; Menzel et al. 2006) and one population of Carolina northern flying squirrels also occurs in hemlock–hardwood habitats (Hughes 2006). However, spruce forests contain a greater abundance of hypogeous fungi than northern hardwoods, which provides greater food abundance (Loeb et al. 2000) and it is unknown if these hardwood forests are sinks or sources.

Various other factors that could be detrimental to Carolina northern flying squirrels may come into play over time as climate changes. In particular, potential interactions with southern flying squirrels may be problematic. Effects of changing climate on the ability of southern flying squirrels to colonize higher elevation areas is unknown, but potential negative interactions between these two species may occur, including transmission of *Strongyloides robustus*. Recent evidence suggests that hybridization may occur between northern and southern flying squirrels in the northeastern states and in southern areas of Canada, where southern flying squirrels have recently expanded their range in response to warming temperatures (Garroway et al. 2010).

Case Study 3: Summer Maternity Range of Indiana Bats

The Indiana bat (*Myotis sodalis*) ranges throughout northeastern and midwestern states, as well as parts of the Southeast. During winter, Indiana bats hibernate in cold caves and mines in 19 states from Massachusetts to Tennessee, with the most important hibernacula (≥10,000 bats) occurring primarily in the Midwest and Southeast (U.S. Department of the Interior Fish and Wildlife Service 2007). In the spring, female bats migrate up to 575 km to their summer range (Winhold and Kurta 2006) where they form maternity colonies that usually contain <100 individuals in the snags of various tree species (Menzel et al. 2001).

In 1967, the Indiana bat was listed as endangered by the U.S. Department of the Interior Fish and Wildlife Service based on destruction and degradation of hibernacula; disturbance during hibernation; and loss and degradation of summer maternity habitat, migratory habitat, and swarming sites (U.S. Department of the Interior Fish and Wildlife Service 2007). Despite protection and recovery efforts, populations continued to decline through the year 2000 when they began to increase. However, beginning in 2007 this trend reversed with the emergence of white-nose syndrome, a disease that affects bats during hibernation (Turner et al. 2011).

Until the late 1990s, the primary summer maternity range of the Indiana bat was assumed to be in the Midwest (U.S. Department of the Interior Fish and Wildlife Service 2007). However, with increased netting efforts and possible range shifts, a number of colonies were found in other areas including the mountainous areas of western North Carolina and eastern Tennessee (Britzke et al. 2003), Pennsylvania (Butchkoski and Hassinger 2002), New York (Britzke et al. 2006; Watrous et al. 2006), and New Jersey, Maryland, and Virginia (U.S. Department of the Interior Fish and Wildlife Service 2007). In contrast, no maternity colonies were found in Alabama even though Indiana bats have been known to hibernate there (Harvey 2002). Nor have they been found in the mountains of South Carolina despite considerable netting in these areas over the past decade (Loeb, S.C. [N.d.] Unpublished data. On file with Southern Research Station, Upland Hardwood Ecology and Management, 233 Lehotsky Hall, Clemson University, Clemson, SC 29634).

In general, female Indiana bats migrate north from their hibernacula to their summer maternity colonies while males often remain in or close to hibernacula during the summer (Gardner and Cook 2002; U.S. Department of the Interior Fish and Wildlife Service 2007). During summer, Indiana bats are not dependent on any particular tree species or forest type for roosting or foraging (Menzel et al. 2001), and suitable forest habitat is not a limiting factor for their potential range—the maximum known migratory distances from the highest priority hibernacula (Gardner and Cook 2002). This suggests that there may be some climatic factor such as temperature or precipitation that restricts Indiana bat summer distribution (Brack et al. 2002).

If temperature, precipitation, or the combination of both limits the summer maternity range of Indiana bats, the result may be a shift in the summer range in response to global climate change. To determine that likelihood, we modeled the potential maternity range of Indiana bats under four climate change scenarios—CSIROMK3.5 A1B, CSIROMK2 B2, HadCM3 B2, and MIROC3.2 A1B (McNulty et al. in press). We were especially interested in understanding whether the Southeast, particularly the Southern Appalachian Mountains, might become a more important area for Indiana bat maternity colonies in the future.

We used a species distribution modeling approach (MAXENT, Phillips et al. 2006) to test whether global climate change may influence the maternity range of Indiana bats. First, we modeled the current summer distribution of Indiana bats based on temperature and precipitation during the maternity season (May to August). Based on the outcome of this model, we modeled the distribution of Indiana bats for the period 2041–2060 using the forecasted temperature and precipitation for each of the four climate scenarios. Locations of maternity colonies were provided by the Fish and Wildlife Service and we used the county center for each occurrence record. The historical climate data were at the 5-arc minute resolution (~10 km grid size) and based on PRISM climatology from 1970 to 1999 (Coulson and Joyce 2010). Projected temperature and precipitation data for each climate scenario were obtained from Coulson et al. (2010a, 2010b). We used elevation, the average maximum summer temperature (May through August), and the precipitation for each summer month as input for our models. We included the entire Eastern United States to determine the potential importance of the Southeast for Indiana bats under the various scenarios.

The models predicted that suitable habitat for Indiana bats will decrease considerably under all climate scenarios during the 2041–2060 period (Figures 11.2 and 11.3). The western portion of the range, which is now considered to be the heart of the summer range, is forecasted to be unsuitable under all climate scenarios (Figure 11.2). Areas that are predicted to be highly suitable are in the Southern Appalachians (scenarios CSIROMK3.5 A1B, CSIROMK2 B2, and HadCM3 B2) and the Northeast. The southeastern states (Kentucky, North Carolina, South Carolina, Tennessee, Virginia, and Georgia) are forecasted to contain a greater proportion of suitable habitat under future conditions than they do now (Figure 11.3).

Our results suggest that warmer climates in the western portion of the Indiana bat's current summer range will force ranges to shift farther north and east as females seek suitable climatic conditions for maternity colonies. Thus, the Northeast and Southern Appalachians could become the heart of the future summer range. However, white-nose syndrome, which is associated with

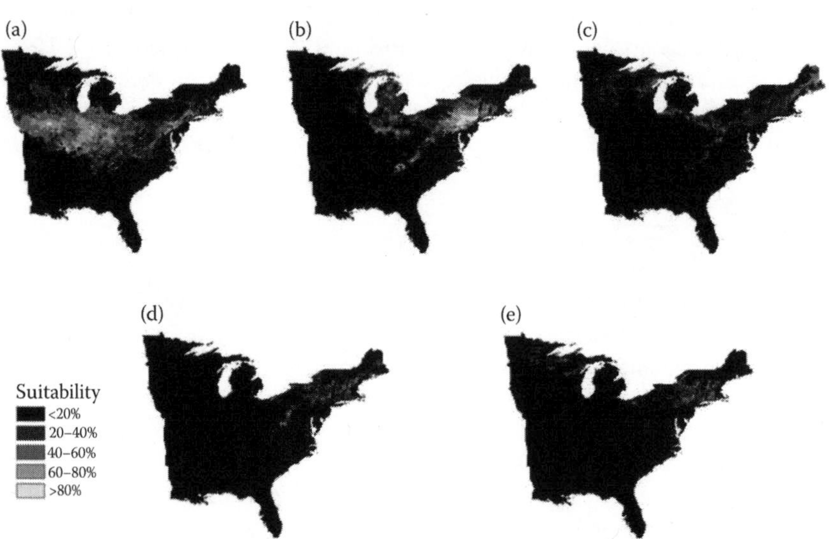

FIGURE 11.2 Suitability of summer maternity habitat of Indiana bats based on historical data, 1970–2000, and predictions from four climate change scenarios, 2040–2060: (a) historical baseline; (b) CSIROMK3.5 A1B prediction of minimal 1.15°C warming combined with moderately increasing 23-mm precipitation; (c) CSIROMK2 B2 prediction of substantial 1.68°C warming and 52-mm drying; (d) HadCM3 B2 prediction of moderate 1.35°C warming and 22-mm drying; and (e) MIROC3.2 A1B prediction of the most severe 2.35°C warming and 219-mm drying. The CSIROMK3.5 A1B, CSIROMK2 B2, HadCM3 B2, and MIROC3.2 A1B climate scenarios (McNulty et al. in press) each combines a general circulation model with an emissions storyline—the A1B storyline representing moderate population growth and high-energy use and the B2 representing lower population growth and energy use.

a fungus that grows primarily in cold temperatures within hibernacula (Gargas et al. 2009), has resulted in the death of approximately 72% of the Indiana bats that hibernate in the Northeast since 2006 (Turner et al. 2011). Studies using radiotelemetry and stable isotopes suggest that female Indiana bats hibernating in the Northeast do not migrate great distances from hibernacula to maternity colonies (Britzke et al. 2006, 2012). Therefore, even though our models forecasted that the Northeast will be important for Indiana bat maternity colonies, low numbers of bats surviving white-nose syndrome in the Northeast may retard growth of summer populations.

Although we did not model winter climate, mortality from white-nose syndrome could potentially be lower in southern hibernacula due to shorter milder winters. Consequently, the forests of the Southern Appalachians may become more important for survival of this species than our models predicted, and maintaining and restoring suitable maternity habitat for Indiana bats in the Southern Appalachians may be critical. Unlike Indiana bats in the rest of the range, Indiana bat maternity colonies in the Southern Appalachians roost almost exclusively in conifer snags in open pine–oak (*Pinus* spp–*Quercus* spp) habitats (Britzke et al. 2003; O'Keefe, J.M., Loeb, S.C. [n.d.] Unpublished data. On file with Southern Research Station, Upland Hardwood Ecology and Management, 233 Lehotsky Hall, Clemson University, Clemson, SC 29634), but there is evidence that pine–oak habitats are declining in this area (Lafon et al. 2007). Therefore, conservation and restoration of mature pine–oak habitats in the Southern Appalachians may be necessary to conserve and recover Indiana bats.

Case Study 4: Black-Throated Blue Warblers in the Appalachian-Cumberland Highlands

Black-throated blue warblers (*Setophaga coerulescens*) winter in Caribbean locales. During breeding season they are restricted to the Southern Appalachian Mountains, where they are most abundant at elevations over 1050 m but can occur down to 750 m (Hamel 1992). The preferred breeding habitat

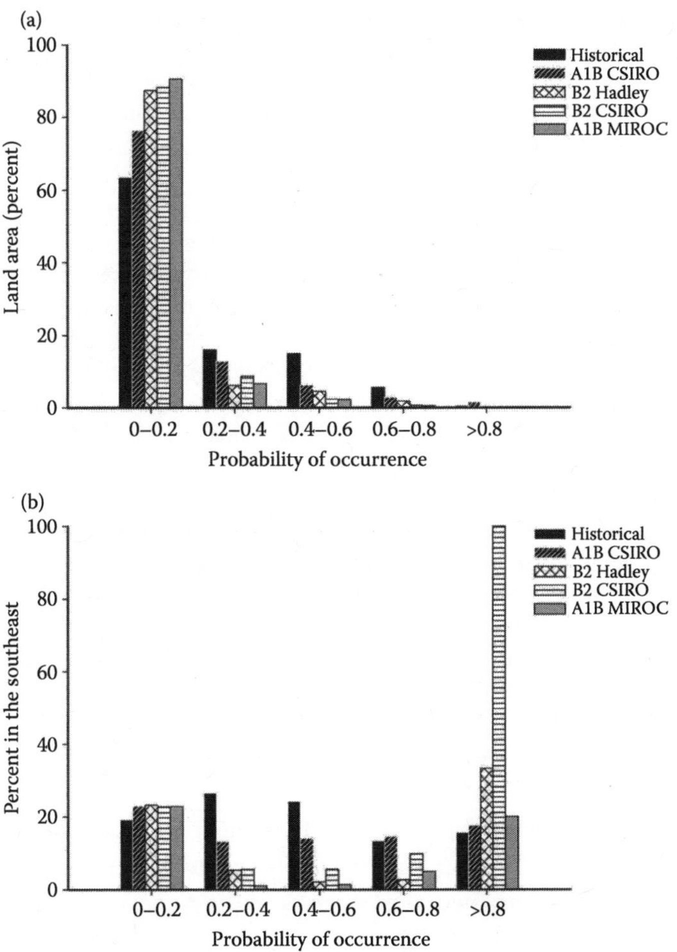

FIGURE 11.3 Land area in five habitat suitability classes for Indiana bats from 1970 to 1999 and under four climate scenarios for 2041–2060: (a) in the Eastern United States and (b) percent of the area that falls within the southeastern United States. The CSIROMK3.5 A1B, CSIROMK2 B2, HadCM3 B2, and MIROC3.2 A1B climate scenarios (McNulty et al. in press) each combines a general circulation model with an emissions storyline—the A1B storyline representing moderate population growth and high-energy use and the B2 representing lower population growth and energy use.

contains large, mature deciduous or mixed deciduous/coniferous forests or spruce–fir forests at higher elevations with a moderate or dense understory (Hamel 1992). Arthropod prey abundance can be affected by weather, which in turn influences nestling survival (Nagy and Holmes 2005; Sherry and Holmes 1992) and the likelihood of double brooding (Sillett et al. 2000). Hence, weather can indirectly affect the future viability of this species. Weather conditions can influence predator abundance and predation rates (Rodenhouse and Holmes 1992), which can in turn affect black-throated blue warblers. Because long-term weather patterns also affect habitat suitability, changes in climate can influence warbler settlement rates, spatial distribution of territories, and abundance. Reproductive success, recruitment rates, and future population size may all eventually reflect changes in weather patterns.

Rodenhouse (1992) suggested that increased precipitation could result in lower reproduction, which may be slightly offset by warmer temperatures with subsequent increases in insect prey and a longer breeding season. He speculated that nest predation could be reduced during periods of higher precipitation, another small potential offset for the predicted negative effects of precipitation on warbler productivity. He also suggested that an annual increase of 2°C could lead to lower reproductive rates and reduced recruitment for this species, but warming temperatures coupled with lower precipitation (≥10%) could result in increased reproduction. Thus, both the change in temperature and in precipitation must be considered before making predictions. Further, not only must the direction of change in precipitation and temperature (increasing or decreasing) be considered, but the intensity of the change must be considered as well.

This study used predicted changes in precipitation and temperature (2000–2060) for the Appalachian-Cumberland highlands under four climate change scenarios—CSIROMK3.5 A1B, MIROC3.2 A1B, CSIROMK2 B2, and HadCM3 B2 (McNulty et al. in press) to predict changes in black-throated blue warbler abundance and range based on temperature and precipitation thresholds that Rodenhouse (1992) determined would affect black-throated blue warbler nest productivity and recruitment. After estimating the 10-year average values for precipitation and temperature for the years 2000–2010, (the baseline), the average value for precipitation and temperature in 20-year intervals was calculated (2000–2020, 2020–2040, and 2040–2060) and these averages were summed to calculate the overall estimate for the entire period.

Under the CSIROMK3.5 A1B model, average annual temperature is predicted to increase by about 1.2°C (~9%) and annual precipitation to increase by about 83.5 mm (~6%) (Table 11.1). Rodenhouse (1992) found that higher temperatures and an average increase in precipitation <10% would result in black-throated blue warblers compensating for potential reductions in reproductive output through increased insect abundance and a longer breeding season, allowing more pairs to raise second broods. Therefore, under the CSIROMK3.5 A1B climate change scenario, black-throated blue warblers would likely maintain their annual reproductive rate.

TABLE 11.1
Predicted Change in Temperature and Precipitation under Four Climate Change Scenarios (2000–2060) Compared to a 10-year Baseline Average (2000–2010), with Probable Consequences to the Black-Throated Blue Warbler in the Appalachian-Cumberland Highlands

	Temperature Change		Precipitation Change		
Scenario[a]	**(°C)**	**(%)**	**(mm)**	**(%)**	**Likely Change in Warbler Reproduction**
MIROC3.2 A1B[b]	3.25	23	–175.27	14	Although compensation is possible, such a high temperature increase could override the potential benefits of reduced rainfall and result in lower productivity
CSIROMK3.5 A1B[c]	1.17	9	83.46	6	This moderate temperature increase and <10% precipitation increase offers the potential for compensation and minimal change in productivity
CSIROMK2 B2[d]	1.98	14	71.43	5	This high temperature increase is likely to result in reduced productivity
HadCM3 B2[e]	1.71	12	17.56	13	This moderate temperature increase and >10% precipitation increase is likely to result in loss of productivity

[a] The MIROC3.2 A1B, CSIROMK2 B2, CSIROMK3.5 A1B, and HadCM3 climate scenarios (McNulty et al. in press) combine a general circulation model (MIROC3.2, CSIROMK2, CSIROMK3.5, HadCM3) with an emissions storyline (A1B storyline representing moderate population growth and high energy use, B2 representing lower population growth and energy use).

[b] Most severe warming (+2.35°C) and drying (–219 mm).

[c] Minimal warming (+1.15°C) and moderately wetter (+23 mm).

[d] Substantial warming (+1.68°C) and drying (–52 mm).

Under the MIROC3.2 A1B scenario, average annual temperature is predicted to increase by approximately 3.3°C (~23%) and annual precipitation to decrease by 175.3 mm (~14%) (Table 11.1). Rodenhouse (1992) determined that a 10–20% precipitation decrease would result in increased productivity and a higher proportion of pairs producing second broods, but that a 2°C temperature increase would result in lower productivity and recruitment. The 3.3°C increase predicted by the MIROC3.2 A1B is over 50% higher than the harmful threshold determined by Rodenhouse et al. (2008). Although the benefits of reduced precipitation could override the harm caused by the higher temperatures in some yet-undetermined way, the magnitude of the temperature increases predicted by MIROC3.2 A1B would likely result in reduced reproduction.

The CSIROMK2 B2 model projected an average annual temperature increase of about 2.0°C (~14) and an annual precipitation decrease of approximately 71.4 mm (~5%) (Table 11.1).

Alone, the predicted decrease in precipitation could increase reproduction; however, the predicted temperature increase is likely to result in reduced productivity and lower recruitment. Because this temperature increase is substantial and the decrease in precipitation is only 5%, the CSIROMK2 B2 climate change scenario would result in reduced reproduction.

For the HadCM3 B2 model, annual temperature is estimated to increase 1.7°C (~12% over baseline), and annual precipitation to increase by 175.6 mm (~13% over baseline) from 2000 to 2060 (Table 11.1). Rodenhouse (1992) found that increased average precipitation greater than 10% over his baseline would likely result in reduced black-throated blue warbler productivity, and that second broods were unlikely to compensate for the reduction. A temperature increase of 1.7°C is higher than the increase that Rodenhouse et al. (2008) predicted would lower productivity and reduce recruitment. Hence, the wetter, hotter projections of the HadCM3 B2 climate change scenario would result in lower reproductive success.

In summary, black-throated blue warbler productivity and recruitment in the Appalachian-Cumberland highlands are unlikely to be affected by the relatively minor changes in temperature and precipitation (2000–2060) predicted under the CSIROMK3.5 A1B climate change scenario. In contrast, reductions in productivity and recruitment are likely under the MIROC3.2 AIB, CSIROMK2 B2, and HadCM3 B2 climate change scenarios, where changes in temperature or precipitation are greater (Table 11.1). These evaluations corroborate predictions of models based on other, similar climate-change scenarios that show a reduction in abundance and a range contraction of black-throated blue warblers in the Appalachian-Cumberland highlands where changes in temperature and (or) precipitation are substantial (Matthews et al. 2007).

Case Study 5: Synergistic Effects of Invasive Species and Climate Change on Aquatic Amphibians

Leaf litter from the nonnative, invasive Chinese tallow tree (*Triadica sebifera*) is known to affect the survival of some aquatic amphibians, likely by lowering dissolved oxygen and pH (Cotten et al. 2012; Leonard 2008). Cotten et al. (2012) reported that the effects from Chinese tallow tree leaf litter may be more severe on winter breeding amphibians than on species that breed later in the spring and summer. A likely reason for this difference in effect is that impacts on water chemistry from the leaf litter diminish over time; therefore, earlier breeding amphibians may be exposed to extremely low dissolved oxygen and pH levels, whereas later breeders will encounter water conditions with more favorable oxygen and pH levels. Local weather conditions can play a significant role in the timing of amphibian breeding activity (Saenz et al. 2006). Thus, climate change may play a significant role in the interactions of amphibians and the Chinese tallow tree.

The goal of this study was to understand the interaction between the Chinese tallow tree and southern leopard frog (*Lithobates sphenocephalus*) larvae when influenced by changes in climate. In Texas, Chinese tallow tree leaves generally fall from November through late December (D. Saenz, research wildlife biologist, Southern Pine Ecology, Southern Research Station, U.S. Forest Service, 506 Hayter St., Nacogdoches, TX 75961, personal observation). When leaves enter a wetland, they are quickly leached of tannins and other soluble materials, which changes the water chemistry. During drought years, leaves may remain unleached until rainfall. Southern leopard frogs breed from November through March, depending primarily on temperature (Saenz et al. 2006). This variation in the timing of leaf fall from tallow tree and the phenology of leopard frog breeding can produce a variety of scenarios, ranging from relatively simultaneous breeding and leaf fall to a four-month lapse between the two. Additionally, rainfall can play an important role in the timing of leaf leaching and breeding (Saenz et al. 2006).

To determine the relationships between invasive species, amphibian survival, and climate change, we raised leopard frog tadpoles in five *in situ* treatments, each treatment representing a different stage of tallow tree leaf decomposition in water. Water chemistry measurements were taken throughout the experiment. Tadpoles in treatments with shorter decomposition times had significantly lower survival and significantly smaller tail muscles. Treatments also had significant differences in water chemistry, supporting the hypothesis that the phenology of amphibian breeding and the timing of leaf leaching are important factors affecting tadpole survival. Because leopard frog breeding phenology and the timing of leaf leaching is regulated by precipitation and temperature (Saenz et al. 2006), climate change would have a profound and predictable impact on tadpole survival in the presence of Chinese tallow trees. Below are some general predictions of how weather may impact amphibian survival.

Hot and wet winter: Early breeding and early leaching of tallow tree leaves. This is probably the worst-case scenario because fresh leaves would fall in the water about the same time that breeding occurs. Dissolved oxygen and pH levels would be very low, causing reduced survival in tadpoles.

Hot and dry winter: Late breeding and late leaching. This scenario is not much of an improvement over hot and wet winters. Although breeding would be delayed because of a lack of water, so would leaf leaching. With the eventual onset of significant rainfall, leaching and breeding would take place about the same time. The likely result would be low dissolved oxygen and pH levels causing reduced survival in tadpoles. However, the potency of the leaves would vary because they would be exposed to the elements for some time before being inundated with water.

Cold and dry winter: Late breeding and late leaching. This scenario is similar to a hot and dry winter because leaching and breeding would be delayed. The minor difference is that both rain and temperature would delay breeding. The likely result would be low dissolved oxygen and pH levels, causing reduced survival in tadpoles.

Cold and wet winter: Late breeding and early leaching. This is the best-case scenario because breeding would take place long after the cessation of leaf leaching. In a wet winter, the leaves would leach soon after leaf drop, but in the event of low temperatures, breeding would not take place at all. Compared to the other scenarios, dissolved oxygen and pH levels would be higher when tadpoles are present, thereby increasing survival.

Shifts in the breeding phenology of many species are believed to have occurred as a result of a changing climate (Parmesan 2007). The major threat to southern leopard frogs and other winter-breeding amphibians that occur in areas invaded by Chinese tallow trees, in a warming climate, is a shift in breeding phenology. Earlier breeding by amphibians would mean greater impacts of the deleterious effects of the Chinese tallow tree.

Because shifts in amphibian breeding phenology are likely out of the control of managers, the most effective way to address negative interactions between Chinese tallow trees and early-breeding aquatic amphibians is to focus on the invasive species side of the equation. An aggressive control program for Chinese tallow tree, especially in or near wetlands, would lessen the interactions between this nonnative invasive species and native amphibians.

Case Study 6: Hydrological Pattern in an Ephemeral Pond—Implications for Amphibians

The timing and duration of wetland hydroperiod (duration of saturation), water and air temperature, and the timing and amount of precipitation have been correlated with breeding behavior, reproductive success, and declines of some amphibian populations (Carey and Alexander 2003). Ephemeral ponds are especially vulnerable to potential changes in climate that could affect water levels, hydroperiods, and the timing of depth changes that are critical for successful amphibian reproduction. Climate change could alter weather patterns, such as the amount of rainfall and temperature, thereby affecting water quality, depth, and the timing and duration of hydroperiods in ephemeral wetlands. Life history attributes, such as dispersal ability, expected life span, breeding cues, and rates of larval development, along with landscape dynamics, would likely mediate amphibian population trends and extinction risk among species.

Small, isolated, ephemeral wetlands play a critical role in sustaining the biological diversity of entire ecosystems and landscapes (Semlitsch and Bodie 1998). For example, one study reported 75,644 juvenile amphibians of 15 species metamorphosed from a 1-ha pond in South Carolina (Pechmann et al. 1989). Several amphibian species, including species of conservation concern such as the Florida gopher frog (*Lithobates capito*) and striped newt (*Notophthalmus perstriatus*), reproduce in fish-free temporary ponds and inhabit surrounding uplands as adults. Many species rely on specific weather events and hydroperiod characteristics for breeding cues and successful metamorphosis.

We used long-term (March 1994–August 2011) monthly measurements of temperature, rainfall, and water depth of an isolated pond to develop a predictive model of pond depth and hydroperiod, and applied the model to the CSIROMK3.5 A1B (best-case) and MIROC3.2 A1B (worst-case) climate change scenarios (McNulty et al. in press) over a 60-year period. Our study area, located in the longleaf pine-wiregrass (*Pinus palustris–Aristida stricta*) sandhills of the Ocala National Forest, consisted of a 0.1-ha isolated, ephemeral pond that was part of a long-term study of amphibian use of ephemeral ponds.

We fit a linear model (PROC REG; SAS Institute Inc. 2000) to the time-series data with pond depth as our dependent variable. We included the total rainfall for the current month, the combined total rainfall for the current and prior month, the date (month/year), the minimum and maximum monthly temperatures, the midrange monthly temperature, and the previous number of months during which the pond was dry. We also included a first-order lag of pond depth (pond depth one month prior) as an independent variable; this accounted for autocorrelation in the data and incorporated seasonal trends.

Our model indicated a close correlation ($R^2 = 0.89$) between rainfall and pond depth the prior month (lag), which were the only significant predictors of pond depth:

$$\text{PD} = -10.2 + 0.922\text{PDL} + 1.05,$$

where PD is pond depth, PDL is pond depth lag (cm), and R is rainfall (cm). The model also calculated belowground pond depths; although we did not measure belowground water levels in the field (dry ponds were recorded as 0 cm depth), we assumed that negative pond depths calculated in our model were acceptable because the linear model was reasonably accurate.

Predictive models indicated that changes to pond hydrological patterns could have more serious impacts on amphibians under the MIROC3.2 A1B climate change scenario than under the CSIROMK3.5 A1B scenario (Figure 11.4). Under CSIROMK3.5 A1B, the number of hydroperiods (about six per decade) and their average duration (about 14 months during most decades) would not change dramatically (Table 11.2, Figure 11.4). In contrast, the MIROC3.2 A1B climate scenariopredicted only five or six hydroperiods per decade for the first 30 years (2000–2029) and only one or two hydroperiods per decade for the last 30 years (2030–2059) modeled (Table 11.2, Figure 11.4). Most hydroperiods were predicted to be very shallow (<5 cm), or short-lived, or both. Hydroperiod length ranged from 1 to 18 months during the first 40 years, and from 1 to 1.5 months during the last 20 years.

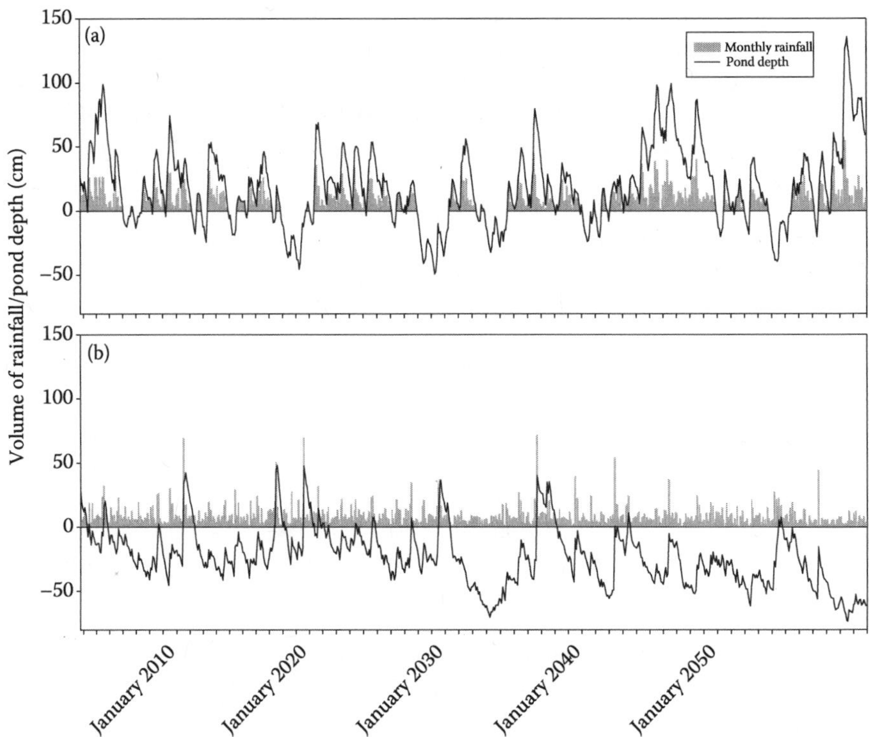

FIGURE 11.4 Predicted water depth and hydroperiod (in months) for an isolated, ephemeral pond on the Ocala National Forest in Florida, based on precipitation data for two climate change scenarios: (a) CSIROMK3.5 A1B prediction of minimal warming with moderately increasing precipitation; and (b) MIROC3.2 A1B prediction of more severe warming and drying. The CSIROMK3.5 A1B and MIROC3.2 A1B climate scenarios (McNulty et al. in press) combine the CSIROMK3.5 and MIROC3.2 general circulation model with the A1B emissions storyline, representing moderate population growth and high-energy use.

Our data represent only one of many isolated, shallow ephemeral ponds in the Ocala National Forest and throughout much of the Coastal Plain. Nonetheless, the severity of altered hydrological pattern under the MIROC3.2 A1B scenario suggests that pond breeding amphibians would be severely impacted at a much greater landscape level. Dispersion of species that are primarily aquatic, such as cricket frogs (*Acris* spp) and pig frogs (*Lithobates grylio*), would likely be severely reduced across the landscape and they could not persist at ponds that were dry most of the time. Most amphibian species that rely on ephemeral ponds for breeding would not live long enough to exploit these less frequent hydroperiods, suggesting that their populations would dramatically shrink or become locally extinct. Even if some species could breed during those infrequent hydroperiods, successful recruitment of juveniles would be unlikely because most hydroperiods would not be sufficiently long-lasting for tadpoles to complete their development to metamorphosis. Thus, Florida gopher frogs, pig frogs, southern leopard frogs (*Lithobates sphenocephalus)* and bullfrogs (*Lithobates catesbeianus*)—all commonly captured at our study pond—would be unlikely to persist, as they have long larval development periods. Populations of summer breeders, such as oak toads, would also likely shrink or become locally extinct because summer hydroperiods under the MIROC3.2 A1B scenario would be rare. The relative abundance of species would likely shift toward species such as the spadefoot toad (*Scaphiopus holbrookii*) that live longer and have fast-developing larvae. However, even persistence of the spadefoot toad is uncertain because the interval between suitable hydroperiods could exceed their lifespan.

TABLE 11.2
Number and Duration (Average and Range) of Actual (2001–2009) and Predicted (2001–2059) Hydroperiods per Decade in an Isolated, Ephemeral Sinkhole Pond in the Ocala National Forest, Based on Two Climate Change Scenarios

	Hydroperiods (Historical)			Hydroperiods (Predicted[a]) CSIROMK3.5 A1B[b]			Hydroperiods (Predicted[a]) MIROC3.2 A1B[c]		
Decade	Number	Average Duration (months)	Range (months)	Number	Average Duration (months)	Range (months)	Number	Average Duration (months)	Range (months)
2001–2009	6	12.0	1 to 51	6	13.5	2 to 32	5	3.6	1 to 6
2010–2019	N/A	N/A	N/A	8	8.6	4 to 22	5	5.0	1 to 10
2020–2029	N/A	N/A	N/A	6	14.0	1 to 39	6	3.3	1 to 11
2030–2039	N/A	N/A	N/A	6	14.0	1 to 42	1	18.0	none
2040–2049	N/A	N/A	N/A	4	26.3	5 to 82	2	1.5	1 to 2
2050–2060	N/A	N/A	N/A	6	14.5	2 to 31	2	1.0	1 to 1

Note: N/A means no data collected for the date range.

[a] The MIROC3.2 A1B, CSIROMK2 B2, CSIROMK3.5 A1B, and HadCM3 climate scenarios (McNulty et al. in press) combine a general circulation model (MIROC3.2, CSIROMK2, CSIROMK3.5, HadCM3) with an emissions storyline (A1B storyline representing moderate population growth and high energy use, B2 representing lower population growth and energy use).

[b] Minimally warmer (+1.15°C) and moderately wetter (+23 mm).

[c] Most severe warming (+2.35°C) and drying (–219 mm).

CONCLUSIONS

Uncertainty associated with climate-change prediction limits the ability of land managers to develop specific management plans for particular species. Also daunting are knowledge gaps in the basic natural history and ranges of tolerance of many wildlife species. Further, the ecological complexities of interactions among species that would occur when vegetation and wildlife communities reassemble are unknown. Clearly, research is needed to fill these gaps, thus enabling land managers and planners to develop strategic plans with a more comprehensive understanding of climate change and likely responses of plants and animals. Proactive strategies include systematic and long-term monitoring across large areas. Programs for monitoring forest status and trends, streams, and air quality at a national level are in place (U.S. Department of Agriculture Forest Service 2011), but programs for systematic long-term monitoring of wildlife are fewer and often piecemeal. The North American Breeding Bird Survey, established in 1966, and the Christmas Bird Count are important efforts that are already proving invaluable in assessing current trends in bird populations (Rodenhouse et al. 2009). A North American bat monitoring program is currently being developed. Similar long-term programs are needed for monitoring amphibians, reptiles, mammals, and other indicator species at a regional or national level. Targeted monitoring of potentially vulnerable populations in specific locales is important for early detection, assessment, and rapid response when climate change threatens population trends or population health (U.S. Department of Agriculture Forest Service 2011). Effectiveness monitoring is also essential to assess the results of management activities designed to increase resilience, reduce stressors, or otherwise benefit potentially vulnerable wildlife species (U.S. Department of Agriculture Forest Service 2011). Most importantly, monitoring can be used as an "early alert" system for adaptive management, so that land managers can be poised and flexible in implementing mitigation measures when needed. Finally, tried and true conservation practices such as restoring land connectivity and ecosystem health would benefit wildlife and help to mitigate multiple stressors on population health regardless of the uncertainties associated with climate change predictions.

REFERENCES

Adams, R.A. 2010. Bat reproduction declines when conditions mimic climate change projections for western North America. *Ecology* 91:2437–2445.

Adams, R.A. and M.A. Hayes. 2008. Water availability and successful lactation by bats as related to climate change in arid regions of western North America. *J. Anim. Ecol.* 77:1115–1121.

Agosta, S.J. 2002. Habitat use, diet and roost selection by the big brown bat (*Eptesicus fuscus*) in North America: A case for conserving an abundant species. *Mammal Rev*. 32:179–198.

Archaux, F. 2004. Breeding upwards when climate is becoming warmer: No bird response in the French Alps. *Ibis* 146:138–144.

Arnett, E.B., W.K. Brown, W.P. Erickson et al. 2008. Patterns of bat fatalities at wind energy facilities in North America. *J. Wildl. Manage*. 72:61–78.

Ashton R.E. Jr. and P.S. Ashton. 1978. Movements and winter behavior of *Eurycea bislineata* (Amphibia, Urodela, Plethodontidae). *J. Herpetol.* 12:295–298.

Ashton, R.E. Jr. and P.S. Ashton. 1988a. *Handbook of Reptiles and Amphibians of Florida: Part One, the Snakes*. Windward Publishing, Inc., Miami, FL. 176 pp.

Ashton, R.E. Jr. and P.S. Ashton. 1988b. *Handbook of Reptiles and Amphibians of Florida: Part Three, the Amphibians*. Windward Publishing Inc., Miami, FL. 191 pp.

Askins, R.A., J.F. Lynch, and R. Greenberg. 1990. Population declines in migratory birds in eastern North America. *Curr. Ornithol.* 7:1–57.

Austen, D.J. 2011. Landscape conservation cooperatives: A science-based network in support of conservation. *Wildl. Prof.* 5:32–37.

Austin, J.D., C.W. Miller, and R.J. Fletcher, Jr. 2012. What role can natural selection and phenotypic plasticity play in wildlife adaptation to climate change? Pp. 38–57 *In* Brodie, J.F., E. Post, and D.F. Doak (eds.). *Wildlife Conservation in a Changing Climate*. University of Chicago Press, IL. xxx p. http://www.press.uchicago.edu/ucp/books/book/chicago/W/bo13770227.html

Badeck, F.W., A. Bondeau, K. Bottcher et al. 2004. Responses of spring phenology to climate change. *New Phytol.* 162:295–309.

Baerwald, E.F. and R.M.R. Barclay. 2011. Patterns of activity and fatality of migratory bats at a wind energy facility in Alberta, Canada. *J. Wildl. Manage*. 75:1103–1114.

Bagne, K.E., M.M. Friggens, and D.M. Finch. 2011. *A System for Assessing Vulnerability of Species (SAVS) to Climate Change*. U.S. For. Serv., Gen. Tech. Rep. RMRS-GTR-257.

Barnosky, A.D., E.A. Hadly, and C.J. Bell. 2003. Mammalian response to global warming on varied temporal scales. *J. Mammal.* 84:354–368.

Beckmann, J.P., G.A. Kaufman, and D.W. Kaufman. 2001. Influence of habitat on distribution and abundance of the eastern woodrat in Kansas. *Gr. Plains Res.* 11:249–60.

Beebee, T.J.C. 1995. Amphibian breeding and climate. *Nature* 374:219–220.

Berteaux, D. and N.C. Stenseth. 2006. Measuring, understanding and projecting the effects of large-scale climatic variability on mammals. *Clim. Res.* 32:95–97.

Berthold, P. 1998. Vogelwelt und Klima: Gegenwartige Veranderungen. *Naturw. Rdsch.* 51:337–346.

Birney, E.C. 1976. An assessment of relationships and effects of interbreeding among woodrats of the *Neotoma floridana* species group. *J. Mammal.* 57:103–132.

Blaustein, A.R., L.K. Belden, D.H. Olson et al. 2001. Amphibian breeding and climate change. *Conserv. Biol.* 15:1804–1809.

Blaustein, A.R., J.M. Romansic, J.M. Kiessecker et al. 2003. Ultraviolet radiation, toxic chemicals and amphibian population declines. *Divers. Distrib.* 9:123–140.

Blois, J.L., J.L. McGuire, and E.A. Hadly. 2010. Small mammal diversity loss in response to late-Pleistocene climate change. *Nature* 465:771–774.

Borchert, J.R. 1950. The climate of the central North American Grassland. *Annals Assoc. Am. Geogr*. 40:1–39.

Bowen, M.S. 1992. Cold weather survival in the northern flying squirrel (*Glaucomys sabrinus*): A comparison of three species of sciurid. MS thesis, Wake Forest University, Winston–Salem, NC.

Bowman, J., G.L. Holloway, J.R. Malcolm et al. 2005. Northern range boundary dynamics of southern flying squirrels: Evidence of an energetic bottleneck. *Can. J. Zool.* 83:1486–1494.

Boyles, J.G., P.M. Cryan, G.F. McCracken et al. 2011. Economic importance of bats in agriculture. *Science* 332:41–42.

Brack, V., Jr., C.W. Stihler, R.J. Reynolds et al. 2002. Effect of climate and elevation on distribution and abundance in the mideastern United States. In: Kurta, A. and J. Kennedy (eds.) *The Indiana Bat: Biology and Management of an Endangered Species*. Pp. 21–28. Bat Conservation International, Austin, TX. 253 pp.

Bradley, S.B. 2010. Salem blowdown roadside salvage environmental assessment, Appendix D. Mark Twain National Forest, biological assessment, Salem, MO. http://a123.g.akamai.net/7/123/11558/abc123/forestservic.download.akamai.com/11558/www/nepa/60738_FSPLT1_026169.pdf. Accessed August 2011.

Britzke, E.R., M.J. Harvey, and S.C. Loeb. 2003. Indiana bat, *Myotis sodalis*, maternity roosts in the southern United States. *Southeast. Nat.* 2:235–242.

Britzke, E.R., A.C. Hicks, S.L. Von Oettingen et al. 2006. Description of spring roost trees used by female Indiana bats (*Myotis sodalis*) in the Lake Champlain Valley of Vermont and New York. *Am. Midl. Nat.* 155:181–187.

Britzke, E.R., S.C. Loeb, C.S. Romanek et al. 2012. Variation in catchment areas of Indiana bat (*Myotis sodalis*) hibernacula inferred from stable hydrogen (δ^2H) isotope analysis. *Can. J. Zool.* 90:1243–1250.

Brown, W.S. 1993. Biology, status, and management of the timber rattlesnake (*Crotalis horridus*): A guide for conservation. Soc. Study Amphib and Rept., *Herpetol. Circ.* 22:1–72.

Burles, D.W., R.M. Brigham, R.A. Ring et al. 2009. Influence of weather on two insectivorous bats in a temperate Pacific Northwest rainforest. *Can. J. Zool.* 87:132–138.

Burns, C.E., K.M. Johnson, and O.J. Schmitz. 2003. Global climate change and mammalian species diversity in U.S. national parks. *Proc. Nat. Acad. Sci.* 100:11474–11477.

Busby, W.H. and W.R. Brecheisen. 1997. Chorusing phenology and habitat associations of the crawfish frog, *Rana areolata* (Anura: Ranidae), in Kansas. *Southwest. Nat.* 42:210–217.

Butchkoski, C.M. and J.D. Hassinger. 2002. Ecology of a maternity colony roosting in a building. In: Kurta, A. and J. Kennedy (eds.) *The Indiana Bat: Biology and Management of an Endangered Species*. Pp 130–142. Bat Conservation International, Austin, TX.

Byers, E. and S. Norris. 2011. Climate change vulnerability assessment of species of concern in West Virginia. Project Report, West Virginia Division of Natural Resources. 69 pp.

Carey, C. and M.A. Alexander. 2003. Climate change and amphibian declines: Is there a link? *Divers. Distrib.* 9:111–121.

Cely, J.E. 1991. Wildlife effects of Hurricane Hugo. *J. Coastal Res.* 8:319–326.

Chapin, F.S., III., G.R. Shaver, A.E. Giblin et al. 1995. Responses of arctic tundra to experimental and observed changes in climate. *Ecology* 76:694–711.

Childs, J.E., T.G. Ksiazek, C.F. Spiropoulou et al. 1994. Serologic and genetic identification of *Peromyscus maniculatus* as the primary rodent reservoir for a new hantavirus in the southwestern United States. *J. Infect. Dis.* 169:1271–1280.

Collatz, G.J., J.A. Berry, and J.S. Clark. 1998. Effects of climate and atmospheric CO_2 partial pressure on the global distribution of C4 grasses: Present, past, and future. *Oecologia* 114:441–454.

Conant, R. and J.T. Collins. 1998. *A Field Guide to the Reptiles and Amphibians: Eastern and Central North America*. Houghton Mifflin Company, Boston, New York, NY.

Cooperative Holocene Mapping Project. 1988. Climate changes of the last 18,000 years: Observations and model simulations. *Science* 241:1043–1052.

Corn, P.S. 2005. Climate change and amphibians. *Anim. Biodiv. Conserv.* 28:59–67.

Cotten, T.B., M.A. Kwiatkowski, D. Saenz, and M. Collyer. 2012. Effects of an invasive plant, Chinese tallow (*Triadica sebiferum*), on development and survivorship of anuran larvae. *J. Herpetol.* 46:186–193.

Coulson, D.P. and L.W. Joyce. 2010. Historical climate data (1940–2006) for the conterminous United States at the 5 arc minute grid spatial scale based on PRISM climatology. Fort Collins, CO: U.S. For. Serv., Rocky Mountain Research Station. http://www.fs.fed.us/rm/data_archive/dataaccess/US_HistClimateScenarios_grid_PRISM.shtml. Accessed August 2010.

Coulson, D.P., L.A. Joyce, D.T. Price et al. 2010a. Climate scenarios for the conterminous United States at the 5 arc minute grid spatial scale using SRES scenario B2 and PRISM climatology. Fort Collins, CO: U.S. Department of Agriculture, Forest Service, Rocky Mountain Research Station. http://www.fs.fed.us/rm/data_archive/dataaccess/US_ClimateScenarios_grid_B2_PRISM.shtml. Accessed August 2010.

Coulson, D.P., L.A. Joyce, D.T. Price et al. 2010b. Climate scenarios for the conterminous United States at the 5 arc minute grid spatial scale using SRES scenarios A1B and A2 and PRISM climatology. Fort Collins, CO: U.S. For. Serv. Rocky Mountain Research Station. http://www.fs.fed.us/rm/data_archive/dataaccess/US_ClimateScenarios_grid_A1B_A2_PRISM.shtml. Accessed August 2010.

Crawford, J.A. and R.D. Semlitsch. 2007. Estimation of core terrestrial habitat for stream-breeding salamanders and delineation of riparian buffers for protection of biodiversity. *Conserv. Biol.* 21:152–158.

Crawford J.A. and R.D. Semlitsch. 2008. Post-disturbance effects of even-aged timber harvest on stream salamanders in southern Appalachian forests. *Anim. Conserv.* doi:10.1111/j.1469-1795.2008.00191.x

Cryan, P.M. and A.C. Brown. 2007. Migration of bats past a remote island offers clues toward the problem of bat fatalities at wind turbines. *Biol. Conserv.* 139:1–11.

Daszak, P., A.A. Cunningham, and A.D. Hyatt. 2001. Anthropogenic environmental change. *Acta Tropica* 78:103–116.

Daszak, P., D.E. Scott, A.M. Kilpatrick et al. 2005. Amphibian population declines at the Savannah River Site are linked to climate, not chytridiomycosis. *Ecology* 86:3232–3237.

Davis, W.B. and D.J. Schmidly. 2004. *The Mammals of Texas*. University of Texas Press, Austin, TX.

Devall, M.S. and B.R. Parresol. 1998. Effects of global climate change on biodiversity in forests of the southern United States. In: Mickler, R.A. and S. Fox (eds.). *The Productivity and Sustainability of Southern Forest Ecosystems in a Changing Environment*. Pp. 663–681. Springer-Verlag, Inc. New York, NY. 892 pp.

Donnelly, N.A. and M.L. Crump. 1998. Potential effects of climate change on two neotropical amphibian assemblages. *Clim. Change* 39:541–561.

Dowling, H.G. 1956. Geographic relations of Ozarkian amphibians and reptiles. *Southwest. Nat.* 1:174–189.

Dyksterhuis, E.J. 1957. The savannah concept and its use. *Ecology* 38:435–442.

Edwards, C.W., C.F. Fulhorst, and R.D. Bradley. 2001. Molecular phylogenetics of the *Neotoma albigula* species group: Further evidence of a paraphyletic assemblage. *J. Mammal.* 82:267–279.

Engelthaler, D.M., D.G. Mosley, J.E. Cheek et al. 1999. Climatic and environmental patterns associated with Hantavirus pulmonary syndrome, Four Corners Region, United States. *Emerg. Infect. Dis.* 5:87–94.

Ensminger, A.B. and L.G. Nichols. 1958. Hurricane damage to Rockefeller Refuge. Proc. Ann. Conf., Southeast. *Assoc. Game and Fish Comm.* 11:52–56.

Epstein, P.R. 2001. Climate change and emerging infectious diseases. *Microbes Infect.* 3:747–754.

Evans, C.A. 1997. Reproductive biology of scaled quail (*Callipepla squamata*) in southern New Mexico. MS thesis, New Mexico State University, Las Cruces.

Ford, W.M., K.N. Mertz, J.M. Menzel et al. 2007. Late winter home range and habitat use by the Virginia northern flying squirrel (*Glaucomys sabrinus fuscus*). U.S. Forest Service, Res. Pap. NRS-4.

Francl, K.E., K. Hayhoe, M. Saunders et al. 2010. Ecosystem adaptation to climate change: Small mammal migration pathways in the Great Lakes states. *J. Gr. Lakes Res.* 36:86–93.

Frazer, N.B., J.L. Greene, and J.W. Gibbons. 1993. Temporal variation in growth rate and age at maturity of male painted turtles, *Chrysemys picta*. *Am. Midl. Nat.* 30:314–324.

Frelich, L.E. and P.B. Reich. 2010. Will environmental changes reinforce the impact of global warming on the prairie-forest border of central North America. *Frontiers Ecol. Environ.* 8:371–378.

Frick, W.F., D.S. Reynolds, and T.H. Kunz. 2010. Influence of climate and reproductive timing on demography of little brown myotis *Myotis lucifugus*. *J. Anim. Ecol.* 79:128–136.

Frost, C.C. 1998. Presettlement fire frequency regimes of the United States: A first approximation. *Proc. Tall Timbers Fire Ecol. Conf.* 20:70–81.

Fucik, E.M. 2011. Interactions between invasive species and climate change: Effect on an East Texas anuran. MS thesis. Stephen F. Austin State University, Nacogdoches, TX.

Galbraith, H., R. Jones, R. Park et al. 2005. *Global Climate Change and Sea Level Rise: Potential Losses of Intertidal Habitat for Shorebirds*. U.S. For. Serv. Gen. Tech. Rep. PSW-GTR-191.

Gardner, J.E. and E.A. Cook. 2002. Seasonal and geographic distribution and quantification of potential summer habitat. In: Kurta, A. and J. Kennedy (eds.) *The Indiana Bat: Biology and Management of an Endangered Species*. Pp. 9–20. Bat Conservation International, Austin, TX.

Gargas, A., M.T. Trest, M. Christensen et al. 2009. *Geomyces destructans* sp. nov. associated with bat white-nose syndrome. *Mycotaxon* 108:147–154.

Garroway, C.J., J. Bowman, T.J. Cascaden et al. 2010. Climate change induced hybridization in flying squirrels. *Glob. Change Biol.* 16:113–121.

Gibbons, J.W., D.E. Scott, T.J. Ryan et al. 2000. The global decline of reptiles, déjà vu amphibians. *Bioscience* 50:653–666.

Gibbons, J.W., R.D. Semlitsch, J.L. Greene et al. 1981. Variation in age and size at maturity of the slider turtle (*Pseudemys scripta*). *Am. Nat.* 117:841–845.

Gibbs, J.P. and A.R. Breisch. 2001. Climate warming and calling phenology of frogs near Ithaca, New York, 1900–1999. *Conserv. Biol.* 15:1175–1178.

Gienapp, P., R. Leimu, and J. Merila. 2007. Responses to climate change in avian migration time—Microevolution versus phenotypic plasticity. *Clim. Res.* 35:25–35.

Glick, P. 2005. *The Waterfowlers' Guide to Global Warming*. National Wildlife Federation, Washington, DC.

Glick, P., B.A. Stein, and N.A. Edelson (eds.). 2011. *Scanning the Conservation Horizon: A Guide to Climate Change Vulnerability Assessment*. National Wildlife Federation, Washington, D.C. www.nwf.org/vulnerabilityguide. Accessed August 2011.

Goheen, J.R. and R.K. Swihart. 2003. Food-hoarding behavior of gray squirrels and North American red squirrels in the central hardwoods region: Implications for forest regeneration. *Can. J. Zool.* 81:1636–1639.

Goldenberg, S.B., C.W. Landsea, A.M. Mestras-Nunez et al. 2001. The recent increases in Atlantic hurricane activity: Causes and implications. *Science* 293:474–479.

Gordo, O. 2007. Why are bird migration dates shifting? A review of weather and climate effects on avian migratory phenology. *Clim. Res.* 35:37–58.

Gosner, K.L. and I.H. Black. 1955. The effects of temperature and moisture on the reproductive cycle of *Scaphiopus h. holbrooki*. *Am. Midl. Nat.* 54:192–203.

Graham, R.W. and E.C. Grimm. 1990. Effects of global climate change on the patterns of terrestrial biological communities. *Tree* 5:289–292.

Graham, R.W., E.L. Lundelius, M.A. Graham et al. 1996. Spatial response of mammals to later Quaternary environmental fluctuations. *Science* 272:1601–1606.

Greenberg, C.H. and G.W. Tanner. 2004. Breeding pond selection and movement patterns by eastern spadefoot toads (*Scaphiopus holbrookii*) in relation to weather and edaphic conditions. *J. Herpetol.* 38:569–577.

Greenberg, C.H. and G.W. Tanner. 2005b. Spatial and temporal ecology of eastern spadefoot toads on a Florida landscape. *Herpetologica* 61:20–28.

Greenberg, C.H. and G.W. Tanner. 2005a. Spatial and temporal ecology of oak toads (*Bufo quercicus*) on a Florida landscape. *Herpetologica* 61:422–434.

Greenberg, C.H., A. Livings-Tomcho, A.D. Lanham et al. 2007. Short-term effects of fire and other fuel reduction treatments on breeding birds in a southern Appalachian hardwood forest. *J. Wildl. Manage.* 71:1906–1916.

Griep, M.T. and B. Collins. In press. Wildlife and forest communities. In: Wear, D. and J. Greis (eds.). *Southern Forest Futures*. Pp. xx–xx. Asheville, NC: U.S. Department of Agriculture, Forest Service, Southern Research Station.

Gubler, D.J., P. Reiter, K.L. Ebi et al. 2001. Climate variability and changes in the United States: Potential impacts on vector- and rodent-borne diseases. *Environ. Health Persp.* 109:223–233.

Gunzburger, M.S., W.B. Hughs, W.J. Barichivich et al. 2010. Hurricane storm surge and amphibian communities in coastal wetlands of northwestern Florida. *Wetl. Ecol. Manage.* 18:651–663.

Haarsma, R.J., J.F.B. Mitchell, and C.A. Senior. 1993. Tropical disturbances in a GCM. *Clim. Dynam.* 8:247–257.

Hamel, P.B. 1992. *The Land Manager's Guide to the Birds of the South*. The Nature Conservancy, Chapel Hill, NC.

Hanski, I. 1987. Pine sawfly population dynamics: Patterns, processes, problems. *Oikos* 50, 327–335.

Harkey, G.A. and R.D. Semlitsch. 1988. Effects of temperature on growth, development, and color polymorphism in the ornate chorus frog *Pseudacris ornata*. *Copeia* 1988(4):1001–1007.

Harris, A.H. 1984. *Neotoma* in the late Pleistocene of New Mexico and Chihuahua. *Spec. Pubs., Carnegie Mus. Nat. Hist.* 8:164–178.

Harvell, C.D., C.E. Mitchell, J.R. Ward et al. 2002. Climate warming and disease risks for terrestrial and marine biota. *Science* 296:2158–2162.

Harvey, M.J. 2002. Status and ecology of the Indiana bat (*Myotis sodalis*) in the southern United States. In: Kurta, A. and J. Kennedy (eds.) *The Indiana Bat: Biology and Management of an Endangered Species*. Pp. 29–34. Bat Conservation International, Austin, TX. 253pp.

Hawkes, L.A., A.C. Broderick, M.H. Godfrey et al. 2007. Investigating the potential impacts of climate change on a marine turtle population. *Glob. Change Biol.* 13:923–932.

Hibbard, C.W., C.E. Ray, D.E. Savage et al. 1965. Quaternary mammals of North America. In: Wright, H.E. Jr. and D. G. Frey (eds.) *The Quaternary of the United States*. Pp. 509–525. Princeton University Press, Princeton, NJ.

Hitch, A.T. and P.L. Leberg. 2007. Breeding distributions of North American bird species moving north as a result of climate change. *Conserv. Biol.* 21:534–539.

Hobbs, R.J. and H.A. Mooney. 1985. Community and population dynamics of serpentine grassland annuals in relation to gopher disturbance. *Oecologia* 67:342–351.

Hoegh-Guldberg, O., L. Hughes, S. Mcintyre et al. 2008. Assisted colonization and rapid climate change. *Science* 321:345–346.

Holt, R.D. 1990. The microevolutionary consequences of climate change. *Trends in Ecol. Evol.* 5:311–315.

Hoying, K.M. and T.H. Kunz. 1998. Variation in size at birth and post-natal growth in the insectivorous bat *Pipistrellus subflavus* (Chiroptera: Vespertilionidae). *J. Zool.* 245:15–27.

Huey, R.B., J.B. Losos, and C. Moritz. 2010. Are lizards toast? *Science* 328:832–833.

Hughes, R.S. 2006. Home ranges of the endangered Carolina northern flying squirrel in the Unicoi Mountains of North Carolina. *Proc. Ann. Conf. Southeast. Assoc. Fish Wildl. Agencies* 60:19–24.

Humphries, M.M., D.W. Thomas, and J.R. Speakman. 2002. Climate-mediated energetic constraints on the distribution of hibernating mammals. *Nature* 418:313–316.

Huntly, N. and R. Inouye. 1988. Pocket gophers in ecosystems: Patterns and mechanisms. *Biosci.* 38:786–793.

Inouye, D.W., B. Barr, K.B. Armitage et al. 2000. Climate change is affecting altitudinal migrants and hibernating species. *Proc. Nat. Acad. Sci.* 97:1630–1633.

Intergovernmental Panel on Climate Change. 2007. Climate Change 2007: Synthesis Report. 52 pp. http://www.ipcc.ch/publications_and_data/publications_ipcc_fourth_assessment_report_synthesis_report.htm. Accessed August 2011.

Janzen, F.J. 1994. Climate change and temperature dependent sex determination in reptiles. *Proc. Natl. Acad. Sci.* 91:7487–7490.

Jarvinen, A. 1982. Effects of exceptionally favorable weather on the breeding of the pied flycatcher *Ficedula hypoleuca* in Finnish Lapland. *Ibis* 124:196–198.

Johnson, C.N. 1996. Interactions between mammals and ectomycorrhizal fungi. *Trends Ecol. Evol.* 11:503–507.

Jones, G., D.S. Jacobs, T.H. Kunz et al. 2009. Carpe noctem: The importance of bats as bioindicators. *Endanger. Species Res.* 8:93–115.

Kalcounis-Ruepell, M.C., M.J. Vonhof, and L.J. Rissler. 2010. Modeling current and future potential for peripheral populations of southeastern bats to mitigate effects of white-nose syndrome in core populations. *Bat Res. News* 51:167.

Kalka, M.B., A.R. Smith, and E.K.V. Kalko. 2008. Bats limit arthropods and herbivory in a tropical forest. *Science* 320:71.

Kim, D.H., D.A. Buehler, E.T. Linder, and K.E. Franzreb. Unpublished report. Effects of climate on avian distributions across an elevational gradient in the southern Appalachian Mountains. Contact: Daniel H. Kim, 4630 NW Woodside Terrace, Portland, OR 97210; texraptor@hotmail.com.

Klempa, B. 2009. Hantaviruses and climate change. *Clin. Microbiol. Infect.* 15:518–523.

Kremen, C. and T. Ricketts. 2000. Global perspectives on pollination disruptions. *Conserv. Biol.* 14:1226–1228.

Kuenzi, A.J., M.L. Morrison, N.K. Madhav et al. 2007. Brush mouse (*Peromyscus boylii*) population dynamics and hantavirus infection during a warm, drought period in southern Arizona. *J. Wildl. Dis.* 43:675–683.

Kurta, A., T.H. Kunz, and K.A. Nagy. 1990. Energetics and water flux of free-ranging big brown bats (*Eptesicus fuscus*) during pregnancy and lactation. *J. Mammal.* 71:59–65.

Kurta, A., G.P. Bell, K.A. Nagy et al. 1989. Water balance of free-ranging little brown bats (*Myotis lucifugus*) during pregnancy and lactation. *Can. J. Zool.* 67:2468–2472.

Lacki, M.J. and L.E. Dodd. 2011. Diet and foraging behavior of *Corynorhinus* in eastern North America. Pp. 39–52 In: Loeb, S.C., M.J. Lacki, and D.A. Miller (eds.). *Gen. Tech. Rep. SRS-145*. Asheville, NC: U.S. Department of Agriculture, Forest Service, Southern Research Station.

Lafon, C.W., J.D. Waldron, D.M. Cairns et al. 2007. Modeling the effects of fire on the long-term dynamics and restoration of the yellow pine and oak forests in the southern Appalachian Mountains. *Restoration Ecol.* 15:400–411.

Lee, D.S. and C. Marsh. 1978. Range expansion of the Brazilian free-tailed bat in North Carolina. *Am. Midl. Nat.* 100:240–241.

Lee, Y.F. and G.F. McCracken. 2005. Dietary variation of Brazilian free-tailed bats links to migratory populations of pest insects. *J. Mammal.* 86:67–76.

Leonard, N.E. 2008. The Effects of the Invasive Exotic Chinese Tallow Tree (*Triadica sebifera*) on Amphibians and Aquatic Invertebrates. PhD dissertation, University of Georgia, Athens, GA.

Licht, P. and P. Leitner. 1967. Physiological responses to high environmental temperatures in three species of microchiropteran bats. *Compar. Biochem. Physiol.* 22:371–387.

Loeb, S.C., F.H. Tainter, and E. Cázares. 2000. Habitat associations of hypogeous fungi in the southern Appalachians: Implications for the endangered northern flying squirrel (*Glaucomys sabrinus coloratus*). *Am. Midl. Nat.* 144:286–296.

Loehle, C. 2012. Relative frequency function models for species distribution modeling. *Ecography* 35:487–498.

Lundy, M., I. Montgomery, and J. Russ. 2010. Climate change-linked range expansion of Nathusius' pipistrelle bat, *Pipistrellus nathusii* (Keyserling & Blasius, 1839). *J. Biogeogr.* 37:2232–2242.

Martin, T.E. 1987. Food as a limit on breeding birds: A life history perspective. *Ann. Rev. Ecol. System.* 18:453–487.

Matthews, S.N., L.R. Iverson, A.M. Prasad et al. 2007. A climate change atlas for 147 bird species of the eastern United States [database]. U.S. For. Serv. Northern Research Station, Delaware, Ohio. http://www.nrs.fs.fed.us/atlas/bird. Accessed March 2012.

McCallum, M.L. 2010. Future climate change spells catastrophe for Blanchard's cricket frog, *Acris blanchardi* (Amphibia: Anura: Hylidae). *Acta Herpetol.* 5:119–130.

McLachlan, J.S., J.J. Hellman, and M.W. Schwartz. 2007. A framework for debate of assisted migration in an era of climate change. *Conserv. Biol.* 21:297–302.

McLaughlin, J.F., J.J. Hellmann, C.L. Boggs et al. 2002. Climate change hastens population extinctions. *Proc. Natl. Acad. Sci.* 99:6070–6074.

McNulty, S.G., J.A. Moore Myers, P. Caldwell et al. In press. Climate change. In: Wear, D. and J. Greis (eds.). *Southern Forest Futures*. Pp. xx-xx. Asheville, NC: U.S. Department of Agriculture, Forest Service, Southern Research Station.

Menzel, J.M., W.M. Ford, J.W. Edwards et al. 2006. Home range and habitat use of the vulnerable Virginia northern flying squirrel (*Glaucomys sabrinus fuscus*) in the central Appalachian Mountains. *Oryx* 40:204–210.

Menzel, M.A., J.M. Menzel, T.C. Carter et al. 2001. Review of the forest habitat relationships of the Indiana bat (*Myotis sodalis*). U.S. For. Serv. Gen. Tech. Rep. NE-284.

Michener, W.K., E.R. Blood, K.L. Bildstein et al. 1997. Climate change, hurricanes and tropical storms, and rising sea level in coastal wetlands. *Ecol. Appl.* 7:770–801.

Milanovich, J.R., W.E. Peterman, N.P. Nibbelink et al. 2010. Projected loss of a salamander diversity hotspot as a consequence of projected global climate change. *PLoS ONE* 5:1–10.

Mills, J.N., T.G. Ksiazek, C.J. Peters et al. 1999. Long-term studies of hantavirus reservoir populations in the southwestern United States: A synthesis. *Emerg. Infect. Dis.* 5:135–142.

Moore, J.A. 1939. Temperature tolerance and rates of development in the eggs of amphibian. *Ecology* 20:459–478.

Moorman, C.E., K.R. Russell, and C.H. Greenberg. 2011. Reptile and amphibian response to hardwood forest management and early successional habitats. In: Greenberg, C.H., B.S. Collins, and F.R. Thompson III (eds.) *Sustaining Young Forest Communities: Ecology and Management of Early Successional Habitats in the Central Hardwood Region*. pp. 191 – 208. Springer, New York. 310p.

Moosman, P.R., Jr., H.H. Thomas, and J.P. Veilleux. 2007. Food habits of eastern small-footed bats (*Myotis leibii*) in New Hampshire. *Am. Midl. Nat.* 158:354–360.

Moritz, C., J.L. Patton, C.J. Conroy et al. 2008. Impact of a century of climate change on small-mammal communities in Yosemite National Park, USA. *Science* 322:261–264.

Murray, S.W. and A. Kurta. 2002. Spatial and temporal variation in diet. In: Kurta, A. and J. Kennedy (eds.) *The Indiana Bat: Biology and Management of an Endangered Species*. pp 182–192. Bat Conservation International, Austin, TX.

Nagy, L.R. and R.T. Holmes. 2005. To double-brood or not? Individual variation in the reproductive effort of black-throated blue warblers (*Dendroica caerulescens*). *Auk* 122:902–914.

NatureServe. 2012. The NatureServe Climate Change Vulnerability Index. http://www.natureserve.org/prodServices/climatechange/ccvi.jsp. Accessed March 2012.

Neuweiler, G. 2000. *The Biology of Bats*. Oxford University Press, New York, NY. 310 p.

Newson, S.E., S. Mendes, H.Q.P. Crick et al. 2009. Indicators of the impact of climate change on migratory species. *Endanger. Species Res.* 7:101–113.

North American Bird Conservation Initiative, U.S. Committee. 2010. The State of the Birds 2010. Report on Climate Change, United States of America. U.S. Department of Interior, Washington, DC, 32 pp.

North American Bird Conservation Initiative, U.S. Committee. 2011. The State of the Birds 2011. Report on Public Lands and Waters, U.S. Department of Interior, Washington, DC, 48 pp.

Oseen, K.L. and R.J. Wassersug. 2002. Environmental factors influencing calling in sympatric anurans. *Oecologia* 133:616–625..

Pank, L.F., 1974. A bibliography of seed-eating mammals and birds that affect forest regeneration. U.S. Department of Interior, Fish and Wildlife Service, Special Science Report. 174.

Parmesan, C. 2007. Influences of species, latitudes and methodologies on estimates of phenological response to global warming. *Glob.Change Biol.* 13:1860–1872.

Parmesan, C. and G. Yohe. 2003. A globally coherent fingerprint of climate change impacts across natural systems. *Nature* 42:37–42.

Parmesan, C., N. Ryrholm, C. Stefanescu et al. 1999. Poleward shifts in geographical ranges of butterfly species associated with regional warming. *Nature* 399:579–583.

Payne, J.L., D.R. Young, and J.F. Pagels. 1989. Plant community characteristics associated with the endangered northern flying squirrel, *Glaucomys sabrinus*, in the southern Appalachians. *Am. Midl. Nat.* 121:285–292.

Pechmann, J.H.K., D.E. Scott, J.W. Gibbons et al. 1989. Influence of wetland and hydroperiod on diversity and abundance of metamorphosing juvenile amphibians. *Wetlands Ecol. Manage.* 1:3–11.

Peppers, K.C. 2004. *Old-growth Forests in the Western Cross Timbers of Texas*. PhD dissertation, University of Arkansas, Fayetteville , AR.

Perry, R.W., S.A. Carter, and R.E. Thill. 2010. Temporal patterns in capture rate and sex ratio of forest bats in Arkansas. *Am. Midl. Nat.* 164:270–282.

Peterman W.E. and R.D. Semlitsch. 2009. Efficacy of riparian buffers in mitigating local population declines and the effects of even-aged timber harvest on larval salamanders. *For. Ecol. Manage.* 257:8–14.

Petranka, J.W. 1998. *Salamanders of the United States and Canada*. Smithsonian Institution Press, Washington, DC.

Phillips, S.J., R.P. Anderson, and R.E. Schapire. 2006. Maximum entropy modeling of species geographic distributions. *Ecol. Model.* 190:231–259.

Piechotowski, I., S.O. Brockmann, C. Schwarz et al. 2008. Emergence of hantavirus in South Germany: Rodents, climate and human infections. *Parasitol. Res.* 103:S131–S137.

Pough, R.H., R.M. Andrews, J.E. Cadle et al. 2001. *Herpetology*. Upper Sadddle River, NJ: Prentice-Hall, Inc. 736 pp.

Price, J.T. and T.L. Root. 2001. Climate change and neotropical migrants. In: Rahm, J. and R. McCabe. (eds.) *Changing Climates of North America: Political, Social, and Ecological*. Trans. Sixty-sixth N. Am. Wildl. Nat. Res. Conf., March 16–20, 2001. Pp. 371–379. Wildlife Management Institute, Washington, DC.

Reading, C.J. 1998. The effect of winter temperatures on the timing of breeding activity in the common toad *Bufo bufo*. *Oecologia* 117:469–475.

Reading, C.J. 2007. Linking global warming to amphibian declines through its effects on female body condition and survivorship. *Oecologia* 151:125–131.

Rebelo, H., P. Tarroso, and G. Jones. 2010. Predicted impact of climate change on European bats in relation to their biogeographic patterns. *Glob. Change Biol.* 16:561–576.

Ricciardi, A. and D. Simberloff. 2009. Assisted colonization is not a viable conservation strategy. *Trends Ecol. Evol.* 24:248–253.

Rice, E.L. and W.T. Penfound. 1959. The upland forests of Oklahoma. *Ecology* 40:593–608.

Robbins, C.S., J.R. Sauer, R.S. Greenberg et al. 1989. Population declines in North American birds that migrate to the neotropics. *Proc. Natl. Acad. Sci.* 86:7658–7662.

Rodenhouse, N.L. 1992. Potential effects of climate change on a Neotropical migrant landbird. *Conserv. Biol.* 6:263–272.

Rodenhouse, N.L. and R.T. Holmes. 1992. Results of experimental and natural food reductions for breeding black-throated blue warblers. *Ecology* 73:357–372.

Rodenhouse, N.L., L.M. Christenson, D. Parry et al. 2009. Climate change effects on native fauna of northeastern forests. *Can. J. For. Res.* 39:249–263.

Rodenhouse, N.L., S.N. Matthews, K.P. McFarland et al. 2008. Potential effects of climate change on birds in the Northeast. *Mitig. Adapt. Strat. Glob. Change* 13:517–540.

Root, T.L. 1988a. Energy constraints on avian distributions and abundances. *Ecology* 69:330–339.

Root, T.L. 1988b. Environmental factors associated with avian distributional boundaries. *J. Biogeogr.* 15:489–505.

Root, T.L. and S.H. Schneider. 2002. Climate change: Overview and implications for wildlife. In: *Wildlife Responses to Climate Change: North American Case Studies*. Pp. 1–54. Island Press, Washington, DC. 437 pp.

Root, T.L., D.P. MacMynowski, M.P. Mastrandrea et al. 2005. Human-modified temperatures induce species changes: Joint attribution. *Proc. Nat. Acad. Sci.* 102:7465–7469.

Root, T.L., J.T. Price, K.R. Hall et al. 2003. Fingerprints of global warming on wild animals and plants. *Nature* 421:57–60.

Roy, D.B. and T.H. Sparks. 2000. Phenology of British butterflies and climate change. *Glob. Change Biol.* 6:407–416.

Rudolph, C.D., S.J. Burgdorf, J. Conner et al. 2002. Prey handling and diet of Louisiana pine snake (*Pituophis ruthveni*) and black pine snake (*Pituophis melanoleucus lodingi*) with comparisons to other selected snakes. *Herpetol. Nat. Hist.* 9:57–62.

Rudolph, D.C. and C.A. Ely. 2000. The influence of fire on lepidopteran abundance and community structure in forested habitats of eastern Texas. *Texas J. Sci.* 53:127–138.

Rudolph, D.C., C.A. Ely, R.R. Schaefer et al. 2006. The Diana fritillary (*Speyeria diana*) and great-spangled fritillary (*S. cybele*): Dependence on fire in the Ouachita Mountains of Arkansas. *J. Lepidopt. Soc.* 60:218–226.

Saenz, D., L.A. Fitzgerald, K.A. Baum et al. 2006. Abiotic correlates of anuran calling phenology: The importance of rain, temperature, and season. *Herpetol. Monogr.* 20:64–82.

Sanz, J.J. 2002. Climate change and birds: Have their ecological consequences already been detected in the Mediterranean region? *Ardeola* 49:109–120.

SAS Institute Inc. 2000. *SAS/STAT User's Guide*, version 8 edition. SAS Institute, Cary, North Carolina.

Schierenbeck, K.A. 2004. Japanese honeysuckle (*Lonicera japonica*) as an invasive species; history, ecology, and context. *Cr. Rev. Plant Sci.* 23:391–400.

Schowalter, T., D. Crossley, and W. Hargrove. 1986. Herbivory in forest ecosystems. *Ann. Rev. Entomol.* 31:177–196.

Schriever, T.A., J. Ramspott, B.I. Crother et al. 2009. Effects of hurricanes Ivan, Katrina, and Rita on a southeastern Louisiana herpetofauna. *Wetlands* 29:112–122.

Schwartz, M.D., R. Ahas, and A. Aasa. 2006. Onset of spring starting earlier across the Northern Hemisphere. *Glob. Change Biol.* 12:343–351.

Scott, J.A. 1986. *The Butterflies of North America.* Stanford University Press, Stanford, CA, 583 pp.

Semlitsch, R.D. 2000. Principles for management of aquatic-breeding amphibians. *J. Wildl. Manage.* 64:615–631.

Semlitsch, R.D. and J.R. Bodie. 1998. Are small, isolated wetlands expendable? *Conserv. Biol.* 12:1129–1133.

Sherry, T.W. and R.T. Holmes. 1992. Population fluctuations in a long-distance neotropical migrant: Demographic evidence for the importance of breeding season events in the American redstart. In: Hagan, J.M. and D.W. Johnston (eds.) *Ecology and Conservation of Neotropical Migrant Landbirds*. Pp. 431–442. Smithsonian Institution Press, Washington, DC.

Sillett, T.S., R.T. Holmes, and T. W. Sherry. 2000. Impacts of a global climate change on the population dynamics of a migratory songbird. *Science* 288:2040–2042.

Sinervo, B., F. Méndez-De-la-Cruz, D.B. Miles et al. 2010. Erosion of lizard diversity by climate change and altered thermal niches. *Science* 328:894–899.

Smith, C.F. and S.E. Aldous. 1947. The influence of mammals and birds in retarding artificial and natural regeneration of coniferous forests of the United States. *J. For.* 45, 361–369.

Smith, J.B. and D.A. Tirpak (eds.). 1989. The potential impact of global climate change on the United States. U.S. Environmental Protection Agency, Washington, DC.

Sparks, T.H. and H.Q.P. Crick. 1999. The times they are a-changing. *Bird Conserv. Int.* 9:1–7.

Stahle, D.W. and M.K. Cleaveland, 1988. Texas drought history reconstructed and analyzed from 1698–1980. *J. Climatol.* 1:59–74.

Stahle, D.W. and J.G. Hehr. 1984. Dendroclimatic relationships of post oak across a precipitation gradient in the southcentral United States. *Annals Assoc. Am. Geogr.* 74:561–573.

Steele, M.A., T. Knowles, K. Bridle et al. 1993. Tannins and partial consumption of acorns: Implications for dispersal of oaks by seed predators. *Am. Midl. Nat.* 130:229–238.

Sumner, E.L. 1935. A life history study of the California quail, with recommendations for its conservation and management. *Part 1. Cal. Fish Game* 21:167–256.

Taulman, J.F. and L.W. Robbins. 1996. Recent range expansion and distribution limits of the nine-banded armadillo (*Dasypus novemcinctus*) in the United States. *J. Biogeogr.* 635–648.

The Nature Conservancy. 2011. Initial estimates of the ecological and economic consequences of sea level rise on the Florida Keys through the year 2100. Arlington, VA. http://www.frrp.org/SLR%20documents/FINAL%20-%20Aug%2021%20-WITH%20COVER.pdf. Accessed July 2011.

Thill, R.E., D.C. Rudolph, and N.E. Koerth. 2004. Shortleaf pine-bluestem restoration for Red-cockaded Woodpeckers in the Ouachita Mountains: Implications for other taxa. In: Costa, R. and S.J. Daniels (eds.) *Red-Cockaded Woodpecker: Road to Recovery*. Pp. 657–671. Hancock House Publishers, Blaine, Washington. 743 pp.

Trani, M.K., W.M. Ford, and B.R. Chapman (eds.). 2007. *The Land Manager's Guide to Mammals of the South.* The Nature Conservancy, Southeastern Region, and U.S. For. Serv., Southern Region, Atlanta, GA.

Trappe, J.M. and C. Maser. 1977. Ectomycorrhizal fungi: Interactions of mushrooms and truffles with beasts and trees. In: Walters, T. (ed.) *Mushrooms and Man, an Interdisciplinary Approach to Mycology*. Pp. 165–179. Linn-Benton Community College, Albany, Oregon.

Trauth, J.B., S.E. Trauth, and R.L. Johnson. 2006. Best management practices and drought combine to silence the Illinois chorus frog in Arkansas. *Wildl. Soc. Bull.* 34:514–518.

Turner, G.G., D.M. Reeder, and J.T.H. Coleman. 2011. A five-year assessment of mortality and geographic spread of white-nose syndrome in North American bats and a look to the future. *Bat Res. News* 52:13–27.

Ursic, S.J. and R.J. Esher. 1988. Influence of small mammals on stormflow responses of pine-covered catchments. *Water Resour. Bull.* 24, 133–139.

U.S. Department of Agriculture, Forest Service. 2011. National Roadmap for Responding to Climate Change. USDA For. Serv. FS-957b.

U.S. Department of the Interior, Fish and Wildlife Service. 1987. Recovery plan for the Alabama beach mouse (*Peromyscus poliontus ammobates*), Perdido Key beach mouse (*Peromyscus poliontus trissyllepsis*), and Choctawhatchee beach mouse (*Peromyscus poliontus allophrys*). Southeast Region, Atlanta, Georgia.

U.S. Department of the Interior, Fish and Wildlife Service. 2007. *Indiana Bat (Myotis sodalis) Draft Recovery Plan: First revision*. U.S. Fish and Wildlife Service, Fort Snelling, Minnesota. 258 pp.

U.S. Department of the Interior, Fish and Wildlife Service. 2010. Recovery Plan for the St. Andrew Beach Mouse (*Peromyscus polionotus peninsularis*). Southeastern Region, Atlanta, Georgia.

U.S. Global Change Research Program. 2012. Regional highlights from global climate change impacts in the United States: The Southeast. http://www.globalchange.gov/publications/reports/scientific-assessments/us-impacts/regional-climate-change-impacts/southeast. Accessed March 2012.

Van Devender, T.R., R.S. Thompson, and J.L. Betancourt. 1987. Vegetation history in the Southwest: The nature and timing of the Late Wisconsin-Holocene transition. In: Ruddiman, W.F. and H.E. Wright, Jr. (eds.) *North America and Adjacent Oceans during the Last Deglaciation*. Pp. 323–352. Geologic Society of America, Boulder, CO.

Vander Wall, S.B. 1990. *Food Hoarding in Animals*. University of Chicago Press, Chicago, IL. 445 p.

Vose, J.M., D.L. Peterson, and T. Patel-Weynand. In press. National climate assessment: Forest sector technical report. Gen. Tech. Rep. PNW-xxx, Portland, Oregon: U.S. Department of Agriculture, Forest Service, Pacific Northwest Research Station. 300 pp.

Walters, B.L., D.W. Sparks, J.O. Whitaker, Jr. et al. 2006. Timing of migration by eastern red bats (*Lasiurus borealis*) through Central Indiana. *Acta Chirop*. 8:259–263.

Walther, G., E. Post, P. Convey et al. 2002. Ecological responses to recent climate change. *Nature* 416:389–395.

Watrous, K.S., T.M. Donovan, R.M. Mickey et al. 2006. Predicting minimum habitat characteristics for the Indiana bat in the Champlain Valley. *J. Wildl. Manage.* 70:1228–1237.

Watson, J.C., R.G. Hooper, D.L. Carlson et al. 1995. Restoration of the red-cockaded woodpecker population on the Francis Marion National Forest: Three years post Hugo. Pp. 172–182. *In*: Kulhavy, D.L., R.G. Hooper, and R. Costa (eds.). Red-cockaded woodpecker: Recovery, ecology, and management. Center for Applied Studies in Forestry, Stephen F. Austin State University, Nacogdoches, TX. 551 p.

Wear, D. and J. Greis (eds.). In press. *Southern Forest Futures*. Asheville, NC: U.S. Department of Agriculture, Forest Service, Southern Research Station.

Webb, S.D. 1974. *Pleistocene Mammals of Florida*. University Press of Florida, Gainesville, FL. 270 p.

Weigl, P.D. 2007. The northern flying squirrel (*Glaucomys sabrinus*): A conservation challenge. *J. Mammal.* 88:897–907.

Weigl, P.D., T.W. Knowles, and A.C. Boynton. 1999. The distribution and ecology of the northern flying squirrel, *Glaucomys sabrinus coloratus*, in the southern Appalachians. North Carolina Wildlife Resources Commission, Raleigh.

Whitaker, J.O., Jr. 2004. Prey selection in temperate zone insectivorous bat community. *J. Mammal*. 85:460–469.

Whittle, A., D.S. Maehr, F. Songlin, and J. Cox. 2008. Global climate change and its effects on large carnivore habitat in Florida. *Proceedings of the Conference on Florida's Wildlife: On the Frontline of Climate Change*, October 1–3, 2008, Orlando, FL (Abstract).

Wiens, J.A., D. Stralberg, D. Jongsomjit, C.A. Howell, and M.A. Snyder. 2009. Niches, models, and climate change: Assessing the assumptions and uncertainties. *Proc. Natl. Acad. Sci.* Sept 17; 106:19729–19736.

Wildlife Management Institute. 2007. Adapting to climate change: Agency science needs to adapt game management to changing global climate. A report to the National Commission on Energy Policy and the Hewlett Foundation. Wildlife Management Institute, Washington, DC.

Wildlife Management Institute, eds. 2008. Season's end: Global warming's threat to hunting and fishing. Bipartisan Policy Center. http://www.seasonsend.org/upland_birds.shtml. Accessed March 2012.

Wiley, R.W. 1980. Neotoma Floridana. *Mammal. Species* 139:1–7.

Wilhide, J.D., B. Baker, and D.A. Saugey. 1998. Arkansas range extension of the Seminole bat (*Lasiurus seminolus*). *J. Ark. Acad. Sci.* 52:140–141.

Williams-Guillén, K., I. Perfecto, and J. Vandermeer. 2008. Bats limit insects in a neotropical agroforestry system. *Science* 320:70.

Winhold, L. and A. Kurta. 2006. Aspects of migration by the endangered Indiana bat, *Myotis sodalis*. *Bat Res. News* 47:1–6.

Withgott, J. 1999. Pollination migrates to top of conservation agenda. *BioSci*. 49(11):857–862.

12 Climate Change and Outdoor Recreation Participation in the Southern United States

J.M. Bowker, Ashley E. Askew, Neelam Poudyal, Stanley J. Zarnoch, Lynne Seymour, and H. Ken Cordell

CONTENTS

In this chapter we begin to assess the potential effects of climate change on future outdoor recreation in the South, a region spanning 13 states from Virginia to Texas (Chapter 1). Our goal is to provide some useful insights about future natural resource-based recreation—an important nontimber product derived from southern forests—in the face of climate change. We develop and present projections of participation and consumption for 10 traditional natural resource-based recreation activities in the South. The work builds on previous outdoor recreation forecasts (Bowker et al. in press) by explicitly incorporating climate, along with population growth, land-use changes, and future socioeconomic conditions into demand models and projections.

The Intergovernmental Panel on Climate Change (2007) predicted that increases in the concentration of greenhouse gases in the atmosphere will further accelerate global warming, thereby resulting in significant climate change across the planet. The potential impacts of climate change have been projected within the context of various natural and biological systems. However, these impacts on human behavior and cultural interaction with nature remain largely unknown. Outdoor recreation opportunities rely heavily on natural settings, thus negative impacts on the quality and availability of forests and water bodies could negatively impact their long-term potential to provide recreational opportunities to humans (Morris and Walls 2009).

Mendelsohn and Markowski (1999) suggested that climate could affect recreation in direct and indirect ways including the effect of severe weather on physical comfort or convenience, the effect that varying season lengths could have on availability and suitability of certain outdoor opportunities, and the degree and pace of alterations to the natural resource base on which outdoor activities depend. Although many expect the impacts of climate change on outdoor recreation to be negative, Gregory (2011) has argued that the reverse could be more likely for some forms of adventure recreation and that rising temperatures could open new opportunities worldwide. For example, enough ice could melt in polar areas to make rowing in the Polar North possible. Similarly, rapid glacial recession in the high Andes of Peru would help open new trekking routes.

A handful of research articles and stories published in popular magazines have speculated on the mixed impacts of climate change on outdoor recreation, which could also go well beyond factors such as opportunity and participation. For example, the many dollars spent by recreation users generate billions in economic impacts, often in small rural economies. So, understanding the future impact of climate change on this industry would help communities take timely action on mitigation and adaptation. A few key questions facing planners and managers today are: (1) how climate change would impact demand for outdoor recreation; (2) whether certain types of activities would be impacted more than others; (3) whether certain places (such as regions or states) would experience higher impacts than others; (4) how the anticipated decrease in outdoor activity would translate into lost public welfare, which would allow mitigation or management decisions to take those costs/benefits into account; and (5) the appropriate analytical framework needed for assessing impacts (both in terms of demand/supply and welfare).

Literature on the relationship between climate change and outdoor recreation has slowly emerged. Available studies can be broadly classified into two types: individual survey-based studies and aggregate modeling studies. The individual survey approach has generally focused on a particular type of recreation, a limited area, or both together. For example, Cato and Gibbs (1973) found that the chance of rain and the expected air temperature could significantly affect the decision to go boating. Ahn et al. (2000) conducted a survey in North Carolina to determine how fishing behavior would be affected by a decline in trout habitat under global warming and found significant welfare loss. Similarly, Richardson and Loomis (2004) surveyed summer tourists at Rocky Mountain National Park in an effort to relate hypothetical climate scenarios to stated recreation trip behavior; they predicted a significant increase in park visitation under all climate change scenarios. Lise and Tol (2002), assessing the impacts of climate on tourist demand, found evidence that under a scenario of global warming, tourists would clearly alter their holiday patterns in Europe and that the effect of climate on tourism demand varied by age and income groups.

Individual survey-based approaches solicit perceptions of climate change and stated recreation behavior under contingent climate scenarios. However, using such survey data to develop a predictive model has some limitations. First, although respondents may accurately remember the number of trips they made or number of days they spent in a particular activity over a year, they might not remember the weather conditions for those days. Second, even if the survey explains the climate scenarios, respondents are still responding hypothetically. And third, linking individual surveys with regional averages of climate data generally means a mismatch on measurements (individual trip data versus state or county level climate data). For these reasons, an indirect approach to demand modeling, which measures observed participation and climate data on a seasonal basis for specific area units, can be more meaningful.

A few studies have adopted aggregate visitation modeling to evaluate the impacts of climate change on outdoor recreation. For example, Wake et al. (2006) combined annual time-series data on annual winter skiing and snowmobiling days with weather data (snow cover days, snowfall, winter temperature) to estimate their correlations. Results suggested a negative relationship with temperature, meaning that warming would have a negative impact on winter recreation. Arbel and Ravid (1985) estimated a time-series model of park visitation and found that weather variables negatively affected visitation in the short run. Mendelsohn and Markowski (1999) used state level data for the conterminous 48 states to assess the impacts of average temperature and precipitation on participation in various outdoor activities. Results revealed a mixed effect, and predicted an overall welfare gain. Loomis and Crespi (1999) also examined state-level data such as total park visits and rounds of golf played in relation to climate variables; they found that many outdoor activities would be negatively affected by climate change, but activities like golf and freshwater recreation would benefit.

In an aggregate modeling study, Whitehead et al. (2009) used data from the National Survey of Recreation and Environment (NSRE) to develop a participation model that he expanded to include some climate variables measured at state level. Their findings showed a significant and negative impact of climate variables, such as monthly temperature and precipitation, for certain months (such as June temperature and January precipitation). More recently, Bowker et al. (2012) also used the NSRE dataset (U.S. Department of Agriculture Forest Service 2009) to model national outdoor recreation participation rates and annual participation days for 17 common activities, and then projected recreation days under various climate change scenarios. They generally found that projections of climate change (U.S. Department of Agriculture Forest Service 2012) had marginally negative effects but that downturns were dramatic for two winter activities, snowmobiling and undeveloped skiing, under some of the climate alternatives.

Aggregated visitation approaches also have limitations. First, the models are usually simple and parsimonious, meaning that important variables are often missing, leading to potential biases in estimation. Second, many models used one-dimensional data, either cross-sectional data for a brief period in time, or time-series data for a limited area; parameters estimated from cross-sectional models are not stable over time, and thus may have limited forecasting accuracy and those estimated with the time-series data are not always applicable to other regions. And third, many studies used standardized or direct measurement of climate data. A variety of indirect but equivalent measurements are available—such as level of thermal comfort, and humidity resistance—that could more effectively capture the climate conditions perceived by people.

PARTICIPATION AND USE

We define participation in an outdoor recreation activity as engaging in that activity at least once in the preceding 12 months. Participation is an indicator of the size of a market and can also be a gauge of public interest. For example, if over 80% of the population engages in day hiking and only 4% engage in snowmobiling, public resource management agencies and private land managers would benefit from knowing that demand for hiking trails could be outpacing that for snowmobiling opportunities. This demonstrates the importance of knowing how many people participate in a given activity, and how this measure could change over time. Participation statistics, either per capita or in absolute numbers of participants, provide the broadest measure of a recreation market.

A second measure of recreation use or quantity demanded is consumption (also known as participation intensity), which can be measured as number of times, days, visits, or trips within a year or other time span; for example, the U.S. Forest Service has used recreation visitor days and national forest visits per year. Consumption measures provide an important additional dimension for resource managers, whose decisions depend on knowing how often and for how long people engage in an activity. This information can be critical to allocating campsites and other existing resources, and is also useful in planning the development of new venues. At the regional level, participation and consumption together provide the broadest measures of an outdoor recreation market. The consumption measure used in this chapter is the number of different days in the previous year that an American adult engaged in a specific activity. Our definition of a day follows the NSRE definition of an activity day: any amount of time spent on an activity on a given day, regardless of the number of hours or whether the activity was the primary reason for the outdoor visit.

The preceding two metrics are origin based—that is, they result from household-level surveying—but they do not specify the location of any activity. Research has shown, however, that the vast majority of outdoor recreation takes place within a few hours' drive of the visitor's residence (Hall and Page 1999). Number of participants and participation rates for 2008, along with total days spent participating, for 10 outdoor recreation activities are reported in Table 12.1. Short- and long-term trends can be important indicators of what could happen with outdoor recreation in the near future (Cordell 2012; Hall et al. 2009). However, simple descriptive statistics or trends do not formally address the underlying factors and associations that could be driving the trends. Thus, a trend could be of limited value if the time horizon is long or if its driving factors are expected to deviate substantially from their historical levels. Trend analysis can be supplemented by projection models that

TABLE 12.1
Outdoor Recreation Activity by Adults in the Southern United States, 2008

Activity	Participation Rate (%)	Participants (millions)	Days
Land based			
Developed site use (family gathering, picnicking, camping)	80	63.2	672
Horseback riding on trails	7	5.7	99
Day hiking	25	20.3	463
Motorized off-road driving	21	16.9	562
Primitive (visiting a wilderness, primitive camping/backpacking)	35	28.2	412
Water based			
Motorized water (motor boating, water skiing, personal water craft)	27	21.3	384
Nonmotorized (canoeing, kayaking, rafting)	15	12.2	80
Wildlife			
Birding (viewing or photographing)	34	27.0	2,862
Fishing	36	28.0	573
Hunting	14	10.8	230

Source: Adapted from National Survey on Recreation and the Environment, 2005 to 2009 ($n = 30{,}394$) (U.S. Department of Agriculture Forest Service 2009).

relate recreation participation directly to the factors that are known to influence behavior. Projection models can then be used with external forecasts of influential factors, including population growth, to simulate future participation. Such modeling allows changes over time to be assessed in light of previously unseen changes in factors that drive the behavior, such as demographics, economic conditions, climactic conditions, and land uses.

Previous research (Bowker et al. 1999, 2006; Cicchetti 1973; Hof and Kaiser 1983b, Leeworthy et al. 2005) has established that race, ethnicity, gender, age, income, and supply or proximity to settings all affect outdoor recreation participation as well as consumption. Similarly, these factors along with others, including distance and quality descriptors, have been used to explain visitation to specific sites (Bowker et al. 2007, 2010; Englin and Shonkwiler 1995). Reliable information about these factors is often available from external sources, like the U.S. Census or parallel research efforts aimed at modeling and simulating influential variables into the future. Such information can thus be available long before results from recreation surveys.

We used a two-step approach to develop projections for participation and consumption of 10 traditional outdoor recreation activities (Table 12.1). The model estimation step focused on developing statistical models of southern adult per capita participation and days-of-participation (conditional on being a participant) for each activity. The models describe the probability of participating in an activity. For those participating, the consumption model describes the number of days. This information provides an understanding of the factors that influence individual recreation choices or behavior and a process for examining individual behavior changes over time in response to changes in underlying factors such as demographics, climate, and resource availability.

The second or simulation step, combined the estimated models with external projections of explanatory variables to generate participation probabilities and days-of-participation for each activity at 10-year intervals to 2060. Per capita estimates for participation and days were combined with population projections to derive estimates of regional adult participants and days-of-participation for each activity. These estimates are then used to create indices by which 2008 estimates of participants and days-of-participation could be scaled.

STORYLINES AND GENERAL CIRCULATION MODELS

Indices of adult participants for each of the 10 activities and days of annual participation are presented across three storylines developed by the Forest Service for the 2010 Resources Planning Act (RPA) assessment. The three storylines, considered equally likely, are globally consistent and well documented by the Intergovernmental Panel on Climate Change (2007). They describe a range of future global and U.S. socioeconomic conditions that are likely to have different effects on future conditions and trends of U.S. forests and grasslands (U.S. Department of Agriculture Forest Service 2010). The global data were scaled to the U.S. national and regional levels and U.S. gross domestic product, and population projections were updated and the updated data were downscaled to county levels for the South (U.S. Department of Agriculture Forest Service 2010; Zarnoch et al. 2010).

As shown in Table 12.2 and Figures 12.1 and 12.2, storyline A1B corresponds to mid-range population growth and the highest household income growth levels. Under these conditions, the South

TABLE 12.2
Key Characteristics of Emissions Storylines Developed by the Intergovernmental Panel on Climate Change (2007)

	Storyline[a]		
Characteristics	**A1B**	**A2**	**B2**
General description	Global economic convergence	Regionalism less trade	Slow change, localized solutions
Global real gross domestic product growth (2010–2060)	High (6.2x)	Low (3.2x)	Low-medium (3.5x)
Global population growth (2010–2060)	Medium (1.3x)	High (1.7x)	Medium (1.4x)
U.S. real gross domestic product growth	High (3.3x)	Low-medium (2.6x)	Low (2.2x)
U.S. population growth	Medium (1.5x)	High (1.7x)	Low (1.3x)
Global expansion of primary biomass energy production	High	Medium	Low

[a] Numbers in parentheses (for example, 6.2x) are factors of change during the projection period.

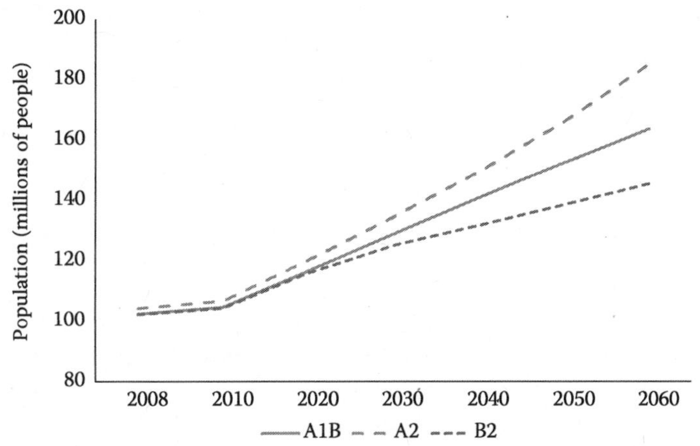

FIGURE 12.1 Projected population growth from 2008 to 2060 in the Southern United States based on an expectation of moderate population growth and high income growth (storyline A1B), high population growth and low income growth (storyline A2), or low population growth and moderate income growth (storyline B2); emissions storylines developed by the Intergovernmental Panel on Climate Change (2007).

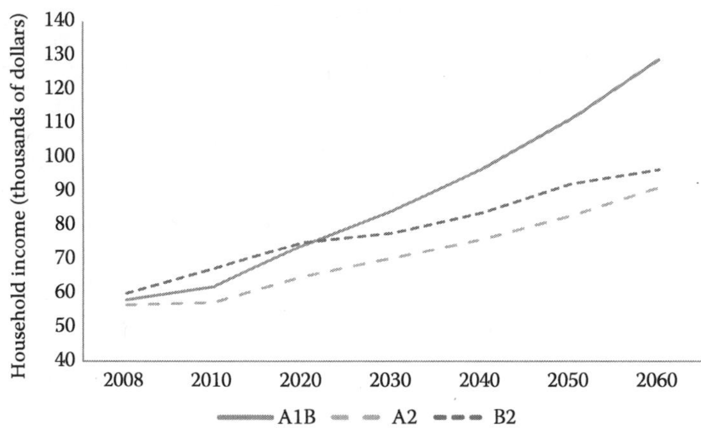

FIGURE 12.2 Projected average household real income (inflation adjusted) growth in the Southern United States based on an expectation of moderate population growth and high income growth (storyline A1B), high population growth and low income growth (storyline A2), or low population growth and moderate income growth (storyline B2); source: National Survey on Recreation and the Environment, emissions storylines developed by the Intergovernmental Panel on Climate Change (2007).

could expect to see about 164 million people (135 million adults) and an average household income of $129,000 by 2060. Storyline A2 projects the highest population growth, reaching about 185 million people (152 million adults) by 2060, and the lowest household income, about $91,000. Storyline B2 projects the lowest population growth and mid-level personal income, predicting a population of 145 million people (120 million adults) with average household income of about $96,000.

Projected land-use changes from Wear (2011) were used to develop the supply variables listed in Table 12.3. Nationally, urban area is expected to increase by 1–1.4 million acres per year from 1997 to 2060, with corresponding decreases of 24–37 million acres in forest area and 19–28 million acres in cropland. About 90% of forecasted losses would be in the Eastern United States with more than half of those losses occurring in the South. For the South, Wear (2011) forecasts forest acreage losses of 11 and 23 million acres (about 7–13%). Based on forecasts of land-use change from 2008 to 2060 by Cordell et al. (2013), southern forest and rangeland per capita is expected to decrease about 45% under A1B, 50% under A2, and about 37% under B2. Federal lands and areas covered by water are assumed static throughout the projection period. Further details about all explanatory variables and their values can be found at: www.forestthreats.org/research/projects/project-summaries/ccammo/Chapter12appendix.

Although not much large-scale work has been done relating climate to outdoor recreation, the general consensus is that long-term changes in climate could affect recreation demand. Walls et al. (2009) assert that the single most important new challenge to recreation supply will be mitigating the adverse effects of climate change, particularly in coastal areas and on western public lands. Disentangling the effects of the climate variables on recreation participation is difficult. Further exploration of these direct and indirect relationships, at both local and macro levels, will be fundamental to improving forecasts of recreation behavior in the future.

Each Intergovernmental Panel on Climate Change storyline had multiple associated climate projections based on levels of greenhouse gas emissions. For this chapter, we linked the three storylines with six general circulation models (Table 12.4) that differ in their approaches to modeling climate dynamics (MIROC3.2, CSIROMK2, CSIROMK3.5, HadCM3, CGCM2, and CGCM3.1), three of which were used to capture a range of future climates for the 2010 RPA assessment (U.S. Department of Agriculture Forest Service 2012).

The Intergovernmental Panel on Climate Change climate projections were first downscaled to the approximately 10 km scale, and then aggregated to the county scale. Detailed documentation

TABLE 12.3
Socioeconomic and Supply Variables for Modeling and Forecasting Outdoor Recreation Participation and Days-of-Participation by Adults in the Southern United States

Variable	Description
Gender	1 = male, 0 = otherwise
American Indian	1 = American Indian, non-Hispanic, 0 = otherwise
Asian/Pacific Islander	1 = Asian/Pacific Islander, 0 = otherwise
Hispanic	1 = Hispanic, 0 = otherwise
Black	1 = African-American, non-Hispanic, 0 = otherwise
Bachelor's	1 = Bachelor degree, 0 = otherwise
Below high school	1 = Less than high school, 0 = otherwise
Post graduate	1 = Post-graduate degree, 0 = otherwise
Some college	1 = Some college or technical school, 0 = otherwise
Age	Respondent age in years
Age squared	Respondent age squared
Income	Respondent household income (2007 dollars)
Population density	County area divided by population (base 1997)
Coastal	1 = County on coast, 0 otherwise
For_ran_pcap	Sum of forest land acres and rangeland acres divided by population at county level and at 50-, 100-, 200-mile radii (base 1997)
Water_pcap	Water acres divided by population at county level and at 50-, 100-, 200-mile radii (base 1997)
Mtns_pcap	Mountainous acres divided by population (base 1997)
Pct_mtns_pcap	Percentage of county acres in mountains divided by population multiplied by 100.000 (base 1997)
Natpark_pcap	Number of nature parks and similar institutions divided by population multiplied by 100,000 (base 1997)
Fed_land_pcap	Sum U.S. Forest Service, National Park Service, U.S. Fish and Wildlife Service, Bureau of Land Management, U.S. Bureau of Reclamation, Tennessee Valley Authority, and U.S. Army Corps of Engineers acreage divided by population (base 1997)
Avg_elev	Average elevation in meters at county level and 50-, 100-, 200-mile radii (base 1997)

TABLE 12.4
Intergovernmental Panel on Climate Change (2007) Emissions Storylines Paired with General Circulation Model Climate Projections

Storyline	General Circulation Model	Model Vintage[a]
	CGCM3.1 (T47)	
A1B	CSIROMK3.5	AR4
	MIROC3.2 (medres)	
	CGCM3.1 (T47)	
A2	CSIROMK3.5	AR4
	MIROC3.2 (medres)	
	CGCM2	
B2	CSIROMK2	TAR
	HadCM3	

[a] AR4 models were downloaded from the Program for Climate Model Diagnosis and Intercomparison Project 3 (www-pcmdi.llnl.gov/), and TAR 47 models were downloaded from the Intergovernmental Panel on Climate Change Data Distribution Centre (www.ipcc-data.org/).

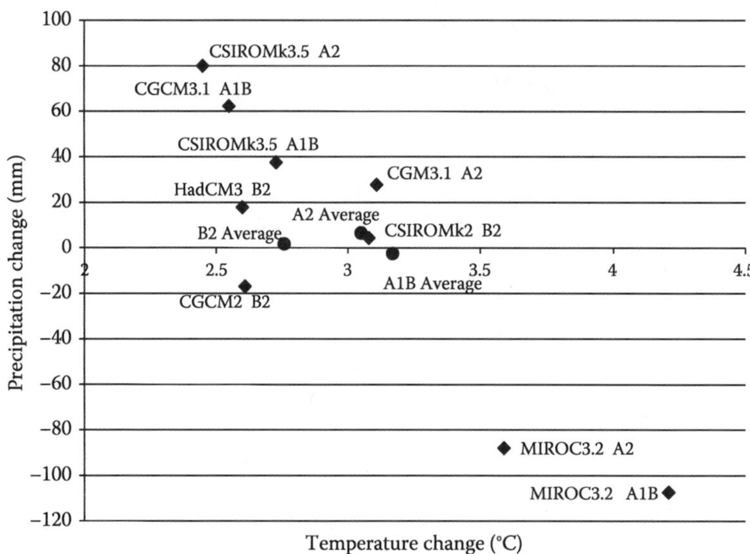

FIGURE 12.3 U.S. temperature and precipitation changes from the reference period (1961–90) to the decade surrounding the year 2060 (2055–64).

of the development of the RPA climate scenario-based projections and downscaling process can be found in U.S. Department of Agriculture Forest Service (2012) and Joyce et al. (in press). At the scale of the conterminous United States, the A1B storyline predicts the warmest and the driest climate of all storylines at 2060 (Figure 12.3), A2 the wettest, and B2 the coolest, although the precipitation changes at the scale of the United States are small to 2060. The individual climate model combinations highlight the variation within each storyline of the individual climate model projections. For example, within the A2 storyline, the CSIROMK3.5 model projects the least warming and the MIROC3.2 model projects the most warming. Although all areas of the United States show increases in temperature, the rate of change varies, and regional precipitation projections vary greatly (U.S. Department of Agriculture Forest Service 2012).

SUMMARY

The objectives of this chapter are to evaluate how population growth, changing demographics and economic conditions, changing land use, and changing climate are likely to affect participant numbers and days-of-participation for an array of 10 natural resource-based recreation activities in the South. The socioeconomic, climate, and land-use projections that are described above were used to develop projections of future resource uses and conditions. Not all of the projected variables are used in all models, but all of the projection models used some subset of these variables. Because the baseline models and orecasts (without climate change) are discussed in detail elsewhere (Bowker et al. in press), the main goal for this chapter is to identify the differences between the baseline recreation forecasts developed for the 2010 RPA assessment and those for which climate futures are explicitly incorporated.

This chapter proceeds as follows. First, we present the statistical methods and previous research on which per capita participation and consumption models were based. Next, we describe the data used in the estimation step—including projections of the various income and population growth factors and relevant assumptions—and present estimation and simulation steps for regional projections of participation and days by activity and climate scenario to 2060. Finally, we discuss some of the key findings within and across activity categories with respect to the factors driving change over the projection period, while focusing particular attention on the effects of climate change and its relation to management options and activities.

METHODS AND DATA

Recreation demand models fall into three categories: site-specific user models, site-specific aggregate models, and population-level models (Cicchetti 1973). Cicchetti (1973) pioneered the use of cross-sectional population-level models with the household-based 1965 National Survey of Recreation. Estimated models and Census Bureau projections of socio-demographic variables and population were then used to forecast recreation participation and use to 2000. Cicchetti's approach has been used to estimate and project participation and use for recreation activities at national and regional levels (Bowker 2001; Hof and Kaiser 1983a; Leeworthy et al. 2005; Walsh et al. 1992) and for previous RPA assessments (Bowker et al. 1999; Hof and Kaiser 1983b). Alternative approaches, wherein population data were combined with individual site-level data or county-level data to project participation or consumption, have also been used to project national or regional recreation demand (Bowker et al. 2006; Cordell and Bergstrom 1991; Cordell et al. 1990; Englin and Shonkwiler 1995; English et al. 1993; Poudyal et al. 2008). A drawback of cross-sectional models is that the estimated model parameters remain constant over the projection period (Bowker et al. in press). A further drawback of these models is that it is difficult to account for future congestion, supply limitations, and relative price changes on growth in participation and use. Moreover, projections of external variables like population and economic growth, used as inputs for simulations across time, may not include the same assumptions as the estimated statistical models.

Logistic models used to describe the probability of adult participation in each of the 10 activities were specified as

$$P_i = \frac{1}{[1 + \exp(-X_i'B_i)]} \tag{12.1}$$

where P_i is the probability that an individual participated in recreation activity i in the preceding year. The vector X_i contains sociodemographic characteristics, supply, and climate variables for activity i, and at least one climate variable related to conditions at or near the individual's residence; B_i represents a vector of parameters that were estimated using NLOGIT 4.0. Models for each activity, based on NSRE data (U.S. Department of Agriculture Forest Service 2009) for the 13 southern states from 1999 to 2008, were combined with 2008 baseline population-weighted sample means for the explanatory variables to create an initial predicted per capita participation rate for each activity. The per capita participation rates were recalculated at 10-year intervals using projected changes in the explanatory variables. Indices were then created for the participation rates by which the NSRE 2005-2009 average population-weighted participation frequencies (baselines) were scaled, leading to indexed per capita participation rates for each of the 10 activities. Indexing the 2005–2009 averages by changes in model-predicted rates was judged to be superior in terms of mitigating potential nonlinearity biases associated with complete reliance on logistic predicted values (Souter and Bowker 1996). The indexed participation rate estimates were then combined with projected changes in population, according to each of the three storylines to yield indexed values for total adult participants across the 10 activities.

Consumption models were similar to the participation models except that an integer metric represented use, for example, the number of times, days, visits, trips, or events is modeled rather than decision to participate. The general specification for the consumption model was

$$Y_i = f(X_i) + u_i \tag{12.2}$$

where Y_i represents the annual number of days that an individual participated in activity i, X_i is a vector of sociodemographic characteristics, supply, and climate variables associated with activity i, and u_i is a random error term. These count data models are often estimated using negative binomial specifications with a semilogarithmic link function (Bowker 2001; Bowker et al. 1999; Zawacki et al. 2000). Variations of these consumption models have been used in onsite applications, where all observed visits are greater than or equal to one, as data are only obtained from actual visitors (Bowker and Leeworthy 1998). Such zero-truncated models have been applied extensively in onsite recreation-demand estimation and valuation research (Ovaskainen et al. 2001). In some situations the estimated models have been extrapolated to general populations (Englin and Shonkwiler 1995). This approach, wherein population data are combined with individual site-level data, was suggested by Cordell and Bergstrom (1991) and used in a previous RPA assessment by Cordell et al. (1990) with linear models to estimate outdoor recreation trips nationally for 31 activities and to project the number of trips by activity from 1989 to 2040. English et al. (1993) extended the Cordell et al. (1990) models and projections to the regional level by combining parameter estimates from national models with regional explanatory variable values. However, others have questioned the efficacy of extrapolating parameter estimates from the onsite demand models to the population at large (Hagerty and Moeltner 2005).

Because household data, like those obtained using the NSRE, may report zero visits, problems related to onsite samples and extrapolating onsite models to general populations are not serious impediments. In a previous RPA assessment analysis, Bowker et al. (1999) used data from the 1994 to 1995 NSRE, the U.S. Census, and the 1997 NORSIS database to project participation and consumption (annual days and trips) for more than 20 natural resource-based outdoor activities, both nationally and for the four geographical regions of the United States, from 2000 to 2050. The scope of his work was broader than participation modeling, including the use of negative binomial count models to estimate consumption (days and trips annually) and the projection of these measures over the same time period. Bowker (2001) followed the same approach using NSRE and state-level SCORP data to project participation and consumption for Alaskans from 2000 to 2020. Moreover, Leeworthy et al. (2005) used NSRE 2000 data to project participation and consumption of marine-related outdoor recreation from 2000 through 2010. Finally, Bowker et al. (2006) applied similar methods with NSRE 2000 and NVUM (National Visitor Use Monitoring) data (English et al. 2002) to project wilderness and primitive-area recreation participation and consumption from 2002 through 2050.

Alternatively, if observed zeros for the dependent variable (days-of-participation) seem excessive or not entirely caused by the same data-generating process as the positive values, a hurdle model structure or a zero-inflated count procedure is recommended (Cameron and Trivedi 1998). The hurdle model, employed in this analysis, combines the probability of participation (threshold) with the estimated number of days for those participating:

$$E\left[Y_i|X_i\right] = Pr\left[Y_i > 0|X1_i\right] * EY_i > 0[Y_i|Y_i > 0, X2_i] \tag{12.3}$$

where Y_i represents days of participation in activity i, and X_i, $X1_i$, and $X2_i$ represent vectors of sociodemographic characteristics, supply, and climate variables associated with activity i. The hurdle model allows different vectors of explanatory variables for the respective products of the expectation in Equation 12.3, with the probability estimated as a logistic and conditional-days portions estimated as a truncated negative binomial, thus leading to two unique sets of estimated parameters. Parameter estimates for each of the 10 regional recreation activity-day hurdle models were estimated with NLOGIT 4.0 (Greene 2009) using NSRE data from southern households from 1999 to 2008 (U.S. Department of Agriculture Forest Service 2009; Cordell 2012), county level climate data (Joyce et al. in press), county land-use data (Wear 2011), and recreation supply data (Cordell et al. 2013).

Similar to the procedure with the participation models and indices, hurdle model parameter estimates are combined with 2008 NSRE baseline participation and days estimates, projected explanatory variables, and projected population changes under each of the storylines to provide indices of projected growth of annual days-of-participation for the activities listed in Table 12.1. Three climate alternatives (Table 12.4) are used for each of the storylines.

Socioeconomic and supply variables for the various models and projections are listed in Table 12.3. The preponderance of these variables was included in the NSRE database (U.S. Department of Agriculture Forest Service 2009; Cordell 2012). Addition variables related to supply were obtained from Cordell et al. (2013). Projections of land-use change variables are from Wear (2011).

Historical as well as projected climate data are from Joyce et al. (in press). As little or no literature was available on linking climate to household participation and consumption of recreation activities, an ad hoc approach was followed during the model estimation stage wherein climate variables were created based on 6-year moving averages and arbitrary distances from county centroids. Climate variables are listed in Table 12.5. Each estimated model was limited to one climate variable to avoid multicollinearity.

Results were estimated for 10 logistic participation models, without and then with climate variables (www.forestthreats.org/research/projects/project-summaries/ccammo/Chapter12-appendix). Reported results for the logistic participation models include parameter estimates for each activity, values for explanatory variables by scenario and year, odds ratios that indicate the odds of participation occurring in one group compared to the odds of occurrence in another group, fit statistics, and graphics of overall participant growth by activity and assessment scenario. Climate variables used in the participation models are reported in Table 12.6.

Parameter estimates were then combined with available projections of explanatory variables to create indexed per capita participation estimates at 10-year intervals through 2060. These indices were in turn combined with population projections for each of the storylines to develop estimated participant indices. The participant indices were then applied to a beginning baseline estimate of participants for each activity based on weighted national averages calculated from 2005 to 2009 NSRE data to yield projection of adult participants. The four-year average around 2008 was chosen to avoid any aberration associated with a single year.

The hurdle model combines probability of participation in an activity with the expected value of days participating, given one actually participated (Equation 12.3). The estimated logistic models (www.forestthreats.org/research/projects/project-summaries/ccammo/Chapter12-appendix) are thus combined with conditional participation-days models to complete the hurdle model. Given that only those participating are included in the conditional days portion of the model, thus eliminating observations of zero for days, a truncated negative binomial model was employed for estimation. Like the participation models above, the days models were estimated for each of the 10 activities, with and without climate variables (www.forestthreats.org/research/projects/project-summaries/ccammo/Chapter12-appendix). Climate variables used in the days models are reported in Table 12.6.

Total days for each activity were estimated following a procedure that is similar to the one used for estimating participants and that uses the same data. First, days-of-participation per participant were regressed on relevant explanatory variables without and then with climate variables (www.forestthreats.org/research/projects/project-summaries/ccammo/Chapter12-appendix). Parameter estimates from the respective negative binomial models were then combined with projected explanatory variables at 10-year intervals to create indexed per capita days-of-participation for each activity. These indices were in turn combined with population projections for each of the storylines to develop estimated per participant-days indices. The participant-days indices were then applied to a beginning baseline estimate of participation days for each activity, based on weighted regional averages calculated from 2005 to 2009 NSRE data, to yield projections of southern adult participation days. Like the participant estimates, the four-year average around 2008 was chosen to avoid any aberration associated with a single year.

TABLE 12.5
Climate Variables Used for Estimating and Forecasting Outdoor Recreation Participation and Days-of-Participation by Adults in the Southern United States

Variable	Description[a]
Ppt_monthly_mean100	Daily average of precipitation for all months for resident county and counties within 100 miles of resident county centroid
Ppt_monthly_mean200	Daily average of precipitation for all months for resident county and counties within 200 miles of resident county centroid
Spring_PET_d200	Spring average daily potential evapotranspiration for resident county and counties within 200 miles of resident county centroid
Tmax_fall50	Average monthly maximum autumn temperature for resident county and counties within 50 miles of resident county centroid
Tmax_geq_25_d200	Percentage of the month when average monthly maximum temperature exceeded 25°C for resident county and counties within 200 miles of resident county centroid
Tmax_geq_35	Percentage of months when average monthly maximum temperature exceeded 35°C in the resident county
Tmax_geq35_d100	Percentage of month when average monthly maximum temperature exceeded 35°C for resident county and counties within 100 miles of resident county centroid
Tmax_geq35_d200	Percentage of month when average monthly maximum temperature exceeded 35°C for resident county and counties within 200 miles of resident county centroid
Tmax_spring	Average of the monthly maximum temperature averages in spring in the resident county
Tmax_spring100	Average of the monthly maximum temperature averages in spring for the resident county and counties within 100 miles of resident county centroid
Tmax_summer	Average of the monthly maximum temperature averages in summer in the resident county
Tmax_summer50	Average of the monthly maximum temperature averages in summer for the resident county and counties within 50 miles of resident county centroid
Tmax_summer100	Average of the monthly maximum temperature averages in summer for the resident county and counties within 100 miles of resident county centroid
Tmax_summer200	Average of the monthly maximum temperature averages in summer for the resident county and counties within 200 miles of resident county centroid
Tmax_winter	Average of the monthly maximum temperature averages in winter in the resident county
Tmax_winter100	Average of the monthly maximum temperature averages in winter for the resident county and counties within 100 miles of resident county centroid
Tmin_leq_0	Percentage of month when average monthly minimum temperature was below 0°C in the resident county
Tmin_leq_neg10	Percentage of month when average monthly minimum temperature was below −10°C in the resident county
Total_ppt100	Monthly average of total monthly precipitation in resident county and counties within 100 miles of resident county centroid
Total_ppt200	Monthly average of total monthly precipitation in resident county and counties within 200 miles of resident county centroid
Tinter_PET_d50	Average of daily potential evapotranspiration averages in winter for resident county and counties within 50 miles of resident county centroid
Tinter_PET_d200	Average of daily potential evapotranspiration averages in winter for resident county and counties within 200 miles of resident county centroid
Tearly_PET_d200	Average of daily potential evapotranspiration averages for resident county and counties within 200 miles of resident county centroid

[a] All averages were calculated over six-year periods, for example, historic data are based on 2000 to 2006 data, 2060 projections are based on averages from 2055 to 2060. Seasons were divided into three-month periods based on the following categories: winter (December, January, and February), spring (March, April, and May), summer (June, July, and August), and autumn (September, October, and November).

TABLE 12.6
Climate Variables for Modeling Activity Participation and Days-of-Participation

Recreation Activity	Model Type	Climate Variable
Land based		
Developed site use (family gathering, picnicking, camping)	Participation	tmax_summer
	Days	tmax_geq_25_d100
Equestrian (horseback riding on trails)	Participation	tmin_leq_0_d200
	Days	tmax_geq_35
Day hiking	Participation	tmax_geq_35_d200
	Days	tmin_leq_0
Off-road driving	Participation	tmax_geq_25
	Days	tmax_geq_35_d200
Primitive area use (visiting wilderness, camping/backpacking)	Participation	tmax_geq25
	Days	tmax_geq_25_d100
Water based		
Motorized water (motorboating, water skiing, jetskiing)	Participation	tmax_geq_25
	Days	tmin_leq_0_d200
Floating (canoeing, kayaking, rafting)	Participation	tmax_summer
	Days	spring_PET_d50
Wildlife		
Birding (viewing or photographing)	Participation	winter_PET_d50
	Days	tmax_geq_35
Hunting	Participation	tmax_winter_d100
	Days	tmin_leq_neg5_d100
Fishing	Participation	tmax_geq_35
	Days	tmax_geq_35

RESULTS

Below, we present per capita and overall changes, from 2008 to 2060, in participation and days-of-participation by storyline for land-based activities (developed site use, hiking, horseback riding on trails, motorized off-road driving, and primitive site use), water-based activities (motorized and nonmotorized), and wildlife-based activities (birding, fishing, and hunting).

Developed site use: Developed site use is the most popular of the land-based outdoor recreation activities, both nationally and in the South. This composite activity includes family gatherings, picnicking, and developed camping. On average, from 2005 to 2009, this activity was practiced by about 80% of southern adults, or more than 63 million people, accounting for 672 million days-of-participation in 2008 (Table 12.1). Moreover, our projections only relate to adults; because many children participate in these activities, participation totals that include all age groups should be much higher than the numbers reported in this chapter. As Table 12.7 indicates, per capita participation growth in this activity is expected to be static over the next 50 years across all storylines; with the moderate population/high income growth-focused A1B—at 2%—showing the most change. This composite activity is already highly popular, and the static participation rate means that overall participant growth is likely to mirror general population increases for all storylines. Thus, under A2, which has the highest expected population growth, participation would increase by nearly 90% to approximately 122 million adults per year. Days-per-participant is projected to remain constant across all storylines. Hence, the total for days is expected to follow growth in participant numbers and is expected to range from 53% to 90% over the next five decades.

TABLE 12.7
Forecasted Developed Site Use (Family Gatherings, Picnicking, or Camping) by Adults in the Southern United States

	Year		Projected Change from 2008 (%)			
Storyline[a]	2008	2060[b]	Baseline[b]	Climate1[c]	Climate2[d]	Climate3[e]
	Per capita participation (%)					
A1B	79.9	81.5	2	1	0	(2)
A2	79.9	80.7	1	(1)	(1)	(2)
B2	79.9	80.7	1	(1)	0	(1)
	Adult participants (millions)					
A1B	63.2	109.9	74	72	70	67
A2	64.2	122.1	90	88	86	84
B2	63.0	96.4	53	50	51	50
	Days per participant					
A1B	10.61	10.61	0	1	3	4
A2	10.61	10.61	0	2	3	3
B2	10.61	10.61	0	2	1	1
	Total days (millions)					
A1B	672	1,170	74	74	76	73
A2	684	1,299	90	91	92	89
B2	670	1,026	53	52	53	52

Note: Based on an expectation of moderate population growth and high income growth (storyline A1B), high population growth and low income growth (storyline A2), or low population growth and moderate income growth (storyline B2) under alternative climate futures derived from general circulation models.

[a] Emissions storylines developed by the Intergovernmental Panel on Climate Change (2007).

[b] Climate variable omitted from model and projection.

[c] Climate1 uses forecast data from CGCM3.1 for storylines A1B and A2; CGCM2 for storyline B2.

[d] Climate2 uses forecast data from CSIROMK3.5 for storylines A1B and A2; CSIROMK2 for storyline B2.

[e] Climate3 uses forecast data from MIROC3.2 for storylines A1B and A2; HadCM3 for storyline B2.

Adding climate projections across the storylines produced only marginal changes in participants and days-of-participation (Table 12.7). Generally, participation rates decreased ≤4 percentage points from the baseline leading to a potential decrease of ≤10 million participants in 2060 under storyline A1B. Alternatively, despite climate change from the baseline, days-per-participant increased from 1 to 4 percentage points across the storylines. This change offset the slight decrease in participant numbers and thus the days total appears largely unaffected by climate change, depending far more on population and income changes.

Hiking: Day hiking is perhaps the single most popular backcountry activity. In 2008 about 33% of adults nationally participated in hiking. In the South, 25.2% of adults participated in hiking, totaling about 20 million participants and 463 million days annually (Table 12.1). For all storylines in the absence of climate change, hiking participation per capita is expected to increase by 12% to 16% by 2060 (Table 12.8). Participant numbers increase the most under A2 at nearly 113% (resulting in about 44 million hikers), followed by A1B at about 96% and B2 at about 70%. Hiking days are expected to increase by slightly more than participants. A notable result for hiking is that it is the only activity for which Hispanic ethnicity is associated with a higher participation rate and higher days-per-participant than other Caucasians (www.forestthreats.org/research/projects/project-summaries/ccammo/Chapter12-appendix).

Projected climate changes appear to have negative impacts across all storylines. For example, relative to the baseline, participation rates decreased by ≤16 percentage points for storyline A1B,

TABLE 12.8
Forecasted Hiking Use by Adults in the Southern United States

	Year		Projected Change from 2008 (%)			
Storyline[a]	**2008**	**2060**[b]	**Baseline**[b]	**Climate1**[c]	**Climate2**[d]	**Climate3**[e]
	Per capita participation (%)					
A1B	25.2	29.2	16	13	10	0
A2	25.2	28.5	13	10	8	1
B2	25.2	28.2	12	5	8	7
	Adult participants (millions)					
A1B	20.3	39.8	96	91	86	70
A2	20.6	44.0	113	107	104	91
B2	20.2	34.4	70	59	63	62
	Days per participant					
A1B	22.93	23.62	3	(5)	(5)	(5)
A2	22.93	23.16	1	(6)	(6)	(7)
B2	22.93	24.08	5	(4)	(3)	(3)
	Total days (millions)					
A1B	463	935	102	82	77	62
A2	471	1,017	116	95	91	78
B2	461	817	77	53	59	57

Note: Based on an expectation of moderate population growth and high income growth (storyline A1B), high population growth and low income growth (storyline A2), or low population growth and moderate income growth (storyline B2) under alternative climate futures derived from general circulation models.

[a] Emissions storylines developed by the Intergovernmental Panel on Climate Change (2007).

[b] Climate variable omitted from model and projection.

[c] Climate1 uses forecast data from CGCM3.1 for storylines A1B and A2; CGCM2 for storyline B2.

[d] Climate2 uses forecast data from CSIROMK3.5 for storylines A1B and A2; CSIROMK2 for storyline B2.

[e] Climate3 uses forecast data from MIROC3.2 for storylines A1B and A2; HadCM3 for storyline B2.

≤12 percentage points for A2, and ≤7 percentage points for B2. The biggest decreases are associated with the MIROC3.2 climate projection, which is characterized by higher temperatures and lower participation. Nevertheless, at the very least, the same percentage of southerners will likely be participating in hiking in 2060 as today, even under the most negative climate alternative. Climate change would also have a downward effect on the annual hiking days-per-participant, accounting for a 7- to 9-percentage point decrease from the baseline across storylines. Despite these decreases, increased population is expected to cause increases, both in participant numbers (59–107%) and days-of-participation (53–95%). On average, projected annual hiking days across storylines will likely be ≤25% points lower in 2060 than if climate remained unchanged.

Horseback riding on trails: Although the least popular of the land based activities, horseback riding is nevertheless enjoyed by 7.1% of southern adults annually (Table 12.9). Unlike developed use and hiking, per capita participation in horseback riding on trails is projected to decrease by 5–8% in B2 and A2 over the next 50 years. In A1B, however, per capita participation is expected to increase by 9%. The number of participants in this activity increases under A1B (a function of high income growth) from about 5.6 million in 2008 to between 10 and 11 million by 2060, followed by a similar increase under A2 (driven by high population growth). Annual riding days-per-participant is static under A2, but increases by 9% under the low population/moderate income growth of B2, and by 26% under A1B. Combined with the participation rate changes and population growth, horseback riding on trails is projected to increase from a total of about 100 million days in 2008 to between 155 and 231 million days annually by 2060.

TABLE 12.9
Forecasted Horseback Riding-on-Trails Use by Adults in the Southern United States

	Year		Projected Change from 2008 (%)			
Storyline[a]	**2008**	**2060[b]**	**Baseline[b]**	**Climate1[c]**	**Climate2[d]**	**Climate3[e]**
	Per capita participation (%)					
A1B	7.1	7.7	9	8	7	8
A2	7.1	6.5	(8)	(9)	(10)	(11)
B2	7.1	6.7	(5)	(7)	(3)	(5)
	Adult participants (millions)					
A1B	5.6	10.5	85	83	82	84
A2	5.7	9.9	73	71	69	67
B2	5.6	5.4	44	41	47	44
	Days per participant					
A1B	17.67	22.26	26	(15)	(18)	(65)
A2	17.67	17.49	(1)	(59)	(40)	(60)
B2	17.67	19.26	9	(36)	(28)	(31)
	Total days (millions)					
A1B	99	231	133	55	31	(35)
A2	101	172	71	(30)	2	(34)
B2	99	155	57	(10)	6	(1)

Note: Based on an expectation of moderate population growth and high income growth (storyline A1B), high population growth and low income growth (storyline A2), or low population growth and moderate income growth (storyline B2) under alternative climate futures derived from general circulation models.

[a] Emissions storylines developed by the Intergovernmental Panel on Climate Change (2007).

[b] Climate variable omitted from model and projection.

[c] Climate1 uses forecast data from CGCM3.1 for storylines A1B and A2; CGCM2 for storyline B2.

[d] Climate2 uses forecast data from CSIROMK3.5 for storylines A1B and A2; CSIROMK2 for storyline B2.

[e] Climate3 uses forecast data from MIROC3.2 for storylines A1B and A2; HadCM3 for storyline B2.

Accounting for associated climate change alternatives has very minor effects on the participation rate and the number of participants by 2060, falling within a few percentage points of the baseline for all storylines (Table 12.9). However, climate seems to have dramatic dampening effects on annual days-per-participant, leading to decreases of 15–65% (2.7–10.6 days). The largest decreases appear for the MIROC3.2 climate projections. Combined with population growth under each storyline, the effect of climate change on total days-of-participation for riding ranges from a 55% increase (A1B with CGCM3.1) to a 35% decrease (A1B with MIROC3.2). Including climate would likely cause substantially fewer riding days per year in 2060 than the baseline. Moreover, in five of nine storyline/climate alternatives, the annual days total decreases compared to 2008.

Motorized off-road driving: Off-road driving increased in popularity among southerners by 42% from 1999 to 2009 (Cordell et al. in press). In 2008, approximately 21% or 17 million adults took part in off-road driving, accounting for more than 560 million days (Table 12.10). This makes motorized off-roading second only to visiting developed sites for days-of-use among the land-based activities. Over the next 50 years, participation rates are projected to decrease by 11–25% across all storylines, meaning that the growth of participant numbers would be lower than the population growth rate, or 26–51%. Annual days-per-participant is expected to decrease by ≤3%; therefore, the total number of days for this activity is expected to grow slightly less than participants, or from 24% to 48%. Although off-roading days will likely increase less than population growth, southerners would nevertheless increase their off-roading days by 135–269 million annually by 2060.

TABLE 12.10
Forecasted Motorized Off-Road Use by Adults in the Southern United States

	Year		Projected Change from 2008 (%)			
Storyline[a]	**2008**	**2060**[b]	**Baseline**[b]	**Climate1**[c]	**Climate2**[d]	**Climate3**[e]
	Per capita participation (%)					
A1B	21.3	19.0	(11)	(12)	(15)	(15)
A2	21.3	16.0	(25)	(25)	(27)	(27)
B2	21.3	17.9	(16)	(18)	(17)	(17)
	Adult participants (millions)					
A1B	16.9	25.5	51	50	45	44
A2	17.2	24.4	42	41	37	37
B2	16.9	21.2	26	25	25	26
	Days per participant					
A1B	33.30	32.63	(2)	(13)	(20)	(39)
A2	33.30	32.30	(3)	(31)	(18)	(32)
B2	33.30	32.63	(2)	(21)	(15)	(32)
	Total days (millions)					
A1B	562	831	48	30	16	(12)
A2	571	788	38	(2)	13	(7)
B2	560	695	24	(1)	6	3

Note: Based on an expectation of moderate population growth and high income growth (storyline A1B), high population growth and low income growth (storyline A2), or low population growth and moderate income growth (storyline B2) under alternative climate futures derived from general circulation models.

[a] Emissions storylines developed by the Intergovernmental Panel on Climate Change (2007).

[b] Climate variable omitted from model and projection.

[c] Climate1 uses forecast data from CGCM3.1 for storylines A1B and A2; CGCM2 for storyline B2.

[d] Climate2 uses forecast data from CSIROMK3.5 for storylines A1B and A2; CSIROMK2 for storyline B2.

[e] Climate3 uses forecast data from MIROC3.2 for storylines A1B and A2; HadCM3 for storyline B2.

Through 2060, climate is expected to have very minor negative effects on participation rates compared to the baseline (Table 12.10). Similarly, the effect on participant numbers is marginal. However, because days-per-participant decreases from 13% to 39% annually compared to 2008, total days-of-participation for off-road driving is expected to be noticeably less than for the no climate change baseline. With four out of nine storyline/climate alternatives, the number of off-road driving days in 2060 is forecasted to be ≤12% lower than in 2008, suggesting that climate change could strongly dampen off-road driving days in the future.

Visiting primitive areas: This activity is an aggregate that consists of activities such as backpacking, primitive camping, and visiting a designated or undesignated wilderness. This composite accounted for over 28 million participants in 2008, or about 35% of all adults in the South (Table 12.11). Participants visited primitive areas on over 411 million days in 2008. Under the baseline with no climate change, annual per capita participation is expected to decrease by ≤7% over the next 50 years. Increased population density, decreases in forest and rangeland per capita, and changing demographics appear to be factors influencing the participation rate decrease (www.forestthreats.org/research/projects/project-summaries/ccammo/Chapter12-appendix). However, overall participation is expected to increase by 44–76% across all storylines by 2060 because population growth would offset the small decrease in participation rates. Annual days-of-participation for visiting primitive areas per participant is projected to remain nearly constant throughout the simulation period; therefore, the growth in total days per year is expected to closely follow adult population growth and range from 43% to 77% across all baselines.

TABLE 12.11
Forecasted Primitive area Use by Adults in the Southern United States

	Year		Projected Change from 2008 (%)			
Storyline[a]	**2008**	**2060[b]**	**Baseline[b]**	**Climate1[c]**	**Climate2[d]**	**Climate3[e]**
	Per capita participation (%)					
A1B	35.3	34.9	(3)	(3)	(4)	(4)
A2	35.3	32.8	(7)	(7)	(8)	(8)
B2	35.3	33.5	(6)	(6)	(6)	(6)
	Adult participants (millions)					
A1B	28.2	47.0	67	65	63	63
A2	28.6	50.4	76	75	74	73
B2	28.1	40.4	44	42	42	43
	Days per participant					
A1B	14.55	14.70	1	(4)	(10)	(11)
A2	14.55	14.70	1	(5)	(9)	(8)
B2	14.55	14.70	1	(5)	(4)	(3)
	Total days (millions)					
A1B	412	697	67	59	47	45
A2	419	751	77	66	58	59
B2	411	592	43	36	37	38

Note: Based on an expectation of moderate population growth and high income growth (storyline A1B), high population growth and low income growth (storyline A2), or low population growth and moderate income growth (storyline B2) under alternative climate futures derived from general circulation models.

[a] Emissions storylines developed by the Intergovernmental Panel on Climate Change (2007).

[b] Climate variable omitted from model and projection.

[c] Climate1 uses forecast data from CGCM3.1 for storylines A1B and A2; CGCM2 for storyline B2.

[d] Climate2 uses forecast data from CSIROMK3.5 for storylines A1B and A2; CSIROMK2 for storyline B2.

[e] Climate3 uses forecast data from MIROC3.2 for storylines A1B and A2; HadCM3 for storyline B2.

Amending the storylines with related climate forecasts would lead to virtually no changes in participation rates or the number of participants visiting primitive areas to 2060. However, climate change would produce a 5- to 22-percentage point decrease from the baseline across the storylines, yielding 36–66% more primitive visit days than in 2008.

Motorized water use: In 2008, 27%, or about 21 million southern adults, engaged in motor boating, waterskiing, and personal watercraft use; and spent approximately 384 million days in this activity. Taken separately, these activities all experienced relatively strong growth in participants from 1999 to 2009, both regionally and nationally (Cordell et al. in press; Cordell 2012). The participation rate for motorized water use is projected to increase by 10% to 2060 under A1B, but decreases by ≤5% under A2 and B2 (Table 12.12). The difference can be attributed to higher growth rate for household income, which is an important driver for this activity (www.forestthreats.org/research/projects/project-summaries/ccammo/Chapter12-appendix). Including population growth yields a 48- to 87-percent increase in total participants by 2060. Days-per-participant is expected to be stable at 18 days per year under A1B (faster than population growth), but decrease slightly under the others for a rate that is somewhat slower than population growth. By 2060, days-of-participation for motorized water use are expected to grow by 38–86%, to between 529 and 715 million days annually.

Climate change would add about a 3 percentage points to the participation rate for A1B and A2, and no change for B2 (Table 12.12). Thus, motor boating participant numbers can be expected to increase by zero to 8 percentage points more than the baselines when climate forecasts are included.

TABLE 12.12

Forecasted Recreational Motorized Water Use (Motor Boating, Waterskiing, Using Personal Watercraft) by Adults in the Southern United States

	Year		Projected Change from 2008 (%)			
Storyline[a]	**2008**	**2060**[b]	**Baseline**[b]	**Climate1**[c]	**Climate2**[d]	**Climate3**[e]
	Per capita participation (%)					
A1B	27.0	29.7	10	10	15	15
A2	27.0	25.7	(5)	(4)	(2)	(2)
B2	27.0	26.5	(2)	(2)	(2)	(2)
	Adult participants (millions)					
A1B	21.3	39.8	87	88	95	95
A2	21.6	38.9	80	80	85	86
B2	21.2	31.4	48	49	48	48
	Days per participant					
A1B	18.21	18.03	(1)	(1)	(1)	(2)
A2	18.21	16.57	(9)	(10)	(9)	(8)
B2	18.21	17.12	(6)	(7)	(8)	(7)
	Total days (millions)					
A1B	384	715	86	85	92	92
A2	391	645	65	63	69	70
B2	383	529	38	39	37	37

Note: Based on an expectation of moderate population growth and high income growth (storyline A1B), high population growth and low income growth (storyline A2), or low population growth and moderate income growth (storyline B2) under alternative climate futures derived from general circulation models.

[a] Emissions storylines developed by the Intergovernmental Panel on Climate Change (2007).

[b] Climate variable omitted from model and projection.

[c] Climate1 uses forecast data from CGCM3.1 for storylines A1B and A2; CGCM2 for storyline B2.

[d] Climate2 uses forecast data from CSIROMK3.5 for storylines A1B and A2; CSIROMK2 for storyline B2.

[e] Climate3 uses forecast data from MIROC3.2 for storylines A1B and A2; HadCM3 for storyline B2.

Annual days-per-participant is expected to remain virtually unchanged from baseline conditions. Hence, for total days for motorized water use can be expected to increase by ≤6 percentage points over the baselines.

Nonmotorized water use: Approximately 15.4% or 12.2 million adults in the South participated in canoeing, kayaking, or rafting in 2008, resulting in 80 million days of use (Table 12.1). Although rafting grew by only 5% from 1999 to 2009, canoeing (39%) and kayaking (154%) grew dramatically (Cordell 2012). Despite rapid growth over the past decade, per capita adult participation is projected to be stable out to 2060, resulting in participant numbers growing at the same rate as the population, or 45–81% (Table 12.13). This activity is less affected by income than its motorized counterpart. Hence, A2 with higher population growth would yield the biggest increase in participants. Days-per-participant is expected to decrease minimally by 2060, meaning that the current 80 million days for this activity is forecasted to increase to 114 to 143 million days by 2060.

Climate change will likely negatively affect participation rates for nonmotorized water activities across all storylines. Participation rates are expected to drop 6–18 percentage points compared to the baseline (Table 12.13). The number of participants is thus expected to grow 33–70% by 2060, or about 10 percentage points less than when climate change is not considered. Conversely, climate change is expected to have a marginally positive effect on annual days-of-participation—2–11 percentage points over the baseline—depending on the particular storyline/climate alternative. Given

TABLE 12.13

Forecasted Recreational Nonmotorized Water Use (Canoeing, Kayaking, Rafting, Tubing) by Adults in the Southern United States

	Year		Projected Change from 2008 (%)			
Storyline[a]	**2008**	**2060**[b]	**Baseline**[b]	**Climate1**[c]	**Climate2**[d]	**Climate3**[e]
	Per capita participation (%)					
A1B	15.4	16.3	6	0	(5)	(12)
A2	15.4	14.8	(4)	(10)	(13)	(17)
B2	15.4	14.8	(4)	(12)	(11)	(12)
	Adult participants (millions)					
A1B	12.2	22.0	80	69	61	49
A2	12.4	22.5	81	70	65	57
B2	12.2	17.6	45	33	35	33
	Days per participant					
A1B	6.58	6.49	(2)	2	7	9
A2	6.58	6.38	(3)	6	5	6
B2	6.58	6.45	(2)	4	2	0
	Total days (millions)					
A1B	80	141	76	73	72	63
A2	81	143	76	80	72	67
B2	80	114	43	38	38	33

Note: Based on an expectation of moderate population growth and high income growth (storyline A1B), high population growth and low income growth (storyline A2), or low population growth and moderate income growth (storyline B2) under alternative climate futures derived from general circulation models.

[a] Emissions storylines developed by the Intergovernmental Panel on Climate Change (2007).

[b] Climate variable omitted from model and projection.

[c] Climate1 uses forecast data from CGCM3.1 for storylines A1B and A2; CGCM2 for storyline B2.

[d] Climate2 uses forecast data from CSIROMK3.5 for storylines A1B and A2; CSIROMK2 for storyline B2.

[e] Climate3 uses forecast data from MIROC3.2 for storylines A1B and A2; HadCM3 for storyline B2.

that the marginal decrease in participation rate is slightly greater than the marginal increase in days-per-participant, on average the total days-of-participation for nonmotorized water activities in 2060 ranges from slightly more (4 percentage points) to somewhat less (13 percentage points) than the baselines when climate change is included.

Birding: This nonconsumptive activity, which consists of viewing or photographing birds, involves 34.2% of the adult population (27 million people) in the South. Among all activities, birding has the highest annual days-per-participant (107) for an annual total of about 2.9 billion days (Table 12.14). This reflects the many levels or intensities of birding, from watching backyard feeders to pursuing sightings in remote forests. Cordell (2012) reports that birding participation increased by nearly 30% from 1999 to 2009. Per capita participation in birding is projected to increase 8–10% through 2060, meaning that birders would increase faster than the general adult population across all storylines, with total participants expected to be 44–56 million. Days-per-participant is expected to decrease 9–13%, meaning that the total number of days per year would increase marginally less than the population, or 47–76%.

Adding climate change to the storylines shows little or no effect on the participation rate or the number of birders. However, the 9–13% decrease in annual days-per-participant under the baselines becomes 20–37% when climate data are included, with the biggest decreases happening under A1B with the MIROC3.2 climate forecast (Table 12.14). Given the shortened participation periods, total

TABLE 12.14
Forecasted Birding Activity (Viewing or Photographing) by Adults in the Southern United States

	Year		Projected Change from 2008 (%)			
Storyline[a]	**2008**	**2060[b]**	**Baseline[b]**	**Climate1[c]**	**Climate2[d]**	**Climate3[e]**
	Per capita participation (%)					
A1B	34.2	37.6	10	11	12	11
A2	34.2	36.9	8	9	9	10
B2	34.2	36.9	8	9	8	8
	Adult participants (millions)					
A1B	26.0	50.4	87	88	90	89
A2	27.4	55.7	103	105	106	108
B2	26.9	43.9	63	65	63	64
	Days per participant					
A1B	106.65	94.92	(11)	(25)	(27)	(37)
A2	106.65	92.79	(13)	(30)	(24)	(31)
B2	106.65	97.05	(9)	(22)	(20)	(20)
	Total days (millions)					
A1B	2862	4752	66	42	39	19
A2	2912	5125	76	43	56	43
B2	2855	4197	47	29	30	30

Note: Based on an expectation of moderate population growth and high income growth (storyline A1B), high population growth and low income growth (storyline A2), or low population growth and moderate income growth (storyline B2) under alternative climate futures derived from general circulation models.

[a] Emissions storylines developed by the Intergovernmental Panel on Climate Change (2007).

[b] Climate variable omitted from model and projection.

[c] Climate1 uses forecast data from CGCM3.1 for storylines A1B and A2; CGCM2 for storyline B2.

[d] Climate2 uses forecast data from CSIROMK3.5 for storylines A1B and A2; CSIROMK2 for storyline B2.

[e] Climate3 uses forecast data from MIROC3.2 for storylines A1B and A2; HadCM3 for storyline B2.

days-of-participation for birding will likely increase 19–56% to 2060; about 17- to 33-percentage points less, on average, than the baselines.

Fishing: Defined here, fishing includes various types of saltwater and freshwater pursuits. Fishing has the second highest participation rate (35.7%) for southerners. In 2008, approximately 28 million anglers accounted for 573 million days-of-participation (Table 12.1). According to Cordell (2012), fishing participants increased by >21% in the past decade. Across all storylines, the fishing participation rate is projected to decrease by 10–18% over the next 50 years (Table 12.15). Thus, the number of anglers is projected to grow somewhat slower than the regional population, with growth rates for participants of 32–54%. Days-per-participant are expected to decrease marginally, remaining at about 20 per year. Therefore, the number of days-of-participation for fishing is expected to grow slightly slower than the number of participants, or 30–51%. Nevertheless, fishing is expected to remain among the top recreation activities in the South, accounting for 742–874 million days annually in 2060.

Adding climate to the fishing forecasts would result in decreases from the baseline for both participation rates and annual fishing days-per-participant (Table 12.15). Across all storylines, participation rates in 2060 are expected to decrease 15–27% from 2008 levels. This implies that participant numbers would increase 24–46%, or about 8 percentage points less than when climate change is not considered. The fishing-days total increases under eight of nine storylines with climate included, but at rates far below the baselines. Increases range from 10% to 33%.

TABLE 12.15
Forecasted Recreational Fishing Activity (Warm-Water, Cold-Water, and/or Saltwater) by Adults in the Southern United States

	Year		Projected Change from 2008 (%)			
Storyline[a]	2008	2060[b]	Baseline[b]	Climate1[c]	Climate2[d]	Climate3[e]
	Per capita participation (%)					
A1B	35.7	32.1	(10)	(15)	(15)	(21)
A2	35.7	29.3	(18)	(23)	(23)	(27)
B2	35.7	31.1	(13)	(18)	(17)	(18)
	Adult participants (millions)					
A1B	28.0	42.9	53	46	44	35
A2	28.5	42.9	54	46	45	38
B2	28.0	36.9	32	24	25	24
	Days per participant					
A1B	20.58	20.17	(2)	(9)	(12)	(26)
A2	20.58	19.96	(3)	(20)	(12)	(21)
B2	20.58	20.37	(1)	(11)	(8)	(9)
	Total days (millions)					
A1B	573	864	51	33	26	(1)
A2	582	874	50	17	27	10
B2	571	742	30	10	15	13

Note: Based on an expectation of moderate population growth and high income growth (storyline A1B), high population growth and low income growth (storyline A2), or low population growth and moderate income growth (storyline B2) under alternative climate futures derived from general circulation models.

[a] Emissions storylines developed by the Intergovernmental Panel on Climate Change (2007).

[b] Climate variable omitted from model and projection.

[c] Climate1 uses forecast data from CGCM3.1 for storylines A1B and A2; CGCM2 for storyline B2.

[d] Climate2 uses forecast data from CSIROMK3.5 for storylines A1B and A2; CSIROMK2 for storyline B2.

[e] Climate3 uses forecast data from MIROC3.2 for storylines A1B and A2; HadCM3 for storyline B2.

Hunting: This activity includes all types of legal hunting including big game, small game, waterfowl, and varmint. Approximately 11 million adults in the South, over 13%, reported hunting in 2008 on a total of 230 million days (Table 12.1). Cordell (2012) reports that small game hunting participants increased by 16%, compared to 25% for big game hunters, from 1999 to 2009. Findings from our models suggest that per capita participation has peaked and is likely to decrease 26–41% over the next 50 years (Table 12.16). A number of factors appear to be driving the decrease including: increasing population density, growth in the Asian and Hispanic components of the general population, increasing levels of education, and declining forest and rangeland per capita (www.forestthreats.org/research/projects/project-summaries/ccammo/Chapter12-appendix). Despite the declining participation rate, the number of southern hunters out to 2060 is expected to increase by 8% under the low population/moderate income growth focused B2, compared to 25% under the moderate population/high income growth focused A1B. Days-of-participation per hunter, currently about 22, is projected to remain constant regardless of storyline. Total days-of-participation for hunting are forecasted to grow at about the same rate as hunter numbers: 8–24% for an annual total of 248–286 million days by 2060.

When climate change projections are included in the hunting forecasts, participation rates and participants remain largely unchanged from the baseline with participant numbers increasing 4–26% by 2060 (Table 12.16). However, climate change appears to have a positive influence on the

TABLE 12.16
Forecasted Recreational Hunting Activity by Adults in the Southern United States

	Year		Projected Change from 2008 (%)			
Storyline[a]	**2008**	**2060**[b]	**Baseline**[b]	**Climate1**[c]	**Climate2**[d]	**Climate3**[e]
	Per capita participation (%)					
A1B	13.7	10.1	(26)	(26)	(27)	(26)
A2	13.7	8.1	(41)	(41)	(42)	(43)
B2	13.7	9.7	(29)	(32)	(27)	(29)
	Adult participants (millions)					
A1B	10,786	13,482	25	26	23	25
A2	10,973	12,180	11	12	10	7
B2	10,758	11,618	8	4	10	7
	Days per participant					
A1B	21.68	21.46	(1)	5	4	1
A2	21.68	21.46	(1)	0	5	2
B2	21.68	21.68	0	12	7	1
	Total days (millions)					
A1B	230	286	24	32	28	26
A2	234	255	9	12	15	9
B2	230	248	8	16	18	9

Note: Based on an expectation of moderate population growth and high income growth (storyline A1B), high population growth and low income growth (storyline A2), or low population growth and moderate income growth (storyline B2) under alternative climate futures derived from general circulation models.

[a] Emissions storylines developed by the Intergovernmental Panel on Climate Change (2007).

[b] Climate variable omitted from model and projection.

[c] Climate1 uses forecast data from CGCM3.1 for storylines A1B and A2; CGCM2 for storyline B2.

[d] Climate2 uses forecast data from CSIROMK3.5 for storylines A1B and A2; CSIROMK2 for storyline B2.

[e] Climate3 uses forecast data from MIROC3.2 for storylines A1B and A2; UKMOHADCM3 for storyline B2.

annual hunting days-per-participant, leading to increases of zero to 12%. Thus, total hunting days increase with climate change across A1B by an average of 29%, and B2 by 14% compared to 2008. The increases average about 5 percentage points higher annually than the no-climate storyline baselines.

CONCLUSIONS

Despite continued losses in forests and rangeland per capita across the South and changing demographics, outdoor recreation activity is expected to continue growing—both in numbers of participants and days-of-participation—at a rate near to or somewhat below population growth rates. Details of participation and consumption forecasts related to population growth and changing demographics are available in Bowker et al. (2012) and at www.forestthreats.org/research/projects/project-summaries/ccammo/Chapter12-appendix. For a few activities, such as developed site use, hiking, and birding, participant numbers as well as days-of-participation are projected to grow faster than population. Other activities typically associated with higher income, such as horseback riding on trails, motorized water use, and non-motorized water use, would grow faster than the population if predictions of higher income eventuate. Otherwise, they would grow at rates slightly less than population. A few activities, such as fishing, hunting, and motorized off-road use, are projected to experience substantial decreases in participation rates; and thus, although increasing, are expected to grow much slower than population. Hunting and motorized off-road use, being relatively land intensive, would be adversely affected by the expected decrease in available forest and rangeland acreage per capita. Moreover, these activities are generally not considered widely popular to the growing numbers of ethnic minorities in the region (Poudyal et al. 2008).

Participant numbers and days-of-participation for southerners were projected for storylines with and without associated climate alternatives (Figure 12.3, Table 12.4). Details about climate effects on recreation participation and use can be found in Tables 12.7 through 12.16 and by examining the models and simulations that support predictions (www.forestthreats.org/research/projects/project-summaries/ccammo/Chapter12-appendix). No specific probabilities were assigned to either the individual storylines or any of the climate alternatives associated with them (U.S. Department of Agriculture Forest Service 2012). However, the general effects of climate change on each of the 10 outdoor recreation activities can be inferred by looking at the range of changes in the participation rates, participants, days-of-participation per participant, and total days-of-participation for each activity compared to the no-climate baseline.

Compared to projections without climate change, participation rates were either marginally decreased by including climate change into the models and simulations, as for developed site use (1–4 percentage points), motorized off-road use (0–4 percentage points), and visiting primitive areas (0–1 percentage points); or the participation rate decreased somewhat more relative to the baseline, as for hiking (3–16 percentage points), nonmotorized water use (6–18 percentage points), and fishing (4–11 percentage points). Motorized water use (0–5 percentage points) and birding (0–2 percentage points) were the only activities for which the participation rate rose as a result of including climate in the forecasts. Changes in the participation rate for hunting and horseback riding on trails were ambiguous, ranging from an increase of 2 percentage points to a decrease of 3 percentage points for both activities.

Adult participant numbers tracked relatively closely to the participation rates with and without climate. For example, activities for which participant numbers decreased slightly compared to the no-climate baseline included developed site use (2–7 percentage points), horseback riding on trails (1–6 percentage points), motorized off-road use (0–7 percentage points), and visiting primitive areas (1–4 percentage points). Larger decreases in participant numbers occurred for hiking (5–26 percentage points), nonmotorized water use (10–31 percentage points), and fishing (7–18 percentage points). For motorized water use (0–8 percentage points) and birding (0–5 percentage points), the number of participants remained constant or rose relative to the baseline. Hunting was the only activity to show mixed results for participant numbers that ranged from an increase of 2 percentage points to a decrease of 4 percentage points relative to the baseline.

Annual days-per-participant would be moderately less for hiking, primitive area use, and motorized water use when climate is considered in the storylines (Tables 12.7 through 12.16). A number of activities, including birding, horseback riding on trails, motorized off-road use, and fishing, would experience very large decreases in annual days-per-participant. For developed site use, hunting, and nonmotorized water use, including climate would slightly increase the average annual days-of-participation. Total activity days per year would generally mirror the effects seen with days-per-participant.

Table 12.17 shows forest and rangeland acres-per-participant, which is a measure of recreation resource availability. In places where congestion is a concern and recreation use can adversely impact theresource, a higher number is preferred. This measure is useful in demonstrating the potential differences between storylines with and without climate change. In the absence of climate change, forest and range acres-per-participant for all activities except hunting are projected to decrease by 24–54% by 2060 as participant numbers increase. For hunting the decrease is expected to be somewhat less, from 11% to 26%, because of slower growth in the number of hunters. Incorporating climate change, all other activities face decreases of 23–54%. The biggest single change is for storyline A1B where hiking is about 4 percentage points less impacted with climate change. Birding is the only land-based activity for which climate change spells a decrease in acres-per-participant by

2060 (about 1 percentage point). Across all activities and storylines the differences in this density of participation measure are minimal.

An alternative measure of congestion or land impact is annual days-of-use per forest or rangeland acre (Table 12.18). This measure is perhaps more accurate for assessing the impact of activities on naturebecause it combines the number of participants with participant intensity per unit of land area. The essential driver for this measure is activity days, or participation intensity. In places where congestion is a concern and recreation use can adversely impact the resource, a lower number is preferred. Over the next 50 years, congestion per unit of land area is expected to be highest for horseback riding on trails (151%) and hiking (130%), and lowest for hunting (13–34%). In general, adding climate change to the storylines would have a more noticeable mitigating effect on annual days-of-use per forest and rangeland acre: 18–36 percentage points lower for birding, 22–31 percentage points for hiking, and similar effects for motorized off-road and primitive area use. The most noticeable difference (nearly 100 percentage points) would be for horseback riding on trails. For hunting and developed site use, the changes would be negligible.

For developed site use and hiking, decreases in acres-per-participant could begin to strain existing infrastructure. Birding and hiking may not require expansive areas for quality experiences as they are often "edge dependent" or along linear corridors. Activities typically considered space intensive—horseback riding on trails, motorized off-road use, and especially hunting—could experience somewhat smaller decreases in acres-per-participant, but could actually "feel" more congested given the

TABLE 12.17
Forest-Based Recreation Acres per Participant Densities in the Southern United States

		Acres per Participant			Percent Increase (Decrease)	
Activity	**Storyline**	**2008**	**2060 Baseline**	**2060 Climate Average**	**No Climate Change**	**With Climate Change**
Birding	A1B	10.5	5.2	5.16	(50)	(51)
	A2	10.4	4.8	4.72	(54)	(54)
	B2	10.5	6.2	6.15	(41)	(42)
Developed site use	A1B	4.5	2.4	2.46	(47)	(45)
	A2	4.4	2.2	2.24	(51)	(49)
	B2	4.5	2.8	2.87	(37)	(36)
Hiking	A1B	14.0	6.6	7.11	(53)	(49)
	A2	13.8	6.1	6.46	(51)	(53)
	B2	14.0	7.9	8.30	(43)	(41)
Horseback riding on trails	A1B	50.2	25.2	25.50	(50)	(49)
	A2	49.4	26.8	27.46	(46)	(44)
	B2	50.2	33.4	33.41	(33)	(33)
Hunting	A1B	26.3	19.5	19.60	(26)	(25)
	A2	25.9	21.9	22.19	(15)	(14)
	B2	26.3	23.3	23.56	(11)	(10)
Motorized off-road	A1B	16.8	10.3	10.65	(38)	(37)
	A2	16.5	10.9	11.21	(34)	(32)
	B2	16.8	12.7	12.85	(24)	(23)
Primitive area use	A1B	10.1	5.6	5.71	(45)	(43)
	A2	9.9	5.3	5.36	(47)	(46)
	B2	10.1	6.7	6.78	(33)	(33)

Note: Based on an expectation of moderate population growth and high income growth (storyline A1B), high population growth and low income growth (storyline A2), or low population growth and moderate income growth (storyline B2) with and without climate change; emissions storylines developed by the Intergovernmental Panel on Climate Change (2007).

TABLE 12.18
Forest-Based Recreation Days per Acre Densities in the Southern United States

		Activity Days per Forest and Range Acre			Percent Increase (Decrease)	
Activity	Storyline	2008	2060 Baseline	2060 Climate Average	No Climate Change	With Climate Change
Birding	A1B	10.1	18.0	14.5	79	43
	A2	10.3	19.2	16.0	87	57
	B2	10.1	15.5	13.7	54	36
Developed site use	A1B	2.4	4.4	4.5	88	88
	A2	2.4	4.9	4.9	102	103
	B2	2.4	3.8	3.8	60	59
Hiking	A1B	1.6	3.6	3.1	118	87
	A2	1.7	3.8	3.3	130	100
	B2	1.6	3.0	2.7	85	63
Horseback riding on trails	A1B	0.3	0.9	0.4	151	26
	A2	0.4	0.6	0.3	82	(15)
	B2	0.3	0.6	0.4	63	3
Hunting	A1B	0.8	1.1	1.1	34	39
	A2	0.8	1.0	1.0	16	19
	B2	0.8	0.9	1.0	13	19
Motorized off-road	A1B	2.0	3.1	2.4	59	20
	A2	2.0	3.0	2.2	47	8
	B2	2.0	2.6	2.1	30	7

Note: Based on an expectation of moderate population growth and high income growth (storyline A1B), high population growth and low income growth (storyline A2), or low population growth and moderate income growth (storyline B2) with and without climate change; emissions storylines developed by the Intergovernmental Panel on Climate Change (2007).

nature of the activity. Notably, for storyline A1B, with or without climate change, the loss in acres-per-participant associated with high income growth, moderate land conversion, and moderate population growth would lead to the most "congestion" for the space intensive activities. Conversely, B2, characterized by the lowest population growth, would lead to the least amount of future congestion or pressure on resources.

Bowker et al. (2012) found that winter activities often done mostly at the local level, such as snowmobiling and undeveloped skiing, were more negatively impacted by projected changes in climate over the next 50 years, going from substantial increases in participants and days-of-participation to a high likelihood of dramatic decreases. Although often enjoyed by many southerners, these activities are obviously not major outdoor recreation activities region-wide.

Finally, although the effects of climate change are summarized as ranges, more often than not the most pronounced differences between the no-climate and climate forecasts occurred under storylines A1B and A2 under the Climate3 alternative (Tables 12.7 through 12.16), which employs the MIROC3.2 climate forecasts (Joyce et al. in press). As discussed in Chapter 2, the MIROC3.2 climate simulations project the highest temperatures and lowest precipitation of all of the models.

MANAGEMENT RESPONSES

In preparing this chapter, we developed models to explain outdoor recreation participation and days-of-participation for residents of the Southern United States. These models—combined with population, socioeconomic, land use, and climate projections from alternative futures—were employed to predict the number of outdoor recreation participants and days-of-participation and to estimate the degree to which projections differ based on the presence of climate change.

The results herein suggest that recreation participant numbers and days for southerners will continue to grow over the next 50 years under nearly all of the considered socioeconomic, land use, and climate conditions. Thus, the general outlook for recreation resources is for opportunities and access per person to decline. Assuming that the public land base for outdoor recreation remains stable into the future, an increasing population would result in decreasing per-person opportunities across most of the United States. Although many other factors are involved in recreation supply, recreation resources (both built and natural) will likely become less "available" as more people compete to use them. For privately owned land, this could mean rising access prices from increased demand relative to supply. On public lands, where access fees cannot be easily adjusted to market or quasi-market conditions, increased congestion and possible declines in the quality of the recreation experience are likely to present important challenges to management.

A major challenge for natural resource managers and planners will be to ensure that recreation opportunities remain viable and grow along with the population. This will probably have to be accomplished through creative and efficient management of site attribute inputs and plans, rather than through major expansions or additions to the natural resource base. Trends toward more flexible work scheduling and telecommuting would allow recreation users to allocate their leisure time more evenly across the seasons and through the week, thus facilitating less concentrated peak demands. In addition, technological innovations like GPS (Global Positioning System) units and light-weight plastic kayaks allow more people to find and get to places more easily and quickly, perhaps leading to overuse pressures not previously considered a threat.

Overall, the infrastructure that supports outdoor recreation opportunities in the South could be severely tested under most foreseeable circumstances. For activities like developed site use and day hiking, fewer acres or trail miles per participant could begin to strain the existing infrastructure as biological and social carrying capacities become strained. Activities like birding and hiking may not require expansive contiguous areas for quality experiences as they are often "edge dependent" or occur along linear corridors. However, activities typically considered space intensive—horseback riding on trails, hunting, and motorized off-road use—are likely to actually "feel" more congested given the nature of the activity, despite relatively slow growth.

Perhaps surprisingly, the effects of climate change on recreation demand by southerners appear to be moderately mitigating insofar as use density measures (like participants per forest and rangeland acre and activity days per acre) are concerned (Tables 12.17 and 12.18). Climate can affect individual willingness to participate in recreation activities, recreation resource availability and quality, or both. The climate variables used in the recreation models were limited to those coming directly from the RPA climate projections (U.S. Department of Agriculture Forest Service 2012), or variables derived from those basic variables. Generally, the climate variables used in these recreation models were presumed to affect willingness to participate and frequency of participation directly. However, despite the lack of existing data, climate change would undoubtedly affect resource availability directly and indirectly. For example, increasing temperatures will likely affect the distribution of plant and animal species, which are fundamental to maintaining fish and game populations. Moreover, changes in regional precipitation would influence stream and reservoir levels, affecting opportunities for fishing and boating. Disentangling the effects of the climate variables on recreation participation is difficult. However, understanding these direct and indirect relationships, at both local and macro levels, will be fundamental to improving forecasts of recreation behavior.

No one can know exactly how changes in income, socioeconomic factors, economic development, and climate change would affect the supply and demand for forest-based outdoor recreation because the assumptions that underlie forecasts are likely to change with time. As well, due to data limitations, the results presented here do not account for detailed interactions among many of the external variables over time. Moreover, people's preferences shift. New technologies alter, and occasionally curtail, enjoyment of the outdoors. Activities like snowboarding, mountain biking, and orienteering were not available options in 1973, the year that Cicchetti published his seminal forecasting work on national recreation use; nor were activities like video gaming and watching movies at home. As ethnic groups continue to acculturate over the course of the next five decades, differences in outdoor recreation and consumption could shift. But it is safe to say that as the population grows, outdoor recreation pressure will increase on the natural environment, public and private; and that management will need to find creative ways to mitigate this pressure, especially on the most pristine and potentially vulnerable areas. What is important to keep in mind is that the effects of climate change in the South are likely to be relatively minor compared to threats resulting from population growth.

REFERENCES

Ahn, S., DeSteiguer, J.E., Palmquist, R.B., Holmes, T.P. 2000. Economic analysis of the potential impact of climate change on recreational trout fishing in the southeastern Appalachian Mountains: An application of a nested multinomial logit model. *Climate Change*. 45: 493–509.

Arbel, A., Ravid, S.A. 1985. On recreation demand: A time-series approach. *Applied Economics*. 17: 979–990.

Bowker, J.M. 2001. Outdoor recreation participation and use by Alaskans: Projections 2000–2020. Gen. Tech. Rep. PNW–GTR–527. Portland, OR: U.S. Department of Agriculture Forest Service, Pacific Northwest Research Station. 28pp.

Bowker, J.M., Leeworthy, V.R. 1998. Accounting for ethnicity in recreation demand: A flexible count data approach. *Journal of Leisure Research* 30(1): 64–78.

Bowker, J.M., Bergstrom, J.C., Gill, J. 2007. Estimating the economic value and impacts of recreational trails: A case study of the Virginia Creeper rail trail. *Tourism Economics*. 13: 241–260.

Bowker, J.M., English, D.B.K., Cordell, H.K. 1999. Outdoor recreation participation and consumption: Projections 2000 to 2050. In: Cordell, H.K., Betz, C.J., Bowker, J.M. et al. (eds.). *Outdoor Recreation in American Life: A National Assessment of Demand and Supply Trends*. Champagne, IL: Sagamore Press, Inc.: 323–350.

Bowker, J.M., Askew, A.E., Cordell, H.K., Bergstrom, J.C. In press. Outdoor recreation in the South: Projections to 2060. In: Wear, D.N. and J.G. Greis (eds). *The Southern Forest Futures Project: Technical Report*. Gen. Tech. Rep. SRS-xxx. Asheville, NC: U.S. Department of Agriculture, Forest Service, Southern Research Station.

Bowker, J.M., Askew, A.E., Cordell, H.K. et al. 2012. *Outdoor Recreation Participation in the U.S.—Projections to 2060: A Technical Document Supporting the Forest Service 2010 RPA Assessment.* Gen. Tech. Rep. SRS-160. Asheville, NC: U.S. Department of Agriculture Forest Service, Southern Research Station. 34pp.

Bowker, J.M., Bergstrom, J.C., Starbuck, C.M. et al. 2010. *Estimating Demographic and Population Level Induced Changes in Recreation Demand for Outdoor Recreation on U.S. National Forests: An Application of National Visitor Use Monitoring Program data.* Fac. Ser. Work. Pap. FS 1001. Athens, GA: The University of Georgia, Department of Agricultural and Applied Economics. 147pp.

Bowker, J.M., Murphy, D., Cordell, H.K. et al. 2006. Wilderness and primitive area recreation participation and consumption: An examination of demographic and spatial factors. *Journal of Agricultural and Applied Economics.* 38(2): 317–326.

Cameron, C.A., Trivedi, P.K. 1998. *Econometric Society Monographs: Regression Analysis of Count Data.* New York: Cambridge University Press. 412pp.

Cato, J. and K. Gibbs. 1973. *An Economic Analysis Regarding the Effects of Weather Forecasts on Florida Coastal Recreationists.* Economics Report No. 50, Gainesville, Food and Resource Economics Department, University of Florida.

Cicchetti, C.J. 1973. *Forecasting Recreation in the United States*. Lexington, MA: D.C. Heath and Co. 200 p.

Cordell, H.K. (ed.) 2012. *Outdoor Recreation Trends and Futures: a Technical Document Supporting the Forest Service 2010 RPA Assessment.* Gen. Tech. Rep. 160. Asheville, NC: U.S. Department of Agriculture Forest Service, Southern Research Station. 167pp.

Cordell, H.K., Bergstrom, J.C. 1991. A methodology for assessing national outdoor recreation demand and supply trends. *Leisure Sciences*. 13(1): 1–20.

Cordell, H.K., Bergstrom, J.C., Hartmann, L.A., English, D.B.K. 1990. *An Analysis of the Outdoor Recreation and Wilderness Situation in the United States: 1989–2040.* Gen. Tech. Rep. RM–189. Fort Collins, CO: U.S. Department of Agriculture Forest Service, Rocky Mountain Forest and Range Experiment Station. 112pp.

Cordell, H.K., Betz, Carter J., Mou, Shela H. In press. Outdoor recreation in a shifting societal setting. In: Wear, D.N. and J.G. Greis (eds.). *The Southern Forest Futures Project: Technical Report*. Gen. Tech. Rep. SRS-xxx. Asheville, NC: U.S. Department of Agriculture, Forest Service, Southern Research Station.

Cordell, H.K, Betz, C.J., Zarnoch, S.J. 2013. *Recreation and Protected Land Resources in the United States: A Technical Document Supporting the Forest Service 2010 RPA Assessment*. Gen. Tech. Rep. SRS-150. Asheville, NC: U.S. Department of Agriculture, Forest Service, Southern Research Station. 198pp.

Englin, J.E., Shonkwiler, J.S. 1995. Estimating social welfare using count data models: An application to long-run recreation demand under conditions of endogenous stratification and truncation. *Review of Economics and Statistics*. 77(1): 104–112.

English, D.B.K., Betz, C.J., Young, J.M. et al. 1993. *Regional Demand and Supply Projections for Outdoor Recreation*. Gen. Tech. Rep. RM–230. Fort Collins, CO: U.S. Department of Agriculture Forest Service, Rocky Mountain Forest and Range Experiment Station. 44pp.

English, D.B.K., Kocis, S.M., Zarnoch, S.J. Arnold, J.R. 2002. *Forest Service National Visitor use Monitoring Process: Research Method Documentation*. Gen. Tech. Rep. SRS–57. Asheville, NC: U.S. Department of Agriculture Forest Service, Southeastern Forest Experiment Station. 14pp.

Greene, W.H. 2009. *NLOGIT 4.0. Plainview*, NY: Econometric Software, Inc.

Gregory, S. 2011. Go with the Floe: Adventure travel's love-hate relationship with climate change. *TIME Magazine*, August, 29, 2011.

Hagerty, D., Moeltner, K. 2005. Specification of driving costs in models of recreation demand. *Land Economics*. 81(1): 127–143.

Hall, C.M., Page, S.J. 1999. *The Geography of Tourism and Recreation*. New York, NY: Routledge. 309pp.

Hall, T.E., Heaton, H., Kruger, L.E. 2009. *Outdoor Recreation in the Pacific Northwest and Alaska: Trends in Activity Participation*. Gen. Tech. Rep. PNW–GTR–778. Portland, OR: U.S. Department of Agriculture Forest Service, Pacific Northwest Research Station. 108pp.

Hof, J.G., Kaiser, H.F. 1983a. Long term outdoor recreation participation projections for public land management agencies. *Journal of Leisure Research*. 15(1): 1–14.

Hof, J.G., Kaiser, H.F. 1983b. *Projections of Future Forest Recreation Use*. Resour. Bull. WO–2. Washington, DC: U.S. Department of Agriculture Forest Service. 12 p.

Intergovernmental Panel on Climate Change. 2007. Climate change 2007, the fourth IPCC assessment report. http//www.ipcc.ch/ipccreports/tar/index.htm [Date accessed: December 10, 2008].

Joyce, L.A., Price, D.T., Coulson, D.P., McKenney, D.W. et al. in press. *Climate Change Projections for the United States: A Technical Document Supporting the Forest Service 2010 RPA Assessment*. Gen. Tech. Rep. RMRS—XXX. Fort Collins, CO: U.S. Department of Agriculture, Forest Service, Rocky Mountain Research Station.

Leeworthy, V.R., Bowker, J.M., Hospital, J.D., Stone, E.A. 2005. *Projected Participation in Marine Recreation: 2005 & 2010*. Report prepared for U.S. Department of Commerce, National Oceanic and Atmospheric Administration, National Ocean Service, Special Projects Division, Silver Spring, MD, March, 152p. http://www.srs.fs.usda.gov/pubs/ja/ja_leeworthy002.pdf [Date accessed: August 31, 2010].

Lise, W., Tol, R.S.J. 2002. Impact of climate on tourist demand. *Climate Change* 55: 429–449.

Loomis, J. B., and Crespi, J. 1999. Estimated effects of climate change on selected outdoor recreation activities in the United States In: *The Impact of Climate Change on the United States Economy*. R. Mendelsohn and J. E. Newmann (eds.). Cambridge, New York, Cambridge University Press. pp. 283–314.

Mendelsohn, R., Markowski, M.1999. The impact of climate change on outdoor recreation. In: R. Mendelsohn and J. E. Newmann (eds.). *The Impact of Climate Change on the United States Economy*. Cambridge, New York, Cambridge University Press. pp. 267–288.

Morris, D., Walls, M. 2009. *Climate Change and Outdoor Recreation Resources*. Backgrounder. Resources for the Future, Washington, DC April. 26pp.

Ovaskainen, V., Mikkola, J., Pouta, E. 2001. Estimating recreation demand with on-site data: An application of truncated and endogenously stratified count data models. *Journal of Forest Economics*. 7(2): 125–144.

Poudyal, N.C., Cho, S.H., Bowker, J.M. 2008. Demand for resident hunting in the southeastern United States. *Human Dimensions of Wildlife*. 13: 154–178.

Richardson, R.B., Loomis, J.B. 2004. Adaptive recreation planning and climate change: A contingent visitation approach. *Ecological Economics*. 50: 83–99.

Souter, R.A., Bowker, J.M. 1996. A note on nonlinearity bias and dichotomous choice CVM: Implications for aggregate benefits estimation. *Agricultural and Resource Economics Review*. 25(1): 54–59.

U.S. Department of Agriculture Forest Service. 2009. National survey on recreation and the environment [Dataset]. www.srs.fs.usda.gov/trends/nsre/nsre2.html. [Date accessed: September 15, 2010].

U.S. Department of Agriculture Forest Service. 2010. National Visitor Use Monitoring Program: FY 2009 NVUM national summary report. www.fs.fed.us/recreation/programs/nvum/. [Date accessed: October 6, 2010].

U.S. Department of Agriculture Forest Service. 2012. Future Scenarios: A technical document supporting the Forest Service 2010 RPA Assessment. USDA Forest Service, Gen. Tech. Rept. RMRS-GTR-272, Fort Collins, CO: Rocky Mountain Research Station. 34pp.

Wake, C., Burakowski, E., Goss, L. 2006. *Winter Recreation and Climate Variability in New Hampshire: 1984–2006*. The Carbon Coalition and Clean Air-Cool Planet. Portsmouth, New Hampshire. October.

Walls, M., Darley, S., Siikamaki, J. 2009. The state of the great outdoors: America's parks, public lands, and recreation resources. Washington, DC: Resources for the Future Report. http://www.rff.org/Publications/Pages/PublicationDetails.aspx?PublicationID=20921. [Date accessed: September 15, 2010].

Walsh, R.G., Jon, K.H., McKean, J.R., Hof, J. 1992. Effect of price on forecasts of participation in fish and wildlife recreation: An aggregate demand model. *Journal of Leisure Research*. 21: 140–156.

Wear, D.N. 2011. *Forecasts of County-Level Land Uses Under three Future Scenarios: A Technical Document Supporting the Forest Service 2010 RPA Assessment*. Gen. Tech. Rep. SRS-141. Asheville, NC: U.S. Department of Agriculture Forest Service, Southern Research Station. 41pp.

Whitehead, J.C., Poulter, B., Dumas, C.F., and Bin, O. 2009. Measuring the economic effects of sea level rise on shore fishing. *Mitigation and Adaptation Strategies for Global Change*. 14: 777–792.

Zarnoch, S.J., Cordell, H.K., Betz, C.J., Langner, L. 2010. *Projecting County-Level Populations Under Three Future Scenarios: A Technical Document Supporting the Forest Service 2010 RPA assessment*. e-Gen. Tech. Rep. SRS–128. Asheville, NC: U.S. Department of Agriculture Forest Service, Southern Research Station. 8pp.

Zawacki, W.T., Marsinko, A., Bowker, J.M. 2000. A travel cost analysis of economic use value of nonconsumptive wildlife recreation in the United States. *Forest Science*. 46(4): 496–505.

13 Summary of Findings, Management Options, and Interactions

James M. Vose, Shelby Gull Laird, Zanethia D. Choice, and Kier D. Klepzig

CONTENTS

Forests in the Southern United States are likely to be very different in the coming decades as a result of climate change. Maintaining resilience and restoring forest ecosystems to ensure a continuing supply of ecosystem services will be a major challenge in the twenty-first century. Fortunately, most forests have inherent resistance and resilience to climatic variability and disturbances. Historically, land managers have been able to leverage these characteristics, shaped by management and on-the-ground experiences, to buffer the effects of climatic stresses and disturbances on forest health and productivity. As noted in the previous chapters, the natural resource community in the Southern

United States is not operating in a vacuum of scientific knowledge or management experience. Clear examples include the use of prescribed fire and tree breeding for insect and disease resistance. Despite this history and decades of practical experience, the rapid pace and magnitude of climate change may exceed the inherent resistance and resiliency of forest ecosystems and pose new management challenges that go beyond current knowledge and experience. Among those challenges will be identifying areas where forests are most vulnerable, determining where the effects of change could be the greatest and the most detrimental, and developing and implementing management activities to increase resilience and resistance or to facilitate a transition to a new condition (Millar et al. 2007).

As discussed in Chapter 7, initial progress for making forest stands more resistant and resilient to climate change will most likely occur as modest modifications of forest management activities currently being practiced on forest industry and timberland investment ownerships, on nonindustrial private forest ownerships already under active management, and on government lands such as national and state forests. For these forests, managing forest stands with attention to the possibility of climate change will not be much different in principle than decisions foresters have been making over the past century, but perhaps with different treatment prescriptions based on stand management objectives that address and emphasize the potential impacts of a changing climate. In many ways, this is essentially a risk-reducing strategy, but where management decisions consider the implications of climate change along with all of the other factors considered when planning and implementing management activities. In other cases, the pace of change, the complexity of highly fragmented landscapes and multiple co-occurring stressors, and the consequences of economic and ecological impacts could be so severe that management actions focusing on historical conditions are likely to fail. In these cases, new management approaches may be required to anticipate and respond to climate change and guide development and adaptation of forest ecosystem structures and function, thereby sustaining desired ecosystems services and values over large landscapes and multiple decades. Examples include facilitating migration of and managing future habitat for impacted species (Chapters 10 and 11), selectively managing for "water-wise" species to maintain streamflow and groundwater recharge (Chapter 9), and developing new breeding programs that favor species resistant to climatic change and variability (Chapter 8). In some cases, management actions will involve facilitating the transition to a different or new mix of species and stand structures that can continue to provided ecosystem services. As such, some of these new management activities may challenge long-held management practices aimed at maximizing productivity per unit area or restoring ecosystems to historic conditions.

To address forest–climate interactions in the Southern Region, chapter authors were provided with a common framework (Chapter 1), consistent definitions (Chapter 3), and down-scaled projections of a range of future temperature and precipitation values (Chapter 2) as starting points. Although the approach to utilizing this information varied from chapter to chapter, all chapters provide a comprehensive analysis and discussion of anticipated climate change impacts on forest threats and values in the Southern Region and discuss potential management options to mitigate those impacts. Most climate change impacts, and subsequent management activities to adapt to or mitigate them, would interact with multiple threats and values simultaneously. Some interactions may occur more quickly than others, and may be either positive or negative, and may be either short- or long term. For example, thinning to reduce the threat of wildfire might also have impacts on water resources, wildlife habitat, recreation experience, forest growth, and carbon storage—suggesting the importance of an interdisciplinary and multifactor approach that identifies and evaluates trade-offs among adaptation and mitigation choices. An important first step is to identify the most important interactions. Here, we highlight the major findings from each of the values and threats chapters, summarize management options, and identify key interactions among effects and management options. For details and scientific references to support the key findings and management options, readers are referred to the specific chapters.

THREATS TO SOUTHERN FORESTS

WILDFIRES

Key Findings

- The South will face the challenge of potentially increased wildfires this century because of projected warmer temperatures and more frequent droughts resulting from climate change.
- Future fuel loading is projected to decrease in the western areas of the South and increase in the eastern areas based on a global dynamic vegetation model and a dynamic downscaled climate change scenario.
- The area with the largest increase in future fire potential is expected to occur along the eastern Gulf of Mexico coastal areas in the early spring, extending to the central area in the late spring, and further to the Atlantic coast in summer and early autumn. The length of the spring and autumn fire seasons is likely to increase and the extent of the drying would be more severe.
- Projected increased dryness during summer may introduce a summer fire season to parts of the South (or at least a later end to the spring season and earlier start of the autumn season). Fire seasons would increase in duration by about 1–5 months in eastern areas of the South with the largest increase in the Appalachian Mountains, and by about 1–3 months in western areas with the largest increase in the Mid-South.
- Wetter weather, which may accompany climate change in some places, would result in less frequent, smaller, and less intense wildfires in those areas.
- The continued population growth of the South increases the potential threat that wildfires pose to life and property.
- Human populations are positive risk factors for wildfire; human-ignited wildfires tend to be clustered around places with human populations, confirming that as human populations grow, wildfire ignitions by people are more frequent, all things being equal. Increased fires in the future will likely increase occurrence of smoke and lead to more severe air quality impacts.
- Public tolerance of smoke has diminished over time, and complaints about smoke impacts from prescribed burning, wildland fire use, and wildfires are frequent.
- The number of days when a prescribed burn is unlikely to become an uncontrolled burn or a wildfire in the South could be reduced by about 40–60% during summer and autumn, 30% during winter, and 10% during spring.

Management Options

- Increase public awareness and education about the potential impacts of climate change on wildfire risk.
- Increase the use of prescribed burning to reduce accumulation of understory fuels and therefore reduce wildfire risk.
- Reduce wildfire risk indirectly by expanding the acreage of fire-resistant species such as longleaf pine.
- Increase focus on arson-prevention efforts.

INSECTS, DISEASES, AND INVASIVES

Key Findings

- Changes in environmental factors including temperature, precipitation, and associated factors, may affect the occurrence and impacts of forest diseases, native and nonnative pest insects, and nonnative plant species in several ways, some of which are difficult to predict.

- Long-term drought events may result in increased host stress and lead to an increase in the severity and mortality attributed to root disease. Conversely, tree mortality associated with drought can reduce stand density, thus reducing the opportunity for root contact, contagion, and demand for nitrogen.
- Increasing air temperature may decrease the incidence of stem rust pathogens because of their exacting requirements for infection and spore survival on plant surfaces.
- Warming from climate change will likely hasten proliferation of potentially harmful insect species such as bark beetles, defoliators, sap-sucking insects, borers, and weevils (all of which would likely see increased numbers of generations per year).
- Climate-change-driven warming could expand the northern ranges for many invasive insect species in the United States.
- In addition to increased insect metabolism during the growing season, warmer temperatures could also reduce insect mortality from the extreme cold winter season, resulting in thriving insect populations through the spring into summer.
- Climate change could also indirectly affect insect populations through impacts on natural enemies, important insect symbionts, host physiology, and host range distributions.
- Future warmer winter temperatures could remove existing range barriers for some native species. This could result in spread into places where hosts are currently abundant and result in competition between native and nonnative insect species.
- Even within the past few decades, invasive species of all kinds have moved higher in latitude and elevation, threatening the native plant species with small population sizes and distribution ranges.
- Climate change will likely influence the establishment of new invasive species and the effectiveness of control strategies. Invasive plants that are fast growing and responsive to resources would be favored by environmental changes that increase resource availability, which would jeopardize the existence of invasive species.

Management Options

- Implement preventive measures such as thinning to reduce stand density—removing infested, damaged, and weakened trees, and harvesting before trees become over mature.
- Replant impacted stands with resistant varieties and species.
- Apply early detection and rapid response systems to prevent the establishment of, or mitigate damage from, new invasive species.
- Plan for monitoring and prompt management of insect outbreaks at higher elevations, and prepare for management of large-scale mortality events and altered fire regimes.
- Promote silvicultural practices that increase seedling regeneration and genetic mixing.
- Reduce homogeneity of stand structure and synchrony of disturbance patterns by promoting diverse age classes, species mixes, and genetic diversity.

VALUES OF SOUTHERN FOREST

Forest Productivity and Carbon Sequestration

Key Findings

- The South's forest sector produces approximately 60% of the total United States' wood production–climate change could have major implications for the nation's wood supply.
- Southern forests play an important role in carbon (C) storage, accounting for 36% of the carbon sequestered in the conterminous United States.
- The primary effects of climate change will likely occur as exogenous or endogenous disturbance events of as-yet unknown duration, frequency, or intensity in southern forest stands and landscapes.

- Climate change may stimulate growth in many areas of the Southern United States; however, lower precipitation and nitrogen (N) availability may alter or negate the growth response.
- Predicted warming over the next century may cause accelerated decomposition of soil C and reduce C storage in southern forests.

Management Options

- Increase the frequency of thinning and manage at lower basal area to maintain individual tree vigor.
- Expand and intensify pine plantation management to increase productivity and C sequestration.
- Plant and manage species less susceptible to extreme weather conditions such as longleaf pine.
- Breed for more climate change resilient ideotypes.
- Manage to create forests with greater genetic diversity and multiple age cohorts.
- Conserve at risk species via *in situ* and *ex situ* genetic conservation.
- In wet or seasonally inundated forests, maintain hydrologic regimes to ensure continued C retention.
- Extend rotations and increase initial planting density to increase net C storage.

Water Resources

Key Findings

- Water resources in the Southern United States are at risk of degradation from climate change, land conversion from forests to urban uses, increasing water demand from growing population, and sea level rise.
- Droughts, floods, and water quality problems are likely to be amplified by climate change.
- As species distributions change in response to climate change, especially if large areas of forests experience catastrophic mortality, streamflow and water quality are also likely to change.
- Decreases in water supply and increases in water demand as a result of population growth would combine to increase water supply stress to humans and ecosystems.
- A warming climate may elevate water temperature, diminishing surface-water value for consumptive and habitat uses.
- Changes in precipitation amount or storm intensity can affect soil erosion and sediment loading by altering the amount of runoff, the kinetic energy of rainfall, or the ability of vegetation cover to resist erosion.
- Future climate change is likely to aggravate the problems of salt-water intrusion in the east coast of the United States by increasing air temperatures, changing regional precipitation regimes, and raising sea level.

Management Options

- Consider the water demands of tree species when planning restoration and afforestation activities. "Water-wise" choices may help maintain streamflow during droughts and reduce flood risk after high precipitation events.
- Natural wetlands should be protected to maintain their hydrologic functions and buffer disturbances.
- The resilience and vulnerability in artificially drained, managed forests should be evaluated based on projected future climatic and hydrologic conditions in the region.
- To mitigate salt-water intrusion, management practices in these areas should include timing of withdrawals to coincide with outgoing tides, increasing storage of raw water, adjusting

the timing of larger releases to move the saltwater–freshwater interface downstream, and blending higher conductance surface water with lower conductance water from an alternative source such as groundwater.

- Mitigating climate change impacts on low flows would require implementation of management practices that help to reduce water use, reclaim wastewater, and enhance infiltration and groundwater discharge to streams.
- Mitigation of increases in stream temperature and greater overland flow would require stable or increased shading from solar radiation through riparian buffer retention, expansion of riparian buffer widths, and restoration of degraded riparian zones.
- To compensate for higher runoff volumes: (1) adjust the relationship between road grade and separation distances for broad-based dips and other water diversion features, (2) increase culvert diameter for a given drainage area, and (3) use hooded inlets on culverts to increase storm water-carrying capacity for a given culvert size.
- Develop and implement new best management practices for road design and construction that more effectively disconnect the road system from the stream network.
- Develop and implement storm water structural best management practices such as rain gardens and other sediment basins, and sustainable urban drainage systems that encourage infiltration, minimize exceeding treatment facility capacity, and ultimately reduce the risk of degraded surface water.

Plant Species and Habitat

Key Findings

- Climate models vary in their forecasts of future climates, but generally predict a warmer and drier environment for much of the South. Vulnerability assessment for common tree species indicates that some, that is, those with relatively high moisture requirements, will decrease their areas of distribution.
- Vulnerability assessments varied considerably among the four climate predictions, particularly when comparing the MIROC3.2 A1B scenario (the general circulation model that projects the warmest and driest future) to the other three model and scenario combinations.
- Vulnerability of tree species to climate change in the Southern Region is potentially highest (measured by number of tree species predicted to decrease their range) in the Mid-Gulf-West of the Coastal Plain, the Central Appalachian Piedmont, and the Ozark-Ouachita Highlands section of the Mid-South. In contrast, the ranges may expand northward for species in the Southern Appalachian Piedmont, Blue Ridge and Northern Ridge and Valley, which are rapidly expanding their northern ranges in response to warming climates.
- The overall effect of climate change on diversity of southern trees was predicted to result in small reductions of species richness in most of the 21 ecological sections studied, although the number of species may remain constant or increase slightly in several sections by the year 2060.
- Although the magnitude of results varied somewhat among climate scenarios, the overall trends were generally consistent: tree diversity could decrease throughout the Southern Region by the year 2060, with the highest risks occurring in certain sections of the Coastal Plain, Piedmont, and eastern zone of the Mid-South.

Management Options

- Landowners with holdings in areas of the Southern Region that face the highest threat to vulnerability for certain species (such as shortleaf pine in western Arkansas) and high risk to species diversity (such as in the Ozark-Ouachita Highlands section of the Mid-South) will be among those who need relevant information on resource management the earliest.

- To maintain the historic oak component in some ecosystems in the likely event of climate change, land managers may need to reduce basal area and stem density of shade-tolerant competitors, such as invading red maple, in all canopy strata.
- Artificial regeneration of preferred tree species by planting may be required as an alternative to natural methods for resource managers concerned about short-term effects of climate change on species composition.
- Genetic selection from natural variability within a species may allow resource managers to respond to climate change when trends of future temperature and precipitation become more apparent for an area.

Wildlife Species and Habitats

Key Findings

- Major factors contributing to population declines of wildlife species include habitat destruction and fragmentation, isolation, small population sizes, low genetic diversity, water diversions, introduction of nonnative invasive species, acid precipitation and other environmental pollutants, commercial development, human disturbance, and exploitation. Stress from climate change may exacerbate the effects of these factors.
- Based on the processes of extinction and colonization and their ability to disperse, species are expected to shift their distributions or move to higher altitudes in response to a warming climate.
- Increased fire frequency or intensity could alter habitat features required by some wildlife species.
- Because of their ability to regulate body temperature, mammals generally respond to climate indirectly through changes in food supplies, predators, parasites, and habitat. However, human-modified landscapes may affect movements of mammals to more favorable climates.
- Increased variability or reductions in precipitation may affect disease outbreaks, survival, physiology, and nutritional state in mammals. Changes in climate could lead to increased zoonotic disease outbreaks in humans.
- Climate modification also can indirectly affect birds by influencing habitat conditions. Various groups of birds, such as waterfowl, shorebirds, game species, and neo-tropical migratory birds may be affected by habitat conditions that change from climate modification.
- Eventually, even highly adaptive bird species may reach their limits and would be forced to employ other measures to deal with climate change effects. Temporal responses may appear as changes in migratory patterns with birds arriving at breeding grounds earlier.
- For some bird species, warming may have a beneficial effect if it results in a widening of the area within which their habitat requirements are met and they utilize this expanded area. The enlarged area may contain food resources that differ from the current range; hence, the species must be capable of adjusting to the new food resources and possibly a different suite of predators.
- Shorebirds that depend on multiple areas of the South for part of their life cycle may be adversely affected by global warming on their breeding habitat, migration routes, and wintering habitat.
- Shifts in climate can have negative effects on amphibian populations. Temperature and precipitation have major influences on the life cycle of amphibians, particularly their breeding activities. Higher temperatures could cause some amphibians to shift their distributions north because their young are unable to develop in warmer waters. Many mountaintop species may already be approaching their thermal maxima and have limited dispersal ability. Increases in temperature would result in lost habitat for many species. Species with small geographic ranges would be at greater risk of extinction.

- As a group, reptiles may be less vulnerable to climate change than amphibians because their scale-covered skin makes them less vulnerable to desiccation and better able to tolerate predicted drier, warmer conditions. Nonetheless, reptiles would be affected by changes to primary habitats, temperature-driven energetic shortfalls, temperature-dependent sex determination, or changes in food availability.
- Butterflies play a significant, critical role in ecosystem function as pollinators. Climate change, to the extent that it alters butterfly and other pollinator populations, would have cascading effects on entire ecosystems. As a consequence of their complex life cycles, diverse larval hosts, and frequent dependence on a particular suite of nectar resources, butterflies would respond to climate change in very complex ways. Their larvae are also important herbivores in terrestrial ecosystems and critical prey for predators (such as insectivorous birds).
- Changes in butterfly populations resulting from climate change would have consequences for both host plant and predator species. Butterfly species would begin to expand northward but their success would depend on concurrent shifts in host plant distribution. Other species, lacking the ability to colonize new habitats rapidly, would be locally or globally extirpated, with resultant loss of diversity and critical ecosystem services.

Management Options

- Increase the amount and connectivity of wildlands and minimally disturbed habitats through acquisitions and conservation programs, with particular attention to interconnected habitats that run from north to south, and from higher to lower elevations.
- Consider wildlife movement and road-kill mortality reduction when planning or improving roads and highways; examples include providing (as needed) elevated sections of highways or building wildlife underpasses to allow uninterrupted migration corridors.
- Where possible, manage ecosystems to restore and maintain historical conditions likely to enhance resilience in associated native wildlife species by promoting optimal habitat and larger populations.
- Protect and conserve coastal habitats through strategic planning and implementation of zoning and building codes, particularly in areas of predicted rising sea levels where development plans may need to provide for inland migration of coastal wetlands.
- Use caution when building barriers for construction or flood control, as they can cause the elimination of existing wetlands.
- Make coastal habitats less susceptible to sea level rise, such as by removing ditches to restore the hydrologic regime and limit saltwater intrusion, assisting development of vegetation by planting salt-tolerant species, and building oyster reefs to buffer shorelines from storm events and wave action.
- Protect and maintain wetland and stream water quality, water quantity, flow, hydrologic processes, and temperatures by restoring riparian zones and watersheds.
- Conserve species and special habitats, paying special attention to restoration and management of rare species or ecosystems.
- Aim for representation, resiliency, redundancy—networks of intact habitats that represent the full range of species and ecosystems in a given landscape, with multiple robust examples of each.
- Reduce existing ecosystem stressors, both from climate and other disturbances, such as habitat loss and alteration, pollution, ozone depletion, invasive species, and pathogens.
- Forge or foster partnerships among agencies, organizations, scientists, and local communities to develop science-driven, landscape scale strategies to maximize the use of scarce resources.
- Monitor, model, and implement adaptive management in response to unforeseen responses or trophic cascades resulting from climate change.

- Increase management for species in areas where they are expected to advance, such as the northern limits of their ranges.
- Consider short-distance, human-facilitated movements of species across artificial and natural impediments to migration, such as areas dominated by intense agriculture and large rivers.
- Consider, *ex situ* gene conservation options for species at risk of extinction. One option could be translocation of species to geographic areas or habitats (assisted migration) where they do not naturally occur.

Recreation

Key Findings

- Outdoor recreation opportunities rely heavily on natural settings and conditions, thus negative impacts on settings like forests and water bodies could negatively impact the quality and availability of these resources, and their long-term potential in providing recreational opportunities.
- Climate may affect recreation in direct and indirect ways including the effects of weather on recreationists' physical comfort or convenience and the effect of varying the length of seasons on the availability/suitability of certain outdoor opportunities. In addition, more frequent and damaging severe weather events could pose risks to recreationist safety.
- Climate change could gradually alter the natural resource base on which outdoor activities depend. For example, increasing temperatures will likely affect the distribution of plant and animal species that are fundamental to maintaining fish and game populations.
- In the coming decades, population growth will be the overriding factor influencing outdoor recreation. The effects of climate change on recreation demand by southerners appear to be moderately mitigating insofar as use density measures (like participants per forest and rangeland acre and activity days per acre) are concerned.
- Increased congestion and possible declines in the quality of the outdoor recreation experience are likely to present important challenges to management.

Management Options

- Management activities that maintain forest settings, provide stable water resources (e.g., streamflow and lake water levels), and provide wildlife habitat will help to ensure quality recreation opportunities.
- While climate change will impact outdoor recreation, managers will need to be more concerned about increased congestion and possible declines in the quality of the outdoor recreation experience. This will probably have to be accomplished through creative and efficient management of site attribute inputs and plans, rather than through any major expansions or additions to the natural resource base for recreation.

INTERACTIONS AMONG MANAGEMENT OPTIONS

In this section, we identify and discuss management options within the context of multiple benefits or conflicts among values and threats (Table 13.1). Management actions can have positive impacts on multiple resource values or threats. For example, increased use of prescribed burning is a management option to reduce wildfire risk; however, it is also recommended for maintaining wildlife habitat and reducing threats from certain insects, diseases, and nonnative plants. On the other hand, expanding the use of prescribed fire into less frequently burned areas in anticipation of future wildfire risk could substantially change current species composition and wildlife habitat. Referring to Table 13.1, other management options with multiple benefits include increased thinning of current stands and managing future stands for lower stocking (recommended by 3 values, 2 threats); increased vegetative, genetic, and age class diversity (recommended by 3 values, 1 threat);

TABLE 13.1
Climate-Change Mitigation Alternatives and Their Potential Impacts (Each Designated by Δ Symbol) on Multiple Forest Values and Threats

	Values					Threats	
Management Options	**Water**	**Species**	**Forest Productivity**	**Carbon Sequestration**	**Recreation**	**Wildfire**	**Invasives**
Improve landscape connectivity, especially south to north		Δ			Δ		
Increase vegetative, genetic age-class diversity		Δ	Δ		Δ		Δ
Increase use of prescribed fire		Δ			Δ	Δ	Δ
Increase thinning, keep stocking levels low	Δ	Δ			Δ	Δ	Δ
Lengthen rotation age				Δ	Δ		
Shorten rotation age			Δ		Δ	Δ	Δ
Redesign road best management practices for larger storms	Δ				Δ		
Increase culvert size for larger storms	Δ				Δ		
Fertilize established stands			Δ	Δ	Δ		
Plant/encourage species that are resistant and resilient to disturbance and stress			Δ		Δ	Δ	Δ
Restore hydrologic function in drained forests and wetlands; maintain or enhance hydrologic function elsewhere	Δ	Δ		Δ	Δ		
Restore and widen riparian buffers	Δ	Δ			Δ		
Manage destinations for species' migrations		Δ			Δ		
Implement facilitated migration		Δ			Δ		
Incorporate water use characteristics of tree species in decision making	Δ				Δ		
Increase and intensify plantation management			Δ	Δ	Δ		
Implement coastal zone water management to abate sea level rise	Δ	Δ			Δ		

restoration, maintenance, and enhancement of hydrologic function (recommended by 4 values); and planting or encouragement of species resistant or resilient to disturbance or stress (recommended by 2 values, 2 threats). As climate change management adaptation and mitigation activities are evaluated and planned for public lands, the identification of management options that provide multiple benefits may help with prioritizing and funding activities that garner the biggest "bang for the buck." This is not to imply that management activities that only benefit a single value or threat are less important or lower priority. Indeed, it is likely that some very specific management activities may be required for activities such as protecting rare and endangered species or addressing high risks such as flood severity or frequency. Management for commercial production will need to translate these threats (Table 13.1) into financial risk profiles for investment alternatives. Increased risks would likely favor shortened rotations and intermediate treatments and possibly increased diversification of management across species and locations.

Specific management options may also have negative impacts on other resource values of threats. For example, maintaining forests at lower stocking levels through thinning and lower initial stand density to increase tree vigor may reduce overall stand productivity and decrease C sequestration potential in the short term. However, if these actions prevent catastrophic wildfire or severe insect outbreaks, then the impacts would be positive in the long term. In contrast, extending rotation lengths to enhance C storage may increase the potential for exposure to insect outbreaks or storm damage that could offset C storage gains. These interactions and trade-offs suggest the need for a broad vision and large-scale approach for implementing "climate smart" management practices. For example, extending rotation ages may not be desirable for all species in all locations. Extending rotation ages in areas where climate change is expected to increase the potential for severe storms or insect outbreaks would increase the risk of tree mortality and reduce productivity and C storage. In areas where future risks are greatest, decisions could be made to shorten rotations and/or favor species more resistant to climate extremes and insect outbreaks.

As noted, evaluating risk (Chapter 3) and landscape scale approaches are at the core of this type of decision making. Risk management has been used by resource managers for many years, but typically not in the context of climate change. The most recent National Climate Assessment (Vose et al. 2012) has advanced a risk-based approach for evaluating climate change impacts and evaluating management options to reduce risk (Yohe and Leichenko 2010), where risk is framed by the likelihood of an impact occurring and the magnitude of the consequence of the impact (Yohe 2010). Figure 13.1 provides an example application of the risk-based framework for a management option to reduce flood risk. Landscape scale planning andmanagement activities will be especially challenging in the Southern United States because of the complex and mixed ownership pattern of the southern forests. Due to the large acreage of private forest land (about 89%), management activities on private forests will be critical in order for many adaptation and mitigation activities to have an impact at meaningful scales. A proactive approach will require working across institutional and ownership boundaries to exchange information and potentially coordinate landscape level approaches.

Robust decision making also requires a full understanding of uncertainty associated with climate change projections (especially for precipitation), climate change impacts on resource values

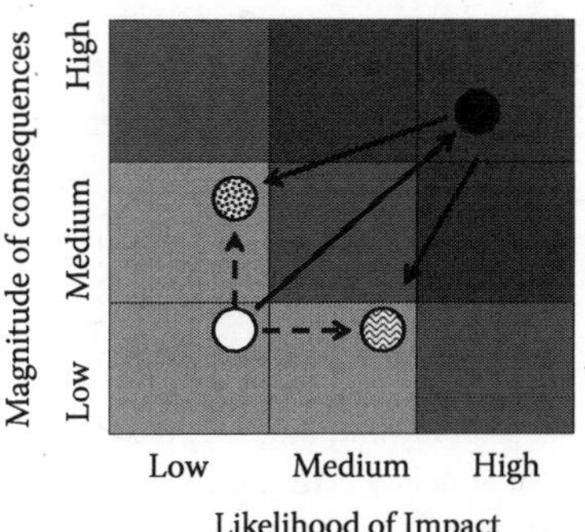

FIGURE 13.1 Example of applying the risk-based framework described in Yohe and Leichenko (2010) to prioritize management decisions and evaluate management options. The *y*-axis is the likelihood of climate change impact (low, medium, high) and the *x*-axis is the magnitude of the climate change impact (low, medium, high). Darker colors indicate greater risk. In this example, we denote the open circle (○) as the flood risk for a watershed under current precipitation regimes. The solid circle (●) represents the future flood risk under a changing precipitation regime that includes more extreme high rainfall events. Management actions can be either proactive (dashed lines) to minimize risk or reactive (solid lines) to reduce risk after changes have occurred. An example of a management action to reduce the likelihood of impacts (◍) would be to increase culvert sizes. An example of a management action to reduce magnitude of impacts (◍) would be to manage for wide riparian zones with species adapted to wet conditions.

and threats, and management actions to reduce impacts or minimize risk. This uncertainty continues to be reduced as new science and observations confirm (or modify) previous predictions and improve model accuracy, and as on-the-ground adaptation activities (such as thinning to reduce wildfire risk) are tested by changes in fire regimes (Vose et al. 2012). Although some level of uncertainty will always exist, sufficient scientific information is available to begin to implement climate smart management activities now, and tools are available to help incorporate climate change science into planning and decision making (Vose et al. 2012). Examples include TACCIMO and ForWarn (see www.forestthreats.org for further information).

Climate change is just one of many considerations that land managers must address when making decisions and choices about specific on-the-ground management activities. These decisions are considered within the context of short- and long-term goals and desired future conditions, the best available science, and the experiences of natural resource professionals. Management actions can be either passive (a decision of no action), reactive, or anticipatory (Carter et al. 1994). The management options advanced by the authors of this book are a mix of reactive and anticipatory actions. Similarly, climate change adaptation management involves a combination of planning (decision making and prioritization in the context of other resource management demands and constraints), strategies (prioritization and decision making of how, where, and when to implement climate change management actions), and tactics (project scale decision making and implementation of climate change management actions). While many of the management options discussed in this book provide guidance for planning, the primary scale of focus is on strategies and tactics.

A few overarching concepts provide broad but useful advice for managers who consider the options described above when developing adaptation strategies and tactics:

- Reduce existing stressors as much as possible early in planning to increase the resilience of the forest ahead of further major climate change impacts. These stressors may include climate impacts, such as more frequent flooding or drought, and others including pollution, habitat loss, alteration, and invasive species. Starting with a healthy ecosystem could provide a greater level of resilience for existing populations.
- Adopt and promote a system of monitoring and implementing adaptive management strategies to accurately assess success and identify areas for improvement. Consistent long-term monitoring will help managers gain valuable information about the ecosystem under a changing climate and adjust their strategies appropriately.
- Develop and enhance existing partnerships with scientists, organizations, agencies and local communities. Such partnerships could enhance the development of science-based management approaches that will provide the essential foundation for managing forests in response to a changing climate.

The rapid pace of new scientific information from observations, models, and experiments; and "on-the-ground" experiences with climate change adaptation activities will place an even greater emphasis on the rapid transfer of new science to land managers and frequent feedback from land managers to scientists. As such, this book serves as a starting point for a dialogue about "climate change adaptation and mitigation and management options" that will be continuously updated and improved through enhanced science–management partnerships.

REFERENCES

Carter, T.R., Parry, M.L., Harasawa, H., Nishioka, S. 1994. *IPCC Technical Guidelines for Assessing Climate Change Impacts and Adaptation*. Working Group 2 of the Intergovernmental Panel on Climate Change. Cambridge, UK: Cambridge University Press: 1–72.

Millar, C.I., Stephenson, N.L., Stephens, S.L. 2007. Climate change and forest of the future: Managing in the face of uncertainty. *Ecological Applications* 17: 2145–2151.

Vose, J.M., Peterson, D.L., Patel-Weynand, T. 2012. Effects of climatic variability and change on forest ecosystems: A comprehensive science synthesis for the U.S. forest sector. Gen. Tech. Rep. PNW-GTR-870. Portland, OR: U.S. Department of Agriculture, Forest Service, Pacific Northwest Station. 265pp.

Yohe, G. 2010. Risk assessment and risk management for infrastructure planning and investment. *The Bridge* 40(3): 14–21.

Yohe, G., Leichenko, R. 2010. Adopting a risk-based approach. *Annals of the New York Academy of Sciences* 1196: 29–40.

Index

D

E

G

H

N

O

P

T

U

V

W

Y